Emerging Modalities in Mitigation of Antimicrobial Resistance

Nadeem Akhtar • Kumar Siddharth Singh
Prerna • Dinesh Goyal
Editors

Emerging Modalities in Mitigation of Antimicrobial Resistance

Springer

Editors
Nadeem Akhtar
Department of Animal Biosciences
University of Guelph
Guelph, Ontario, Canada

Prerna
Department of Biotechnology
Thapar Institute of Engineering and Technology
Patiala, India

Kumar Siddharth Singh
Institute for Microbiology
Leibniz University
Hannover, Germany

Dinesh Goyal
Department of Biotechnology
Thapar Institute of Engineering and Technology
Patiala, Punjab, India

ISBN 978-3-030-84125-6 ISBN 978-3-030-84126-3 (eBook)
https://doi.org/10.1007/978-3-030-84126-3

This Springer imprint is published by the registered company Springer Nature Switzerland AG
The registered company address is: Gewerbestrasse 11, 6330 Cham, Switzerland

About This Book

Antimicrobial resistance (AMR) is one of the deadliest threats to global public health. This book focuses on dynamics in the landscape of AMR while informing about the latest technologies and strategies to mitigate it. The menace of AMR in different niches, routes of penetration across various domains, socio-economic impact, and the need for a "One Health" approach in mitigating AMR have been emphasized. Factors involved in AMR, underlying mechanisms, and pharmacometrics in developing antimicrobials are highlighted. Emphasis is given to emerging technologies that are sustainable, scalable, and applicable to the global community, such as big data analytics, bioactive agents, phage therapy, and nanotechnology. The book also explores current and alternative treatment strategies to combat AMR, emphasizing the use of nanoparticles to target pathogens and as a viable alternative to antibiotics.

Contents

About the Editors

Nadeem Akhtar, PhD currently works at the University of Guelph, Canada, as a Research Associate. His current research focuses on biotechnology in animal nutrition, where the main objective is to improve animals' health for a better and disease-free lifestyle. Before the Research Associate role, he has worked as a Postdoctoral Researcher in the Department of Chemical Engineering, University of Waterloo, Canada (2019–2021), and in the Department of Animal Biosciences, University of Guelph, Canada (2015–2019). Dr. Akhtar was awarded a Ph.D. from the Department of Biotechnology, Thapar University, India. He has qualified National Eligibility Test (NET-2010) and Graduate Aptitude Test (GATE) in Life Sciences (2010) and Biotechnology (2011). His scientific contributions include more than 23 articles in peer-reviewed international journals, 10 book chapters, and a patent. He reviewed several research articles for reputed journals, including *RSC Advances*, *Journal of Applied Microbiology*, *Scientific Reports*, *Poultry Science*, and many more. Department of Animal Biosciences, University of Guelph, Guelph, Ontario, Canada

Kumar Siddharth Singh is a trained biotechnologist interested in host-microbe interactions, protein science, molecular microbiology, and omics technologies. He has a Master's degree in Biotechnology from Thapar University, Patiala, India, and a Bachelor's degree from Anugrah Narayan College, Patna, India. He obtained a Ph.D. in Animal Biotechnology from ICAR-National Dairy Research Institute, Karnal, India, and developed a potential nutraceutical of probiotic origin to protect against gastrointestinal infections. He is also trained in food safety and quality management, emphasizing regulatory standards relevant to the food and biotechnology industry. He has teaching experience at graduate and postgraduate levels and has many years of mentoring experience. He has industry exposure, more than 4 years of postdoctoral research experience, and has been awarded many national and international academic awards at different levels of his career from different agencies. He has many scientific articles and reviews to his name and serves as an editorial board member of two journals. Institute for Microbiology, Leibniz University, Hannover, Germany

Prerna is a Ph.D. Candidate at the Department of Biotechnology, Thapar Institute of Engineering and Technology, India, since July 2013. She received a scholarship under the Government of India scheme TEQIP-II (July 2013–June 2016) and served as a teaching assistant in Biotechnology. Her current research focuses on developing green novel strategies to fabricate various nanoparticles and commercial interests, including nanobiotechnology in medicine, industry, and agriculture. She has optimized gram-scale synthesis of nanoparticles from various plants and alga. She is involved in evaluating strategies employed to remediate industrial wastes with a vision to conserve soil and water, promote water reuse efficiency, and protect soil and water quality in diverse agro-ecosystems. As a member of the interdisciplinary research group, she used different plants, algae, fungi, and bacteria to fabricate ZnO nanoparticles. She has extensive experience in characterizing nanoparticles using different spectroscopic, diffractographic, crystallographic, and advanced microscopic analyses. She served as a resource person for different training programs and workshops at STEP (Science and Technology Entrepreneur's Park), Thapar Institute of Engineering and Technology. She has published two research articles, two book chapters, and presented her research in many conferences and symposia. Thapar Institute of Engineering and Technology, Patiala, Punjab, India

Dinesh Goyal is a microbiologist graduated from the Division of Microbiology, Indian Agricultural Research Institute, New Delhi, in 1990. He served at the University of Delhi, New Delhi, and Thapar Centre for Industrial Research, Patiala, before joining Thapar Institute of Engineering and Technology (Deemed to be University), Patiala, as a faculty in 2000 and is presently the Professor in the Department of Biotechnology. He did Postdoctoral Research at the Department of Bioengineering, Tokyo Institute of Technology, Tokyo, Japan, during 1993 to 1995 and later underwent 2 weeks training on Technology Entrepreneurship and Education at Walter A Haas Business School, University of California, Berkeley, online training on Marketing for Incubator Managers and Entrepreneurs organized by Asia Pacific Incubator's Network (APIN), and accomplished many orientation courses on higher education, human values, IPR, and commercialization of technology and innovation.His current interests are in applied microbiology and biotechnology, and he is focusing on nano-bioremediation and nano-antimicrobials. He has published more than 100 articles, two patents, 14 book chapters, compiled three manuals, and guided the research work of 20 Ph.D. students, 10 project fellows, B.Tech, 60 M.Sc., and 08 M.Tech. students.He has organized more than 35 training programs and workshops in microbiology, biotechnology, and entrepreneurship and eight national and one international conference. He was nominated four times (2006–08) as an expert member of National Board of Accreditation (NBA) by All India Council for Technical Education (AICTE) for evaluation and accreditation of B.Tech Biotechnology program.He is also the recipient of prestigious Japanese Govt (Mombusho) Scholarship (1993–95), Fellow of International Congress of Chemistry and Environment (2009), Fellow of Association of Biotechnology and Pharmacy (2010), Shiksha Rattan Puraskar (2011), Fellow of Association of Microbiologists of India (2012), and Fellow of Institute of Water and Environment (2014). Thapar Institute of Engineering and Technology, Patiala, Punjab, India

Part I
Current Status and Active Dynamics of AMR

Chapter 1
Antimicrobials in Livestock Production and Its Cross-Domain Dynamics

Bishwo Pokharel and Sandeep Raj Karna

1.1 Introduction

Antimicrobials are natural, seminatural (semisynthetic), or synthetic substances that kill or inhibit the growth of microorganisms (Page and Gautier 2012). Antimicrobials are effective against various classes of microorganisms such as virus, bacteria, protozoa, and fungi; however, the antimicrobials of common interest are those that are effective against bacteria. This is because bacteria can mutate to variants that are resistant to the antimicrobials used against them. These bacteria, after the mutation, can become a public health concern and jeopardize livestock and human health. The coexisting nature of livestock and human and their dependency on each other provide a greater host range to these resistant bacteria. Fear is growing among the scientific community that such resistance could result in another costly pandemic. Therefore, this chapter focuses on the use of antimicrobials in livestock production; dynamics of antimicrobial flow between humans, animals, and the environment; antimicrobial resistance; and potential alternatives to antimicrobials. This chapter also discusses One Health approach to antimicrobial resistance and provides information on antimicrobial stewardship to provide guidelines to stakeholders involved in the use of antimicrobials.

B. Pokharel (✉)
Department of Animal Biosciences, University of Guelph, Guelph, ON, Canada

S. R. Karna
School of Veterinary Science, Massey University, Palmerston North, New Zealand

N. Akhtar et al. (eds.), *Emerging Modalities in Mitigation of Antimicrobial Resistance*,
https://doi.org/10.1007/978-3-030-84126-3_1

1.2 Antimicrobial Use in Livestock

1.2.1 History

Tracing the date when antimicrobials started to have a dramatic impact on livestock farming can be daunting. Kirchhelle (2018), in their review, mentioned that antimicrobials started to play a bigger role in food production since the 1930s when synthetic sulfonamides came into existence. Sulfonamides were found to be effective against streptococcal infection providing a therapeutic effect on agricultural animals. Food products coming from livestock became even more important during World War II when there was a need to optimize livestock production to meet the increasing demand for food products. Researchers started to study alternatives in the form of antimicrobials to produce more meat at a cheaper cost; however, the practice started to come under scrutiny with the emergence of antibiotic residues in food products and antibiotic resistance (Table 1.1).

Table 1.1 Summary of major events on antimicrobial use in livestock

1920–1930	Discovery of penicillin (1928)
1930–1940	Development of the first agricultural antibiotic (1935): sulfochrysoidine (prontosil), first-time use of sulfonamides in animals in Britain (1938)
1940–1950	A rapid surge in the production of antibiotics during world war II, first-time use of penicillin/gramicidin against mastitis, antibiotics to control fish infection (1947), sulfaquinoxaline in poultry feed against coccidiosis (1948)
1950–1960	For the first time, concerns started to emerge on the use of antimicrobials in food animals, antimicrobial use boomed in several European countries, and nearly all piglets had access to food with tetracycline in the late 1950s
1960–1970	Agricultural antibiotics became widespread in Japan; the Food and Drug Administration (FDA) started monitoring programs against antibiotic residue in milk and meat; *animal machines*—a revolutionary book published backlashing antimicrobials (1964); nearly 80% of animal feed in Germany had some form of antimicrobials
1970–1980	Antimicrobial use in livestock boomed in the United States, South Africa, and several other countries
1980–1990	Sweden banned antimicrobial use as a growth promoter (1986) and prophylactic medications (1988)
1990–2000	Denmark banned the prophylactic use of antimicrobials (1994), initiation of European lobby to ban antimicrobial growth promoters, ban of several antimicrobials in EU; the World Health Organization (WHO) concluded that the use of antimicrobials as growth promoter should be stopped
2000–2010	Ban of all antibiotic growth promoters by the EU (2006); the WHO, Food and Agriculture Organization (FAO), and Office International des Epizooties (OIE) held a workshop that coined the term "critically important antimicrobials" to tackle antibiotic resistance
2010–2020	Substantial publications on antimicrobial use in livestock and antimicrobial resistance, accelerated discussion on one health approach and antimicrobial stewardship

Sources: Cogliani et al. (2011) and Kirchhele (2016, 2018)

1.2.2 Numbers Behind Antimicrobial Use in Livestock

Globally, more than 27 billion chickens, 1.7 billion cattle and buffaloes, 850 million pigs, and 2.3 billion sheep and goats are farmed (FAOSTAT 2020). Also, other groups of livestock share a significant proportion among the total livestock population. This suggests that a significant portion of the global population relies on some forms of livestock farming directly or indirectly. For those directly dependent on livestock, poor health and productivity of their animals could be devastating with a serious negative impact on their economy for years. Thus, many of them knowingly or unknowingly use substances such as antimicrobials that enhance the productivity of their livestock and prevent/protect their livestock from diseases.

In recent years, the use of antimicrobials is growing at an unprecedented rate in food animals. This is expected to grow similarly for some time as demand for animal protein is growing rapidly (Tilman et al. 2011). An estimated 63 thousand tons of antimicrobials were used in 2010 in livestock, which doubled in 2013 (131,109 tons). This use is further expected to rise up to 67% by 2030 (Van Boeckel et al. 2015). Figures are even more alarming in Asia, where antimicrobial use in chicken and pigs are expected to rise by 129% and 124%, respectively, by 2030 (Van Boeckel et al. 2015). In India, industrial poultry production is expected to grow by 312% by 2030, further increasing the demand for antimicrobials. Developed countries such as Denmark, Sweden, and Norway have been cautious in using antimicrobials; however, the developing countries have not shown any signs of reducing the use of antimicrobials for agricultural purposes (for instance, 8 mg/PCU of antimicrobial use in Norway compared to 318 mg/PCU in China) (Van Boeckel et al. 2017).

In 2010, the top five countries that shared the largest proportion of global antimicrobial use in livestock were China, the United States, Brazil, India, and Germany (Van Boeckel et al. 2015). The more alarming data from BRICS (Brazil, Russia, India, China, and South Africa) suggest that the use of antimicrobials is expected to grow by more than 99% in those countries in the next 10 years, making them future hotspot of global antimicrobial use. The global rise in the use of antibiotics is attributed to the shift toward more intensified livestock farming where a large group of animals is kept in an enclosed environment, increasing antimicrobial pressure to maintain and improve health and productivity. Also, livestock farming has seen significant changes in the last few decades owing mainly to the genetic advancements. Genetic selection has been practiced heavily, and the focus is mostly laid on improving productivity, which has unintentionally given rise to undesirable side effects such as increased frequency of rare recessive alleles. As a result, immune incompetence is more common leading to increased occurrence of pathologies and compromised animal welfare (Rauw et al. 1998; Hocking 2014). Compromised immune system is also one of the reasons that has caused increase in prophylactic and therapeutic use of antimicrobials, possibly giving rise to increased antimicrobial resistance.

Most of the abovementioned data come from poultry, cattle, and pig. There is hardly any accurate data available on antimicrobial use from fish farming. However, fish farming may already be contributing to the major proportion of antimicrobial use globally. The data from South Asia and South America already suggest an extremely high rate of antimicrobial use up to 1400 mg/kg (Van Boeckel et al. 2015) in fish farming. Like any other livestock farming, aquaculture is also shifting toward more efficient and intensive farming with the potential to become a major shareholder of global antimicrobial use.

1.3 Why Are Antimicrobials Used in Livestock?

1.3.1 Antimicrobials as a Growth Promoter

The single most controversial use of antimicrobials in livestock is its use as a growth promoter, which dates to the 1950s in the United States, Australia, and some European countries (Dibner and Richards 2005). Studies have reported improved feed conversion and growth in cattle, pigs, poultry, and other animals (Gallo and Berg 1995; Cromwell 2002; Castanon 2007; Chattopadhyay 2014) with some of these studies reporting productivity improvement of up to 10% after the use of antimicrobial growth promoters. The interaction between gut, microbiota, and antimicrobials is thought to be the reason behind growth-promoting effects of antimicrobials, more specifically the reduction of microbial metabolites that cause growth reduction in animals (Visek 1978; Anderson et al. 1999). Antimicrobials reduce the population of opportunistic pathogens and subclinical infections, limiting competition for food and thereby improving growth (Visek 1978). Antimicrobials also increase nutrient availability and absorption by maintaining gut microflora compositions, thereby thinning the barrier in the small intestine, and assisting in the digestion of high-energy diets (Peng et al. 2014).

Although antimicrobials are being used as a growth promoter for decades, there is a lack of reliable recent data on the effect of antimicrobials as growth promoters. Most of the studies on antimicrobials as growth promoter were conducted in the decades of the 1980s and 2000s (Teillant 2015). With the readily available antimicrobials to be used for growth promotion, it is often ignored by farmers that similar results could be achieved by selecting high growing lines, good hygiene, nutrition, and health management. Focusing on these things rather than just relying on antimicrobials for growth-promoting effects could dramatically reduce the use of antimicrobials in livestock.

Additionally, it is important to understand the economic aspects of using antimicrobials as a growth promoter and the potential economic effect of banning antimicrobials as growth promoters. There is limited knowledge on these; however, studies from the countries such as Denmark and Sweden, where antimicrobials have already been banned as a growth promoter, suggest that there is minimal impact on economy (Graham et al. 2007; Sneeringer et al. 2015). The most likely cost after banning antimicrobials as a growth promoter will be to improve hygiene and management, which is significant, but with a long-term positive effect both on animals and humans. Developing countries are a major concern where the production is less controlled, and the impact of the ban is likely higher compared to that of developed countries. The ban could become counterproductive if not handled properly as it could lead to more therapeutic use of antimicrobials to keep animals healthier and more productive. Additionally, to meet the demand of increasing world population, more animals need to be raised if growth-promoting antimicrobials are prohibited, which may subsequently lead to negative impacts on environment and other areas (Hao et al. 2014). Therefore, this is an extraordinarily complex issue requiring intervention from each country to make a common alliance with common goal.

Compared to very few positive effects (such as improved growth and improved feed efficiency) of antimicrobials used as growth promoters, there are numerous negative effects (Edqvist and Pedersen 2001; Hao et al. 2014). They are summarized below:

- Increases the pool of antimicrobial-resistant genes.
- Camouflages bad feed, subsequently discouraging improvement in feed development and its alternatives.
- Helps to hide the subclinical diseases and associated stress.
- Promotes intensive farming that is less animal-friendly.
- Disrupts disease treatment by increasing antimicrobial resistance.
- Provides the best possible environment to bacteria that are mutating to become antimicrobial-resistant.
- Indirectly impacts human health due to the transfer of antimicrobial resistance.

Based on the above, it is of utmost importance to identify alternatives of growth-promoting antimicrobials and implement those alternatives as soon as possible. Some of the alternatives to antimicrobials are discussed later in this chapter.

1.3.2 Prophylactic Use of Antimicrobials in Livestock

Farmers do not have any other choices but to use antimicrobials when animals are sick. The use depends on the animal species, stage of production, and disease risk. Similarly, when only a few individuals are sick, farmers choose to use antimicrobials to prevent the spread of disease to other animals. Usually, such antimicrobials are administered at critical points during the livestock production cycle to prevent diseases.

When antimicrobials are used as a prophylactic agent against certain diseases, they are generally used for a short duration and administered via feed or water to a group of animals. For example, most feedlot cattle in the United States (~83%) are administered with at least one antimicrobial in feed and water to control different disease outbreaks such as diarrhea and pneumonia (Animal and Plant Health

Inspection Service 1999). Similarly, broilers are usually administered with bacitracin and sulfonamides via feed to prevent necrotic enteritis and coccidiosis, respectively (McEwen and Fedorka-Cray 2002). In pigs, several antimicrobials such as tiamulin, sulfonamides, tetracyclines, and ceftiofur are used to prevent pneumonia. Additionally, most pigs receive antimicrobials during weaning to prevent them from infectious disease as weaning is one of the most stressful periods in a pig's life (McEwen and Fedorka-Cray 2002).

1.3.3 Therapeutic Use of Antimicrobials in Livestock

Antimicrobial use as a therapeutic agent is a common practice throughout the world and is the least controversial among the three uses of antimicrobials. Usually, antimicrobials are administered to a targeted individual(s) via feed and water or through direct injection. During disease outbreaks, especially in large pig and poultry farms, antimicrobials are administered through the water as a disease can depress feed intake in animals and it is usually believed that animals continue to drink water despite reducing the feed intake during sickness.

Gentamicin, apramycin, and neomycin are used to treat bacterial diarrhea in pigs caused by *E. coli* and *C. perfringens* (McEwen and Fedorka-Cray 2002). Similarly, nearly all weaned piglets have access to some form of antimicrobials to control disease outbreaks because of stress during weaning (Dewey et al. 1999). Fluoroquinolones are used to treat *E. coli* infections in poultry, and it is a common practice to use ionophores and sulfonamides to control coccidiosis. Hatchery use of antimicrobials is also common to control omphalitis in day-old chicks (Ouckema and Phillipe 2009). In dairy cattle, antimicrobials such as penicillins, cephalosporins, and erythromycins are used to treat mastitis (Erskine 2000). Such drugs are a routine practice in cattle, which are usually administered to the entire herd during nonlactating periods (Erskine 2000).

1.4 Antimicrobial Resistance

Bacteria are referred to as resistant to antimicrobials when they become nonsusceptible to one or more antimicrobials. When they become resistant to three or more antimicrobials, they become multidrug-resistant bacteria and then called pan drug-resistant if they are immune to any antimicrobials (Magiorakos et al. 2011).

Many antimicrobials (especially antibiotics) that are used in livestock are also essential for human use. When antimicrobials are used in livestock to prevent disease, it suppresses and eliminates bacteria that are susceptible to the antimicrobials. However, such antimicrobials cannot eliminate those bacteria that are resistant to them. Bacteria have an extraordinary potential to be adaptive to the new environment including the environment with antimicrobials. Those bacterial that are tolerant to antibiotics can multiply within the host and likely become the dominant bacterial population. Such bacteria are also able to transfer the resistant genes to other bacteria. When humans consume food products coming from animals, such bacteria can enter human being and subsequently colonize in the intestine. Once these tolerant bacteria are widespread within the human population and the antibiotics stop working against those bacteria, the treatment strategies can fail and lead to the devastating outcome (Hall et al. 2011; Marshall and Levy 2011).

1.5 How Antimicrobials and Antimicrobial Resistance Flow Between Humans, Livestock, and the Environment?

The dynamics behind the movement of antimicrobials and antimicrobial resistance from food animals to humans and vice versa is a complex phenomenon. The emergence of antimicrobial resistance and dissemination across and within different species has been summarized below:

1.5.1 Agricultural Production Method

Housing is one of the major drivers increasing the rate of emergence and dissemination of antimicrobials and antimicrobial resistance. In modern housing systems, a large group of animals is confined within a building or closed space (e.g., battery housing, feedlot cattle, pig barns, etc.). Hundreds of animals share food, water, air, and bedding for a long period. Animals are exposed to their own and other wastes containing antimicrobials and resistant bacteria (Gormaz et al. 2014). Additionally, workers get exposed to a large group of animals and resistant pathogens, who further transmit these pathogens to communities through contaminated clothing, shoes, and surfaces (Fey et al. 2000; Rinsky et al. 2013). Humans are not only exposed to these pathogens in farms but can also get these pathogens from a slaughterhouse. In slaughterhouses, workers are in close contact with animal bodies and equipment used in slaughtering, handling, cutting, processing, and storage of carcasses (Madden 1994; Sammarco et al. 1997). Besides, cross-contamination of pathogens is also linked to the trucks and other vehicles, when such vehicles are not thoroughly cleaned and decontaminated after their use in transporting other animals and food products (Hennessy et al. 1996; Pell 1997). Overcrowding, lack of appropriate sanitary measures, and cross-contamination during handling, transport, and slaughterhouse operations can amplify the dissemination of resistant pathogens, further worsening the situation.

Housing (especially intensive) is the major stressors to animals compromising their immune function. A compromised immune system leads to increased shedding of different kinds of pathogens. Animals are exposed to a series of stressors throughout their life from housing, handling, transport, and lairage at a slaughterhouse. Studies have suggested that stress can result in an increased prevalence of infections (Hayes et al. 2004; Verbrugghe et al. 2012), leading to an increase in the demand for antimicrobials.

1.5.2 Livestock Waste

Livestock farming results in a large volume of waste products often bigger than the carrying capacity of the environment. Livestock waste may contain resistant pathogens and genes, feed wastes, and spilled antimicrobials (from the feed, water, and excreta). In many countries, these waste products are largely unregulated, which means they are not treated before going into solid and water. This can lead to the release of a large number of antimicrobials and antimicrobial-resistant pathogens to the environment. From the environment, other animals and humans can get exposed to it, which might create an uncontrolled and widespread transfer of resistant pathogens across different species.

1.5.3 Exposure to Other Animals and Insects

Often livestock buildings are intruded by rodents, birds, insects, and other animals, mostly due to poor biosecurity. Nazni et al. (2005) reported similar pathogens to that found in poultry houses in the flies found in the poultry barn. Rodents in poultry and swine barn have been found to carry antimicrobial-resistant pathogens and disseminate them to the environment (Backhans and Fellstrom 2012). There is a potential transfer of such pathogens and antimicrobials from domesticated animals to wild animals.

1.5.4 Movement of Animals and Food

There is extensive movement of live animals across the different parts of the world, for example, the movement of poultry breeding stock from Europe to Asia and within Europe and live sheep export from Australia to the Middle East. If the use of antimicrobials is permitted in exporting countries but not in the importing countries, there is a likelihood of antimicrobial-resistant pathogen transfer from the exporting country to the importing one.

In addition to live animals, there is an extensive export and import of food products throughout the world. Major producers of pork, poultry, fish, and beef extensively export these products to other countries (Silbergeld and Dailey 2017). This extensive trading makes it impossible for countries to assess the flow of pathogens through food products between the countries. Food can be contaminated with resistant bacteria through several routes, i.e., from bacteria present in animals, from bacteria added during culture, and from bacterial cross-contamination during the processing of foods (Verraes et al. 2013).

Especially in developing countries, antimicrobials are misused due to poor regulations in the supply chain. Moreover, a large population in such countries is in close contact with the animals. It hence burgeons the chances of transmissions of resistant microorganisms from animals to humans from handling of the animals.

1.5.5 Environment

The environment is not only a significant reservoir of many pathogens but also facilitates their dissemination by forming a cycle of pathogen contamination. In addition to getting pathogens from livestock wastes, antimicrobials used as crop pesticides also lead to soil and water contamination (Bhandari et al. 2019), subsequently leading to the emergence of resistant bacteria. Moreover, globalization and urbanization have led to environmental pollution, further compromising livestock and human health and increasing the demand for antimicrobials (Balakrishnan et al. 2019). Antimicrobials used in agriculture, human, and veterinary medicine are partially metabolized by animals and humans and end up being released into the environment through sewage systems. Antimicrobials used in aquaculture are directly added into the water, leading to a high antimicrobial concentration in water and the sediments. Studies in various countries have detected a low concentration of antimicrobials in different environmental compartments such as municipal wastewater, sewage plant effluent, and even groundwater (Kümmerer 2004; Kolpin et al. 2002; Sacher et al. 2001). Most of the commonly used antimicrobials are not biodegradable and persist in the aquatic ecosystem (Kümmerer 2003). These antibiotics may have direct effects upon the resident microbial community of sediments in the ecosystem (Nygaard et al. 1992). The presence of active antibiotic compounds in the environment exerts a selective pressure which might create the occurrence of antibiotic-resistant phenotypes that may spread in the environment through the microbial species (Thanner et al. 2016). In addition to the release of antibiotics leading to the development of resistant bacteria, bacteria themselves are also excreted by humans and animals which end up in the ecosystems.

Humans and animals are a part of a complex environmental phenomenon. Several human activities such as traveling, contact with livestock, and contact with wild animals lead to the dissemination of antimicrobials and pathogens that are resistant to antimicrobials. The environment in which both human and animal live completes the cycle of this dissemination. Therefore, we must reduce the release of antimicrobials to the environment to disrupt this cycle and to slow down the zoonotic transmission of antimicrobial resistance (Fig. 1.1).

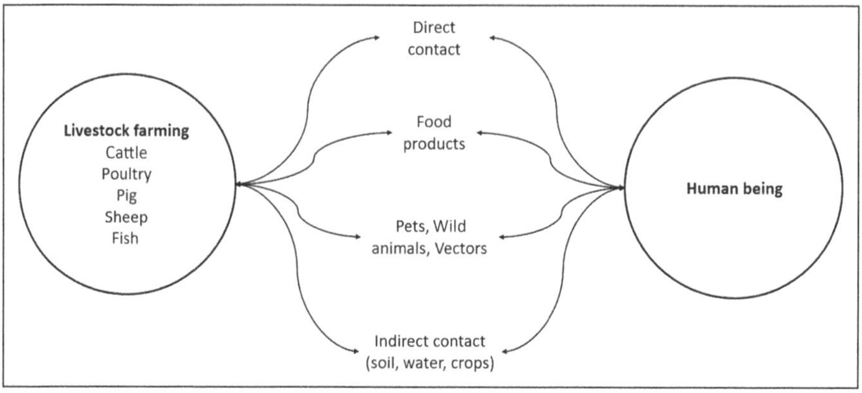

Fig. 1.1 Potential route of exchange of antimicrobials and antimicrobial-resistant pathogens between animals, humans, and the environment

1.6 Zoonosis of Antimicrobial Resistance

It is estimated that more than 50% of pathogens that can infect human beings can also infect other animals (Taylor et al. 2001). Therefore, there is a huge potential for transfer of antimicrobial resistance from animals to human beings and vice versa. The earliest documented evidence of animal-human transmission of antimicrobial resistance was in the 1970s during the *Salmonella* epidemic in a human hospital that was traced back to the calves infected by *Salmonella* (Labro and Bryskier 2014). Since then, documentation of antimicrobial resistance in livestock and humans and the spread of resistant bacteria between animals and humans is large and readily available (Woolhouse and Ward 2013).

In the United States alone, more than 2.8 million cases of illnesses are due to some form of antimicrobial-resistant infections leading to more than 35,000 deaths per year (Centers for disease control and prevention (CDC) 2019). Antimicrobial resistance contributes to 700,000 deaths annually with estimated 214,000 neonatal deaths attributed to resistant sepsis infections (Pokharel et al. 2020). The data on livestock deaths due to antimicrobial resistance is scarce; however, studies have reported antimicrobial resistance to antibiotics in *E. coli*, *Salmonella* spp., *Campylobacter* spp., and *Enterococcus* spp. that are responsible for most infections in livestock (McEwen and Fedorka-Cray 2002; WHO 2003; Aarestrup et al. 2008).

Increased resistance to antimicrobials usually coincides with the use of such antimicrobials in livestock that are used for food production. For example, in poultry, fluoroquinolones are heavily used to treat respiratory diseases, and it is no surprise that increased resistance to fluoroquinolones has been heavily documented in humans, mostly linked to poultry consumption (Endtz et al. 1991; Nelson et al. 2007). In a more recent study, approximately 90% of isolates from poultry showed some form of resistance to antimicrobials such as sulfonamides, tetracyclines, fluoroquinolones, and third-generation cephalosporins (Kaesbohrer et al. 2012).

Similarly, methicillin-resistant *Staphylococcus aureus* (MRSA) is a growing concern among people that are in contact with animals, both livestock and pets (Labro and Bryskier 2014). *Enterococcus* is another commensal bacteria found in both human and animal guts, which are intrinsically resistant to cephalosporins and can also acquire resistance to quinolones, macrolides, and tetracyclines (Murray 1990).

As a global public health threat affecting both humans and animals, antimicrobial resistance has warranted several national and international communities to work together on implementing policies to preserve the efficacy of medically important antimicrobials. The concept of critically important antimicrobials was developed in a second workshop held between the WHO, FAO, and OIE in 2004. The WHO classified antimicrobials into five groups based on their importance to human medicine and released a guideline in 2018, which recommended that the highest priority critically important antimicrobials (HPCIA) should not be used in food-producing animals (WHO 2017). The HPCIA includes five classes of antimicrobials: quinolones; third-, fourth-, and fifth-generation cephalosporins; macrolides and ketolides; glycopeptides; and polymyxins. Study in some European countries has shown that it is possible to maintain health and productivity with no use of cephalosporins and fluoroquinolones in livestock; however, total exclusion of macrolides is difficult as they are critically important in managing respiratory disease in pigs, poultry, cattle, and other animals. This makes it more complex as respiratory diseases in livestock are associated with significant economic losses in most countries (Lhermie et al. 2020).

1.7 One Health and Antimicrobial Resistance

Antimicrobial resistance is a multifaceted global issue (Pokharel et al. 2020). Both human and veterinary medicine are the major contributor to the emergence of antimicrobial resistance. The issue is not going to affect one single species in the world; it could well become the most widespread pandemic in the future, affecting the largest number of species throughout the world. Therefore, there is a need for a multidisciplinary approach involving humans, animals, and the environment, which is referred to as One Health.

The WHO defines One Health as a "concept and approach to designing and implementing programs, policies, legislation and research in which multiple sectors communicate and work together to achieve better public health outcomes" (WHO 2017). The origin of One Health is centuries old and recognizes both human and animal health. More recently, this concept recognizes environmental health too. In summary, there are three domains in this approach: human health, animal health, and environmental health.

Among the three domains, human health takes a major emphasis. Antimicrobial resistance genes have been reported to be highly prevalent in some common pathogens in humans such as *E. coli*, *K. pneumonia*, and *S. aureus* (Robinson et al. 2016). Livestock has played a major role in the transmission of antimicrobial resistance,

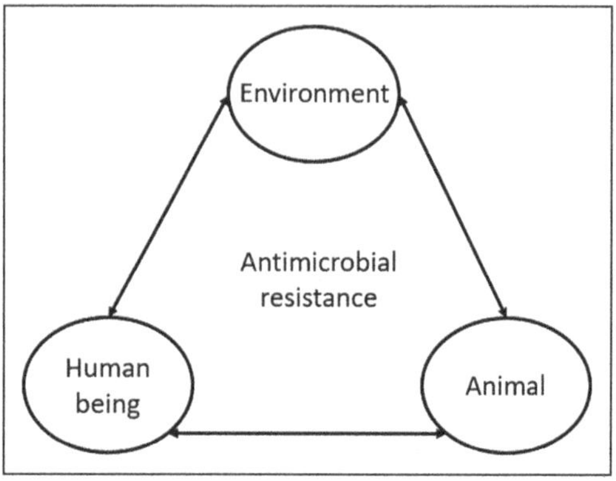

Fig. 1.2 Relationship between human being, animal, and environment in a One Health concept

which was already discussed previously in this chapter. Livestock and associated products will continue to play a significant role in the dissemination of antimicrobial residue and antimicrobial resistance in the future. At present, there is a lack of knowledge transfer between human and veterinary medicine, causing inconsistencies in the use of antimicrobials in humans and animals. The collaborative approach between human and veterinary medicine can mitigate this and provide sustainable solutions (Fig. 1.2).

The third domain, environment, is getting considerable recognition in recent years. As discussed earlier, the environment is a significant transmission reservoir for most of the pathogens in humans and animals, without which the disease cycle cannot be mostly completed (Pornsukarom and Thakur 2017). Soil and water contamination of antimicrobials can lead to the emergence of antimicrobial-resistant bacteria that are already in soil and water (Grenni et al. 2018). Similarly, other aspects of the environment such as air pollution have led to increased infections in humans and animals, subsequently increasing the demand for antimicrobials that further aids in the emergence of antimicrobial resistance.

1.8 Third-Generation Cephalosporins: A One Health Example

Third-generation cephalosporins are widely used in humans and animals. These are classified as critically important antimicrobials by the WHO (2017). Third-generation cephalosporins have a broad-spectrum activity, and some of their uses include controlling bovine respiratory disease in cow, preventing *E. coli* infections in chicks, and treating pneumonia, arthritis, and other conditions in humans (Greko et al. 2009). Resistance to third-generation cephalosporins has been reported in *E. coli* and *K. pneumonia* (Park 2014; de Kraker et al. 2011). Several studies reported that voluntary withdrawal of third-generation cephalosporin use in chicks was followed by the drop in its resistance in *E. coli* (Hiki et al. 2015; Dutil et al. 2010).

Countries such as Denmark, Australia, and Canada have placed a voluntary ban on the use of these drugs recognizing the resulting human health risk with their use in animals (Collignon and McEwen 2019).

1.8.1 Antimicrobial Stewardship

Antimicrobial resistance is a one-world issue. Therefore, it is in everyone's interest to preserve the efficacy of antimicrobials by properly using them, following the guidelines, monitoring their use and resistance, and implementing good stewardship programs. Antimicrobial stewardship is a set of actions that promote the responsible use of antimicrobials and can be summarized with 5Rs: responsibility, review, reduce, refine, and replace (Page et al. 2014). The 5R approach guides livestock farmers, veterinarians, physicians, and other relevant stakeholders who are involved in antimicrobial use to adopt best practice and management of antimicrobial use. With regard to good stewardship, prevention of disease in livestock is more important than the treatment, which means vaccination and good husbandry are critical in putting antimicrobial use in check (Table 1.2).

1.9 Alternatives to Antimicrobial Use in Livestock

As suggested by good antimicrobial stewardship, we can identify and implement the practices that can either replace or reduce antimicrobial use and also reduce the likelihood of infections in animals. Such practices can include early intervention long time before the infections such as vaccinations. There are several vaccines available that can help prevent several infections in livestock (e.g., cattle, *E. coli*, *Salmonella* vaccine; pigs, *E. coli* vaccine, vaccine against bacterial pneumonia; and poultry, vaccine against pasteurellosis, *Salmonella* vaccine). Another important

Table 1.2 The 5R approach to tackle misuse of antimicrobials

Responsibility	Everyone using antimicrobials need to understand that antimicrobial use can be a risk to both human and livestock. Therefore, responsible use of antimicrobials should be practiced to reduce public health risk
Review	Everyone using antimicrobials should review the use regularly and make strategies to reduce the use of antimicrobials
Reduce	Whenever possible, there should be an attempt to look for the ways to reduce antimicrobial use
Refine	Right drugs at the right time and right dose should be used for the right amount of time
Replace	Whenever possible, strategies should be implemented to consider replacing antimicrobials with non-antimicrobial products such as probiotics, herbal medicines, vaccines, and immune modulators

strategy that can reduce the antimicrobial load includes good husbandry practices. Good sanitation in and around the farm can reduce bacterial load around the farm, good air and water quality can prevent horizontal transmission of diseases, and good feed can help protect animals against many conditions such as salmonellosis and mycotoxins. Good air quality and appropriate ventilation in the animal farm can help control high gaseous levels (e.g., ammonia level in poultry houses) subsequently reducing several bacterial infections.

Good husbandry practices also involve farmers following appropriate biosecurity measures. Controlling what goes into the farm can help prevent a lot of diseases in animals. For example, the use of appropriate clothing and foot baths, control of vectors, control of birds and rodents, and use of *Salmonella* free food can be easily practiced on the farm. Another less common practice involves the use of beneficial bacteria in the form of probiotics, which can act as an antibiotic growth promoter in animals (Reid and Friendship 2000); however, more studies are yet to be conducted to understand more about probiotics and their role in the farm as an alternative to antimicrobials. In addition to probiotics, prebiotics and organic acids can also provide health benefits to animals by stimulating growth, metabolism, and composition of beneficial bacteria in the gastrointestinal tract and eliminating the harmful one (Solis-Cruz et al. 2019).

Genetic selection is another avenue that could provide potential solution to the widespread use of antimicrobials. Herds that are resistant to certain diseases can be selected that could possibly eliminate the use of antimicrobials for that disease. Studies are scarce on the use of genetic selection to achieve pathogen-resistant animals but can be food for thought for animal scientists to tackle the issue. More recently, bacteriophages have emerged as a potential alternative to antimicrobials, which works by specifically attacking bacteria; however, lack of regulatory guidance and clinical trials has hindered the possibility of using bacteriophages in large scale (Romero-Calle et al. 2019). Different alternatives to antimicrobials and how they work have been summarized in Table 1.3.

Table 1.3 Summary of alternatives to antimicrobials in livestock

Alternatives to antimicrobials	Mechanism of action
Vaccines	Preparing immune system to recognize and combat pathogens
Good husbandry practices	Reducing microbial load in the farms and thereby lowering exposure to microbes
Prebiotics, probiotics, and organic acids	Promoting growth, selectively stimulating beneficial bacteria, and eliminating harmful ones
Genetic selection	Selecting animals that are resistant to certain diseases
Bacteriophage	Attacking and killing bacteria

1.10 Conclusion

Antimicrobials are the most important discoveries of human and animal health, and ironically, antimicrobial resistance is one of the greatest crises to public health. The use of antimicrobials in the livestock sector in different parts of the globe is indiscriminate and unregulated. There is a lack of data about the scale of their use, and more studies are required to understand the fate of these antimicrobials in the environment and their consequences on human health. Livestock farming should urgently be recognized as a major contributor to the development of antimicrobial resistance, and countries need to develop legislation regulating prophylactic use of antimicrobials in farming. The evidence presented across countries indicates that it is possible to reduce antimicrobial use and gain highly intensive and productive production systems (Cogliani et al. 2011). A coordinated effort between governments, industry, and scientists is required for effective action on antimicrobial resistance. An immediate step to tackle the problem would be to develop strategies for improved antimicrobial stewardship involving both human medicine and livestock industry and develop alternative approaches to combat microbial disease and improve livestock production.

Key Notes

- Antimicrobial is a complex subject.
- Antimicrobial use in livestock is rising at an alarming rate, driven by increasing demand for animal protein globally.
- Data on the antimicrobial use is not sufficient, which warrants more study on the topic.
- Antibiotic resistance is a public health crisis.
- Livestock farming is a major contributor to the development of antimicrobial resistance.
- Keeping animals healthy is important in reducing the use of antimicrobials.
- Antimicrobial resistance is a One Health issue. More than that, it is a one-world issue.
- 5R approach can help become a good antimicrobial steward and help tackle antimicrobial resistance.
- Strategies that can help reduce the use of antimicrobials include good farm management, vaccination, biosecurity, probiotics, and genetic selection.

References

Aarestrup FM, Wegener HC, Collignon P (2008) Resistance in bacteria of the food chain: epidemiology and control strategies. Expert Rev Anti Infect Ther 6:733–750

Anderson DB, McCracken VJ, Aminovi RI et al (1999) Gut microbiology and growth-promoting antibiotics in swine. Pig News Inform 20:115–122

Animal and Plant Health Inspection Service (1999) Feedlot '99. Part 3: health management and biosecurity in US feedlots. US Department of Agriculture, Washington, DC

Backhans A, Fellstrom C (2012) Rodents on pig and chicken farms—a potential threat to human and animal health. Infect Ecol Epidemiol 2:10

Balakrishnan K, Dey S, Gupta T et al (2019) The impact of air pollution on deaths, disease burden, and life expectancy across the states of India: the global burden of disease study. Lancet Planet Health 3:e26–e39

Bhandari G, Zomer P, Atreya K et al (2019) Pesticide residues in Nepalese vegetables and potential health risks. Environ Res 172:511–521

Castanon JIR (2007) History of the use of antibiotic as growth promoters in European poultry feeds. Poult Sci 86:2466–2471

Centers for disease control and prevention (CDC) (2019) 2019 AR Threats Report. https://www.cdc.gov/drugresistance/biggest-threats.html. Accessed 25 Jan 2021

Chattopadhyay MK (2014) Use of antibiotics as feed additives: a burning question. Front Microbiol 5:334

Cogliani C, Goossens H, Greko C (2011) Restricting antimicrobial use in food animals: lessons from Europe. Microbe 6:274–279

Collignon PJ, McEwen SA (2019) One health—its importance in helping to better control antimicrobial resistance. Trop Med Infect Dis 4:22

Cromwell GL (2002) Why and how antibiotics are used in swine production. Anim Biotechnol 13:7–27

de Kraker MEA, Wolkewitz M, Davey PG et al (2011) Burden of antimicrobial resistance in European hospitals: excess mortality and length of hospital stay associated with bloodstream infections due to Escherichia coli resistant to third-generation cephalosporins. J Antimicrob Chemother 66:398–407

Dewey CE, Cox BD, Straw BE et al (1999) Use of antimicrobials in swine feeds in the United States. J Swine Health Prod 7:19–25

Dibner JJ, Richards JD (2005) Antibiotic growth promoters in agriculture: history and mode of action. Poult Sci 84:634–643

Dutil L, Irwin R, Finley R et al (2010) Ceftiofur resistance in salmonella enterica serovar Heidelberg from chicken meat and humans, Canada. Emerg Infect Dis 16:48

Edqvist L, Pedersen KB (2001) Antimicrobials as growth promoters: resistance to common sense. In: Harremoës P, Gee D, MacGarvin M (eds) Late lessons from early warnings: the precautionary principle 1896–2000. Environment issue report, no 22. European Environment Agency, Copenhagen, pp 93–100

Endtz HP, Ruijs GJ, van Klingeren B et al (1991) Quinolone resistance in campylobacter isolated from man and poultry following the introduction of fluoroquinolones in veterinary medicine. J Antimicrob Chemother 27:199–208

Erskine RJ (2000) Antimicrobial drug use in bovine mastitis. In: Prescott JF, Baggot JD, Walker RD (eds) Antimicrobial therapy in veterinary medicine, 3rd edn. Iowa State University Press, Ames, pp 712–734

FAOSTAT (2020). http://www.fao.org/faostat/en/#data/QA. Accessed 25 Jan 2021

Fey PD, Safranek TJ, Rupp ME et al (2000) Ceftriaxone-resistant salmonella infection acquired by a child from cattle. N Engl J Med 342:1242–1249

Gallo GF, Berg JL (1995) Efficacy of a feed-additive antibacterial combination for improving feedlot cattle performance and health. Can Vet J 36:223

Gormaz JG, Fry JP, Erazo M et al (2014) Public health perspectives on aquaculture. Curr Environ Health Rep 1:227–238

Graham JP, Boland JJ, Silbergeld E (2007) Growth promoting antibiotics in food animal production: an economic analysis. Public Health Rep 122:79–87

Greko C, Badiola JI, Catry B et al (2009) Reflection paper on the use of third and fourth generation cephalosporins in food producing animals in the European Union: development of resistance and impact on human and animal health. J Vet Pharmacol Ther 32:515–533

Grenni P, Ancona V, Caracciolo AB (2018) Ecological effects of antibiotics on natural ecosystems: a review. Microchem J 136:25–39

Hall MAL, Dierikx CM, Stuart JC et al (2011) Dutch patients, retail chicken meat and poultry share the same ESBL genes, plasmids and strains. Clin Microbiol Infect 17:873–880
Hao H, Cheng G, Iqbal Z et al (2014) Benefits and risks of antimicrobial use in food-producing animals. Front Microbiol 5:288
Hayes JR, English LL, Carr LE et al (2004) Multiple-antibiotic resistance of enterococcus spp. isolated from commercial poultry production environments. Appl Environ Microbiol 70:6005–6011
Hennessy TW, Hedberg CW, Slutsker L et al (1996) A national outbreak of salmonella enteritidis infections from ice cream. N Engl J Med 334:1281–1286
Hiki M, Kawanishi M, Abo H et al (2015) Decreased resistance to broad-spectrum cephalosporin in Escherichia coli from healthy broilers at farms in Japan after voluntary withdrawal of ceftiofur. Foodborne Pathog Dis 12:639–643
Hocking PM (2014) Unexpected consequences of genetic selection in broilers and turkeys: problems and solutions. Br Poultry Sci 55:1–12
Kaesbohrer A, Schroeter A, Tenhagen BA et al (2012) Emerging antimicrobial resistance in commensal Escherichia coli with public health relevance. Zoonoses Publ Health 59(Suppl 2):158–165
Kirchhelle C (2016) Toxic confusion: the dilemma of antibiotic regulation in west German food production (1951–1990). Endeavour 40:114–127
Kirchhelle C (2018) Pharming animals: a global history of antibiotics in food production (1935–2017). Palgrave Commun 4:1–13
Kolpin DW, Furlong ET, Meyer MT et al (2002) Pharmaceuticals, hormones, and other organic wastewater contaminants in US streams, 1999–2000: a national reconnaissance. Environ Sci Technol 36:1202–1211
Kümmerer K (2003) Significance of antibiotics in the environment. J Antimicrobiol Chem 52:5–7
Kümmerer K (2004) Resistance in the environment. J Antimicrobiol Chem 54:311–320
Labro M, Bryskier J (2014) Antibacterial resistance: and emerging "zoonosis"? Expert Rev Anti Infect Ther 12:1441–1461
Lhermie G, La Ragione RM, Weese JS et al (2020) Indications for the use of highest priority critically important antimicrobials in the veterinary sector. J Antimicrobiol Chem 75:1671–1680
Madden RH (1994) Microbial hazards in animal products. Proc Nutr Soc 53:309–316
Magiorakos AP, Srinivasan A, Carey RB et al (2011) Multidrug resistant, extensively drug-resistant and pandrug-resistant bacteria: an international expert proposal for interim standard definitions for acquired resistance. Clin Microbiol Infect 18:268–281
Marshall BM, Levy SB (2011) Food animals and antimicrobials: impacts on human health. Clin Microbiol Rev 24:718–733
McEwen SA, Fedorka-Cray PJ (2002) Antimicrobial use and resistance in animals. Clin Infect Dis 34(Suppl 3):S93–S106
Murray BE (1990) The life and times of the enterococcus. Clin Microbiol Rev 3:46–65
Nazni WA, Seleena B, Lee HL et al (2005) Bacterial fauna from the house fly, Musca domestica (L.). Trop Biomed 22:225–231
Nelson JM, Chiller TM, Powers JH et al (2007) Fluoroquinolone-resistant campylobacter species and the withdrawal of fluoroquinolones from use in poultry: a public health success story. Clin Infect Dis 44:977–980
Nygaard K, Lunestad BT, Hektoen H et al (1992) Resistance to oxytetracycline, oxolinic acid and furazolidone in bacteria from marine sediments. Aquaculture 104:31–36
Ouckema R, Phillipe C (2009) Salmonella isolations: historical OHSFP trends. In: Proceedings of salmonellosis, antimicrobial use and antimicrobial resistance symposium; May 6, Guelph, Ontario
Page SW, Gautier P (2012) Use of antimicrobial agents in livestock. Rev Sci Tech OIE 31:145–188
Page S, Prescott J, Weese S (2014) The 5Rs approach to antimicrobial stewardship. Vet Rec 175:207–209

Park SH (2014) Third-generation cephalosporin resistance in gram-negative bacteria in the community: a growing public health concern. Korean J Intern Med 29:27
Pell AN (1997) Manure and microbes: public and animal health problem? J Dairy Sci 80:2673–2681
Peng M, Salaheen S, Biswas D (2014) Animal health: global antibiotic issues. Encycl Agric Food Syst 1:346–357
Pokharel S, Shrestha P, Adhikari B (2020) Antimicrobial use in food animals and human health: time to implement 'one health' approach. Antimicrob Resist Infect Control 9:181
Pornsukarom S, Thakur S (2017) Horizontal dissemination of antimicrobial resistance determinants in multiple salmonella serotypes following isolation from the commercial swine operation environment after manure application. Appl Environ Microbiol 83:1–14
Rauw WM, Kanis E, Noordhunizen-Stassen EN et al (1998) Undesirable side effects of selection for high production efficiency in farm animals: a review. Livest Prod Sci 56:15–33
Reid G, Friendship R (2000) Alternatives to antibiotic use: microbiological perspective. In: Pork industry conference on addressing issues of antibiotic use in livestock production, 16–17 October 2000, Urbana, Illinois
Rinsky JL, Nadimpalli M, Wing S et al (2013) Livestock-associated methicillin and multidrug resistant Staphylococcus aureus is present among industrial, not antibiotic-free livestock operation workers in North Carolina. PLoS One 8:e67641
Robinson TP, Bu DP, Carrique-Mas J et al (2016) Antibiotic resistance is the quintessential one health issue. Trans R Soc Trop Med Hyg 110:377–380
Romero-Calle D, Benevides RG, Góes-Neto A et al (2019) Bacteriophages as alternatives to antibiotics in clinical care. Antibiotics 8:138
Sacher F, Lange FT, Brauch HJ et al (2001) Pharmaceuticals in groundwaters: analytical methods and results of a monitoring program in Baden-Württemberg, Germany. J Chromatogr A 938:199–210
Sammarco ML, Ripabelli G, Ruberto A et al (1997) Prevalence of salmonellae, listeriae, and Yersiniae in the slaughterhouse environment and on work surfaces, equipment, and workers. J Food Prot 60:367–371
Silbergeld EK, Dailey JL (2017) Biological plausibility of associations between antimicrobial use in food-producing animals and increased risks of human exposures to, and infections by, antimicrobial resistant zoonotic pathogens. In: WHO guidelines on use of medically important antimicrobials in food-producing animals. World Health Organization, Geneva
Sneeringer S, MacDonald J, Key N et al (2015) Economics of antibiotic use in U.S. livestock production. https://www.ers.usda.gov/webdocs/publications/45485/err-200.pdf?v=0. Assessed 24 Jan 2021
Solis-Cruz B, Harnandez-Patlan D, Hargis BM et al (2019) Use of prebiotics as an alternative to antibiotic growth promoters in the poultry industry. In: Franco-Robles E (ed) Prebiotics and probiotics: potential benefits in nutrition and health. IntechOpen, London, p 89053
Taylor LH, Latham SM, Woolhouse ME (2001) Risk factors for human disease emergence. Philos Trans R Soc Lond B Biol Sci 356:983–989
Teillant A (2015) Cost and benefits of antimicrobial use in livestock. In: AMR Control, pp 116–112
Thanner S, Drissner D, Walsh F (2016) Antimicrobial resistance in agriculture. MBio 7:e02227–e02215
Tilman D, Balzer C, Hill J et al (2011) Global food demand and the sustainable intensification of agriculture. Proc Natl Acad Sci 108:20260–20264
Van Boeckel TP, Brower C, Gilbert M et al (2015) Global trends in antimicrobial use in food animals. Proc Natl Acad Sci 112:5649–5654
Van Boeckel TP, Glennon EE, Chen D et al (2017) Reducing antimicrobial use in food animals. Science 357:1350–1352
Verbrugghe EM, Boyen F, Gaastra W et al (2012) The complex interplay between stress and bacterial infections in animals. Vet Microbiol 155:2–4

Verraes C, Van Boxstael S, Van Meervenne E et al (2013) Antimicrobial resistance in the food chain: a review. Int J Environ Res Public Health 10:2643–2669
Visek WJ (1978) The mode of growth promotion by antibiotics. J Anim Sci 46:1447–1469
WHO (2003) Impacts of antimicrobial growth promoter termination in Denmark: the WHO international review panel's evaluation of the termination of the use of antimicrobial growth promoters in Denmark: Foulum, Denmark 6–9 November 2002
WHO (2017) One health. https://www.who.int/news-room/q-a-detail/one-health Assessed 30 Jan 2021
Woolhouse ME, Ward MJ (2013) Sources of antimicrobial resistance. Science 341:1460–1461

Chapter 2
Antibiotics and Resistance in Environment

Rahil Changotra, Atul Chalotra, and Himadri Rajput

2.1 Introduction

In 1928, the penicillin discovery by Alexander Fleming is often considered to be one of the most vital discoveries in the field of medical technology (Levy and Bergman 2003). After that, in between the 1940s and 1980s, the worldwide growth of pharmaceutical industries led to the production of over 160 antibiotics and semi-synthetic pharmaceutical compounds which soon became the foundation for the treatment of several infectious diseases (Davies 2006). Despite great success in the reduction of morbidity and mortality due to infectious diseases, these pharmaceuticals have been regarded as emerging pollutants and invoked increasing concern worldwide in the past few decades (Rivera-Utrilla et al. 2013). Even at subtherapeutic concentrations in the environment, they have adverse effects on human health and ecosystems. However, the risk associated with the exposure of pharmaceuticals is essentially vital in areas that practice indirect water reuse, where pharmaceutical effluent is discharged to rivers and streams that are in turn used as source of drinking water for the nearby human communities. Pharmaceutical manufacturing activity has been classified into "red category" owing to its production of huge volume, complexity, and hazardous waste. The process involved in the pharmaceutical manufacturing industry includes the manufacture, extraction, processing, purification, and packaging of solid and liquid materials to be utilized for the ailments of humans and animals. In the manufacturing unit, the synthesis of bulk drugs involves several reactants with or without fermented natural antibiotics depending on the type of bulk drugs being synthesized. Moreover, the increasing rate of unused medicines

R. Changotra · H. Rajput (✉)
School of Energy and Environment, Thapar Institute of Engineering and Technology, Patiala, Punjab, India

A. Chalotra
Bhojia Dental College and Hospital, Baddi, Himachal Pradesh, India

N. Akhtar et al. (eds.), *Emerging Modalities in Mitigation of Antimicrobial Resistance*,
https://doi.org/10.1007/978-3-030-84126-3_2

from hospitals and households and its disposal malpractices have also been considered as a source of pharmaceutical waste (Tong et al. 2011; Vellinga et al. 2014). The effective disposal and management of pharmaceutical waste pose a serious challenge for the scientific research community, particularly due to the occurrence of pharmaceuticals and their metabolites in the outlets of wastewater treatment plants, which may find their ways to enter the aquatic environment. Knowing the existence of pharmaceutical compounds in the aquatic environment, there are growing concerns to explore their fate, transport, effects, and toxicity as well as their resistance in the environment.

2.2 Treatment Systems and Antibiotics in the Environment

The pharmaceuticals comprise of nonsteroidal anti-inflammatory drugs, antibiotics, analgesics, antiepileptics, hormones, steroids, toxic substances, volatile organic compounds (VOCs) and surfactants, etc. Among these pharmaceuticals, antibiotics are the major class of pharmaceuticals with extensive demand and use with the growing attention. These antibiotics despite being intended to treat certain ailments in animals and humans have exhibited adverse impacts on the ecosystem due to their existence in wastewater, surface water, groundwater, and drinking water. The previous years have witnessed significant and wide research on the detection of more than 160 different antibiotics and their metabolites in the various environment samples ranging from few ng L^{-1} to several μg L^{-1} (Kümmerer 2010; Kanakaraju et al. 2017), and as a result, these pharmaceuticals are now categorized as an emerging pollutant in various water bodies. Antibiotics enter into the environment from the pharmaceutical production industries and direct and improper disposal of unused/expired medicines by humans and through human urine or feces and hospital wastes and household sewage which enter into the influent of wastewater treatment plants (WWTPs) in an unmetabolized form or as moderately active metabolites (Mompelat et al. 2009; Changotra et al. 2019). It has been reported that about 10 to 90% of the administrated dose of antibiotic drug is excreted from the human body as a parent compound, whereas the rest is excreted as metabolites and/or in transformed forms (Kümmerer 2009; Changotra et al. 2017). The recent research showed that numerous pharmaceutical manufacturing units were found to be sources of surpassing antibiotic concentrations into the aquatic environment than those caused by the usage of drugs (Kessler 2010). The excreted antibiotics reach the WWTPs and are finally discharged as raw or treated wastewater into the rivers, surface water, groundwater, oceans, and soil (Fig. 2.1).

A major pathway of antibiotic mobilization in the environment is considered to be from WWTPs as they accumulate discharges from pharmaceutical industries, hospitals, households, veterinaries, and pharmacies (Tarpani and Azapagic 2018; Changotra et al. 2018). This is due to the inefficient removal of antibiotics in the conventionally used treatment technologies which are intended to treat biodegradable, nonpolar, and large compounds, as opposed to polar tendency, small

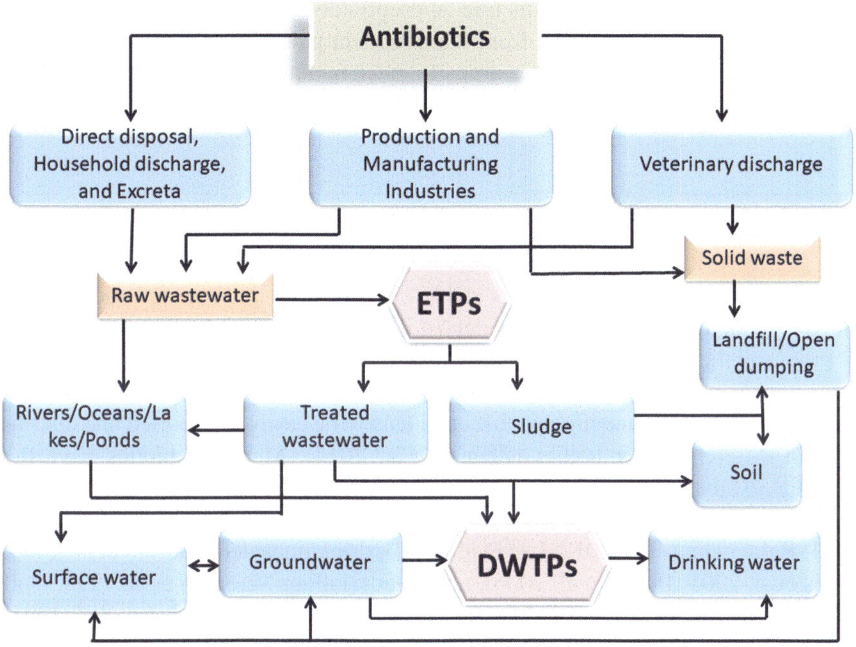

Fig. 2.1 Fate and transport of the antibiotics in the environment. *ETPs* effluent treatment plants, *DWTPs*, drinking water treatment plants

compounds, and nonbiodegradable nature of antibiotic compounds (Ratola et al. 2012; Changotra et al. 2020). It is notable that antibiotics enter in the soil and surface waters and eventually in the ground and drinking water after their excretion (as active metabolites or in unmetabolized form) in urine or feces from humans and animals (Klavarioti et al. 2009). Besides excretion, antibiotics may find their pathways into water bodies due to the disposal of medicines used in industry, agriculture, households, and hospitals. Veterinary pharmaceuticals may directly contaminate soil through manure and surface water and groundwaters by surface runoff (Khetan and Collins 2007). Moreover, numerous transformation products and metabolites are also likely to form in WWTPs as a consequence of numerous chemical reactions in the course of treatment (Zuccato et al. 2000) that could facilitate synergetic effects and become more bioactive when compared to parent compounds. In order to treat the wastewater containing antibiotics, most of the pharmaceutical industry utilized general approaches like recovery of antibiotic drugs which are expected to be present in wash waters and solvents, physicochemical treatments like sedimentation or floatation, conventional treatment technologies like aerobic/anaerobic biological treatment, inactivation of active compounds using UV oxidation, sterilization or inactivation of bioactive substances, and a new hybrid technology specific to the pharmaceutical industry (Gadipelly et al. 2014). Due to the inadequate design, complex operations, lack of technical manpower, and frequent electricity breakdowns, the treatment plants do not function appropriately (CPCB 2007). As a result, improper disposals of effluents from the treatment plants are considered to be the major pathways for the antibiotic drugs to enter in aquatic environment (Gothwal and Shashidhar 2017).

2.3 Impacts of Antibiotics on Environment

Chemical structure of antibiotics is stable enough to produce therapeutic activity, and due to their constant input, they persist in the environment for a longer duration of time; and their existence at trace, low, and high concentration is dangerous to the environment and ecosystem (Klavarioti et al. 2009). These antibiotics are characterized by low volatility and high polarity and tend to be easily transported and distributed in different water bodies (Zhang et al. 2014). Many antibiotics have the potential to bioaccumulate and are highly persistent in the environment. Antibiotics have been detected in various bodies like surface water and groundwater (WHO 2011; Balakrishna et al. 2017; Lee et al. 2017), drinking water (Zuccato et al. 2000; Jones et al. 2005; Buffle et al. 2006), tap water (Halling-Sørensen et al. 1998; Doll and Frimmel 2003), soil and sediments, ocean water (Halling-Sørensen et al. 1998; Klavarioti et al. 2009), and sewage influents and effluents (Heberer 2002; Arlos et al. 2015). Additionally, few studies confirmed the presence of pharmaceuticals in Indian rivers (Fick et al. 2009; Ramaswamy et al. 2011; Mutiyar and Mittal 2014; Subedi et al. 2015; Archana et al. 2016). Studies on the treatment of wastewater comprising antibiotics revealed that antibiotics are not readily biodegradable and their genotoxicity was not eliminated by biological treatment (Alexy et al. 2001). The most commonly used activated sludge treatment process cannot remove antibiotics efficiently that are persistent to biodegradation (Celiz et al. 2009). In Japan, Matsuo et al. (2011) detected the presence of 100–2000 ng L^{-1} of amoxicillin concentration in an activated sludge-based biological treatment of wastewater. A similar presence of amoxicillin was reported in the influent of WWTP involving activated sludge treatment in Australia by Watkinson et al. (2007), whereas 20 times higher concentration was found in the effluent of WWTP comprising biological treatment in Hong Kong (Minh et al. 2009). The concentrations of amoxicillin, ofloxacin, sulfamethoxazole, ciprofloxacin, and metoprolol were found to be 40 times higher in the treated wastewater of Indian WWTPs when compared to the treated effluents of WWTPs in Europe, Japan, North America, and Australia (Balakrishna et al. 2017). Although biological treatment is considered to be effective in the abundance of industrial wastewater treatment applications, its efficacy decreases in the presence of complex and tough antibiotic compounds in wastewater. Moreover, the complex design, overall cost, and economic return are the major problems associated with the treatment of antibiotic-laden wastewater at an industrial scale.

Antibiotics discharged in the environment may impose serious toxicity at many levels of life, i.e., humans, animals, microorganisms, ecosystems, or the ecosphere. They are resistant to biodegradation processes and frequently escape from the conventional treatment plants. Moreover, antibiotics may cause irreversible change and lasting effects on the microorganisms, making them resistant to antibiotics, even at trace concentrations (Bredhult et al. 2007; Kristiansson et al. 2011). The antibiotic's presence in the environment has effects such as retardation of nitrite oxidation and methanogenesis, disruption of the human endocrine system, and increased toxicity of chemical combinations and metabolites (Gadipelly et al. 2014). The scientific community is more concerned about the development of antibiotic-resistant bacterial strains and the endocrine-disrupting effects on aquatic organisms exposed to antibiotics. Hence, the presence of antibiotics or their metabolites in the aquatic environment in particular poses a serious environmental problem on a local and a global scale.

2.4 Antibiotic Resistance in Environment

At present, the emergence of pathogenic microbes that are able to resist antibiotic treatments is regarded as one of the most concerning public health issues (CDCP 2013). According to the report by the CDCP, nearly 25,000 people die per year in Europe due to antibiotic-resistant bacterial infections with a worldwide mortality of half a million people (Davies et al. 2013). Antibiotic resistance also poses a significant impact on the world economies, with an approximate investment of 35 billion USD per annum by the United States alone for the treatment of antibiotic-resistant infections (CDCP 2013). To further worsen the situation, the worldwide discoveries of new antibiotics have declined in the past few years due to economic and technical challenges, thereby leading to the "antibiotic crises" (Livermore 2011). The existence of an increased variety of bacteria harboring antibiotic resistance genes in different environmental samples is one of the most prominent consequences of antibiotic pollution. At an early stage, the antibiotics were found to be effective in treating bacterial infections, leading to short and intense research on developing new antibiotics. However, the prolonged and larger-scale use of antibiotics led to the emergence of antibiotic-resistant bacteria. For instance, an example of the remarkable speed of resistance could be observed when 70% of *S. aureus* strains became resistant to erythromycin within 6 months after the beginning of treatment (Finland 1979). Antibiotic resistance can be disseminated through a wide range of mechanisms, such as inactivation of antibiotics (e.g., cleavage of beta-lactams, penicillin, through beta-lactamases), transportation of antibiotics from the bacterial cells through efflux pumps (e.g., tetracyclines pumped out of the cell through *TetA* proteins), or modifications of antibiotic's target cell (e.g., ciprofloxacin binding to cell is prevented by point mutations in *gyrA*) (Gullberg et al. 2011).

Due to the complex and multifaceted ecological role of antibiotics, it is believed that antibiotic resistance is an ancient phenomenon. It has been investigated that antibiotic resistance genes have been found in a range of nonhuman impacted and pristine environments, from permafrost soils to caves, and have enlightened our understanding of the natural occurrence of antibiotic resistance in microbial

Table 2.1 List of antibiotic-resistant "priority pathogens" (WHO 2017)

Priority 1	Critical • *Acinetobacter baumannii*, carbapenem-resistant • *Pseudomonas aeruginosa*, carbapenem-resistant • *Enterobacteriaceae*, carbapenem-resistant, ESBL-producing
Priority 2	High • *Enterococcus faecium*, vancomycin-resistant • *Staphylococcus aureus*, methicillin-resistant, vancomycin-intermediate and vancomycin-resistant • *Helicobacter pylori*, clarithromycin-resistant • *Campylobacter* spp., fluoroquinolone-resistant • *Salmonella*, fluoroquinolone-resistant • *Neisseria gonorrhoeae*, cephalosporin-resistant, fluoroquinolone-resistant
Priority 3	Medium • *Streptococcus pneumoniae*, penicillin-non-susceptible • *Haemophilus influenzae*, ampicillin-resistant • *Shigella* spp., fluoroquinolone-resistant

ecosystems (D'Costa et al. 2011; Perron et al. 2015). For instance, 70% of Gram-positive and 65% of Gram-negative bacterial isolates from a pristine cave carried several functional antibiotic resistance genes (Bhullar et al. 2012). As a human, the important concern is not the presence of resistance genes in the pathogens, but the main concern is the horizontal transfer of the resistance genes in the environmental microbes, thereby limiting our ability to fight against infectious diseases with antibiotics. There are different mechanisms for the transfer of resistance genes into the microbes including phage-supported transmission (transduction) (Xu et al. 2011), uptake of DNA from the environment (transformation) (Springman et al. 2009), and transmission through conjugative plasmids (Palmer et al. 2010).

From the perspective of public health, environmental microbes carrying the resistance genes against the variety of antibiotics, also known as "superbugs," are an important concern. In 2017, the World Health Organization (WHO) released a list of "superbugs" comprising of 13 families of bacteria that pose the greatest threat to the public health (WHO 2017). The list is divided into three categories according to the urgency of need for new antibiotics: critical, high, and medium priority as shown in Table 2.1. The list highlights in particular the threat of Gram-negative bacteria that are resistant to multiple antibiotics. The high consumption and anthropogenic use of antibiotics have triggered the experimental evolution of the antibiotics in each environment. In the subsequent sections, we shall discuss how the resistance genes are selected and flow through different environments and different vectors that contribute to the speed of gene transfer.

2.4.1 Resistance Genes in the Environment

Due to the increasing awareness of the risks posed by the use of antibiotics and the availability of a wider range of detection techniques, there is a substantial increase in the research focusing on the detection of antibiotic resistance in the environment through qPCR, conventional PCR, culture-dependent techniques, or mutagenomic methods (Graham et al. 2016; Ju et al. 2019). The possibility of the environment affected with antibiotic resistance genes is based on the flow of antibiotic resistance genes between the different environments as outlined based on a schematic model in Fig. 2.2. Extensive investigation has been done to explore the hotspot environments of antibiotic resistance genes, where microbes could grow rapidly due to the availability of nutrients and are able to quickly encounter heavy and repetitive doses of antibiotics (Berendonk et al. 2015). The major hotspot environments of antibiotic resistance genes include animal feeding operations, hospitals, aquaculture operations, and WWTPs.

With the emergence of hospital-acquired infections due to antibiotic-resistant microbes, the hospital facilities and intensive care units are important to explore the evolution and distribution of antibiotic resistance genes. The various members of human-associated microbes existing in hospital water and airflow systems are the major microbial communities in hospitals with diverse groups of antibiotic resistance genes (Kizny Gordon et al. 2017). The extended use of a wide variety of antibiotics in hospitals could drive de novo evolution of resistance, for instance, during the long-term treatment of chronic infectious diseases in patients (Folkesson

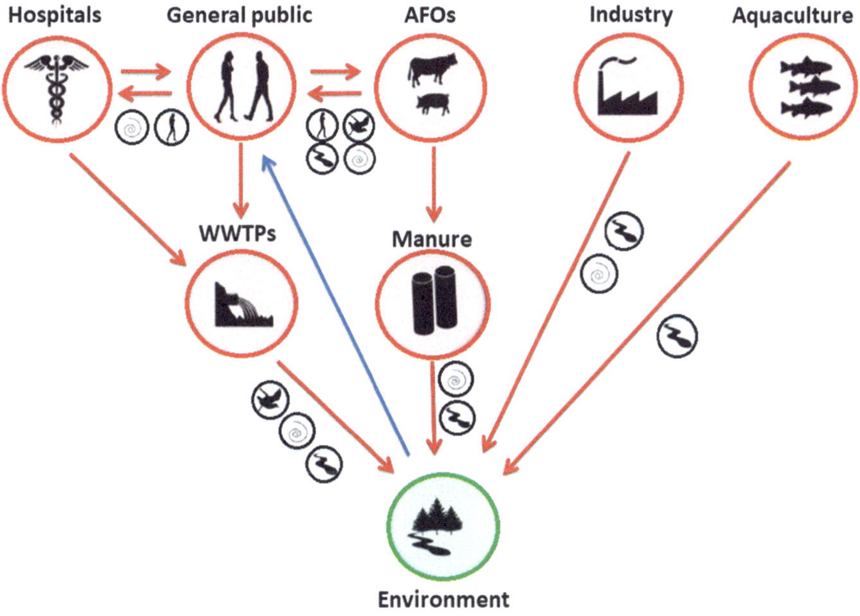

Fig. 2.2 Representation of flow of antibiotic resistance genes and antibiotic-resistant bacteria to the environment (green circle) from hotspots of transmission and evolution (red circles). Red arrows designate the known flows of antibiotic resistance genes and antibiotic-resistant bacteria, blue arrow designates the probable route of transmission from the polluted environments back to the general public. Black circles indicate the possible vectors that might assist in the transmission of infection between environments including waters, air, humans, surface, and animals. (Kraemer et al. 2019)

et al. 2012). In addition, the microbes with newly evolved antibiotic resistance genes could spread epidemically in patients, or the direct gene transmission may occur in different genetic hosts through horizontal gene transfer. Thus, in addition to resistance evolution, the microbes carrying the antibiotic resistance genes could enter the hospital environments through infected humans, where they could enter the new genetic host or could spread epidemically. The antibiotic-resistant microbes that have been found in hospital effluents are vancomycin-resistant enterococci (VRE), *P. aeruginosa*, and extended-spectrum beta-lactamases (ESBL) carrying *E. coli* (Hocquet et al. 2016). Even though the hospital encompasses a variety of antibiotic-resistant microbes, however, the overall role of the hospital as a source of environmental resistance is highly uncertain due to their relative low contribution to the total wastewater in comparison to the total population ($\approx$1%) (Kümmerer 2009). Even though hospitals are under extensive surveillance, they offer controlled environments for the use of antibiotics, and resistance evolution is easily tracked due to the testing of hospital effluents, thus eliminating the problem. Contrary, the inappropriate use of in-home antibiotics by the general population always remains unsupervised. Moreover, even if the antibiotics are correctly utilized, approximately 70% of antibiotics are excreted in the environment unaltered via the human digestive system (Changotra et al. 2019, 2020). In comparison to hospital-derived sewage, the small quantity of hospital independently generated municipal sewage is the major contributor of antibiotic resistance genes from the general public (Kümmerer 2009).

Municipal and hospital sewage are combined and transferred to WWTPs, which employs various physicochemical and biological treatments to treat the waste and diminish the amount of pathogens before their discharge. Sewage comprises of different types of antibiotics, seeded with new antibiotic-resistant bacteria and antibiotic resistance genes, and contains a nutrient-rich environment to support the high quantities of bacteria. Since the wide varieties of antibiotics are present at sub-minimum inhibitory concentrations (MIC) in the sewage, thus they could easily encourage horizontal gene transfer mechanisms to increase the transmission of antibiotic resistance genes (Szczepanowski et al. 2009; Ju et al. 2019). Due to these reasons, WWTPs and sewage are regarded as the hotspot environments for the evolution, recombination, and transmission of antibiotic resistance (Rizzo et al. 2013). While the abundances of antibiotic-resistant bacteria and antibiotic resistance genes are greatly reduced in the water fraction in sewage passage through WWTPs, the WWTPs are ineffective in the removal of antibiotic resistance genes in biosolids (Chen and Zhang 2013; Yang et al. 2014). The current policies allow the use of biosolids as fertilizers for crops, which were subjected to bury in landfills in the past. Thus, the potential contamination of crops, grazing livestock, and soils with resistant microbes and antibiotic resistance genes needs to be examined. Sewage, hospitals, and WWTPs are enriched with human-associated antibiotic-resistant bacteria and antibiotics. As it is very difficult to control the usage of antibiotics in the general public, WWTPs may provide an essential role in the intervention to curb the transmission and evolution of resistance genes.

Independent of the application of antibiotics on the general public, antibiotics are extensively utilized in animal feeding operations for livestock production and aquacultures. Generally, the amount of antibiotics used in such facilities is much higher than that being utilized by hospitals (Chee-Sanford et al. 2009). In Canada, the amount of antibiotics used by biomass in animals is twice the amount used by the general public, and in the United States, approximately 80% of antibiotics sold are used for veterinary purposes (Ventola 2015; PHAC 2016). Waste manure containing antibiotic-resistant bacteria from animal feeding operations is generally collected and treated in waste lagoons and is independent of WWTPs and municipal wastes (Chee-Sanford et al. 2009). Although manure storage could decrease the concentration of culturable bacteria, still manure and lagoons have been found to contain an extensive amount of antibiotic-resistant bacteria and resistance genes (Pruden et al. 2006; Peak et al. 2007). As a result, manure is being used as a fertilizer for crops and soil, where the surviving antibiotic-resistant bacteria and resistance genes come in contact with the soil environment. Subsequently, antibiotic-resistant bacteria and resistance genes from manure enter the surface water and groundwater via surface runoff. In addition to animal feeding operations, the installations of aquaculture, where fish are farmed, are an example of livestock production in which an extensive amount of antibiotics are frequently used (Cabello et al. 2016). Among the open- and closed-water aquaculture, the closed ones are of superior interests, in which animal or human waste is fed to the fish, thereby promoting the resistance gene transfer between the systems (Watts et al. 2017). In contrast, in open-water aquaculture, antibiotics are directly added to open waters to promote the growth and prevent the fishes from disease, leading to the transmission and evolution of antibiotic resistance genes and antibiotic-resistant bacteria in open waters and sediments (Shah et al. 2014). This method of administration of antibiotics leads to their affecting both diseased and healthy fish (metaphylaxis) in the aquaculture (Sørum 2006). Antibiotics not only are utilized in animals for therapeutic use to treat infectious diseases and prevent postsurgery infections but also can be added in low concentrations to the animal feed as antibiotic growth promoters (AGPs) or as prophylaxis to increase the farm productivity and profitability (Turnidge 2004; Hao et al. 2014). In 2006, AGPs were banned in the many European Union (EU) countries as well as in the United Kingdom, leading to subsequent lower antibiotic consumption especially for colistin and tetracycline (Cogliani et al. 2011). Till date, an amended regulation on the veterinary medication in the EU proposed the limit of antibiotic use in the metaphylaxis and prophylaxis was confirmed by the European Parliament and EU Council in 2018 (Council of the European Union 2018). In 2017, subtherapeutic antibiotics in water and feed to enhance feed efficiency and animal growth became illegal in the United States (U.S. Food and Drug Administration 2018). In the rest of the world (particularly low- or middle-income countries), the usage of antibiotic remains partially regulated provoking significant risks for the environmental management and transfer of resistance genes due to the release of antibiotics to water bodies and soils affected by animal farming operations (Van Boeckel et al. 2019; Moore 2019). Based on a One Health framework, a global action plan was launched by the World Organisation for Animal Health (OIE) in 2015 on antibiotic resistance genes and antibiotic-resistant bacteria, and over 100 countries have now established their national action plans against antibiotic resistance genes and antibiotic-resistant bacteria in animals (Góchez et al. 2019). Following the One Health framework policy by the OIE, the Chinese government published "The National Action Plan on controlling antibiotic-resistant bacteria on animal origins (2016–2020)" integrated with the One Health concept for food safety and public health. In the end, the OIE has built a global database on antibiotic agents proposed for the use in animals and published the methodology for data collection in 2019 for the global development of monitoring and surveillance in line with OIE international standards (Góchez et al. 2019).

2.4.2 Transmission Vectors

The transport of antibiotic resistance genes and antibiotic-resistant bacteria to the pristine environments from hotspots of evolution resistance takes place through a variety of vectors (as outlined in Fig. 2.2). Vectors can either allow the transportation or occur as reservoirs where antibiotic resistance genes and antibiotic-resistant bacteria could persist, evolve, or multiply. In most cases, the antibiotic resistance genes are transported within the antibiotic-resistant bacteria between environments. Thus, the genetic framework of most of the existing antibiotic resistance genes in the environment is still anonymous.

2.4.2.1 Air and Surface Water

In the past, the dissemination of antibiotic-resistant bacteria through air has received considerable attention. For instance, antibiotic resistance genes against the seven frequently utilized antibiotics have been identified with high relative abundances in particulate matter in the air throughout the major cities in the world (Li et al. 2018). Similarly, the antibiotic resistance genes were found to be enriched in particulate matter from smog (Hu et al. 2018). Most of the antibiotic resistance genes are concomitant with the wide range of bacteria bounded to the particulate matter floating in the air and disperse in the air through fog, wind, and precipitation (Dueker et al. 2017). Some species of *Pseudomonas*, such as *P. syringae* and *P. aeruginosa*, have been identified to contain multiple resistance genes associated with precipitation (Hwang et al. 2005; Christner et al. 2008). Antibiotic resistance genes have also been identified in the seed air originating during the animal feeding operations (Gao et al. 2017) and in close vicinity and downwind of WWTPs (Li et al. 2016), thereby posing the health risk to the workers. At last, the airborne transmissions of antibiotic-resistant bacteria from hospitals have not been extensively explored; however, airborne antibiotic resistance genes have been found there (Gilbert et al. 2010). On the other hand, surface waters receive industrial and WWTP effluents as well as runoff from animal feeding operations and manure-fertilized fields and hence located at a central hub for the dissemination and transportation of the antibiotic-resistant bacteria (Pei et al. 2006; Graham et al. 2011). Profiles of antibiotic-resistant bacteria and antibiotic resistance genes have been found in rivers through surface runoff and were strongly impacted by the contaminations from animal feeding operations and WWTP effluents.

2.4.2.2 Animal Vectors

In comparison to air and water vectors, animal vectors including humans provide nutrient-rich environments to antibiotic-resistant bacteria to amplify or change their genetic context within the animal vectors (Langelier et al. 2019). Labors and farmers have shown to develop augmented resistance on direct contact with livestock (Hanselman et al. 2006; Price et al. 2007). Antibiotic-resistant bacteria originating from animals or animal feeding operations have to cross the species barrier for their transmission in humans, whereas human-originated antibiotic-resistant bacteria from sewage and hospitals can easily be transmitted to humans. Certainly, hospital workers are more vulnerable to increased transmission risk of antibiotic-resistant bacteria and thus develop increased antibiotic resistance (Lis et al. 2009). Interestingly, the antibiotic-resistant bacteria could easily be transferred to the family members of hospital workers, thereby promoting the possible chain of transmission.

In addition to workers from hospitals and farming, antibiotic-resistant bacteria have been found to spread through international travel. Studies have confirmed that antibiotic-resistant bacteria and resistance genes could persist for a longer period in a traveler's microbiome and thus could remain there even after the return of travelers to their origin country (Bengtsson-Palme et al. 2015; Langelier et al. 2019). Apart from humans, insects such as cockroaches and houseflies have been identified as potential antibiotic-resistant bacteria-carrying vectors for the transmission of pathogens from WWTPs (Doud et al. 2014), animal feeding operations (Zurek and Ghosh 2014), and hospitals (Tilahun et al. 2012). A wide range of vertebrates, specifically small rodents, have been associated to carry and transmit antibiotic-resistant bacteria from animal feeding operations, water feeding birds, and even WWTPs (Allen et al. 2011); however, a direct link between the hotspots of resistance evolution and resistance gene carriage has been rarely established. Recent research has established that birds exposed to wastewater, especially ducks, are characterized by high diversity and abundance of antibiotic resistance genes. As such, the ducks are regarded as a potential vector for the transmission of resistance genes due to their habitats and feeding habits (Marcelino et al. 2018). Another group of birds that is responsible for carrying the antibiotic-resistant bacteria (for instance, pathogenic bacteria such as ESBL carrying *E. coli*) from the contaminated water is seagulls (Poeta et al. 2008; Marcelino et al. 2018).

Although the hotspots (such as WWTPs, animal feeding operations, and hospitals) for the transmission and evolution of antibiotic-resistant bacteria have been extensively evaluated, the risk posed by the transmission vectors between the environments has not been explored thoroughly. As transmission vectors provide the selective conditions for the recombination of resistance genes and multiplication of resistant bacteria, they are regarded as the central hub for understanding the flow of resistant genes between the environments. Indeed, the risks associated with the combination of resistance genes from animals and humans and the passage of species barrier must be assessed.

2.5 Cytotoxic Effects of Antibiotics in the Environment

Even though the primary role of antibiotics is to prevent and treat microbial infections, research is still underway to know whether antibiotics could harbor inhibitory activities under the natural condition (Davies and Davies 2010). Nevertheless, the resistance genes, for example, β-lactamases, are anticipated to have originated millions of years ago, signifying that antibiotics have been affecting the roles of pathogens long before their use in clinical medicines (Hall and Barlow 2004; Aminov and Mackie 2007). The quantities of antibiotics discharged and accumulated into the environment as a result of human activities have negatively affected the local communities of animals and microorganisms.

2.5.1 Toxic Effects of Antibiotic Pollution on Microbial Communities

To favor the transmission and evolution of resistance genes in environments, the selective pressures build up by antibiotic pollution could adversely affect the dynamics of evolution in microbes in different ways (Martínez 2017). Diverse populations of bacteria present different tolerance levels to antibiotics because of variations in physiological traits or changes in gene expression (El Meouche and Dunlop 2018). Under strong selective pressures, the antibiotics could decrease the microbial population's diversity by favoring the growth of tolerant or resistant microbial lineages. Under weak selective pressures, the lower concentrations of antibiotics could increase the genotypic and phenotypic diversity by selectively favoring the growth of resistant microbial lineages. In addition to the fluctuations in microbial population's diversity, exposure of microbial population to even low concentration of antibiotics could also increase the genetic diversity in microbes via the direct mutagenic effect on the DNA (Andersson and Hughes 2014) and via activation of bacterial SOS response, causing an augmented mutation rate in the bacterial genome (Foster 2007). Antibiotics have been found to enhance the horizontal gene transfer between bacterial populations either by augmenting the competence leading to uptake of extracellular DNA (Slager et al. 2014) or by conjugation (Maiques et al. 2006). At last, antibiotics could also affect the regulation of the gene at the transcription level

(Davies et al. 2006), either via regulatory mechanisms such as quorum sensing (Rémy et al. 2018) and riboswitches (Blount and Breaker 2006) or via direct binding, leading to increase virulence and phenotypic variability (Rémy et al. 2018). As a whole, these mechanisms could increase the existing pond of phenotypic and genetic diversity in microbial populations with the exposure to antibiotics. Consecutively, if selective pressure increases, these mechanisms may promote the evolution of antibiotic resistance according to Fisher's Fundamental Theorem of Evolution by Natural Selection (Lee et al. 2018). Selective pressures related to antibiotic pollution could also act on the composition of the overall microbial community by shifting microbial composition or by decreasing taxa diversity. For instance, exposure to antibiotics tends to favor a decrease in Gram-positive bacteria as opposed to Gram-negative bacteria. The former displayed increased susceptibility to disinfectants and antibiotic exposure due to the lack of outer cell membrane. Antibiotic exposure to microbial populations may also lead to the loss of critical ecological roles. For example, the presence of antibiotics in the aquatic environments tends to reduce the overall diversity of microbial populations, including primary productivity and microbial taxa accountable for carbon cycling (Grenni et al. 2018; Eckert et al. 2019). Likewise, the antibiotic pollution in soil environments could alter the physiology of microbes leading to reduction in microbial activity such as respiration, denitrification, nitrification, loss of biomass, and loss of enzyme activity of bacteria (Thiele-Bruhn and Beck 2005; Cycoń et al. 2019). Lastly, the disruption of microbial communities by antibiotics could also increase the plenty of toxic pathogens and parasites in soil and water environments, for example, *Cyanobacteria* species, triggering eutrophication in freshwater and soil environments and causing health threats to humans (Drury et al. 2013).

2.5.2 *Toxic Effects of Antibiotic Pollution on Higher Organisms*

2.5.2.1 Physiological Effects

In addition to the adverse effects of antibiotics on microbial communities, these antibiotics have a strong ability to affect higher organisms in aquatic environments where these organisms are exposed to pollutants even at low concentrations. It has been reported that the concentration of antibiotics in environmental samples is quite low and exhibits negligible risks to humans (Mojica and Aga 2011). Nevertheless, the presence of low concentrations of antibiotics in the environment could also accumulate in humans via long-term exposure to food, drinking water, or other consumer goods inducing some unknown health hazards. For instance, quinolones and macrolides have been identified in chlorinated drinking water (Ye et al. 2007). Triclosan, an antimicrobial agent used in consumer goods, for example, clothes and soap, has been detected in human urine and serum which are not using this antimicrobial and was found to pose health hazards from muscle weakness to reproductive problems (Weatherly and Gosse 2017; Bever et al. 2018). Approximately 70% of the US population has been exposed to various antibiotics and triclosan via consumer goods (Calafat et al. 2008). As a result, triclosan has been detected in rivers and streams across the world (Halden and Paull 2005).

Exposure to low concentrations of antibiotics (such as erythromycin and streptomycin) has been found to negatively impact the behaviors and survival of smaller organisms and micro-invertebrates such as *Artemia* (Migliore et al. 1997) and *Daphnia magna* (Flaherty and Dodson 2005) in laboratory conditions. The presence of subinhibitory concentrations of macrolides in the aquatic environments has also been found to affect the vertebrates such as zebra fish by causing malformations, such as uninflated swim bladder and yolk sac edema, and manipulating movement frequency of the embryo (Wang et al. 2014). Similar malformations have been observed in the case of other fish species on the exposure to antibiotics, including quinolone (Wang et al. 2009), sulfonamide (Lin et al. 2014), and tetracycline (Zhang et al. 2015). Besides, quinolones and their metabolites could persist in the human body for a longer time, thereby contributing toward bioaccumulation risk as well as chronic toxic effects (Liu et al. 2018). The presence of antibiotic tetracycline in aquatic environments has also affected the amphibians, thereby inducing malformations such as pericardial edema and shortened body length in *Xenopus tropicalis* (Liu et al. 2018).

Besides the physiological effects, the exposure of organisms to antibiotics could also lead to interfere with behavior and development changes, possibly resulting from the modifications in gene regulations (Zhang et al. 2016). In 2017, Kim et al. found that exposure to antibiotic tetracycline of zooplankton species *D. magna* caused a gene regulation in the species, impacting the carbohydrate and protein metabolism as well as promoting general stress response. Additionally, these alterations in gene regulation can carry over in subsequent generations of organisms even in the absence of tetracycline, thereby impacting the organism population even after the removal of antibiotic compounds from the environment (Kim et al. 2017).

2.5.2.2 Effect on Host Microbiome

It has been established that antibiotics could adversely affect the higher organisms by disrupting the microbial populations related to the host animals. Host microbiomes in animals accomplish different roles for the host animals, ranging from bone formation to effective nutrient metabolism (Raymann and Moran 2018). Thus, the disruption of host's microbiomes, also known as dysbiosis, could promote health consequences, such as metabolic diseases, allergies, developmental defects, or augmented susceptibility to pathogens (Jie et al. 2017). Recent investigations have shown that the low concentrations of antibiotics in the aquatic environment could also impact other local organisms, such as the fish microbiome, which is highly susceptible to variations in the aquatic environment (Schmidt et al. 2016). In aquaculture, a lower concentration of antibiotics can increase mortality and decrease microbial diversity in the fish gut (Navarrete et al. 2008). Even exposure to the high concentration of certain chemicals and antibiotics in waters contaminated with WWTP effluent could impact the fishes by increasing mortality. Previous research also demonstrated that the adult fish can recover from the short-term exposure to antibiotics, whereas, in some cases, even the lower concentration of antibiotics (tetracycline and streptomycin) as low as 1 μg mL^{-1} could increase mortality and cause dysbiosis in zebra fish embryos (Pindling et al. 2018). Another negative impact of antibiotic exposure on higher organisms includes reduction in fertility and disruption of the growth cycle of host microbiomes (Yan et al. 2016). Thus, there is an imperative need to investigate the extent of negative impact of antibiotic pollutions on aquatic organisms, as antibiotic pollution is only providing additional threats to aquatic organisms in the world due to anthropogenic activity.

2.6 Gaps in Current Policies to Tackle Antibiotic Pollution

With the aim of tackling the global crisis due to antibiotic resistance and pollution, the WHO launched the Global Antimicrobial Resistance Surveillance System (GLASS) under the global action plan in 2015, hence providing a framework for action plans for individual nations to develop regulations and policies to combat antibiotic resistance and pollution problems (WHO 2019). GLASS is a platform for global data-sharing worldwide that aimed to support countries in establishing national antimicrobial resistance (AMR) surveillance systems as part of national action plans to enable collection, analysis, and sharing of data. Till date, just 82 countries (less than half of the WHO member states) have enrolled in GLASS and 66 have submitted AMR data. Technically, this means that national AMR data is either unavailable in a shareable format or not readily accessible, thus hindering the comparison of local or global trends in AMR. Another surveillance system, namely, the National Antimicrobial Resistance Monitoring System (NARMS), was established by the United States in 1996 (FDA 2020). The NARMS is a collaborative program of state and local public health universities and departments, the Centers for Disease Control and Prevention (CDCP), the FDA, and the US Department of Agriculture (USDA). This surveillance system tracks changes in the antimicrobial susceptibility of enteric (intestinal) bacteria found in retail meats (FDA), ill people (CDCP), and food animals (USDA) in the United States. NARMS data are used by the FDA to make regulatory decisions to preserve the effectiveness of antibiotics for animals and humans. In 2009, the Global Antibiotic Resistance Partnership (GARP) was started to create a platform for developing actionable policy proposals on antibiotic resistance in low-income and middle-income countries including India, South Africa, Kenya, Vietnam, Mozambique, Tanzania, Nepal, Uganda, and Bangladesh (CDDEP, New Delhi n.d.). GARP is a project of the Center for Disease Dynamics, Economics and Policy (CDDEP), funded by the Bill & Melinda Gates Foundation.

According to the report of the European Commission, the increasing extent of antibiotic resistance and pollution crisis leads to approximately 70,000 deaths per year on a global scale, making it a global problem due to economic and social fallout (European Commission 2017). With the aim to collect comparable and reliable data on antibiotic use in Europe, the European Surveillance of Antimicrobial Consumption Network (ESAC-Net) project was granted by the European Commission in 2001 (ECDC 2020). The international network continues collecting reference data on the consumption of antimicrobials for systemic use in the community and in the hospital sector in EU countries through the European Surveillance System. Another surveillance network in EU, namely, European Antimicrobial Resistance Surveillance Network (EARS-Net), is the largest publicly funded system for AMR surveillance in Europe. EARS-Net is based on routine clinical antimicrobial susceptibility data from local and clinical laboratories reported to the European Centre for Disease Prevention and Control by appointed representatives from the EU Member States. In South Africa, the South African Society for Clinical Microbiology (SASCM) was implemented with the objectives to include the monitoring of antimicrobial resistance patterns in the public and private medicine sector in South Africa including the publication of antibiotic resistance patterns in selected publications (Singh-Moodley et al. 2018). Due to the lack of reliable and accurate information in African countries, there is an urgent need for the implementation of standardized surveillance programs. In view of estimates of AMR burden, the Indian Council of Medical Research (ICMR), in 2013, initiated the Antimicrobial Resistance Surveillance and Research Network (AMRSN) to enable compilation of data on pathogenic groups on antimicrobial resistance from the country (Walia et al. 2019). The ICMR has developed a real-time online AMR data entry system for its network and will have AMR data analysis capacity. It is a user-friendly web-based informatics solution/suite for collection, storage, and analysis of surveillance data.

The major benefit of having a surveillance system is the collection of real-time accurate data on AMR including the mechanisms of resistance and representativeness to community; sustaining the current effort and expanding the current activities to next levels of healthcare settings are the major challenges. As per the UNEP report, present research emphasized the investigation of health threats associated with the accumulation of antibiotics, antibiotic-resistant bacteria, and resistance genes in the environmental reservoirs (UNEP 2017). However, most of the regulations and action plans laid by different nations fail to address this critical issue and do not aim to combat the pollution problems associated with antibiotics and resistance in the natural environments. Similarly, all national action plans do not include any regulations and policies to control the environmental release of antibiotics by the pharmaceutical manufacturing industries. Considering the magnitude of antibiotics released from industries and accumulation in the local environments, the negligence of such issue might facilitate the localized hotspots of evolution and transmission of antibiotic-resistant bacteria and resistance genes that may consequently spread worldwide. Lastly, there are no policies and regulations in any national action plans to address the problems of antibiotic pollution of WWTPs. Currently, the design of WWTPs is not entitled to remove antibiotics, antibiotic-resistant bacteria, and resistance genes; thus, policies need to be amended to establish WWTP technologies to allow an acceptable load of antibiotics and resistance genes in biosolids and WWTP effluents.

2.7 Conclusion

The presence of antibiotics and resulting antibiotic resistance is a critical problem, leading to serious health problems worldwide. The environmental constituents of this issue, such as evolution and transmission of resistant genes between bacteria, and environmental hotspots of resistant bacteria and genes have received much attention in the past years. However, many elements of antibiotic resistance and pollution are still unknown and need further research. Hospitals, veterinary use, and WWTPs are the major sources of antibiotic pollution in the natural environments. Studies need to be focused on linking the abundance and presence of antibiotic resistance genes directly to the general public's risk of becoming infected on exposure to antibiotic-resistant bacteria. Besides the predominance of antibiotic resistance in natural environments, antibiotic pollution has negatively impacted the ecosystem's functions in the animals and subsequently affected the human health directly. The antibiotic's presence in the environment has effects such as retardation of nitrite oxidation and methanogenesis, disruption of the human endocrine system, and increased toxicity of chemical combinations and metabolites. For such reasons, along with the contribution of antibiotic pollution toward antibiotic resistance, it is vital to consider the overall impacts of antibiotic pollutions on natural environments as well as on human health. Current policies and regulations are based on the surveillance data, but more descriptive studies are required to develop efficient action plans and policies to combat the global problems of antibiotic pollution. Antimicrobial resistance surveillance in healthcare-associated facilities, e.g., hospitals, dialysis clinics, long-term care facilities, etc., is vital in order to understand the prevalent healthcare-associated pathogenic organisms, the antimicrobial resistance patterns of critical organisms, and their mechanisms of resistance. National laboratory systems should support the surveillance for antimicrobial resistance and implement standardized practices, manuals, and quality systems. Besides, this information will support the upliftment and development of national treatment guidelines, regulating policies and strategies for antimicrobial infection prevention and control.

References

Alexy R, Kümpel T, Dörner M, Kümmerer K (2001) Effects of antibiotics against environmental bacteria studied with simple tests. In: Proceedings of the 11th annual meeting of SETAC Europe, Madrid, pp 6–11

Allen SE, Boerlin P, Janecko N et al (2011) Antimicrobial resistance in generic *Escherichia coli* isolates from wild small mammals living in swine farm, residential, landfill, and natural environments in southern Ontario, Canada. Appl Environ Microbiol 77:882–888

Aminov RI, Mackie RI (2007) Evolution and ecology of antibiotic resistance genes. FEMS Microbiol Lett 271:147–161

Andersson DI, Hughes D (2014) Microbiological effects of sublethal levels of antibiotics. Nat Rev Microbiol 12:465–478

Archana G, Dhodapkar R, Kumar A (2016) Offline solid-phase extraction for preconcentration of pharmaceuticals and personal care products in environmental water and their simultaneous determination using the reversed phase high-performance liquid chromatography method. Environ Monit Assess 188:1–10

Arlos MJ, Bragg LM, Parker WJ, Servos MR (2015) Distribution of selected antiandrogens and pharmaceuticals in a highly impacted watershed. Water Res 72:40–50

Balakrishna K, Rath A, Praveenkumarreddy Y et al (2017) A review of the occurrence of pharmaceuticals and personal care products in Indian water bodies. Ecotoxicol Environ Saf 137:113–120

Bengtsson-Palme J, Angelin M, Huss M et al (2015) The human gut microbiome as a transporter of antibiotic resistance genes between continents. Antimicrob Agents Chemother 59:6551–6560

Berendonk TU, Manaia CM, Merlin C et al (2015) Tackling antibiotic resistance: the environmental framework. Nat Rev Microbiol 13:310–317

Bever CS, Rand AA, Nording M et al (2018) Effects of triclosan in breast milk on the infant fecal microbiome. Chemosphere 203:467–473

Bhullar K, Waglechner N, Pawlowski A et al (2012) Antibiotic resistance is prevalent in an isolated cave microbiome. PLoS One 7:e34953

Blount KF, Breaker RR (2006) Riboswitches as antibacterial drug targets. Nat Biotechnol 24:1558–1564

Bredhult C, Bäcklin BM, Olovsson M (2007) Effects of some endocrine disruptors on the proliferation and viability of human endometrial endothelial cells in vitro. Reprod Toxicol 23:550–559

Buffle MO, Schumacher J, Salhi E et al (2006) Measurement of the initial phase of ozone decomposition in water and wastewater by means of a continuous quench-flow system: application to disinfection and pharmaceutical oxidation. Water Res 40:1884–1894

Cabello FC, Godfrey HP, Buschmann AH, Dölz HJ (2016) Aquaculture as yet another environmental gateway to the development and globalisation of antimicrobial resistance. Lancet Infect Dis 16:e127–e133

Calafat AM, Ye X, Wong LY et al (2008) Urinary concentrations of triclosan in the U.S. population: 2003–2004. Environ Health Perspect 116:303–307

CDCP, Centers for Disease Control and Prevention (2013) Antibiotic resistance threats in the United States. Centers for Disease Control and Prevention, Atlanta, GA

CDDEP, New Delhi (n.d.). https://cddep.org/projects/global-antibiotic-resistance-partnership/

Celiz MD, Perez S, Barcelo D, Aga DS (2009) Trace analysis of polar pharmaceuticals in wastewater by LC-MS-MS: comparison of membrane bioreactor and activated sludge systems. J Chromatogr Sci 47:19–25

Changotra R, Rajput H, Dhir A (2017) Natural soil mediated photo Fenton-like processes in treatment of pharmaceuticals: batch and continuous approach. Chemosphere 188:345–353

Changotra R, Guin JP, Dhir A, Varshney L (2018) Decomposition of antibiotic ornidazole by gamma irradiation in aqueous solution: kinetics and its removal mechanism. Environ Sci Pollut Res 25:32591–32602

Changotra R, Rajput H, Paul Guin J et al (2019) Hybrid coagulation, gamma irradiation and biological treatment of real pharmaceutical wastewater. Chem Eng J 370:595–605

Changotra R, Rajput H, Guin JP et al (2020) Techno-economical evaluation of coupling ionizing radiation and biological treatment process for the remediation of real pharmaceutical wastewater. J Clean Prod 242:118544

Chee-Sanford JC, Mackie RI, Koike S et al (2009) Fate and transport of antibiotic residues and antibiotic resistance genes following land application of manure waste. J Environ Qual 38:1086–1108

Chen H, Zhang M (2013) Occurrence and removal of antibiotic resistance genes in municipal wastewater and rural domestic sewage treatment systems in eastern China. Environ Int 55:9–14

Christner BC, Morris CE, Foreman CM et al (2008) Ubiquity of biological ice nucleators in snowfall. Science 319:1214–1215

Cogliani C, Goossens H, Greko C (2011) Restricting antimicrobial use in food animals: lessons from Europe. Microbe Mag 6:274–279
Council of the European Union (2018) Veterinary medicines: new EU rules to enhance availability and fight against antimicrobial resistance. Council of the EU, Brussels
CPCB, Central Pollution Control Board (2007) Advance methods for treatment of textile industry effluents, resource recycling series: RERES/&/2007. India
Cycoń M, Mrozik A, Piotrowska-Seget Z (2019) Antibiotics in the soil environment—degradation and their impact on microbial activity and diversity. Front Microbiol 10:338
D'Costa VM, Osta VM, King CE et al (2011) Antibiotic resistance is ancient. Nature 477:457–461
Davies J (2006) Where have all the antibiotics gone? Can J Infect Dis Med Microbiol 17:287–290
Davies J, Davies D (2010) Origins and evolution of antibiotic resistance. Microbiologia 74:417–433
Davies J, Spiegelman GB, Yim G (2006) The world of subinhibitory antibiotic concentrations. Curr Opin Microbiol 9:445–453
Davies SC, Fowler T, Watson J et al (2013) Annual report of the chief medical officer: infection and the rise of antimicrobial resistance. Lancet 381:1606–1609
Doll TE, Frimmel FH (2003) Fate of pharmaceuticals—photodegradation by simulated solar UV-light. Chemosphere 52:1757–1769
Doud CW, Scott HM, Zurek L (2014) Role of house flies in the ecology of enterococcus faecalis from wastewater treatment facilities. Microb Ecol 67:380–391
Drury B, Scott J, Rosi-Marshall EJ, Kelly JJ (2013) Triclosan exposure increases triclosan resistance and influences taxonomic composition of benthic bacterial communities. Environ Sci Technol 47:8923–8930
Dueker ME, O'Mullan G, Martínez JM et al (2017) Onshore wind speed modulates microbial aerosols along an urban waterfront. Atmosphere (Basel) 8:215
ECDC (2020). https://www.ecdc.europa.eu/en/about-us/partnerships-and-networks/disease-and-laboratory-networks/ears-net
Eckert EM, Quero GM, Di Cesare A et al (2019) Antibiotic disturbance affects aquatic microbial community composition and food web interactions but not community resilience. Mol Ecol 28:1170–1182
El Meouche I, Dunlop MJ (2018) Heterogeneity in efflux pump expression predisposes antibiotic-resistant cells to mutation. Science 362:686–690
European Commission (2017) A European one health action plan against antimicrobial resistance (AMR). European Comission, Brussels
FDA (2020). https://www.fda.gov/animal-veterinary/antimicrobial-resistance/national-antimicrobial-resistance-monitoring-system
Fick J, Söderström H, Lindberg RH et al (2009) Contamination of surface, ground, and drinking water from pharmaceutical production. Environ Toxicol Chem 28:2522
Finland M (1979) Emergence of antibiotic resistance in hospitals, 1935–1975. Clin Infect Dis 1:4–21
Flaherty CM, Dodson SI (2005) Effects of pharmaceuticals on daphnia survival, growth, and reproduction. Chemosphere 61:200–207
Folkesson A, Jelsbak L, Yang L et al (2012) Adaptation of Pseudomonas aeruginosa to the cystic fibrosis airway: an evolutionary perspective. Nat Rev Microbiol 10:841–851
Foster PL (2007) Stress-induced mutagenesis in bacteria. Crit Rev Biochem Mol Biol 42:373–397
Gadipelly C, Pérez-González A, Yadav GD et al (2014) Pharmaceutical industry wastewater: review of the technologies for water treatment and reuse. Ind Eng Chem Res 53:11571–11592
Gao M, Jia R, Qiu T et al (2017) Size-related bacterial diversity and tetracycline resistance gene abundance in the air of concentrated poultry feeding operations. Environ Pollut 220:1342–1348
Gilbert Y, Veillette M, Duchaine C (2010) Airborne bacteria and antibiotic resistance genes in hospital rooms. Aerobiologia 26:185–194
Góchez D, Raicek M, Pinto Ferreira J et al (2019) OIE annual report on antimicrobial agents intended for use in animals: methods used. Front Vet Sci 6:317

Gothwal R, Shashidhar T (2017) Occurrence of high levels of fluoroquinolones in aquatic environment due to effluent discharges from bulk drug manufacturers. J Hazard Toxic Radioact Waste 21:05016003

Graham DW, Olivares-Rieumont S, Knapp CW et al (2011) Antibiotic resistance gene abundances associated with waste discharges to the Almendares river near Havana, Cuba. Environ Sci Technol 45:418–424

Graham DW, Knapp CW, Christensen BT et al (2016) Appearance of β-lactam resistance genes in agricultural soils and clinical isolates over the 20th century. Sci Rep 6:1–8

Grenni P, Ancona V, BarraCaracciolo A (2018) Ecological effects of antibiotics on natural ecosystems: a review. Microchem J 136:25–39

Gullberg E, Cao S, Berg OG et al (2011) Selection of resistant bacteria at very low antibiotic concentrations. PLoS Pathog 7:e1002158

Halden RU, Paull DH (2005) Co-occurrence of triclocarban and triclosan in U.S. water resources. Environ Sci Technol 39:1420–1426

Hall BG, Barlow M (2004) Evolution of the serine β-lactamases: past, present and future. Drug Resist Update 7:111–123

Halling-Sørensen B, Nors Nielsen S et al (1998) Occurrence, fate and effects of pharmaceutical substances in the environment—a review. Chemosphere 36:357–393

Hanselman BA, Kruth SA, Rousseau J et al (2006) Methicillin-resistant Staphylococcus aureus colonization in veterinary personnel. Emerg Infect Dis 12:1933–1938

Hao H, Cheng G, Iqbal Z et al (2014) Benefits and risks of antimicrobial use in food-producing animals. Front Microbiol 5:288

Heberer T (2002) Occurrence, fate, and removal of pharmaceutical residues in the aquatic environment: a review of recent research data. Toxicol Lett 131:5–17

Hocquet D, Muller A, Bertrand X (2016) What happens in hospitals does not stay in hospitals: antibiotic-resistant bacteria in hospital wastewater systems. J Hosp Infect 93:395–402

Hu J, Zhao F, Zhang XX et al (2018) Metagenomic profiling of ARGs in airborne particulate matters during a severe smog event. Sci Total Environ 615:1332–1340

Hwang MSH, Morgan RL, Sarkar SF et al (2005) Phylogenetic characterization of virulence and resistance phenotypes of pseudomonas syringae. Appl Environ Microbiol 71:5182–5191

Jie Z, Xia H, Zhong SL et al (2017) The gut microbiome in atherosclerotic cardiovascular disease. Nat Commun 8:1–12

Jones OA, Lester JN, Voulvoulis N (2005) Pharmaceuticals: a threat to drinking water? Trends Biotechnol 23:163–167

Ju F, Beck K, Yin X et al (2019) Wastewater treatment plant resistomes are shaped by bacterial composition, genetic exchange, and upregulated expression in the effluent microbiomes. ISME J 13:346–360

Kanakaraju D, Ravichandar S, Lim YC (2017) Combined effects of adsorption and photocatalysis by hybrid TiO_2/ZnO-calcium alginate beads for the removal of copper. J Environ Sci (China) 55:214–223

Kessler R (2010) Industry issues: pharmaceutical factories as a source of drugs in water. Environ Health Perspect 118:A383

Khetan SK, Collins TJ (2007) Human pharmaceuticals in the aquatic environment: a challenge to green chemisty. Chem Rev 107:2319–2364

Kim HY, Asselman J, Jeong TY et al (2017) Multigenerational effects of the antibiotic tetracycline on transcriptional responses of Daphnia magna and its relationship to higher levels of biological organizations. Environ Sci Technol 51:12898–12907

Kizny Gordon AE, Mathers AJ, Cheong EYL et al (2017) The hospital water environment as a reservoir for Carbapenem-resistant organisms causing hospital-acquired infections—a systematic review of the literature. Clin Infect Dis 64:1435–1444

Klavarioti M, Mantzavinos D, Kassinos D (2009) Removal of residual pharmaceuticals from aqueous systems by advanced oxidation processes. Environ Int 35:402–417

Kraemer SA, Ramachandran A, Perron GG (2019) Antibiotic pollution in the environment: from microbial ecology to public policy. Microorganisms 7:180

Kristiansson E, Fick J, Janzon A et al (2011) Pyrosequencing of antibiotic-contaminated river sediments reveals high levels of resistance and gene transfer elements. PLoS One 6:e17038

Kümmerer K (2009) Antibiotics in the aquatic environment: a review—part I. Chemosphere 75:417–434

Kümmerer K (2010) Pharmaceuticals in the environment. Annu Rev Env Resour 35:57–75

Langelier C, Graves M, Kalantar K et al (2019) Microbiome and antimicrobial resistance gene dynamics in international travelers. Emerg Infect Dis 25:1380–1383

Lee CM, Palaniandy P, Dahlan I (2017) Pharmaceutical residues in aquatic environment and water remediation by TiO_2 heterogeneous photocatalysis: a review. Environ Earth Sci 76:1–19

Lee L, Savage VM, Yeh PJ (2018) Intermediate levels of antibiotics may increase diversity of Colony size phenotype in bacteria. Comput Struct Biotechnol J 16:307–315

Levy SB, Bergman MM (2003) The antibiotic paradox: how the misuse of antibiotics destroys their curative powers. In: Clinical infectious diseases, 2nd edn. Perseus Publishing, Boston, MA

Li J, Zhou L, Zhang X et al (2016) Bioaerosol emissions and detection of airborne antibiotic resistance genes from a wastewater treatment plant. Atmos Environ 124:404–412

Li J, Cao J, Zhu YG et al (2018) Global survey of antibiotic resistance genes in air. Environ Sci Technol 52:10975–10984

Lin T, Yu S, Chen Y, Chen W (2014) Integrated biomarker responses in zebrafish exposed to sulfonamides. Environ Toxicol Pharmacol 38:444–452

Lis DO, Pacha JZ, Idzik D (2009) Methicillin resistance of airborne coagulase-negative staphylococci in homes of persons having contact with a hospital environment. Am J Infect Control 37:177–182

Liu L, Wu W, Zhang J et al (2018) Progress of research on the toxicology of antibiotic pollution in aquatic organisms. Acta Ecol Sin 38:36–41

Livermore DM (2011) Discovery research: the scientific challenge of finding new antibiotics. J Antimicrob Chemother 66:1941–1944

Maiques E, Úbeda C, Campoy S et al (2006) β-Lactam antibiotics induce the SOS response and horizontal transfer of virulence factors in *Staphylococcus aureus*. J Bacteriol 188:2726–2729

Marcelino VR, Wille M, Hurt AC et al (2018) High levels of antibiotic resistance gene expression among birds living in a wastewater treatment plant. In: BioRxiv, p 462366

Martínez JL (2017) Effect of antibiotics on bacterial populations: a multi-hierachical selection process. F1000Res 6:51

Matsuo H, Sakamoto H, Arizono K, Shinohara R (2011) Behavior of Pharmaceuticals in Waste Water Treatment Plant in Japan. Bull Environ Contam Toxicol 87(1):31–35. https://doi.org/10.1007/s00128-011-0299-7

Migliore L, Civitareale C, Brambilla G, Dojmi Di Delupis G (1997) Toxicity of several important agricultural antibiotics to Artemia. Water Res 31:1801–1806

Minh TB, Leung HW, Loi IH et al (2009) Antibiotics in the Hong Kong metropolitan area: ubiquitous distribution and fate in Victoria harbour. Mar Pollut Bull 58:1052–1062

Mojica E-RE, Aga DS (2011) Antibiotics pollution in soil and water: potential ecological and human health issues. In: Encyclopedia of environmental health. Elsevier, Burlington, NJ, pp 97–110

Mompelat S, Le Bot B, Thomas O (2009) Occurrence and fate of pharmaceutical products and by-products, from resource to drinking water. Environ Int 35:803–814

Moore CE (2019) Changes in antibiotic resistance in animals. Science 365:1251–1252

Mutiyar PK, Mittal AK (2014) Occurrences and fate of selected human antibiotics in influents and effluents of sewage treatment plant and effluent-receiving river Yamuna in Delhi (India). Environ Monit Assess 186:541–557

Navarrete P, Mardones P, Opazo R et al (2008) Oxytetracycline treatment reduces bacterial diversity of intestinal microbiota of Atlantic salmon. J Aquat Anim Health 20:177–183

Palmer KL, Kos VN, Gilmore MS (2010) Horizontal gene transfer and the genomics of enterococcal antibiotic resistance. Curr Opin Microbiol 13:632–639

Peak N, Knapp CW, Yang RK et al (2007) Abundance of six tetracycline resistance genes in wastewater lagoons at cattle feedlots with different antibiotic use strategies. Environ Microbiol 9:143–151

Pei R, Kim SC, Carlson KH, Pruden A (2006) Effect of river landscape on the sediment concentrations of antibiotics and corresponding antibiotic resistance genes (ARG). Water Res 40:2427–2435

Perron GG, Whyte L, Turnbaugh PJ et al (2015) Functional characterization of bacteria isolated from ancient Arctic soil exposes diverse resistance mechanisms to modern antibiotics. PLoS One 10:e0069533

PHAC, Public Health Agency of Canada (2016) Canadian Antimicrobial Resistance Surveillance System—Report 2016

Pindling S, Azulai D, Zheng B et al (2018) Dysbiosis and early mortality in zebrafish larvae exposed to subclinical concentrations of streptomycin. FEMS Microbiol Lett 365:188

Poeta P, Radhouani H, Igrejas G et al (2008) Seagulls of the Berlengas natural reserve of Portugal as carriers of fecal *Escherichia coli* harboring CTX-M and TEM extended-spectrum beta-lactamases. Appl Environ Microbiol 74:7439–7441

Price LB, Graham JP, Lackey LG et al (2007) Elevated risk of carrying gentamicin-resistant Escherichia coli among U.S. poultry workers. Environ Health Perspect 115:1738–1742

Pruden A, Pei R, Storteboom H, Carlson KH (2006) Antibiotic resistance genes as emerging contaminants: studies in northern Colorado. Environ Sci Technol 40:7445–7450

Ramaswamy BR, Shanmugam G, Velu G et al (2011) GC-MS analysis and ecotoxicological risk assessment of triclosan, carbamazepine and parabens in Indian rivers. J Hazard Mater 186:1586–1593

Ratola N, Cincinelli A, Alves A, Katsoyiannis A (2012) Occurrence of organic microcontaminants in the wastewater treatment pocess. Hazard Mater 239–240:1–18

Raymann K, Moran NA (2018) The role of the gut microbiome in health and disease of adult honey bee workers. Curr Opin Insect Sci 26:97–104

Rémy B, Mion S, Plener L et al (2018) Interference in bacterial quorum sensing: a biopharmaceutical perspective. Front Pharmacol 9:203

Rivera-Utrilla J, Sánchez-Polo M et al (2013) Pharmaceuticals as emerging contaminants and their removal from water. a review. Chemosphere 93:1268–1287

Rizzo L, Manaia C, Merlin C et al (2013) Urban wastewater treatment plants as hotspots for antibiotic resistant bacteria and genes spread into the environment: a review. Sci Total Environ 447:345–360

Schmidt V, Amaral-Zettler L, Davidson J et al (2016) Influence of fishmeal-free diets on microbial communities in Atlantic salmon (Salmo Salar) recirculation aquaculture systems. Appl Environ Microbiol 82:4470–4481

Shah SQA, Cabello FC, L'Abée-Lund TM et al (2014) Antimicrobial resistance and antimicrobial resistance genes in marine bacteria from salmon aquaculture and non-aquaculture sites. Environ Microbiol 16:1310–1320

Singh-Moodley A, Perovic O, Ismail H (2018) An overview of antimicrobial resistance surveillance among healthcare-associated pathogens in South Africa. Afr J Lab Med 7:1–6

Slager J, Kjos M, Attaiech L, Veening JW (2014) Antibiotic-induced replication stress triggers bacterial competence by increasing gene dosage near the origin. Cell 157:395–406

Sørum H (2006) Antimicrobial drug resistance in fish pathogens. In: Aarestrup FM (ed) Antimicrobial resistance in bacteria of animal origin. ASM Press, Washington, DC, pp 213–238

Springman AC, Lacher DW, Wu G et al (2009) Selection, recombination, and virulence gene diversity among group B streptococcal genotypes. JBacteriol 191:5419–5427

Subedi B, Balakrishna K, Sinha RK et al (2015) Mass loading and removal of pharmaceuticals and personal care products, including psychoactive and illicit drugs and artificial sweeteners, in five sewage treatment plants in India. J Environ Chem Eng 3:2882–2891

Szczepanowski R, Linke B, Krahn I et al (2009) Detection of 140 clinically relevant antibiotic-resistance genes in the plasmid metagenome of wastewater treatment plant bacteria showing reduced susceptibility to selected antibiotics. Microbiology 155:2306–2319

Tarpani RRZ, Azapagic A (2018) A methodology for estimating concentrations of pharmaceuticals and personal care products (PPCPs) in wastewater treatment plants and in freshwaters. Sci Total Environ 622–623:1417–1430

Thiele-Bruhn S, Beck IC (2005) Effects of sulfonamide and tetracycline antibiotics on soil microbial activity and microbial biomass. Chemosphere 59:457–465

Tilahun B, Worku B, Tachbele E et al (2012) High load of multi-drug resistant nosocomial neonatal pathogens carried by cockroaches in a neonatal intensive care unit at Tikur Anbessa specialized hospital, Addis Ababa. Ethiopia Antimicrob Resist Infect Control 1:12

Tong AYC, Peake BM, Braund R (2011) Disposal practices for unused medications around the world. Environ Int 37:292–298

Turnidge J (2004) Antibiotic use in animals-prejudices, perceptions and realities. J Antimicrob Chemother 53:26–27

U.S. Food and Drug Administration (2018) Summary report on antimicrobial sold or distributed for use in food-producing animals. U.S. Food and Drug Administration, Silver Spring, MD

UNEP (2017) United Nations Environment Programme. Frontiers 2017: emerging issues of environmental concern. United Nations Environment Programme, Nairobi

Van Boeckel TP, Pires J, Silvester R et al (2019) Global trends in antimicrobial resistance in animals in low- and middle-income countries. Science 365:1944

Vellinga A, Cormican S, Driscoll J et al (2014) Public practice regarding disposal of unused medicines in Ireland. Sci Total Environ 478:98–102

Ventola CL (2015) The antibiotic resistance crisis: causes and threats. Pharm Ther J 40:277–283

Walia K, Madhumathi J, Veeraraghavan B et al (2019) Establishing antimicrobial resistance surveillance & research network in India: journey so far. Indian J Med Res 149:164

Wang N, Noemie N, Hien NN et al (2009) Adverse effects of enrofloxacin when associated with environmental stress in Tra catfish (Pangasianodonhypophthalmus). Chemosphere 77:1577–1584

Wang H, Che B, Duan A et al (2014) Toxicity evaluation of β-diketone antibiotics on the development of embryo-larval zebrafish (Danio rerio). Environ Toxicol 29:1134–1146

Watkinson AJ, Murby EJ, Costanzo SD (2007) Removal of antibiotics in conventional and advanced wastewater treatment: implications for environmental discharge and wastewater recycling. Water Res 41:4164–4176

Watts J, Schreier H, Lanska L, Hale M (2017) The rising tide of antimicrobial resistance in aquaculture: sources. Sinks Solutions Mar Drugs 15:158

Weatherly LM, Gosse JA (2017) Triclosan exposure, transformation, and human health effects. J Toxicol Environ Health Pt B Crit Rev 20:447–469

WHO (2017). https://www.who.int/news/item/27-02-2017-who-publishes-list-of-bacteria-for-which-new-antibiotics-are-urgently-needed

WHO (2019). https://www.who.int/glass/en/

WHO, World Health Organization (2011) Pharmaceuticals in Drinking-water. 1–49

Xu Z, Li L, Shirtliff ME et al (2011) Resistance class 1 integron in clinical methicillin-resistant Staphylococcus aureus strains in southern China, 2001–2006. Clin Microbiol Infect 17:714–718

Yan Z, Lu G, Ye Q, Liu J (2016) Long-term effects of antibiotics, norfloxacin, and sulfamethoxazole, in a partial life-cycle study with zebrafish (Danio rerio): effects on growth, development, and reproduction. Environ Sci Pollut Res 23:18222–18228

Yang Y, Li B, Zou S et al (2014) Fate of antibiotic resistance genes in sewage treatment plant revealed by metagenomic approach. Water Res 62:97–106

Ye Z, Weinberg HS, Meyer MT (2007) Trace analysis of trimethoprim and sulfonamide, macrolide, quinolone, and tetracycline antibiotics in chlorinated drinking water using liquid chromatography electrospray tandem mass spectrometry. Anal Chem 79:1135–1144

Zhang D, Gersberg RM, Ng WJ, Tan SK (2014) Removal of pharmaceuticals and personal care products in aquatic plant-based systems: a review. Environ Pollut 184:620–639

Zhang Q, Cheng J, Xin Q (2015) Effects of tetracycline on developmental toxicity and molecular responses in zebrafish (*Danio rerio*) embryos. Ecotoxicology 24:707–719

Zhang Y, Wang X, Yin X et al (2016) Toxicity assessment of combined fluoroquinolone and tetracycline exposure in zebrafish (*Danio rerio*). Environ Toxicol 31:736–750
Zuccato E, Calamari D, Natangelo M, Fanelli R (2000) Presence of therapeutic drugs in the environment. Lancet 355:1789–1790
Zurek L, Ghosh A (2014) Insects represent a link between food animal farms and the urban environment for antibiotic resistance traits. Appl Environ Microbiol 80:3562–3567

Chapter 3
Antimicrobial Agents in Agriculture and Their Implications in Antimicrobial Resistance

Shweta Singh and Arun Goyal

3.1 Introduction

The spread of antimicrobial resistance genes, among microbial pathogens, is becoming a global threat to environment, human, and animal health (Fig. 3.1). Antibiotics are the chemotherapeutic compounds used to treat bacterial diseases (Amábile-Cuevas 2016). Its development is the greatest achievement in the history of chemotherapy. The first antimicrobial compound discovered by Sir Alexander Fleming was penicillin: a β-lactam antibiotic (Fleming 1929). Antibiotics treat patients with an active infection and permit clinicians to execute medical techniques safely. The use of antibiotics allows organ transplantation, myelosuppressive chemotherapy, basic surgery, and implantation of cardiac devices successfully without infection (Amábile-Cuevas 2016). The misuse and overuse of antibiotics have led to the selection of resistant bacterial strains to every antibiotic which has been introduced in the past 70 years and is a major threat to human health. The center for disease control and prevention in the USA reported that nearly two million illnesses and 23,000 deaths annually are due to antibiotic resistance (Chang et al. 2015).

In agriculture, the use of antibiotics is carried out: (1) to treat bacterial infections in sick animals, (2) for prophylactic use when there is the risk of infection, and (3) in small quantities in water and feed to promote animal growth (Marshall and Levy 2011; Economou and Gousia 2015). The flocks and herds can be given antimicrobial agents in early feeding period when there is a risk of an outbreak of infectious disease, and this mass medication is known as metaphylaxis (Checkley et al. 2010).

S. Singh
North East Centre for Technology Application and Reach, Shillong, Meghalaya, India

A. Goyal (✉)
Department of Biosciences and Bioengineering, Indian Institute of Technology Guwahati, Guwahati, Assam, India
e-mail: arungoyl@iitg.ac.in

N. Akhtar et al. (eds.), *Emerging Modalities in Mitigation of Antimicrobial Resistance*,
https://doi.org/10.1007/978-3-030-84126-3_3

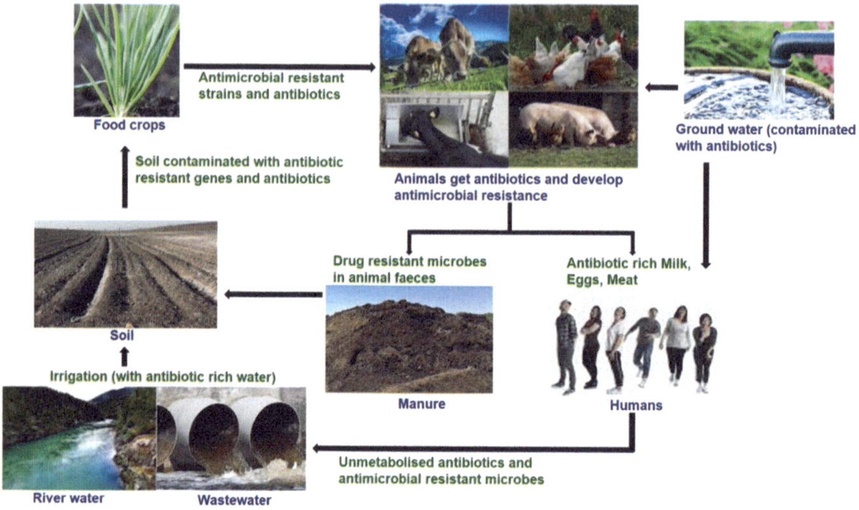

Fig. 3.1 Transfer pathway of antimicrobial resistance gene within microbial community via agriculture, environment, and food chain

Feed prophylaxis and injectable metaphylaxis in cattle improve the feed efficiency and daily weight gain. Some important antibiotics used in the treatment of animals in their food source are amoxicillin, bacitracin, cephalosporins, erythromycin, fluoroquinolones, gentamicin, lincomycin, neomycin, penicillin, streptomycin, tetracycline, and sulfonamides (Shea 2003). Similarly, antifungal drugs used in agriculture served as the environmental drivers for the evolution of resistance in pathogenic fungal strains. Therefore, to decrease the development of resistance in fungal strains for use in agriculture, a mixture of antifungal compounds with different mechanisms of action is supported (Thomson 1982). Antimicrobial agents used for veterinary and human as growth promoters are not fully metabolized and therefore excreted by animals and end up in manure. This manure is used as agricultural fertilizer because of which antimicrobial agents penetrate the soil and enters the groundwater (Thanner et al. 2016). In many countries, the unmetabolized antimicrobial agents are disposed of in the sewage system. If these drugs are not metabolized during sewage treatment, then they seep into the soil or in other environments, and then they can mix with surface water and groundwater (Sanseverino et al. 2018). For example, tetracyclines have been detected in concentrations of up to 0.2 μg/kg in soil (Hamscher et al. 2002). The antibiotics detected in sewage effluent, municipal sewage, and surface water and groundwater include quinolones such as sulfonamides, ciprofloxacin, roxithromycin, erythromycin, and others (Kümmerer 2003). Ciprofloxacin and ampicillin were found in concentrations between 0.7 and 124.5 μg/L and 20–80 μg/L at the hospital effluent which was equal to the minimum inhibitory concentration for susceptible pathogenic bacteria (Kümmerer 2001). Most of the time this untreated sewage water is used for irrigation purposes, and thus it contaminates the soil with antibiotics even above minimum inhibitory concentrations of antimicrobial agents (Kümmerer 2003). In addition to this, the antibiotic concentration in soil at subminimum inhibitory concentrations also selects the antimicrobial-resistant soil microflora via transfer of antimicrobial resistance genes and associated mobile genetic elements (transposons and plasmids) and brings about the genetic modifications in bacterial genome (Cycoń et al. 2019). Gullberg et al. (2011) reported that 1/4 and 1/10 of minimum inhibitory concentration of streptomycin and ciprofloxacin, respectively, select resistant *Salmonella typhimurium* LT2 strain and *Escherichia coli* strain, respectively. This leads to the transmission of antimicrobial drug resistance genes into the soil microflora and as a result to the plants. Therefore, the use of these plant products can transmit drug resistance to humans and animals. In 2013, the US Food and Drug Administration marked a change in its stance, calling on the industry to eliminate the use of essential medical antibiotics.

Various studies have shown that the treatment of foodborne infections in human caused by multidrug-resistant *Campylobacter*, *Salmonella*, *Vibrio*, *Shigella*, *Helicobacter*, and *E. coli* is difficult and needs last-resort antibiotics (Karunasagar et al. 1994; Mousavi et al. 2014; Giacometti et al. 2021; Zachariah et al. 2021). Young children and infants are at higher risk of developing foodborne gastrointestinal infections (Shea 2003). Bacterial populations acquire resistance against antibiotics through either horizontal transfer, vertical gene transfer, or chemical modification of antibiotics. For example, the β-lactamase enzymes were expressed from plasmids; therefore, the horizontal gene transfer is the important mechanism of drug resistance in β-lactam antibiotics (Barlow and Hall 2002). Li et al. (2019) showed the antimicrobial resistance against sulfonamides in *E. coli* via vertical gene transfer. The resistance against β-lactam and glycopeptide antibiotics via chemical modification of penicillin-binding proteins and structure of peptidoglycan precursor inhibits the binding of the drug to the target in *Staphylococcus aureus* (Reygaert 2018).

The human gastrointestinal tract provides an environment for antibiotic resistance genes to spread in normal gut microflora (Thanner et al. 2016). Therefore, to control the acceleration of antibiotic resistance in the microbial community, an adequate understanding of molecular biology of the mobile genetic elements is needed. The knowledge of interactions between the health system and the food chain will reduce the infections caused by antimicrobial resistance microbes and thus improve the patient health. It also guides policymakers and consumers to minimize the nontherapeutic uses of antimicrobial agents (Van Hoek et al. 2011).

3.2 Antibiotics in Agriculture

3.2.1 Antibiotics

The word antibiotic comes from the word "antibiosis" which means "against life." Antibiotics produced from microorganisms or synthesized wholly or partly via synthetic means can inhibit bacterial growth (bacteriostatic) or completely able to kill bacteria (bactericidal). They are differentiated as antibacterials, antivirals, and antifungals on the basis of group of microorganisms they act upon. Antibiotics are classified on the basis of chemical or molecular structures, mode of action, and spectrum of activity (Etebu and Arikekpar 2016).

3.2.1.1 On the Basis of Chemical or Molecular Structures

On the basis of chemical or molecular structures, antibiotics are grouped as β-lactams, tetracyclines, macrolides, quinolones, sulfonamides, aminoglycosides, oxazolidinones, and glycopeptides (van Hoek et al. 2011; Frank and Tacconelli 2012).

3.2.1.1.1 β-Lactams

The member of β-lactam antibiotics contained highly reactive one-nitrogen and three-carbon ring. The β-lactam antibiotics are penicillins, monobactams, cephalosporins, and carbapenems (Etebu and Arikekpar 2016). These antibiotics inhibit the protein essentially required for the synthesis of peptidoglycan layer of bacterial cell wall. Penicillin was the first antibiotic discovered by Alexander Fleming in 1929 from *Penicillium notatum* (McGeer et al. 2001). Penicillin group of antibiotics contained a common nucleus, 6-animopenicillanic acid. This class included several antibiotics such as penicillin G, oxacillin, penicillin V, methicillin, ampicillin, carbenicillin, nafcillin, amoxicillin, piperacillin, ticarcillin, and mezlocillin (Boundless 2016). The monobactam group of antibiotics was produced by bacterium *Chromobacterium violaceum*, and in this group of β-lactam antibiotics, the β-lactam is not fused with other ring (Bonner and Sykes 1984). Aztreonam antibiotic of this group is the only antibiotic which is commercially available. It has narrow range of antimicrobial activity against aerobic Gram-negative bacteria (Sykes et al. 1981). Cephalosporin group of antibiotics are produced by fungus *Cephalosporium acremonium* that contains a common nucleus of 7-aminocephalosporanic acid and a side chain 3,6-dihydro-2 H-1,3-thiazane ring. The cephalosporins are subdivided into five generations depending on their target organism (Abraham 1987). Carbapenem group of β-lactam antibiotics displayed resistance against β-lactamase enzyme (Papp-Wallace et al. 2011). Therefore, these antibiotics are administered to the patients, when they have acquired the infection with resistant microbial strains. The two members of this group are clavulanic acid and thienamycin isolated from *Streptomyces clavuligerus* and *Streptomyces cattleya*, respectively (Brown et al. 1976; Butterworth et al. 1979). They have a broad-range antimicrobial activity against both Gram-positive and Gram-negative bacteria.

3.2.1.1.2 Tetracyclines

Tetracycline produced by *Streptomyces* was discovered by Benjamin Duggar in 1945 (Sánchez et al. 2004). The members of this class of antibiotics have four hydrocarbon rings, and on the basis of the method of synthesis, they are grouped into three different generations (Fuoco 2012). First-generation tetracyclines are produced from biosynthesis, which includes tetracycline, oxytetracycline, chlortetracycline, and demeclocycline. Second-generation tetracyclines are derivatives of semi-synthesis, which includes doxycycline, meclocycline, lymecycline, rolitetracycline, minocycline, and methacycline. The third-generation tetracyclines are produced from total synthesis which includes tigecycline. The target of tetracyclines is ribosome in bacteria, which inhibits the antimicrobial activity by the disruption of the addition of amino acids to polypeptide chains during protein synthesis. They are used in the treatment of amoebic infections, malaria, rickettsia, and elephantiasis (Sánchez et al. 2004).

3.2.1.1.3 Quinolones

Quinolones were first discovered as nalidixic acid. The members of quinolones have been dived into two major groups, i.e., quinolones and naphthyridones. These include nalidixic acid, cinoxacin, ofloxacin, norfloxacin, temafloxacin, ciproxacin, sparfloxacin, temafloxacin, and sparfloxacin (Domagala 1994). The chemical structure of quinolones consists of two rings, but the recent generations also have an added ring which enables them, the broad-spectrum antimicrobial activity. Quinolones interfere the transcription and DNA replication in bacteria.

3.2.1.1.4 Aminoglycosides

In this class of antibiotics, streptomycin was the first drug isolated in 1943 (Mahajan and Balachandran 2012). The members of this class were produced from soil: actinomycetes. They have a broad-spectrum antimicrobial activity. The aminoglycosides consist of 3-amino sugars which are connected by glycosidic bonds. These include streptomycin, neomycin, gentamicin, amikacin, and tobramycin. The members of this class inhibited the protein synthesis by binding with one of the subunits of ribosome. These drugs have been extensively used against *Mycobacterium tuberculosis* (Etebu and Arikekpar 2016).

3.2.1.1.5 Sulfonamides

Sulfonamides are the first group of antimicrobial agents used as therapeutic medicine (Eyssen et al. 1971). They inhibit Gram-positive and Gram-negative bacteria and some protozoa (Eyssen et al. 1971). The antimicrobial sulfonamides are synthetic and bacteriostatic (Henry 1943). They also demonstrated that at higher concentrations they may become bactericidal.

3.2.1.1.6 Glycopeptides

Glycopeptide antibiotics are obtained as natural products or as semisynthetic derivatives. The glycopeptides are consisted of a cyclic peptide having seven amino acids and two bound sugars (Kang and Park 2015). They bind to its target via formation of five hydrogen bonds from the peptidic backbone of the glycopeptide. In few glycopeptides, an additional sugar or chlorine is attached to the backbone of glycopeptide (oritavancin) at the time of synthesis. Similarly, the lipophilic side chain prolongs the half-life of glycopeptides (Etebu and Arikekpar 2016).

3.2.1.1.7 Oxazolidinones

These antimicrobial agents are a group of synthetic antibiotics. Linezolid is the first member of this group and approved for medical application. Earlier studies showed that these drugs inhibited antimicrobial activity via interfering with protein synthesis. The members of this class have a broad-spectrum activity against Gram-positive bacteria (Shinabarger et al. 1997; Bozdogan and Appelbaum 2004). Oxazolidinones are given in surgical infections as they can easily penetrate in the tissue, i.e., the lung, bone and vegetations (plant-like growth in tissues), and cerebrospinal fluid (Bozdogan and Appelbaum 2004).

3.2.1.1.8 Macrolides

The first macrolide, erythromycin A, was isolated from fungus *Streptomyces erythraeus* (Moore 2015). The members of this class have 14-, 15-, or 16-membered macrocyclic lactose rings along with unusual deoxy sugars D-desosamine and L-cladinose (Moore 2015). These antibiotics have broad-range spectrum for antimicrobial activity. Macrolides can inhibit growth or kill microorganisms by inhibiting the protein synthesis. The members of this class are erythromycin, clarithromycin, and azithromycin. These antibiotics showed broad-range spectrum antimicrobial activity. These antibiotics generally cause inflammation in the body; therefore, clinicians recommended low-dose administration (Hamilton-Miller 1973).

3.2.1.2 On the Basis of Mode of Action

In this mode of classification, antibiotics are divided into groups as follows: (a) inhibits the synthesis of bacterial cell wall, (b) disrupts the function or structure of cell membrane, (c) inhibits the function or structure of nucleic acid, (d) inhibits the protein synthesis, and (e) alters the metabolic pathways.

3.2.1.2.1 Antibiotics Inhibits the Synthesis of Bacterial Cell Wall

The members of this class of antibiotics are cephalosporins, carbapenems, and penicillins (Josephine et al. 2004). Antibiotics of glycopeptide class such as vancomycin also comes in this class (Kahne et al. 2005). These antibiotics inhibit the synthesis of peptidoglycan, a main component of bacterial cell wall. The rigid layer of peptidoglycan protects the bacterial cell by maintaining the high osmotic pressure in the cell (Bugg and Walsh 1992).

3.2.1.2.2 Antibiotics Disrupts the Function or Structure of Cell Membrane

The members of this class of antibiotics are daptomycin (Alborn et al. 1991) and polymyxins (Falagas et al. 2010). On the basis of differences in the type of lipids in cell membrane, the above class of antibiotics are grouped. Daptomycin depolarizes calcium-dependent membrane, whereas polymyxins binds to the lipopolysaccharide, i.e., lipid moiety of the cell membrane, and resulted in the disintegration of bacterial cell membrane (Falagas et al. 2010).

3.2.1.2.3 Antibiotics Inhibits the Synthesis of Nucleic Acid

The members of this class of antibiotics are quinolones (Etebu and Arikekpar 2016). The synthesis of nucleic acid is essential for the survival of bacterial cell. This group of antibiotics interferes the synthesis of nuclei acid by stopping replication or transcription. In DNA replication, helicase enzyme unwinds the double helical structure of DNA (Gale et al. 1981). Antibiotic quinolones disrupt the functionality of the helicase enzyme because of which it no longer unwinds the DNA in bacteria. In bacteria, this group of antibiotics also inhibits the functionality of topoisomerase II and topoisomerase IV because of which they prevent the RNA synthesis (Etebu and Arikekpar 2016).

3.2.1.2.4 Antibiotics Inhibiting the Protein Synthesis

Antibiotics inhibiting the protein synthesis are divided into two subclasses: (a) 30S inhibitors and (b) 50S inhibitors. The members of 30S ribosome inhibitors of antibiotic subclass are spectinomycin, streptomycin, and tetracycline (Hong et al. 2014). These antibiotics inhibit the entry of aminoacyl-tRNAs to the ribosome. The members of 50S ribosome inhibitors of antibiotic subclass are chloramphenicol, clindamycin, erythromycin, lincomycin, and linezolid (Douthwaite 1992; Katz and Ashley 2005). These antibiotics inhibit the initiation or the elongation phase of protein synthesis (Patel et al. 2001).

3.2.1.2.5 Antibiotics Blocking the Key Metabolic Pathway

The members of this class of antibiotics are sulfonamides and trimethoprim. These antibiotics mimic the structure of substrate required in cellular metabolism of bacteria. Due to this, the antibiotic attaches to the bacterial enzyme instead of the original substrate. The sulfonamide antibiotics resembles tetrahydrofolate required in the synthesis of folic acid in bacteria (Talaro and Chess 2008). This folic acid plays a vital role in the nucleic acid and amino acid metabolism. Therefore, these sulfonamide drugs inhibit the nucleic acid synthesis (Talaro and Chess 2008).

3.2.2 Crop Protection

In agriculture, streptomycin was the first antibiotic used against an enterobacterium *Erwinia amylovora* for controlling fire blight of pome fruits (McManus et al. 2002). The annual economic loss due to fire blight and control cost in the USA was estimated to be more than USD 100 million (Norelli et al. 2003). It was estimated to be USD 42 million and USD 68 million in Michigan and Washington, respectively (Aćimović et al. 2015). Therefore, the antibiotic streptomycin was registered in 1959 in the USA for its use in crop protection (McManus et al. 2002). Blasticidin S is the first antibiotic developed selectively for crop protection by Japan. It was isolated from *Streptomyces griseochrogenes* and used against rice blast disease (Kidd and James 1991). The rice blast causes nearly 30% of global rice production loss which was equivalent to the feeding of nearly 60 million people (Nalley et al. 2016). The control on rice blast motivated other researchers to develop antifungal antibiotics such as aureofungin (Nene and Thapliyal 1993), kasugamycin (Tomlin 2006), and validamycin (Liao et al. 2009) for crop protection. Aureofungin is a broad-spectrum fungicide registered in India against gummosis in citrus. The disease gummosis is caused by *Phytophthora* species which causes major economic losses to the citrus industry. The epidemic outbreak of gummosis in California caused 46% decrease in the annual yield of citrus plants which resulted in 12.9 million dollars of annual economic loss to the citrus industry (Savita and Nagpal 2012). Validamycin is a non-systemic fungicidal antibiotic used against soilborne disease caused by *Rhizoctonia solani* in potatoes, rice, and vegetables. In addition, validamycin is also used for curing infection in vegetable seedlings such as sugar beets, cotton, and rice (Thomson 1982). In India, tetracycline and streptomycin are registered against many plant diseases (Vidaver 2002). Earlier report of Vidaver (2002) showed that gentamicin, an aminoglycoside antibiotic, is used to control various bacterial diseases of vegetable crops caused by *Pseudomonas*, *Pectobacterium*, *Ralstonia*, and *Xanthomonas*. Oxolinic acid is a quinolone antibiotic used to treat fiber blight of pear and blight of rice (Maeda et al. 2004). So, to avoid the global economic losses, antibiotics were used in the plant agriculture for controlling plant pathogens. Most of the above antibiotics used in controlling plant pathogens are also used for treating human diseases. The antibiotic streptomycin is used in the treatment of tuberculosis caused by *Mycobacterium* in humans (Grosset and Singer 2013). It is also used in the treatment of bacterial infections in pig (Luk'aš et al. 1963), poultry (Barbour et al. 1985), and sheep and goat (Radwan et al. 1992) and as growth promoters. The tetracyclines are used in the treatment of cholera, acne, malaria, plague, brucellosis, pneumonia, and syphilis in humans (Chopra and Roberts 2001) and as growth promoter and in bacterial infection treatment in animals (Di Cerbo et al. 2019). The antibiotic gentamicin is used in the treatment of meningitis (Mccracken et al. 1980) and infections in the lungs (Pines et al. 1967), abdomen (Gomez et al. 1999), urinary tract (Chong et al. 2003), and skin (Gemeinder et al. 2021) in humans, and it is a common medicine used in treating bacterial infections in veterinary practices (Jensen et al. 2006).

3.2.3 Animal Husbandry

The antibiotics are used since the late 1940s in veterinary medicine to treat animals from preventing infections in flocks or herds. In the early 1950s, the Food and Drug Administration (FDA) in the USA approved the use of antibiotics in animal feed (Mellon et al. 2001; Amábile-Cuevas 2016). This practice improves animal growth with a better economy to farmers in a shorter period of time (Mellon et al. 2001). The antibiotics used in livestock are classified as therapeutic agents, prophylactic agents, and metaphylactic agents (growth promoters) (Sarkar et al. 2018). Antimicrobials as therapeutic agents are used in relatively high doses for treating infected animals in a shorter period of time. Antimicrobial agents as prophylactic agents are given to animal at sub-therapeutic doses via drinking water or feed to prevent disease, when symptoms of infection are not present but suspected. Antimicrobial agents as growth promoters are given to animals at a very low dose on regular basis over the lifetime through fodder to increase productivity and growth rate. In animal husbandry, the worldwide use of antibiotics was 63,000 tons in 2010, out of which 3%, 3%, 9%, 13%, and 23% were consumed in Germany, India, Brazil, the USA, and China, respectively (Van Boeckel et al. 2015; Amábile-Cuevas 2016). This use of antibiotics is anticipated to increase by two-third to 105,600 tons by 2030. Most of the antibiotics used in animal husbandry are also used for humans. The key classes of in-use antibiotics shared by humans and animals are shown in Table 3.1. The antibiotics from classes, macrolide, penicillin, and tetracycline are mostly used for treating infected animals, each in the annual order of USD 500 million (Laxminarayan et al. 2015). Earlier study of Sarkar et al. (2018) reported that the worldwide average annual consumption of antibiotics for pig, chicken, and cattle is estimated to be 172 mg/kg, 148 mg/kg, and 45 mg/kg, respectively. The World Health Organization (WHO) has given a list of critically important, highly important, and important antibiotics that must be well preserved for human use (Table 3.2) to prevent the spread of antimicrobial resistance (Scott et al. 2019).

Table 3.1 Key classes of antibiotics shared by humans and animals

Antibiotic class	Used in humans	Used in animals	References
β-lactams	Amoxicillin, penicillin G	Amoxicillin	Neu (1974), Sarkar et al. (2018)
Macrolides	Erythromycin, azithromycin	Erythromycin, azithromycin	Noli and Boothe (1999)
Tetracyclines	Doxycycline, chlortetracycline, oxytetracycline	Doxycycline, oxytetracycline	Di Cerbo et al. (2019)
Fluoroquinolones	Lomefloxacin, ciprofloxacin, ofloxacin	Ciprofloxacin, enrofloxacin, sarafloxacin, lomefloxacin, ofloxacin	Ihrke et al. (1999)
Streptogramins	Quinupristin-dalfopristin	Virginiamycin	Schwarz et al. (2016)
Glycopeptide	Vancomycin, teicoplanin, avoparcin	Avoparcin	Pyörälä et al. (2009), Sarkar et al. (2018)
Phenicol	Chloramphenicol, thiamphenicol	Florfenicol	Schwarz et al. (2016)
Polypeptides	Bacitracin, polymyxin B	Enramycin	Dowling (2013a, b)
Pleuromutilins	Retapamulin	Tiamulin, valnemulin	Novak and Shlaes (2010), Schwarz et al. (2016)

3.2.4 Aquaculture

The Food and Agriculture Organization estimated that the annual supply of fish in fisheries is around 110 million tons. However, the pathogenesis in fisheries causes unpredictable mortalities and hinders fish production in aquaculture (Cabello 2006; Defoirdt et al. 2011). The Food and Drug Administration has approved the use of antibiotics such as florfenicol, oxytetracycline, and sulfadimethoxine/ormetoprim in aquaculture to prevent this loss (Romero et al. 2012). In India, antibiotics such as erythromycin, streptomycin, chloramphenicol, and co-trimoxazole have been used in aquaculture (Sarkar et al. 2018). Antibiotics are given to fish after mixing them in the formulated feed. However, the fish subsequently passed the unmetabolized antibiotics into the environment via feces. In India, the Marine Products Export Development Authority has listed the banned pharmacologically active substances and antibiotics for use in aquaculture such as nitrofurans, chloramphenicol, nalidixic acid, neomycin, sulfamethoxazole, chloroform, colchicine, chlorpromazine, dapsone, dimetridazole, sulfonamides, fluoroquinolones, nitroimidazoles, ipronidazole, and ronidazole (Sarkar et al. 2018).

Table 3.2 Class of antibiotics used in the treatment of infection in humans and animals

Antibiotic class	Animal use	Human medicine importance	References
Cephalosporin	Treatment of infection in cattle and swine	Critically important	Doyle et al. (2013), Asai et al. (2011)
Macrolide	Treatment of disease in cattle and swine	Critically important	Doyle et al. (2013), Pyörälä et al. (2014)
Fluoroquinolones	Treatment of infection in cattle, swine, and poultry	Critically important	Huq (2006), Doyle et al. (2013)
Tetracycline	Treatment of infection in cattle and swine	Highly important	Sengeløv et al. (2003), Doyle et al. (2013)
Penicillins	Treatment of infection in cattle, poultry, and swine	Highly important	Doyle et al. (2013), De Briyne et al. (2014)
Lincosamide	Treatment of infection in cattle and swine	Highly important	Doyle et al. (2013), Pyörälä et al. (2014)
Streptogramin	Treatment and prevention of infection in cattle, swine, and poultry	Highly important	Giguère (2013), Doyle et al. (2013)
Phenicol	Treatment of infection in cattle and swine	Not	Dowling (2013a, b)
Sulfonamide	Treatment of infection in swine	Not	Stahl et al. (2016)
Pleuromutilin	Treatment of infection in swine	Not	Alban et al. (2017)
Bambermycin	Treatment in cattle and chickens	Not	Dealy and Moeller (1977), Letellier et al. (2000)
Carbadox	Treatment of infection in swine	Not	Looft et al. (2014)

3.3 Routes of transmission of Antimicrobial-Resistant Bacteria in the Environment

According to the methods of waste management released from information for all program, the United Nations Educational, Scientific and Cultural Organization (UNESCO), the antimicrobial-resistant bacteria are released into the environment via air, water, and soil through disposals from animal houses and feedlots (Islam et al. 2019). The earlier report of the US Department of Agriculture found that food animals nearly produce 335 million tons of manure annually which is 40 times more than the mass of human biosolids (7.6 million) (Silbergeld et al. 2008; Linville et al. 2015). The unmetabolized antimicrobial agents and resistant microbial strains are excreted by the animals into the manure, and its applications in soil are the main cause of spread of antimicrobial resistance in soil microflora. The surface water and groundwater get contaminated by the animal waste and sewage disposal which leads antimicrobial resistance pathogenic strains in drinking water and irrigated soil (Islam et al. 2004). Food crops grown in wastewater-irrigated soil and manure application soil get contaminated with antibiotic-resistant bacteria and are the major source of transmission of multidrug resistance in humans. Mass gatherings in religious congregations, sports events, beach sports, and water parks are another source of transmission of antibiotic-resistant bacteria and antibiotic-resistant genes in the water bodies and resulted in the spread of waterborne diseases, i.e., endemic cholera (Mutreja et al. 2011; Fouz et al. 2020). Recently, antimicrobial resistance genes corresponding to different classes of antibiotics, including β-lactams, fluoroquinolones, aminoglycosides, macrolide-lincosamide-streptogramins (MLS), sulfonamides, and rifampicin, were identified in samples collected from the Ganges-Yamuna river confluence site (Samson et al. 2019). Hajj is a religious gathering in Saudi Arabia where pilgrims stayed in tents in Mina for 3–5 days. The wastewater from septic tanks of Hajj pilgrim camps showed the antibiotic resistance genes for β-lactam antibiotics and aminoglycoside (Fouz et al. 2020). In addition to this, water sports in water and sea parks cause the spread of antibiotic resistance genes by direct ingestion of water by many people (Leonard et al. 2015; Williams et al. 2016). The wastewater from antibiotic manufacturing plant or pharmaceuticals disposed antimicrobials in aquatic environment which is directly associated with the spread of antibiotic resistance genes in the community. β-Lactams, tetracyclines, aminoglycosides, glycopeptides, fluoroquinolones, macrolides, sulfonamides, and trimethoprim have been detected in wastewater from pharmaceutical industries (Felis et al. 2020). Hospital wastewater is the hotspot of antibiotic resistance genes as it contains the mixture of antibiotics, metabolized antibiotic from the patient excreta, and multidrug-resistant bacteria having resistance to extended-spectrum β-lactams, tetracyclines, carbapenems, and sulfonamides (Zhang et al. 2020).

3.4 Consequences of Antibiotic Use in Agriculture

3.4.1 Consequences of Antimicrobial Resistance in Crop Disease

The antibiotics have been used against crop disease since the 1950s. After the continuous use of antibiotics in agriculture for years, the plant pathogens became resistant to antibiotics (Sarkar et al. 2018). Streptomycin was used against plant pathogen, *Xanthomonas campestris*, for controlling disease in tomato and pepper plants (Minsavage et al. 1990). Nowadays, *Xanthomonas campestris* strain has acquired resistance against antibiotic streptomycin. This is due to the single point mutation in *rpsl* chromosomal gene and thus leads to the inhibition of the binding of streptomycin to ribosome (Sarkar et al. 2018). Similarly, streptomycin-resistant *Erwinia amylovora*, the causative agent of fire blight in apple and pears, and *Pseudomonas syringae*, the causative agent of necrosis in plants, evolved (Sundin and Bender 1993; Russo et al. 2008). *Erwinia amylovora* showed resistance against oxolinic acid (Chen et al. 2019). *Pectobacterium carotovorum*, the causative agent of soft rot potato tubers and bacterial canker, showed multidrug resistance against penicillin, polymyxin B, erythromycin, and vancomycin (Koh et al. 2012). *Acidovorax avenae*, the causative agent of bacterial leaf blight, showed resistance against kasugamycin (Chen et al. 2019). Tetracycline resistance was reported in plant pathogenic bacteria, *Pseudomonas syringae* and *Agrobacterium tumefaciens* (Chen et al. 2019). The bacterium *Pseudomonas syringae* pandemically affected *Actinidia chinensis* and *Actinidia deliciosa* in China, Italy, New Zealand, Spain, Portugal, France, Japan, and South Korea (Scortichini et al. 2012). Therefore, countries having kiwifruit industries suffered from grievous economic losses. The symptoms of the disease are brown leaf spots along with chlorotic haloes, discoloration of cankers, and bud damage fruits (Cameron and Sarojini 2014). This report showed that the development of antibiotic resistance in *Pseudomonas syringae* is a major threat to plant ecosystem. *Agrobacterium tumefaciens* is the most important plant pathogen known to cause crown gall disease eudicots (De Cleene and De Ley 1976). Antimicrobial resistance in plant pathogenic bacteria is a major threat to pathosystems because it decreases the productivity of the agrosystem and transfers resistance to human pathogenic strains (Williams-Nguyen et al. 2016; Sundin and Wang 2018; Raman et al. 2020).

3.4.2 Medical Consequences of Antibiotic Use in Agriculture

Several studies showed increasing multiple drug resistance among key enteric pathogens such as *Escherichia coli*, *Salmonella* spp., *Vibrio cholera*, *Shigella* spp., *Campylobacter* spp., and *Helicobacter* sp. to nearly all commonly available antibiotics. Therefore, it is imperative that this trend should be reversed.

3.4.2.1 *Escherichia coli*

Escherichia coli is the most predominant facultative anaerobic bacteria in the gastrointestinal tract of animals and humans. It is the most common enteric pathogen that also has harmful strains which causes serious urinary tract infection (Varughese and Beniwal 2015). So, multidrug resistance in *Escherichia coli* is of great concern. Rasheed et al. (2014) found that *Escherichia coli* isolates from raw chicken, raw meat, vegetable salad, unpasteurized milk, and raw egg surface showed multidrug resistance. The isolates from raw chicken possessed multidrug resistance against ampicillin, amoxicillin, aztreonam, cefotaxime, ceftazidime, chloramphenicol, ciprofloxacin, co-trimoxazole, gentamicin, ofloxacin, streptomycin, and tetracycline (Rasheed et al. 2014). Isolates from unpasteurized milk showed drug resistance against ampicillin, amoxicillin, amoxyclav, cefotaxime, chloramphenicol, ciprofloxacin, ofloxacin, streptomycin, and tetracycline. Food chain is the most common mode of cycling of pathogens among humans and animals (Laxminarayan et al.

2015). Therefore, good hygienic practices are required to avoid contamination from poultry and cattle products in humans (Marshall and Levy 2011; Meng et al. 2017).

3.4.2.2 *Salmonella* spp.

Salmonella enterica causes salmonellosis, the most commonly reported foodborne bacterial disease (Shang et al. 2021). Chicken meat is the second largest source of animal protein for humans. So, the dissemination of antimicrobial resistance *Salmonella* sp. through the food chain, by chicken, has important consequences for the failure of salmonellosis treatment (Shang et al. 2021). A recent report showed multidrug resistance monophasic *Salmonella typhimurium* against streptomycin, ampicillin, and tetracyclines (Giacometti et al. 2021). The *Salmonella* serotype isolates from humans were found to be identical to *Salmonella typhimurium*, *Salmonella enteritidis*, and *Salmonella infantis* isolated from food (Rospotrebnadzor 2018). All *Salmonella* serotypes were resistant to amoxicillin-clavulanic acid, 93.8% to ceftazidime and cefuroxime, 68.8% to tetracycline, 47.9% to gentamicin, 39.6% to ciprofloxacin, 41.7% to nalidixic acid, 31.3% to ofloxacin, 27.1% to cefixime, and 52.1% to multidrug resistance against all these antibiotics (Rospotrebnadzor 2018).

3.4.2.3 *Vibrio* spp.

Vibrio sp. are the causative agent of diarrheal and systemic diseases in humans. They grow at temperatures from 10 to 30 °C in aquatic environment. Several *Vibrio* sp. such as *Vibrio harveyi*, *Vibrio splendidus*, *Vibrio vulnificus*, *Vibrio parahaemolyticus*, *Vibrio anguillarum*, *Vibrio alginolyticus*, and *Vibrio parahaemolyticus* induce infections in the aquatic ecosystem. *Vibrio vulnificus* cause wound infections of soft tissues, gastroenteritis, and septicemia (Armstrong et al. 1983). *Vibrio cholerae*, *Vibrio parahaemolyticus*, *Vibrio vulnificus*, and *Vibrio alginolyticus* strains are human pathogens. *Vibrio cholerae* is the most common causative agent of cholera in humans. Earlier report of Pan et al. (2008) showed that *Vibrio cholera* has resistance against various antibiotics, ampicillin, tetracycline, streptomycin, chloramphenicol, gentamicin, and trimethoprim-sulfamethoxazole. Other *Vibrio* sp. showed resistance against antibiotics: ampicillin, cefazolin, ciprofloxacin, cefotaxime, streptomycin, and cefuroxime sodium (Yano et al. 2014).

3.4.2.4 *Shigella* spp.

Shigella are Gram-negative bacteria and cause acute gastrointestinal infection, shigellosis, or bloody diarrhea (Shahin et al. 2019). *Shigella dysenteriae* type 1 was first isolated in Japan in 1896 by Kiyoshi Shiga, as it caused severe dysentery epidemic. The global report on mortalities due to shigellosis was more than 700,000 and 60% of them occurred in children below the age of 5 years (Puzari et al. 2018). Shigella strains are transmitted through fresh vegetables (Jo et al. 2019), contaminated meat and dairy products (Ahmed and Shimamoto 2014), contaminated food and water, and person-to-person contact in human beings (Niyogi 2005). Shigellosis can occur in the pandemic, epidemic, and sporadic forms. Epidemics have been reported from India, Myanmar, Bhutan, Nepal, Maldives, Sri Lanka, Bangladesh, and Central American countries (Muthuirulandi Sethuvel et al. 2017). In the year 2017, the World Health Organization (WHO) released a report that for treating *Shigella* infections, there is an urgent need for new antibiotics or the alternative approaches to be explored. The National Antimicrobial Resistance Monitoring System reported that in the year 1999, only 57%, 56%, 1.6%, 0.3%, and 1% of *Shigella* isolates were resistant to tetracycline, streptomycin, nalidixic acid, ciprofloxacin, and ceftriaxone, respectively, whereas in recent years these rates increased to 87%, 57%, 98.3%, 23.1%, and 10.8%, respectively (Shahin et al. 2020). Zachariah et al. (2021) reported that *Shigella* spp. showed 91.7%, 83.3%, 82.1%, 73.1%, 66.7%, and 54.2% against erythromycin, doxycycline, ampicillin, co-trimoxazole, minocycline, and cefuroxime, respectively.

3.4.2.5 *Campylobacter* spp.

Campylobacter spp. are microaerophilic curved comma or s-shape Gram-negative bacteria. They are part of the normal intestinal flora of domestic ruminants. Food products from ruminants such as meat and milk predominantly spread this bacterium to humans. *Campylobacter coli* and *Campylobacter jejuni* cause the most commonly detected enteric infections, campylobacteriosis in humans (Samie et al. 2007). Earlier studies of Hong et al. (2007) showed that *Campylobacter* isolates from chicken raw meat showed resistance to doxycycline, ciprofloxacin, nalidixic acid, tetracycline, enrofloxacin, and erythromycin. Kinana et al. (2006) reported the predominance of macrolide- and fluoroquinolone-resistant *Campylobacter jejuni* and *Campylobacter coli* is increasing globally. In tropical developing countries, *Campylobacter* infection among children is hyperendemic. The community-based studies assessed the occurrence rate of *Campylobacter* infection in children below 5 years of age which is between 40,000 and 60,000 notifications per 100,000 population (Coker et al. 2002). According to a recent report of Zachariah et al. (2021), *Campylobacter jejuni* showed 87.5%, 75%, 73.7%, 73.3%, and 68.8% resistance against erythromycin, doxycycline, ampicillin, co-trimoxazole, and minocycline, respectively.

3.4.2.6 *Helicobacter* spp.

Helicobacter pylori is a curved spiral shape, microaerophilic Gram-negative bacterium. It is the causative agent of peptic ulcer disease, gastric adenocarcinoma, type B gastritis, and mucosa-associated lymphoid tissue lymphoma and the fourth common type of cancer (Mousavi et al. 2014). Earlier studies showed that *Helicobacter pylori* is the natural host in the gut of domestic ruminants, i.e., cow, sheep, buffalo, goat, and camel species without any gastritis (Rahimi and Kheirabadi 2012). It was found that the milk and feces of these domestic species showed the presence of *Helicobacter pylori* (Rahimi and Kheirabadi 2012). *Helicobacter pylori* can survive in water, milk, fresh meat, fresh fruit and vegetables, and some dairy products at temperatures below 30 °C and provide optimal conditions for its transmission to humans (Fan et al. 1998). *Helicobacter pylori* infects about 50% of persons above the age of 60 years and 20% of persons below the age of 40 years (Sosa et al. 2010). Metronidazole and clarithromycin belong to nitroimidazole and macrolide class of antibiotics, respectively, and are used with amoxicillin for the treatment of gastritis and peptic ulcer caused by *Helicobacter pylori*. They are the first-line treatment antibiotics for infections caused by *Helicobacter pylori*. Therefore, resistance in Helicobacter pylori against these antibiotics hampers the treatment (Abadi 2017). Resistance in *Helicobacter pylori* against these antibiotics is low in Asia and Europe (0.8–9.1%) and higher (nearly 18%) in developing countries (Lwai-Lume et al. 2005).

3.4.2.7 Urinary Tract Infection Causing Pathogens

In urinary tract infection, bacteria colonize in the urinary tract. The lower urinary tract infection is known as cystitis (bladder infection) and in the upper urinary tract infection is known as pyelonephritis. The common symptoms of lower urinary tract infection are pain with frequent urination and feeling the need to urinate and are flank pain, fever, and sometimes bloody urine. The urinary tract infections are caused by Gram-positive and Gram-negative bacteria, some fungi, viruses, and protozoa. The most common bacteria causing urinary tract infection are *Escherichia coli*, *Klebsiella pneumoniae*, *Proteus mirabilis*, *Enterococci*, *Pseudomonas aeruginosa*, *Staphylococcus aureus*, and *Enterobacter* spp. (Ziółkowski et al. 2021). Mechal et al. (2021) reported that *Escherichia coli* and *Staphylococcus aureus* acquired high level of resistance against ampicillin and tetracycline. The antimicrobial resistance in urinary tract infection causing microbes causes recurrent infections and fails to antibiotic therapy (Vachvanichsanong et al. 2021).

3.5 Mechanisms of Antibiotic Resistance

Bacteria have different mechanisms to survive antibiotics, which in turn are dependent on the mode of action, biochemical properties, and structures of different antibiotics (Barker 1999). The mechanisms of antimicrobial resistance can be divided into five major groups:

1. Prevention of the antimicrobial agent from reaching its cellular target by decreasing active export or uptake of the antimicrobial agents, e.g., a decrease in the

number of porin channels decreases the antibiotic entry into the cell, and efflux pump decreases the concentration of antibiotics in cell by exporting them outside the cell (Fernández and Hancock 2012),
2. Alterations of the antimicrobial agents, e.g., resistant microbes convert penicillin into penicilloic acid (Cole et al. 1973).
3. Microorganism protects the target site of antimicrobials via phosphorylation, acetylation, or nucleotidylation, e.g., in aminoglycosides (Mingeot-Leclercq et al. 1999).
4. Formation of alternative metabolic pathways, which bypass some steps which are inhibited by antimicrobial agents (Reynolds 1989).
5. Alteration of the target protein through mutation (Barker 1999; Morar and Wright 2010; Nguyen et al. 2014; Munita and Arias 2016).

3.5.1 Transfer of Antibiotic Resistance in Clinical Isolates via Agricultural Sources

Humans are at high risk of exposure to resistant pathogens through direct contact during handling, contact with infected humans, and ingestion of contaminated food and water (Chang et al. 2015). The mechanisms by which antibiotic resistance through agricultural microbes is transferred to human microbes are mainly via horizontal transfer of mobile genetic elements such as broad host range plasmid DNA, integrons and transposable elements, and vertical gene transfer.

3.5.1.1 Horizontal Gene Transfer

Horizontal gene transfer is the process of transmission of the genetic material between unicellular and multicellular organisms from donor to recipient. It is carried out via three main mechanisms such as conjugation, transformation, and transduction (Stearns and Hoekstra 2005). It occurs in the digestive system of humans and animals, food, soil, and water. Most of the antimicrobial resistance genes are carried on plasmids, integrons, or transposons. These can act as vectors for transferring antimicrobial resistance genes to other members of the same species or other genus or species (Keeling and Palmer 2008). The frequency of horizontal gene transfer mainly depends on the characteristics of the donor and recipient populations, properties of mobile genetic elements, and the environment (Verraes et al. 2013). Moreover, the high intrinsic recombination ability in certain microbes such as *Escherichia coli*, *Salmonella enterica*, *Bacillus subtilis*, and *Streptococcus pneumoniae* increases the rate of horizontal gene transfer (Dutta and Pan 2002). *Salmonella typhimurium* transferred ampicillin resistance gene in *Escherichia coli* in milk, and *Enterococcus faecalis* transferred tetracycline and erythromycin resistance in *Leuconostoc monocytogenes* (Von Wintersdorff et al. 2016).

3.5.1.1.1 Conjugation

Conjugation is the process of transfer of genetic material (DNA) between the live bacteria that requires a direct contact between the recipient and the donor cells (Fig. 3.2). The antimicrobial resistance genes are present on the mobile elements, such as transposons, plasmids, insertion sequence, and integrons. The transposons and insertion sequences can translocate the antimicrobial resistance genes only within the bacterial cells (Lyras et al. 2004). An integron is an immobile element that expresses and integrates or releases the gene cassettes. It is divided into two groups: (1) the chromosomal integrons which are associated with the bacterial chromosome and (2) the mobile integrons associated with plasmids and transposons (Cambray et al. 2010). In Gram-negative bacteria, the production of extended-spectrum beta-lactamase (ESBL) enzymes commonly imparts resistance against β-lactam antimicrobial agents. The genes encoding ESBL enzymes are usually acquired or transferred in microbes belonging to *Enterobacteriaceae* by conjugation (Liu et al. 2019). A recent study by Johansson et al. (2021) showed that an isolate of *Salmonella enterica* from animal origin has translocatable mobile resistance elements for tetracycline, sulfonamide, and aminoglycoside.

3.5.1.1.2 Transformation

Transformation is the process of uptake, integration, and expression of foreign genetic material from the environment by a recipient cell as shown in Fig. 3.3 (Stearns and Hoekstra 2005). *Bacillus subtilis*, *Campylobacter* spp., *Helicobacter pylori*, *Neisseria gonorrhoeae*, *Haemophilus influenzae*, *Acinetobacter baylyi*, and *Streptococcus* spp. have natural tendency of competence (Lorenz and Wackernagel 1994; Verraes et al. 2013). Natural transformation is observed in various environments, such as in soil, water, foodstuff, and human fluids, and contributed to the

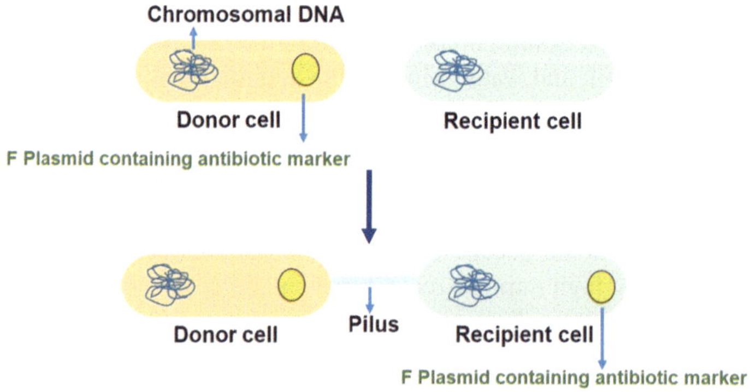

Fig. 3.2 Conjugation by direct contact between two bacterial plasmids by forming a mating bridge between the bacteria which results in DNA transfer for antibiotic resistance genes to a recipient cell

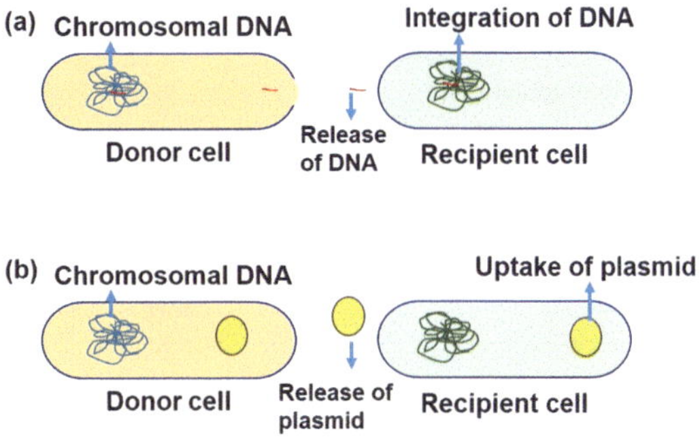

Fig. 3.3 (**a**) Transformation by transfer of antibiotic resistance gene from donor to a recipient cell and its integration into its chromosome. (**b**) Uptake of a plasmid DNA by the recipient cell from a donor cell

genetic variations in human pathogens (Domingues et al. 2012). The report of Domingues et al. (2012) showed that *Pseudomonas aeruginosa* acquired resistance against ampicillin and sulfamethoxazole and *Acinetobacter* sp. acquired resistance against ampicillin, ceftazidime, gentamicin, cefotaxime, kanamycin, and sulfamethoxazole via transformation. A recent report of Jin et al. (2020) showed that the anthropogenic activity, i.e., the chlorination of drinking water by sodium hypochlorite, enhanced the transmission of antibiotic resistance genes encoding tetracycline, ampicillin, kanamycin, and nalidixic acid markers in enteric pathogens *Enterococcus faecalis* and *Escherichia coli*.

3.5.1.1.3 Transduction

Transduction is the process in which bacterial DNA is transferred from one bacterium to another by a bacteriophage. The mobilization of antibiotic resistance genes by bacteriophages has been reported in strains of *Streptococcus pyogenes* for tetracycline, in enterococci for tetracycline and gentamicin, in *Escherichia coli* for β-lactamase, and in *Salmonella* for multidrug resistance (Von Wintersdorff et al. 2016).

3.5.1.2 Vertical Gene Transfer

In vertical gene transfer, the genetic material is inherited by the transmission of genes from parents to offspring (Fig. 3.4). A recent report of Li et al. (2019) showed that transconjugants of *E. coli* have tetracycline, trimethoprim, and amoxicillin resistance genes by vertical gene transfer. Meng et al. (2017) found that amoxicillin resistance gene was transferred through vertical gene transfer in *Phocoenobacter*, *Ignavibacterium*, *Aminobacterium*, *Spirochaeta*, and *Cloacibacillus*, which are present in wastewater. The BLAST analysis of homologous sequences of methicillin-resistant gene showed that it was transferred through vertical gene transfer in *Staphylococcus aureus*, which is a common pathogenic species of skin infection (Ito et al. 2003).

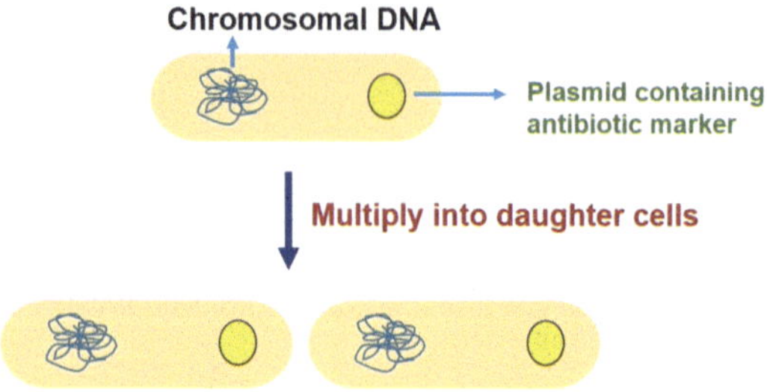

Fig. 3.4 Vertical gene transfer in bacterial species from parent to daughter cells

3.6 Antimicrobial Resistance and Health Risks to Humans

The severity of infections in humans has increased worldwide due to the antimicrobial-resistant bacteria. This is because of the following reasons:

1. *Failure or delay in treatment*: Antibiotic administration to patients, mainly in severe cases, which required intensive care units (ICU) is usually done before the antibiogram results. This results in the increase of the multidrug resistance in microbial strain. This leads to worsening of the patient's condition during relapse time. This condition in patients is avoided by increasing the awareness in people regarding antibiotic use (Khilnani et al. 2019). Further, the switching from broad-spectrum antibiotics to narrow-spectrum antibiotics, i.e., de-escalation, is recommended in the guidelines of antibiotic use to avoid increase in multidrug resistance in microbial strains (Leone et al. 2014). The recent studies showed that the de-escalation reduces the number of mortalities in patients (Guo et al. 2016; Tabah et al. 2016).
2. *Choice of antimicrobial agents is limited*: The antimicrobial agents' categories are limited due to the appearance of antimicrobial-resistant strains (Khilnani et al. 2019).
3. *Selection of antimicrobial-resistant strains*: The antibiotic treatment in patients selects the suppressed resistant pathogenic strains. This is because the prolonged exposure of microbial strains to antimicrobial agents helps them to evolve resistance against antimicrobial agents. Every year in the USA, nearly 35,000 deaths of human were reported due to multidrug-resistant microbial strain (Jordt et al. 2020).

3.7 Antibiotic Stewardship and Public Awareness to Mitigate the Health Risks to Humans

Antimicrobial stewardship is a coordinated program that promotes the appropriate use of antibiotics by selection of optimum dosage and time of antibiotic treatment, and it improves patient outcomes, reduces microbial resistance, and decreases the spread of infections caused by multidrug-resistant organisms (Gerding 2001). In this the first aim of medical practitioners is to prescribe the correct dose and duration of antimicrobial agents for the patients. The second aim is the prevention of misuse and overuse of antimicrobial agents, and the third aim is the minimization of the development of antimicrobial agent resistance in individual patient and in community level (Doron and Davidson 2011). The increase in antimicrobial resistance in community can be avoided or decreased via public awareness regarding antibiotic use and misuse. This can be done by increasing awareness in school and college students by adding lectures on the subject. The campaign regarding the use of antibiotics should be done to increase the knowledge of rural community and less educated people (Manyi-Loh et al. 2018). The consumer associations and groups can increase the awareness of antibiotic uses in food animals, which increases the demand of antibiotic smart food. They are also guiding farmers to do antibiotic free farming to avoid spread of antimicrobial resistance (Mathew et al. 2019).

3.8 Alternatives to Antimicrobial Agents in Agriculture

One of the most important alternatives for reducing the use of antimicrobial agents is immunization by vaccination (Hoelzer et al. 2018). The use of prebiotics, probiotics, and synbiotics also reduces the usage of antibiotics because they improve the health of human gut microflora (Hume 2011). The health of human digestive tract microbiota generally contributes to the immune system functionality and offers less area for pathogen colonization. In poultry, the major decrease in *Salmonella* colonization was observed after the administration of anaerobic bacterial commensal. Phages are also used as an alternative of antimicrobial agents, and they neutralize the selective pathogen in lytic cycle (Lin et al. 2017). Moreover, phages can also be used with other antimicrobials without any side effects. The oral administration of virus in humans reduces the symptoms of inflammatory bowel disease caused by *Butyricicoccus pullicaecorum*. The use of phages is limited to neutralization of foodborne pathogens in animal husbandry, curing of tropical diseases in humans, and control of phytopathogens.

Other useful agents are antimicrobial peptides which can act as alternative to antibiotics (Baltzer and Brown 2011). Bacteriocins are the antimicrobial peptides which are used as food preservative. The US Food and Drug Administration has approved the daily uptake of nisin A (bacteriocin) to 2.9 mg per person per day (Delves Broughton et al. 1996; Joo et al. 2012). It is used in food products and recognized as safe ferments. Bacteriocins have been used in food industries to reduce the growth of *Listeria*. Predatory bacteria are also used as an alternative to antimicrobial agents (Allende et al. 2007). *Micavibrio aeruginosavorus* and *Bdellovibrio bacteriovorus* have been used as prey against multidrug-resistant pathogens of genus *Escherichia coli*, *Acinetobacter baumannii*, *Pseudomonas aeruginosa*, *Klebsiella pneumonia*, *Aeromonas*, *Salmonella*, *Shigella*, *Serratia*, *Yersinia*, *Vibrio*, *Enterobacter*, *Proteus*, *Bordetella*, *Burkholderia*, and *Pseudomonas putida* (Dashiff et al. 2011). They act as both probiotics and antibiotics. *Bdellovibrio bacteriovorus* has been used for treating ocular diseases in rabbits caused by *Shigella flexneri* and in cows caused by *Moraxella bovis* (Shanks et al. 2013; Galvão and Angelos 2010).

3.9 Regulatory Agencies and Their Role in Regulating Antibiotic Use and Global Partnerships in Antimicrobial Resistance

The World Health Organization introduced the Global Action Plan so that every country must develop the national action plans in an attempt to reduce the spread of antibiotic resistance (World Health Organisation 2015). Further, the monitoring and surveillance of antibiotic use and antibiotic resistance are important strategy against spread of antibiotic resistance. Surveillance systems regarding consumption of antimicrobial agents are South African National Veterinary Surveillance and Monitoring Programme for Resistance to Antimicrobial Drugs (SANVAD), Pretoria, South Africa; Monitoring of Antimicrobial Resistance and Antibiotic Usage in Animals in Wageningen, Netherlands (MARAN); and Danish Integrated Antimicrobial Resistance Monitoring and Research Programme (DANMAP), Kongens Lyngby, Denmark (Schaekel et al. 2017). The antimicrobial stewardship is implemented in agriculture sector as well as in clinical practices in order to maintain the efficiency of available antimicrobial agents and to reduce the spread of antibiotic resistance from animal- and plant-derived food products (Moudgil et al. 2017). The stakeholders involved in reducing the spread of antibiotic resistance must report the antimicrobial stewardship from the stand point of surveillance, regulations, treatment guidelines, research, education, infection control, and awareness (Sharma et al. 2018). The National Farmed Animal Health and Welfare Council stated that the antimicrobial stewardship can be implemented in agriculture with the following perspectives such as clinical microbiology, biosecurity, surveillance on antibiotic use and antimicrobial resistance, alternatives to antibiotics, and animal management (National Farmed Animal Health and Welfare Council 2016). Some European countries banned the use of antimicrobial agents as growth promoters (Cogliani et al. 2011), whereas in New Zealand and Australia the partial ban is implemented. The tripartite alliance among the World Health Organization, Food and Agricultural Organization of the United Nations, and World Organisation for Animal Health was formed in 2003. This alliance helps in the categorization of medicines used in veterinary practices into important drugs for human health, highly important and critically important. Therefore, it is helpful in guiding the use of antimicrobial agents in animal agriculture and thus helps in reducing the spread of antimicrobial resistance (Manyi-Loh et al. 2018).

3.10 Conclusions

The continued increase of antimicrobials in the environment from agriculture and other sources has disrupted the natural balance between the antimicrobial agents and microbes. This has resulted in the dissemination of antimicrobial resistance in the microbial communities via horizontal gene transfer and vertical gene transfer. The multidrug resistance genes from commensal or beneficial microbes of human or plant origin are often transferred to pathogens and ultimately create superbugs. Therefore, to control the overuse and misuse of antimicrobial agents, alternative approaches have been used such as vaccination, prebiotics, probiotics, phages, and predatory bacteria. In addition to this, awareness should be spread against indiscriminate antibiotic usage, and strict regulatory policies should be implemented for nonhuman usage of antibiotics.

References

Abadi ATB (2017) Resistance to clarithromycin and gastroenterologist's persistence roles in nomination for *Helicobacter pylori* as high priority pathogen by World Health Organization. World J Gastroenterol 23(35):6379

Abraham E (1987) Cephalosporins. Drugs 4(2):1–4

Aćimović SG, Zeng Q, McGhee GC, Sundin GW, Wise JC (2015) Control of fire blight (*Erwinia amylovora*) on apple trees with trunk-injected plant resistance inducers and antibiotics and assessment of induction of pathogenesis-related protein genes. Front Plant Sci 6:16

Ahmed AM, Shimamoto T (2014) Isolation and molecular characterization of *Salmonella enterica*, *Escherichia coli* O157: H7 and *Shigella* spp. from meat and dairy products in Egypt. Int J Food Microbiol 168:57–62

Alban L, Ellis-Iversen J, Andreasen M, Dahl J, Sönksen UW (2017) Assessment of the risk to public health due to use of antimicrobials in pigs-an example of pleuromutilins in Denmark. Front Vet Sci 4:74

Alborn WE, Allen NE, Preston DA (1991) Daptomycin disrupts membrane potential in growing *Staphylococcus aureus*. Antimicrob Agents Chemother 35(11):2282–2287

Allende A, Martínez B, Selma V, Gil MI, Suárez JE, Rodríguez A (2007) Growth and bacteriocin production by lactic acid bacteria in vegetable broth and their effectiveness at reducing *Listeria monocytogenes* in vitro and in fresh-cut lettuce. Food Microbiol 24(7–8):759–766

Amábile-Cuevas CF (2016) Antibiotics and antibiotic resistance in the environment. CRC Press, Boca Raton, FL

Armstrong CW, Lake JL, Miller GB (1983) Extraintestinal infections due to halophilic vibrios. 76(5):571–4.

Asai T, Masani K, Sato C, Hiki M, Usui M, Baba K, Sawada T (2011) Phylogenetic groups and cephalosporin resistance genes of *Escherichia coli* from diseased food-producing animals in Japan. Acta Vet Scand 53(1):1–5

Baltzer SA, Brown MH (2011) Antimicrobial peptides promising alternatives to conventional antibiotics. J Mol Microbiol Biotechnol 20(4):228–235

Barbour EK, Nabbut NH, Al-Nakhli HM (1985) Use of epidemiologic markers to identify the source of *Escherichia coli* infections in poultry. Am J Vet Res 46(4):989–991

Barker KF (1999) Antibiotic resistance: a current perspective. J Clin Pharmacol 48:109–124

Barlow M, Hall BG (2002) Phylogenetic analysis shows that the OXA β-lactamase genes have been on plasmids for millions of years. J Mol Evol 55:314–321
Bonner DP, Sykes RB (1984) Structure activity relationships among the monobactams. J Antimicrob Chemother 14(4):313–327
Boundless (2016) Antibiotic classifications. In: Boundless microbiology. https://www.boundless.com/microbiology/textbooks/boundlessmicrobiology-textbook/antimicrobial-drugs-13/overview-ofantimicrobial-therapy-153/antibiotic-classifications-775-4905/
Bozdogan B, Appelbaum PC (2004) Oxazolidinones: activity, mode of action, and mechanism of resistance. Int J Antimicrob Agents 23(2):113–119
Brown AG, Butterworth D, Cole M, Hanscomb G, Hood JD, Reading C, Rolinson GN (1976) Naturally-occurring betalactamase inhibitors with antibacterial activity. J Antibiot 29(6):668–669
Bugg TDH, Walsh CT (1992) Intracellular steps of bacterial cell wall peptidoglycan biosynthesis: enzymology, antibiotics, and antibiotic resistance. Nat Prod Rep 9(3):199–215
Butterworth D, Cole M, Hanscomb G, Rolinson GN (1979) Olivanic acids, a family of beta-lactam antibiotics with beta-lactamase inhibitory properties produced by *Streptomyces* species. Detection, properties and fermentation studies. J Antibiot 32(4):287–294
Cabello FC (2006) Heavy use of prophylactic antibiotics in aquaculture: a growing problem for human and animal health and for the environment. Environ Microbiol 8(7):1137–1144
Cambray G, Guerout AM, Mazel D (2010) Integrons. Annu Rev Genet 44:141–166
Cameron A, Sarojini V (2014) *Pseudomonas syringae* pv. actinidiae: chemical control, resistance mechanisms and possible alternatives. Plant Pathol 63(1):1–11
Chang Q, Wang W, Regev-Yochay G, Lipsitch M, Hanage WP (2015) Antibiotics in agriculture and the risk to human health: how worried should we be? Evol Appl 8(3):240–247
Checkley SL, Campbell JR, Chirino-Trejo M, Janzen ED, Waldner CL (2010) Associations between antimicrobial use and the prevalence of antimicrobial resistance in fecal *Escherichia coli* from feedlot cattle in western Canada. Can Veterin J 51(8):853
Chen QL, Cui HL, Su JQ, Penuelas J, Zhu YG (2019) Antibiotic resistomes in plant microbiomes. Trends Plant Sci 24(6):530–541
Chong CY, Tan AL, Ng W, Tan-Kendrick A, Balakrishnan A, Chao SM (2003) Treatment of urinary tract infection with gentamicin once or three times daily. Acta Paediatr 92(3):291–296
Chopra I, Roberts M (2001) Tetracycline antibiotics: mode of action, applications, molecular biology, and epidemiology of bacterial resistance. Microbiol Mol Biol Rev 65(2):232–260
Cogliani C, Goossens H, Greko C (2011) Restricting antimicrobial use in food animals: lessons from Europe. Microbe 6(6):274–279
Coker AO, Isokpehi RD, Thomas BN, Amisu KO, Obi CL (2002) Human campylobacteriosis in developing countries. Emerg Infect Dis 8:237–243
Cole M, Kenig MD, Hewitt VA (1973) Metabolism of penicillins to penicilloic acids and 6-aminopenicillanic acid in man and its significance in assessing penicillin absorption. Antimicrob Agents Chemother 3(4):463–468
Cycoń M, Mrozik A, Piotrowska-Seget Z (2019) Antibiotics in the soil environment-degradation and their impact on microbial activity and diversity. Front Microbiol 10:338
Dashiff A, Junka RA, Libera M, Kadouri DE (2011) Predation of human pathogens by the predatory bacteria *Micavibrio aeruginosavorus* and *Bdellovibrio bacteriovorus*. J Appl Microbiol 110(2):431–444
De Briyne N, Atkinson J, Borriello SP, Pokludová L (2014) Antibiotics used most commonly to treat animals in Europe. Vet Rec 175(13):325–325
De Cleene M, De Ley J (1976) The host range of crown gall. Bot Rev 42(4):389–466
Dealy J, Moeller MW (1977) Effect of bambermycins on *Escherichia coli* and antibiotic resistance in calves. J Anim Sci 45(6):1239–1242
Defoirdt T, Sorgeloos P, Bossier P (2011) Alternatives to antibiotics for the control of bacterial disease in aquaculture. Curr Opin Microbiol 14(3):251–258

Delves Broughton J, Blackburn P, Evans RJ, Hugenholtz J (1996) Applications of the bacteriocin, nisin. Antonie Van Leeuwenhoek 69(2):193–202

Di Cerbo A, Pezzuto F, Guidetti G, Canello S, Corsi L (2019) Tetracyclines: insights and updates of their use in human and animal pathology and their potential toxicity. Open Biochem J 13(1):1

Domagala JM (1994) Structure-activity and structure-side-effect relationships for the quinolone antibacterials. J Antimicrob Chemother 33(4):685–706

Domingues S, Nielsen KM, da Silva GJ (2012) Various pathways leading to the acquisition of antibiotic resistance by natural transformation. Mob Genet Elem 2(6):257–260

Doron S, Davidson LE (2011) Antimicrobial stewardship. Mayo Clin Proc 86(11):1113–1123

Douthwaite S (1992) Interaction of the antibiotics clindamycin and lincomycin with *Escherichia coli* 23S ribosomal RNA. Nucleic Acids Res 20(18):4717–4720

Dowling PM (2013a) Chloramphenicol, thiamphenicol, and florfenicol. Antimicrob Ther Vet Med:269–277

Dowling PM (2013b) Peptide antibiotics: polymyxins, glycopeptides, bacitracin, and fosfomycin. Antimicrob Ther Vet Med 189

Doyle MP, Loneragan GH, Scott HM (2013) Antimicrobial resistance: challenges and perspectives. Compr Rev Food Sci Food Saf 12:234–248

Dutta C, Pan A (2002) Horizontal gene transfer and bacterial diversity. J Biosci 27(1):27–33

Economou V, Gousia P (2015) Agriculture and food animals as a source of antimicrobial-resistant bacteria. Infect Drug Resist 8:49

Etebu E, Arikekpar I (2016) Antibiotics: classification and mechanisms of action with emphasis on molecular perspectives. Int J Appl Microbiol Biotechnol Res 4(2016):90–101

Eyssen HJ, Van den Bosch JF, Janssen GA, Vanderhaeghe H (1971) Specific inhibition of cholesterol absorption by sulfaguanidine. Atherosclerosis 14(2):181–192

Falagas ME, Rafailidis PI, Matthaiou DK (2010) Resistance to polymyxins: mechanisms, frequency and treatment options. Drug Reresist Updat 13(4–5):132–138

Fan XG, Chua A, Li TG, Zeng QS (1998) Survival of *Helicobacter pylori* in milk and tap water. J Gastroenterol Hepatol 13(11):1096–1098

Felis E, Kalka J, Sochacki A, Kowalska K, Bajkacz S, Harnisz M, Korzeniewska E (2020) Antimicrobial pharmaceuticals in the aquatic environment-occurrence and environmental implications. Eur J Pharmacol 866:172813

Fernández L, Hancock REW (2012) Adaptive and mutational resistance: role of porins and efflux pumps in drug resistance. Clin Microbiol Rev 25(4):661–681

Fleming A (1929) On the antibacterial action of cultures of a penicillium, with special reference to their use in the isolation of *B. influenza*. Br J Exp Pathol 10:226–236

Fouz N, Pangesti KN, Yasir M, Al-Malki AL, Azhar EI, Hill-Cawthorne GA, Ghany MAE (2020) The contribution of wastewater to the transmission of antimicrobial resistance in the environment: implications of mass gathering settings. Trop Med Infect Dis 5(1):33

Frank U, Tacconelli E (2012) The Daschner guide to in-hospital antibiotic therapy: European standards, vol 10. Springer Science and Business Media, Berlin, p 176

Fuoco D (2012) Classification framework and chemical biology of tetracycline-structure-based drugs. Antibiotics 1:1–13

Gale E, Cundliffe E, Reynolds PE, Richmond MH, Waring MJ (1981) The molecular basis of antibiotic action, 2nd edn. John Wiley & Sons, New York, NY, p 670

Galvão KN, Angelos JA (2010) Ulcerative blepharitis and conjunctivitis in adult dairy cows and association with *Moraxella bovoculi*. Can Veterin J 51(4):400

Gemeinder JLP, Barros NRD, Pegorin GSA, Singulani JDL, Borges FA, Arco MCGD, Giannini MJSM, Almeida AMF, Salvador SLDS, Herculano RD (2021) Gentamicin encapsulated within a biopolymer for the treatment of *Staphylococcus aureus* and *Escherichia coli* infected skin ulcers. J Biomater Sci Polym Ed 32(1):93–111

Gerding DN (2001) The search for good antimicrobial stewardship. Joint Comm J Qual Improv 27(8):403–404

Giacometti F, Pezzi A, Galletti G, Tamba M, Merialdi G, Piva S, Serraino A, Rubini S (2021) Antimicrobial resistance patterns in *Salmonella enterica* subsp. enterica and *Escherichia coli* isolated from bivalve molluscs and marine environment. Food Control 121:107590

Giguère S (2013) Lincosamides, pleuromutilins, and streptogramins. In: Giguère S, Prescott JF, Dowling PM (eds) Antimicrobial therapy in veterinary medicine, 5th edn. Wiley, Hoboken, NJ, pp 199–210

Gomez GGV, Guerrero TS, Lluck MC, Delgado FJ (1999) Effectiveness of collagen-gentamicin implant for treatment of "dirty" abdominal wounds. World J Surg 23(2):123

Grosset JH, Singer T (2013) Streptomycin. Brenner's Encycl Genet 2:568–569

Gullberg E, Cao S, Berg OG, Ilbäck C, Sandegren L, Hughes D, Andersson DI (2011) Selection of resistant bacteria at very low antibiotic concentrations. PLoS Pathog 7(7):e1002158

Guo Y, Gao W, Yang H, Ma C, Sui S (2016) De-escalation of empiric antibiotics in patients with severe sepsis or septic shock: a meta-analysis. Heart Lung 45:454–459

Hamilton-Miller JM (1973) Chemistry and biology of the polyene macrolide antibiotics. Bacteriol Rev 37(3):166

Hamscher G, Sczesny S, Höper H, Nau H (2002) Determination of persistent tetracycline residues in soil fertilized with liquid manure by high-performance liquid chromatography with electrospray ionization tandem mass spectrometry. Anal Chem 74(7):1509–1518

Henry RJ (1943) The mode of action of sulfonamides. Bacteriol Rev 7(4):175

Hoelzer K, Bielke L, Blake DP, Cox E, Cutting SM, Devriendt B, Van Immerseel F (2018) Vaccines as alternatives to antibiotics for food producing animals. Part 1: challenges and needs. Vet Res 49(1):1–10

Hong J, Kim JM, Jung WK, Kim SH, Bae W, Koo HC, Park YH (2007) Prevalence and antibiotic resistance of *Campylobacter* spp. isolated from chicken meat, pork, and beef in Korea, from 2001 to 2006. J Food Prot 70(4):860–866

Hong W, Zeng J, Xie J (2014) Antibiotic drugs targeting bacterial RNAs. Acta Pharm Sin B 4(4):258–265

Hume ME (2011) Historic perspective: prebiotics, probiotics, and other alternatives to antibiotics. Poult Sci 90(11):2663–2669

Huq F (2006) Molecular modelling analysis of the metabolism of enrofloxacin. Asian J Biochem 1(4):316–324

Ihrke PJ, Papich MG, Demanuelle TC (1999) The use of fluoroquinolones in veterinary dermatology. Vet Dermatol 10(3):193–204

Islam M, Doyle MP, Phatak SC, Millner P, Jiang X (2004) Persistence of enterohemorrhagic *Escherichia coli* O157:H7 in soil and on leaf lettuce and parsley grown in fields treated with contaminated manure composts or irrigation water. J Food Prot 67:1365–1370

Islam S, Aldstadt J, Aga D (2019) Global antimicrobial resistance: a complex and dire threat with few definite answers. Trop Med Int Health 24:658–662

Ito T, Okuma K, Ma XX, Yuzawa H, Hiramatsu K (2003) Insights on antibiotic resistance of *Staphylococcus aureus* from its whole genome: genomic island SCC. Drug Resist Updat 6(1):41–52

Jensen VF, Jakobsen L, Emborg HD, Seyfarth AM, Hammerum AM (2006) Correlation between apramycin and gentamicin use in pigs and an increasing reservoir of gentamicin-resistant Escherichia coli. J Antimicrob Chemother 58(1):101–107

Jin M, Liu L, Wang DN, Yang D, Liu WL, Yin J, Yang ZW, Wang HR, Qiu ZG, Shen ZQ, Shi DY, Li HB, Guo JH, Li JW (2020) Chlorine disinfection promotes the exchange of antibiotic resistance genes across bacterial genera by natural transformation. ISME J 14(7):1847–1856

Jo SH, Lee J, Park E, Kim DW, Lee DH, Ryu CM, Choi D, Park JM (2019) A human pathogenic bacterium *Shigella* proliferates in plants through adoption of type III effectors for shigellosis. Plant Cell Environ 42(11):2962–2978

Johansson MH, Bortolaia V, Tansirichaiya S, Aarestrup FM, Roberts AP, Petersen TN (2021) Detection of mobile genetic elements associated with antibiotic resistance in *Salmonella*

enterica using a newly developed web tool: mobile element finder. J Antimicrob Chemother 76(1):101–109
Joo NE, Ritchie K, Kamarajan P, Miao D, Kapila YL (2012) Nisin, an apoptogenic bacteriocin and food preservative, attenuates HNSCC tumorigenesis via CHAC 1. Cancer Med 1(3):295–305
Jordt H, Stalder T, Kosterlitz O, Ponciano JM, Top EM, Kerr B (2020) Coevolution of host-plasmid pairs facilitates the emergence of novel multidrug resistance. Nat Ecol Evol 4(6):863–869
Josephine HR, Kumar I, Pratt RF (2004) The perfect penicillin? Inhibition of a bacterial DD-peptidase by peptidoglycan-mimetic β-lactams. J Am Chem Soc 126(26):8122–8123
Kahne D, Leimkuhler C, Lu W, Walsh C (2005) Glycopeptide and lipoglycopeptide antibiotics. Chem Rev 105(2):425–448
Kang HK, Park Y (2015) Glycopeptide antibiotics: structure and mechanisms of action. J Bacteriol Virol 45(2):67–78
Karunasagar I, Pai R, Malathi GR, Karunasagar I (1994) Mass mortality of *Penaeus monodon* larvae due to antibiotic-resistant *Vibrio harveyi* infection. Aquaculture 128(3–4):203–209
Katz L, Ashley GW (2005) Translation and protein synthesis: macrolides. Chem Rev 105(2):499–528
Keeling PJ, Palmer JD (2008) Horizontal gene transfer in eukaryotic evolution. Nat Rev Genet 9(8):605–618
Khilnani GC, Zirpe K, Hadda V, Mehta Y, Madan K, Kulkarni A, Mohan A, Dixit S, Guleria R, Bhattacharya P (2019) Guidelines for antibiotic prescription in intensive care unit. Ind J Critic Care Med 23(Suppl 1):S1–S63
Kidd H, James DR (1991) The agrochemicals handbook, 3rd edn. Royal Society of Chemistry, Cambridge, p 1500
Kinana AD, Cardinale E, Tall F, Bahsoun I, Sire JM, Garin B, Perrier-Gros-Claude JD (2006) Genetic diversity and quinolone resistance in *Campylobacter jejuni* isolates from poultry in Senegal. Appl Environ Microbiol 72(5):3309–3313
Koh YJ, Kim GH, Lee YS, Sohn SH, Koh HS, Kwon S, Heu S, Jung JS (2012) *Pectobacterium carotovorum* subsp. actinidiae subsp. a new bacterial pathogen causing canker-like symptoms in yellow kiwifruit, Actinidia chinensis. N Z J Crop Hortic Sci 40(4):269–279
Kümmerer K (2001) Drugs in the environment: emission of drugs, diagnostic aids and disinfectants into wastewater by hospitals in relation to other sources-a review. Chemosphere 45:957–969
Kümmerer K (2003) Significance of antibiotics in the environment. J Antimicrob Chemother 52(1):5–7
Laxminarayan R, Van Boeckel T, Teillant A (2015) The economic costs of withdrawing antimicrobial growth promoters from the livestock sector. OECD Food, Agriculture and Fisheries Papers. OECD, Paris, p 78
Leonard AF, Zhang L, Balfour AJ, Garside R, Gaze WH (2015) Human recreational exposure to antibiotic resistant bacteria in coastal bathing waters. Environ Int 82:92–100
Leone M, Bechis C, Baumstarck K, Lefrant JY, Albanèse J, Jaber S, Lepape A, Constantin J, Papazian L, Bruder N, Allaouchiche B, Bézulier K, Antonini F, Textoris J, Martin C (2014) De-escalation versus continuation of empirical antimicrobial treatment in severe sepsis: a multicenter non-blinded randomized noninferiority trial. Intensive Care Med 40(10):1399–1408
Letellier A, Messier S, Lessard L, Quessy S (2000) Assessment of various treatments to reduce carriage of *Salmonella* in swine. Can J Vet Res 64(1):27
Li B, Qiu Y, Song Y, Lin H, Yin H (2019) Dissecting horizontal and vertical gene transfer of antibiotic resistance plasmid in bacterial community using microfluidics. Environ Int 131:105007
Liao Y, Wei Z, Bai L, Deng Z, Zhong J (2009) Effect of fermentation temperature on validamycin A production by *Streptomyces hygroscopicus*. J Biotechnol 142:271–274
Lin DM, Koskella B, Lin HC (2017) Phage therapy: an alternative to antibiotics in the age of multidrug resistance. World J Gastrointest Pharmacol Therapeut 8(3):162
Linville JL, Shen Y, Wu MM, Urgun-Demirtas M (2015) Current state of anaerobic digestion of organic wastes in North America. Curr Sustain Renew Energy Rep 2(4):136–144

Liu G, Bogaj K, Bortolaia V, Olsen JE, Thomsen LE (2019) Antibiotic-Induced, increased conjugative transfer is common to diverse naturally occurring ESBL plasmids in *Escherichia coli*. Front Microbiol 10:2119

Looft T, Allen HK, Casey TA, Alt DP, Stanton TB (2014) Carbadox has both temporary and lasting effects on the swine gut microbiota. Front Microbiol 5:276

Lorenz MG, Wackernagel W (1994) Bacterial gene transfer by natural genetic transformation in the environment. Microbiol Rev 58:563–602

Luk'aš B, Hejzlar M, Libich J (1963) Effect of streptomycin and kanamycin on the course of experimental tularaemia in guinea-pigs. Folia Microbiol 8(2):80–88

Lwai-Lume L, Ogutu EO, Amayo EO, Kariuki S (2005) Drug susceptibility pattern of *Helicobacter pylori* in patients with dyspepsia at the Kenyatta National Hospital, Nairobi. East Afr Med J 82(12):603–608

Lyras D, Adams V, Lucet I, Rood JI (2004) The large resolvase TnpX is the only transposon encoded protein required for transposition of the Tn4451/3 family of integrative mobilizable elements. Mol Microbiol 51:1787–1800

Maeda Y, Kiba A, Ohnishi K, Hikichi Y (2004) Implications of amino acid substitutions in GyrA at position 83 in terms of oxolinic acid resistance in field isolates of *Burkholderia glumae*, a causal agent of bacterial seedling rot and grain rot of rice. Appl Environ Microbiol 70:5613–5620

Mahajan GB, Balachandran L (2012) Antibacterial agents from actinomycetes-a review. Front Biosci 4(4):240–253

Manyi-Loh C, Mamphweli S, Meyer E, Okoh A (2018) Antibiotic use in agriculture and its consequential resistance in environmental sources: potential public health implications. Molecules 23(4):795

Marshall BM, Levy SB (2011) Food animals and antimicrobials: impacts on human health. Clin Microbiol Rev 24(4):718–733

Mathew P, Sivaraman S, Chandy S (2019) Communication strategies for improving public awareness on appropriate antibiotic use: bridging a vital gap for action on antibiotic resistance. J Fam Med Prim Care 8(6):1867

Mccracken JRG, Mize S, Threlkeld N (1980) Intraventricular gentamicin therapy in gram-negative bacillary meningitis of infancy. Lancet 315(8172):787–791

McGeer A, Fleming CA, Green K, Low DE (2001) Antimicrobial resistance in Ontario: are we making progress. Lab Profic Test Prog Newsl 293:1–4

McManus PS, Stockwell VO, Sundin GW, Jones AL (2002) Antibiotic use in plant agriculture. Annu Rev Phytopathol 40:443–465

Mechal T, Hussen S, Desta M (2021) Bacterial profile, antibiotic susceptibility pattern and associated factors among patients attending adult OPD at Hawassa University Comprehensive Specialized Hospital, Hawassa, Ethiopia. Infect Drug Resist 14:99

Mellon M, Benbrook C, Benbrook KL (2001) Hogging it! Estimates of antimicrobial abuse in livestock. Union of Concerned Scientists, Washington, DC

Meng L, Li X, Wang X, Ma K, Liu G, Zhang J (2017) Amoxicillin effects on functional microbial community and spread of antibiotic resistance genes in amoxicillin manufacture wastewater treatment system. J Environ Sci 61:110–117

Mingeot-Leclercq MP, Glupczynski Y, Tulkens PM (1999) Aminoglycosides: activity and resistance. Antimicrob Agents Chemother 43(4):727–737

Minsavage GV, Canteros BI, Stall RE (1990) Plasmid-mediated resistance to streptomycin in *Xanthomonas campestris* pv. vesicatoria. Phytopathology 80:719–723

Moore D (2015) Antibiotic classification and mechanism. http://www.orthobullets.com/basic-science/9059/antibioticclassification-and-mechanism

Morar M, Wright GD (2010) The genomic enzymology of antibiotic resistance. Annu Rev Genet 44:25–51

Moudgil P, Bedi JS, Moudgil AD, Gill JPS, Aulakh RS (2017) Emerging issue of antibiotic resistance from food producing animals in India: perspective and legal framework. Food Rev Int 34(5):447–462

Mousavi S, Dehkordi FS, Rahimi E (2014) Virulence factors and antibiotic resistance of *Helicobacter pylori* isolated from raw milk and unpasteurized dairy products in Iran. J Venom Anim Toxins Incl Trop Dis 20:51

Munita JM, Arias CA (2016) Mechanisms of antibiotic resistance. Microbiol Spect 4(2):VMBF-0016-2015

Muthuirulandi Sethuvel DP, Devanga Ragupathi NK, Anandan S, Veeraraghavan B (2017) Update on: *Shigella* new serogroups/serotypes and their antimicrobial resistance. Lett Appl Microbiol 64(1):8–18

Mutreja A, Kim DW, Thomson NR, Connor TR, Lee JH, Kariuki S, Croucher NJ, Choi SY, Harris SR, Lebens M, Niyogi SK, Jin E, Holmgren J, Parkhill J, Dougan G (2011) Evidence for several waves of global transmission in the seventh cholera pandemic. Nature 477(7365):462–465

Nalley L, Tsiboe F, Durand-Morat A, Shew A, Thoma G (2016) Economic and environmental impact of rice blast pathogen (*Magnaporthe oryzae*) alleviation in the United States. PLoS One 11(12):e0167295

National Farmed Animal Health and Welfare Council (2016) Antimicrobial stewardship in food animals in Canada. National Farmed Animal Health and Welfare Council, Bluevale, ON

Nene YL, Thapliyal RN (1993) Evaluation of fungicides in fungicides for plant disaease control, vol 3. Oxford, IHB Pub. Co, New Delhi, p 331

Neu HC (1974) Antimicrobial activity and human pharmacology of amoxicillin. J Infect Dis 129:S123–S131

Nguyen F, Starosta AL, Arenz S, Sohmen D, Donhofer A, Wilson DN (2014) Tetracycline antibiotics and resistance mechanisms. Biol Chem 395:559–575

Niyogi SK (2005) Shigellosis. J Microbiol 43(2):133–143

Noli C, Boothe D (1999) Macrolides and lincosamides. Vet Dermatol 10(3):217–223

Norelli JL, Jones AL, Aldwinckle HS (2003) Fire blight management in the twenty-first century: using new technologies that enhance host resistance in apple. Plant Dis 87(7):756–765

Novak R, Shlaes DM (2010) The pleuromutilin antibiotics: a new class for human use. Curr Opin Investig Drugs 11(2):182–191

Pan JC, Ye R, Wang HQ, Xiang HQ, Zhang W, Yu XF, Meng DM and He ZS (2008) *Vibrio cholerae O139 multiple-drug resistance mediated by Yersinia pestis pIP1202-like conjugative plasmids Antimicrob Agents Chemother, 52(11):3829–3836*

Papp-Wallace KM, Endimiani A, Taracila MA, Bonomo RA (2011) Carbapenems: past, present, and future. Antimicrob Agents Chemother 55(11):4943–4960

Patel U, Yan YP, Hobbs FW Jr, Kaczmarczyk J, Slee AM, Pompliano DL, Kurilla MG, Bobkova EV (2001) Oxazolidinones mechanism of action: inhibition of the first peptide bond formation. J Biol Chem 276(40):37199–37205

Pines A, Raafat H, Plucinski K (1967) Gentamicin and colistin in chronic purulent bronchial infections. Br Med J 2(5551):543

Puzari M, Sharma M, Chetia P (2018) Emergence of antibiotic resistant *Shigella* species: a matter of concern. J Infect Publ Health 11:451–454

Pyörälä S, Greko C, Pomba MCMF (2009) Use of glycopeptides in veterinary medicine. Vet Rec 165(18):543

Pyörälä S, Baptiste KE, Catry B, Van Duijkeren E, Greko C, Moreno MA, Törneke K (2014) Macrolides and lincosamides in cattle and pigs: use and development of antimicrobial resistance. Vet J 200(2):230–239

Radwan AI, Bekairi SI, Mukayel AA (1992) Treatment of *Brucella melitensis* infection in sheep and goats with oxytetracycline combined with streptomycin. Revue Scientifique et Technique 11:845–845

Rahimi E, Kheirabadi EK (2012) Detection of *Helicobacter pylori* in bovine, buffalo, camel, ovine, and caprine milk in Iran. Foodborne Pathog Dis 9(5):453–456

Raman NM, Easwaran M, Kaul R, Bharti J, Motelb KFA, Kaul T (2020) Antimicrobial resistance with special emphasis on pathogens in agriculture. In: Antimicrobial resistance. IntechOpen, London

Rasheed MU, Thajuddin N, Ahamed P, Teklemariam Z, Jamil K (2014) Antimicrobial drug resistance in strains of *Escherichia coli* isolated from food sources. Rev Inst Med Trop Sao Paulo 56(4):341–346
Reygaert WC (2018) An overview of the antimicrobial resistance mechanisms of bacteria. AIMS Microbiol 4(3):482
Reynolds PE (1989) Structure, biochemistry and mechanism of action of glycopeptide antibiotics. Eur J Clin Microbiol Infect Dis 8(11):943–950
Romero J, Feijoo CG, Navarrete P (2012) Antibiotics in aquaculture - use, abuse and alternatives. In: Health and environment in aquaculture, p 159
Rospotrebnadzor (2018) The Russian Federal Service for Surveillance on Consumer Rights Protection and Human Wellbeing, State Report. On the State of Sanitary and Epidemiological Welfare of the Population in the Russian Federation in 2017, 6
Russo NL, Burr TJ, Breth DI, Aldwinckle HS (2008) Isolation of streptomycin-resistant isolates of *Erwinia amylovora* in New York. Plant Dis 92(5):714–718
Samie A, Ramalivhana J, Igumbor EO, Obi CL (2007) Prevalence, haemolytic and haemagglutination activities and antibiotic susceptibility profiles of *Campylobacter* spp. isolated from human diarrhoeal stools in Vhembe District, South Africa. J Health Popul Nutr 25(4):406
Samson R, Shah M, Yadav R, Sarode P, Rajput V, Dastager SG, Dharne MS, Khairnar K (2019) Metagenomic insights to understand transient influence of Yamuna River on taxonomic and functional aspects of bacterial and archaeal communities of River Ganges. Sci Total Environ 674:288–299
Sánchez AR, Rogers RS III, Sheridan PJ (2004) Tetracycline and other tetracycline-derivative staining of the teeth and oral cavity. Int J Dermatol 43(10):709–715
Sanseverino I, Navarro Cuenca A, Loos R, Marinov D, Lettieri T (2018) State of the art on the contribution of water to antimicrobial resistance. JRC Technical Reports. European Union, Brussels. ISSN 1831-9424
Sarkar DJ, Mukherjee I, Shakil NA, Rana VS, Kaushik P, Debnath S (2018) Antibiotics in agriculture: use and impact. Ind J Ethnophytopharmaceut 4:4–19
Savita GSV, Nagpal A (2012) Citrus diseases caused by *Phytophthora* species. GERF Bull Biosci 3(1):18–27
Schaekel F, May T, Seiler J, Hartmann M, Kreienbrock L (2017) Antibiotic drug usage in pigs in Germany-are the class profiles changing? PLoS One 12(8):e0182661
Schwarz S, Shen J, Kadlec K, Wang Y, Michael GB, Feßler AT, Vester B (2016) Lincosamides, streptogramins, phenicols, and pleuromutilins: mode of action and mechanisms of resistance. Cold Spring Harb Perspect Med 6(11):a027037
Scortichini M, Marcelletti S, Ferrante P, Petriccione M, Firrao G (2012) *Pseudomonas syringae* pv. actinidiae: a re-emerging, multi-faceted, pandemic pathogen. Mol Plant Pathol 13:631–640
Scott HM, Acuff G, Bergeron G, Bourassa MW, Gill J, Graham DW, Kahn LH, Morley PS, Salois MJS, Simjee S, Singer RS, Smith TC, Storrs C, Wittum TE (2019) Critically important antibiotics: criteria and approaches for measuring and reducing their use in food animal agriculture. Ann N Y Acad Sci 1441(1):8
Sengeløv G, Halling-Sørensen B, Aarestrup FM (2003) Susceptibility of *Escherichia coli* and *Enterococcus faecium* isolated from pigs and broiler chickens to tetracycline degradation products and distribution of tetracycline resistance determinants in *E. coli* from food animals. Vet Microbiol 95(1–2):91–101
Shahin KM, Bouzari R, Wang MR, Khorasgani (2019) Distribution of antimicrobial resistance genes and integrons among *Shigella* spp. isolated from water sources. J Glob Antimicrob Resist 19:122–128
Shahin K, Bouzari M, Komijani M, Wang R (2020) A New Phage Cocktail against multidrug, ESBL-producer isolates of *Shigella sonnei* and *Shigella flexneri* with highly efficient bacteriolytic activity. Microb Drug Resist 26(7):831–841

Shang K, Wei B, Cha SY, Zhang JF, Park JY, Lee YJ, Kang M (2021) The occurrence of antimicrobial-resistant *Salmonella enterica* in hatcheries and dissemination in an integrated broiler chicken operation in Korea. Animals 11(1):154

Shanks RM, Davra VR, Romanowski EG, Brothers KM, Stella NA, Godboley D, Kadouri DE (2013) An eye to a kill: using predatory bacteria to control Gram-negative pathogens associated with ocular infections. PLoS One 8(6):e66723

Sharma C, Rokana N, Chandra M, Singh BP, Gulhane RD, Gill JPS, Ray P, Puniya AK, Panwar H (2018) Antimicrobial resistance: its surveillance, impact, and alternative management strategies in dairy animals. Front Vet Sci 4:237

Shea KM (2003) Antibiotic resistance: what is the impact of agricultural uses of antibiotics on children's health? Pediatrics 112(1):253–258

Shinabarger DL, Marotti KR, Murray RW, Lin AH, Melchior EP, Swaney SM, Dunyak DS, Demyan WF, Buysse JM (1997) Mechanism of action of oxazolidinones: effects of linezolid and eperezolid on translation reactions. Antimicrob Agents Chemother 41(10):2132–2136

Silbergeld EK, Graham J, Price LB (2008) Industrial food animal production, antimicrobial resistance and human health. Annu Rev Public Health 29:151–169

Sosa ADJ, Byarugaba DK, Amábile Cuevas CF, Hsueh PR, Kariuki S, Okeke IN (2010) Antimicrobial resistance in developing countries. Springer, New York, NY, p 554

Stahl J, Zessel K, Schulz J, Finke JH, Müller-Goymann CC, Kietzmann M (2016) The effect of miscellaneous oral dosage forms on the environmental pollution of sulfonamides in pig holdings. BMC Vet Res 12(1):1–8

Stearns SC, Hoekstra RF (2005) Evolution: an introduction, 2nd edn. Oxford University Press, Oxford, NY, pp 38–40

Sundin GW, Bender CL (1993) Ecological and genetic analysis of copper and streptomycin resistance in *Pseudomonas syringae* pv. syringae. Appl Environ Microbiol 59:1018–1024

Sundin GW, Wang N (2018) Antibiotic resistance in plant-pathogenic bacteria. Annu Rev Phytopathol 56:161–180

Sykes RB, Cimarusti CM, Bonner DP, Bush K, Floyd DM, Georgopapadakou NH, Koster WH, Liu WC, Parker WL, Principe PA, Rathnum ML, Slusarchyk WA, Trejo WH, Wells JS (1981) Monocyclic β-lactam antibiotics produced by bacteria. Nature 291(5815):489–491

Tabah A, Cotta MO, Garnacho-Montero J, Schouten J, Roberts JA, Lipman J (2016) A systematic review of the definitions, determinants, and clinical outcomes of antimicrobial de-escalation in the intensive care unit. Clin Infect Dis 62:1009–1017

Talaro KP, Chess B (2008) Foundations in microbiology, 8th edn. McGraw Hill, New York, NY

Thanner S, Drissner D, Walsh F (2016) Antimicrobial resistance in agriculture. MBio 7(2):e02227

Thomson WT (1982) Agricultural chemicals book IV fungicides. Thomson Publications, Fresno, CA

Tomlin C (2006) The pesticide manual. British Crop Protection Council, Hampshire, p 14

Vachvanichsanong P, McNeil EB, Dissaneewate P (2021) Extended-spectrum beta-lactamase *Escherichia coli* and *Klebsiella pneumoniae* urinary tract infections. Epidemiol Infect 149:e12

Van Boeckel TP, Brower C, Gilbert M, Grenfell BT, Levin SA, Robinson TP, Teillant A, Laxminarayan R (2015) Global trends in antimicrobial use in food animals. Proc Natl Acad Sci U S A 112:5649–5654

Van Hoek AH, Mevius D, Guerra B, Mullany P, Roberts AP, Aarts HJ (2011) Acquired antibiotic resistance genes: an overview. Front Microbiol 2:203

Varughese LR, Beniwal V (2015) High quinolone resistance pattern among enteric pathogens isolated from patients with urinary tract infection. Indian J Biotechnol 14(1):167–171

Verraes C, Van Boxstael S, Van Meervenne E, Van Coillie E, Butaye P, Catry B, Herman L (2013) Antimicrobial resistance in the food chain: a review. Int J Environ Res Public Health 10(7):2643–2669

Vidaver AK (2002) Uses of antibiotics in plant agriculture. Clin Infect Dis 34:S107–S110

Von Wintersdorff CJ, Penders J, Van Niekerk JM, Mills ND, Majumder S, Van Alphen LB, Savelkoul PHM, Wolffs PF (2016) Dissemination of antimicrobial resistance in microbial ecosystems through horizontal gene transfer. Front Microbiol 7:173

Williams MR, Stedtfeld RD, Guo X, Hashsham SA (2016) Antimicrobial resistance in the environment. Water Environ Res 88(10):1951–1967

Williams-Nguyen J, Sallach JB, Bartelt-Hunt S, Boxall AB, Durso LM, McLain JE, Singer RS, Snow DD, Zilles JL, Zilles JL (2016) Antibiotics and antibiotic resistance in agroecosystems: state of the science. J Environ Qual 45(2):394–406

World Health Organisation (2015) Global action plan on antimicrobial resistance. World Health Organization, Geneva

Yano Y, Hamano K, Satomi M, Tsutsui I, Ban M, Aue-Umneoy D (2014) Prevalence and antimicrobial susceptibility of *Vibrio* species related to food safety isolated from shrimp cultured at inland ponds in Thailand. Food Control 38:30–36

Zachariah OH, Lizzy MA, Rose K, Angela MM (2021) Multiple drug resistance of *Campylobacter jejuni* and *Shigella* isolated from diarrhoeic children at Kapsabet Country referral hospital, Kenya. BMC Infect Dis 21(1):1–8

Zhang S, Huang J, Zhao Z, Cao Y, Li B (2020) Hospital wastewater as a reservoir for antibiotic resistance genes: a meta-analysis. Front Public Health 8:574968

Ziółkowski G, Pawłowska I, Stasiowski M, Jachowicz E, Wójkowska-Mach J, Bielecki T (2021) Multidrug-resistant micro-organisms associated with urinary tract infections in orthopedic patients: a retrospective laboratory-based study. Antibiotics 10(1):7

Chapter 4
Modern Diagnostic Tools for Rapid Detection of Multidrug Resistance

Nimisha Tehri, Saurabh Kadyan, Tejinder P. Singh, Piyush Tehri, and Amit Vashishth

4.1 Introduction

In the past few decades, there has been a consistent rise in cases exposing the prevalence of multiple drug resistance throughout the world. The emergence of antimicrobial-resistant pathogens is the greatest epidemiological menace in this twenty-first century (Prestinaci et al. 2015). It has become a major challenge for calculated antimicrobial therapies to treat various types of infectious diseases and thus presents a serious threat to patient's treatment. Various clinically important bacteria including Gram-positive and Gram-negative such as methicillin-resistant *Staphylococcus aureus* (MRSA), vancomycin-resistant *Enterococci* (VRE), *Acinetobacter baumannii*, *Pseudomonas aeruginosa*, and members of family Enterobacteriaceae like *Klebsiella pneumoniae*, *Escherichia coli*, and *Proteus* sp. rapidly develop and transmit antimicrobial resistance in hospital environment

N. Tehri (✉)
Centre for Biotechnology, Maharshi Dayanand University, Rohtak, Haryana, India

S. Kadyan
Dairy Microbiology Division, ICAR-National Dairy Research Institute, Karnal, Haryana, India

T. P. Singh
Dairy Microbiology Department, College of Dairy Science and Technology, Lala Lajpat Rai University of Veterinary and Animal Science, Hisar, India

P. Tehri
Department of Medicinal Chemistry, National Institute of Pharmaceutical Education and Research, Guwahati, Assam, India

A. Vashishth
Department of Biochemistry, International Institute of Veterinary Education and Research (LUVAS), Rohtak, Haryana, India

N. Akhtar et al. (eds.), *Emerging Modalities in Mitigation of Antimicrobial Resistance*,
https://doi.org/10.1007/978-3-030-84126-3_4

(Basak et al. 2016; Lee et al. 2012; Livermore 2012). Further, it results in increased rate of morbidity and mortality with long stay of patients in hospitals.

Keeping in view, the WHO declared "combat drug resistance: no action today, no cure tomorrow" in 2011 (Sharma 2011). In the twenty-first century, MDR represents the largest threat to health globally according to the WHO and needs immediate actions in place. Annually, death of more than 700,000 people results from drug-resistant infections, and this number is speculated to reach ten million by 2050 (UN 2016; Vasala et al. 2020).

Treatment of common infections being a challenge due to an exponential rise in antimicrobial drug resistance worldwide has led to the development of several novel antimicrobials in recent years. However, it also needs to focus on using the currently available antimicrobials. This highlights the importance of detection of susceptibility profile of targeted pathogens toward different drugs for particular infection. The use of appropriate diagnostic methods to determine drug resistance holds great potential to play significant role in initiating the right type of treatment (Frickmann et al. 2014).

Methods used for evaluating the multiple drug resistance and susceptibility profile mainly include phenotypic and genotypic types. Classical AST systems chiefly include the use of culture-based approaches which require more detection time due to long incubation steps usually requiring 48–72 h for completion (van Belkum et al. 2020). This contributes toward the indiscriminate use of antibiotics that further pave the way to drug resistance.

Therefore, various attempts have been made to develop methods offering rapid detection of drug resistance in the past few years. According to most clinical microbiologists, a testing can be defined as rapid if it is feasible during a single working shift, i.e., within 8 h or less (van Belkum et al. 2020).

This chapter presents the origin and transmission of multiple drug resistance with major emphasis on the current progress toward different types of AST systems including microscopy; molecular, i.e., PCR-based, RNA-based, isothermal amplification, and DNA microarrays; flow cytometry; biosensing; immunodiagnostic; mass spectrometry; colorimetric imaging; and optical approaches for rapid detection of MDR. In addition, we discuss the existing challenges and future prospects of new techniques for rapid detection of antimicrobial resistance.

4.2 Origin and Transmission of Multidrug Resistance

The development of AMR is considered as a phenomenon of natural selection that happens usually through genetic manipulations naturally over time due to the selective pressure posed by unchecked applications of drugs. The resistant strains may develop and be found in humans, animals, plants, and environment, but what makes the condition worse is the possibility of transmission of such AMR strains from one carrier to another, and the carrier may be a person, animals, or food of animal origin. Several factors include misuse of drugs as prophylactic, metaphylactic, or growth promoter (Woolhouse et al. 2015), unhygienic practices, ignorance, improper health-care facilities, unaffordable and inaccessible quality treatments, lack of awareness and knowledge, as well as negligence and/or lack of law enforcement.

The scientific literature has shown an existence of a positive correlation between resistance to single drug and resistance to multiple drugs within a bacterial population. Numerous mechanisms have been proposed to explain the multidrug resistance among different pathogens but still need to be assessed and tested critically. Several unanswered questions still exist like whether mechanisms are taxon-specific, do pathogens share the pathways to accumulation of MDR to some extent, and why do resistance determinants aggregate in certain strains of bacteria. The increasing cases of MDR may be attributed to unexpected high rates of origin, high rates of spread of MDR strains or determinants, or both. Horizontal gene transfer is believed to be a key factor that can disseminate resistance to multiple drugs in a single step. However, it is better to understand the MDR concept in terms of origin and transmission.

Various possible mechanisms explained so far for the emergence of resistant strains include biochemical mechanisms, genetic linkage, highly mutable or recombinogenic bacterial lineages, and multidrug therapy with accelerated treatment failure in resistant infections. The biochemical mechanisms employed by resistant strains are modification of the target molecules, inactivation of the drug, bypassing of the target molecule, or efflux of drugs. Each biochemical mechanism mentioned can confer resistance to more than one drug. Also, the resistant strain may employ one or many different biochemical mechanisms to resist one or multiple drugs (Fig. 4.1).

Other reports have mentioned that the determinants of resistance to multiple drugs are genetically linked. They are arranged in a sequence on bacterial

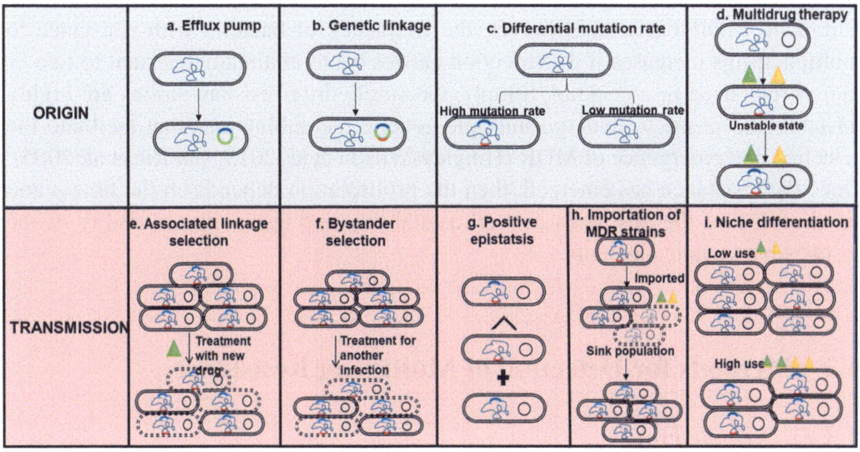

Fig. 4.1 Mechanisms for the origin and transmission of the MDR strains. Green and yellow triangles indicate the treatment with drug A and drug B. Red and blue marks represent drug A and drug B resistance determinants, respectively

chromosome or on the horizontally transmittable elements such as plasmid and conjugative transposons. Hence, the determinants can be co-inherited vertically and potentially co-transformed when shared horizontally. Researchers have believed that once the strain has acquired one resistance phenotype, then it can acquire many. The elements like plasmids, transposons, and integrons that transmit horizontally are highly extensible in nature and hence can incorporate new elements with time (Hall 2012; Roberts and Mullany 2009). Such elements do carry genes encoding for the specific enzymes like transposases and other site-specific recombinases that facilitate genetic incorporations. These elements may be organized on the chromosome or plasmid and may get transferred among interspecies or intraspecies through conjugation, transformation, and/or transduction (Alekshun and Levy 2007). The molecular mechanisms involved in transfers are well known, but how the resistance elements stay together under selective pressure is still unclear.

The rate of mutational changes varies among bacterial lineages which are attributed to variability in DNA proofreading mechanisms such as the mismatch repair system and also their ability to accept and integrate transforming DNA (Croucher et al. 2013). Therefore, the highly recombinogenic lineages possess higher frequencies of acquiring MDR determinants. An exposure to sublethal concentrations of antibiotics is considered as a selective pressure that can induce the mutation in bacteria (Baquero et al. 1998; Kohanski et al. 2010). Under antibiotic stress, the bacteria generate the reactive oxygen species that elevates the rate mutational changes irrespective of their specificity to the AMR (Kohanski et al. 2010). Acclimatization to the environmental conditions with antibiotics may provide second-order selection for high mutation rates (Mao et al. 1997).

Multidrug therapy is commonly recommended to treat infections, which may be involving strains resistant to single drug, with a belief that a combination of multiple drugs may control the resistance strain and minimize the probability of the emergence of resistance. Such belief was based on the assumptions that the host will harbor no MDR bacteria. However, the frequency of bacteria with resistance to multiple drugs increases if the infection carries bacteria already resistant to two or more drugs used in a cocktail. Simply, the single drug-resistant strains are highly unstable and prone to undergo multiple genetic recombinations that facilitate the selection and emergence of MDR (Hingley-Wilson et al. 2013; van Rie et al. 2005). Once the resistance has emerged, then the proliferation depends on the fitness and survivability of the host cells as well as stability and transmittance ability of the responsible genetic elements.

4.3 Methods for Detection of Multidrug Resistance

4.3.1 Microscopy

Microscopy has been used as a rapid means of detecting drug resistance in many bacteria. In March 2010, the WHO approved the microscopic observation drug susceptibility (MODS) assay to test susceptibility of mycobacteria against isoniazid and rifampicin. The rapid growth of *Mycobacterium* in liquid medium than solid medium forms the basis of this assay. Rope-like growth/cord formation in liquid medium is characteristic of *M. tuberculosis* which is detected by an inverted microscope (Caviedes et al. 2000; Moore et al. 2006; Bwanga et al. 2010). This is an easy, simple, sensitive, rapid, and inexpensive assay for detection and drug resistance testing of mycobacteria. Sanogo et al. (2017) in their study conducted in West Africa evaluated the performance of this technique by comparing the results with other validated methods, i.e., mycobacteria growth indicator tube/antimicrobial susceptibility testing/streptomycin, isoniazid, rifampicin, and ethambutol. Caviedes et al. (2000) also analyzed sputum samples known to be positive for *M. tuberculosis* for susceptibility to isoniazid and rifampin using MODS and microwell alamar blue assay (MABA) in concordance with 89% of cases. MODS showed results rapidly with a median of 9 days and 9.5 days for diagnostic and susceptibility testing, respectively. Thus, the findings of these studies show the great potential of MODS for better management of multidrug-resistant TB cases throughout the world.

Accelerate Diagnostics (USA) has commercialized a multiplexed automated digital microscopy (MADM) based on fluorescence in situ hybridization (FISH) as a rapid online technique for AST (Metzger et al. 2014; Chantell 2015). Accelerate Pheno uses a preliminary step of electrophoresis to separate impurities like blood and urine from clinical samples. Subsequent change of electric field polarity leads to repel the target microorganism again toward the liquid. Here, the sample representing microbial culture growing in Mueller-Hinton medium is used to detect fluorescence signal after every 10 min (Charnot-Katsikas et al. 2017). Till date, this method is the only FDA-approved rapid diagnostic AST known which is based on microbial growth (Doern 2018).

4.3.2 *Molecular Approaches*

Molecular-based approaches have been increasingly used to trace the antibiotic resistance determinants (ARDs) in clinically relevant pathogens by targeting their specific sequences of deoxyribonucleic acid (DNA) and ribonucleic acid (RNA). Polymerase chain reaction (PCR)- and RNA-based detection, loop-mediated isothermal amplification (LAMP), and DNA microarrays are some of the molecular techniques used for early detection of antibiotic resistance which are discussed below.

4.3.2.1 PCR-Based Approach

The usage of PCR in clinical settings for detection of ARDs was first described to detect β-lactam resistance genes among *Enterobacteriaceae*. Since then, PCR and real-time PCR (qPCR) have proven to be a great diagnostic asset not only for precise identification/quantification of infectious agents but also for profiling their ARDs. Conventional PCR usually takes 12 h to present result and is composed of three steps. Firstly, the genetic material has to be extracted from the specimen (blood, urine, etc.) followed by amplification of target DNA in a thermocycler. Secondly, thermocycler allows continuous amplification of DNA in around 25–35 cycles of denaturation, annealing, and extension. Lastly, the amplified target is detected through agarose gel electrophoresis, southern blotting, DNA fingerprinting, and other sequencing-based techniques (March Rosselló and Bratos Pérez 2016). To further shorten the turnaround time (TAT) of diagnosis to few hours, real-time PCR (qPCR) allows simultaneous amplification and detection of amplicons via different methods. PCR had also been adopted for detection of many ARDs in Gram-negative bacteria such as carbapenemase and cephalosporinase encoding genes like NDM, VIM, KPC, IMP, SHV, TEM, and OXA (Monteiro et al. 2012; Bogaerts et al. 2012).

qPCR can be used to differentiate antibiotic-susceptible strains from resistant ones by monitoring their growth under different antibiotic concentrations. For example, antibiotic susceptibility of *A. baumannii* against ciprofloxacin, imipenem, and colistin has been concluded within 6 h after culturing for isolated colonies wherein highly conserved region of *ompA* gene was used as bacterial growth indicator (Martín-Peña et al. 2013). Recently, a 2-D multiplex qPCR approach was used to detect four cephalosporin and five carbapenemase genes simultaneously in a single tube using a combination of probe colors and Tm values with high sensitivity of 30–3000 copies per reaction (Li et al. 2020). However, the limitation of qPCR is time-consuming culturing protocols and overestimation of viable cells, i.e., amplification of DNA from nonviable bacteria leading to increased false positives.

As the PCR technology refined in terms of sound DNA extraction protocols, multiplexing, and high-throughput automation, several commercial PCR instruments and/or kits came into force for enhanced antibiotic stewardship. Examples include Cepheid Xpert, Roche SeptiFast, Check-Points, Check-Direct CPE, Molzym's SepsiTest, Seegene MagicPlex, and BD GeneOhm MRSA (Li et al. 2017a). BD GeneOhm MRSA is suitable for antibiotic assessment from less complex samples like nasal swabs. These assays deliver results in the form of sample to answer and have short TAT (1–8 h). The GeneXpert® system (Cepheid) is one such example of fully automated qPCR-based point-of-care diagnostic tool to detect several carbapenemase genes directly from rectal samples without the need for skilled technicians. The Xpert® MRSA/SA BC and MRSA/SA SSTI cartridge could detect methicillin-resistant *S. aureus* (MRSA) in grown blood bottles and skin-soft tissue infections, respectively, in an hour with high sensitivity (93%) and specificity (close to 100%) (March Rosselló and Bratos Pérez 2016). This instrument proved useful in clinical laboratory settings as evident by saving 25.4 h of time for optimal antibiotic treatment (Carver et al. 2008). Similarly, Xpert® MTB/RIF detects *Mycobacterium tuberculosis* and its rifampicin resistance within 2 h with 86.8% sensitivity and 93.1% specificity in sputum samples (March Rosselló and Bratos Pérez 2016). SeptiFast test (Roche) is another rapid culture-independent amplification assay studied extensively for diagnosis of bloodstream infections directly from EDTA whole blood samples. It uses dual FRET probes for identification plus quantification and can detect around 25 bacteria and fungi in about 6 h with limit of detection of 3–30 cfu/mL (Mongell et al. 2015). Although SeptiFast possesses specificity and sensitivity values in the range of 74–100% and 60–95%, respectively, the technology is costly and labor-intensive and provides no information on antibiotic susceptibility testing (AST) except for MRSA (Jordana-Lluch et al. 2015). FilmArray (bioMérieux) Blood Culture Identification Panel is another closed and fully automated culture-independent system which is approved by the FDA and CE. The instrument could detect 24 etiological agents (5 *Candida* spp. and 11 Gram-positive and eight Gram-negative bacteria) and three resistance genes (*mecA*, *KPC*, and *vanA/B*) with a TAT of 1 h (Salimnia et al. 2016). It can detect contaminant species also and consists of automated steps of cell lysis, DNA purification, multiplex PCR, nested PCR, and lastly melting curve analysis. The instrument demonstrated overall sensitivity and specificity in the range of 50–100% and 77–100%, respectively (Maugeri et al. 2019). The Curetis Unyvero system employs comprehensive resistance panels for diagnosing many ASDs per test in pneumonia, bloodstream infection, implant and tissue infection, and intra-abdominal infection (Gadsby et al. 2019).

Compared to conventional and real-time PCR, digital PCR (dPCR) has the ability to improve sensitivity as well as robustness in detecting low-abundance targets by partitioning the extracted nucleic acids into many individual picoliter droplets containing either one or zero molecules per droplet reaction. This system allows absolute quantification of target DNA or RNA in digital format (1 or 0) without the need for normalization leading to enhanced precision and reproducibility (Zhang et al. 2015). Besides, the partitioning also offers unique characteristics of determining ASTs by evaluating the delay in DNA replication. This property had been adopted to determine the antibiotic susceptibility of clinical UTI isolates after 15 min of exposure to four clinically relevant antibiotics (Schoepp et al. 2016).

However, PCR-based approaches are rapid, specific, and sensitive and could possibly reduce the turnaround times in antibiotic stewardship but suffer from drawbacks such as expensive, poor correlation between phenotypic and genotypic resistance especially for ESBLs and CREs, and their inability to detect novel and uncharacterized genes or the coincidental mutations thereof especially for continuous emerging variants of carbapenemases in gram negative bacteria (Pulido et al. 2013). Finally, their complicated upstream sample processing makes them more difficult to automate than the conventional phenotypic-based culture methods. Recently, a rapid sample processing-free microfluidic diagnostic platform based on single-step blood droplet digital PCR coupled with 3D particle counter system for bacterial identification and ARD determination directly from whole blood had been reported (Abram et al. 2020). This sample-to-answer device can detect CTX-M-9 family ESBLs with high sensitivity of 10 cfu/mL within 1 h assay time.

4.3.2.2 RNA-Based Approach

This approach of measurement of gene transcripts (RNA) instead of gene itself (DNA) could resolve the inherent disadvantages of DNA-based detection techniques, i.e., the lack of differentiation between viable and nonviable bacteria and between silent and expressed genes (Mancini et al. 2014). NucliSENS EasyQ KPC Test® (bioMérieux) is based on detection of RNA levels of blaKPC gene variants with TAT of 2 h (McEwan et al. 2013). Fluorescence in situ hybridization (FISH) is a non-amplification technique typically utilizing bacterial and fungal rRNA target-specific oligonucleotide probes to capture naturally abundant rRNA genes circumventing the amplification step (Wellinghausen et al. 2006). PNA-FISH is the modified version to detect clinical bacterial isolates using peptide nucleic acid (PNA) probes (Pellestor et al. 2008). In PNA probes, negatively charged phosphate backbone of nucleic acids is replaced by neutral *N*-(2-aminoethyl)-glycine scaffold allowing robust hybridization and less susceptibility to inhibition by impurities.

Furthermore, direct targeting the mRNA could provide both genotypic and phenotypic information via its unique sequence and abundance, respectively. Bhattacharyya et al. (2019) developed a novel assay GoPhAST-R combining the phenotypic and genotypic AST via hybridization-based multiplexed RNA detection for improving molecular epidemiology and identifying emerging resistance mechanisms (Bhattacharyya et al. 2019). The assay presented with 94–99% accuracy in strain classification through machine learning analysis and determined key antibiotic resistance markers based on differential transcriptional responses to antibiotic exposure in <4 h from positive blood cultures compared to 24–36 h by conventional phenotypic ASTs.

4.3.2.3 Isothermal Amplification

Loop-mediated isothermal amplification (LAMP) is another rapid technique for real-time evaluation of antibiotic susceptibility testing (ASTs). Based on the strand displacement activity of Bst DNA polymerase, the gene of interest is amplified to detectable levels at isothermal temperature of 60–65 °C, thereby circumventing the need for thermocycler (Li et al. 2017b). LAMP had been evaluated for detection of carbapenem resistance in *Acinetobacter baumannii* through amplification of OXA-type genes (Vergara et al. 2014). The eazyplex system (Amplex Biosystems GmbH) utilizes LAMP reagents in lyophilized form in a tube and amplifies extracted DNA when added to tube within 30 min at constant temperature maintained in Genie II apparatus (OptiGene) followed by detection of amplicons in real time (March Rosselló and Bratos Pérez 2016). Notably, eazyplex SuperBug CRE had been developed to detect genes encoding for ESBL (CTX-M) and carbapenemases (NDM, KPC, VIM, and OXA-48) directly either from urine samples or from isolated colonies. Hicke et al. (2012) utilized helicase-dependent isothermal amplification with DNA array for multiplexing of 64 distinct targets on a single platform. This technology had been evaluated for detection of *C. difficile* on 130 patient samples with sensitivity and specificity of 97% and 100%, respectively. The ultrafast real-time digital LAMP for phenotypic antibiotic susceptibility measurements of *E. coli* from urine samples based on quantification of transcriptional responses upon antibiotic exposure had been developed which completes its entire workflow within 30 min (Schoepp et al. 2017).

4.3.2.4 DNA Microarrays

DNA microarrays are promising technologies for rapid screening of antibiotic susceptibility as these are capable of detecting broad range of resistance markers in a single step. Multiple cDNA probes on nylon membranes are closely attached to a glass or silicon platform, and specific hybridization of labeled probes with target antibiotic gene marker helps in determining the resistance (Khan et al. 2019). DNA microarrays had been used successfully for evaluating rifampin and isoniazid resistance in *M. tuberculosis* (Huang et al. 2014). An automated system VERIGENE (Luminex) for multiple detection of individual targets is composed of two panels. One panel (BC-GN) detects 8 Gram-negative bacterial targets and their six ARDs (*IMP*, *OXA*, *NDM*, *VIM*, *KPC*, and *CTX-M*), while the other (BC-GP) detects 12 Gram-positive targets and their three ARDs (*mecA*, *vanA*, and *vanB*) with turnaround time of around 2.5 h (Minejima and Wong-Beringer 2016). The cartridges of this system composed of glass slide printed with several capture probes for target DNA sequence-specific binding. A mediator probe binds to target DNA region and has polyA tail for further attachment to polyT region of probe labeled with gold nanoparticles. Afterward, silver nanoparticles surround the gold nanoparticles, and their aggregation is detected by light scattering (Ledeboer et al. 2015). BC-GP/BC-GN have sensitivity and specificity of 92.6–100%/97.1% and 95.4–100%/99.5%, respectively, for bacterial identification (Bork et al. 2014; Buchan et al. 2013). The Check assay (Check-Points) is based on ligation-mediated multiplex qPCR and microarray hybridization using different probes specific to genetic variants of β-lactamases and carbapenemases. It can provide results within 2–6 h; however, DNA extraction has to be performed outside of this platform (Lau et al. 2015). Examples include Check-MDR, Check-MD, and Check-CPE kits. After whole-cell DNA extraction, multiplex ligation reaction produces DNA molecules followed by PCR amplification using universal primers labeled with biotin. The amplicons are hybridized with resistance gene-specific probes, and image-acquired results in the form of presence or absence of marker gene are automatically interpreted by software (March Rosselló and Bratos Pérez 2016).

Microarray offers potential of high-throughput screening of resistance genes unlike PCR which can detect only handful of genes at a time, but it suffers from the same limitations as PCR such as poor correlation between phenotypic and genotypic markers, no information about MIC, and failure to detect novel or uncharacterized mechanisms of resistance. Moreover, the technology is also cost-intensive. Lately, a proof of concept for development of low-cost microarray platform for detection of highly clinically relevant antibiotic genes was developed (Wolff et al. 2019). It is based upon cost-effective in-house elongation of biotin-labeled oligonucleotide probes using terminal deoxynucleotidyl transferase reaction and has comparable sensitivity and specificity at one-tenth the cost of commercial-labeled probes.

4.3.3 Flow Cytometry

Flow cytometry (FC) has also been studied for AST by monitoring the multiple characteristics (cell membrane abnormality, cell size, enzyme activity, and microbial concentration) of microbial populations exposed to different concentrations of antibiotics over time in the presence of relevant fluorescent stains (Huang et al. 2018). After extensive sample preparation and labeling with fluorescent dyes, FC conducts automated data interpretation for over millions of cells (Leonard et al. 2018). For instance, carbapenem susceptibility profiling of *Klebsiella pneumonia* could be qualitatively and quantitatively estimated within 1 h and 3 h, respectively, by flow cytometry (Mulroney et al. 2017). The Luminex xTAG® assay (Luminex) is a well-established technology used for detection of different ESBLs, carbapenemases, and cephalosporinases in bacterial strains in TAT of 5 h and is based on combination of multiplex ligation PCR and colored microsphere-based flow cytometry (Ceyssens et al. 2016). The histograms produced by 90 °C side scattering of light to fluorescent signal enabled differentiation of methicillin-sensitive and methicillin-resistant strains of *S. aureus* upon 4 h exposure to antibiotic (Shrestha et al. 2011). Similarly, Faria-Ramos et al. (2013) differentiated ESBL- and non-ESBL-producing strains on the basis of antibiograms depicting the viability status of bacteria in the presence of antibiotics (ceftazidime, cefotaxime, and clavulanic acid), using the fluorochrome $DiBAC_4$ due to its ability to penetrate into nonviable cells.

FC is also a promising approach to study the MIC in clinical specimens. The MIC of piperacillin, gentamicin, and amoxicillin in pathogens *P. aeruginosa*, *E. coli*, and *S. aureus* has been calculated based on the antibiogram obtained in 4 h, and results were found to be 100% correlated with other systems like broth macrodilution, *E*-test, and VITEK2 (Broeren et al. 2013). Similarly, comparing the decrease in intensity of fluorescent signal obtained by hydrolysis of fluorescein diacetate substrate by the enzyme esterase present in mycobacteria can be used to determine the sensitivity of *M. tuberculosis* to antituberculosis drugs (rifampicin, isoniazid, and ethambutol) (Kirk et al. 1998). However, standardized protocols and data interpretation are limited for AST at present which in turn varies between pathogen and antibiotic combination used (Schumacher et al. 2018). Recently, FC-based AST protocol for early detection and quantification of uropathogenic *E. coli* strains against antibiotics, namely, trimethoprim-sulfamethoxazole, nitrofurantoin, ceftriaxone, and ciprofloxacin, had been optimized, thereby giving results within 4 h in urine samples (Velican et al. 2020). Thus, FC approach could play a pivotal role in enhancing the therapeutic success rate by providing early interpretation on AST against 24 h time taken by most conventional AST methods.

4.3.4 MALDI-TOF Mass Spectrometry

MALDI-TOF mass spectrometry is a technique which analyzes cellular protein reflecting gene products and metabolic products in order to identify different types of microorganisms including bacteria, yeasts, and molds (Vrioni et al. 2018; March Rosselló and Bratos Pérez 2016). This technique usually analyzes highly abundant, chiefly ribosomal proteins of microbes lying in the mass range of 2000–20,000 Da (Caroll and Patel 2015). The use of this technique for detection of drug resistance offers a major advantage of filling the gap between identification and resistance status of a species.

MALDI-TOF offers rapid AST using mainly four types of method (Kostrzewa et al. 2013; Sparbier et al. 2016) including (1) detection of resistance peak pattern by comparing analysis of mass spectra of susceptible and resistant strains, (2) studying bacterial-induced hydrolysis of antibiotic by observing mass shifts during incubation of pathogen with antibiotic, (3) examination of newly synthesized proteins by detection of the amount of isotope-labeled amino acids in pathogen incubated in the presence of antibiotic (Sparbier et al. 2013), and (4) measurement of bacterial growth in the presence and in the absence of antibiotics wherein an internal standard is used to compare the mass spectra and results in low peak intensities indicating susceptibility, i.e., lack of growth and intense bacterial protein peaks indicating resistance, viz., normal growth in the presence of antibiotics (Lange et al. 2014).

Expression of enzymes, namely, β-lactamase and carbapenemases, is the most common mechanism providing resistance to pathogens against β-lactam antibiotics and carbapenems by making them inactive. MALDI-TOF has successfully been used to detect drug resistance against β-lactams (Kostrzewa et al. 2013) and carbapenems (Hrabák et al. 2011) by detecting the hydrolyzed antibiotic that remains after the β-lactamase and carbapenemase activity, respectively.

Genes vanA and vanB are the keynote players in coding for proteins providing resistance to vancomycin and teicoplanin. Presence of van A gene provides resistance against vancomycin as well as teicoplanin and vanB against vancomycin only. Many strains of *Enterococcus faecium* are vanB-positive. MALDI-TOF has been used to differentiate between vancomycin-susceptible and vancomycin-resistant enterococci on the basis of the presence or absence of some specific peaks. This showed a sensitivity of 96.7% and a specificity of 98.1% when compared to PCR, the control method (Griffin et al. 2012; Kasper 1976).

4.3.5 Biosensors

Biosensors are rapid analytical devices which consist of two components, namely, biorecognition element for sensing the presence of analyte and transducer for capturing the signal. Recently, biosensing technology has also been used to test drug resistance in various microorganisms. The presence of antibiotics results in changes in membranes, morphology, metabolism, movements, mass, heat production, and nucleic acid content of bacteria that can be targeted for detection by biosensing technology (Vasala et al. 2020). Several attempts have been made to develop different types of biosensors using various types of biosensing elements like bacteriophages, aptamers, and transducers like electrochemical, piezoelectric, etc.

Guntupalli et al. (2013) developed a lytic phage spheroid-based sensor for detection of methicillin-resistant (MRSA) and methicillin-sensitive (MSSA) *S. aureus* species. Bacteriophage-spheroid interactions were evaluated by a quartz crystal microbalance with dissipation tracking. Bacteria-spheroid interactions result in low resonance frequency and increase in dissipation energy for both MRSA and MSSA. Another study reported the development of a label-free electrochemical DNA biosensor for determination of MDR1 gene. The working of this sensor included the hybridization of DNA by differential pulse voltammetry in terms of monitoring the changes in peak currents based on Au nanoparticles/toluidine blue-graphene oxide (Au NPs/TB-GO) modified electrode. The developed sensor was simple, rapid, and cost-effective with high selectivity and stability and showed a detection limit of 2.95×10^{-12} M (at an *S/N* of 3) (Peng et al. 2015). Chen et al. (2020) developed a novel multiplex loop-mediated isothermal amplification linked to a nanoparticle-based lateral flow biosensor (m-LAMP-LFB). This sensor was used to detect all *S. aureus* species and identify MRSA and MSSA strains from clinical samples. The working of LFB was based on the use of two sets of primers specific for femA gene (*S. aureus*-specific gene) and mecA gene (encoding penicillin-binding protein 2a) and analysis of products produced. A function-based biosensing approach for detection of antibiotic resistance has been developed by Mecklenburg et al. (2017). The working of this sensor was based on the identification of resistance enzymes in terms of measurement of thermal signal generated upon enzyme-mediated degradation of antibiotics. This biosensing system was able to differentiate between two AMR β-lactamase enzymes, namely, penicillinase and metallo-β-lactamase. Enzyme thermistor technology was used to create the resistance profile by analyzing a panel of antibiotics. Till date, most of the biosensing technologies reported to detect drug resistance present only proof-of-concept studies. Evaluation of their performance on clinical samples and subsequent progress for faster commercialization still remains a challenge. Currently, the exploration of biosensing technology for their application in rapid detection of drug resistance is further in progress in order to felicitate its faster commercialization.

4.3.6 Immunodiagnostics

Immunodiagnostics constitute one of the highly specific and sensitive approaches in order to felicitate identification of pathogenic bacteria (Verma et al. 2013). In the past decades, several attempts have been made by different researchers to apply immunodiagnostic concept for development of rapid methods for drug resistance detection. Poetz et al. (2018) reported the development of a highly sensitive, reproducible, and reliable sandwich immunoassay for detection of multidrug resistance protein 1 (MDR1). MDR1 presents a prototypic 12 transmembrane domain transporter. In order to develop this assay, antibodies were produced specifically to target a proteotypic peptide derived from MDR1. The performance of this assay was cross-validated using mass spectrometry. Another study also reported the development of a strip-based immunoassay based on the principle of detection of enzyme conferring drug resistance. The working of this assay included the application of diluent containing bacteria representing antigen on the test position that has previously been loaded with specific antibody. Interaction of antigen and antibody results in the development of a colored band, thus indicating positive test. One such test has been marketed by BinaxNOW PBP2a test (Alere) to detect methicillin resistance in *S. aureus*. This test is highly sensitive and specific and has successfully been applied to blood culture with practically a sensitivity and specificity of 95.4% and 100%, respectively (Pence et al. 2013; March Rosselló and Bratos Pérez 2016). In this line, Volland et al. (2019) also developed a rapid, highly sensitive, and cost-effective lateral flow immunoassay using monoclonal antibodies selectively designed to detect colistin resistance. Colistin resistance has mainly been reported in bacteria due to a plasmid-encoded colistin resistance gene mcr-1. The working of this assay included suspension of bacterial colony in extraction buffer with subsequent dispensing on the cassette and allowing migration for 15 min. The positive results were recorded in terms of appearance of a pink color band. This assay was successfully applied to detect colistin resistance in human and animal enterobacterial isolates showed a sensitivity of 100% and specificity of 98% for detection of MCR-1.

4.3.7 Chemiluminescence and Bioluminescence

In chemiluminescence, emission of light occurs upon return of excited molecules to ground state in some chemical reactions. On the other hand, bioluminescence is a form of chemiluminescence which results from chemical reactions occurring inside living organisms such as fireflies (March Rosselló and Bratos Pérez 2016). Maity et al. (2020) developed a chemiluminescent probe for detection of beta-lactamase, the expression which is known as a major cause of drug resistance. The sensitivity reported with this probe for beta-lactamases has been found to be four times higher when compared with fluorocillin, a commercially available fluorogenic substrate. Similarly, another study carried out by Das et al. (2020) showed in their work development of a carbapenemase-sensitive chemiluminescent probe. This probe facilitates enzyme-mediated breakdown of carbapenem core and subsequently followed by a facile 1,8-elimination process leading to green light emission resulting from faster chemical excitation. Moreover, this probe has successfully been used to detect several clinically relevant carbapenemases in their work.

Chemiluminescence approach has also been applied for direct analysis of drug resistance in microbial culture. Here, menadione being permeable to bacterial membrane is added to microbial culture. After menadione enters into the microbial cells, it gets reduced and generates several compounds that undergo self-oxidation, resulting in emission of light (March Rosselló and Bratos Pérez 2016). The working principle of this method includes determination of viability of microbial cultures incubated in the presence and absence of antibiotics by comparing their chemiluminescence signal. This approach has been used to detect intermediate and vancomycin-heteroresistant strains of *S. aureus* in 8 h. Moreover, this technique can also be used to detect mycobacteria sensitivity in 4 days (Yamashoji 2002).

Like chemiluminescence, bioluminescence approach compares the bioluminescent signal of microbial cultures incubated in the presence and absence of antibiotics. Adenosine triphosphate (ATP) bioluminescence has been used for rapid AST. Bioluminescent signal is measured in terms of extracellular, intracellular, or total ATP levels of microbial culture. In the past decades, this approach has successfully been used to test drug resistance using direct sample. This method has successfully been used on urine sample. Here, antibiogram using urine sample was made by comparing the ATP signal obtained from two different aliquots of urine sample incubated with and without antibiotic (Ivancic et al. 2008). Application of ATP bioluminescence method has also been attempted on blood culture specimens to test levofloxacin susceptibility with no false-susceptible results in a pilot study (Matsui et al. 2019). Blood corpuscles are known to have abundant intracellular ATP possibly leading to false-susceptible results. Keeping in view, some new interventions including specimen centrifugation with subsequent dilution of supernatant with broth and an ATP-eliminating reagent before addition to bacterial culture have been made to minimize background ATP in this study.

4.3.8 Colorimetric Methods

Various types of colorimetric methods chiefly employing redox indicators, tetrazolium salts, and nitrate reduction principle have been developed to rapidly detect drug resistance. Till date, various types of tetrazolium salts have been evaluated for their use in rapid detection of multiple drug resistance in causal agent of TB. Out of various salts, the compound, namely, 3-(4,5-dimethylthiazol-2-yl)-2,5-diphenyl tetrazolium bromide (MTT), has widely been used for testing of drug resistance. Mshana and colleagues first suggested its use in development of a colorimetric assay for detection of rifampicin resistance (Mshana et al. 1998; Abate et al. 1998). The working principle of this test includes the reduction of MTT which is usually of yellow color in its oxidized state to purple-colored MTT formazan crystals. This test has also been used earlier to determine the viability of cells (Mosmann 1983). In a study conducted by Mohammadzadeh et al. (2006), 2,3,5-triphenyltetrazolium chloride was used to test antibiotic susceptibility of multidrug-resistant *M. tuberculosis* isolates against isoniazid and rifampicin at a critical concentration of 0.2 μg/mL and 2.0 μg/mL, respectively. Using this method, results were obtained and recorded in terms of color change after 4.9 days.

Colorimetric methods based on the use of indicator dyes record the change in color of the indicator dye in medium with and without antibiotic. The change in color of the dyes occurs as a result of metabolic activity of bacteria during growth (Palomino et al. 2007). This concept of using redox indicators has successfully been reported to offer rapid detection of resistance. In a study carried out by Coban (2012), the crystal violet decolorization assay (CVDA) has successfully been used to test and compare susceptibility of *Mycobacterium tuberculosis* to isoniazid and rifampicin. Similarly, some other studies have shown the determination of susceptibility of pathogens against different antibiotics using resazurin. Coban et al. (2006; Coban 2012) tested the susceptibility of *S. aureus* for methicillin and vancomycin in 6 h using resazurin. The major drawback associated with resazurin is the inability of non-fermenting Gram-negative bacilli to metabolize it.

Another colorimetric method that has widely been used to evaluate multiple drug resistance is nitrate reductase assay. The working of this assay includes the measurement of metabolic activity in terms of nitrate reductase in bacterial culture in the presence of nitrate ions and antibiotics. It is a cost-effective, rapid phenotypic method and has also been recommended by the WHO for rapid detection of drug-resistant TB (Angeby et al. 2002; WHO 2009). It has also been used to test methicillin resistance in *Staphylococcus aureus* (Coban et al. 2014) and rifampin, isoniazid, and streptomycin resistance in *Mycobacterium tuberculosis* (Angeby et al. 2002).

4.4 Conclusions and Future Outlook

The methodologies covered in this chapter are the modern approaches which offer rapid detection of antimicrobial drug resistance when compared to traditional methods. The standard methods including disc diffusion and broth dilution used for determination of antimicrobial susceptibility are mainly based on microbial growth and need lengthy protocols and pure cultures. Keeping in view, in the past decades, several attempts have been made by researchers and are currently in progress to develop rapid AST. Development of new ASTs and possible improvements of existing technologies offering satisfactory performance need to well recognize the limitations of conventional and modern AST methods in order to get their faster approval and marketing. Clinical samples usually are of polymicrobial nature, and only a few growth-based ASTs can work efficiently with such samples. This indicates the need for highly sensitive measurement of microbial growth by online analysis of samples taken after short intervals using techniques like FISH and immunoassays. The WHO-approved MODA and FDA-approved Accelerate Pheno system based on FISH methods should further be validated using diversity of other clinical samples. Techniques like MALDI-TOF offer rapid formation of antibiogram. However, the interpretation of antibiogram further requires the information about bacterial species. Therefore, it is important to increase the number of strains and antibiotics under study to determine the sensitivity and specificity of such methods. Many new AST methods directly detect all types of resistance phenotypes and/or genotypes present within a clinical sample without culturing, identification, and subsequent study of mechanism underlying resistance. This is not of much significance. Moreover, regulatory bodies like the EUCAST, the Clinical and Laboratory Standards Institute, and the FDA should lay down guidelines for developers and manufacturers of AST and give MIC breakpoints for compliance by diagnostic laboratories. The development of new AST should strictly fulfill the need of some desirable features chiefly including reduced detection time, less cost for faster implementation, and high multiplexing capacity. Approaches focusing on the prevention of infection and antibiotic stewardship should also be given attention.

References

Abate G, Mshana RN, Miorner H (1998) Evaluation of a colorimetric assay based on 3-(4,5-dimethylthiazol-2-yl)- 2,5-diphenyl tetrazolium bromide (MTT) for rapid detection of rifampicin resistance in Mycobacterium tuberculosis. Int J Tuberc Lung Dis 2:1011–1016

Abram TJ, Cherukury H, Ou CY, Vu T, Toledano M et al (2020) Rapid bacterial detection and antibiotic susceptibility testing in whole blood using one-step, high throughput blood digital PCR. Lab Chip 20:477–489

Alekshun MN, Levy SB (2007) Molecular mechanisms of antibacterial multidrug resistance. Cell 128:1037–1050

Angeby KA, Klintz L, Hoffner SE (2002) Rapid and inexpensive drug susceptibility testing of Mycobacterium tuberculosis with a nitrate reductase assay. J Clin Microbiol 40:553–555

Baquero F, Negri MC, Morosini MI, Blazquez J (1998) Antibiotic selective environments. Clin Infect Dis 27(Suppl 1):S5–S11

Basak S, Singh P, Rajurkar M (2016) Multidrug resistant and extensively drug resistant bacteria: a study. J Pathog 2016:4065603. 1–5

van Belkum A, Burnham CAD, Rossen JWA et al (2020) Innovative and rapid antimicrobial susceptibility testing systems. Nat Rev Microbiol 18:299–311

Bhattacharyya RP, Bandyopadhyay N, Ma P, Son SS, Liu J, He LL et al (2019) Simultaneous detection of genotype and phenotype enables rapid and accurate antibiotic susceptibility determination. Nat Med 25:1858–1864

Bogaerts P, Hamels S, de Mendonca R (2012) Analytical validation of a novel high multiplexing real-time PCR array for the identification of key pathogens causative of bacterial ventilator-associated pneumonia and their associated resistance genes. J Antimicrob Chemother 68:340–347

Bork JT, Leekha S, Heil E, Zhao L, Badamas R, Johnson JK (2014) Rapid testing using Verigene® BC-GN microarray in combination with antimicrobial stewardship intervention in gram-negative bacteremia. Antimicrob Agents Chemother 59:1588–1595

Broeren MA, Maas Y, Retera E, Arents NL (2013) Antimicrobial susceptibility testing in 90 min by bacterial cell count monitoring. Clin Microbiol Infect 19:286–291

Buchan BW, Ginocchio CC, Manii R, Cavagnolo R, Pancholi P, Swyers L et al (2013) Multiplex identification of gram-positive bacteria and resistance determinants directly from positive blood culture broths: evaluation of an automated microarray-based nucleic acid test. PLoS Med 10:e1001478

Bwanga F, Joloba ML, Haile M, Hoffner S (2010) Evaluation of seven tests for the rapid detection of multidrug-resistant tuberculosis in Uganda. Int J Tuberc Lung Dis 14:890–895

Caroll K, Patel R (2015) Systems for identification of bacteria and fungi. In: Jorgensen J, Pfaller M, Carroll K et al (eds) Manual of clinical microbiology, 11th edn. ASM Press N.W, Washington, DC, pp 29–43

Carver PL, Lin SW, DePestel DD et al (2008) Impact of mecA gene testing and intervention by infectious disease clinical pharmacists on time to optimal antimicrobial therapy for Staphylococcus aureus bacteremia at a university hospital. J Clin Microbiol 46:2381–2383

Caviedes L, Lee TS, Gilman RH, Sheen P, Spellman E, Lee EH et al (2000) Rapid, efficient detection and drug susceptibility testing of Mycobacterium tuberculosis in sputum by microscopic observation of broth cultures. The Tuberculosis Working Group in Peru. J Clin Microbiol 38:1203–1208

Ceyssens PJ, Garcia-Graells C, Fux F et al (2016) Development of a Luminex xTAG® assay for cost-effective multiplex detection of β-lactamases in Gram-negative bacteria. J Antimicrob Chemother 71:2479–2483

Chantell C (2015) Multiplexed automated digital microscopy for rapid identification and antimicrobial susceptibility testing of bacteria and yeast directly from clinical samples. Clin Microbiol Newsl 37:161–167

Charnot-Katsikas A, Tesic V, Love N, Hill B, Bethel C, Boonlayangoor S et al (2017) Use of the accelerate pheno system for identification and antimicrobial susceptibility testing of pathogens in positive blood cultures and impact on time to results and workflow. J Clin Microbiol 56:e01166–e01117

Chen X, Ma K, Yi X, Xiong L, Wang Y, Li S (2020) The rapid and visual detection of methicillin-susceptible and methicillin-resistant Staphylococcus aureus using multiplex loop-mediated isothermal amplification linked to a nanoparticle-based lateral flow biosensor. Antimicrob Resist Infect Control 9:111

Coban AY (2012) Rapid determination of methicillin resistance among Staphylococcus aureus clinical isolates by colorimetric methods. J Clin Microbiol 50:2191–2193

Coban AY, Bozdogan B, Cihan CC, Cetinkaya E, Bilgin K, Darka O et al (2006) Two new colorimetric methods for early detection of vancomycin and oxacillin resistance in Staphylococcus aureus. J Clin Microbiol 44:580–582

Coban AY, Demirpek U, Ciftçi A, Bozdoğan B (2014) Staphylococcus aureus metisilin direncinin hızlı saptanmasında nitrat redüktaz testi: Bir sınır değer duyarlılık test yöntemi [Nitrate reductase assay for the rapid detection of Staphylococcus aureus methicillin resistance: a breakpoint susceptibility testing method]. Mikrobiyol Bul 48:40–47. Turkish. PMID: 24506714

Croucher NJ, Finkelstein JA, Pelton SI, Mitchell PK, Lee GM, Parkhill J, Bentley SD, Hanage WP, Lipsitch M (2013) Population genomics of post-vaccine changes in pneumococcal epidemiology. Nat Genet 45:656–663

Das S, Ihssen J, Wick L, Sitz U, Shabat D (2020) Chemiluminescent carbapenem-based molecular probe for detection of carbapenemase activity in live bacteria. Chem Eur J 26:3647–3652

Doern CD (2018) The slow march toward rapid phenotypic antimicrobial susceptibility testing: are we there yet? J Clin Microbiol 56:e01999–e01917

Faria-Ramos I, Espinar MJ, Rocha R, Santos-Antunes J, Rodrigues AG, Cantón R et al (2013) A novel flow cytometric assay for rapid detection of extended-spectrum beta-lactamases. Clin Microbiol Infect 19:E8–E15

Frickmann H, Masanta WO, Zautner AE (2014) Emerging rapid resistance testing methods for clinical microbiology laboratories and their potential impact on patient management. Biomed Res Int 2014:375681. 1–19

Gadsby NJ, McHugh MP, Forbes C, MacKenzie L, Hamilton SK, Griffith DM, Templeton KE (2019) Comparison of unyvero P55 pneumonia cartridge, in-house PCR and culture for the identification of respiratory pathogens and antibiotic resistance in bronchoalveolar lavage fluids in the critical care setting. Eur J Clin Microbiol 38:1171–1178

Griffin PM, Price GR, Schooneveldt JM et al (2012) The use of MALDI-TOF MS to identify vancomycin resistant Enterococci and investigate the epidemiology of an outbreak. J Clin Microbiol 50:2918–2931

Guntupalli R, Sorokulova I, Olsen E, Globa L, Pustovyy O, Vodyanoy V (2013) Biosensor for detection of antibiotic resistant *Staphylococcus* bacteria. J Vis Exp 2013:e50474

Hall RM (2012) Integrons and gene cassettes: hotspots of diversity in bacterial genomes. Ann N Y Acad Sci 1267:71–78

Hicke B, Pasko C, Groves B et al (2012) Automated detection of toxigenic clostridium difficile in clinical samples: isothermal Tcdb amplification coupled to array-based detection. J Clin Microbiol 50:2681–2687

Hingley-Wilson SM, Casey R, Connell D, Bremang S, Evans JT, Hawkey PM, Smith GE, Jepson A, Philip S, Kon OM, Lalvani A (2013) Undetected multidrug-resistant tuberculosis amplified by first-line therapy in mixed infection. Emerg Infect Dis 19:1138–1141

Hrabák J, Walkova R, Studentova V et al (2011) Carbapenemase activity detection by matrix-assisted laser desorption ionization-time of flight mass spectrometry. J Clin Microbiol 49:3222–3227

Huang WL, Hsu ZJ, Chang TC, Jou R (2014) Rapid and accurate detection of rifampin and isoniazid-resistant Mycobacterium tuberculosis using an oligonucleotide array. Clin Microbiol Infect 20:O542–O549

Huang TH, Tzeng YL, Dickson RM (2018) FAST: rapid determinations of antibiotic susceptibility phenotypes using label-free cytometry. Cytom A 93:639–648

Ivancic V, Mastali M, Percy N, Gornbein J, Babbitt JT, Li Y et al (2008) Rapid antimicrobial susceptibility determination of uropathogens in clinical urine specimens by use of ATP bioluminescence. J Clin Microbiol 46:1213–1219

Jordana-Lluch E, Giménez M, Quesada MD, Rivaya B, Marcó C, Domínguez MJ et al (2015) Evaluation of the broad-range PCR/ESI-MS technology in blood specimens for the molecular diagnosis of bloodstream infections. PLoS One 10:e0140865

Kasper DL (1976) The polysaccharide capsule of Bacteroides fragilis subspecies fragilis: immunochemical and morphologic definition. J Infect Dis 133:79–87

Khan ZA, Siddiqui MF, Park S (2019) Current and emerging methods of antibiotic susceptibility testing. Diagnostics 9:49

Kirk SM, Schell RF, Moore AV, Callister SM, Mazurek GH (1998) Flow cytometric testing of susceptibilities of Mycobacterium tuberculosis isolates to ethambutol, isoniazid, and rifampin in 24 hours. J Clin Microbiol 36:1568–1573

Kohanski MA, DePristo MA, Collins JJ (2010) Sublethal antibiotic treatment leads to multidrug resistance via radical-induced mutagenesis. Mol Cell 37:311–320

Kostrzewa M, Sparbier K, Maier T et al (2013) MALDI-TOF MS: an upcoming tool for rapid detection of antibiotic resistance in microorganisms. Proteomics Clin Appl 7:767–778

Lange C, Schubert S, Jung J et al (2014) Quantitative matrix assisted laser desorption ionization-time of flight mass spectrometry for rapid resistance detection. J Clin Microbiol 52:4155–4162

Lau AF, Fahle GA, Kemp MA, Jassem AN, Dekker JP, Frank KM (2015) Clinical performance of Check-Direct CPE, a multiplex PCR for direct detection of bla KPC, bla NDM and/or bla VIM, and bla OXA-48 from perirectal swabs. J Clin Microbiol 53:3729–3737

Ledeboer NA, Lopansri BK, Dhiman N, Cavagnolo R, Carroll KC, Granato P et al (2015) Identification of Gram-negative bacteria and genetic resistance determinants from positive blood culture broths by use of the Verigene Gram-negative blood culture multiplex microarray-based molecular assay. J Clin Microbiol 53:2460–2472

Lee YL, Chen YS, Toh HS, Huang CC, Liu YM, Ho CM, Lu PL, Ko WC, Chen YH, Wang JH, Tang HJ, Yu KW, Liu YC, Chuang YC, Liu CE, Hsueh PR (2012) Antimicrobial susceptibility of pathogens isolated from patients with complicated intra-abdominal infections at five medical centers in Taiwan that continuously participated in the Study for Monitoring Antimicrobial Resistance Trends (SMART) from 2006 to 2010. Int J Antimicrob Agents 40:29–36

Leonard H, Colodner R, Halachmi S, Segal E (2018) Recent advances in the race to design a rapid diagnostic test for antimicrobial resistance. ACS Sens 3:2202–2217

Li Y, Yang X, Zhao W (2017a) Emerging microtechnologies and automated systems for rapid bacterial identification and antibiotic susceptibility testing. SLAS Technol 22:585–608

Li Y, Fan P, Zhou S, Zhang L (2017b) Loop-mediated isothermal amplification (LAMP): a novel rapid detection platform for pathogens. Microb Pathog 107:54–61

Li Y, Xu P, Wan Z, Du H, Jin X, Zhang C (2020) Single-tube detection of nine bacterial antibiotic-resistance genes by a 2-dimensional multiplex qPCR assay based on fluorescence and melting temperature. Mol Biol Rep 47:7341–7348

Livermore DM (2012) Current epidemiology and growing resistance of Gram-negative pathogens. Korean J Intern Med 27:128–142

Maity S, Wang X, Das S, He M, Riley LW, Murthy N (2020) A cephalosporin-chemiluminescent conjugate increases beta-lactamase detection sensitivity by four orders of magnitude. Chem Commun (Camb) 56:3516–3519

Mancini N, Infurnari L, Ghidoli N, Valzano G, Clementi N, Burioni R, Clementi M (2014) Potential impact of a microarray-based nucleic acid assay for rapid detection of Gram-negative bacteria and resistance markers in positive blood cultures. J Clin Microbiol 52:1242–1245

Mao EF, Lane L, Lee J, Miller JH (1997) Proliferation of mutators in a cell population. J Bacteriol 179:417–422

March Rosselló GA, Bratos Pérez MA (2016) Antibiograma rápido en Microbiología Clínica. Enferm Infecc Microbiol Clin 34:61–68

Martín-Peña R, Domínguez-Herrera J, Pachón J, McConnell MJ (2013) Rapid detection of antibiotic resistance in *Acinetobacter baumannii* using quantitative real-time PCR. J Antimicrob Chemother 68:1572–1575

Matsui A, Niimi H, Uchiho Y et al (2019) A rapid ATP bioluminescence-based test for detecting levofloxacin resistance starting from positive blood culture bottles. Sci Rep 9:13565

Maugeri G, Lychko I, Sobral R, Roque AC (2019) Identification and antibiotic-susceptibility profiling of infectious bacterial agents: a review of current and future trends. Biotechnol J 14:1700750

McEwan AS, Derome A, Meunier D et al (2013) Evaluation of the NucliSENS EasyQ KPC assay for detection of *Klebsiella pneumoniae* carbapenemase-producing Enterobacteriaceae. J Clin Microbiol 51:1948–1950

Mecklenburg M, Chen Q, Andersson A, Xie B (2017) A biosensing strategy for fast profiling of antibiotic resistance. Proc Technol 27:33–34

Metzger S, Frobel RA, Dunne WM (2014) Rapid simultaneous identification and quantitation of *Staphylococcus aureus* and *Pseudomonas aeruginosa* directly from bronchoalveolar lavage specimens using automated microscopy. Diagn Microbiol Infect Dis 79:160–165

Minejima E, Wong-Beringer A (2016) Implementation of rapid diagnostics with antimicrobial stewardship. Expert Rev Anti-Infect Ther 14:1065–1075

Mohammadzadeh A, Farnia P, Ghazvini K, Behdani M, Rashed T, Ghanaat J (2006) Rapid and low-cost colorimetric method using 2,3,5-triphenyltetrazolium chloride for detection of multidrug-resistant Mycobacterium tuberculosis. J Med Microbiol 55(Pt 12):1657–1659

Mongell G, Romeo MA, Denaro C, Gennaro M, Fraggetta F, Stefani S (2015) Added value of multi-pathogen probe-based real-time PCR SeptiFast in the rapid diagnosis of bloodstream infections in patients with bacteraemia. J Med Microbiol 64:670–675

Monteiro J, Widen RH, Pignatari AC et al (2012) Rapid detection of carbapenemase genes by multiplex real-time PCR. J Antimicrob Chemother 67:906–909

Moore DAJ, Evans CAW, Gilman RH, Caviedes L, Coronel J, Vivar A et al (2006) Microscopic-observation drug-susceptibility assay for the diagnosis of TB. N Engl J Med 355:1539–1550

Mosmann T (1983) Rapid colorimetric assay for cellular growth and survival: application to proliferation and cytotoxicity assays. J Immunol Methods 65:55–63

Mshana RN, Tadesse G, Abate G, Miorner H (1998) Use of 3-(4,5- dimethylthiazol-2-yl)-2,5-diphenyl tetrazolium bromide for rapid detection of rifampin-resistant Mycobacterium tuberculosis. J Clin Microbiol 36:1214–1219

Mulroney KT, Hall JM, Huang X, Turnbull ES, Bzdyl NM, Chakera A, Naseer U, Corea EM, Ellington MJ, Hopkins KL et al (2017) Rapid susceptibility profiling of carbapenem-resistant Klebsiella pneumoniae. Sci Rep 7:1903

Palomino JC, Martin A, Portaels F (2007) Rapid drug resistance detection in Mycobacterium tuberculosis: a review of colourimetric methods. Clin Microbiol Infect 13:754–762

Pellestor F, Paulasova P, Hamamah S (2008) Peptide nucleic acids (PNAs) as diagnostic devices for genetic and cytogenetic analysis. Curr Pharm Des 14:2439–2444

Pence MA, McElvania TeKippe E, Burnham CA (2013) Diagnostic assays for identification of microorganisms and antimicrobial resistance determinants directly from positive blood culture broth. Clin Lab Med 33:651–684

Peng HP, Hu Y, Liu P, Deng YN, Wang P, Chen W, Liu AL, Chen YZ, Lin XH (2015) Label-free electrochemical DNA biosensor for rapid detection of mutidrug resistance gene based on Au nanoparticles/toluidine blue–graphene oxide nanocomposites. Sensors Actuators B Chem 207(Part A):269–276

Poetz O, Dieze T, Hammer H, Wei F, Sommersdorf C, Templin MF, Esdar C, Zimmermann A, Stevanovic S, Bedke J, Stenzl A, Joos TO (2018) Anal Chem 90:5788–5794

Prestinaci F, Pezzotti P, Pantosti A (2015) Antimicrobial resistance: a global multifaceted phenomenon. Pathog Glob Health 109:309–318

Pulido MR, García-Quintanilla M, Martín-Peña R, Cisneros JM, McConnell MJ (2013) Progress on the development of rapid methods for antimicrobial susceptibility testing. J Antimicrob Chemother 68:2710–2717

van Rie A, Victor TC, Richardson M, Johnson R, van der Spuy GD, Murray EJ, Beyers N, Gey van Pittius NC, van Helden PD, Warren RM (2005) Reinfection and mixed infection cause changing Mycobacterium tuberculosis drug-resistance patterns. Am J Respir Crit Care Med 172:636–642

Roberts AP, Mullany P (2009) A modular master on the move: the Tn916 family of mobile genetic elements. Trends Microbiol 17:251–258

Salimnia H, Fairfax MR, Lephart PR, Schreckenberger P, DesJarlais SM, Johnson JK et al (2016) Evaluation of the FilmArray blood culture identification panel: results of a multicenter controlled trial. J Clin Microbiol 54:687–698

Sanogo M, Kone B, Diarra B, Maiga M, Baya B, Somboro AM, Sarro YS, Togo ACG et al (2017) Performance of microscopic observation drug susceptibility for the rapid diagnosis of tuberculosis and detection of drug resistance in Bamako, Mali. Clin Microbiol Infect 23:408

Schoepp NG, Khorosheva EM, Schlappi TS, Curtis MS, Humphries RM, Hindler JA, Ismagilov RF (2016) Digital quantification of DNA replication and chromosome segregation enables determination of antimicrobial susceptibility after only 15 minutes of antibiotic exposure. Angew Chem Int Ed 55:9557–9561

Schoepp NG, Schlappi TS, Curtis MS, Butkovich SS, Miller S, Humphries RM, Ismagilov RF (2017) Rapid pathogen-specific phenotypic antibiotic susceptibility testing using digital LAMP quantification in clinical samples. Sci Transl Med 9:eaal3693

Schumacher A, Vranken T, Malhotra A, Arts JJC, Habibovic P (2018) In vitro antimicrobial susceptibility testing methods: agar dilution to 3D tissue-engineered models. Eur J Clin Microbiol 37:187–208

Sharma A (2011) Antimicrobial resistance: no action today, no cure tomorrow. Indian J Med Microbiol 29:91–92

Shrestha NK, Scalera NM, Wilson DA, Procop GW (2011) Rapid differentiation of methicillin-resistant and methicillin-susceptible Staphylococcus aureus by flow cytometry after brief antibiotic exposure. J Clin Microbiol 49:2116–2120

Sparbier K, Lange C, Jung J et al (2013) MALDI biotyper-based rapid resistance detection by stable-isotope labeling. J Clin Microbiol 51:3741–3748

Sparbier K, Schubert S, Kostrzewa M (2016) MBT-ASTRA: a suitable tool for fast antibiotic susceptibility testing? Methods 104:48–54

UN (2016) United Nations meeting on antimicrobial resistance. Bull World Health Organ 94:638–639

Vasala A, Hytönen VP, Laitinen OH (2020) Modern tools for rapid diagnostics of antimicrobial resistance. Front Cell Infect Microbiol 10:308

Velican AM, Măruţescu L, Kamerzan C, Cristea VC, Banu O, Borcan E, Chifiriuc MC (2020) Rapid detection and antibiotic susceptibility of uropathogenic Escherichia coli by flow cytometry. Microorganisms 8:1233

Vergara A, Zboromyrska Y, Mosqueda N, Morosini MI, García-Fernández S, Roca I, Cantón R, Marco F, Vila J (2014) Evaluation of a loop-mediated isothermal amplification-based methodology to detect carbapenemase carriage in acinetobacter clinical isolates. Antimicrob Agents Chemother 58:7538–7540

Verma J, Saxena S, Babu SG (2013) ELISA-based identification and detection of microbes. Springer, Berlin, pp 169–186

Volland H, Dortet L, Bernabeu S et al (2019) Development and multicentric validation of a lateral flow immunoassay for rapid detection of MCR-1-producing enterobacteriaceae. J Clin Microbiol 57:e01454–e01418

Vrioni G, Tsiamis C, Oikonomidis G, Theodoridou K, Kapsimali V, Tsakris A (2018) MALDI-TOF mass spectrometry technology for detecting biomarkers of antimicrobial resistance: current achievements and future perspectives. Ann Transl Med 6:240

Wellinghausen N, Nockler K, Sigge A, Bartel M, Essig A, Poppert S (2006) Rapid detection of Brucella spp. in blood cultures by fluorescence in situ hybridization. J Clin Microbiol 44:1828–1830

WHO (2009) Strategic and Technical Advisory Group for Tuberculosis (STAG-TB). Report of the Ninth Meeting. WHO, Geneva. http://www.who.int/tb/advisory_bodies/stag_tb_report_2009.pdf. Accessed 12 Jan 2021

Wolff N, Hendling M, Schönthaler S, Geiss AF, Barišić I (2019) Low-cost microarray platform to detect antibiotic resistance genes. Sens Biosens Res 23:100266

Woolhouse M, Ward M, vanBunnik B, Farrar J (2015) Antimicrobial resistance in humans, livestock and the wider environment. Philos Trans R Soc Lond Ser B Biol Sci 370:20140083

Yamashoji S (2002) Menadione-catalyzed luminol chemiluminescent assay for viability of Mycobacterium bovis. Microbiol Immunol 46:571–573

Zhang K, Kang DK, Ali MM, Liu L, Labanieh L, Lu M et al (2015) Digital quantification of miRNA directly in plasma using integrated comprehensive droplet digital detection. Lab Chip 15:4217–4226

Chapter 5
The Use of Antimicrobials in Agriculture and Socioeconomic Considerations in Global Perspective

Himadri Rajput, Swati Dhiman, and Rahil Changotra

5.1 Introduction

The use of antimicrobials in agriculture at large scale is a growing concern globally. The precise measurement of antimicrobials used in agriculture worldwide is difficult; however, the studies suggest that the extent of their use in agriculture is as great as the quantity used by humans. It has been estimated that 40 million pounds of antimicrobials are used every year in the United States, out of which around 0.1% is utilized in the field of plant agriculture (Levy 1992). Also, in the United States, the use of antimicrobials for animals is far greater than in humans, as for animals 70% or even more antibiotics are used. A report by the Food and Drug Administration (FDA) in 2011 stated that in the United States, 93% of the human medically important antibiotics are administered in agriculture via feed or water (FDA 2014). It has been also studied that 75–90% of tested antibiotics are excreted unmetabolized from animals which then enter the water resources (Marshall and Levy 2011). The waste generated from animals possesses resistant bacteria strains and antibiotics that could then foster resistance in bacteria that may pose a greater risk to humans. The manure is often utilized for fertilization of crops which comes from animal farms, a major cause of resistance (Sengeløv et al. 2003). The utilization of antimicrobials in animals for meat production in the BRICS (Brazil, Russia, India, China, South Africa) countries alone could probably double between the years 2010 and 2030. Prolonged exposure to antimicrobials could cause development of drug resistance, and high-risk factors are associated with the huge consumption of

H. Rajput · R. Changotra (✉)
School of Energy and Environment, Thapar Institute of Engineering and Technology, Patiala, India

S. Dhiman
Department of Agriculture, DAV College of Education, Hoshiarpur, India

N. Akhtar et al. (eds.), *Emerging Modalities in Mitigation of Antimicrobial Resistance*,
https://doi.org/10.1007/978-3-030-84126-3_5

antimicrobials such as transfer of drug-resistant strains by direct contact and transfer of these strains to humans by the food chain. Fungal diseases are more common and frequent in crops and possess huge threat. Hence, fungicides are employed in considerable quantities for crop protection. In general, they are used on grapes, cereals, and tulip production. Antifungals are widely used for the treatment of human diseases; hence, their increased use in agriculture could impose serious threat of resistance (Azevedo et al. 2015). Approximately three quarters of a million people die every year due to fungal infections. Azole-based fungicides majorly face the problem of resistance and pose threat to humans. Infections due to *Aspergillus* fungus are cured by azole-based oral treatments. If the azole resistance increases, it will highly affect the human health (Kleinkauf et al. 2013).

In the United States, antimicrobial agents of a minimum of 17 classes, including penicillin, lincomycin, tetracyclines, virginiamycin, and macrolides, have been authorized for their potential application in growth promotion or for the improvement of feed efficiency. For better understanding of the consequences due to the utilization of antimicrobials in agriculture on human health, it is important to quantify the used amount of antimicrobials in food animals. However, in the United States, there is no public health reporting system available for the quantification of antimicrobials used for food animals. The Animal Health Institute in the United States has estimated that in 1998, around 18 million pounds of antimicrobials is generated by its member companies for its application in food animals (Animal Health Institute 2000). The Union of Concerned Scientists has provided an alternative estimation in 2001 for the United States and estimated that around 31 million pounds of antimicrobials have been utilized for food animals annually. This organization also estimated that 93% of the antimicrobials which is approx. 28 million pounds used in the field of agriculture is exploited unnecessarily even if there is no disease (Mellon et al. 2001). These initial estimations provide a preliminary perspective on the huge antimicrobial quantities being used for food animals in the United States. However, more precise and accurate data is needed on the amount and motive of antimicrobials used for food animals.

For the case of human medicine, utilization of antimicrobials for agriculture possesses a selective pressure due to the problem of antimicrobial-resistant bacteria, including human pathogens, animal pathogens which have food animal reservoirs, and bacteria that are also present in food animals (Cohen and Tauxe 1986; Levy 1997). The transfer of these bacteria to humans may occur either by the food supply or through direct animal contact (Witte 1998). Once antimicrobials enter into the environment, prediction of their degradation time is very difficult as their chemical structure is very diverse, whether they come from human/animal use or from manufacturing. Some of the antimicrobials may degrade very easily, while others may react with organic matter and persist in the environment in their active form for longer periods. This opens the scope and need for further advanced studies on this serious issue.

5.2 Antimicrobials in Agriculture

The US Environmental Protection Agency (US-EPA) regulates the use of antimicrobials in agriculture, and Food and Drug Administration regulates all the other uses of antimicrobial. Streptomycin and oxytetracycline are the only two EPA-approved antimicrobials for their use in agriculture (Vidaver 2002). These two antimicrobials are grouped with fungicides and are used as prophylactic treatments. Oxytetracycline is approved for its use on four fruit crops, while streptomycin is approved for its use on 12 fruits, ornamental fruit crops, and vegetables as shown in Table 5.1.

Erwinia amylovora, a relative specie of *E. coli* and enteric bacteria, causes a major plant disease called fire blight. For the treatment of fire blight, spray treatments have been employed for 3–4 days by streptomycin and/or 4–6 days by oxytetracycline as prophylactic treatment to prevent damage during the blossom time (Johnson and Stockwell 1998). Around 53,000 hectares of antimicrobials are sprayed annually for the protection of crops (Vidaver 2001). Depending upon the type of specie, the blossom time might be extended to 6 weeks or even more by employing prophylactic treatment. The studies conducted in the past have shown that there are no streptomycin residues in detectable amount in the fruits at the time of harvest; however, the activity of streptomycin was detectable in the leaves (Shaffer and Goodman 1969). In 1992, a fact sheet was published by the EPA on streptomycin which indicated that the drug is nontoxic to freshwater invertebrates, honeybees, and birds while showing slight toxicity toward fish (warm-water and cold-water species) (US EPA 1992). In 1993, a fact sheet was published by the EPA on oxytetracycline which indicated that the drug has nontoxic effects on fishes, honeybees, birds, and aquatic invertebrates (US EPA 1993). The concentration recommended for streptomycin ranges from 50 to 200 mg L^{-1}, based on the type of crop and severity of infection. In treating fire blight-infected apples and pears, streptomycin application dose of ~2–4 L $hectare^{-1}$ is suggested. The recommended concentration of oxytetracycline is 150–200 mg L^{-1}. For the treatment of nectarines and peaches, the application dose of 240 gallons $acre^{-1}$ can be employed. This dose can increase for large-sized trees but shall not exceed 500 gallons $acre^{-1}$ per application. For treating pears, 50–100 gallons $acre^{-1}$ of solution is recommended.

For agriculture, the use of gentamicin in Latin America is of increasing concern as the antimicrobial quantity used in this region is unknown and the extent of human exposure is also not known. The US-EPA along with the American Society for Microbiology have instructed that the gentamicin-treated food products (vegetables/fruits) must not be imported as this drug plays a very important role in the human medicine. These measures have been taken as the unnecessary residue of this drug on food could cause resistance to this antimicrobial drug which is among the few economically available drugs used against bacterial infections in human. Till now, there is no data available on the use of this drug in the areas of Latin America and on the antimicrobial resistance (AMR) (Vidaver 2002). Table 5.2 represents the region wise use of drug gentamicin for agriculture. The relative utilization of antimicrobials in the field of agriculture without the implementation of better and

Table 5.1 Registered antibiotics for agricultural utilization in the United States (Agri-Mycin 17 2001; Mycoshield 2001; US EPA 1992, 1993; Vidaver 2002)

Crop	Disease	Disease-causing agent	Approved treatment	
			Streptomycin	Oxytetracycline
Food and/or feed crops				
Apple	Fire blight	*Erwinia amylovora*	√	√
Celery	Bacterial blight	*Pseudomonas cichorii*	√	–
Pear	Fire blight	*E. amylovora*	√	√
Tomato	Bacterial spot	*X. campestris* pv. *Vesicatoria*	√	–
Bean	Halo blight	*Pseudomonas syringae* pv. *phaseolicola*	√	–
Pepper	Bacterial spot	*X. campestris* pv. *vesicatoria*	√	–
Crabapple	Fire blight	*E. amylovora*	√	–
Peach	Bacterial leaf, fruit spot	*X. campestris* pv. *pruni*	–	√
Quince	Fire blight	*E. amylovora*	√	–
Nectarine	Bacterial leaf, fruit spot	*Xanthomonas campestris* pv. *pruni*	–	√
Potato	Bacterial soft rot	*E. carotovora* subspecies *carotovora*, *E. chrysanthemi*	√	–
Nonfood crops				
Tobacco	Wildfire	*Pseudomonas syringae* pv. *tabaci*	√	–
Sugar beets	Bacterial rot/ blight	*Erwinia* spp.	√	√
Ornamental herbaceous plants, *vines*, *shrubs*, and *greenhouse ornamentals*				
Roses	Crown gall	*Agrobacterium tumefaciens*	√	–
Anthurium	Bacterial blight	*X. campestris* pv. *dieffenbachiae*	√	–
Quince	Fire blight	*E. amylovora*	√	–
Cotoneaster	Fire blight	*E. amylovora*	√	√
Pyracantha	Fire blight	*E. amylovora*	√	–
Elm	Lethal yellows	*Phytoplasma*	–	√
Hawthorn	Fire blight	*E. amylovora*	√	–
Philodendron	Bacterial leaf spot	*X. campestris* pv. *dieffenbachiae*	√	√
Dieffenbachia	Bacterial stem rot	*Erwinia* spp.	√	–
Palm	Lethal yellows	*Phytoplasma* sp.	–	√

Table 5.2 Utilization of drug gentamicin for agriculture in different countries (Vidaver 2002)

Country and crop	Disease	Disease-causing agent
Mexico		
Watermelon	Bacterial spot	*Xanthomonas* sp.
Potato	Black leg	*E. carotovora* subspecies *atroseptica*
Tomato and chili	Bacterial spot	*X. campestris* pv. *vesicatoria*
Apple, pear, and ornamentals	Fire blight	*E. amylovora*
Agave	Bland rottenness of the heart of agave	*Erwinia* spp.
Chile		
Pear	Fire blight	*Erwinia amylovora*
Tomato	Bacterial canker	*Clavibacter michiganensis* subspecies *michiganensis*
Central America (Honduras, Guatemala, Costa, Rica, El Salvador)		
Cauliflower and broccoli	Bacterial soft rot	*Erwinia* spp.
Potato	Blackleg	*Erwinia carotovora* subspecies *atroseptica*
Chili	Bacterial spot	*Xanthomonas campestris* pv. *vesicatoria*
Cabbage	Bacterial black rot	*X. campestris* pv. *campestris*
Tomato	Bacterial speck	*Pseudomonas syringae* pv. tomato

appropriate policies is likely to expand sooner due to the increasing economic growth and food demand.

5.2.1 AMR due to Manure

High frequency of resistance has been observed toward several metals such as nickel, zinc, cobalt, and cadmium and toward the antibiotics like streptomycin, spiramycin, ampicillin, and olaquindox in copper-resistant isolates as compared to the copper-sensitive bacteria isolated from agricultural lands where pig manure with copper contamination had been used (Huysman et al. 1994). Pig manure is a public health hazard as it is a major source of resistance genes. In addition to the utilization of antibiotic agents, utilization of heavy metals in aquaculture and animal farming might further promote the AMR via co-selection. When the soil is fertilized with antibiotic-contaminated manure and sewage sludge, the transfer of antibiotics occurs to the soil and water (Heuer et al. 2011). The utilization of metal-contaminated fertilizers, liquid manure, and sewage sludge is a common practice followed in European countries and in other parts of the world. Due to these applications, heavy metals such as Cd, Pb, Cu, Hg, Ni, Zn, and Cr are transferred to the arable soil (Han et al. 2002). Pesticides containing Cu have been used in organic and conventional agriculture due to the bactericidal and fungicidal properties of Cu (Nemecek et al.

2011). Also, metals like iron, cobalt, manganese, Zn, and Cu have been used as nutritional supplements in animal feed for fish production and livestock farming in European countries. The European Union demonstrated that the major contributor of Zn and Cu metal pollution in the environment is the aquaculture and agriculture activities. It was estimated that the no effect amount of Zn and Cu will be exceeded in some of the water and soil systems within the coming 10–50 years (Monteiro et al. 2010). Agriculture has been also identified as a major source of Zn and Cu contamination of arable soil in Wales and England (Nicholson et al. 2003). Moreover, 30% of the Cd identified in the agricultural soil comes from inorganic fertilizers. Manure is one of the major antibiotic pollution sources existing in the environment, and China annually produces almost 618 billion kg of swine manure (Wang et al. 2006). Most of the veterinary antibiotics used for animals are poorly absorbed by them and are excreted and dispersed in soil when the manure is used in farming land as fertilizers (Alcock et al. 1999). The utilization of subtherapeutic levels of antibiotics for animal feeds leads to an increased antibiotic resistance traits in manure (Looft et al. 2012), manure-amended soils (Ghosh and LaPara 2007), and river waters and sediments (Pruden et al. 2012). Moreover, several metals are added to swine feeds for disease control and as a growth promoter which may impart a long-term pressure for antibiotic resistance (Baker-Austin et al. 2006).

5.2.2 *Fate of Antimicrobials in Agriculture*

Pharmaceutical drugs used for humans, animals, fragrances, sunscreens, and cosmetics are discharged in raw sewage water after their regular use. These drugs are inefficiently treated in the sewage treatment plants and hence make their entry in water resources (Verlicchi et al. 2012). It has been observed that intensive livestock activities and farming practices are majorly responsible for the entry of pharmaceuticals in the environment either in their active form or in their unmetabolized form (Bottoni et al. 2010). Worldwide, the application of recycled organic manure is increasing which also has its environmental implications (Motoyama et al. 2011). Manure is widely used in agriculture as a valuable fertilizer as it possesses essential nutrients needed by plants like phosphorus, nitrogen, potassium, organic carbon, etc. The excessive utilization of manure obtained from medicated animals in the farming lands is a major route through which pharmaceuticals enter the environment and ultimately to water reservoirs through runoff and percolation, respectively. The main concern related to the entry of these pharmaceuticals in the environment is better understanding of their final fate as once they enter in the system their fate is governed by various factors (Kemper et al. 2008). The major concern related to pharmaceuticals entered into the environment is that they may have potential biological effects on aquatic life and human health (Howard and Muir 2011). The presence of antibiotics has been observed in German soils where swine liquid manure has been applied as fertilizer, and 15 $\mu g\ kg^{-1}$ of sulfadimidine was also detected suggesting the stability of sulfadimidine in the soil (Christian et al. 2003). In China, the presence of pharmaceuticals has also been detected in the soils where cattle, swine, and chicken manures were employed as organic fertilizers (Wang et al. 2014). The highest concentrations of pharmaceuticals detected in the soils where pig manure was used were oxytetracycline, chlortetracycline, and sulfamethazine. Oxytetracycline was detected in the soils where cattle manure was used. However, high concentration of quinolones and tetracyclines was observed in the soils where chicken manure was used (Wang et al. 2014). Sulfachloropyridazine was detected in the soils where poultry manure was used indicating that this can be the main source of soil contamination by such antimicrobial compound (Karcı and Balcıoğlu 2009).

5.3 Antimicrobials as Growth Promoters

Since the early 1950s, antimicrobials have been utilized in the agriculture field including livestock and poultry for the treatment of infections and to promote growth and feed efficiency. There has been always a developing controversy toward the utilization of pharmaceutical drugs as growth promoters. These pharmaceuticals have been used at low concentration in animal feeds which improves the product quality with low-fat percentage and higher protein content in meat. When antibiotics are used as growth promoters, there are several other associated advantages such as control of zoonotic pathogens like *Escherichia coli*, *Salmonella*, *Enterococcus*, and *Campylobacter*. The utilization of drugs as growth promoters imparts a selection pressure for bacteria which are resistant to antibiotics and may be used in human or animal treatment. It has been observed in the 1940s that when pigs and chickens were treated with the broad-spectrum antibiotics, the size and amount of meat produced increased. Afterward, antibiotics were widely applied as growth promoters even after the detection of antibiotic resistance in farm animals in the year 1951. In countries where this practice is followed including the United States, more than half of the antibiotics administered to the animals are for increasing the growth. The utilization of antibiotics for growth promotion purpose is most common in pigs. Almost 30% of the world's pork is produced by China, and its farmers utilize antibiotics four times more than their US counterparts for producing the same amount of meat (Cully 2014). It has been reported that resistance genes have been increased up to 28,000-fold in soil and manure at the Chinese pig farms (Zhu et al. 2013). These resistance genes can easily get transferred from bacterial specie that infects animals to the species that infect humans. It was observed that overuse of antibiotics in livestock could lead to larger contribution toward antibiotic resistance. In the United States, pigs are often exposed to antibiotics like lincosamides, teracyclines, penicillins, macrolides, and erythromycin. All these pharmaceuticals are also employed in humans for the treatment of infections. Pigs in the United States are exposed to a range of compounds such as pleuromutilins, bacitracin, virginiamycin, flavophospholipol, and arsenical compounds for the purpose of growth promotion. In the United States, flavophospholipol and virginiamycin are employed as cattle growth promoters, which are also used as poultry growth promoters. Cattle are also treated with ionophores like monensin for growth promotion purpose, and poultry are exposed to arsenical compounds for growth promotion.

The accurate quantity of used antimicrobials in agriculture is difficult to estimate; however, a considerable amount of subtherapeutic doses has been provided to food animals for disease protection and growth promotion (Mellon et al. 2001). The World Health Organization (WHO) has recommended stopping the use of antimicrobial growth promoters in agriculture, which are used in human medicine (World Health Organization 1997, 1999). The Institute of Medicine (USA) has made the same recommendation in the year 2003 (IOM 2003). Various European countries have also taken measures toward the restricted use of antimicrobial growth promoters in agriculture which are used for human treatment. Due to the problem of antimicrobial resistance, in the year 1999, farmers from Denmark willingly stopped the usage of antimicrobials as growth promoters (Aarestrup et al. 2001). This voluntary action has led to the 60% reduction (from 206 to 81 tons) in the amount of antimicrobials used for food animals annually (DANMAP 2001; Sørensen et al. 2002). The antimicrobial agents used as growth promoters which are also used for human medicine such as bacitracin, tylosin, virginiamycin, avoparcin, and spiramycin are banned by the European Union in the year 2001 (European Commission, Belgium). Also, in the European Union, the Health Ministries have agreed to stop the utilization of all antimicrobial growth promoters by the year 2006.

Several studies have been performed to assess the effect of this ban, and no negative impact has been observed in the profit of farmers and on animal health in broiler chicken. The similar conclusions were made in the case of fattening pigs; however, incidences of diarrhea in weaned piglets were observed to increase which required other modifications like changes in the feeding and weaning processes (WHO report). In 1986, Sweden completely banned the utilization of antimicrobials as growth promoters which lead to 55% decrease in the total usage without affecting the long-term productivity. This showed that competitive production can be achieved without using antimicrobial agents as growth promoters (Wierup 1998; Greko 1999). The complete elimination of antimicrobial use as growth promoters has led to a significant decrease in the resistance in humans, animals, and food products (Bager et al. 1999; Klare et al. 1999).

The National Antimicrobial Resistance Monitoring System (NARMS) has been launched in the year 1996 for monitoring the resistance in foodborne pathogens against the use of antimicrobials. The NARMS is collaboration between the US-FDA Center for Veterinary Medicine and Centers for Disease Control and Prevention (CDC) and state/local health departments. Moreover, the Foodborne Diseases Active Surveillance Network (FoodNet) performed studies in ten states on the basis of population for estimating the sources and impact of specific foodborne infections (Angulo et al. 2004). The occurrence of antimicrobial resistance in non-Typhi *Salmonella* has been monitored using NARMS since 1996. In the year 1996, the study was done on *Salmonella* isolates collected from 14 sites against 14 antimicrobials, and 164 out of 1527 isolates showed resistance toward five or more antimicrobial agents (Centers for Disease Control and Prevention 2003). The multidrug resistance (MDR) strains have posed a threat to the world which is growing day by day. In the United States and other countries, MDR *Salmonella* serotype (DT 104) has shown its resistance toward various drugs like streptomycin, ampicillin, sulfonamides, chloramphenicol, and tetracycline (ACSSuT). This shows the increasing development of *Salmonella*-resistant clone having the ability to spread among animals and subsequently to humans. Another MDR strain of *Salmonella* which is commonly emerging is *Salmonella* Newport, which is resistant to several drugs such as chloramphenicol, tetracycline, ampicillin, sulfamethoxazole, cephalothin, streptomycin, amoxicillin/clavulanate, ceftiofur, and cefoxitin. The MDR strain possesses a decreased susceptibility toward the drug ceftriaxone, i.e., MIC $\geq 16\ \mu g\ mL^{-1}$, which is termed as MDR *S.* Newport Amp-C. The NARMS has been employed to monitor the extent of resistance among *Campylobacter jejuni*, which is the most common strain of *Campylobacter* and *Campylobacter coli* in the United States since 1997. In the monitoring process, isolates from five different sites were considered, and 28 isolates out of 217 were found to be resistant toward the drugs like ciprofloxacin and fluoroquinolone (Centers for Disease Control and Prevention 2003).

The use of antimicrobials in agriculture can significantly affect the treatment employed to cure human diseases. In the United States, increasing resistance has been noticed in the foodborne pathogens like *Campylobacter* and *Salmonella* (Centers for Disease Control and Prevention 2003) which limited the use of therapeutic agents and increases the probability of treatment failures and undesirable clinical outcomes (Travers and Michael 2002). Also, patients who are on antimicrobial drugs for their treatment are at high risk of developing antimicrobial-resistant foodborne infection (Barza and Travers 2002). To maintain the effectiveness of antimicrobial agents, their appropriate utilization in food animals and humans is of extreme importance. The fact cannot be denied that the utilization of antimicrobials in food animals is important for their health; however, the effectiveness of antimicrobial agents which are used in human medicine must also be preserved for their long-term effectiveness.

5.4 Clinical Implications

One of the major consequences of increased antimicrobial resistance in foodborne bacteria on human health is the increase in cases of foodborne illness. The antimicrobial-resistant foodborne pathogens may cause human infection due to the interaction between the antimicrobial treatment, the antimicrobial-resistant *Salmonella*, and the host flora. The resistant bacteria might get a short-term advantage as the treatment might suppress the activity of normal protective flora. If the infection causing pathogen strain is resistant to the antimicrobial drug used for its treatment, it may lower down the infectious dose for *Salmonella* (Barza and Travers 2002). An antimicrobial-resistant *Salmonella* outbreak analysis has suggested that the parallel subjection of antimicrobial agents could lead to increase in the number of cases if the outbreak is caused by sensitive strain (Cohen and Tauxe 1986). A study conducted in the early 1960s showed that mice with "undisturbed" normal intestinal flora have approx. 106 organism *Salmonella* infectious doses (Bohnhoff and Miller 1962). However, when the normal intestinal flora was "disturbed" by administering streptomycin, the streptomycin-resistant *Salmonella* infectious dose was decreased to merely ten organisms. In the *Salmonella* outbreak studies, it was concluded that the antimicrobial treatment which is unrelated to *Salmonella* infections can cause human infections with either susceptible (Pavia et al. 1990) or resistant *Salmonella* (Holmberg et al. 1984; Spika et al. 1987). When the sporadic salmonellosis was studied, it was suggested that the previous treatment using antimicrobial agent imposes the risk for acquisition of antimicrobial-resistant infections, when compared to the susceptible infections (Riley et al. 1984; Lee et al. 1994).

Personnel in the field of medicine must be aware that as the resistance of foodborne pathogens increases, the treatment of patients using antimicrobial agents enhances the possibilities for patients to develop infection soon after the treatment. Based on resistance frequency and antibiotic administration frequency, the public health impact could increase illness cases and larger outbreaks (Travers and Michael 2002). In addition to causing increased illnesses in humans, the increasing resistance may lead to unsuccessful treatments when a resistant antimicrobial agent is employed for the treatment. Hence, efforts have been made to increase the optimum utilization of antimicrobial agents in human medicine, for example, the American Academy of Pediatrics suggested very conscious and selective utilization of antimicrobials for treating upper respiratory infections in children (Dowell et al. 1998). The problematic issue of resistance is faced by the antimicrobial agents generally utilized for treating major *Salmonella* infections, such as fluoroquinolones for adults. An example of treatment failure due to drug resistance has been reported in Denmark, where an outbreak of *S. typhimurium* DT104 leads to contaminated pork (Mølbak et al. 1999). The *Salmonella* isolates from pork and human samples exhibited reduced response toward the drug fluoroquinolones which lead to the death of two patients where fluoroquinolones were used for their treatment.

5.5 High-Throughput Methods in the Study of AMR in Agriculture

There is very limited knowledge available on the transmission modes of antimicrobial resistance in between the agricultural sites and to humans through food chain. Also, limited information is there concerning the health risks to humans due to agricultural release of antimicrobial resistance genes (ARG), antimicrobial agents (AMA), and antimicrobial-resistant bacteria (ARB). Food animals are one of the main modes of AMR transfer and are the endpoints in the selection, spread, and dissemination of ARG and ARB. In determining the ARG and ARB profiles of pathogenic bacteria, surveillance laboratories are using culture- and PCR-based techniques. The major limitation associated with these methods is the identification is possible of those bacteria which are capable to grow under laboratory conditions, and prior knowledge of resistant genes is required for the primer designing. There are a large number of resistance genes present in the environment; hence, PCR runs needed to test the sample for all genes are also large which is not practically feasible and is not cost-effective either. However, with time the advancement in molecular biology has enabled the utilization of other methods for screening the total DNA of a sample for ARG, AR-carrying plasmids, and antibiotic resistance proteins. Nowadays, chip-based AR gene detection is also feasible which has significantly reduced the DNA sample screening cost for several numbers of resistance genes. This also eliminated the need of bacterial isolation resulting in reduced detection time and cost (Zhu et al. 2013). The next-generation sequencing (NDS) is capable of detecting millions of resistance genes present in a DNA sample. However, there are few limitations associated with this method such as requirement of minimum copy number of a gene needed to be detected. Pacific Biosciences has applied a method known as single-molecule real-time sequencing (SMRT) for plasmid genome sequencing which can perform the sequencing of long fragments (40 kb) in one read. PLACNET (plasmid constellation networks) tool has solved the issue of plasmid construction and has made it feasible. Other methods utilize functional metagenomics for the identification of novel resistance genes. This method needs longer time than that of metagenomics but has the advantage of identification of resistance phenotype at start. Microarray expression analysis has helped in the expression profiles of bacteria which are under antibiotic stress. The more recent and advanced versions of transcriptomics and meta-transcriptomics tools enable the identification of gene transcripts present in the sample. However, these methods are not common as whole-genome sequencing and metagenomics, but they are capable in the identification of function genes present in the sample.

5.6 Agriculture, AMR, and Socioeconomic Aspects

AMR has created a very significant economic and health burden on the society, and due to AMR, bacterial infections are untreatable (Stewardson et al. 2016). The threat of AMR is increasing due to inappropriate utilization of antibiotics (Cheng et al. 2012), poverty (Alividza et al. 2018), poor sanitation practices (Hendriksen et al. 2019), international travel (Frost et al. 2019), and increased healthcare interventions (Chatterjee et al. 2018). Efforts and future research are required to assess the contribution of AMR toward the occurrence of diseases and to develop efficient strategies. For effectively tackling the rapidly increasing issue of AMR, estimation and understanding of global burden of disease are most important. On the basis of these estimations' strategies could be decided for evaluation of interventions, decisions on resource allocation, comparisons between countries, driving research priorities, and comparisons with other diseases (Hay et al. 2018). Various studies have been reported on the national and regional burden of AMR including the estimations from the US CDC (Centre for Disease Control and Prevention (CDC) 2013) and the

European Centre for Disease Prevention and Control (ECDC) (Cassini et al. 2019) and from Thailand (Pumart et al. 2012). However, the direct comparison of these estimations is not possible as each of the study utilizes different data source, case definitions, and methods to calculate the burden of disease. A report from the UK government (2014) has estimated that approximately 700,000 annual deaths may occur globally due to AMR bacterial infections, MDR, and drug-resistant tuberculosis (Hay et al. 2018). Various good quality surveillance AMR networks are present in low- to middle-income countries (LMICs); however, there is a need to widen and deepen the surveillance. The WHO has launched the Global Antimicrobial Resistance Surveillance System (GLASS) for developing a standard approach for the collection, analysis, and sharing of data on AMR globally. This initiative taken by the WHO has established a national AMR monitoring system which provides a reliable platform for the collection of data which allows the monitoring of followed trends and estimation of future interventions. An international collaborative effort has been made with the Global Research on AntiMicrobial resistance (GRAM) study combined with Oxford University's tropical medicine and Big Data Institute expertise with the Global Burden of Disease (GBD) framework at the Institute for Health Metrics and Evaluation at the University of Washington to map and determine the burden of AMR globally (Hay et al. 2018). The main challenge is production of reliable estimates of global AMR burden which greatly depends upon the selection of methodology (Limmathurotsakul et al. 2019). Disability-adjusted life years (DALYs), excess length of stay, annual mortality, and years of life lost are all important matrices which have several calculation approaches. The global burden of disease framework utilizes a single cause of death mentioned in the death certificate for the estimations of mortality (GBD 2017 Causes of Death Collaborators 2018). This method has the advantage of being easily understandable; it can utilize data from international death certificate and also avoids double counting of deaths. However, this method is not very suitable for the estimation of AMR as the cause of death in the certificate may be different from the actual cause of death. The best approach for the estimation of AMR burden may vary for community-acquired versus hospital-acquired infection and high-income countries versus LMICs.

5.7 Conclusions

Antimicrobial resistance is an emerging public health issue in front of the world. The widespread presence of drug resistance is rapidly rising worldwide which can have serious medical implications. To avoid the problem of resistance, appropriate utilization of antimicrobials for humans and animals must be practiced which will also need the collaborative effort by partners from different fields such as public health communities, veterinary, farming, and medical. To monitor the preventive efforts, increased surveillance is needed for antimicrobial use and resistance problems. The promotion of veterinarian's education for the appropriate utilization of antimicrobial agents has been suggested. In the United States, Public Health Action

Plan outlined collaborative federal actions for the antimicrobial resistance in agriculture and published the report in 2001 (Interagency Task Force on Antimicrobial Resistance). The suggested actions mentioned in the plan includes increased surveillance rate of antimicrobial utilization and drug resistance, enhanced research and education, and refining and implementing the FDA's Framework being on the top priority item. This also proposed an improvised approval process for the utilization of antimicrobial drugs in animals (Food and Drug Administration–Centre for Veterinary Medicine). The actions were taken for ensuring the safety of humans toward antimicrobial drugs used for animals by prioritizing the utilization of these antimicrobials on the basis of their importance in human treatment. The large-scale utilization of antimicrobial drugs for food animals may lead to increased drug resistance in microorganisms, which ultimately can be transferred in humans. The potential increase in the treatment failures is the major adverse health consequence of transfer of resistant bacteria in humans. For controlling and avoiding this public health issue, the appropriate use of antimicrobials for food animals and humans must be promoted and regulated. Adherence to the guidelines is an important factor for the appropriate utilization of antimicrobials in food animals for reduction in the resistance.

References

Aarestrup FM, Seyfarth AM, Emborg HD, Pedersen K, Hendriksen RS, Bager F (2001) Effect of abolishment of the use of antimicrobial agents for growth promotion on occurrence of antimicrobial resistance in fecal enterococci from food animals in Denmark. Antimicrob Agents Chemother 45(7):2054–2059

Agri-Mycin 17 (2001) Crop protection reference, 17th edn. Chemical and Pharmaceutical Press, New York, NY, pp 1928–1930

Alcock RE, Sweetman A, Jones KC (1999) Assessment of organic contanhnant fate in waste water treatment plants I: selected compounds and physicochemical properties. Chemosphere 38(10):2247–2262

Alividza V, Mariano V, Ahmad R, Charani E, Rawson TM, Holmes AH, Castro-Sanchez E (2018) Investigating the impact of poverty on colonization and infection with drug-resistant organisms in humans: a systematic review. Infect Dis Pov 7(1):1–11

Angulo FJ, Baker NL, Olsen SJ, Anderson A, Barrett TJ (2004) Antimicrobial use in agriculture: controlling the transfer of antimicrobial resistance to humans. Semin Pediatr Infect Dis 15(2):78–85

Animal Health Institute (2000) Survey indicates most antibiotics used in animals are used for treating and preventing disease. AHI, Chennai. http://www.ahi.org/news%20room/press%20release/2000/feb/antibiotic%20usage%20data.htm

Azevedo MM, Faria-Ramos I, Cruz LC, Pina-Vaz C, Goncalves Rodrigues A (2015) Genesis of azole antifungal resistance from agriculture to clinical settings. J Agric Food Chem 63(34):7463–7468

Bager F, Aarestrup FM, Madsen M, Wegener HC (1999) Glycopeptide resistance in Enterococcus faecium from broilers and pigs following discontinued use of avoparcin. Microb Drug Resist 5(1):53–56

Baker-Austin C, Wright MS, Stepanauskas R, McArthur JV (2006) Co-selection of antibiotic and metal resistance. Trends Microbiol 14(4):176–182

Barza M, Travers K (2002) Excess infections due to antimicrobial resistance: the "Attributable Fraction". Clin Infect Dis 34(Suppl 3):S126–S130
Bohnhoff M, Miller CP (1962) Enhanced susceptibility to Salmonella infection in streptomycin-treated mice. J Infect Dis 111:117–127
Bottoni P, Caroli S, Caracciolo AB (2010) Pharmaceuticals as priority water contaminants. Toxicol Environ Chem 92(3):549–565
Cassini A, Högberg LD, Plachouras D, Quattrocchi A, Hoxha A, Simonsen GS, Colomb-Cotinat M, Kretzschmar ME, Devleesschauwer B, Cecchini M, Ouakrim DA (2019) Attributable deaths and disability-adjusted life-years caused by infections with antibiotic-resistant bacteria in the EU and the European Economic Area in 2015: a population-level modelling analysis. Lancet Infect Dis 19(1):56–66
Centers for Disease Control and Prevention (2003) National Antimicrobial Resistance Monitoring System for Enteric Bacteria (NARMS): 2001 annual report. US Department of Health and Human Services, CDC, Atlanta, GA
Centre for Disease Control and Prevention (CDC) (2013) Antibiotic resistance threats in the United States report. U.S. Department of Health and Human Services, CDC, Atlanta, GA. https://www.cdc.gov/drug resistance/threat-report-2013/index.html
Chatterjee A, Modarai M, Naylor NR, Boyd SE, Atun R, Barlow J, Holmes AH, Johnson A, Robotham JV (2018) Quantifying drivers of antibiotic resistance in humans: a systematic review. Lancet Infect Dis 18(12):e368–e378
Cheng AC, Turnidge J, Collignon P, Looke D, Barton M, Gottlieb T (2012) Control of fluoroquinolone resistance through successful regulation, Australia. Emerg Infect Dis 18(9):1453
Christian T, Schneider RJ, Färber HA, Skutlarek D, Meyer MT, Goldbach HE (2003) Determination of antibiotic residues in manure, soil, and surface waters. Acta Hydrochim Hydrobiol 31(1):36–44
Cohen ML, Tauxe RV (1986) Drug-resistant Salmonella in the United States: an epidemiologic perspective. Science 234(4779):964–969
Cully M (2014) Public health: the politics of antibiotics. Nature 509(7498):S16–S17
DANMAP (2001) 2000: Consumption of antimicrobial agents and resistance to antimicrobial agents in bacteria from food animals, food and humans in Denmark: report from Statens Serum Institut, Danish Veterinary and Food Administration. Danish Medicines Agency and Danish Veterinary Laboratory, Glostrup. http://www.keepantibioticsworking.com/library/uploaded-files/Danmap 2000.pdf
Dowell SF, Marcy SM, Phillips WR, Gerber MA, Schwartz B (1998) Principles of judicious use of antimicrobial agents for pediatric upper respiratory tract infections. Pediatrics 101(Suppl 1):163–165
FDA (2014) 2011 Summary report on Antimicrobials sold or distributed for use in Food producing animals. Department of Health and Human Services, FDA, Silver Spring, MD
Frost I, Van Boeckel TP, Pires J, Craig J, Laxminarayan R (2019) Global geographic trends in antimicrobial resistance: the role of international travel. J Trav Med 26(8):taz036
GBD 2017 Causes of Death Collaborators (2018) Global, regional, and national age-sex-specific mortality for 282 causes of death in 195 countries and territories, 1980–2017: a systematic analysis for the Global Burden of Disease Study 2017. Lancet 392:1736–1788
Ghosh S, LaPara TM (2007) The effects of subtherapeutic antibiotic use in farm animals on the proliferation and persistence of antibiotic resistance among soil bacteria. ISME J 1(3):191–203
Greko C (1999) Antibiotics as growth promoters. Acta Vet Scand Suppl 92:87–100
Han FX, Banin A, Su Y, Monts DL, Plodinec JM, Kingery WL, Triplett GE (2002) Industrial age anthropogenic inputs of heavy metals into the pedosphere. Naturwissenschaften 89(11):497–504
Hay SI, Rao PC, Dolecek C, Day NP, Stergachis A, Lopez AD, Murray CJ (2018) Measuring and mapping the global burden of antimicrobial resistance. BMC Med 16(1):1–3

Hendriksen RS, Munk P, Njage P, Van Bunnik B, McNally L, Lukjancenko O, Röder T, Nieuwenhuijse D, Pedersen SK, Kjeldgaard J, Kaas RS (2019) Global monitoring of antimicrobial resistance based on metagenomics analyses of urban sewage. Nat Commun 10(1):1–12
Heuer H, Schmitt H, Smalla K (2011) Antibiotic resistance gene spread due to manure application on agricultural fields. Curr Opin Microbiol 14(3):236–243
Holmberg SD, Osterholm MT, Senger KA, Cohen ML (1984) Drug-resistant Salmonella from animals fed antimicrobials. N Engl J Med 311(10):617–622
Howard PH, Muir DC (2011) Identifying new persistent and bioaccumulative organics among chemicals in commerce II: pharmaceuticals. Environ Sci Technol 45(16):6938–6946
Huysman F, Verstraete W, Brookes PC (1994) Effect of manuring practices and increased copper concentrations on soil microbial populations. Soil Biol Biochem 26(1):103–110
IOM (2003) Microbial threats to health: emergence, detection, and response. Report of the Institute of Medicine. IOM, Nepal. http://www.nap.edu/books/030908864X/html
Johnson KB, Stockwell VO (1998) Management of fire blight: a case study in microbial ecology. Annu Rev Phytopathol 36(1):227–248
Karcı A, Balcıoğlu IA (2009) Investigation of the tetracycline, sulfonamide, and fluoroquinolone antimicrobial compounds in animal manure and agricultural soils in Turkey. Sci Total Environ 407(16):4652–4664
Kemper N, Färber H, Skutlarek D, Krieter J (2008) Analysis of antibiotic residues in liquid manure and leachate of dairy farms in Northern Germany. Agric Water Manag 95(11):1288–1292
Klare I, Badstübner D, Konstabel C, Böhme G, Claus H, Witte W (1999) Decreased incidence of VanA-type vancomycin-resistant enterococci isolated from poultry meat and from fecal samples of humans in the community after discontinuation of avoparcin usage in animal husbandry. Microb Drug Resist 5(1):45–52
Kleinkauf N, Verweij PE, Arendrup MC, Donnelly PJ, Cuenca-Estrella M, Fraaije B, Melchers WJ, Adriaenssens N, Kema GH, Ullmann A, Bowyer P (2013) Risk assessment on the impact of environmental usage of triazoles on the development and spread of resistance to medical triazoles in Aspergillus species. European Centre for Disease Prevention and Control technical report. ECDC, Stockholm
Lee LA, Puhr ND, Maloney EK, Bean NH, Tauxe RV (1994) Increase in antimicrobial-resistant Salmonella infections in the United States, 1989–1990. J Infect Dis 170(1):128–134
Levy S (1992) The antibiotic paradox: how miracle drugs are destroying the miracle. Plenum Press, New York, NY
Levy SB (1997) Antibiotic resistance: an ecological imbalance. CIBA Found Symp 207:1–9
Limmathurotsakul D, Dunachie S, Fukuda K, Feasey NA, Okeke IN, Holmes AH, Moore CE, Dolecek C, van Doorn HR, Shetty N, Lopez AD (2019) Improving the estimation of the global burden of antimicrobial resistant infections. Lancet Infect Dis 19(11):e392–e398
Looft T, Johnson TA, Allen HK, Bayles DO, Alt DP, Stedtfeld RD, Sul WJ, Stedtfeld TM, Chai B, Cole JR, Hashsham SA (2012) In-feed antibiotic effects on the swine intestinal microbiome. Proc Natl Acad Sci 109(5):1691–1696
Marshall BM, Levy SB (2011) Food animals and antimicrobials: impacts on human health. Clin Microbiol Rev 24(4):718–733
Mellon M, Benbrook C, Benbrook KL (2001) Hogging it. Estimates of antimicrobial abuse in livestock, pp 7–9
Mølbak K, Baggesen DL, Aarestrup FM, Ebbesen JM, Engberg J, Frydendahl K, Gerner-Smidt P, Petersen AM, Wegener HC (1999) An outbreak of multidrug-resistant, quinolone-resistant Salmonella enterica serotype Typhimurium DT104. N Engl J Med 341(19):1420–1425
Monteiro SC, Lofts S, Boxall ABA (2010) Pre-assessment of environmental impact of zinc and copper used in animal nutrition. European Food Safety Authority, Parma
Motoyama M, Nakagawa S, Tanoue R, Sato Y, Nomiyama K, Shinohara R (2011) Residues of pharmaceutical products in recycled organic manure produced from sewage sludge and solid waste from livestock and relationship to their fermentation level. Chemosphere 84(4):432–438

Mycoshield (2001) Crop protection reference, 17th edn. Chemical and Pharmaceutical Press, New York, NY, pp 2086–2087
Nemecek T, Dubois D, Huguenin-Elie O, Gaillard G (2011) Life cycle assessment of Swiss farming systems: I. Integr Org Farm Agric Syst 104:217–232
Nicholson FA, Smith SR, Alloway BJ, Carlton-Smith C, Chambers BJ (2003) An inventory of heavy metals inputs to agricultural soils in England and Wales. Sci Total Environ 311(1–3):205–219
Pavia AT, Shipman LD, Wells JG, Puhr ND, Smith JD, McKinley TW, Tauxe RV (1990) Epidemiologic evidence that prior antimicrobial exposure decreases resistance to infection by antimicrobial-sensitive Salmonella. J Infect Dis 161(2):255–260
Pruden A, Arabi M, Storteboom HN (2012) Correlation between upstream human activities and riverine antibiotic resistance genes. Environ Sci Technol 46(21):11541–11549
Pumart P, Phoda T, Thamlikitkul V, Riewpaiboon A, Prakongsai P, Limwattananon S (2012) Health and economic impacts of antimicrobial resistance in Thailand. J Health Serv Res Pol 358:352–360
Riley LW, Cohen ML, Seals JE, Blaser MJ, Birkness KA, Hargrett NT, Martin SM, Feldman RA (1984) Importance of host factors in human salmonellosis caused by multiresistant strains of Salmonella. J Infect Dis 149(6):878–883
Sengeløv G, Agersø Y, Halling-Sørensen B, Baloda SB, Andersen JS, Jensen LB (2003) Bacterial antibiotic resistance levels in Danish farmland as a result of treatment with pig manure slurry. Environ Int 28(7):587–595
Shaffer WH, Goodman RN (1969) Effectiveness of an extended Agri-Mycin-17 spray schedule against fireblight. Plant disease reporter. Bureau of Plant Industry, U.S. Dept. of Agriculture, Washington, DC
Sørensen TL, Wegener HC, Fridmodt-Møller N (2002) Resistant bacteria in retail meats and antimicrobial use in animals. N Engl J Med 346(10):779–779
Spika JS, Waterman SH, Hoo GWS, St. Louis ME, Pacer RE, James SM, Bissett ML, Mayer LW, Chiu JY, Hall B, Greene K (1987) Chloramphenicol-resistant Salmonella newport traced through hamburger to dairy farms. N Engl J Med 316(10):565–570
Stewardson AJ, Allignol A, Beyersmann J, Graves N, Schumacher M, Meyer R, Tacconelli E, De Angelis G, Farina C, Pezzoli F, Bertrand X (2016) The health and economic burden of bloodstream infections caused by antimicrobial-susceptible and non-susceptible Enterobacteriaceae and Staphylococcus aureus in European hospitals, 2010 and 2011: a multicentre retrospective cohort study. Eurosurveillance 21(33):30319
Travers K, Michael B (2002) Morbidity of infections caused by antimicrobial-resistant bacteria. Clin Infect Dis 34(Suppl 3):S131–S134
US EPA (1992) RED facts: streptomycin and streptomycin sulfate. US Environmental Protection Agency, Washington, DC
US EPA (1993) RED facts: hydroxytetracycline monohydrochloride and oxytetracycline calcium. US Environmental Protection Agency, Washington, DC
Verlicchi P, Al Aukidy M, Zambello E (2012) Occurrence of pharmaceutical compounds in urban wastewater: removal, mass load and environmental risk after a secondary treatment—a review. Sci Total Environ 429:123–155
Vidaver AK (2001) Horticultural and other uses of antibiotics. In: International Congress and Symposium Series-Royal Society of Medicine, vol 247. Royal Society of Medicine Press, London, pp 125–130
Vidaver AK (2002) Uses of antimicrobials in plant agriculture. Clin Infect Dis 34(Suppl 3):S107–S110
Wang FH, Ma WQ, Dou ZX, Ma L, Liu XL, Xu JX, Zhang FS (2006) The estimation of the production amount of animal manure and its environmental effect in China. China Environ Sci 26(5):614–617
Wang N, Guo X, Xu J, Kong X, Gao S, Shan Z (2014) Pollution characteristics and environmental risk assessment of typical veterinary antibiotics in livestock farms in Southeastern China. J Environ Sci Health B 49(7):468–479

Wierup M (1998) Preventive methods replace antibiotic growth promoters: ten years experience from Sweden. APUA Newsl 16:1–4

Witte W (1998) Medical consequences of antibiotic use in agriculture. Science 279:996

World Health Organization (1997) The medical impact of the use of antimicrobials in food animals: report of a WHO Meeting, Berlin, Germany, 13-17 October 1997 (No. WHO/EMC/ZOO/97.4). World Health Organization, Geneva

World Health Organization (1999) Containing antimicrobial resistance: review of the literature and report of a WHO Workshop on the Development of a Global Strategy for the Containment of Antimicrobial Resistance, Geneva, Switzerland, 4-5 February 1999. WHO, Geneva

Zhu YG, Johnson TA, Su JQ, Qiao M, Guo GX, Stedtfeld RD, Hashsham SA, Tiedje JM (2013) Diverse and abundant antibiotic resistance genes in Chinese swine farms. Proc Natl Acad Sci 110(9):3435–3440

Chapter 6
Epidemiology of Microbial Infections

Kanika, Aneesh Ali, Md. Meraj Ansari, and Rehan Khan

6.1 Introduction

The emergence of infectious disease and even re-emergence of deadly infectious disease are found to be a threat to public health and welfare. Microbes and hosts that include humans are interdependent for their survival. They also represent a close connection between microbes and humans as half of human DNA originated from viruses. The skin, gut, and mucous membranes are occupied by microbes.

The germ theory of disease was found to be essential in identifying the disease that is caused by microbes and also aids in finding the ways that prevent the infection through sanitation, immunization, etc. Human civilization has a major role in the spread of microbial infection, and it has become the major cradle of humans travailing from both morbidity and mortality. Urbanization and large settlements lead to play a major role in the spread of microbial infection. Large pandemics such as smallpox, cholera, plague, and other diseases like tuberculosis and syphilis have been affecting the human throughout the ages.

The spectrum of occurrence of diseases in a particularly specified population comprises the presence of sporadic methods of occurrence: endemic indicates that there is continuous occurrence, epidemic implies that there is a considerable increase of disease, and ultimately pandemic means that occurs in several nations.

In a particular defined population, the study of determinants, distribution of health and disease, and occurrence is known as epidemiology. The study of

Kanika · A. Ali · R. Khan (✉)
Department of Nano-Therapeutics, Institute of Nano Science and Technology, Mohali, Punjab, India
e-mail: rehankhan@inst.ac.in

M. M. Ansari
Centre for Pharmaceutical Nanotechnology, Department of Pharmaceutics, National Institute of Pharmaceutical Education and Research, Mohali, Punjab, India

N. Akhtar et al. (eds.), *Emerging Modalities in Mitigation of Antimicrobial Resistance*,
https://doi.org/10.1007/978-3-030-84126-3_6

epidemiology can be descriptive, analytic, and experimental. It exploits the problem-solving by developing a case; hypothesizing; analyzing the data by time, place, and person; preparing the public report; and lastly evaluating control and taking different preventive measures (Hennekens and Mayrent 1987; Beaglehole et al. 1993).

An epidemiology is a science that gives descriptive information regarding rate determination and measurement of the diseases occurring in certain populations. The major goal of this is to find out the effective measures for control of disease by defining the various parameters including risk factors, etc. MacMahon and Pugh 1970).

Microbial agents can grow into two main compartments that are intracellular and extracellular. The intracellular pathogens invade the host cell to replicate and can be divided into two parts: one that replicates freely in the cell such as viruses and some bacteria and another one that replicates in cellular vesicle such as mycobacteria. There are major unique features of a pathogen that include its mechanism of replication, its pathogenesis, and its modes of transmission that majorly affect the pathogenicity.

6.1.1 Chain of Infection

The chain of infection, i.e., how infections initiate and further spread to the human body, includes three main agents, which are the etiologic agent; the transmission method that can be through air, vector, etc.; and lastly the host:

6.1.1.1 Etiologic Agent

The agents that come under this category are microorganisms that have the ability to cause a disease. The microorganism's virulence and invasiveness depict its pathogenicity. The other characteristic that describes the agent is the infecting dose site of entrance of the organism into the host and host species. The other is its agent specificity. The site where the organism resides, metabolizes, and multiplies is known as the reservoir of an organism.

6.1.1.2 Transmission Methods

How agents spread through the source to the host is known as the method of transmission. The methods include via contact, by air, or vector:

6.1.1.2.1 Contact Transmission

In this method, the spread is through air droplets directly or indirectly. The transmission that used to take place when the transmission is via from source to host without the involvement of intermediate object also known as the person-to-person transmission is called direct contact. An example is the hepatitis virus that is transmitted by hand contact. The transmission that takes place from a source to a host via a means of an inanimate object is known as indirect transmission. An example is the transmission of *Pseudomonas* organism via shaving brush.

6.1.1.2.2 Common Vehicle Transmission

The transmission that takes place through the inanimate vehicle leads to multiple cases resulting from such exposure. This category of transmission includes the food or water that acts as the vehicle of infection. An example is waterborne *shigellosis* that results from the use of intravenous fluid that was contaminated with Gram-negative bacteria.

6.1.1.2.3 Airborne Transmission

The spread of infection is by dust or droplet nuclei. They are found to be light enough to be transmitted by more than 3 feet from the source. The major airborne disease is tuberculosis in which the source is a coughing patient who creates aerosols that contain tubercle bacilli.

6.1.1.2.4 Vector-Borne Transmission

In this transmission, arthropods act as vectors and these can be external or internal, for example, *Salmonella* organisms and malaria parasites in the mosquito vector. The transmission of infection can be by more than one route like *Salmonella* transmitted by food or human carrier.

6.1.1.3 Host

The entrance of microorganisms is through the skin, lungs, and gastrointestinal tract or may enter the fetus through the placenta. The entrance of organisms through the skin, mucous membrane, genetic factors, nutrition, and the presence of other diseases comes under the category of nonspecific defense mechanism. The other crucial factor that affects the chain of infection is temperature, air velocity, low humidity, and ultraviolet radiation that can kill the microorganism (Mandell et al. 1990; Bennett and Brachman 1992; Benenson 1995).

The major microbial causes of infection can be explained as infection can be caused by different sets of vectors that include bacteria, viruses, fungi, or parasites. The particular infection can be endogenous which means that it occurs when a microorganism penetrates the skin or mucosal barrier as a result of surgery. In contrast, where the microorganism is acquired from the environment is known as exogenous infection (Koneman et al. 1997).

6.2 Mechanism of Microbial Infection

6.2.1 Mechanism of Bacterial Infection

Many bacterial pathogens have many similar mechanisms of infection like having the ability to adhere, invade, and cause damage to host cells and establish an infection. There are various characteristics that indicate the pathogenicity of bacteria including the adherence to host cell, transmissibility, invasion of the host cell and tissues, and ability to evade the host' immune system. Numerous bacteria are found to be the vital cause of the disease which include *Streptococcus pneumoniae* and *Staphylococcus aureus*, but in contrast bacteria like *Salmonella typhi* are pathogens, but their infection became latent and the host acts as the carrier. Hence, several mechanisms have been used by the bacteria to cause disease in human hosts. The various mechanisms of bacterial infection are as follows:

6.2.1.1 A Bacterial Pathogen After Encountering A Human Host Initiates Several Mechanisms To Evade the Host defenses:

1. Various components present on the bacteria are capsules that will interact with the host even though it acts as frustrated phagocytosis that causes enhanced inflammatory response leading to more tissue damage; hence, more WBCs are recruited to the infection site. It aids in protecting the pathogen from neutrophil engulfment and macrophage. *Streptococcus pneumonia* (pneumococcus), *Neisseria meningitides* (meningococcus), and *Pseudomonas aeruginosa* are found to be capsules producing bacteria (Liu et al. 1971; García et al. 1999).
2. The cell wall of the bacteria and its toxic components play a crucial role in the pathogenesis of bacterial septic shock. Indeed, lipopolysaccharide, a large amphiphilic molecule present in the outer membrane of the Gram-negative bacteria, aids in triggering the event of septic shock that is found to be the combined action of complement components, cytokines, and coagulation cascade. Gram-negative bacteria, like *E. coli*, *P. aeruginosa*, and *Meningococcus*, and Gram-positive bacteria, like *Staphylococcus aureus*, *Staphylococcus epidermidis*, and *Streptococcus*, are commonly involved in septic shock (Walker 1998).

3. Bacteria also produce toxins that act as biological weapons to destroy or damage the host cell. Toxins can be non-proteinaceous also known as endotoxins that are lipopolysaccharide; another one is proteinaceous toxins known as exotoxins that are delivered to eukaryotic cells. Several different species of bacteria contain A-B toxins including *P. aeruginosa*, *E. coli*, *Vibrio cholera*, *Corynebacterium diphtheriae*, and *Bordetella pertussis* (Finlay and Falkow 1997).
4. Adhesins are microbial adherence factors and can be made from polypeptides or polysaccharides. The main step in host-pathogen interaction is the adherence of the pathogen to the host surfaces like the skin, mucous membranes, and deeper tissues. Hence, after the adherence, pathogen would be able to start its biochemical process that includes toxin secretion, host cell invasion, and activation of host cell-signaling cascades that lastly result in disease. Gram-negative bacterial pathogens, in particular, rely on fimbriae for adherence. Examples include *E. coli* (for both urinary tract infections and gastroenteritis), *P. aeruginosa*, and *Neisseria* species (Hahn 1997; Donnenberg 2000; Merz and So 2000).

6.2.1.2 Invasion

To further perpetuate the infection cycle, pathogens gain deeper access into the host termed as an invasion. This invasion can be both extracellular and intracellular. Extracellular invasion happens while remaining outside of host cells in which the microbe breaks down the barrier of the tissue to disseminate in the host. For example, this strategy is used by group A b-hemolytic *Streptococcus* and *S. aureus*.13. In intracellular invasion, the microbe survives within the environment and penetrates the cells of host tissue. The bacteria that show this intracellular nature are Gram-negative, Gram-positive, and mycobacterial pathogens. Some pathogens have an obligate intracellular life cycle which requires a mammalian cell for growth. These include *Chlamydia* spp., *Rickettsia* spp., and *Mycobacterium leprae*. This specific biochemical cross talk has become vital for host and pathogen interaction and meant essential for the penetration of host cells. A type 3 secretion mechanism, which injects bacterial signaling proteins into the host cell, is a common technique used by pathogens to induce uptake. The pathogen may be present in the host cell cytosol, phagolysosomes (phagosomes that have fused with lysosomes), or phagosomes that have not fused with lysosomes (Walker 1998) Fig. 6.1.

6.2.2 Mechanism of Viral Infection

The process by which infection through virus leads to disease is called pathogenicity that comprises of various steps:

The steps include the entry of virus followed by replication and ultimately virus shedding where spread and multiplication to target organs take place. This mechanism of pathogenicity is affected by so many factors that are its determinant,

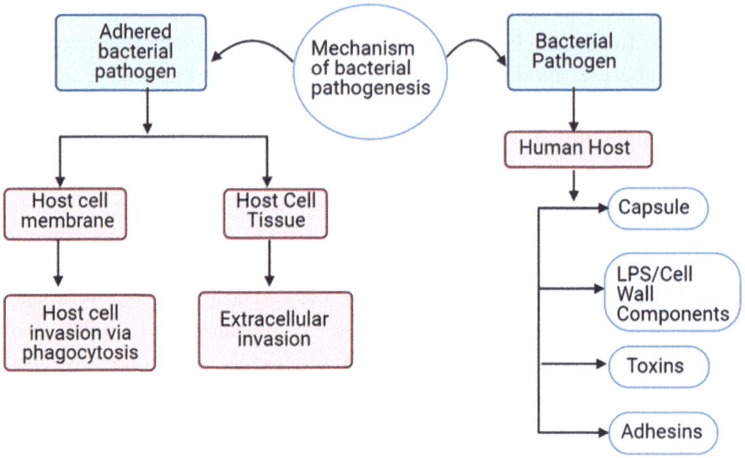

Fig. 6.1 Schematic representation of the mechanism of bacterial infection

accessibility which is affected by physical barriers and cellular susceptibility, and characteristic of the virus that makes it virulence, hence making it susceptible to get spread and target in large numbers to various specific organs (Strayer et al. 1990). The mechanism includes the following steps:

Penetration at the Portal of Entry The possible routes like air, food, bites, and contaminated objects help in carrying the virus to the body. The highest implantation is observed when the virus contacts the living cell directly.

Replication and Spread of the Virus The first step is followed by the local replication within the first initiated cell and that it may further spread to the adjacent cells can be possible by extracellular (almost all virus) or intracellular method (include paramyxoviruses that spread by both routes). Therefore, infection establishment at entry further leads to shedding of virus, causing disease.

Dissemination The most prevalent and common route for the spread of the virus is through circulation that can be in the bloodstream and nerves although the latter one was found to be less common but has been considered as the main source in the spread of many important diseases. It mainly occurs in rabies virus, herpesvirus, and, occasionally, poliomyelitis virus infections. This step is further followed by the incubation period in which the virus travels only a short distance to reach the target organ.

Multiplication in Specific and Targeted Organs During the time of virus progression, the various recovery mechanisms like local immunity, local inflammation, and interferon are activated. Nevertheless, these systemic defenses can diffuse in various degrees into target organs and thereby help retard virus replication and disease.

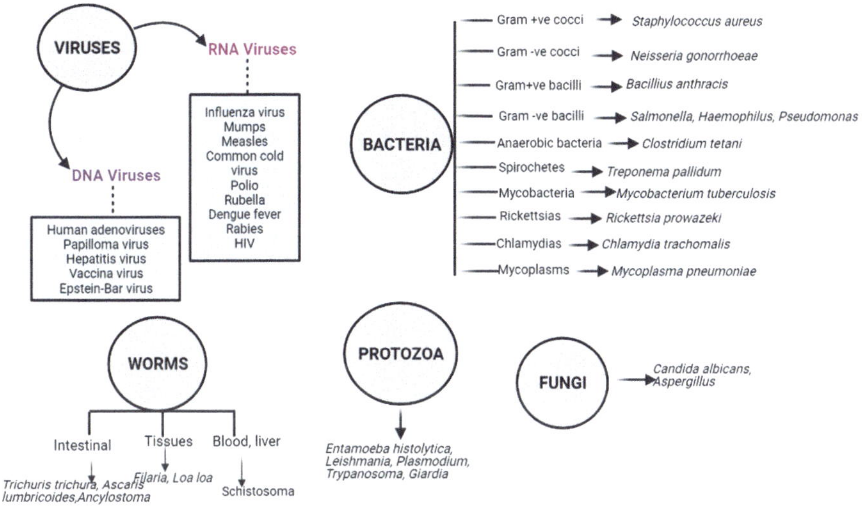

Fig. 6.2 Different types of microbial infection

Shedding of Virus Since biting insects become infected along this path, the blood and lymph are sites of shedding for arboviruses. The disease is transmitted through blood the and sperm. Viruses such as certain RNA tumor viruses (retroviruses) and cytomegalovirus shed their coats in milk. Several viruses (e.g., cytomegaloviruses) are shed around the same time from the urinary tract and other shedding locations. Herpesvirus type 2 shedding is normal in the genital tract, and the virus may be spread to sexual partners via this path (Fields 1983; Coen 1994; Grieder et al. 1995) (Fig. 6.2).

6.3 Types of Microbial Infection and Its Reservoirs

6.3.1 Bacterial Classification

Being ubiquitous in nature, bacteria play a crucial role in maintaining the environment. There has been seen a huge effect on health of public by small percentage of bacteria. Although having double-stranded DNA, they considered as prokaryotic organism. Giving the bacteria two categories named as Gram-negative and Gram-positive bacteria makes them differ in susceptibilities toward specific antibiotics. They further classified as obligate aerobes that require oxygen to sustain and another one facultative organism that can sustain in the absence or presence of oxygen (Engelkirk et al. 2020).

According to Koch's postulates, bacteria have been classified as pathogens, opportunistic pathogens, or nonpathogens. *Mycobacterium tuberculosis* and *Yersinia pestis* are pathogens, yet their presence is found to be abnormal. Other species of bacteria are known as *Escherichia coli* that are part of the normal flora of the human but cause disease like they act as the gastrointestinal flora of normal humans but act as a major cause of urinary tract infection because of the virulence in animals or models of infection and genetic makeup that is specifically associated with the production of disease. Other bacteria mainly termed as opportunistic pathogens are *Stenotrophomonas maltophilia*, *Pseudomonas* species, and yeasts and molds and therefore cause disease in immunosuppressed persons (Falkow 1988; Bannister et al. 1996). Meningitis is caused by the bacteria *Neisseria meningitidis* that infects the meninges and further can also infect the lungs and cause pneumonia.

The reservoir of bacterial infection: The site where the survival of the pathogen is possible until it gets transfer to the host is known as its reservoir that can comprise of birds, humans, animals, etc., but the reservoirs that come under nonliving are soil, water, and air. Some infections are caused even without making the person conscious of it especially through bacteria *Salmonella typhi*. *Salmonella* diarrhea may be caused by touching turtles and contaminating one's hands with their feces, ingesting undercooked chicken infected with the bacteria, or eating undercooked or raw chicken eggs, among other items. There are also nonliving reservoirs like air, food, soil, etc. It includes various diseases, for example, the various species of *Clostridium* can be acquired from exposure of a wound to dirt or soil. Food may be contaminated by feces, or the animal itself may be infected, such as in the case of chickens with *Campylobacter* or *Salmonella*. Food can also be contaminated with the ubiquitous spores of *botulinum*, which can cause a form of paralysis called botulism. Water generally becomes a reservoir for infection when it is contaminated by soil microbes or animal or human feces and causes a disease when contaminated by *Shigella* and *Legionella*. (Evans 1976).

6.3.2 Viral Classification

The genome of the virus can be either RNA or DNA which are packed with proteins that are encoded by the genome of the virus. The virus is categorized into an enveloped and non-enveloped virus. The earlier one has a lipid membrane, and the latter one does not have a membrane. The steps included in infection are the attachment to the cell surface receptors.

Acute viral diseases including respiratory, diarrheal, exanthematous, and neurological infections will overlap and manifest as seasonal epidemics that peak in incidence every few years and overlap with the aggregation of a sufficient number of nonimmune hosts in the young population. In Southeast Asia, arboviral disease infection also correlates with arthropod vector activity, such as mosquito breeding during hot rainy seasons, and is linked to an uptick in hemorrhagic fever or neurological diseases like dengue hemorrhagic fever, West Nile virus, or Japanese encephalitis. Many lifelong viral infections, such as the human immunodeficiency virus (HIV), hepatitis A, and hepatitis B, are spread via the bloodstream. Some of these chronic viral infections such as HBV, HCV, HIV, polyomaviruses, and papillomaviruses are also linked to the genesis of cancers. Over 70% of emerging viral infections such as the Ebola virus, severe acute respiratory syndrome (SARS) coronavirus.

Major viral families	Virus within family
Herpesviridae	HSV-1 HSV-2 VZV
Orthomyxoviridae	Influenza A Influenza B Influenza C
Filoviridae	Ebola virus Marburg virus
Picornaviridae	Poliovirus Rhinovirus
Rhabdoviridae	Rabies virus
Flaviviridae	Dengue virus, Hepatitis C virus, Yellow fever, Tick-borne virus
Paramyxoviridae	Measles, mumps virus
Togaviridae	Rubella virus, Chikungunya
Retroviridae	HIV-1, HIV-2
Hepadnaviridae	Hepatitis B virus
Hepeviridae	Hepatitis E virus
Reoviridae	Rotavirus
Coronaviridae	Human corona virus
Papillomaviridae	Human papillomavirus

Cheng et al. (2017)

6.3.3 Parasitic Classification

Intestinal parasitic infections are present nearly everywhere on the globe, with elevated rates of occurrence in many countries. The ten most common diseases on the planet include amoebiasis, ascariasis, hookworm infection, and trichuriasis. While these diseases have a low mortality rate, complications do occur often, and certain cases require hospitalization. Malabsorption, diarrhea, blood loss, decreased job ability, and reduced growth rate caused by intestinal parasitic infections are serious health and social concerns in many countries. *Giardia lamblia*, which causes giardiasis, is the most common pathogenic protozoan found in the human duodenum and jejunum. *Leishmania donovani* spread from the site of inoculation to multiply in reticuloendothelial cells which causes kala-azar in humans, and this is usually fatal (Lange Medical Microbiology 2007).

6.3.4 Fungal Classification

Being beneficial to humankind, fungi reside in nature and are helpful in breaking down and recycling organic matter. Fungi are aerobes that can be obligate or facultative and are eukaryotic. Infection caused by fungi is called mycoses. These mycoses are classified based on the initial site of involvement and their usual portal of entry into superficial, cutaneous, subcutaneous, opportunistic, and systemic:

Superficial Mycoses It includes various infections like pityriasis versicolor. The infection caused by lipophilic yeast that requires lipid in the medium for growth is known as *Malassezia globosa* that cause chronic mild infection of the stratum corneum.

Tinea Nigra The infection caused by the dematiaceous fungus *Hortaea werneckii* that causes infection in the stratum corneum.

Cutaneous Mycoses They are the types of fungi that only infect the superficial keratinized tissues, which are the skin, hair, and nails. The fungi that participate in this type of infection are *Trichophyton*, *Epidermophyton*, and *Microsporum* but are restricted to the nonviable skin. Tinea pedis is found to be the most prevalent of all dermatophytoses.

Subcutaneous Mycoses They normally reside in soil or on vegetation and are only confined to subcutaneous tissues. Examples of this are *Sporothrix schenckii*.

6.4 Epidemiology of Microbial Infections

Epidemiology is the analysis of the spread and variables of health-related conditions, diseases, or incidents in special communities, as well as the interpretation of the findings to the management of medical issues. As we are aware, the microbial infection takes place when immune-compromised individuals are exposed to pathogens and to the host via a portal insertion site, which typically includes recognizable immune responses to the specific microbial agents. Morbidity and mortality are detected based on the incidence rate of ill persons per 100,000 persons at risk and incidences of deaths per 100,000 persons at risk. The study of the geographic spread of different microbial diseases has become a priority in public and global health. These unique infectious disease trends are critical in resolving the upcoming infectious disease threats. It was observed in the geographic distributions of human infectious disease that mainly depict that specific rise in the disease was found in the tropics relative to higher latitude and even more diseases were found on the larger islands rather than smaller ones. It was observed in the geographical ranges of human infectious disease that a particular increase in disease was seen in the tropical regions compared to the high latitudes and even more diseases were found on the larger islands rather than smaller ones (Cliff and Haggett 1995; Guernier et al. 2004; Cliff and Haggett 2004). These biogeographical patterns have long been used to maintain, discover, and track the world's biodiversity (Diamond et al. 1976; Channell and Lomolino 2000; Myers et al. 2000). A better understanding of the huge range of biogeographical patterns of various human microbial diseases has considerable potential in managing and monitoring the risk faced by the global health community (Scheiner 2009).

6.4.1 Types of Epidemiological Study

Mainly epidemiological studies are descriptive and analytical as it purely depends on the available data. The descriptive study describes the pattern of the disease based on place and personal factors; on the contrary, analytical studies describe the hypothesis involved in the development of descriptive study. Furthermore, analytical studies are classified into three subclasses which are as follows:

1. *Cohort Study*: These experiments are used to illustrate how an infection develops over time in the absence or presence of risk factors that are assessed at the start of the study (Song and Chung 2010).
2. *Case-Control Study*: These are the observational experiments in which the researcher investigates the relationship between exposure and outcome in study subjects.
3. *Cross-Sectional or Prevalence Study*: In these types of studies, the investigator looks for the exposures which are followed by the results of the study participants simultaneously. The participant's selection for these studies is done based on exclusion and inclusion criteria (Setia 2016).

6.5 Prevalence and Temporal Patterns of Microbial Infections

In the following paragraph, we have included the few major epidemiological stats of microbial infection which are as follows:

Infectious diseases are also a major cause of the extreme poverty troubles so much of the world. These diseases kill about nine million individuals per year, most of whom are children under the age of 5, and they often inflict severe economic burdens due to lifelong disability. In 2018, 104,017 new cases have been reported including tuberculosis, salmonella infection, Lyme disease, and meningococcal disease in the United States alone (Villano and Ogden 2017; Ilyas and Alowibdi 2018).

According to the US CDC report, from 2013 to 2018, there have been 195,203 human immunodeficiency virus (HIV) cases registered among people between the ages of 13 to 65 and above. In contrast to other age groups, we find that Black or African Americans are the main focus of the spread of the disease. In 2019, 5044 mortality was recorded per 100,000 population as per the same report. In 2019, approximately 690,000 mortalities were recorded with AIDS-related illness globally, whereas 1.7 million and 1.1 million cases were recorded in 2004 and 2010. The mortality incidence rate due to AIDS has been declined by 39% since 2010.

According to the WHO TB report of 2020, around 10.0 million people suffered from TB worldwide. India, Indonesia, China, the Philippines, Pakistan, Nigeria, Bangladesh, and South Africa account for approximately two-thirds of the global number. In 2019, 1.4 million people died of tuberculosis, and the worldwide TB incidence rate dropped by 9% from 2015 to 2019 per 100,000 population (Ahmad et al. 2021).

As per the WHO Global hepatitis report 2017, the viral hepatitis pandemic is responsible for around 1.4 million deaths per year from acute infection and hepatitis-related liver cirrhosis and cancer; these stats are very much comparable to tuberculosis and HIV. Hepatitis B virus causes approximately 47% of deaths, hepatitis C virus causes 48% of deaths, and hepatitis A virus and hepatitis E virus cause the remaining 3% of deaths. Viral hepatitis is also a rising cause of death among HIV-positive people. Hepatitis C and B coinfected almost 2.9 million HIV-positive individuals and 2.6 million with hepatitis B. Worldwide, lower than 5% of people with chronic viral hepatitis are aware of their condition.

According to the WHO COVID-19 22 March 2021 report, coronavirus disease (COVID-19) was first reported from Wuhan, China, on 31 December 2019 which is, later on, declared as a pandemic. COVID-19 confirmed cases increased for the fourth week in a row as only 3.3 million new cases reported in the previous week. Interestingly, the mortality rate of COVID-19 has been plateaued after 6 weeks as there were only 60,000 new death cases reported. Western Pacific, Southeast Asia, and European and Mediterranean regions recorded a marked spike in new cases in the last few weeks. The morbidity and mortality rate in the European region and region of the Americas is individually reported to be around 80% of all cases.

As per the WHO Global leishmaniasis surveillance (2017–2018), leishmaniasis is one of the major diseases in the following world regions: the Americas, North Africa, East Africa, Southeast Asia, and West Asia. In 2018, 253,435 cutaneous leishmaniasis cases were reported, and along with that, 17,223 new visceral leishmaniasis cases were reported by the WHO. Visceral leishmaniasis (VL) which is also known as kala-azar is a severe condition and leads to death if not treated in over 95%. The majority of VL cases are reported in Brazil, East Africa, and India. Roughly approximately 50,000 to 90,000 new cases of VL are reported globally and very less around 25–45% recorded by the WHO between 25 and 45%. It is one of the parasitic threads which can be a pandemic outbreak and causes more mortalities.

According to the 2019 CDC report, malaria took the lives of approximately 409,000 people in sub-Saharan Africa, mostly children under the age of 5. In 2019, the WHO recorded 94 percent of all deaths in the African region alone. Over the past decades, intensive work has been undertaken to eradicate malaria, which has saved millions of lives around the world. Malaria is mainly found in rural tropical and subtropical regions of the world. From 2010 to 2019, mortality decreased by 44%, increasing expectations and plans for reduction and, finally, abolishment by the joint efforts of various agencies. Young children, pregnant women, and travelers or migrants from areas with little or no malaria transmission are prone to malaria since they are lacking immunity or have compromised immunity.

6.6 Prevention and Cure of Microbial Infections

6.6.1 Prevention and Cure of Bacterial Infections

Lower respiratory infection and diarrhea which are the third and sixth most common causes of mortality in the world are caused by bacteria. In this age of increasing antibiotic resistance, prevention is found to be crucial for survival. Interruption in chain of infection, protection of host against infection, and elimination of the main source that causes the infection were found to be the three main causes of infections. Primary prevention includes prevention via vaccine, and prevention of symptomatic infection is primarily named as secondary prevention, and to prevent further transmission, treatment is given to infected people, which is known as tertiary prevention. Using personal equipment, the use of pesticides and proper disposal of animal waste are measures of infection that help in preventing the infection. Untreated water can cause various parasitic and viral diseases (Detels et al. 2011).

In developing countries, improving water quality has been critical in reducing the burden of infection in cities. Filtration, settling, and coagulation to kill bacteria-carrying particles, aeration, and chlorination or treatment with another reactive halogen are all part of the current water treatment procedure. It's incredibly difficult to keep bacteria from spreading across the air. It is difficult to sterilize the weather. Laminar flow units are used in hospitals to ensure that air polluted by patients with airborne bacterial diseases such as tuberculosis does not spread within the facility. Quarantine, as well as the application of ionizing radiation, can be used to prevent infectious transmission from person to person (Benenson 1995).

Treatment: Antibiotics can target four major sites, which are the cell wall, the cell membrane, the nucleic acid synthetic pathway, and the ribosome. The antibacterial agent can be classified based on its chemical structure, its target site, and whether the agent is bactericidal or bacteriostatic. Some bacteria are innately resistant to certain classes of antibiotics, because they either lack the target or are impermeable to the drug. The mechanism of resistance includes alteration in the target site, by decreasing permeability of the cell an alteration can do in target site or lastly by the production of enzymes that inactivate the antibiotic. Antibiotics can target various sites: nucleic acid synthetic pathway, cell membrane, ribosome, and cell wall. On the basis of target site and its chemical structure, the antibacterial agent can be classified as bacteriostatic or bactericidal. Resistance mechanism can work by doing target site alteration and ultimately by enzyme production that makes the inactive antibiotic (http://www.cdc.gov – Centre for Disease Control and Prevention).

6.6.2 Prevention and Cure of Viral Infections

Vaccination, which is used to avoid infections by developing immunity to a virus or virus family, is the most effective way of controlling infectious disease. Vaccines can be made of live or killed viruses, as well as viral molecular subunits. Virus vaccines that have been killed and subunit viruses that have not been killed are both unable to cause disease. The threat of using live vaccines, which are always more successful than killed vaccines, is that these viruses will return to their disease-causing state due to back mutations. The "wild-type" (disease-causing) virus is normally attenuated (weakened) in the laboratory by growing it in tissues or at a certain temperature (Ada 2001).

Antiviral agents have the potential to selectively block viral activities, making drug production impossible without harming the host. Viruses affect a wide range of infections in animals and humans, from the common cold to potentially deadly illnesses like meningitis. Antiviral medications and vaccines may cure these infections, but certain viruses, such as HIV, are capable of evading the immune system and mutating to become antiviral drug-resistant (Hawley and Eitzen Jr 2001).

6.7 Conclusions

The study of the determinants, prevalence, spread, and control of health and disease in a given population is the subject of epidemiology, which is discussed in this chapter. The incidence rate is the number of disease cases separated by the population in which the cases occurred. Before the origin of a disease is discovered, epidemiology can reliably explain it and all of the causes that influence its incidence. The relationship between the host and the microorganism is reflected in how an infectious disease presents clinically. Host immune status and microbial virulence factors affect this relationship. The signs and symptoms differ depending on where the infection is located and how severe it is. Infectious diseases disproportionately impact poorer countries in the developing world, becoming a global burden as a result. Infectious diseases are the leading cause of death for children and young adults in developed countries, accounting for one out of every two deaths. An hour, 1500 people are expected to die from an infectious disease, with more than half of them being children under the age of 5. Respiratory infections, an acquired immunodeficiency syndrome (AIDS), diarrheal diseases, tuberculosis (TB), malaria, and measles account for 90% of all infectious disease deaths worldwide.

References

Ada G (2001) Vaccines and vaccination. N Engl J Med 345(14):1042–1053

Ahmad I, Jadhav H, Shinde Y, Jagtap V, Girase R, Patel H (2021) Optimizing Bedaquiline for cardiotoxicity by structure-based virtual screening, DFT analysis, and molecular dynamic simulation studies to identify selective MDR-TB inhibitors. In Silico Pharmacol 9(1):1–15

Bannister BA, Begg NT, Gillespie SH (1996) *Structure and classification of pathogens*, 2nd edn. Blackwell Science Ltd, Oxford, pp 23–34. (Infectious Disease). ch. 2

Beaglehole R, Bonita R, Kjellström T (1993) *Basic epidemiology*. World Health Organization, Geneva, pp 133–142

Benenson AS (1995) Control of communicable diseases manual. The Association, Washington

Bennett JV, Brachman PS (1992) Hospital infections, 3rd edn. Little Brown & Company, Boston

Channell R, Lomolino MV (2000) Dynamic biogeography and conservation of endangered species. Nature 403(6765):84–86

Cheng VCC, Chan JFW, Hung IFN, Yuen KY (2017) Viral infections, an overview with a focus on prevention of transmission. International Encyclopedia of Public Health 2017:368

Cliff A, Haggett P (2004) Time, travel, and infection. Br Med Bull 69(1):87–99

Cliff AD, Haggett P (1995) The epidemiological significance of islands. Health Place 1(4):199–209

Coen DM (1994) Acyclovir-resistant, pathogenic herpes viruses. Trends Microbiol 2(12):481–485

Detels R, Beaglehole R, Lansang MA, Gulliford M (2011) Oxford textbook of public health. Oxford University Press, Oxford

Diamond JM, Terborgh J, Whitcomb RF, Lynch JF, Opler PA, Robbins CS et al (1976) Island biogeography and conservation: strategy and limitations. Science 193(4257):1027–1032

Donnenberg MS (2000) Pathogenic strategies of enteric bacteria. Nature 406(6797):768–774

Engelkirk PG, Duben-Engelkirk J, Fader RC (2020) Burton's microbiology for the health sciences. Jones & Bartlett Publishers

Evans AS (1976) Epidemiological concepts and methods. In: Viral infections of humans. Springer, Boston, pp 1–32

Falkow S (1988) Molecular Koch's postulates applied to microbial pathogenicity. *Rev Infect Dis* 10:S274–S276

Fields BN (1983) How do viruses cause different diseases? JAMA 250(13):1754–1756

Finlay BB, Falkow S (1997) Common themes in microbial pathogenicity revisited. Microbiol Mol Biol Rev 61(2):136–169

García E, Llull D, López R (1999) Functional organization of the gene cluster involved in the synthesis of the pneumococcal capsule. Int Microbiol 2(3):169–176

Grieder FB, Davis NL, Aronson JF, Charles PC, Sellon DC, Suzuki K, Johnston RE (1995) Specific restrictions in the progression of Venezuelan equine encephalitis virus-induced disease resulting from single amino acid changes in the glycoproteins. Virology 206(2):994–1006

Guernier V, Hochberg ME, Guégan JF (2004) Ecology drives the worldwide distribution of human diseases. PLoS Biol 2(6):e141

Hahn HP (1997) The type-4 pilus is the major virulence-associated adhesin of Pseudomonas aeruginosa–a review. Gene 192(1):99–108

Hawley RJ, Eitzen EM Jr (2001) Biological weapons—a primer for Microbiologists1. Annu Rev Microbiol 55:235–253

Hennekens CH, Mayrent SL (1987) *Epidemiology in medicine* (No. 614.4 H4). Little, Brown, Boston
http://www.cdc.gov n.d. – Centers for Disease Control and Prevention
https://www.cdc.gov/nchs/data/hus/2019/010-508.pdf. Hepatitis 2019
https://www.cdc.gov/nchs/data/hus/2019/011-508.pdf. 2019 HIV
https://www.who.int/docs/default-source/hq-tuberculosis/global-tuberculosis-report2020/fact-sheet.pdf).TB 2020
Ilyas MU, Alowibdi JS (2018, April) Disease tracking in GCC region using Arabic language tweets. In: *Companion Proceedings of the Web Conference 2018*, pp 417–423
Koneman EW, Allen SD, Janda WM, Schreckenberger PC, Winn WC (1997) Diagnostic microbiology. In: The non fermentative gram-negative bacilli. Lippincott-Raven Publishers, Philadelphia, pp 253–320
Liu TY, Gotschlich EC, Jonssen EK, Wysocki JR (1971) Studies on the meningococcal polysaccharides: I. composition and chemical properties of the group a polysaccharide. J Biol Chem 246(9):2849–2858
MacMahon B, Pugh TF (1970) Epidemiology: principles and methods. Epidemiology: principles and methods
Mandell GL, Bennett JE, Dolin R (1990) Principles and practice of infectious diseases. Churchill Livingstone, London, pp 1755–1775
Merz AJ, So M (2000) Interactions of pathogenic neisseria with epithelial cell membranes. Annu Rev Cell Dev Biol 16(1):423–457
Myers N, Mittermeier RA, Mittermeier CG, da Fonseca GAB, Kent J (2000) Biodiversity hotspots for conservation priorities. Nature 403(6772):853–858
Scheiner SM (2009) The intersection of the sciences of biogeography and infectious disease ecology. EcoHealth 6(4):483–488
Setia MS (2016) Methodology series module 2: case-control studies. Indian J Dermatol 61(2):146
Song JW, Chung KC (2010) Observational studies: cohort and case-control studies. Plast Reconstr Surg 126(6):2234
Strayer DS, Laybourn KA, Heard HK (1990) Determinants of the ability of malignant fibroma virus to induce immune dysfunction and tumor dissemination in vivo. Microb Pathog 9(3):173–189
Villano JS, Ogden BE (2017) Infectious disease research: animal models and risk management. Comp Med 67(3):189–191
Walker T (1998) Microbiology. WB Saunders Company, Philadelphia

Chapter 7
Socioeconomic Impact of Antimicrobial Resistance and Their Integrated Mitigation by One Health Approach

Md. Meraj Ansari, Kaushik Kuche, Rohan Ghadi, Tushar Date, Dasharath Chaudhari, Rehan Khan, and Sanyog Jain

7.1 Introduction

Microbes are generally microscopic in size and we cannot see it by our naked eye. They are about 1/10th the size of a typical animal cell and generally measured in micrometer scale. Antimicrobial agents are group of chemical compound that have been used against microbes for combating infectious disease (Becker 2013; Uchil et al. 2014; McEwen and Collignon 2018). The antimicrobial agents generally inhibit bacterial cell cycle by acting on the cell wall, DNA replication, and protein synthesis. In early times, antimicrobial agent was synthesized from plant extracts. Whereas microorganisms become resistant to drugs due to mutation in chromosomes, bacteria become resistant to drugs due to decrease in cell membrane permeability, poor influx of drugs, rapid efflux of drugs, modification in drug binding site, drug inactivation, and formation of 3D biofilm of extracellular polymeric matrix. Nowadays, pharmaceutical industries are involved in the synthesis of antimicrobial agent on a large scale. Synthesis of new antimicrobial agent is a long-term process and very expensive, whereas development to resistance takes hardly few years. Nowadays, antimicrobial resistance (AMR) has become a global threat for public health (Prestinaci et al. 2015; Founou et al. 2017). Development of drug resistance by microorganism is an evolutionary survival process, and human activities are also responsible for the development of AMR (Moo et al. 2020). Humans have major contribution in the development of antimicrobial resistance (AMR). Using

M. M. Ansari · K. Kuche · R. Ghadi · T. Date · D. Chaudhari · S. Jain (✉)
Centre for Pharmaceutical Nanotechnology, Department of Pharmaceutics, National Institute of Pharmaceutical Education and Research, Mohali, Punjab, India
e-mail: sanyogjain@niper.ac.in

R. Khan
Department of Nano-Therapeutics, Institute of Nano Science and Technology, Mohali, Punjab, India

N. Akhtar et al. (eds.), *Emerging Modalities in Mitigation of Antimicrobial Resistance*,
https://doi.org/10.1007/978-3-030-84126-3_7

antimicrobial agent to overcome microbial infection without proper guidelines, without advice of expert physician, and without proper diagnosis of infection and not taking antimicrobial drugs as prescribed by physician are major causes of resistance development in microorganism. Once resistance is developed, then microorganisms are able to adapt and grow in the presence of medication (Founou et al. 2017). Day by day, human practices such as misuse or overuse of antimicrobial drugs in health care, poor sanitary conditions, inappropriate food handling, and poor infection control and prevention practices in hospitals is making AMR situation worse. The World Health Organization (WHO) is extensively working on AMR, and in 2001, the WHO explains that AMR infection is most common threats for human health. In 2020, according to the WHO, AMR infection has become one of the top ten global public health threats to humanity. According to expert opinion on AMR infection impact on health, it is estimated that there would be about ten million human deaths annually by 2050 (O'Neill 2014; Garau and Bassetti 2018).

Day by day, AMR has become a challenge for human health globally. It has become a big hurdle for developing country to tackle AMR due to their social and economic impacts. According to researcher's opinion, it has expected that 7% of the world's economic GDP will lose due to AMR infection (O'Neill 2014). Infection with AMR strain causes serious health problem and prolonged hospital care that increase health-care cost, due to expensive second-line drugs and sometime treatment failures (ECDC. Surveillance of Antimicrobial Resistance in Europe 2017, Shrestha et al. 2018). For example, the European region is estimated to spend about 9 billion euro annually due to AMR infection (Llor and Bjerrum 2014; Prestinaci et al. 2015). Furthermore, the Centers for Disease Control and Prevention (CDC) has calculated about 20 billion dollars additional in direct health-care cost in the United States which may cause 35 billion dollars loss in annual productivity (Antibiotic resistance threats in the United States; 2013). Therefore, the health-care sector recently considered AMR infection as a serious health problem and needs crucial treatment with new-generation drugs which are expensive. AMR infection is responsible for public health problem in terms of rate of mortality as well as economic loss (Taneja and Sharma 2019).

The health-care system is also affected by country's economic condition as well as patient individual conditions. Generally rich or developed countries have more health budget as compared to poor or developing countries. The people of developed countries are mostly economically rich, and they can spend more money on health and hygiene, whereas people in poor countries are not economically rich and they are not able to spend more money on health and hygiene because these people are struggling for their basic needs. Therefore, these rich countries generally have less mortality rate as compared to poor countries from AMR infection. For the assessments of socioeconomic impact on AMR infection, some viewpoints are very important such as patient perspectives, physician perspectives, drug industry perspectives, health-care perspectives, and societal perspectives. These viewpoints will be helpful in exact assessments of socioeconomic impact on AMR.

"One Health" approach is essential for the integrated mitigation of AMR. One Health approach combines the efforts of multiple sector and agency action plans on national and international level to ensure optimum health for animals, humans, and the environment. In 2001, the World Health Organization (WHO) has described AMR as a global problem. The Global Strategy for Containment of Antimicrobial Resistance has provided a framework of interventions to slow the emergence and reduce the spread of AMR bacteria (World Health Organization 2001). However, there is a lack of sufficient financial and human resources for the implementation of the strategy for the intended goal. In 2015, the Global Action Plan (GAP) for the purpose of endorsing to tackle AMR was adopted at the 68th World Health Assembly in Geneva (World Health Organization 2015). In 2020, AMR condition has become one of the top ten global public health problems (World Health Organization 2020), and till date, it has failed to gain proper attention. The proper and global implementation of the One Health approach will be relevant in the mitigation of the AMR. If actions are not taken as early as possible for integrated mitigation of AMR, their bad impact will be on socioeconomic conditions as well. The social inequality will increase, the economy will shrink, and the public health-care sector will be less sustainable. Even more serious will be the increase in rate of mortality from microbial infection. Lord Jim O'Neill and his team have published a review entitled "Antimicrobial Resistance: Tackling a crisis for the health and wealth of nations" in 2014, and the review has estimated that there would be ten million deaths every year caused by AMR condition by 2050 (Review on Antimicrobial Resistance 2014). The European Commission (EC) is also working on the broad vision of One Health. The EC has claimed that AMR condition attributed costs approximately €1.5 billion every year (European Commission 2011).

The WHO is eagerly concerned on AMR and ways of tackling it by One Health approach; in spite of mentioning the importance of establishing AMR surveillance system at the local, national, and global level (Grundmann et al. 2011), the lack of capacity, the surveillance gap, and the integration remain to be the major challenge. The US National Action Plan (NAP) view for One Health concept is appreciable. The NAP has recommended the enhancement of a "One Health" national surveillance system, for humans, animals, and the environment, with the upgraded international association and capacity (White House 2015). However, the One Health goal also refers to extensive, system-based access to complex problems (Zinsstag et al. 2011). The socioeconomic factors are directly connected to the use and misuse of antimicrobial agents; therefore, AMR has major impact on the global socioeconomic conditions. Hence, the One Health goal is appropriate for AMR surveillance because it acknowledges the underlying structural factors which influence AMR, including the material, biological, socioeconomic, and political factors (Kock 2015). The One Health approach also introduces the analysis of context across all levels of society (Rushton 2015). The main objective of One Health approach is to make sure that the entire community has sound health. Therefore, public awareness of One Health goal and proper implementation of this approach will be useful in integrating mitigation of AMR.

7.2 Antimicrobial Agent

Antimicrobial agent is a general term for drugs, chemical substance, and other plant derivatives that either kill or inhibit the growth of microbes and most frequently has been used against infections and infectious disease (Punjataewakupt et al. 2019). Antimicrobial agents have been used since ancient times (Shukla et al. 2016) such as some specific molds and plant extracts for treatment of infections. Antimicrobials are grouped according to their action against microbe. For example, antimicrobial agents used against bacteria are called antibiotics, and those acting against fungal infection are called antifungal. Penicillin was the first antibiotic accidently discovered by Sir Alexander Fleming in 1928, and now it becomes the most widely used antibiotics in the world. Penicillin was first naturally occurring antibiotic and most frequently used therapeutically to fight against microbial infections. Penicillin was served as a gift from God during World War II because it was used to reduce the number of deaths from wound infection. In the meantime, Sir Alexander Fleming warned the misuse of antibiotics due to its unexpected adverse effect. Nowadays, people are facing unexpected adverse effect in the form of development of bacterial resistance from multiples of antibiotics, leading cause of deaths of humanity from antimicrobial resistance.

7.3 Antimicrobial Resistance

Antimicrobial resistance is a major public health concern (Bishop 2016). When a microbe developed defense mechanism against antimicrobial medicine, it is called antimicrobial resistance. Nowadays, antimicrobial resistance is a major health threat globally, and for this all we are responsible. Sir Alexander Fleming warned at the time of penicillin discovery that misuse and overuse of antibiotic will give an opportunity to microorganism to develop resistance for their survival. All we know is that the famous phrase "survival of the fittest" of Herbert Spencer describes the idea; in nature, there is very tough competition to survive and reproduce. Therefore, only the fittest can survive in any condition. The development of antimicrobial resistance is an evolutionary process for microbes to survive and reproduce. Due to the development of resistance, day by day classical drugs become noneffective against infectious disease and cause severe illnesses and deaths. Nowadays, antimicrobial resistance has become major cause of deaths globally. The World Health Organization described the development of antimicrobial resistance as a global threat for humanity in 2001 and predicted it as a leading cause of mortality by infection through antimicrobial resistance strains in the future. According to the World Health Organization, antimicrobial resistance has become one of the top ten worldwide threats to humanity in 2020. It's time to take action against the unfair means that are leading cause of development of antimicrobial resistance.

7.4 Mechanism of Resistance

Currently, we are using antimicrobial agent against infections which have mainly three sites of action to inhibit bacterial cell cycle in the course of combating infections: the bacterial cell wall, DNA replication, and translation of proteins. However, bacteria become resistant against antimicrobial agent due to induction of mutation in chromosomes or some other intrinsic factors. After resistance development, older strain of bacteria may terminate with the use of classical drugs, and mutant strain can survive and reproduce. The mutant strain become resistant to multiple drugs by some basic mechanisms such as poor drug influx by porin due to the modification in their cell wall permeability by mutation, rapid and active drug efflux from drug action site due to mutation regulatory protein, inactivation of drug molecule by covalent binding, resistance plasmid and modification in drug target site. Biofilm formation is another way of bacterial resistance against drugs. In biofilm formation, bacteria develop extracellular polymeric matrix around it and prevent from high drug concentration around it. Therefore, these resistance mechanisms of microorganism are leading cause of hindering the antimicrobial drug effectiveness against harmful microorganism.

7.5 Human Practice Accelerating Rate of AMR

From first antibiotics to latest new antibiotics, human is responsible for their synthesis as well as major contributor in drug resistance development. Some microorganism becomes resistant against antibiotics by evolutionary process for their survival (Davies and Davies 2010), and some becomes resistant by human practices as summarized in Fig. 7.1:

1. *Misuse or overuse of antimicrobial drugs in health care*: From early day of discovery of antibiotics, Sir Alexander Fleming warned the people that the high demand of antibiotic could cause their misuse or overuse (Ventola 2015; Langford and Morris 2017; Zaman et al. 2017). Misuse or overuse of antimicrobial drugs is a major problem in developing AMR. Use of antimicrobial agent against infections without proper diagnosis or without recommendation of expert physician, use of antibiotics in healthy animal for growth and production and unnecessary use of antibacterial accelerating AMR.
2. *Inappropriate food handling*: Inappropriate food handling is responsible for major infectious disease. Inappropriate food handling is responsible for foodborne infectious disease and may promote drug resistance as well as acceleration in AMR. We should take care about hygiene for combating AMR.
3. *Poor sanitary conditions*: Poor sanitation is a major cause of many infectious diseases. Open defecation and unsafe drinking water are responsible for transmission of many infectious diseases. Human practices such as poor sanitation may be responsible in the acceleration of AMR. For combating AMR, we should focus on health and hygiene.

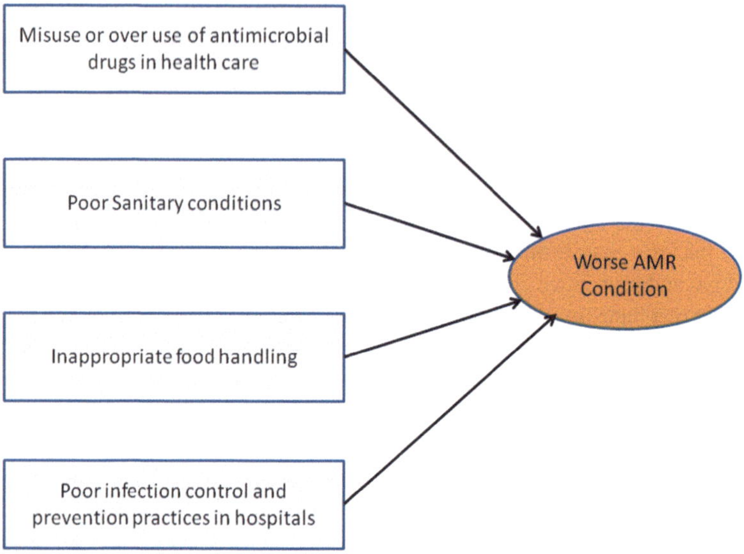

Fig. 7.1 Practices by human that are making AMR worse

4. *Poor infection control and prevention practices in hospitals*: Nowadays, healthcare system is not efficient in infection control and prevention in hospital. Human practices like this may accelerate the AMR. For combating AMR, people should observe proper hand-washing and clean the hospital surface with disinfectant, and special infected people must be at isolated place and use gloves as required.

7.6 Socioeconomic Impact

The costs of new-generation drug are generally high (Gronde et al. 2017). Treatment of AMR-infected patient day by day becomes very expensive. Nowadays, AMR infection has become the leading cause of loss in the global GDP. There are many variables that have contributed in the assessments of socioeconomic impact on AMR infection. The different viewpoints are involved in assessments of socioeconomic impact on AMR infection and very well-represented in Fig. 7.2. Understanding socioeconomic perspective on AMR infection can be better explained by another subheading such as social impact and economic impact.

7.6.1 Economic Impact

We can say that every disease and infection has their impact on global economy. AMR infection is a major health threat to galvanize political and financial investment (Chandler 2019). AMR has their global impact on economic loss. A recent

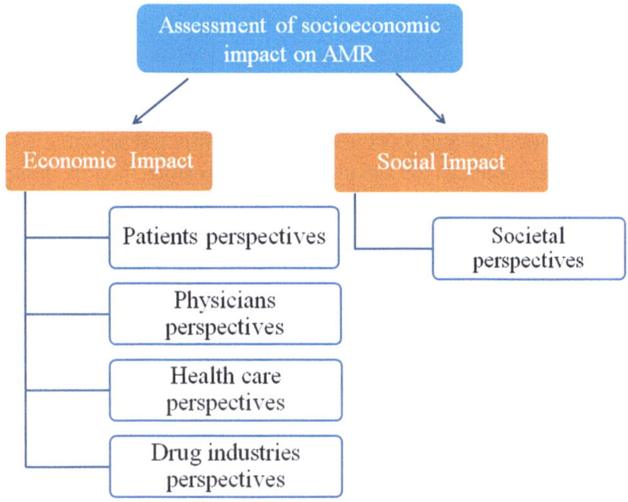

Fig. 7.2 Different viewpoints for the assessment of socioeconomic impact on AMR infection

Table 7.1 Estimated economic loss by 2050 by different medical conditions

Medical condition	Percentage loss of the world's GDP
Caesarean sections	2%
Joint replacements	0.65%
Cancer drugs	0.75%
Organ transplants	0.1%
AMR infections	7%

O'Neill (2014)

review on "Antimicrobial Resistance: Tackling a crisis for the health and wealth of nations" by Jim O'Neill in 2014 estimated the loss in global GDP by 2050 as summarized in Table 7.1. It has been estimated the global percentage loss in GDP by 2050 with different medication conditions such as 2% loss by caesarean sections, 0.65% by joint replacements, 0.75% by cancer drugs, 0.1% by organ transplants, and 7% by AMR infection. We can understand from these data that antimicrobial resistance strain infections are the major contributor in the loss of global GDP by 2050. There will be great impact of antimicrobial resistance on global GDP, and we should take action as early as possible. There are many variables that have significant role to address economic perspectives. Therefore, exact evaluation will be very tough, but proper evaluation of economic perspectives on AMR infection, several viewpoints, and its impact include those of patient, physician, drug industry, and health-care perspectives.

7.6.1.1 Patient Perspectives

Patients suffering from AMR infection have their views and observations that may be valuable in addressing the economic perspective. There are some important consequences such as mortality and morbidity in AMR-infected patients. AMR-infected patients have double chances of development of serious health complication as compared to normal infection and triple chances of mortality (Cecchini et al. 2015). Economic impact can also be measured in terms of consequences faced by patients from day of infection till death by taking into account added cost of treatment by AMR infection because patients pay drug price and services. Patients have their own experience in paying such charges directly or indirectly. If patients have healthcare coverage, then they have to increase their premiums for treatments of such type of infections. As we know that there is three times mortality from AMR infection, this may create fear in susceptible patients that leads to more severity in AMR-infected patients (Friedman et al. 2016). From a review article by Jim O'Neill on "Antimicrobial resistance: tackling a crisis for the health and wealth of nations," the comparison between expected mortality by AMR infection and cause by other conditions by 2050 is summarized in Table 7.2. Also, in Table 7.2, the global mortality rate of AMR infection by 2050 will be ten million, and it is a matter of concern. This high mortality rate of AMR infection may have great impact on the global economy.

Nowadays, about 700,000 patients lose their lives annually due to AMR infection (Porooshat Dadgostar 2019). The mortality rate due to AMR infection in different regions of the world is summarized in Table 7.3. These data may create fears in people suffering from AMR infection and consequently cause more complication.

The United States have reported that annually two million individuals are affected due to AMR infection that leads to cause about 23 thousand deaths annually (Davis et al. 2017). In the same pattern, the European Union has about 25 thousand annual mortality due to AMR infection (WHO 2016). In the meantime, researchers have estimated that about ten million individual will die due to AMR infection globally if strong action is not taken against AMR infection (Laxminarayan et al. 2013).

Table 7.2 Mortality rates by AMR related infections and other conditions by 2050

Condition	Mortality rate
Cholera	0.1 - 0.12 million
Measles	0.13 million
Diabetes	1.5 million
Diarrheal disease	1.4 million
Tetanus	0.06 million
Cancer	8.2 million
Road traffic accidents	1.2 million
Antimicrobial resistance	10 million

O'Neill (2014)

Table 7.3 Mortality rates due to AMR infection in different regions of the world by 2050

Regions	Mortality rates
Asia	4,730,000
Europe	390,000
Africa	4,150,000
Latin America	392,000
North America	317,000
Oceania	22,000

O'Neill (2014)

7.6.1.2 Physician Perspectives

Physicians are directly linked to the infected patient, and their point of view will be very valuable during assessment of economic impact due to AMR infections. When a person gets infected by AMR and consults to the physician for the treatment, then financial condition may be affected by treatment. Physicians are highly professional and they charge money for diagnosis and treatment. AMR-infected patients take time to cure. Treatment for AMR-infected patient is very costly; latest drugs required for the treatment are very expensive for patients. In the absence of suitable drug agent, it is very difficult for physician to treat AMR-infected patient; then, the main economic problem start from this point. Therefore, the economic impact diminishing the effectiveness of a given drug depends on the availability of other potent drugs.

7.6.1.3 Drug Industry Perspectives

The pharmaceutical firms are extensively involved in the production of drugs for the treatment of and prevention from infectious disease. The production of antimicrobial agent by pharmaceutical firm is health-care business. The purposes of these industries are production and making money directly from patients or some time through the health-care sector and government. The product sales are their desire outcome. For the industries, development of bacterial resistance becomes a challenge. Their old antibacterial product day by day is becoming useless or noneffective, and production of new viable drugs for the treatment of AMR infection is very expensive process for them. The production of new-generation drug by extensive research will be very costly, and their impact will be directly on the infected patient as well as on country economic conditions. However, development of antimicrobial resistance gives opportunity to develop new-generation drugs for the treatment.

7.6.1.4 Health-Care Perspectives

Nowadays, health-care sector financial resources are more frequently controlled by administrators and financial manager as compared to doctors and nurses. These peoples are highly professional and see reduced illness and death as a reasonable goal, and their objective is to achieve the goal with fiscal efficiency (McGowan Jr 1991). AMR infections have disastrous impacts on health-care costs. According to CDC reports, in the United States alone, AMR infection treatment could add about $1400 to the treatment bill with any bacterial infection (Antibiotic resistance threats in the United States 2013; Thorpe et al. 2018). According to the researcher, AMR infection could cost from $300 to more than $1 trillion every year globally by 2050 (Chokshi et al. 2019). Extensive treatment requiring intensive care units and isolation beds to prevent the spread of infection against AMR-infected people (Friedman et al. 2016) may cause economic burden on health-care sectors. Researchers have also reported indirect health-care burden as secondary effect from AMR infection (Naylor et al. 2018). These secondary effects make more expensive treatment from AMR infection. Furthermore, organ transplants will be challenging because patients may be exposed to different infections (Li and Webster 2018). Therefore, we can say that AMR infections have economic burden on health-care sector, leading cause of loss in the global economy.

7.6.2 Social Impact

The social impact or societal perspectives are a final view and to be considered for the betterment of public health. The societal perspectives are fuelled by social good conclusively the entire population of town, village, cities, and even the world (McGowan Jr 2001). The goal of societal perspective is maximum heath for entire the population; then it will take long time. Misuse or overuse of antimicrobial drug causes drug resistance and mean time inappropriate decrease in resources. The societal perspectives have great impact on AMR. When we are treating one person causes to decrease in effectiveness in treating to another the person by same drugs, society is adversely affected. The antimicrobial resistance infection prevalence has been increased and become major health problem for the people. Nowadays, mortality rate has increased due to AMR infection. Furthermore, the global loss in economy has been observed due to AMR infection. The recovery rate in developed counties from AMR infection is generally high as compared to developing countries. The treatment cost and mortality rate of AMR infection are high which may create anxiety and psychological problem in poor people leading to complications and poor recovery rate from AMR infection.

7.7 Importance of One Health Approach

The One Health term was first used in 2003–2004, and it was related with the severe acute respiratory syndrome (SARS) in early 2003 (Mackenzie and Jeggo 2019). Nowadays, One Health approach has become a global vision for tackling several infectious diseases. The main objective of the One Health approach is to maintain best health for animals, humans, and the environment (McEwen and Collignon 2018), and it plays a very significant role to tackle most of the contagious disease (zoonotic in nature) that are affecting human health. The zoonotic contagious disease represents a significant threat to global health, economic growth, and security (Allen et al. 2017). The One Health objective can be effectively implemented for combating AMR as well as zoonotic threats. The main motive of One Health goal is represented in Fig. 7.3.

Over the past three decades, it has been observed that majority of the novel, emerging zoonotic infectious diseases originated from animals, especially the wildlife (Taylor et al. 2001). However, the key drivers of such emergence are thought to be related with human activities like changes in habitats, use of land for agricultural intensification, urbanization, trade, and travel (Jones et al. 2008; Karesh et al. 2012; Jones et al. 2013). Collaborative and multidisciplinary access, cutting across boundaries of animals, humans, and the environmental health, is necessary to explain the ecology of each budding zoonotic disease in order to start the assessment of risk and develop strategies for response and control. These assumptions are very useful to recognize the demand and importance of collaborative cross-disciplinary access for the acknowledgment of budding and resurging diseases and in particular for the addition of wildlife health as a crucial component of global disease prevention, surveillance, mitigation, and control.

The One Health definition most commonly used was shared by the US Centers for Disease Control and Prevention and is defined as a collaborative, multisectoral, and transdisciplinary approach – functioning at the local, national, and global levels – with an ultimate goal of accomplishing optimal health outcomes, perceived interconnection between plants, animals, and their shared environment (Mackenzie and Jeggo 2019). One Health Global Network suggested a definition: One Health recognizes that the health of humans, animals, and ecosystems are interconnected.

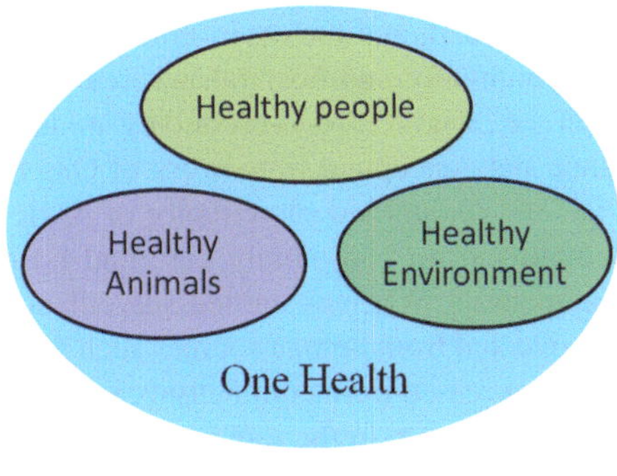

Fig. 7.3 Pictorial representation of One Health goal

It associates applying a collaborative, coordinated, multidisciplinary, and cross-sectoral approach with forwarding potential risk that originates from animals, humans, and ecosystem interface (Mackenzie and Jeggo 2019). According to One Health Institute of the University of California, Davis; One Health is a match to ensure the well-being of people, animals, and the ecosystem through combining problem-solving locally, nationally, and globally. However, the opportunity of One Health as conceptualized by many international organizations such as the WHO, UNICEF, and other organizations also clearly grasps other disciplines and domains, including environment, social sciences, ecosystem health, wildlife, land use, and biodiversity.

The American Medical Association is also working on One Health objective. The American Medical Association, WHO, and Public Health England are engaging in the incorporation of One Health objectives into the medical school curricula as a necessary part of the perspective of public health and infectious disease (Rabinowitz et al. 2017). The celebration of 3 November as One Health Day every year is important for global awareness of One Health views in public, particularly among young generation and students. The One Health Platform Foundation, the One Health Commission, and the One Health Initiative have started in 2016 to celebrate "One Health Day" by organizing One Health education sessions and public awareness events around the globe. In nutshell, we can say that the implementation of One Health approach in surveillance of AMR might be relevant in the integrated mitigation of AMR.

7.8 AMR Surveillance: Role of One Health Framework

For the integrated mitigation of AMR, the One Health approach role is significant. For integrated mitigation of AMR, we have to emphasize the way by which microorganisms develop resistance against certain types of antimicrobial agents and unify the AMR surveillance. In this section, we have focused on a collective data analysis approach from unified AMR surveillance and unified antimicrobial consumption surveillance. In the surveillance of AMR, the One Health approach framework plays a significant role. At a time, it unified and assimilates surveillance of antibiotic usage as well as utilization for humans and animals with AMR data from animals, humans, food, and the environment. The human consumption data are reuniting in the community and hospital, whereas in animals they are collected at the species level, i.e., wildlife, food-producing animals, and companion animals. AMR surveillance and conceptual framework of One Health is proposed here and illustrated in Fig. 7.4. The unified surveillance of AMR in humans include invasive clinical case sampling from community care and hospitals while noninvasive sampling from commensals. Whereas unified surveillance of antimicrobial consumption data can be collected from human sources such as hospitals and community care, the isolates can be accessed from commensals and clinical cases from the healthy animals in the food processing. In the same way, clinical case samples from wildlife can easily be

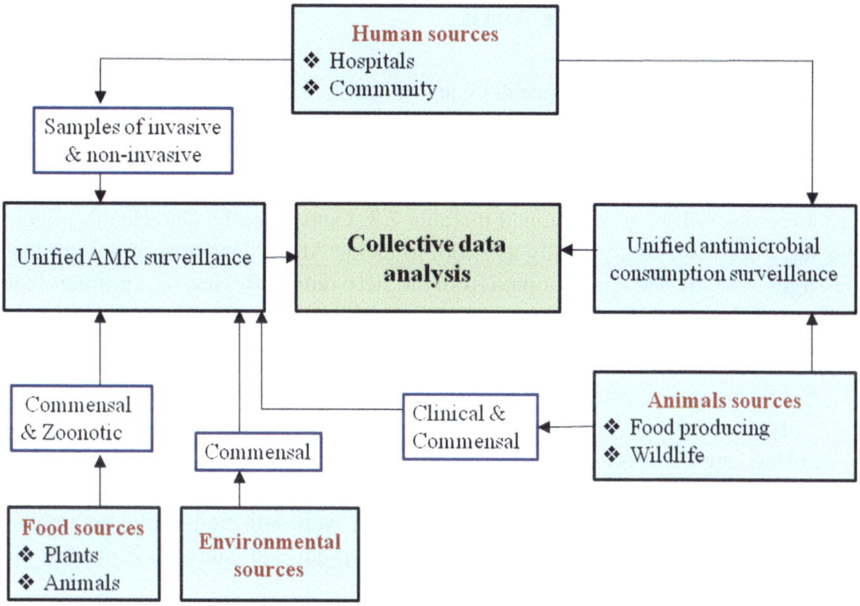

Fig. 7.4 Schematic representation of One Health framework in AMR surveillance

accessed through veterinarians and commensals from food production and wildlife. AMR surveillance in organisms isolated from food includes plants and animals, while the surveillance of zoonotic pathogen is from the commensals. However, samples from water sources and soil can be taken for the surveillance of the environment. After the collection of data from such categories, it requires a unified analysis method facilitated by a company experienced in analyzing such various sectors. In the same way, clarification of analysis requires the company of sectoral professionals who also maintain intersectoral understanding and knowledge which unify skills to work with those from other sectors, social scientists, and behavior change experts. This is to assure that the One Health concept is developing the subsequent recommendations for AMR surveillance. For surveillance purpose, a centralized program is required to set standards for the data collection from various sources for analysis and interpretation. Meantime, this approach will also improve communications and networking between sectors and disciplines via shared meetings, discussions, and report editing.

For integrated mitigation of AMR, the goal set by the One Health is global. For proper and unified AMR surveillance, guidelines given by different global agencies for the fulfillment of the One Health objective might be implemented at the global level. The antimicrobial agent consumption surveillance might be implemented at the global level for the achievement of One Health goal in a way of mitigation of AMR. A strong network recommended for global agencies are dealing with unified AMR surveillance, unified antimicrobial consumption surveillance, and their collective analysis. In nutshell, we can say that public awareness and implementation of One Health approach aim can resolve AMR.

7.9 Ten Ways to Tackle AMR

AMR is a major global health problem, and if action is not taken against AMR, then it will be the leading cause of mortality of almost ten million peoples annually by 2050 (Review on antimicrobial resistance 2016). Recently, researchers have reported that AMR is a global health problem and the expected mortality rate by 2050 in different regions will be as mentioned in Table 7.3. Consequently, One Health principles have been working globally as a tool to tackle AMR. The most important key messages to tackle AMR are adopted from the 2016 report (Review on Antimicrobial Resistance) by Jim O'Neill. This review has mentioned ten different perspectives to tackle AMR:

1. *Public awareness program*

   It is a very tough task to educate our entire societies about the side effects, misuse, and overuse of antimicrobials or antibiotics. This is because patients directly demand antibiotics from doctors or buy them directly from pharmacists for any infectious diseases. An efficient and well-delivered public awareness program can definitely improve public viewpoint and can reduce the sale of unprescribed antibiotics.
2. *Prevent the spread of infection by improving hygiene and sanitation*

   Improving sanitation and hygiene is essential to counter infectious diseases. In the twenty-first century, this is true and is very crucial to reduce the rise of multiple drug resistance. If fewer people get infected by microorganisms, the demand for antibiotics will automatically reduce, and fewer cases for drug resistance will arise. Countries must work on this common aim. The economic condition of all developing countries in the world is not good; these countries should give emphasis on improving the basics first, i.e., by expanding access to sanitation and clean water. The developed countries should focus on reducing the infections among the healthy, by preventive superbugs in hospitals. We all can contribute to this goal of controlling the infection by simply following proper sanitary measures.
3. *Restrict unusual use of antimicrobial in the agriculture sector and their dissemination into the environment*

   Nowadays, in agriculture and aquaculture to maintain the animal welfare and food security, antimicrobials are used frequently. However, the use of antimicrobial agent is not for treating the diseased animal, but it is to prevent infections or simply to promote growth and production. The majority of antimicrobial agent consumption across the world comes from aquaculture and agriculture. In the United States, around 70% (by total volume annually) of antimicrobials that are medically important are frequently used in the agriculture sector. The use of such antimicrobials for infection prevention and growth promotion should be considered unnecessary and dangerous.
4. *Improve worldwide surveillance of drug resistance and antimicrobial consumption*

   To improve global surveillance of AMR as well as antimicrobial consumption, surveillance is essential. For scientists and general physicians to explain the mechanisms of AMR development of new drugs by monitoring the already present cases and to predict the future health threats, they need to have proper insights into the present and past data of AMR. Therefore, three main areas are required for better information and structure, i.e., (1) consumption of antimicrobial among animals and humans, (2) rate of resistance for the available drugs, and (3) scientific knowledge on AMR molecular foundations.

5. *To cut unnecessary use of antimicrobials and promote new and rapid diagnostic*

   Rapid diagnosis of any disease plays a significant role in rapid management and accurate treatment. Therefore, the development of new rapid diagnostics could transform the way of antimicrobial drug use in humans and animals, reducing the unnecessary use of antimicrobials, so that, existing drugs can last longer. Annually, about 27 million patients in the United States are taking antimicrobials unnecessarily due to misdiagnosis of the disease. Therefore, only the development of rapid and accurate diagnostic tests would be helpful for doctors to target disease with suitable antimicrobials.

6. *Encourage development and use of vaccines and other alternatives*

   Development of vaccine for infectious disease can prevent infection and can therefore decrease the demand of antimicrobial drugs automatically. This will inevitably reduce the cases of drug resistance. Although, at present, there is a scarcity of licensed vaccines against the urgent pathogens, there are some promising clinical candidates coming up against *Pseudomonas aeruginosa* and *Clostridium difficile*. However, greater investment is recommended for early-stage research of such new vaccines. Further, alternatives to antibiotics can also be opted which includes probiotics, antibodies, and phage therapy.

7. *Improve the number and recognition of people working with infectious disease*

   A qualified workforce is required for proper implementation and management of AMR. Currently, there is a shortage of expert physicians and microbiologists to deal with such infectious diseases. Therefore, countries need to invest in rewarding and training such specialists.

8. *A worldwide innovation fund for noncommercial and early stage*

   Nowadays, there is no sufficient public and private investment in research and development which focuses on tackling AMR. We need to emphasize on early-stage extensive research in the area of pharmacology or diagnostic. Therefore, we should generate a global innovation fund to support attractive research in this field.

9. *Better incentives for improving the existing ones and promote investment for new drugs*

   Developments of new antimicrobial agent for pharmaceutical companies are not attractive until widespread AMR has emerged against previous generations of drugs. It is very difficult to predict exactly when and how AMR will develop, leaving the pharmaceutical companies in an uncertainty in making business decisions. Developing new antibiotics is an expensive process for companies. The clinical trial networks and harmonized regulations can play a significant role in the regulatory aspects of new drugs for reducing the development costs.

10. *Build a global coalition for real action*

    AMR is not a problem of any one country or even anyone region, so this problem cannot be solved by any one country. AMR is a global problem, and we should plan a global action to make meaningful progress in tackling AMR. The One Health objectives are impacting in tackling AMR on the international political agenda.

7.10 Strategies of One Health in Mitigation of AMR

A strategy of One Health approach is important for integrated mitigation of AMR. In this section, we are highlighting the significance of One Health strategies in combating AMR, and the same is pictorially represented in Fig. 7.5. At present, the emphasis of One Health strategy is to reduce the usage of antimicrobial agent in food animals (White and Hughes 2019). Recently, a meta-analysis of multiple scientific studies and systematic review observed an association between reduction in antimicrobial use in food of animals and decrease in AMR in animals with limited evidence of decline in humans. Thus, by recognizing the complexity of this system, it prevents expression of highlighting research gap and clearing causal pathway (Hoelzer et al. 2017; Tang et al. 2017; Scott et al. 2018). In a report to the president in 2014 on combating AMR (President's Council of Advisors on Science and Technology 2014; World Health Organization 2017), this complexity was highlighted. The WHO is very actively involved in addressing AMR via One Health

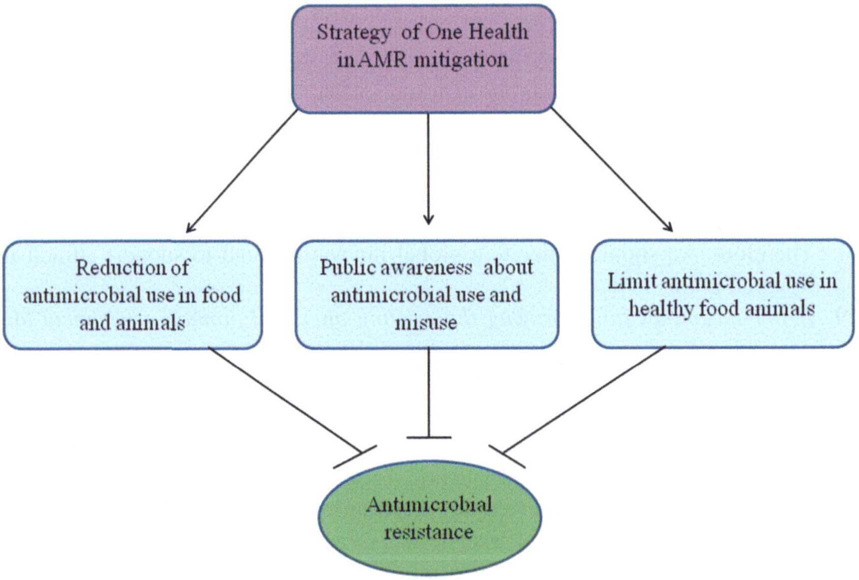

Fig. 7.5 Pictorial representation of One Health strategy in AMR mitigation

approach. The WHO has a role in making guidelines from time to time for the mitigation of AMR. Recently, the WHO published a new guideline to limit or reduce the use of antimicrobial in healthy food animals (World Health Organization 2017). This guideline on antimicrobial use in food and humans is important in combating AMR, a comprehensive approach that addresses aquaculture, wildlife, and the environment (Aga et al. 2018; Thakur and Gray 2019). The wildlife serves as a resistant gene as well as a resistant organism reservoir. For example, drug-resistant *E. coli* have been found in many wildlife species, which could spread among human and livestock (Furness et al. 2017; Dolejska and Papagiannitsis 2018; Weiss et al. 2018). The unnecessary use of antimicrobial agent is the leading cause of AMR, and One Health research gives stress on public awareness about antimicrobial use and misuse for combating AMR. The increasing AMR has an impact on wildlife too. One study revealed that the human strain of drug-resistant *Staphylococcus aureus* indicated the antimicrobial resistance in chimpanzees as well and this indicates the impact humans have on nonhuman primates (Schaumburg et al. 2012). However, there is a paucity of available data on the public health impact of the use of antimicrobial agent in aquaculture, and therefore, the administration of antimicrobial in aquatic systems may promote antimicrobial resistance development (Cabello et al. 2016). Other studies have revealed that antibacterial-resistant genes in aquatic systems can transfer the gene from aquatic microorganisms to human pathogens (Jiang et al. 2012; Tomova et al. 2015). Due to the unnecessary use of antimicrobials which are widely dispersed in aquatic system, it is becoming a containment challenge (Heuer et al. 2009). Environmental contamination related to antimicrobial administration both in food production (aquaculture, cropland, and livestock) and in humans via runoff has been identified (Kumar et al. 2005; Boxall 2012). However, a recent review on the One Health has concluded that environmental health expertise was underrepresented (Khan et al. 2018), and therefore, inclusion of environmental issues becomes necessary in the action plan for mitigation of AMR. Environmental pollution from the production of antimicrobial has been connected to contamination of the surrounding environment (Khan et al. 2013; Larsson 2014). So we can say that the unnecessary use and misuse of antimicrobial agent promote its resistance development, and for the integrated mitigation of AMR, we should limit the use of antimicrobial agent. One Health research is very focused on some basic and easy steps in the way of mitigation of AMR such as public awareness regarding the use and misuse of antimicrobial agent, reduction of their use in food and animals, and limiting their use in healthy animals and the environment.

7.11 Conclusions

Antibiotics are lifesaving drugs and have been used for treatment of infectious disease. The prevalence of antimicrobial resistance has increased. Nowadays, AMR has become global health threats. Day by day, human practices such as misuse or overuse of antimicrobial agent in health care, inappropriate food handling, poor sanitary conditions, and poor infection control and prevention practices in hospitals make AMR worse. Due to AMR infection, the global mortality rate has increased. Therefore, for extensive treatment of AMR-infected patients, the health-care sector needs new-generation drugs. Researchers are extensively working on the synthesis of new-generation drugs. New-generation drugs are very expensive, which is leading to the cause of global economic loss. The socioeconomic perspectives have great impact on AMR infection. For exact evaluation of socioeconomic perspectives on the impact of AMR, some viewpoints such as patient perspectives, physician perspectives, health-care perspective, and societal perspectives are very useful. Generally, the annual budget for health and family welfare of developed counties is higher as compared to developing countries. Therefore, people of economically developed countries have better life and recovery rate from AMR infection compared to developing countries, whereas the rate of AMR infection in people of urban area is higher than the rural area due to sewage problem, supply of contaminated water, and biomedical waste disposal. Social stigma has great impact on AMR infection. Poor people know that AMR infection treatment is very expensive and patient becomes panic that may hamper recovery rate. Scientists, the WHO, and different health-care organizations have focused on the integrated approach for mitigation of antimicrobial resistance, and the One Health approach is one of them. The concept of One Health approach is to ensure best health for animals, humans, and the environment. The WHO and these organizations have developed guidelines to limit the misuse of antimicrobial agents for agriculture, animals, and human. Furthermore, the food chain and environment can spread antimicrobial resistance. Thus, understanding the importance of environmental, food, and agriculture sectors in developing AMR can lead us in preventing the spread and development of AMR. The integrated mitigation of AMR may reduce loss of global GDP. Therefore, we can say that AMR has great impact on global socioeconomic condition and implementation of One Health concept may reduce global economic burden.

Acknowledgments This work was supported by the Department of Health Research (DHR), Ministry of Health and Family Welfare, Government of India. Author Md.Meraj Ansari is thankful to the DHR for the Young Scientist Award under DHR Young Scientist Scheme (YSS/2019/000052/PRCYSS).

Conflict of Interest There is no conflict of interest among the authors.

References

Aga D, Davies J, Gandra S, Kasprzyk-Hordern B, Larsson J, McLain J, Singer A, Snape J, Slijkhuis H, Sweetman A, Voulvoulis N (2018) Initiatives for addressing antimicrobial resistance in the environment: current situation and challenges. Wellcome Trust, London. https://wellcome.ac.uk/sites/default/files/antimicrobial-resistance-environment-report.pdf

Allen T, Murray KA, Zambrana-Torrelio C, Morse SS, Rondinini C, Di Marco M, Breit N, Olival KJ, Daszak P (2017) Global hotspots and correlates of emerging zoonotic diseases. Nat Commun 8(1):1124. https://doi.org/10.1038/s41467-017-00923-8

Antibiotic resistance threats in the United States (2013). https://www.cdc.gov/drugresistance/pdf/ar-threats-2013-508.pdf. Accessed 16 Sept 2019

Becker DE (2013) Antimicrobial drugs. Anesth Prog 60(3):111–123. https://doi.org/10.2344/0003-3006-60.3.111

Bishop BM (2016) Antimicrobial Stewardship in the Emergency Department: Challenges, Opportunities, and a Call to Action for Pharmacists. J Pharm Pract 29(6):556–563. https://doi.org/10.1177/0897190015585762

Boxall ABA (2012) New and Emerging Water Pollutants arising from Agriculture, United Kingdom. Organisation for Economic Co-Operation and Development University of York, York. https://www.oecd.org/tad/sustainable-agriculture/

Cabello FC, Godfrey HP, Buschmann AH, Dölz HJ (2016) Aquaculture as yet another environmental gateway to the development and globalisation of antimicrobial resistance. Lancet Infect Dis 16(7):e127–e133. https://doi.org/10.1016/S1473-3099(16)00100-6

Cecchini M, Langer J, Slawomirski L (2015). Antimicrobial resistance in G7 countries and beyond: economic issues, policies and options for action. Available from: https://www.oecd.org/els/healthsystems/Antimicrobial-Resistance-in-G7-Countries-and-Beyond.pdf. Accessed 16 Sept 2019

Chandler C (2019) Current accounts of antimicrobial resistance: stabilisation, individualisation and antibiotics as infrastructure. Palgrave Commun 5(1):53. https://doi.org/10.1057/s41599-019-0263-4

Chokshi A, Sifri Z, Cennimo D, Horng H (2019) Global Contributors to Antibiotic Resistance. J Global Infect Dis 11(1):36–42. https://doi.org/10.4103/jgid.jgid_110_18

Dadgostar P (2019) Antimicrobial resistance: implications and costs. Infection Drug Resist 12:3903–3910. https://doi.org/10.2147/IDR.S234610

Davies J, Davies D (2010) Origins and evolution of antibiotic resistance. Microbiol Mol Biol Rev 74(3):417–433. https://doi.org/10.1128/MMBR.00016-10

Davis ME, Liu TL, Taylor YJ, Davidson L, Schmid M, Yates T, Scotton J, Spencer MD (2017) Exploring patient awareness and perceptions of the appropriate use of antibiotics: a mixed-methods study. Antibiotics 6(4):23. https://doi.org/10.3390/antibiotics6040023

Dolejska M, Papagiannitsis CC (2018) Plasmid-mediated resistance is going wild. Plasmid 99:99–111. https://doi.org/10.1016/j.plasmid.2018.09.010

ECDC (2017) Surveillance of antimicrobial resistance in Europe. Available from: https://www.ecdc.europa.eu/sites/default/files/documents/EARS-Net-report-2017-update-jan.2019.pdf. Accessed 16 Sept 2019

European Commission (2011) Action plan against the rising threats from antimicrobial resistance. In Communication from the Commission to the European Parliament and the Council. Available at: https://ec.europa.eu/health/sites/health/files/antimicrobial_resistance/docs/communication_amr_2011_748_en.pdf

Founou RC, Founou LL, Essack SY (2017) Clinical and economic impact of antibiotic resistance in developing countries: A systematic review and meta-analysis. PLoS One 12(12):e0189621. https://doi.org/10.1371/journal.pone.0189621

Friedman ND, Temkin E, Carmeli Y (2016) The negative impact of antibiotic resistance. Clin Microbiol Infect 22(5):416–422. https://doi.org/10.1016/j.cmi.2015.12.002

Furness LE, Campbell A, Zhang L, Gaze WH, McDonald RA (2017) Wild small mammals as sentinels for the environmental transmission of antimicrobial resistance. Environ Res 154:28–34. https://doi.org/10.1016/j.envres.2016.12.014

Garau J, Bassetti M (2018) Role of pharmacists in antimicrobial stewardship programmes. Int J Clin Pharm 40(5):948–952. https://doi.org/10.1007/s11096-018-0675-z

Gronde TV, Uyl-de Groot CA, Pieters T (2017) Addressing the challenge of high-priced prescription drugs in the era of precision medicine: A systematic review of drug life cycles, therapeutic

drug markets and regulatory frameworks. PLoS One 12(8):e0182613. https://doi.org/10.1371/journal.pone.0182613
Grundmann, H., Klugman, K. P., Walsh, T., Ramon-Pardo, P., Sigauque, B., Khan, W., Laxminarayan, R., Heddini, A., & Stelling, J. (2011). \. Drug Res Updat, 14(2), 79–87. doi: https://doi.org/10.1016/j.drup.2011.02.007
Heuer OE, Kruse H, Grave K, Collignon P, Karunasagar I, Angulo FJ (2009) Human health consequences of use of antimicrobial agents in aquaculture. Clin Infect Dis 49(8):1248–1253. https://doi.org/10.1086/605667
Hoelzer K, Wong N, Thomas J, Talkington K, Jungman E, Coukell A (2017) Antimicrobial drug use in food-producing animals and associated human health risks: what, and how strong, is the evidence? BMC Vet Res 13(1):211. https://doi.org/10.1186/s12917-017-1131-3
Jiang HX, Tang D, Liu YH, Zhang XH, Zeng ZL, Xu L, Hawkey PM (2012) Prevalence and characteristics of β-lactamase and plasmid-mediated quinolone resistance genes in Escherichia coli isolated from farmed fish in China. J Antimicrob Chemother 67(10):2350–2353. https://doi.org/10.1093/jac/dks250
Jones BA, Grace D, Kock R, Alonso S, Rushton J, Said MY, McKeever D, Mutua F, Young J, McDermott J, Pfeiffer DU (2013) Zoonosis emergence linked to agricultural intensification and environmental change. Proc Natl Acad Sci U S A 110(21):8399–8404. https://doi.org/10.1073/pnas.1208059110
Jones KE, Patel NG, Levy MA, Storeygard A, Balk D, Gittleman JL, Daszak P (2008) Global trends in emerging infectious diseases. Nature 451(7181):990–993. https://doi.org/10.1038/nature06536
Karesh WB, Dobson A, Lloyd-Smith JO, Lubroth J, Dixon MA, Bennett M, Aldrich S, Harrington T, Formenty P, Loh EH, Machalaba CC, Thomas MJ, Heymann DL (2012) Ecology of zoonoses: natural and unnatural histories. Lancet 380(9857):1936–1945. https://doi.org/10.1016/S0140-6736(12)61678-X
Khan GA, Berglund B, Khan KM, Lindgren PE, Fick J (2013) Occurrence and abundance of antibiotics and resistance genes in rivers, canal and near drug formulation facilities--a study in Pakistan. PLoS One 8(6):e62712. https://doi.org/10.1371/journal.pone.0062712
Khan MS, Rothman-Ostrow P, Spencer J, Hasan N, Sabirovic M, Rahman-Shepherd A, Shaikh N, Heymann DL, Dar O (2018) The growth and strategic functioning of One Health networks: a systematic analysis. Lancet 2(6):e264–e273. https://doi.org/10.1016/S2542-5196(18)30084-6
Kock R (2015) Structural One Health--are we there yet? Vet Rec 176(6):140–142. https://doi.org/10.1136/vr.h193
Kumar K, Gupta SC, Chander Y, Singh AK (2005) Antibiotic use in agriculture and its impact on the terrestrial environment. Adv Agron 87:1–54. https://doi.org/10.1016/S0065-2113(05)87001-4
Langford BJ, Morris AM (2017) Is it time to stop counselling patients to "finish the course of antibiotics"? Canad Pharm J 150(6):349–350. https://doi.org/10.1177/1715163517735549
Larsson DG (2014) Pollution from drug manufacturing: review and perspectives. Philos Trans Royal Soc Lond Ser B Biol Sci 369(1656):20130571. https://doi.org/10.1098/rstb.2013.0571
Laxminarayan R, Duse A, Wattal C, Zaidi AK, Wertheim HF, Sumpradit N, Vlieghe E, Hara GL, Gould IM, Goossens H, Greko C, So AD, Bigdeli M, Tomson G, Woodhouse W, Ombaka E, Peralta AQ, Qamar FN, Mir F, Kariuki S et al (2013) Antibiotic resistance-the need for global solutions. Lancet Infect Dis 13(12):1057–1098. https://doi.org/10.1016/S1473-3099(13)70318-9
Li B, Webster TJ (2018) Bacteria antibiotic resistance: New challenges and opportunities for implant-associated orthopedic infections. J Orthop Res 36(1):22–32. https://doi.org/10.1002/jor.23656
Llor C, Bjerrum L (2014) Antimicrobial resistance: risk associated with antibiotic overuse and initiatives to reduce the problem. Therapeut Adv Drug Safety 5(6):229–241. https://doi.org/10.1177/2042098614554919
Mackenzie JS, Jeggo M (2019) The One Health Approach-Why Is It So Important? Tropical medicine and infectious disease 4(2):88. https://doi.org/10.3390/tropicalmed4020088

McEwen SA, Collignon PJ (2018) Antimicrobial resistance: a one health perspective. Microbiol Spect 6(2). https://doi.org/10.1128/microbiolspec.ARBA-0009-2017
McGowan JE Jr (1991) Cost and benefit of perioperative antimicrobial prophylaxis: methods for economic analysis. Rev Infect Dis 13(Suppl 10):S879–S889. https://doi.org/10.1093/clinids/13.supplement_10.s879
McGowan JE Jr (2001) Economic impact of antimicrobial resistance. Emerg Infect Dis 7(2):286–292. https://doi.org/10.3201/eid0702.010228
Moo CL, Yang SK, Yusoff K, Ajat M, Thomas W, Abushelaibi A, Lim SH, Lai KS (2020) Mechanisms of Antimicrobial Resistance (AMR) and Alternative Approaches to Overcome AMR. Curr Drug Discov Technol 17(4):430–447. https://doi.org/10.2174/1570163816666190304122219
Naylor NR, Atun R, Zhu N, Kulasabanathan K, Silva S, Chatterjee A, Knight GM, Robotham JV (2018) Estimating the burden of antimicrobial resistance: a systematic literature review. Antimicrob Resist Infect Control 7:58. https://doi.org/10.1186/s13756-018-0336-y
Jim O'Neill (2014) Antimicrobial resistance: tackling a crisis for the health and wealth of nations. Available from: https://amr-review.org/sites/default/files/AMR.Review. Paper-Tackling a crisis for the health and wealth of nations_1.pdf. Accessed 17 Sept 2019
President's Council of Advisors on Science and Technology (2014) Report to the President on Combating Antibiotic Resistance. Executive Office of the President, President's Council of Advisors on Science and Technology, Washington
Prestinaci F, Pezzotti P, Pantosti A (2015) Antimicrobial resistance: a global multifaceted phenomenon. Pathogens Global Health 109(7):309–318. https://doi.org/10.1179/2047773215Y.0000000030
Punjataewakupt A, Napavichayanun S, Aramwit P (2019) The downside of antimicrobial agents for wound healing. Eur J Clin Microbiol Infect Dis 38(1):39–54. https://doi.org/10.1007/s10096-018-3393-5
Rabinowitz PM, Natterson-Horowitz BJ, Kahn LH, Kock R, Pappaioanou M (2017) Incorporating one health into medical education. BMC Med Educ 17(1):45. https://doi.org/10.1186/s12909-017-0883-6
Review on Antimicrobial Resistance (2014) Antimicrobial resistance: tackling a crisis for the health and wealth of nations. Available from: https://amr-review.org/sites/default/files/AMR%20Review%20Paper%20%20Tackling%20a%20crisis%20for%the%20health%20and%20wealth%20of%20nations_1.pdf
Review on Antimicrobial Resistance (2016) Antimicrobial resistance: tackling drug-resistant infections globally: final report and recommendations. Available from: https://www.biomerieuxconnection.com/wp-content/uploads/2018/04/Tackling-Drug-Resistant-Infections-Globally_-Final-Report-and-Recommendations.pdf
Rushton J (2015) Anti-microbial use in animals: how to assess the trade-offs. Zoonoses and public health, 62 Suppl 1(Suppl 1):10–21. https://doi.org/10.1111/zph.12193
Schaumburg F, Mugisha L, Peck B, Becker K, Gillespie TR, Peters G, Leendertz FH (2012) Drug-resistant human Staphylococcus aureus in sanctuary apes pose a threat to endangered wild ape populations. Am J Primatol 74(12):1071–1075. https://doi.org/10.1002/ajp.22067
Scott AM, Beller E, Glasziou P, Clark J, Ranakusuma RW, Byambasuren O, Bakhit M, Page SW, Trott D, Mar CD (2018) Is antimicrobial administration to food animals a direct threat to human health? A rapid systematic review. Int J Antimicrob Agents 52(3):316–323. https://doi.org/10.1016/j.ijantimicag.2018.04.005
Shrestha P, Cooper BS, Coast J, Oppong R, Do Thi Thuy N, Phodha T, Celhay O, Guerin PJ, Wertheim H, Lubell Y (2018) Enumerating the economic cost of antimicrobial resistance per antibiotic consumed to inform the evaluation of interventions affecting their use. Antimicrob Resist Infect Control 7:98. https://doi.org/10.1186/s13756-018-0384-3
Shukla R, Sharma DC, Baig MH, Bano S, Roy S, Provazník I, Kamal MA (2016) Antioxidant, Antimicrobial Activity and Medicinal Properties of Grewia asiatica L. Med Chem (Shariqah (United Arab Emirates)) 12(3):211–216. https://doi.org/10.2174/1573406411666151030110530
Taneja N, Sharma M (2019) Antimicrobial resistance in the environment: The Indian scenario. Indian J Med Res 149(2):119–128. https://doi.org/10.4103/ijmr.IJMR_331_18

Tang KL, Caffrey NP, Nóbrega DB, Cork SC, Ronksley PE, Barkema HW, Polachek AJ, Ganshorn H, Sharma N, Kellner JD, Ghali WA (2017) Restricting the use of antibiotics in food-producing animals and its associations with antibiotic resistance in food-producing animals and human beings: a systematic review and meta-analysis. Lancet 1(8):e316–e327. https://doi.org/10.1016/S2542-5196(17)30141-9

Taylor LH, Latham SM, Woolhouse ME (2001) Risk factors for human disease emergence. *Philosophical transactions of the Royal Society of London*. Series B, Biol Sci 356(1411):983–989. https://doi.org/10.1098/rstb.2001.0888

Thakur S, Gray GC (2019) The mandate for a global "One Health" Approach to Antimicrobial Resistance Surveillance. Am J Trop Med Hyg 100(2):227–228. https://doi.org/10.4269/ajtmh.18-0973

Thorpe KE, Joski P, Johnston KJ (2018) Antibiotic-resistant infection treatment costs have doubled since 2002, now exceeding $2 billion annually. Health Affairs (Project Hope) 37(4):662–669. https://doi.org/10.1377/hlthaff.2017.1153

Tomova A, Ivanova L, Buschmann AH, Rioseco ML, Kalsi RK, Godfrey HP, Cabello FC (2015) Antimicrobial resistance genes in marine bacteria and human uropathogenic Escherichia coli from a region of intensive aquaculture. Environ Microbiol Rep 7(5):803–809. https://doi.org/10.1111/1758-2229.12327

Uchil RR, Kohli GS, Katekhaye VM, Swami OC (2014) Strategies to combat antimicrobial resistance. J Clin Diagn Res 8(7):ME01–MME4. https://doi.org/10.7860/JCDR/2014/8925.4529

Ventola CL (2015) The antibiotic resistance crisis: part 1: causes and threats. P & T 40(4):277–283

Weiss D, Wallace RM, Rwego IB, Gillespie TR, Chapman CA, Singer RS, Goldberg TL (2018) Antibiotic-Resistant Escherichia coli and Class 1 Integrons in Humans, Domestic Animals, and Wild Primates in Rural Uganda. Appl Environ Microbiol 84(21):e01632–e01618. https://doi.org/10.1128/AEM.01632-18

White A, Hughes JM (2019) Critical Importance of a One Health Approach to Antimicrobial Resistance. EcoHealth 16(3):404–409. https://doi.org/10.1007/s10393-019-01415-5

White House (2015) National action plan for combating antibiotic-resistant bacteria. White House, Washington. https://obamawhitehouse.archives.gov/sites/default/files/docs/national_action_plan_for_combating_antibotic-resistant_bacteria.pdf

WHO report on surveillance of antibiotic consumption (2016). https://www.who.int/medicines/areas/rational_use/who-amr-amc-report-20181109.pdf. Accessed 17 Sept 2019

World Health Organization (2001) WHO global strategy for containment of antimicrobial. Available at: http://www.who.int/drugresistance/WHO_Global_Strategy_English.pdf?ua =1

World Health Organization (2015) Global action plan on antimicrobial resistance. Available at: http://www.who.int/drugresistance/global_action_plan/en/

World Health Organization (2017) WHO guidelines on use of medically important antimicrobials in food-producing animals. Licence: CC BY-NC-SA 3.0 IGO. Available at: https://apps.who.int/iris/bitstream/handle/10665/258970/9789241550130-eng.pdf?sequence=1

World Health Organization (2020) Global action plan on antimicrobial resistance. Available at: https://www.who.int/news-room/fact sheets/detail/antimicrobial-resistance

Zaman SB, Hussain MA, Nye R, Mehta V, Mamun KT, Hossain N (2017) A Review on Antibiotic Resistance: Alarm Bells are Ringing. Cureus 9(6):e1403. https://doi.org/10.7759/cureus.1403

Zinsstag J, Schelling E, Waltner-Toews D, Tanner M (2011) From "one medicine" to "one health" and systemic approaches to health and well-being. Prev Vet Med 101(3–4):148–156. https://doi.org/10.1016/j.prevetmed.2010.07.003

Part II
Mode of Action of Antimicrobials in Mitigation of AMR

Chapter 8
Regulations in Antimicrobial Drug Development: Challenges and New Incentives

Sanjana Kumariya, Richa Mehra, and Rashmi Kumariya

8.1 Introduction

Recent decline in the developmental rate of antimicrobial agents resulted in less frequent approval of novel antimicrobial drugs (AMDs). Due to the slow development of AMDs, the microbial population developing the resistance to existing drugs could not be counterbalanced, therefore restricting the choices of treatments. Several reasons account for this decline in developmental rate including monetary issues, trouble in designing clinical trials, and hindrances in examining the medication for acute infections. The approval of new drugs for their use in particular treatment is decided by regulatory bodies like the US Food and Drug Administration (FDA) and European Medicines Agency (EMA). Though irrespective of the regulatory bodies, the enrolment and approval of drugs are exorbitant. Several measures have been taken by the FDA to encourage the development of antimicrobial medications so that medical practitioners have access to more number of treatment choices. The collaboration of the FDA and EMA is a good initiative to exchange information and in taking decisions regarding clinical trial designs, considering that most of the drug advancement programs are delivered to both the USA and European controllers (Nambiar et al. 2014). From the last 35 years, the regulatory standards set by the FDA have been improved significantly, and newly developed regimens are more focused to accelerate the regulatory procedure to achieve the required medication for treating serious conditions (Darrow et al. 2014, 2020).

S. Kumariya
Pharmaceutics and Pharmacokinetics Division, Central Drug Research Institute, Lucknow, Uttar Pradesh, India

R. Mehra · R. Kumariya (✉)
Advanced Technology Platform Centre, Regional Centre for Biotechnology, Faridabad, Haryana, India

N. Akhtar et al. (eds.), *Emerging Modalities in Mitigation of Antimicrobial Resistance*,
https://doi.org/10.1007/978-3-030-84126-3_8

In comparison to the pre-antibiotic era, the discovery of AMDs dramatically decreased the mortality rate resulting from acute microbial infections. In the pre-antibiotic era, 70–90% of the mortality rate was accounted for acute meningococcal meningitis, which was reduced to approximately 10% by subcutaneous use of sulfanilamides (Schwentker et al. 1937). Later, the use of antimicrobials was expanded to less severe diseases by clinicians, and this expansion was based on the selective and specific target of antimicrobials without any adverse effects on the host. However, this expansion did not consider the fact that several diseases are self-limited and get cured on their own. Due to insufficient data on the placebo-controlled trials of these antimicrobials, their efficacies for the self-limited disease are still questionable (Powers 2004).

Another question is about the safety of these antimicrobials in terms of public health. In the past, there were safety issues associated with the use of AMDs, like the use of sulfanilamides resulted in the loss of several lives due to side effects (Ballentine 1981). This incidence resulted in the enactment of the Food, Drug, and Cosmetic (FD&C) Act in 1938, under which, for the first time, the new drugs were subjected to pre-market safety evaluation (Temple 1995). Later in 1962, FD&C Act was amended due to the detrimental effects of thalidomide that was used for leprosy treatment. The amendment clearly stated that the FDA's decision on the drug's approval would rely on scientific testing, proof of safety and "substantial evidence" of a drug's efficacy. However, the majority of the antimicrobials were introduced in the market for use before this amendment. Nonetheless, this amendment led to the recognition of the fact that the adverse effects of the drugs should be compensated by their beneficial effects; otherwise, there is no point in their use as antimicrobials (Powers 2004).

Since the 1960s, new antimicrobial classes were evidently absent from the FDA-approved list of antimicrobial agents. In the 1980s and 1990s, the FDA approved 29 and 22 new AMDs, respectively (Powers 2004). Only 3% of the 225 anti-infective new molecular entities approved by the FDA were new antimicrobial agents (Spellberg et al. 2004), which showed that the AMD development would soon reach its saturation point. This decline in the development despite huge expenditure on the research and development (R&D) of AMDs suggests the inherent challenges in this field and the rigorous regulatory guidelines that pharmaceutical industries have to go through for approval.

The FDA, later, took up the testing standards for drugs previously established voluntarily by the American Medical Association's (AMA) Council on Drugs, the US Pharmacopeia, and the National Formulary. Since 1962, the FDA has overseen substantial refinements to the broad legal requirement that post-1962 new drugs be approved on the basis of "adequate and well-controlled" studies (Daemmrich 2007). In the 1980s, the FDA and Infectious Diseases Society of America (IDSA) formulated the first ever guidelines for conducting the clinical trials of AMDs for their subsequent approval (Shlaes 2010). Currently, the FDA along with the EMA, International Council for Harmonisation (ICH), and World Health Organization (WHO) are the leading international institutions guiding the evaluation of controlled clinical trials and market release of medicinal drugs (White 2008).

8.2 Drug Development Pipeline

Drug development is a complex and lengthy process involving several steps that usually take years and decades to complete. The drug development pipeline comprises five major steps comprising drug discovery, preclinical research, clinical trials, approval, and post-market surveillance (Fig. 8.1) (FDA 2021a). In the first step, researchers identify the most promising potential candidate from possibly hundreds or thousands of compounds and develop an assay for further evaluation and target optimization. The second step involves extensive preclinical studies that generally comprise in vitro, in vivo, and ex vivo screening complying with good laboratory practices guidelines led by drug regulatory agencies like the FDA and EMA. Also, detailed information about the target, ADME-Tox (absorption, distribution, metabolism, excretion, and toxicity), and dose profiles of the drugs is calculated. Based on dose and toxicity profiles obtained in the preclinical studies, the clinical trials are planned. This step may be considered as the most crucial step, as most of the drugs do not qualify the clinical trials owing to its stringent criteria and ethical issues involved (Chow and Liu 2008). The clinical trials are planned after acquiring information on the qualification criteria of participants, number of participants, duration of the study, control and target group to limit any bias, dose administration, assessment criteria, type of review, and analyses. The successful clinical trials are followed by approval of regulatory authorities if it is considered safe and effective for the proposed use in the specified target population. It is important to acknowledge that despite the rigorous drug development pipeline, few limitations exist in the process. Thus, the true scenario about the drug's safety and optimal usage evolves over a period of months or maybe years that makes up the lifetime of the product in the market. The problems and/or side effects of over-the-counter drugs are always under surveillance, and regulatory bodies like FAERS (FDA Adverse Event Reporting System) may direct to add cautions and usage information or may ask to retract the product in case of serious issues (FDA 2021a).

The developmental pipeline faces several challenges of cost and uncertainties of positive outcomes at each step. Besides, additional hurdles, like reproducibility, result validations, insufficient knowledge of the mode of action, lack of specific targets and biomarkers, high failure rate, patient heterogeneity, and regulatory challenges, pose a burden on the researchers (Norris et al. 2012). Each year, a few tens of drugs are approved for use, while several thousands of drug candidates fall by the wayside. The drug development journey of the licensed drugs begins from the

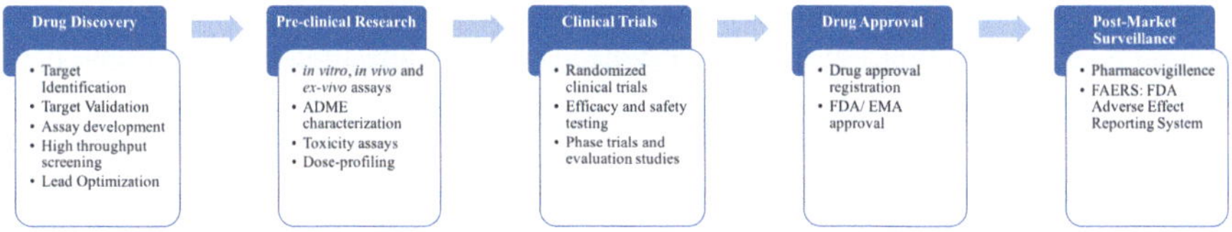

Fig 8.1 Overview of the drug development pipeline

laboratory bench and undergoes an extensive analysis at the cellular and molecular level that might have taken around 10–30 years and cost around billions of dollars (Torjesen 2015).

The modern drug history dates back to the early 1800s when J V Liebig began synthesizing organic molecules and synthesized the first sleeping drug "chloral hydrate" in 1832 (Gauillard et al. 2002). The historical drug development process took several decades for drug commercialization, for example, quinine (isolation drug) was discovered in 1820, but its synthesis mechanism was deciphered in 1944. In the early twentieth century, drug development picked up the pace, and later, toward the end of the twentieth century, more stringent norms were introduced for drug development and approval process. Several drugs like bevacizumab (anticancer), valdecoxib (anti-inflammatory drugs for rheumatoid arthritis), etoricoxib (cox-II inhibitor for rheumatoid arthritis), and alirocumab (whole antibody against high cholesterol) have been approved after 10–15 years since their discoveries (CentreWatch 2021). This time period is cut down to 1–6 years recently, and constant efforts are being made to further streamline the process.

8.3 Clinical Trials

Clinical trial is an important and specialized form of a bioassay to test the therapeutic efficacy, effectiveness, and safety of a new AMD against a specific disease (Chow and Liu 2008). Other trials like investigational trials and observational trials are conducted to define the safety as well as the effectiveness of AMDs under controlled conditions and the health-related concerns of the population in natural conditions, respectively. However, clinical trials are considered most important as these measure the therapeutic effectiveness of AMDs. For a successful clinical trial, every AMD has to pass preclinical trials that involve in vitro and animal studies subjected to a wide range of AMD doses to obtain the preliminary data on efficacy, toxicity, and pharmacokinetics of AMDs (Dudley and Ambrose 2000; MacGowan 2004; MacGowan and Bowker 2002).

The clinical trials are conducted in different phases constituting extensive treatment and observational strategies. Phase I trials are carried out to determine the pharmacokinetics, safety, and gross effects of AMDs on healthy individuals. After successful completion of phase I, phase II trials are conducted wherein pharmacokinetics, safety, and the therapeutic efficiency of AMDs are determined on selected patients. In phase III, the safety and therapeutic efficiency of AMDs are studied on hundreds of selected patients. Once a drug qualifies all three phases with good potential of minimizing the disease condition without any major side effects, it is approved by the regulatory bodies and then commercialized. Lastly, in phase IV, clinicians and medical practitioners review and evaluate the effectiveness and adverse reactions of the test drugs (Stolberg et al. 2004). In 2006, the FDA incorporated one more phase, phase 0, before the phase I trial. Phase 0 trial is meant for speeding up the promising drug development. It involves the testing of AMDs on a comparatively smaller number of subjects to obtain the preliminary pharmacokinetic and pharmacodynamic data (Martinez and Silley 2010).

To reduce the effect of any imbalance or bias, the clinical trials are randomized and controlled. The randomized and controlled clinical trials are the simplest, yet the strongest tools for clinical studies and are considered as the gold standard for establishing the efficacy of new drugs. In the randomized controlled trials, random allocation of participants is done to receive the test AMDs, and this allocation is not decided by the clinicians or the investigators; thus, all the participants have an equal chance of being allocated to any of the test groups, minimizing the chances of imbalance and biasness (Altman and Champman and Hall 1990; Jadad 1998). Randomization of the participants rules out the possibility of any other variations except the AMDs; thus, any significant differences among the groups can be directly attributed to the AMDs. Despite being an effective tool, randomized controlled trials have some drawbacks. Due to financial constraints, low compliance expectations, or high dropout rates, random clinical trials (RCTs) are not feasible. To conduct a successful RCT, the sample size plays a key role. In order to get statistically significant differences between the groups under study, the difference in the effects should be large; otherwise, the sample size should be increased to get reliable data (Chan et al. 2001). The RCTs can be classified into several categories based on the needs of results as shown in Table 8.1.

In order to follow the strict regulatory rules and/or guidelines to reduce the imbalance and biasness in the outcomes, even the seemingly promising AMDs failed to get approval, resulting in the approval of very few drugs. Therefore, several investigators struggle to get approval for many AMDs after successful clinical trials. Thus, there is a need of few amendments in the regulatory guidelines that relax a few parameters in the trials without compromising the quality, safety, efficacy, and effectiveness of the AMD trials.

8.4 Challenges in Antimicrobial Drug Development

The decline in the development of new AMDs started in the 1980s that may be credited to several setbacks like recruitment of patients for clinical trials, spiraling costs to develop new AMDs, acceptance of the new AMDs in the market, and also the regulatory requirements.

The most frequent challenge faced in the AMD development is the slow recruitment process during the clinical trials which are mainly due to low number of eligible patients (Kadam et al. 2016). Even if sufficient patients are available to conduct the clinical trials, other factors, such as resources and stringent policies for trials, come into the picture. Regardless of every attempt to figure out and tackle the problems associated with patient recruitment, the challenge persists. To overcome these issues, Clinical Trials Transformation Initiative (CTTI) came up with effective clinical trial recruitment planning which includes (1) trial design and protocol development that focus on the information source, design elements, and activities

Table 8.1 Classification of randomized clinical trials (RCTs)

Basis of classification	Classification of trials	Description of trials	References
Treatment efficacy	Superiority trials	It shows that the new AMD is superior or better than the control or the existing AMDs	(Sackett 2014; Bokai et al. 2017; Sackett 2004; Lesaffre 2008)
	Non-inferiority (NI) trials	It shows that the new AMD is not better but also not worse than the control or the existing AMDs	(Bokai et al. 2017; Sackett 2004; Lesaffre 2008)
	Equivalence trials	It shows that the new AMD is not too different clinically than the control or the existing AMDs	(Bokai et al. 2017; Sackett 2004; Lesaffre 2008)
Evaluation of different aspects of AMDs	Explanatory	It shows whether AMD works and its mode of action	(Stolberg et al. 2004)
	Pragmatic	It shows whether AMD works and its side effects	(Stolberg et al. 2004)
	Efficacy	It shows the efficacy of AMDs, i.e., effect of AMDs under ideal conditions	(Stolberg et al. 2004)
	Effectiveness	It shows the effectiveness of AMDs, i.e., effect of AMDs under daily life conditions	(Stolberg et al. 2004)
	Phases 1, 2, 3, and 4	These trials are done for the market release of AMDs	(Stolberg et al. 2004)
Participant's exposure and response to AMDs	Parallel	Each group of participants is treated with only one AMD	(Louis et al. 1984)
	Crossover	Each group of participants is treated with all the AMDs under study in succession	(Louis et al. 1984)
	Factorial	Each group of participants is treated with one or more AMDs separately as well as in combination and against a control	(Jadad 1998)
Number of participants	N-of-one or individual patient	It considers an individual patient as the sole unit of observation to study the AMD treatment.	(Stolberg et al. 2004)
	Mega-trials	Thousands of patients from multiple centers and different countries are treated with AMD, but the data collection is limited	(Charlton 1995; Woods 1995)
	Sequential trials	The number of participants is added sequentially and continuously in the study until a clear effect of the AMD is observed or until the investigators get convinced that there is no effect of the AMD	(Stolberg et al. 2004; Altman and Champman and Hall 1990)
	Fixed trials	Either arbitrarily or statistically, investigators decide the sample size to study the effect of AMD. This method is more commonly used	(Stolberg et al. 2004)

(continued)

Table 8.1 (continued)

Basis of classification	Classification of trials	Description of trials	References
Level of blinding	Single-blinded trials	It involves blinding of participants only, i.e., they do not know which AMD treatment is given to them	(Stolberg et al. 2004)
	Double-blinded trials	It involves the blinding of participants as well as clinicians	(Stolberg et al. 2004)
	Triple-blinded trials	It involves blinding of participants, clinicians, and investigators	(Stolberg et al. 2004)
	Quadruple-blinded trials	It involves blinding of patients, clinicians, investigators, and statisticians, but this method is rarely used	(Stolberg et al. 2004)

that drive recruitment, (2) trial feasibility and site selection that mainly consider practicability of trial- and site-related issues, and (3) recruitment communication planning to keep the study team informed about the stakeholder's requirements since communication is necessary to complete trial enrolment successfully. Therefore, to improve the clinical trial recruitment process, CTTI suggested initial stage strategies for recruitment design that produces a positive impact in reducing later-stage hurdles (Huang et al. 2018). Ethical and regulatory obstacles are also important barriers in the recruitment process. Enormous time required for the ethical and regulatory review of the implementing grant sometimes leads to the expiry of grants even before the recruitment process started (Alemayehu et al. 2018).

Sample size is an important factor in the clinical trials, which if not met may incur an increase in type 2 error (erroneously concluding that there is no significant difference observed between treatment groups) (Kadam et al. 2016). In order to show the efficacy of AMDs among different races, it becomes necessary to increase the sample size with the heterogeneous population. Bigger sample size with a heterogeneous population may help in getting AMD globally accepted. However, to launch the AMD globally or across different countries, it is necessary to get the AMD patented individually in the targeted country since the rights granted by a country's patent are extendable only to that particular country and do not affect a foreign country (Saha and Bhattacharya 2011). So, companies willing to protect patents in other countries must apply for a patent in each of the other countries or regional patent offices.

Developing a new AMD and successfully launching it into the market are risky, lengthy, and expensive ventures (Atkinson and Jones 2009). Irrespective of the AMDs, the regulatory cost to launch them in the market was rising. For instance, it globally costs approximately $1581 million to develop a new AMD, excluding the postlaunch surveillance (Towse et al. 2017). Profitability is also a major challenge in the development of new AMDs. The new AMDs are often considered as a last resort in the treatment of any infectious disease where already existing AMDs failed to work, thus resulting in the sale of fewer units. Due to this, several large pharmaceutical companies have already exited the market, and the "small- and medium-sized enterprises" (SMEs) are struggling to get finances for their efforts (Årdal et al. 2020).

Second perception was that there are already enough antibiotics in the market to tackle infectious diseases and there is no requirement of the new AMDs (Duval et al. 2019). However, there was a concern about antimicrobial resistance (AMR), but the numbers of resistance development cases were not enough to encourage the pharmaceutical and biotechnology companies to invest. At the same time, there are extensive regulatory barriers for the new AMDs which require strict validation for the potential of AMDs in fighting against the AMR. Regulatory bodies, however, relaxed some rules in order to address the issues, but they are still bound by the overarching intent of enabling legislation that is designed to protect patient safety. Also, due to the lack of regulatory guidelines for pediatric drug development and evidence-based treatments, infectious diseases are still a major cause of morbidity and death in children. The regulatory plans need to be developed to carefully evaluate the dose regimen and formulation of the new AMDs for the treatment of infectious diseases in children (Årdal et al. 2020).

In order to ensure the patient's safety, regulatory bodies have devised several guidelines. However, these guidelines pose exhaustive regulatory barriers which are big hurdles in the development of new AMDs, resulting in decreased progress in the drug development pipeline of new AMDs. In addition, the possibility of adverse outcomes from the trials is also hampering the R&D of novel AMDs (Alvan et al. 2011). It is necessary to understand the problems hindering the development of new AMDs to get effective resolutions to be executed at the R&D level. However, to address these problems, the perspectives of the R&D professionals are required (Bettiol et al. 2015). The difficulties encountered by the pharmaceutical companies are multifactorial that include (1) monetary issues, (2) the challenging research to study the therapeutics for an acute infectious disease, and (3) scientific issues with regulatory guidelines for AMDs.

There is a need for quick diagnostic tests, updated regulatory norms, as well as advanced trial designs specifically in the case of multidrug-resistant pathogens. Mittra et al. proposed some factors that impede the development of AMDs such as market feasibility and the interplay between identification of rapid pathogens and their impact on trial expenses, specifically, for narrow-spectrum AMDs (Mittra et al. 2019). It has also been reported that previously the capacity to produce commercially viable AMDs was hindered by the regulatory guidelines set for the clinical trials, but now unappealing business prospects are the primary boundary.

As per the FDA regulatory norms, the efficacy and safety drug profile of the new AMDs has to be determined before patients are exposed to them. According to the FDA, the review of AMDs under investigations should have abstained until sufficient data is obtained to justify the risk-benefit profile of investigational AMDs (Darrow et al. 2014). Conclusively, it seemed difficult to fulfill the high standards of regulatory guidelines regarding the trial conduct. To allow the successful conduct of practical and ethical trials, the FDA and EMA have tailored regulatory guidelines. To enable small trials, there is a need to modify the statistical demands together with advances in enrolment conditions. An effort has also been made to decrease the regulatory weights for new AMDs for patients with restricted treatment alternatives. Therefore, revised guidelines are now no longer an obstruction to the pharmaceutical companies in the path of discovering new AMDs (Mittra et al. 2019).

For the first time, the clinical trials on the patients infected with the same infectious agent but at different anatomical places could be pooled. These pooled trials can show the positive profile of the drugs for the different infections caused by the same infection agent and thus made the trials more economical for such pathogens (Rex et al. 2013). To ascertain the safety and efficacy of commercialized medicines, the FDA established new strategies in July 2012, and a new class of "breakthrough therapy" was established by the Congress in the FDA Safety and Innovation Act (FDASIA). The drugs included under the breakthrough therapy were the new molecular entities that are used for the treatment of serious diseases and whose preclinical evidences showed their substantial superiority over the existing drugs (FDASIA 2012). To promote the supporters of antimicrobials and to seek after creative statistical and clinical improvement schemes, the FDA had proclaimed directions involving usage of (1) minor, shorter, or lesser clinical trials, (2) greater NI margins, (3) external controls (nonrandomized design), and (4) enhanced reliability on preclinical researches (animal studies) (FDA 2019; FDA 2017). The regulatory bodies are working hard to streamline and harmonize the regulatory policies to work toward globalization.

8.5 Harmonization of Regulatory Policies: A Collaborative Effort

The recent COVID-19 pandemic has made us realize the worth of globalization of healthcare products irrespective of physical and political boundaries. Despite rapid globalization in the development, production, and distribution of various technologies, the globalization of regulatory policies in medicines, especially AMDs, needs to be addressed on an immediate basis. The pharmaceutical industries that tend to expand in the international markets are required to obtain separate approvals for a single product from different countries. The different regulatory guidelines may require additional data, and even the cost and efforts of putting the same data in several different formats add to the total cost and delay in the release of the product. Thus, globalization and convergence of regulatory policies are necessary for early access of AMDs to patients (Bassetti et al. 2019).

The process of harmonization usually falls under three major categories, viz., regional, bilateral, and global. As the name indicates, the regional initiatives are carried out in a confined region only. For instance, the Europe Union (EU) can be considered as an example of a regional initiative; however, there exist several subgroups including the Gulf Cooperation Council, the Pan American Network for Drug Regulatory Harmonization, the Asia-Pacific Economic Cooperation (APEC), etc. Likewise, bilateral agreements are usually made between two individual countries or groups of countries. For example, in the bilateral cooperation among the EU and the USA in April 2007, the agenda for promoting trans-Atlantic economic integration was signed which particularly targeted the simplification of administrative procedures in the regulation of medicinal drugs. The third major harmonization initiative is global and involves several establishments and countries (Lezorte 2014; EMA 2021a).

World Health Organization (WHO) and International Council for Harmonisation (ICH) To expand the horizons of global harmonization, many initiatives are being taken to achieve convergent regulatory policies like international collaborations in pharmaceutical projects, mutual recognition agreements (MRAs), and the establishment of the organizations like the WHO and ICH. Besides the WHO's key role in the pandemics and treatments across the globe, it has also given an internationally recognized classification system of medicinal drugs – anatomical therapeutic chemical/defined daily dose. The WHO has developed its own prequalification system for drug evaluation which comprises a team of evaluators from WHO staff and professionals from the national regulatory authorities (NRAs) (WHO 2013). Another international body, the ICH, was introduced in 1990 which united the pharmaceutical regulatory agencies from three countries, Europe, Japan, and the USA, to build consensual scientific and technical standards to achieve approval for medicinal products (Louis et al. 1984). At the first ICH meeting, the main criteria for the approval of new medicinal products were decided which were based on the safety, quality, and efficacy of the testing product. The ICH brought regulatory bodies and pharmaceutical companies together to analyze and review the technical and scientific issues related to pharmaceutical industries to frame comprehensive ICH guidelines addressing overall problems. The primary objective of the ICH is to accomplish better alignment in implementing necessary technical guidelines for the registration of the pharmaceutical product, which offers various profits to both regulatory agencies and pharmaceutical companies such as (1) minimizing repeated testing of the new medicinal products on humans and reducing animal testing along with maintaining the required safety standards and effectiveness, (2) decreasing the duration and resources for drug development, and (3) optimizing the regulatory evaluation for the application of the new medicinal products. Since its origin, ICH guidelines have gradually improved. Besides quality, safety, and efficacy, it extends its guideline to multidisciplinary topics like drug interactions, mutagenic impurities, etc. Recent ICH guidelines cover broad categories including clinical safety for the drug used in long-term treatment, pharmacovigilance, dose-response studies, general considerations for clinical trials, statistical principles for clinical trials, choice of control groups in clinical trials, clinical evaluation by therapeutic category, and safety data collection (I ICH 1990).

European Medicines Agency (EMA) The EMA performs a crucial role to combat AMR by promoting new developmental and treatment approaches, encouraging appropriate usage of existing AMDs, and gathering the necessary data to further formulate policies and research. The EMA, FDA, and Japanese Pharmaceuticals and Medical Devices Agency (PMDA) are making collaborative efforts to promote a single development program for new AMDs fulfilling the regulatory standards of all three regulatory bodies. In successive meetings in the years 2016 and 2017, the decisions to harmonize data requirements and the process of clinical trials in specific ailments like gonorrhea and urinary tract infections were made. Consequently, they agreed upon comprehensive and multidirectional approaches to facilitate the common regulatory strategies (EMA 2021b). The major points addressed in these guidelines are (1) explanation of minimal inhibitory concentration of the test AMD, (2) pharmacokinetic (PK)/pharmacodynamic (PD) indices and PK-PD targets from nonclinical data using in vitro and/or in vivo PD models, (3) clinical PK data to support PK-PD analysis, (4) identifying the probability of target attainment (PTA) in order to decide dosage regimen, and (5) assessment of clinical exposure-response (E-R) relationships (EMA 2015a). Besides, several newer instructions are formulated by the EMA that emphasize on the PK/PD effects of antibiotics which provides the guidance to evaluate the development of new antibiotics especially drugs against multiple drug-resistant (MDR) bacteria. These guidelines delineate the regulatory outcomes for the application dossier and describe the scientific progress in pharmacometrics allowing better developmental programs for antimicrobial agents. Consequently, PK-PD analysis has become a key player in deciding the efficient dosage regimen for the AMDs, e.g., beta-lactamase inhibitors, and drugs against MDR pathogens (EMA 2015b).

In the addendum to the guidelines on the evaluation of medicinal products for treating bacterial infections, EMA provided meticulous guidance on the best suited study design for conducting studies in human infections. The guidance regarding the clinical evaluation of the drug administered directly at the site of infection to exert local antimicrobial action remains inadequate, for instance, topical remedy for superficial skin infection, inhalational therapy, and oral administration to exert an effect on the gut. Thus, the following indications in this addendum have presented guidance on clinical data demands: (1) indications for which NI trials are admissible, (2) indications which might need superiority study designs, (3) circumstances in which limited clinical data may be accepted, and (4) other indications requiring special attention to design the clinical trials and data interpretation, e.g., bacteremia, treatment of acute bacterial infections in neutropenic patients, eradication of carriage, and oral medication acting on the gut (EMA 2013).

Food and Drug Administration (FDA) The FDA plays a pivotal role in promoting public health and outracing MDR. The FDA is a member of several international regulatory establishments like the ICH, Pharmaceutical Inspection Co-operation Scheme, and APEC and has always encouraged the implementation of common regulatory guidelines worldwide. The FDA had identified priority areas in the years 2012–2013 and labeled them as Focus Areas of Regulatory Science (FARS). These FARS are mainly based on the following initiatives: (1) emergency preparedness and response in public health, (2) innovation strategies, (3) data sciences, and (4) patient and consumer empowerment. The FDA also developed a report on Advancing Regulatory Science: FARS with a tendency to accommodate frequent updates and revisions. The FDA's strategic plan against AMR announced in 2019 includes comprehensive approaches to coordinate policy and external collaborations in AMR. It addresses the following four key areas: (1) product development to combat AMR with a robust pipeline, (2) promotion of antibiotic stewardship to slow down the AMR, (3) surveillance for the adequate use of AMDs and resistance against them, and (iv) scientific initiatives to support stakeholders in case of resistance development (F FDA 2021). Besides these efforts, the FDA is also actively engaged in the R&D of novel/products and methods involved in animal research with the 3Rs strategy: replace, reduce, and refine 2021 (FDA 2021b).

Central Drugs Standard Control Organisation (CDSCO) The CDSCO is the national regulatory authority of the Government of India which has its headquarter in New Delhi. It has six zonal offices, four sub-zonal offices with associated ports and laboratories that carry out pre-approval as well as post-approval assessments, and post-market surveillance. The CDSCO performs regulatory tasks for the regulation of drugs and cosmetics in the country as conferred by "Drugs and Cosmetics Act," 1940 (Press Information Bureau 2021). It is the key authority preparing national guidelines that specify the regulatory pathways for approval of AMDs and emergency drugs, as in the case of COVID. The Drugs Controller General of India, CDSCO, granted emergency approval to "Covishield" and "Covaxin" developed by Serum Institute of India and Bharat Biotech, respectively (Press Information Bureau 2021). Under the 12th 5-year plan, the Center had allocated budget support of Rs. 1750 crores to strengthen the drug regulatory system in the country, and rapid measures are being taken to divert enough funds during the public health crisis of the COVID pandemic (Hirodkar 2021).

National Medical Products Administration (NMPA) NMPA, previously known as China Food and Drug Administration, is China's national body streamlining drug regulatory policies and has its headquarter in Beijing. Under the State Council of the People's Republic of China, the responsibilities of NMPA include supervision of the safety of drugs and medical devices, developing national regulatory guidelines, encouraging drug development, maintaining a quality management system, and conducting pre-approval and post-approval surveillance (National Medical Products Administration 2021). The NMPA has also approved the use of CoronaVac developed by Sinovac as an emergency measure against COVID (Xinhua 2021).

Besides coordinated and consistent efforts toward harmonization at the international level, authorities are also reviewing and updating existing policies and are receptive for formulating new ones. This is evident from several new incentives that are being considered for implementation.

8.6 New Incentives in Regulations and Policies

The scarcity of effective AMDs in the impending health crisis calls for newer incentives to encourage R & D in the pharmaceuticals. The regulatory norms for the R & D of AMDs are constantly being updated from the past few years to meet the global health implications caused by MDR organisms. In the wake of this global health challenge, the FDA and EMA have revised their regulatory guidelines for the development of AMDs (EMA 2013). Both the FDA and EMA are working toward antimicrobial stewardship; however, this, in turn, poses an obstacle in the path. It is critically important to maintain an equilibrium between supporting general well-being by authorizing the development of novel medicines and safeguarding community health. These regulatory bodies have faced huge criticism from pharmaceutical industries because of posing restrictions and implementing strict guidelines which led to huge financial losses. Thus, several strategies are being implemented to safeguard the interests of stakeholders.

The "push and pull incentives" are designed wherein government "pushes" SMEs by providing financial injections at the entry level, thereby lowering the financial risk and reduction in cost inputs. Such incentives usually come in terms of contracts, grants, public-private collaborations, tax credits, and clinical trial target networks (Morel et al. 2020). For example, CARB-X (Combating Antibiotic-Resistant Bacteria Biopharmaceutical Accelerator) has invested $500 million for the year 2016–2021 for the development of novel AMDs, diagnostics of multiple drug-resistant (MDR) pathogens, vaccines, and related innovations. On the other hand, pull incentives ensure a significant reward only toward the end of the pipeline, once an effective product/process is developed. These incentives may include indirect benefits like reduction in time of regulatory review, premium pricing, and extension of patent terms. However, the pull incentives take much longer implementation times with uncertain rewards and, thus, are not of immediate interest for SMEs (Morel et al. 2020).

In addition, the licensing authorities are constantly reviewing their policies to achieve a balance between the development of innovative AMDs and public health safety. The regulators are also displaying a better understanding toward the acceptance of patients for orphan drugs (lifesaving drugs) with rigorous scientific and legal implications. Thus, alternative pathways for early-limited registration and early-patient access are also proposed for such drugs. The FDA has recommended the enlistment of more patients with more stringent enlistment measures to ensure a better assessment of treatment impacts and trial risks. Mostly, the licensing process is very bureaucratic, time-consuming, and globally inconsistent that further adds on to the cost and timeline of the drug production. Consequently, drugs are released at much economical rates into low-cost generic markets leading to a decline in profit (Wise et al. 2011; Piddock 2013). The licensing authorities conceptualized the adaptive licensing processes that incorporate scientific advice, market sanctions under normal as well exceptional circumstances, risk management plans, and regulatory measures within the legal framework of a particular region (Bax and Green 2014).

The EMA introduced two addenda in 2012 and 2013 to its guidelines for evaluation of the medicinal products against bacterial infections. These addenda permit flexibility in regulatory measures to enable the collection of satisfactory information and hence guarantee the balance between risk and benefit ratios. These addenda specified five main indications for acceptability of NI trials and emphasized the clinical development program for antibiotics to address the requirements of AMDs especially in the case of treatment of MDR pathogens. Additionally, other studies requiring superiority study models and special considerations in study designs are also enlisted (EMA 2013, 2011).

The EMA guidelines emphasize more over the PK/PD standards, wherein only a single key study may be accepted in support of an indication. In fewer cases, data extrapolation from adults to kids may be allowed. In the case of rare diseases, distinct targeted studies and/or standard randomized clinical trials can be used to gather efficacy data. However, in the case of MDR pathogens, a minimal number of treated cases may be required to support any claim on a case-to-case basis. The phase 1 investigations in human participants are carried out only if satisfactory results are achieved by the preclinical and early safety studies. Thus, all the data derived from PK/PD predictions based on in vitro/in vivo studies and preclinical and clinical studies provides better conclusions about the drug efficacy (Bax and Green 2014).

The tiered framework with categorical indications for either disease-based or pathogen-based labels is also in consideration for internationalized standards. The adaptive strategies are more convenient for pharmaceutics in obtaining early approvals and providing adequate risk-benefit data in later stages. Several pilot projects with "safe harbor" sessions among regulators, pharmaceutics, and stakeholders are already being conducted. Also, advanced approval may also be granted to conduct studies on restricted human volunteers if promising risk-benefit ratios are attained (EMA 2017). Besides, the pharmaceutical industries have welcomed the approach of post-marketing allegiances between the industries and the regulators which will result in efficient data collection and observational studies. However, cost-determining concerns are still an issue in such cases. The regulators have also acknowledged the evidence-based performance of "controlled clinical trials" which are the gold standard till date. As per the available data, NI trials for antibiotics don't produce reliable results (Bax and Green 2014). The customizable study designs allow pharmaceutical industries to acquire limited endorsement of new treatments much quicker during the drug development process with the aim to further deliver sufficient data on risk-benefit analysis (EMA 2007).

In exceptional cases, special provisions are provided by regulatory bodies in life-threatening diseases. During the current COVID pandemic, the FDA has created Coronavirus Treatment Acceleration Program (CTAP) to accelerate and assess the possible treatment strategies for COVID patients (FDA n.d.-a). The approval of the first antiviral drug, remdesivir, was done under an Emergency

Use Authorization (EUA) plan issued in May 2020 (FDA n.d.-b). Likewise, in January 2021, under President's Emergency Plan for AIDS Relief (PEPFAR), the FDA launched an interactive database offering crucial information about antiretroviral drugs to combat HIV owing to the global pandemic (FDA n.d.-c). The guidelines are also available for emergency use of unapproved investigational drugs/test drugs against life-threatening and/or severely debilitating diseases which may be exempted from approvals by institutional review boards (IRBs) and investigators (FDA n.d.-d). The "Right to Try Act" was signed in May 2018 to get access to unapproved drugs for patients who have exhausted all approved treatment options available and can't participate in clinical trials (Shlaes 2010). In this regard, informed legal consent and legal authorization are mandatory, and the drug should meet the qualification criteria set by Right to Try Act (FDA n.d.-e).

Despite making the required adaptations, the current regulations still pose challenges in many cases and, hence, require further modifications. However, several useful impacts are observed after modifying the regulatory guidelines. As per a survey, 74% of the respondents believe that significant positive outcomes would be achieved on harmonizing worldwide regulatory needs. Additionally, 50% of respondents suggested the four following modifications: (1) establishment of completely new regulatory pathways, (2) provision of regulatory guidelines based on the pathogen-specific indication in MDR pathogens, (3) restricted use of approved existing and new antibiotics, and (4) regulatory approval of external control data. It is believed that gradual revisions of the guidelines with additional rooms for special and emergency cases will be more beneficial (IDSA 2012).

8.7 Conclusions

The AMD development pipeline has been facing many challenges from the past two decades; many of them are identified and addressed while evaluation of others is in progress. It is very important to maintain a strict balance between drug development and its safety, efficacy, and accessibility which consequently leads to equally stringent regulatory guidelines by authorities. While disease-causing pathogens see no boundaries, the availability of drugs is many a times restricted by jurisdictional boundaries, thereby posing challenges in AMD development. Coordinated global efforts toward conceptualizing the common regulatory policies are being made. The extensive debates at regular conferences and meetings during collaborations of international regulatory bodies provide room for new updates and revisions. A thorough understanding, technological advancements, strong leadership, and willingness are all that take to frame and make changes in legal policies. It is believed that harmony among drug development, drug surveillance, and market release will pave the way for early access to novel, effective, and cost-effective drugs in the future.

References

Alemayehu C, Mitchell G, Nikles J (2018) Barriers for conducting clinical trials in developing countries-a systematic review. Int J Equity Health. 17(1):1–11

Altman DG (1990) In: Champman and Hall (ed) Practical statistics for medical research. CRC press, Boca Raton

Alvan G, Edlund C, Heddini A (2011) The global need for effective antibiotics - A summary of plenary presentations. Drug Resist Updates. 14(2):70–76

Årdal C, Balasegaram M, Laxminarayan R, McAdams D, Outterson K, Rex JH et al (2020) Antibiotic development — economic, regulatory and societal challenges. Nat Rev Microbiol 18(5):267–274

Atkinson JDM, Jones R (2009) Intellectual property rights: An overview and implications in pharmaceutical industry. Future Med Chem. 1(9):1547–1500

Ballentine C (1981) Sulfanilamide disaster. FDA Consumer Magazine

Bassetti M, Giacobbe DR, Vena A, Brink A (2019) Challenges and research priorities to progress the impact of antimicrobial stewardship. Drugs Context. 8:212600

Bax R, Green S (2014) Antibiotics: The changing regulatory and pharmaceutical industry paradigm. J. Antimicrob Chemother. 70(5):1281–1284

Bettiol E, Wetherington JD, Schmitt N, Harbarth S (2015) Challenges and solutions for clinical development of new antibacterial agents: Results of a survey among pharmaceutical industry professionals. Antimicrob Agents Chemother 59(7):3695–3699

Bokai W, Hongyue W, Xin M, Changyong F (2017) Comparisons of superiority, non-inferiority, and equivalence trials. Shanghai Arch Psychiat 29(6):385

CentreWatch (2021) FDA approved drugs. https://www.centerwatch.com/directories/1067-fda-approved-drugs. Accessed 1 May 2021

Chan KBY, Man-Son-Hing M, Molnar FJ, Laupacis A (2001) How well is the clinical importance of study results reported? An assessment of randomized controlled trials. Cmaj. 165(9):1197–1202

Charlton BG (1995) Mega-trials: methodological issues and clinical implications. J R Coll Physicians Lond. 29(2):96–100

Chow S-C, Liu J (2008) Design and analysis of clinical trials: concepts and methodologies. John Wiley & Sons, New York, p 507

Daemmrich A (2007) AIHP Kremers Award Address 2006: Pharmacovigilance and the Missing Denominator: The Changing Context of Pharmaceutical Risk Mitigation. Pharm Hist. 49(2):61–75

Darrow JJ, Avorn J, Kesselheim AS (2014) New FDA Breakthrough-Drug Category — Implications for Patients. N Engl J Med 371:1252–1258

Darrow JJ, Avorn J, Kesselheim AS (2020) FDA Approval and Regulation of Pharmaceuticals, 1983–2018. JAMA. 323(2):164–176

Dudley MN, Ambrose PG (2000) Pharmacodynamics in the study of drug resistance and establishing in vitro susceptibility breakpoints: ready for prime time. Curr Opin Microbiol. 3(5):515–521

Duval RE, Grare M, Demoré B (2019) Fight against antimicrobial resistance: We always need new antibacterials but for right bacteria. Molecules. 24(17):3152

EMA (2007) Reflection paper on methodological issues in confirmatory clinical trials planned with an adaptive design. CHMP/EWP/2459/02. https://www.ema.europa.eu/en/documents/scientific-guideline/reflection-paper-methodological-issues-confirmatory-clinical-trials-planned-adaptive-design_en.pdf. Accessed 14 Feb 2021

EMA (2011) Guideline on the evaluation of medicinal products indicated for treatment of bacterial infections. CPMP/EWP/558/95 rev 2. https://www.ema.europa.eu/en/documents/scientific-guideline/guideline-evaluation-medicinal-products-indicated-treatment-bacterial-infections-revision-2_en.pdf. Accessed 14 Feb 2021

EMA (2013) Addendum to the guideline on the evaluation of medicinal products indicated for treatment of bacterial infections. EMA/CHMP/351889/2013. https://www.ema.europa.eu/

en/documents/scientific-guideline/addendum-guideline-evaluation-medicinal-products-indicated-treatment-bacterial-infections_en.pdf. Accessed 14 Feb 2021
EMA (2015a) Guidelines on the use of pharmacokinetics and pharmacodynamics in the development of antibacterial medicinal products. EMA/CHMP/594085/2015. 2015. https://www.ema.europa.eu/en/documents/scientific-guideline/draft-guideline-use-pharmacokinetics-pharmacodynamics-development-antibacterial-medicinal-products_en.pdf. Accessed 14 Feb 2021
EMA (2015b) New guidance to speed up development of antibiotics. https://www.ema.europa.eu/en/news/new-guidance-speed-development-antibiotics. Accessed 14 Feb 2021
EMA (2017) Guideline on good pharmacovigilance practices (GVP) Module V – Risk management systems (Rev 2). EMA/838713/2011 Rev. 2*. https://www.ema.europa.eu/en/documents/scientific-guideline/guideline-good-pharmacovigilance-practices-module-v-risk-management-systems-rev-2_en.pdf. Accessed 14 Feb 2021
EMA (2021a) International Council on Harmonisation of Technical Requirements for Registration of Pharmaceuticals for Human Use (ICH). https://www.ema.europa.eu/en/partners-networks/international-activities/multilateral-organisations-initiatives/international-council-harmonisation-technical-requirements-registration-pharmaceuticals-human-use. Accessed 14 Feb 2021
EMA (2021b) Antimicrobial resistance. https://www.ema.europa.eu/en/human-regulatory/overview/public-health-threats/antimicrobial-resistance. Accessed 14 Feb 2021
F FDA (2021) Advancing Regulatory Science at FDA: Focus Areas of Regulatory Science (FARS). https://www.fda.gov/media/145001/download. Accessed 14 Feb 2021
FDA (2017) Antibacterial therapies for patients with an unmet medical need for the treatment of serious bacterial diseases. https://www.fda.gov/regulatory-information/search-fda-guidance-documents/antibacterial-therapies-patients-unmet-medical-need-treatment-serious-bacterial-diseases. Accessed 14 Feb 2021
FDA (2019) FDA public meeting: limited population pathway for antibacterial and antifungal drugs. https://www.fda.gov/news-events/fda-meetings-conferences-and-workshops/fda-public-meeting-limited-population-pathway-antibacterial-and-antifungal-drugs-07122019-07122019. Accessed 14 Feb 2021
FDA (2021a) The drug development process. https://www.fda.gov/patients/learn-about-drug-and-device-approvals/drug-development-process. Accessed 1 May 2021
FDA (2021b) Antimicrobial resistance information from FDA. https://www.fda.gov/emergency-preparedness-and-response/mcm-issues/antimicrobial-resistance-information-fda. Accessed 14 Feb 2021
FDA (n.d.-a) Coronavirus Treatment Acceleration Program (CTAP). https://www.fda.gov/drugs/coronavirus-covid-19-drugs/coronavirus-treatment-acceleration-program-ctap. Accessed 1 May 2021
FDA (n.d.-b) FDA Approves First Treatment for COVID-19. https://www.fda.gov/news-events/press-announcements/fda-approves-first-treatment-covid-19. Accessed 1 May 2021
FDA (n.d.-c) President's Emergency Plan for AIDS Relief (PEPFAR). https://www.fda.gov/international-programs/presidents-emergency-plan-aids-relief-pepfar. Accessed 1 May 2021
FDA (n.d.-d) Emergency use of an investigational drug or biologic. https://www.fda.gov/regulatory-information/search-fda-guidance-documents/emergency-use-investigational-drug-or-biologic. Accessed 1 May 2021
FDA (n.d.-e) Right to try. https://www.fda.gov/patients/learn-about-expanded-access-and-other-treatment-options/right-try. Accessed 1 May 2021
FDASIA (2012) Public Law 112–144. https://www.congress.gov/112/plaws/publ144/PLAW-112publ144.pdf. Accessed 14 Feb 2021
Gauillard J, Cheref S, Vacherontrystram MN, Martin JC (2002) Chloral hydrate: a hypnotic best forgotten? Encephale. 28(3 Pt 1):200–204

Hirodkar SN (2021). Centre allots Rs. 1750-cr to CDSCO to strengthen drug regulatory system across the country. https://web.archive.org/web/20151001040940/http://pharmabiz.com/ArticleDetails.aspx?aid=82785&sid=1. Accessed 1 May 2021

Huang GD, Bull J, Johnston McKee K, Mahon E, Harper B, Roberts JN (2018) Clinical trials recruitment planning: A proposed framework from the Clinical Trials Transformation Initiative. Contemp Clin Trials 66:74–79

I ICH: Harmonization for better health (1990). https://www.ich.org. Accessed 14 Feb 2021.

IDSA (2012) White paper: recommendations on the conduct of superiority and organism-specific clinical trials of antibacterial agents for the treatment of infections caused by drug-resistant bacterial pathogens. Clin Infect Dis. 55(8):1031–1046

Jadad AR (1998) Randomised controlled trials: a user's guide. Health Technol Assess. 2(13):214

Kadam RA, Borde SU, Madas SA, Salvi SS, Limaye SS (2016) Challenges in recruitment and retention of clinical trial subjects. Perspect Clin Res. 7(3):137

Lesaffre E (2008) Superiority, equivalence, and non-inferiority trials. Bull NYU Hosp Jt Dis. 66(2):150–154

Lezorte P-L (2014) International cooperation, convergence and harmonization of pharmaceutical regulations: a global perspective, 1st edn. Academic press, Cambridge

Louis TA, Lavori PW, Bailar JC, Polansky M (1984) Crossover and self-controlled designs in clinical research. N Engl J Med. 310(1):83–104

MacGowan A, Bowker K (2002) Developments in PK/PD: optimising efficacy and prevention of resistance. A critical review of PK/PD in in vitro models. Int J Antimicrob Agents. 19(4):291–298

MacGowan AP (2004) Elements of design: the knowledge on which we build. Clin Microbiol Infect. 10(Suppl 2):6–11

Martinez M, Silley P (2010) Antimicrobial drug resistance. Handb Exp Pharmacol. 199(2):227–264

Mittra J, Bruce A, Scannell JW, Tait J (2019) Regulatory and market influences on innovation pathways for the development of new antimicrobial drugs. Tech Anal Strat Manag. 33:1–13

Morel CM, Lindahl O, Harbarth S, de Kraker MEA, Edwards S, Hollis A (2020) Industry incentives and antibiotic resistance: an introduction to the antibiotic susceptibility bonus. J. Antibiot. 73(7):421–428

Nambiar S, Laessig K, Toerner J, Farley J, Cox E (2014) Antibacterial drug development: challenges, recent developments, and future considerations. Clin Pharmacol Ther. 96(2):147–149

National Medical Products Administration (2021). http://english.nmpa.gov.cn/. Accessed 1 May 2021

Norris SMP, Pankevich DE, Davis M, Altevogt BM (2012) Improving and Accelerating therapeutic development for nervous system disorder. The National Academies Press, Washington, D.C., pp 9–24

Piddock LJV (2013) Antibiotic action: helping deliver action plans and strategies. Lancet Infect Dis 13(12):1009–1011

Powers JH (2004) Antimicrobial drug development - The past, the present, and the future. Clin Microbiol Infect 10(Suppl 4):23–31

Press Information Bureau (2021) Press Statement by the Drugs Controller General of India (DCGI) on Restricted Emergency approval of COVID-19 virus vaccine. https://pib.gov.in/Pressreleaseshare.aspx?PRID=1685761. Accessed 1 May 2021

Rex JH, Eisenstein BI, Alder J, Goldberger M, Meyer R, Dane A et al (2013) A comprehensive regulatory framework to address the unmet need for new antibacterial treatments. Lancet Infect Dis. 13(3):269–275

Sackett BDL (2014) Superiority trial. Wiley, Hoboken, pp 1–11

Sackett DL (2004) Superiority trials, non-inferiority trials, and prisoners of the 2-sided null hypothesis. BMJ Evid Bas Med 9(2):38–39

Saha CN, Bhattacharya S (2011) Intellectual property rights: An overview and implications in pharmaceutical industry. J Adv Pharm Technol Res. 2(2):88–93

Schwentker FF, Gelman S, Long PH (1937) The treatment of meningococcic meningitis with sulfanilamide: preliminary report. JAMA. 108(17):1407–1408
Shlaes DM (2010) The perfect storm. Antibiotics:1–7
Spellberg B, Powers JH, Brass EP, Miller LG, Edwards JE (2004) Trends in antimicrobial drug development: Implications for the future. Clin Infect Dis. 38(9):1279–1286
Stolberg HO, Norman G, Trop I (2004) Fundamentals of clinical research for radiologists. AJR 183:1539–1544
Temple R (1995) Development of drug law, regulations, and guidance in the United States. In: Principles of pharmacology: basic concepts and clinical applications, 1st edn. Chapman and Hall, New York, pp 1643–1663
Torjesen I (2015) Drug development: the journey of a medicine from lab to shelf. Pharm J. https://pharmaceutical-journal.com/article/feature/drug-development-the-journey-of-a-medicine-from-lab-to-shelf. Accessed 1 May 2021
Towse A, Hoyle CK, Goodall J, Hirsch M, Mestre-Ferrandiz J, Rex JH (2017) Time for a change in how new antibiotics are reimbursed: Development of an insurance framework for funding new antibiotics based on a policy of risk mitigation. Health Policy. 121(10):1025–1030
White JS (2008) FDA and clinical drug trials: a short history. In: A quick guide to clinical trials. U.S. Food and Drug Administration, Silver Spring. https://www.fda.gov/media/110437/download. Accessed 14 Feb 2021
WHO (2013) Prequalification of medicines by WHO. https://www.who.int/news-room/fact-sheets/detail/prequalification-of-medicines-by-who. Accessed 14 Feb 2021
Wise R et al (2011) The urgent need for new antibacterial agents. J Antimicrob Chemother. 66(9):1939–1940
Woods KL (1995) Mega-trials and management of acute myocardial infarction. Lancet. 346(8975):611–614
Xinhua (2021) Sinovac vaccine prevents 67% of symptomatic COVID-19 infections: Chilean study. https://covid-19.chinadaily.com.cn/a/202104/17/WS607a99e6a31024ad0bab636e.html. Accessed 1 May 2021

Chapter 9
New Insights into and Updates on Antimicrobial Agents

Vagish Dwibedi and Santosh Kumar Rath

9.1 Introduction

Worldwide, microorganisms and their contribution towards sustainable development are obliging for advanced research in microbiology and microbial drug discovery (Kuhad 2012; Koehn and Carter 2005). Natural products and their semisynthetic analogous have played a crucial role in the identification and development of antimicrobial drug innovation programme (Wright et al. 2014; Atanasov et al. 2021; Moloney 2016). In spite of the notable impact on wellbeing, nature-derived compounds have achieved specific attention for their potential activities against various pathogens. Undoubtedly, antimicrobial agents have saved the human race from piles of microbial infectious disease burden and remain one of the most significant discoveries in the twenty-first century (Moloney 2016). However, at present, the most crucial health trouble is widely seen due to the rise and spread of antimicrobial resistance among the different microorganisms (bacteria, fungi, virus, and parasites). The mechanisms for survival of the bacterial resistance under various unfavourable and toxic environmental conditions include (i) enzymatic alteration or degradation of drug, (ii) variation or modification in target, and (iii) reduced uptake or increased efflux. These mechanisms when act together are responsible for enhanced resistance (Abreu et al. 2012; Lambert 2005). Efflux-mediated resistance is an important mechanism for bacteria to expel the

V. Dwibedi
University Institute of Biotechnology, Chandigarh University, Mohali, Punjab, India

S. K. Rath (✉)
Department of Pharmaceutical Chemistry, Danteswari College of Pharmacy, Borpadar, Jagdalpur, Chhattisgarh, India
e-mail: santosh@pharmadanteswari.org

N. Akhtar et al. (eds.), *Emerging Modalities in Mitigation of Antimicrobial Resistance*,
https://doi.org/10.1007/978-3-030-84126-3_9

chemotherapeutic agent out of cell to render them ineffective. Inhibition of efflux is regarded as an efficient strategy for the rejuvenation of old antibiotics again to market as finding new antibiotics is a much time-consuming and costly affair. Many microorganisms are the major sources of precious bioactive molecules considered as useful secondary metabolites to stand and fight against various microbial resistant strains (Singh et al. 2017a). Many pure natural isolates along with newly developed scattered synthetic analogues have proved their eligibility as the best alternatives as antimicrobial agents against resistant pathogens (Abdel-Razek et al. 2020; Martelli and Giacomini 2018). Furthermore, natural antimicrobial agents have gained extensive interest among young and established researchers to reinstate the potency of ineffective antibiotics. Thereby, re-evaluation approach of existing drugs with a combination of newer pharmacophore as efflux pump inhibitors (EPIs) is now considered as best alternatives against multidrug resistance strains (P Tegos et al. 2011; Lamut et al. 2019). Many heterocyclic natural alkaloids are now well accepted along with known antibacterial due to their significant role as efflux pump inhibitory activity against many infectious diseases (Y Mahmood et al. 2016). A natural piperidine-type alkaloid, piperine, isolated from *Piper nigrum* and *Piper longum* and berberine, an isoquinoline alkaloid, isolated from roots and rhizomes of *Berberis vulgaris*, *Rhizoma coptidis* and *Cortex phellodendri* were identified as effective natural EPI to overcome the multidrug-resistant pathogens and also can improve the clinical performance of various other antibiotics when co-administered (Jin et al. 2011). Piperine and many of its analogues when co-administered with ciprofloxacin were observed to inhibit the growth of a mutant *S. aureus* strain by reducing MIC values noticeably (Rath et al. 2019). Palmatine, a newer natural alkaloid, acts as EPIs in *P. aeruginosa* by lowering the MIC-MBC level of ciprofloxacin (Aghayan et al. 2017). Reserpine, another plant alkaloid, is a known inhibitor of the Bmr efflux pump of *Bacillus subtilis* used to accelerate the action of tetracycline in *Staphylococcus aureus* strains and also observed reversing NorA-conferred multidrug resistance in *S. aureus* (Shaheen et al. 2019; Rath et al. 2019). Microorganisms are considered as useful drug targets for various widespread diseases. Though the fundamental life path of microorganisms, their responses to antimicrobials and concerned biochemical pathways seem to be quite complex they need to be understood and explored using modern tools of molecular biology.

Foodborne illness due to fungal or bacterial growth is another major issue in recent times. The widespread microorganisms can easily reach food, grow by utilizing nutritious materials and produce metabolites which are the major cause for the spoilage of plentiful food and food products (Pitt and Hocking 2009; Petruzzi et al. 2017). They can survive even in adverse conditions like low temperature, vacuum packing, processing, and modified atmosphere (Carpena et al. 2021). Thereby, considering the food safety and improving shelf life of foods, many significant efforts have been made by food industries and researchers to find existing or new natural antimicrobials as food preservatives (Gutiérrez-del-Río et al. 2018;

Carpena et al. 2021). Plants, bacteria, fungus, and animals are different sources of the production and recognition of antimicrobials. Plants, the major source of natural products, have been largely used in the domain of the antimicrobial drug finding process. The plant extracts, crude drugs and different class of secondary metabolites are now considered as major opportunities to identify newer antimicrobial medicines and food preservatives. Many recently identified extracts/compounds which are showing antimicrobial activity belong to the families of Asphodelaceae, Solanaceae, Rutaceae, Berberidaceae, Anacardiaceae, Rhamnaceae, Euphorbiaceae, Myrtaceae, Zygophyllaceae, Asteraceae, Erythroxylaceae, Lamiaceae, Colchicaceae, Amaryllidaceae, Verbenaceae, Lythraceae, Podocarpaceae, Salicaceae, Apocynaceae, Zingiberaceae, etc. (Singh 2018; VasudhaUdupa et al. 2021; Swain and Rautray 2021) Altogether, several class of compounds such as alkaloids, glycosides, terpenoids, flavonoids, tannins, and phenolic or polyphenolics isolated from natural sources especially plants are now taken in major consideration towards to development of newer antimicrobials (Takó et al. 2020). Natural crude extracts of ginger, mustard, garlic, cinnamon, basil, sage, and other herbal products are typically high in terpenes such as carvacrol, eraniol, linalool, and several other phenolic compounds, which serve as food additives and antimicrobials against broad spectrum of Gram-positive and Gram-negative bacteria (Makroo et al. 2021). Citral, a main component of lemongrass essential oil, has demonstrated important antioxidant and antimicrobial activity against a variety of food pathogens (Moumni et al. 2021). Furthermore, numerous extracts from Chinese chives and cassia have been documented to dramatically reduce the proliferation of *Escherichia coli* and other bacteria during the preparation and storage of foods, juices, and dairy products. Understanding the process of antimicrobial activity of medicinal plant extracts is therefore needed for their optimum use as natural antimicrobial agents to extend shelf life and maintain food safety (Makroo et al. 2021).

9.2 Antimicrobial Agents from Natural Origin

Natural antimicrobial agents are getting major attention of researchers due to their structural diversity, safety, and nontoxic status. Plants, microbes, and fungal sources are considered as best possible alternatives in finding natural preservatives to avoid or control microbial food spoilage (Saeed et al. 2019). Majorly, plants are having rich sources of many bioactive scaffolds bearing secondary metabolites which are now the primary focus of scientists to explore them to any particular target site to prevent/cure ailments.

9.3 Plant-Derived Antimicrobial Agents

Various phenolic compounds, terpenoids, volatile oils, flavonoids, and sulphur-containing compounds have been detected in seeds, herbs, and spices. These bioactive compounds can be present in plant leaves, branches, seeds, roots, flowers, bulbs, and other pieces. Many herbal and medicinal plants have been recognized for centuries for their preservative and antimicrobial effects (Tuyen and Le 2021). The rich sources of essential oils and different classes of secondary metabolites like terpenes, flavones, aromatic and aliphatic compounds bearing functional groups alcohols, esters, ethers, aldehydes, ketones, and lactones in plants can most effectively destroy several bacterial, fungal, or microbial pathogens (Hyldgaard et al. 2012; Orey 2019). Since times of yore, essential oils like peppermint oil, eucalyptus, and lemongrass are mostly used widely in tribal areas as natural antibacterial and antimicrobial agents due to their benefic application for myriad of cleaning and cleansing function (Sarkic and Stappen 2018; Orey 2019; Desam and Al-Rajab 2021). Traditional use of peppermint essential oil for mouthwash, tea tree essential oil as jewellery cleaner, cedarwood oil for flu and cold, and lavender oil as cleaner are most customary treatments usually followed (Chouhan et al. 2017; Sarkic and Stappen 2018; Desam and Al-Rajab 2021). The essential oils like 1,8-cineole, camphor, borneol, α-pinene, oleanolic acid, β-bisabolene, longicyclene, β-pinene, limonene, β-pinene, eugenol, β-isoeugenol, caryophyllene, α-humulene, p-cymene, γ-terpinene, thymol, and methyl chavicol in many plant species are responsible for antimicrobial activity (Chouhan et al. 2017; Martelli and Giacomini 2018; Ju et al. 2020; Orey 2019). Plant-derived antimicrobials like thymol, eugenol, carvone, citral, carvacrol, linalool, etc. were identified active against *L. monocytogenes* in food model systems (Kawacka et al. 2021; Ju et al. 2020). Alongside many naturally isolated flavonoids like quercetin, kaempferol, apigenin, chrysin, epicatechin gallate, naringenin, myricetin, phloretin, genistein, luteolin, etc. are responsible for promising antimicrobial/antibacterial activity (Manzoor et al. 2020). Many of these substances have a protective function and are effective for inactivating or inhibiting a wide variety of microorganisms. Coumarins and its analogues are widely accepted among various classes of natural bioactive agents for the treatment against diverse diseases related to inflammation (Sharma et al. 2019), cancer (Küpeli Akkol et al. 2020), and additionally these are also useful to control, prevent, and destruct various microbial pathogens (Gouda et al. 2020). Cinnamic acids and coumarins are examples of a large class of phenylpropane-derived compounds with the maximum oxidation state (Gupta and Pandey 2020; Sharma et al. 2018). The increase in hydroxylation of phenolic compounds might be a cause of better effectiveness against microbial pathogenic bacteria. It was proved that hydroxylated phenolic catechol and pyrogallol, which are having two and three hydroxyl groups respectively, are found lethal to microorganisms (Lima et al. 2019; Leontopoulos et al. 2017). Phloretin, a natural bioactive flavonoid, isolated from *Malus sylvestris* has shown antimicrobial activity against a variety of microbial pathogens. Withaferin A, isolated from *Withania somnifera,* is a potential drug lead itself considered strong

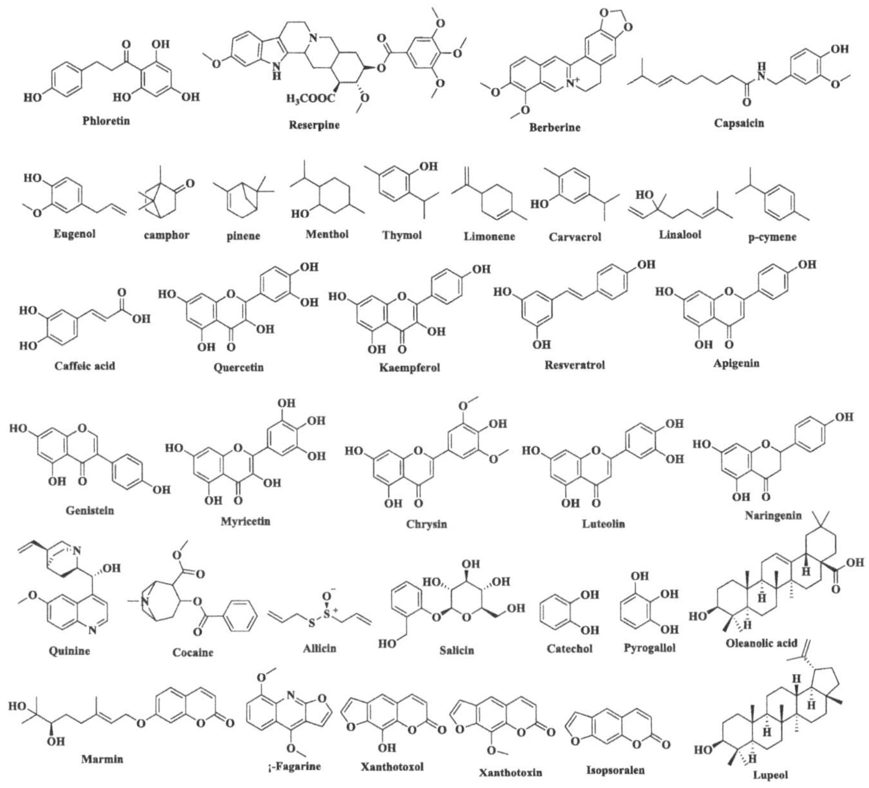

Fig. 9.1 Plant-derived antimicrobial agents

antimicrobial and useful starting material for the development of newer antimicrobials due to the presence of lactone ring and ketone containing unsaturation. Marmin, xanthotoxol, xanthotoxin, lupeol, γ-fragarin, and isopsoralen are class of alkaloids, flavonoids, and terpenoids in *Aegle marmelos* with many reported antimicrobial effects in different in vitro and in vivo assay methods (Reiter et al. 2020). Allicin, a diallylthiosulphinate bioactive defence molecule isolated from *Allium sativum* L., is useful as a broad spectrum antimicrobial agent. However, the instability issue of the molecule retards its effectiveness against microbes in normal or raise in temperature. Allicin's antimicrobial role is largely related to the thiosulphinate functional group (Leontiev et al. 2018) of the molecule. Resveratrol, a naturally occurring phenolic phytoalexin belonging to the stilbene family, has antibacterial activity against diverse Gram-positive and Gram-negative pathogens found in fruit (Dwibedi et al. 2021) (Fig. 9.1).

9.4 Bacterial Origin Antimicrobial Agents

Bacterial infectious diseases are most common in today's time especially in infants and a major cause of paediatric mortality. The antibiotics are the most widely used drugs as powerful therapeutics against various pathogenic bacterial infections (Berkley 2021). Antibacterial drugs, such as ertapenem, erythromycin, gentamycin, tobramycin (*Staphylococcus sp.*), *Aloe vera* (Ghani et al. 2019), retapamulin, periconicins A and ß-resorcyclic acid (*Staphylococcus aureus*), were all identified from natural products and are effective in treating many microbial infections (Suresh and Sona n.d.; Alter and Reich 2021). As a consequence, extensive and injudicious use of antibiotics can be a cause of development for multidrug-resistant microorganisms. The issue of resistance necessitates a renewed attempt to find antibacterial agents from natural sources that are selective against pathogenic bacteria. The 'penicillin' was discovered by Alexander Fleming in 1928. But industrial production of this antibiotic was performed only in 1940 by Howard Florey et Ernst Chain, using *Penicillium chrysogenum* (Gould 2016). Discovery of penicillin made the era of

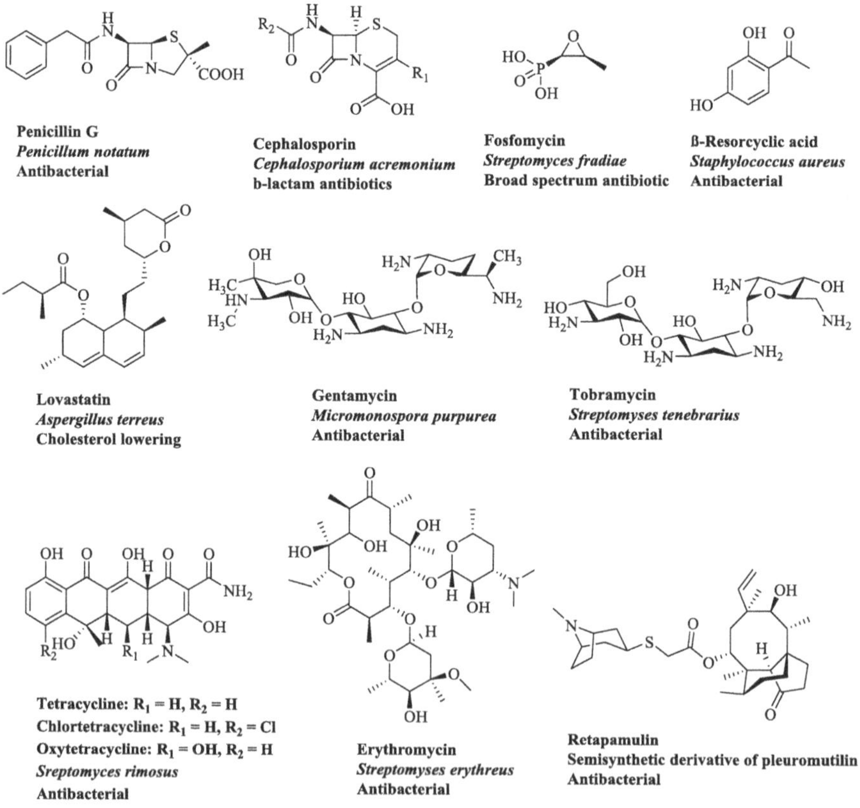

Fig. 9.2 Antimicrobial agents from bacterial source

antibiotics possible, as well as drove the modernization of new methodologies for penicillin discovery (Gould 2016). Many antibiotics used today are derived from microbial classes like β-lactams (penicillin), aminoglycins (gentamicins) and macrolides (erythirosyms), as per instructions (Fig. 9.2). Chlortetracycline, the first antibiotic of tetracycline class, was discovered in 1945 by Benjamin Minge Duggar from Streptomyces (Fig. 9.2). The oxytacycline collaboration between Pfizer and Harvard was called Terramycine (Hochstein et al. 1953; Aminov 2010).

9.5 Fungal-Derived Antimicrobial Agents

Fungi have the ability to produce toxic secondary metabolite mycotoxins which can withstand various harsh/unfavourable conditions at different stages of food chain causing many unavoidable severe health issues and also death in both humans and other animals. Even though fungi are considered as a major cause of food spoilage, still they can have the ability to produce many effective and successful antimicrobials against various ailments. The discovery of first antibiotic Penicillin, a beta-lactum drug that targets the cell wall of bacteria, was derived from the fungus *Penicillium notatum* (Guzmán-Chávez et al. 2018). For many years, *Penicillium notatum* has undergone a program of classical strain improvement (CSI) to improve antibiotics titters. This achievement allowed the lower than normal quantities of BGC-expressed natural products to be generated, resulting in a considerable reduction in the scope of BGC-associated natural product output, or, thus reduced the abundance of a diverse array of these items, which resulted in a significant increase in penicillin enzyme capability alongside the downregulation of a variety of biosynthetic gene clusters (DGCs), resulting in a smaller than usual volume of BGC-encoded DGC-enriched products (NPs). Similarly, edible fungi, such as mushrooms, have possible nutraceutical and inhibitory action against pathogenic microbes (Guzmán-Chávez et al. 2018). From the other side, the fungus *Acremonium fusidioides* (formerly *Fusidium coccineum*) produces steroidal antibiotic fusidine (fusidic acid), the biosynthetic pathway of which is quite close to cholesterol synthesis throughout the human body (Trenin 2013).

Several microbes from the aquatic ecosystem have been shown to secrete secondary metabolites, such as *E. coli*, Proteus genus, and others (Valente et al. 2020). Echinocinocandin, a particular antimycotic, was extracted from *Aspergillus nidulans* using lipo-hexosides as the carbon source (Hu et al. 2020). Interest in the pleuromutilin class has exploded in the modern century, as shown by the production of new human derivatives. Patients with impetigo and untreated superficial lacerations, abrasions, or sutured wounds caused by *Staphylococcus aureus* and *Streptococcus pyogenes* were given Retapamulin, a medicinal antibiotic (Paukner and Riedl 2017).

Trichothecium cinnamon was found to be stable in the organic fraction of the fungus and was also tested for antifungal activity against filamentous fungi. It also reported anti-tumour activity against breast cancer cells, MDA-MB-231 and HeLa, and against MDA- lines B10F and MDA-MB231. (Taware et al. 2015; Silva et al.

2017). Silva et al. isolated and characterized three new isoaigialones, A, B, and C, as well as an aigialone from the endophytic fungus *Phaeoacremonium sp.*, and measured them against the phytopathogenic fungi *Cladosporium cladosporioides* and *C. sphaerospermum* (Silva et al. 2017). Curvularine was found in a leaf of the Murraycian tree (*Hyloomsantia myrmexella*) and produced the antifungalecycol products: murolide A, murolide A, and murop acid, along with six previously unknown compounds: mupiranol A, murasin, mursan, and muran-6, all together with the well-known components muracidin, and murin (Mondol et al. 2017). Acetolide compound (2-amino-1-1-acetapregnadicapramide) 3-ben-β-ol-C and ergosylan-7,22(2,5,6), obtained from ethyl acetate extract of *Anvillearcium chasteriinense*, was used to characterize ethyl acetate extract a novel fusario acetohydide (2-AG:ET:1O) while three known compounds (1-acetolan, 8:3β-diol:7, 6:6-d) and epichlororenone (6:3β-dihydroxymide) were used to complete the characterization of the fused amide. Disc diffusion assay was used to monitor the antibacterial and antifungal efficacy of compound fusarithioamide A against various microorganisms. It demonstrated antibacterial activity against *B. cereus*, *S. aureus*, and *E. coli*, with inhibition zone diameters (IZDs) of 19.0, 14.1, and 22.7 mm, and MICs of 3.1, 4.4, and 6.9 g ml1, respectively.

One additional antifungal toxin out of three new examples of triovirabacinolides and three new trioviriridines from the endophytic fungus *Penicillium raccum* (Kajula et al. 2016) found in the literature by expounding upon three ways of looking at this genus of fungus was noted for antifungostase and anti-inflammation, the terms described above (Kajula et al. 2016). The recognized compounds of seven different species of fungi ((R)-3-Hydroxybutynonine isolated from the endophyte fungus *Aspergillus* sp.) have bioactive dianospermine as the seventh in their list of isolated bioactive compounds. These secondary metabolites were screened against fungi that are phytopathogenic (*Botrytis cinerea*, *Gibberella saubberti*, *Colletrix gloeoproides*, and *Magnipeniella grossi*). A test was performed against *Phytophthora capsici, Escherichia coli, Rhizoctonia solani.* Compounds R-3-hydroxybutanonil were effective against all of the phytopathogenic fungi studied, with the exception of those with a minimum inhibitory concentration (6.25–50 μM) and below MIC level of 6.25 μM, although less active on viruses, antimicrobial compounds less than 25-fold inhibitory concentrations, outosaminic acid had a MIC in the range of 25–100 μM, but was ineffective, or outamin C was active against all pathogenic bacteria, but active down to the low MIC values (of 25-fold) less than 25-to-mM concentrations were ineffective (Xiao et al. 2014). Similarly Trichodermin is a strong antifungal bioactive compound isolated from endophytic fungus *Trichoderma brevicompactum* with EC_{50} of 25 μg/mL fraction possessed significant ability to hinder growth of the plant pathogen *Phenacoccus solani*. Also, it had minimal ability to influence *B. cinerea* at EC_{50} of 25 mg/L (Shentu et al. 2014). A fifth mammalian antifungal that was sourced from *Bignonia magnifica*, evaluated for their anti-pityriasis properties on walnut and mediterrane fungus species *O. fragariaeola*, *O. cinereoxys*, *C. glosum* were also tested. (Silva-Hughes et al. 2015).

It was found that four compounds in Wang's study include cladosin, isocladoside, and 5-hydroxyaspeona. Additionally, Wang et al. (2013) discovered an additional one that foundcladoside, isocladoside, and 5-hydroxyaspeona could be extracted from the endophytic fungus *Cladosioquinidium*. In the presence of this weed, both the synthetic growth inhibitors were found to be effective against *Colletodothis viti* (weed) and the natural relatives (the synthesized and natural kinase inhibitors) (Wang et al. 2013). Altenusin showed activity against clinical strains of *Aspergillus* fungi, and some other *Aspergillus* and *Penicillium* molds. Endophytic *Alternaria alternata* extract shows strong antifungal activity against *Staphylococcus aureus, Escherichia coli* and *Chlamydia trachomatis*. (Johann et al. 2012). Two amides called trimethynilic [also known as tetramic] and tetraethlyinic were obtained from the endophytic (or graminophilic) fungus *Bimucidula* MU34. IC50 µg/mL against plant pathogen, 1.6 mg per gm/ml, 3.2 mg/ml, and 1.6 mg/gm per millilitre of bacteria, which proved to be useful antifungal compounds, in addition to the beneficial for the fungi *Cladosporina*, *Gylezymea*, which has a MIC of 16 mg/mL, and *Gylezymea* which has a MIC of 32 mg/gm and bacteria, that has a 1.5 mg/g 3 g Tiyzin, which can be utilized as anti fungicide (Siriwach et al. 2014). An endophyte-derived phioprothine (an inhibitory one, phorbininophoreinorein compound, used for Giberella root rot), with an IC50 of 15.9 mm was found for *Pestalopsis* sp. a new PC 50-82, also from the root system of an endophytic fungus (Liu et al. 2013).

Chemical investigation of an acetonitrile fraction led to isolation of novel product 2-hydroxyethylol and monoglycolate, along with cytochalasins J and H and 5'-epialtenuene, and the mycotoxins alternariol monomethyl ether, alternariol and cytosporone C from the endophytic fungus *Phomopsis* sp. Furthermore, the antioxidant, anti-inflammatory, antifungal and cytotoxic activities of these compounds, which were isolated from *Phomopsis* sp., *were calculated. C. globosum* and *clostridium extracts* was proven to be strong antimicrobial activity against the human pathogenic bacteria such as *Salmonella* sp., *Staphylococcus* sp. and *Streptococcus* sp. (Chapla et al. 2014). The novel marine bacterium CN-328 grows a novel fungal extract made in coculture was treated with an antifungaliotic, *Potia* sp., as commonly found in this medium. It showed a strong antimicrobial potency (or activity) in the human microdilution assay against methicillin-resistant *Staphylococcus* with a MIC of 37 ng/mL and against vancomyceicm *Antophysomonas endocarditium* with a MIC of 78 ng (Cueto et al. 2001). Strong antimicrobial activity was found in Emerrla red from *Proteus fragilasens*, which had a MIC of 12.5, and bioactive collagen extracts from *S. geliferum* produced excellent antimicrobial activity against S. a that had a MIC of 12 µg (Bugni and Ireland 2004). Periconin-forming diterpenic A and B, which was created by endophytic bacteria *Klebsiella*, *Staphylococcus*, and *Salmonella*, tested in the same range (measured in millilitres per litre), had bacteriocins *Klebsiella*, *Staphylococcus*, and *Salmonell*a typhi with a 3.12–12.5 micomol per litre resulting in measurable bacterization, *Staphylococcus microclo* and *Salmonella* (Kim et al. 2004).

9.6 Animal Origin Antimicrobial Agents

Studies into the antibiotic resistance of animals (whether terrestrial or marine) have been made considerably less in comparison to their use in plants and microorganisms, as well as in smaller animals, which includes amphibians and molluscs (Wang et al. 2018). Quite a small number of the roughly 7.77 million animal species living in different habitats have been evaluated for their antimicrobial efficacy (Jiravanichpaisal et al. 2007). The incredible competency of complex fauna to thrive in difficult environments provides a road on which to discover their survival causes for decades. Because several groups of animals, such as fish, amphibians, reptiles, and rodents, are exposed to changing habitats, it is believed that they have built-in defence against pathogenic threats. Many animals, for the development of new antimicrobial drugs, are ubiquitous and have a significant and mostly underutilized supply (Wang et al. 2018; Jiravanichpaisal et al. 2007).

A novel cecropins B-derived peptide with potent antimicrobial activity against Grame-negative bacteria such as *Micrococcus luteus*, *Aerococcus viridans*, *Bacillus megaterium*, and *Bacillus subtilis*, as well as low toxicity in human cells (Wang et al. 2018). This particular compound generally found in insects was isolated from the *Musca domestica* (Wang et al. 2018). It is also the duty of them to safeguard the crayfish against in the marine world, diverse fish pathogens. The antimicrobial peptide astacidin was derived from the freshwater crayfish *Pacifastacus leniusculus*, and it has a large spectrum of bactericidal potential against both Gram-positive and Gram-negative bacteria (Jiravanichpaisal et al. 2007; Ennaas et al. 2016). In an In vitro study Ennaas et al. (2016) extracted and characterized Collagencin, a bactericidal peptide with good action against multidrug-resistant *Staphylococcus aureus* (Ennaas et al. 2016). Dermaseptin is a brand-new linear peptide with antimicrobial effects. It was first discovered in amphibian skin secretions. Dermaseptin was produced by Ying et al. (2019), and it demonstrated high antimicrobial potential against planktonic bacteria *M. luteus*, *S. aureus*, *S. epidermidis*, *S. enterica*, *Aeromonas hydrophila*, and *E. coli*, which were extracted from cystic fibrosis (CF) patients (Ying et al. 2019). Squalamine is a compound polycationic aminosterol obtained from the shark *Squalus acanthias*. Squalamine has shown to be successful against multidrug-resistant Gram-negative and Gram-positive bacteria. Squalamine's membranolytic efficacy and outstanding biocompatibility render it one of the most powerful antibiotics against nosocomial pathogens including *Acinetobacter baumannii* (Nicol et al. 2019).

Crocodiles and alligators are recognized for their longest life span, and they experience many infectious agents, toxicants, contaminants, carcinogens, etc., during their lives, but they survive under these circumstances (Leelawongtawon et al. 2010). *Siamese crocodile* (*Crocodylus siamensis*) serum has been purified into different antimicrobial agents and has been shown to be effective against so many pathogenic bacteria, including *S. typhi*, *E. coli*, *S. aureus*, *S. epidermidis*, *K. pneumoniae*, *P. aeruginosa* and *V. cholerae* (Leelawongtawon et al. 2010). Birds, such as crows, chicken, ostrich, vulture, turkeys with antimicrobial peptides that regularly feed on tainted food. Janecko et al. (2018) reported in their study a strong antimicrobial peptide with MDR activity against *Escherichia coli* and *Klebsiella* sp. isolated from *Corvus corax* in Canada (Janecko et al. 2018). Pancreatic juice, which is present in the intestine of mammals such as rabbits, guinea goats, pigs, dogs, and livestock, was discovered to have antibacterial action against *Micrococcus pyogenes*, *E. coli*, *Shigella* sp., *Salmonella* sp., *K. pneumoniae*, *Staphylococci*, and *Pseudomonus aeruginosa* (Pierzynowski et al. 1993). L-lysophillic peptides have antimicrobial efficacy against Gram-positive and Gram-negative microbes, such as *Streptococcus* and *Pseudomonas* (Szponder et al. 2018) (Table 9.1).

9.7 Mechanism of Antimicrobials

Due to the immense chemical diversity available in bioactive compounds, the mode of action of all these molecules are not well understood (Wink 2015). Numerous studies have shown that different bioactive molecules target different levels of organization, varying through cellular to individual scale and population scale, and also in some instances, including such biofilms (Wink 2015; Singh et al. 2017b; Abushaheen et al. 2020). The complexity of mode of action posed by bioactive natural products appears to become very encouraging in combating the development of multidrug resistance often seen in pathogens responsible for various infectious diseases. At the cellular level, different antimicrobial phytochemicals react with different biomolecules present at the target site and thus alter themselves chemically and physically to the degree whereby they drop their bio functionality whether partially or fully. During these interactions, bioactive natural compounds bind to different biomolecules, including such protein and nucleic acid, through various bond formation. Many of these bioactive components contain very active sites, like C=O and R-S-R′, RCO_3H, double bonds with anion configuration, and triple bonds in their framework, which can form covalent bonds with proteins and sometimes the DNA of microorganisms (Abushaheen et al. 2020; Singh et al. 2017b; Wink 2015). For example, during defined circumstances, the reactive aldehyde group of these molecules may create a Schiff base with amino/imino groups that occur in amino acid residues and protein and DNA nucleotide bases, respectively.

On the one side, a number of bioactive compounds such as polyphenols have the potential to minimize ROS generation via their strong antioxidant potential, whereas on the other hand, some bioactive compounds induce ROS generation. ROS tends to contribute significantly in the inducing of programmed cell death. After which the O_2-generated in mitochondria by aerobic cellular respiration is changed to H_2O_2 by superoxide dismutase, which then in turn reacts with ferrous ions and produces highly reactive OH-radicals. OH-radicals interact wantonly with various macromolecules, such as unsaturated fatty acids, proteins and DNA, and thus induce apoptosis induction (Le et al. 2017; Memar et al. 2018). The mechanism of the antimicrobial

Table 9.1 Antimicrobials from natural sources

S.N.	Sources	Antibiotic compound	Tested system	Biological activity response	References
(A) Plant-derived antimicrobial agents					
1.	Aloe vera (*Aloe barbadensis*)	Latex (Complex mixture)	*Staphylococcus aureus* (*S. aureus*) and *Escherichia coli* (*E. coli*)	Inhibition zone on *S. aureus* bacterium (26.33 mm) was larger than *E. coli* (21 mm)	Hilmi et al. (2019)
2.	Apple (*Malus sylvestris*)	Phloretin (Flavonoid derivative)	Oral-toxicity test in Kunming mice	The total bacterial and *Pseudomonas* sp. counts suppressed by 2 and 1.5 logarithms	Wei et al. (2020)
3.	Ashwagandha (*Withania somniferum*)	Withaferin A (Lactone)	*Leishmania donovani*-infected peritoneal macrophages and BALB/c mice	Withaferin-A (1.5 μM) reduce amastigote count in peritoneal macrophages	Chandrasekaran et al. (2017)
4.	Bael tree (*Aegle marmelos*)	Essential oil (Terpenoid)	*S. aureus, C. diphtheriae, B. cereus* and *C. diphtheriae*	Inhibition zone was 21.4 mm, 17.2 mm, lowest MIC was against *B. cereus* and *C. diphtheriae* (MIC = 125 μg/ml)	Mahomoodally et al. (2018)
5.	Barberry (*Berberis vulgaris*)	Berberine (Alkaloid)	Influenza A/FM1/1/47 (H1N1) in vivo and in vitro	Berberine strongly suppressed viral replication in A549 cells and in mouse lungs	Yan et al. (2018)

S.N.	Sources	Antibiotic compound	Tested system	Biological activity response	References
6.	Brazilian pepper tree (*Schinus terebinthifolius*)	Terebinthone (Terpenoids)	*Candida albicans*; *Candida krusei*; *Candida glabrata*; and *Candida tropicalis*	There were no significant differences regarding the different strains of *Candida* tested	Torres et al. (2016)
7.	Buchu (*Agathosma betulina*)	Essential oil (Terpenoid)	*Trichophyton rubrum* and *Trichophyton mentagrophytes*	Fungal growth index of 2.3% against *Trichophyton rubrum*	Fajinmi et al. (2019)
8.	Cascara sagrada (*Rhamnus purshiana*)	Tannins (Polyphenols)	*Staphylococcus aureus*, *Bacillus subtilis* and *Streptococcus mutans*	Biofilm formation up to 99.9% and reduced planktonic cell growth up to 10 log units relative to untreated controls	Campbell et al. (2019)
9.	Cashew (*Anacardium pulsatilla*)	Anacardic acid (Polyphenols)	*E. multilocularis*, *E. granulosuss*	Anacardic acid exerted a better efficacy against both pathogen in vitro, and in vivo compared to positive control	Yuan et al. (2019)
10.	Castor bean (*Ricinus communis*)	Volatile organic compounds (VOCs)	Plant-parasitic nematode *Meloidogyne incognit*a	Immobility (>97.3%) and death (>96.9%) of M. *incognita*	Pedroso et al. (2019)
11.	Chamomile (*Matricaria chamomilla*)	Anthemic acid (Phenolic acid)	*M. tuberculosis*, *S. typhimurium*, *S. aureus*	— —	Khameneh et al. (2019)

(continued)

Table 9.1 (continued)

S.N.	Sources	Antibiotic compound	Tested system	Biological activity response	References
12.	Chapparal (*Larrea tridentata*)	Nordihydroguaiaretic acid (Lignan)	Larvae of *Haemonchus contortus*	The effective concentration of the *L. tridentata* extract for 50% larvae mortality (EC_{50}) was 36 mg/mL	García et al. (2018)
13.	Chili peppers, Paprika (*Capsicum annuum*)	Capsaicin (Terpenoid)	Clinical MRSA strains	MIC values ranging from 8 to 256 mg/L against effluxing MRSA strains SA1199B (NorA), XU212 (TetK) and RN4220 (MsrA)	Oyedemi et al. (2019)
14.	Clove (*Syzygium aromaticum*)	Eugenol (Terpenoid)	Several proinflammatory biomarkers VCAM-1, IP-10, I-TAC, MIG	Eugenol significantly inhibited VCAM-1 and collagen III at both protein and gene expression levels	Han and Parker (2017)
15.	Coca (*Erythroxylum coca*)	Cocaine (Alkaloid)	Gram-negative and Gram-positive cocci	— —	Tsuchiya (2017)
16.	Coriander, Cilantro (*Coriandrum sativum*)	Essential oil (Terpenoid)	*Staphylococcus aureus*	Essential oils have excellent activity against both growing and stationary phase *S. aureus*	Xiao et al. (2020)

S.N.	Sources	Antibiotic compound	Tested system	Biological activity response	References
17.	Eucalyptus (*Eucalyptus globulus*)	Tannin (Polyphenol)	Various Gram-negative and Gram-positive bacteria	*E. camaldulensis* is active against many Gram-positive (0.07–1.1%) and Gram-negative bacteria (0.01–3.2%)	Sabo and Knezevic (2019)
18.	Fava bean (*Vicia faba*)	Fabatin (Thionin)	Food spoilage caused by pathogenic and nonpathogenic bacteria	— —	Kraszewska et al. (2016)
19.	Garlic (*Allium sativum*)	Allicin, Ajoene (Sulphoxide)	Inhibits DNA gyrase activity in *E. coli*	Cys_{433} in DNA gyrase subunit A (GyrA) was 6% oxidized in untreated bacteria. After allicin treatment the degree of Cys_{433} oxidation increased to 55% in susceptible Pf0–1 but only to 10% in tolerant PfAR-1.	Reiter et al. (2020)
20.	Glory lily (*Gloriosa superba*)	Colchicine (Alkaloid)	MRSA, *S. pyogenes*, and DR extended-spectrum beta-lactamase (ESBL)	The results suggest that the nanoemulsion of colchicine effective against DR bacteria, and acts by inhibiting the drug efflux mechanism of DR strains.	Krishnamoorthy et al. (2018)

(continued)

Table 9.1 (continued)

S.N.	Sources	Antibiotic compound	Tested system	Biological activity response	References
21.	Goldenseal (*Hydrastis Canadensis*)	Berberine, Hydrastine (Alkaloids)	Bacteria, *Giardia* duodenale, Trypanosomes	— —	Mandal et al. (2020)
22.	Green tea (*Camellia sinensis*)	Catechin (Alavonoid)	*Shigella*, Vibrio, *S. mutans*, Viruses	Catechins, especially epigallocatechin-3-gallate (EGCG), have antiviral effects against diverse viruses and microbes	Xu et al. (2017)
23.	Hemp (*Cannabis sativa*)	ß-Resercyclic acid (Organic acid)	*Staphylococcus aureus* (MRSA)	— —	Karas et al. (2020)
24.	Henna (*Lawsonia inermi*)	Gallic acid (Phenolic acid)	*S. aureus*	— —	Karas et al. (2020)
25.	Hops (*Humulus lupulus*)	Lupulone, Humulone (Phenolic acids)	*Bacteroides fragilis* or *Clostridium perfringens*	MIC and MBC values ranging from 15–107 μg/mL	Cermak et al. (2017)
26.	Lantana (*Lantana camara*)	Squalene, β-ionone, Caryophyllene oxide, β-caryophyllene	*Leishmania amazonensis*; *Leishmania mexicana*	IC_{50} value ranging from 12.02 ± 0.36 μM	Delgado-Altamirano et al. (2019)
27.	Lavender-cotton (*Santolina chamaecyparissus*)	Essential oil (Terpenoid)	Nine strains of bacteria and fungi	IC_{50} value ranging from 0.10/0.30 μl/ml for lavender	Mesic et al. (2021)
28.	Onion (*Allium cepa*)	Allicin (Sulphoxide)	*S. aureus*, *Candida albicans*	— —	Fuchs et al. (2018)
29.	Oregon grape (*Mahonia aquifolia*)	Berberine (Alkaloid)	*Plasmodium falciparum*, *Trypanosoma cruzi*	— —	Karas et al. (2020)
30.	Periwinkle (*Vinca minor*)	Reserpine (Alkaloid)	*S. aureus*, *Candida*	— —	Fuchs et al. (2018)

S.N.	Sources	Antibiotic compound	Tested system	Biological activity response	References
31.	Peppermint (*Mentha piperita*)	Menthol (Terpenoid)	ATCC 25922, ATCC 27853, ATCC 14452, ATCC 29213, ATCC 6633, *S. typhimurium*, *B. cereus*	MIC ranging from 0.062% to 0.5% (v/v)	Marwa et al. (2017)
32.	Quinine (*Cinchona sp.*)	Quinine (Alkaloid)	(MDR) Gram-negative bacteria in vitro and in vivo	Significant improvement in the inactivation of MDR *P. aeruginosa* and *A. baumannii* (planktonic cells and biofilms) when aBL was illuminated during Q-HCL exposure	Leanse et al. (2020)
33.	Chandra (*Rauvolfia serpentina*)	Reserpine (Alkaloid)	*Pseudomonas aeruginosa* PAO1 biofilms	Reserpine reduced biofilm formation, cell motility, virulence factor production, and QS-controlled gene expression	Parai et al. (2018)
34.	Rosemary (*Rosmarinus officinalis*)	Essential oil (Terpenoid)	*Candida* sp. and *Streptococcus pneumoniae*	The results showed that the essential oil from *Rosmarinus officinalis* was the most active on all the tested microorganisms	Akroum (2020)

(continued)

Table 9.1 (continued)

S.N.	Sources	Antibiotic compound	Tested system	Biological activity response	References
35.	Tarragon (*Artemisia dracunculus*)	Caffeic acids (Terpenoid)	Nine strains of bacteria, both gram-negative and Gram-positive	MIC ranging from 0.09 mg/mL to 47 mg/mL	Majdan et al. (2020)
36.	Thyme (*Thymus vulgaris*)	Caffeic acids (Terpenoid), thymol (phenolic alcohol)	ATCC 25922, ATCC 25923	The highest inhibition zone was obtained by Tv extracts at 20 μL loading (22 mm)	Akin and Saki (2019)
37.	Tree bard (*Podocarpus nagi*)	Totarol (Flavonol)	*P. acnes*, other gram-positive bacteria	— —	Hou et al. (2020)
38.	Tua-Tua (*Jatropha gossyphiifolia*)	— —	Different strains of bacteria and fungi	The seeds and fruits can be used against influenza and as a sedative, analgesic or anti-diarrheal agents	Wu et al. (2019)
39.	Turmeric (*Curcuma longa*)	Curcumin (Terpenoids)	*Escherichia coli* O157:H7 and *Listeria innocua*	Enhancement in antimicrobial activity reduced the time required for the inactivation of 5 log CFU mL^{-1} of *E. coli* O157:H7 from 10 min to 2 min of treatment	de Oliveira et al. (2018)
40.	Willow (*Salix alba*)	Salicin (Phenolic glucoside)	Human peripheral leucocyte cells and human hepatoma cell line HepG2	The results using trypan blue staining test showed viability decreases (viability less than 70%) for concentrations of SA extract equal and higher to 200 μg/ml.	Maistro et al. (2019)

S.N.	Sources	Antibiotic compound	Tested system	Biological activity response	References
(B) Bacterial-derived antimicrobial agents					
1.	*Streptomyces antibioticus*	Actinomycin	Antibacterial activities against *Sarcinalutea*, *Bacillus mycoides*, *Bacillus subtilis*, *E. coli*, *Aerobacter aerogenes* and *Brucella abortus*	Transcription inhibitor which preventing RNA polymerase elongation	Genilloud (2014)
2.	*Streptomyces lavendulae*	Streptothricin	Various Gram-positive and Gram-negative bacteria	A protein synthesis inhibitor with miscoding activity	Mander and Liu (2010)
3.	*Streptomyces erythraeus*	Erythromycin	*Streptococcus*, *Staphylococcus*, *Haemophilus* and *Corynebacterium*	Inhibits protein synthesis by binding to the 23S rRNA molecule of the bacterial ribosome and hence blocking the exit of the growing peptide chain	Majer et al. (1977)
4.	*Streptomyces aureofaciens*	Tetracycline	*E. coli*, *Haemophilus influenzae*, *M. tuberculosis*, *Pseudomonas aeruginosa*	Inhibits protein synthesis by binding to the 30S ribosomal subunit in the mRNA translation complex	Nelson and Levy (2011)

(continued)

Table 9.1 (continued)

S.N.	Sources	Antibiotic compound	Tested system	Biological activity response	References
5.	*Micromonospora*	Aminoglycosides	Aerobic, Gram-negative bacteria and *M. tuberculosis*	Binding specifically to the 30S ribosome of the bacteria, preventing attachment of the aminoacyltRNA to the RNA-ribosome complex	Mingeot-Leclercq et al. (1999)
6.	*Streptomyces lincolnensis*	Lincomycin	Gram-positive bacteria	Inhibits protein synthesis	Spížek and Řezanka (2004)
7.	*Amycolatopsis orientalis*	Vancomycin	Gram-positive bacteria particularly methicillin-resistant *Staphylococci*	Inhibition of peptidoglycan synthesis	Levine (2006)
8.	*Bacillus* sp.	Mersacidin	MRSA	— —	Kruszewska et al. (2004)
9.	*Bacillus amyloliquefaciens*	Surfactin, Iturin, Fengycin, Plipastatin, Bacitracin, Amylolysin	*M. luteus*, *Micrococcus roseus*, *Bacillus anthracis*, *Bacillus mycoides*, *Corynebacterium pseudodiphthericum*, *S. aureus*, *L. monocytogenes*, *B. cereus*, *Serratia marcescens*, and *Pasteurella haemolytica*, *Proteus vulgaris*	Inhibitory activity	Benitez et al. (2010); Awais et al. (2007)
10.	*Amycolatopsis rifamycinica*	Rifamycin	*Mycobacterium intracellulare* and *Mycobacterium avium*	Inhibition of bacterial DNA-dependent RNA synthesis	Lin et al. (2017)

S.N.	Sources	Antibiotic compound	Tested system	Biological activity response	References
11.	*Dactylosporangium aurantiacum*	Tiacumicins	Gram-positive bacteria	— —	Hochlowski et al. (1987)
12.	*Streptomyces roseosporus*	Daptomycin	MRSA	— —	Baltz (2009)
13.	*Streptomyces tenebrarius*	Tobramycin	Gram-positive *Staphylococcus aureus* and various Gram-negative bacteria.	<0.25 μg/mL - 92 μg/mL and 0.5 μg/mL - >512 μg/mL	Meylan et al. (2017)
14.	*Streptomyces cattleya*	Thienamycin	Have a high specificity for PBP2 of gram-positive and Gram-negative microorganisms	Hydrolysis by metallo β-lactamases and other β-lactamases	Bennett et al. (2014)
15.	*Streptomyces clavuligerus*	Clavulanic acid	Three mammalian species, hamsters, rats and cotton-top tamarin monkeys in a series of behavioural assays	Enhance the activity of antibiotics by blocking bacterial beta-lactamases	Kim et al. (2009)
16.	*Burkholderia pseudomallei*	Malleobactin A	*B. pseudomallei*	MICs in the range 0.004–0.016 μg/mL	Mima et al. (2011)
17.	*Burkholderia cepacia*	Burkholderic acid	MRSA	MIC of 0.125–0.5% (v/v)	Vasireddy et al. (2018)
18.	*Janthinobacterium lividum*	Janthinocin A	Gram positive bacteria in vitro and in vivo	— —	Johnson et al. (1990)

(continued)

Table 9.1 (continued)

S.N.	Sources	Antibiotic compound	Tested system	Biological activity response	References
19.	*Janthinobacteriumag aricidamnosum*	Jagaricin	*Candida albicans*	Jagaricin shows haemolytic effects, at which the growth of prevalent phytopathogenic fungi is inhibited	Fischer et al. (2019)
20.	*Pseudomonas syringae*	Pseudomycins	*Candida albicans*	— —	Kumari Chikkode Narayana et al. (2017)
21.	*Pseudomonas viridiflava*	Ecomycins	*Cryptococcus neoformans* and *C. Albicans*	— —	Kumari Chikkode Narayana et al. (2017)
(C) Fungal-derived antimicrobial agents					
1.	Penicillin	*Penicillium notatum*	Mainly against gram-positive bacteria	Cell wall of bacteria	Guzmán-Chávez et al. (2018)
2.	Fusidic acid	*Acremonium fusidioides*	Methicillin-resistant *S. aureus* and Vancomycin-resistant *Enterococcus faecium*	Sterol biosynthesis	Trenin (2013)
3.	Griseofulvin	*Penicillium griseofulvum*	Mainly against different species of ringworm	Griseofulvin binds to tubulin, interfering with microtubule function thus inhibiting mitosis	Valente et al. (2020)

S.N.	Sources	Antibiotic compound	Tested system	Biological activity response	References
4.	Echinocandin B	*Aspergillus nidulans*	Antifungal drug	Inhibits the synthesis of glucan, a major component of the fungal cell wall, via noncompetitive inhibition	Hu et al. (2020)
5.	Retapamulin	*Pleurotusmutilus* sp.	Gram-positive and fastidious Gram-negative pathogens as well as against mycoplasmas	Inhibit bacterial translation	Paukner and Riedl (2017)
6.	Cephalosporin	*Cephalosporium acremonium*	Gram-positive cocci, *Citrobacter*, *Enterobacter*, *Haemophilus influenzae*	Inhibit synthesis of the bacterial cell wall, causing cell lysis	Hu and Zhu (2016)
7.	*Trichothecium* sp.	Trichothecin	*S. cerevisiae*, *C. albidus* var diffluens (NCIM 3371), *C. albidus* var *diffluens* (NCIM 3372), *F. oxysporum*, *P. expansum*, *T. viride*, *P. varioti* and *A. niger*	MIC of 6.0, 20.0, 12.0, 10.0, 30.0, 40.0, 20.0 and 12.0 μg/mL respectively	Taware et al. (2015)
8.	*Phaeoacremonium* sp.	Isoaigialone B and C, aigialone	*C. cladosporioides* and *C. sphaerospermum*	7 exhibited antifungal activity, with a detection limit of 5 μg nystatin, positive control, showing a detection limit of 1 μg	Silva et al. (2017)
9.	*Curvularia* sp., strain M12	Murranolide A, Murranopyrone, Curvularin, Pyrenolide A, Modiolide A	*Phytophthora capsici*	IC_{50}: 50–100 μg/mL	Mondol et al. (2017)

(continued)

Table 9.1 (continued)

S.N.	Sources	Antibiotic compound	Tested system	Biological activity response	References
10.	*Fusarium chlamydosporium*	Fusarithioamide A	*C. albicans*	MIC 2.6 μg/mL	Ibrahim et al. (2016)
11.	*Penicillium raciborski*	Outovirin C	*F. oxysporum*, *B. cinerea*, and *V. dahlia*	Compound active against *B. cinerea* (57% inhibition) and slightly less effective against *V. dahliae* (45% inhibition)	Kajula et al. (2016)
12.	*Aspergillus sp.* KJ-9	Fonsecinone A	*G. saubinetti*, *M. grisea*, *B. cinerea*, *C. gloeosporioides* and *A. solani*	MIC range of 6.25–50 μM	Xiao et al. (2014)
13.	*Trichoderma brevicompactum* 0248	Trichodermin	*R. solani*, *B. cinereal*, *C. lindemuthianum*	EC_{50} of 0.25, 2.02 and 25.60 μg/mL respectively	Shentu et al. (2014)
14.	*Biscogniauxia mediterraneaOhu* 19B	5-methylmellein	*C. acutatum*, *C. fragariae*, *C. gloeosporioides*, *F. oxysporum*, *B. cinerea*, *P. obscurans*, and *P. viticola*	——	Silva-Hughes et al. (2015)
15.	*Cladosporium cladosporioides*	Cladosporin, Isocladosporin	*C. acutatum*, *C. fragariae*, *C. gloeosporioides* and *P. viticola*	At 30 μM compound exhibited 92.7, 90.1, 95.4, and 79.9% growth inhibition	Wang et al. (2013)
16.	*Alternaria sp.* UFMGCB 55	Altenusin	Eleven strain of *P. brasiliensis*	MIC values ranging between 1.9 and 31.2 μg/mL	Johann et al. (2012)
17.	*Bipolaris sp.* MU34	Bipolamide B	*C. cladosporioides*, *C. cucumerinum*, *S. cerevisiae*, *A. niger* and *R. oryzae*	MIC values of 16, 32, 32, 64 and 64 μg/mL, respectively	Siriwach et al. (2014)
18.	*Pestalotiopsis fici*	Ficipyrone A	*G. zeae*	IC_{50} 15.9 μM	Liu et al. (2013)

S.N.	Sources	Antibiotic compound	Tested system	Biological activity response	References
19.	*Phomopsis sp.*	Cytochalasin H	*C. cladosporioides* and *C. sphaerosphermum*	MIC 10.0 and 25.0 μg, respectively	Chapla et al. (2014)
20.	*Pestalotiopsis sp.* when cocultured with marine bacterium, strain CNJ-328	Pestalone	Methicillin-resistant *S. aureus* and vancomycin-resistant *Enterococcus faecium*	MIC = 37 ng/mL and 78 ng/mL	Cueto et al. (2001)
21.	*Emericella variecolor*	Varixanthone	*E. coli*	MIC values of 12.5 μg mL^{-1}	Bugni and Ireland (2004)
22.	*Periconia sp.*	Periconicins A and B	*Bacillus subtilis*, *Staphylococcus aureus*, *Klebsiella pneumoniae*, and *Salmonella typhimurium*	MIC in the range of 3.12–12.5 μg/mL	Kim et al. (2004)
(D)animal origin antimicrobial agents					
1.	*Musca domestica*	Cecropins	*M. luteus*, *Aerococcus viridians*, *Bacillus megaterium*, *B. subtilis*	Cecropin DH has potential as a therapeutic agent for both antibacterial and anti-inflammatory applications	Wang et al. (2018)
2.	Crayfish	Astacidin	Gram-positive and Gram-negative bacteria	Bactericidal activities	Jiravanichpaisal et al. (2007)
3.	*Larimichthys crocea*	Collagencin	*S. aureus*	The peptide completely inhibited the growth of *S. aureus* at 1.88 mM and non-toxic at 470 μM	Ennaas et al. (2016)

(continued)

Table 9.1 (continued)

S.N.	Sources	Antibiotic compound	Tested system	Biological activity response	References
4.	*Phyllomedusa distincta*	Dermaseptin	*M. luteus*, *S. aureus*, *S. epidermidis*, *S. enterica*, *Aeromonashydrophila* and *E. coli*	Rapidly killed planktonic bacteria isolated from cystic fibrosis (CF) patient	Ying et al. (2019)
5.	*Squalus acanthias*	Squalamine	Anti-persister activity against *Acinetobacterbaumannii*	Squalamine at 100 mg/L (below its haemolytic concentration) was able to kill dormant cells	Nicol et al. (2019)
6.	*Crocodylus siamensis*	Antimicrobial peptides	ATCC-registered strains of nine bacterial species and two fungal species	MIC was in the range of 25.00–100.00 mg/ml	Leelawongtawon et al. (2010)
7.	*Corvus corax*	Peptides	*Escherichia coli* and *Klebsiella*spp	Specific target site not determined	Janecko et al. (2018)
8.	Pancreatic juice of rabbit, Guinea pig, rats, pig, sheep and cattle	Heat-stable proteinaceous substance with molecular weight of <4000	*Micrococcus pyogenes*, *E. coli*, *Shigella sp.*, *Salmonella sp.*, *K. pneumoniae*, *Staphylococci* and *Pseudomonus aeruginosa*	The antibacterial activity remained unchanged after heating to 65 degrees C and upon dilution to 1:10	Pierzynowski et al. (1993)
9.	Rabbit granulocytes	Antimicrobial peptides	Gram-positive and Gram-negative bacteria	These peptides exhibit microbicidal activity due to increased acidity and ionic strength	Szponder et al. (2018)

agent is primarily due to two pathways, namely chemical interaction with the synthesis or function of essential bacterial components and/or circumvention of traditional antibacterial resistance mechanisms. Multiple targets for antimicrobial agents include microbial protein biosynthesis; microbial cell-wall biosynthesis; microbial cell membrane destruction; microbial DNA replication and repair; and metabolic pathway inhibition. Cell wall is an ultra-dynamic structure in some microbes, such as fungi and bacteria, which protects the body from environmental osmotic shocks which are also essential for the distinctive phenotypes of different species. Any alteration triggered by an antimicrobial triggering an organizational or functional disruption of the cell wall will lead to the death of the microorganism (Timofeeva and Kleshcheva 2011; Le et al. 2017; Memar et al. 2018) (Fig. 9.3).

In the case of microbial antibiotics such as penicillin which inhibit cell synthesis, the mechanism of cell wall disintegration is well understood. Two types of family enzymes, including transglycosylases and transpeptidases, have critical roles in the creation of this sheet, while their functionality has been defined previously. Bifunctional enzymes containing both the transpeptidase and transglycosylase

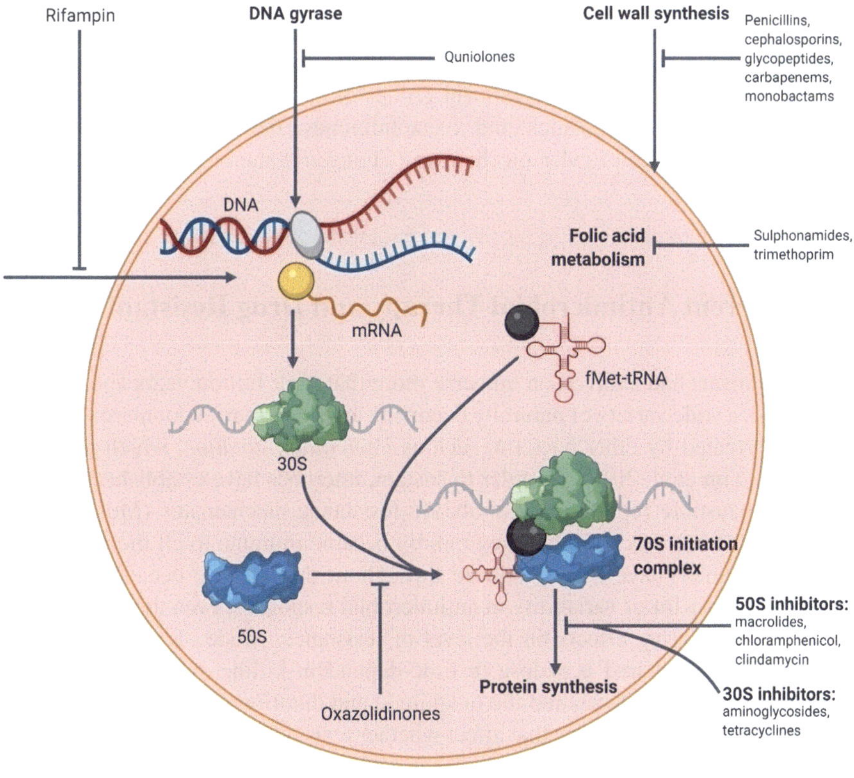

Fig. 9.3 Antimicrobial drug target; in microbes, there can be five major antimicrobial drugs targets: cell-wall synthesis, DNA gyrase, metabolic enzymes, RNA polymerase directed by DNA, and protein synthesis

domains are appropriate targets for bactericidal antibiotics such as penicillins and cephalosporins (Memar et al. 2018; Le et al. 2017; Timofeeva and Kleshcheva 2011). The glycopeptide antibiotics group, like vancomycin, has often been identified to attack the peptidoglycan layer in some other way inside the cell-wall assembly. These antibiotics are capable of binding to the peptide substrate of the peptidoglycan layer, thus preventing enzyme reactions from occurring. However, the net result is very similar, reducing peptidoglycan cross-linkage and thereby weakening the cell wall (Wink 2015; Singh et al. 2017b).

The cell membrane is an essential element of the lipid bilayers that includes integrated extrinsic and intrinsic proteins that serve the roles of enzymes, signalling protein and transport proteins. Owing to their lipophilic nature or bonding to some particular membrane part, numerous bioactive compounds can trigger membrane degradation, leading to loss of membrane stability and functionality (Ibrahim et al. 2000; Chongsiriwatana et al. 2008). Multiple antibiotics including polymyxins may bound to the lipid a constituent of lipopolysaccharide and thus cause substantial modifications through phospholipid interchange, which might lead in osmotic disturbance and, eventually, lead to microbial death. In the case of microbial biosynthesis, there seem to be a significant number of molecular steps involved in the initiation, elongation and termination of microbial ribosome protein assembly. Inhibiting protein synthesis by targeting ribosomal subunits is also an efficient way to fight microbial infections. Significant groups of antibiotics, such as macrolides, tetracycline's, aminoglycosides and oxazolidinones, demonstrate antimicrobial activity through this particular mechanism (Chongsiriwatana et al. 2008; Ibrahim et al. 2000).

9.8 Current Antimicrobial Therapy and Drug Resistant

Microorganisms had evolved on universe more than four billion years ago. During that period, a wide variety of naturally occurring antibiotics are encountered, including those created by other bacteria, such as *Penicillium notatum*, which produces penicillin (Yim et al. 2017). In order to sustain, microbes have established a seemingly inexhaustible repertoire of antibiotic resistance mechanisms (Mulani et al. 2019). This is not shocking that they rapidly became immune to all the antimicrobial agents which have been produced throughout the last five decades. For this reason, there is a lot of variability in antimicrobial responses; even the best of antibiotics have differing effects on the level of resistance. Mode of operation, if an antimicrobial compound is a dose or time-dependent killing agent, effectiveness against pathogenic bacteria, and the magnitude and duration of the available serum concentration are all variables that affect whether resistance arises (Petchiappan and Chatterji 2017). For example, the resistance of β-lactam within streptococci class a still has not been established. But at the other hand, certain antimicrobial agents, like rifampicin, are easily selected for resistance. Antimicrobials that target single enzymes, such as rifampicin, are thought to be the most resistant to resistance production, while agents like penicillin, which irreversibly inactivates several targets, may build resistance more steadily. Because pathogens have been exposed to natural antibiotics such as ß-lactams and macrolides in the environment, it is rational to believe that susceptibility determinants to natural products have formed and spread horizontally. While it was anticipated that resistance to synthetic antimicrobial agents such as fluoroquinolones and linezolid will be sluggish to develop, resistance to synthetic agents developed rather rapidly. It seems that if an antibacterial agent is widely employed in the human community, tolerance can develop rapidly, at least in some microbe populations (Buehrle et al. 2017; Laws et al. 2019).

The development and dissemination of resistant pathogens is a significant concern as the main trigger of antimicrobial drug resistance (Juárez-Verdayes et al. 2012; Iino et al. 2012). The pathways entail modification of drug targets or enzymatic inactivation of antimicrobial agents like ß-lactams, macrolides, tetracyclines and fluoroquinolones. Many antibiotics were discovered to be efflux pump substrates, resulting in medication extrusion from cells. Problem becomes more serious due to intensive use of antibiotics which result in clonal selection of efflux pump overexpressing strains for which chemotherapeutic agents are good substrate. Moreover, hyper expression of naturally occurring multidrug efflux transporters plays an ubiquitous type of resistant element which could use chemical energy (e.g. ATP, Na + or H+ gradients) to expel a set of dissimilar molecules or antibiotics from the cytoplasm through an antiport mechanism (Campion et al. 2004; Stavri et al. 2007; Abdali et al. 2017). Protein architecture has distinguished between five families, i.e. Multidrug, Multid, MATE, ABC, the resistance-family, and the main facilitator superfamily. Secondaryly these have been studied at present in significant amounts including NorA, NorB, MdeA, and LmrP pump. NorA among these has been found overexpressed in nearly half of resistant clinical isolates as compared to other efflux pumps (Abdali et al. 2017; Jang 2016). As a consequence of the intense battle against MDR pathogens, efflux pump inhibitors (EPIs) are potentially effective as adjunctive therapies with an antibiotic to obstruct the activity of such efflux proteins and could be a better approach to improve antibacterial potency at low concentration and help in decreased virulence of bacterial infection (Patkari and Mehra 2013; German et al. 2008). Capsaicin is shown to alter fluoroquinolone pump tolerance in clinical isolates of *Staphylococcus aureus*. Similarly, polystyryladines, for example, dihydropanamidic polyamine esters with amino acid esters, have recently been discovered as antibacterial agents against NorA-overexp bacterium strains. It's worth exploring whether these drug-intermediate infections can even be treated with non-EPIs, which could have new therapeutic benefit for obsolete antibiotics (Fig. 9.4).

Ampuse from available drugs may come from the organism's intrinsic properties, or due to genetic transformation. Resistance is likely to occur in the commensal microflora as well, and the more likely it is (Buehrle et al. 2017; Laws et al. 2019).

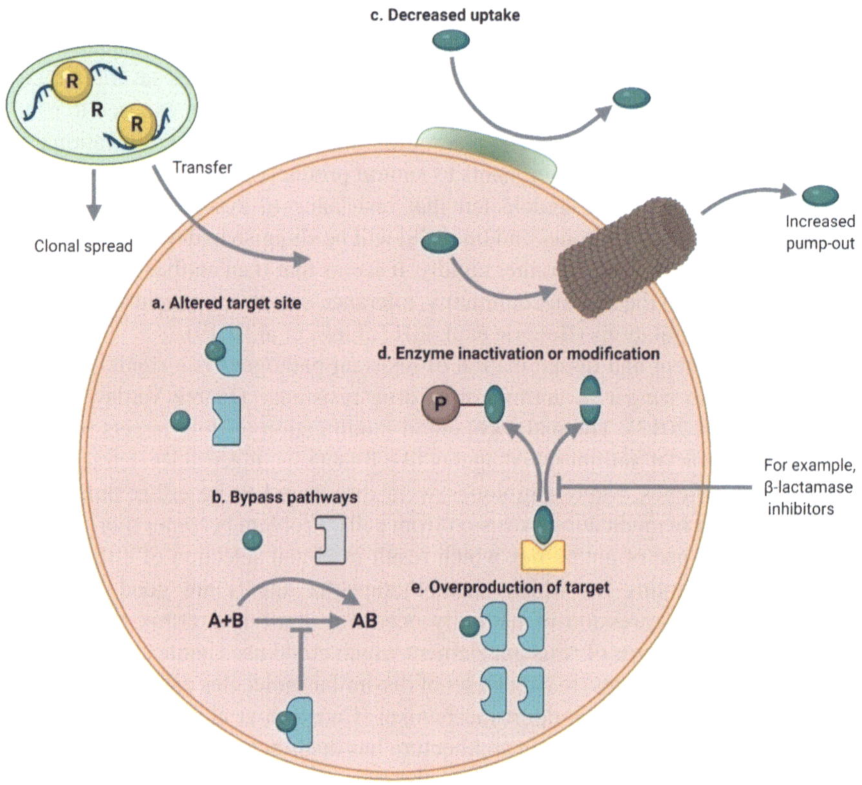

Fig. 9.4 Mechanisms of genetic resistance to antimicrobial agent

9.9 Future Opportunities

There have been a change in the way the drugs/lead molecules used in experimental trials to clinical studies, as researchers began to use advancement in the techniques of synthesizing them from the results of in vitro study. Bioavailability is a challenge because certain bacteria can not only move through the skin but also because of tissue penetration, so when using bioactive products is mixed with the natural antimicrobials. According to that theory, phenolic compounds are said to profoundly influence the body's ability to enter both the liver and the blood. A significant challenge to effective therapy of pathogenic microorganisms has been the emergence of antibiotic-resistant microorganisms. As of now, there is an urgent need to establish a new drug resistance strategy. Bioactive moieties with different chemical structures and modes of action are promising therapeutic platforms for the discovery of novel bioactive compounds in the years to come. However, more study should be done to properly completely comprehend mechanisms as well as the pharmacokinetic and pharmacodynamics characteristics of the bioactive compounds. Although conducting more research on combinations of antibiotics to improve their duration of action would further the duration of these compounds, This class of multidrug-resistant bacteria is a true to life origins, so more research on them must also be performed to reduce resistance in normal flora. Currently, checks are needed to ensure the efficacy of any pathogens that are still in the sample. Since many antibiotics in modern treatments lack specificity, this could yield medications that are less effective when combined with the conventional antimicrobials that can mitigate environmental pathogens that do not have established resistance to these days. If these potential advantages are combined, then a more compliant patient-friendly and cost-conscious approach to antibiotic therapy is taken, such resistance could be prevented, longer durations of use could be achieved, and so less resistance to medications could be developed.

References

Abdali N, Parks JM, Haynes KM, Chaney JL, Green AT, Wolloscheck D, Walker JK, Rybenkov VV, Baudry J, Smith JC (2017) Reviving antibiotics: efflux pump inhibitors that interact with AcrA, a membrane fusion protein of the AcrAB-TolC multidrug efflux pump. ACS infectious diseases 3(1):89–98

Abdel-Razek AS, El-Naggar ME, Allam A, Morsy OM, Othman SI (2020) Microbial natural products in drug discovery. PRO 8(4):470

Abreu AC, McBain AJ, Simoes M (2012) Plants as sources of new antimicrobials and resistance-modifying agents. Nat Prod Rep 29(9):1007–1021

Abushaheen MA, Fatani AJ, Alosaimi M, Mansy W, George M, Acharya S, Rathod S, Divakar DD, Jhugroo C, Vellappally S (2020) Antimicrobial resistance, mechanisms and its clinical significance. Dis Mon 66(6):100971

Aghayan SS, Mogadam HK, Fazli M, Darban-Sarokhalil D, Khoramrooz SS, Jabalameli F, Yaslianifard S, Mirzaii M (2017) The effects of berberine and palmatine on efflux pumps inhibition with different gene patterns in Pseudomonas aeruginosa isolated from burn infections. Avicenna J Med Biotechnol 9(1):2

Akin M, Saki N (2019) Antimicrobial, DPPH scavenging and tyrosinase inhibitory activities of Thymus vulgaris, Helichrysum arenarium and Rosa damascena mill ethanol extracts by using TLC bioautography and chemical screening methods. J Liquid Chromat Relat Technol 42(7–8):204–216

Akroum S (2020) Antimicrobial activity of Rosmarinus officinalis and Zingiber officinale extracts on the species of the genus Candida and on streptococcus pneumonia. In: Annales pharmaceutiques francaises, 2020

Alter T, Reich F (2021) Management strategies for prevention of campylobacter infections through the poultry food chain: a European perspective. Curr Top Microbiol Immunol 431:79–102

Aminov RI (2010) A brief history of the antibiotic era: lessons learned and challenges for the future. Front Microbiol 1:134

Atanasov AG, Zotchev SB, Dirsch VM, Supuran CT (2021) Natural products in drug discovery: advances and opportunities. Nat Rev Drug Discov:1–17

Awais M, Shah AA, Hameed A, Hasan F (2007) Isolation, identification and optimization of bacitracin produced by bacillus sp. Pak J Bot 39(4):1303

Baltz RH (2009) Daptomycin: mechanisms of action and resistance, and biosynthetic engineering. Curr Opin Chem Biol 13(2):144–151

Benitez LB, Velho RV, Lisboa MP, da Costa Medina LF, Brandelli A (2010) Isolation and characterization of antifungal peptides produced by bacillus amyloliquefaciens LBM5006. J Microbiol 48(6):791–797

Bennett JE, Dolin R, Blaser MJ (2014) Mandell, Douglas, and bennett's principles and practice of infectious diseases: 2-volume set, vol 2. Elsevier Health Sciences
Berkley JA (2021) Bacterial infections and nutrition: a primer. In: Nutrition and infectious diseases. Springer, pp. 113–131
Buehrle DJ, Shields RK, Clarke LG, Potoski BA, Clancy CJ, Nguyen MH (2017) Carbapenem-resistant Pseudomonas aeruginosa bacteremia: risk factors for mortality and microbiologic treatment failure. Antimicrob Agents Chemother 61(1)
Bugni TS, Ireland CM (2004) Marine-derived fungi: a chemically and biologically diverse group of microorganisms. Nat Prod Rep 21(1):143–163
Campbell M, Zhao W, Fathi R, Mihreteab M, Gilbert ES (2019) Rhamnus prinoides (gesho): a source of diverse anti-biofilm activity. J Ethnopharmacol 241:111955
Campion JJ, McNamara PJ, Evans ME (2004) Evolution of ciprofloxacin-resistant Staphylococcus aureus in in vitro pharmacokinetic environments. Antimicrob Agents Chemother 48(12):4733–4744
Carpena M, Nuñez-Estevez B, Soria-Lopez A, Garcia-Oliveira P, Prieto MA (2021) Essential oils and their application on active packaging systems: a review. Resources 10(1):7
Cermak P, Olsovska J, Mikyska A, Dusek M, Kadleckova Z, Vanicek J, Nyc O, Sigler K, Bostikova V, Bostik P (2017) Strong antimicrobial activity of xanthohumol and other derivatives from hops (Humulus lupulus L.) on gut anaerobic bacteria. APMIS 125(11):1033–1038
Chandrasekaran S, Veronica J, Sundar S, Maurya R (2017) Alcoholic fractions F5 and F6 from Withania somnifera leaves show a potent Antileishmanial and immunomodulatory activities to control experimental visceral Leishmaniasis. Front Med 4:55
Chapla VM, Zeraik ML, Ximenes VF, Zanardi LM, Lopes MN, Cavalheiro AJ, Silva DHS, Young MCM, Fonseca LMd, Bolzani VS (2014) Bioactive secondary metabolites from Phomopsis sp., an endophytic fungus from Senna spectabilis. Molecules 19 (5):6597–6608
Chongsiriwatana NP, Patch JA, Czyzewski AM, Dohm MT, Ivankin A, Gidalevitz D, Zuckermann RN, Barron AE (2008) Peptoids that mimic the structure, function, and mechanism of helical antimicrobial peptides. Proc Natl Acad Sci 105(8):2794–2799
Chouhan S, Sharma K, Guleria S (2017) Antimicrobial activity of some essential oils—present status and future perspectives. Medicines 4(3):58
Cueto M, Jensen PR, Kauffman C, Fenical W, Lobkovsky E, Clardy J (2001) Pestalone, a new antibiotic produced by a marine fungus in response to bacterial challenge. J Nat Prod 64(11):1444–1446
Delgado-Altamirano R, López-Palma RI, Monzote L, Delgado-Domínguez J, Becker I, Rivero-Cruz JF, Esturau-Escofet N, Vázquez-Landaverde PA, Rojas-Molina A (2019) Chemical constituents with leishmanicidal activity from a pink-yellow cultivar of Lantana camara var. aculeata (L.) collected in Central Mexico. Int J Mol Sci 20(4):872
Desam NR, Al-Rajab AJ (2021) The importance of natural products in cosmetics. Bioactive Natural Products for Pharmaceutical Applications. Springer, In, pp 643–685
Dwibedi V, Rath SK, Prakash R, Saxena S (2021) Response surface statistical optimization of fermentation parameters for resveratrol production by the endophytic fungus Arcopilus aureus and its tyrosinase inhibitory activity. Biotechnol Lett 43(3):627–644
Ennaas N, Hammami R, Gomaa A, Bédard F, Biron É, Subirade M, Beaulieu L, Fliss I (2016) Collagencin, an antibacterial peptide from fish collagen: activity, structure and interaction dynamics with membrane. Biochem Biophys Res Commun 473(2):642–647
Fajinmi O, Kulkarni M, Benická S, Zeljković SĆ, Doležal K, Tarkowski P, Finnie J, Van Staden J (2019) Antifungal activity of the volatiles of Agathosma betulina and Coleonema album commercial essential oil and their effect on the morphology of fungal strains Trichophyton rubrum and T. mentagrophytes. S Afr J Bot 122:492–497
Fischer D, Gessner G, Fill TP, Barnett R, Tron K, Dornblut K, Kloss F, Stallforth P, Hube B, Heinemann SH (2019) Disruption of membrane integrity by the bacterium-derived antifungal jagaricin. Antimicrob Agents Chemother 63(9)

Fuchs AL, Weaver AJ Jr, Tripet BP, Ammons MCB, Teintze M, Copié V (2018) Characterization of the antibacterial activity of Bald's eyesalve against drug resistant Staphylococcus aureus and Pseudomonas aeruginosa. PLoS One 13(11):e0208108

García JE, Gómez L, Mendoza-de-Gives P, Rivera-Corona JL, Millán-Orozco J, Ascacio JA, Medina MA, Mellado M (2018) Anthelmintic efficacy of hydro-methanolic extracts of Larrea tridentata against larvae of Haemonchus contortus. Trop Anim Health Prod 50(5):1099–1105

Genilloud O (2014) The re-emerging role of microbial natural products in antibiotic discovery. Antonie Van Leeuwenhoek 106(1):173–188

German N, Kaatz GW, Kerns RJ (2008) Synthesis and evaluation of PSSRI-based inhibitors of Staphylococcus aureus multidrug efflux pumps. Bioorg Med Chem Lett 18(4):1368–1373

Ghani U, Rehman HU, Ghani AM, Gerdazi SA, Kamil M, Ullah W (2019) Aloe vera plant products as antimicrobial agents

Gouda MA, Abu-Hashem AA, Salem MA, Helal MH, Al-Ghorbani M, Hamama WS (2020) Recent progress on coumarin scaffold-based anti-microbial agents (part III). J Heterocyclic Chem 57(11):3784–3817

Gould K (2016) Antibiotics: from prehistory to the present day. J Antimicrob Chemother 71(3):572–575

Gupta A, Pandey AK (2020) Antibacterial lead compounds and their targets for drug development. In: Phytochemicals as lead compounds for new drug discovery. Elsevier, pp. 275–292

Gutiérrez-del-Río I, Fernández J, Lombó F (2018) Plant nutraceuticals as antimicrobial agents in food preservation: Terpenoids, polyphenols and thiols. Int J Antimicrob Agents 52(3):309–315

Guzmán-Chávez F, Zwahlen RD, Bovenberg RA, Driessen AJ (2018) Engineering of the filamentous fungus Penicillium chrysogenum as cell factory for natural products. Front Microbiol 9:2768

Han X, Parker TL (2017) Anti-inflammatory activity of clove (Eugenia caryophyllata) essential oil in human dermal fibroblasts. Pharm Biol 55(1):1619–1622

Hilmi B, Bustami Y, Trongsatitkul T, Hamid ZAA (2019) The effect of natural antimicrobial agents on Staphylococcus aureus and Escherichia coli growth. J Phys Sci 30:55–63

Hochlowski J, Swanson S, Ranfranz L, Whittern D, Buko A, McAlpine J (1987) Tiacumicins, a novel complex of 18-membered macrolides II. Isolation and structure determination. J Antibiot 40(5):575–588

Hochstein F, Stephens C, Conover L, Regna P, Pasternack R, Gordon P, Pilgrim F, Brunings K, Woodward R (1953) The structure of terramycin1, 2. J Am Chem Soc 75(22):5455–5475

Hou Y, Zhang X, Wang C, Guo M (2020) Formulation and functional properties of whey protein-based tissue adhesive using Totarol as an antimicrobial agent. PRO 8(4):496

Hu Y, Zhu B (2016) Study on genetic engineering of Acremonium chrysogenum, the cephalosporin C producer. Synth Syst Biotechnol 1(3):143–149

Hu Z-C, Li W-J, Zou S-P, Niu K, Zheng Y-G (2020) Mutagenesis of echinocandin B overproducing aspergillus nidulans capable of using starch as main carbon source. Prep Biochem Biotechnol 50(8):745–752

Hyldgaard M, Mygind T, Meyer RL (2012) Essential oils in food preservation: mode of action, synergies, and interactions with food matrix components. Front Microbiol 3:12

Ibrahim HR, Sugimoto Y, Aoki T (2000) Ovotransferrin antimicrobial peptide (OTAP-92) kills bacteria through a membrane damage mechanism. Biochimica et Biophysica Acta (BBA)-General Subjects 1523(2–3):196–205

Ibrahim SRM, Elkhayat ES, Mohamed GAA, Fat'hi SM, Ross SA (2016) Fusarithioamide a, a new antimicrobial and cytotoxic benzamide derivative from the endophytic fungus fusarium chlamydosporium. Biochem Biophys Res Commun 479(2):211–216

Iino R, Hayama K, Amezawa H, Sakakihara S, Kim SH, Matsumono Y, Nishino K, Yamaguchi A, Noji H (2012) A single-cell drug efflux assay in bacteria by using a directly accessible femtoliter droplet array. Lab Chip 12(20):3923–3929

Janecko N, Halova D, Jamborova I, Papousek I, Masarikova M, Dolejska M, Literak I (2018) Occurrence of plasmid-mediated quinolone resistance genes in Escherichia coli and Klebsiella

spp. recovered from Corvus brachyrhynchos and Corvus corax roosting in Canada. Lett Appl Microbiol 67(2):130–135
Jang S (2016) Multidrug efflux pumps in Staphylococcus aureus and their clinical implications. J Microbiol 54(1):1–8
Jin J, Zhang J, Guo N, Feng H, Li L, Liang J, Sun K, Wu X, Wang X, Liu M (2011) The plant alkaloid piperine as a potential inhibitor of ethidium bromide efflux in mycobacterium smegmatis. J Med Microbiol 60(2):223–229
Jiravanichpaisal P, Lee SY, Kim Y-A, Andrén T, Söderhäll I (2007) Antibacterial peptides in hemocytes and hematopoietic tissue from freshwater crayfish Pacifastacus leniusculus: characterization and expression pattern. Develop Comparat Immunol 31(5):441–455
Johann S, Rosa LH, Rosa CA, Perez P, Cisalpino PS, Zani CL, Cota BB (2012) Antifungal activity of altenusin isolated from the endophytic fungus Alternaria sp. against the pathogenic fungus Paracoccidioides brasiliensis. Revista iberoamericana de micologia 29(4):205–209
Johnson JH, Tymiak AA, Bolgar MS (1990) Janthinocins a, B and C, novel peptide lactone antibiotics produced by Janthinobacterium lividum. J Antibiot 43(8):920–930
Ju J, Xie Y, Yu H, Guo Y, Cheng Y, Qian H, Yao W (2020) Synergistic interactions of plant essential oils with antimicrobial agents: a new antimicrobial therapy. Crit Rev Food Sci Nutr:1–12
Juárez-Verdayes MA, Parra-Ortega B, Hernández-Rodríguez C, Betanzos-Cabrera G, Rodríguez-Martínez S, Cancino-Diaz ME, Cancino-Diaz JC (2012) Identification and expression of nor efflux family genes in Staphylococcus epidermidis that act against gatifloxacin. Microb Pathog 52(6):318–325
Kajula M, Ward JM, Turpeinen A, Tejesvi MV, Hokkanen J, Tolonen A, Häkkänen H, Picart P, Ihalainen J, Sahl H-G (2016) Bridged epipolythiodiketopiperazines from Penicillium raciborskii, an endophytic fungus of Rhododendron tomentosum Harmaja. J Nat Prod 79(4):685–690
Karas JA, Wong LJ, Paulin OK, Mazeh AC, Hussein MH, Li J, Velkov T (2020) The antimicrobial activity of cannabinoids. Antibiotics 9(7):406
Kawacka I, Olejnik-Schmidt A, Schmidt M, Sip A (2021) Natural plant-derived chemical compounds as listeria monocytogenes inhibitors in vitro and in food model systems. Pathogens 10(1):12
Khameneh B, Iranshahy M, Soheili V, Bazzaz BSF (2019) Review on plant antimicrobials: a mechanistic viewpoint. Antimicrob Resist Infect Control 8(1):1–28
Kim S, Shin D-S, Lee T, Oh K-B (2004) Periconicins, two new fusicoccane diterpenes produced by an endophytic fungus Periconia sp. with antibacterial activity. J Nat Prod 67(3):448–450
Kim DJ, King JA, Zuccarelli L, Ferris CF, Koppel GA, Snowdon CT, Ahn CH (2009) Clavulanic acid: a competitive inhibitor of beta-lactamases with novel anxiolytic-like activity and minimal side effects. Pharmacol Biochem Behav 93(2):112–120
Koehn FE, Carter GT (2005) The evolving role of natural products in drug discovery. Nat Rev Drug Discov 4(3):206–220
Kraszewska J, Beckett MC, James TC, Bond U (2016) Comparative analysis of the antimicrobial activities of plant defensin-like and ultrashort peptides against food-spoiling bacteria. Appl Environ Microbiol 82(14):4288–4298
Krishnamoorthy R, Athinarayanan J, Periasamy VS, Adisa AR, Al-Shuniaber MA, Gassem MA, Alshatwi AA (2018) Antimicrobial activity of nanoemulsion on drug-resistant bacterial pathogens. Microb Pathog 120:85–96
Kruszewska D, Sahl H-G, Bierbaum G, Pag U, Hynes SO, Ljungh Å (2004) Mersacidin eradicates methicillin-resistant Staphylococcus aureus (MRSA) in a mouse rhinitis model. J Antimicrob Chemother 54(3):648–653
Kuhad R (2012) Microbes and their role in sustainable development. Springer, Cham
Kumari Chikkode Narayana K, Bai Aswathanarayan J, Rai Vittal R (2017) Endophytic peptides-a source of therapeutic agents. Curr Protein Peptide Sci 18(3):284–290
Küpeli Akkol E, Genç Y, Karpuz B, Sobarzo-Sánchez E, Capasso R (2020) Coumarins and coumarin-related compounds in pharmacotherapy of cancer. Cancers 12(7):1959

Lambert PA (2005) Bacterial resistance to antibiotics: modified target sites. Adv Drug Deliv Rev 57(10):1471–1485
Lamut A, Peterlin Mašič L, Kikelj D, Tomašič T (2019) Efflux pump inhibitors of clinically relevant multidrug resistant bacteria. Med Res Rev 39(6):2460–2504
Laws M, Shaaban A, Rahman KM (2019) Antibiotic resistance breakers: current approaches and future directions. FEMS Microbiol Rev 43(5):490–516
Le C-F, Fang C-M, Sekaran SD (2017) Intracellular targeting mechanisms by antimicrobial peptides. Antimicrob Agents Chemother 61(4)
Leanse LG, Dong P-T, Goh XS, Lu M, Cheng J-X, Hooper DC, Dai T (2020) Quinine enhances photo-inactivation of gram-negative bacteria. J Infect Dis 221(4):618–626
Leelawongtawon R, Siruntawineti J, Chaeychomsri W, Sattaponpan C (2010) Antibacterial and antifungal activities from Siamese crocodile blood. J Med Assoc Thail 93(Suppl 7):S58–S64
Leontiev R, Hohaus N, Jacob C, Gruhlke MC, Slusarenko AJ (2018) A comparison of the antibacterial and antifungal activities of thiosulfinate analogues of allicin. Sci Rep 8(1):1–19
Leontopoulos S, Skenderidis P, Kalorizou H, Petrotos K (2017) Bioactivity potential of polyphenolic compounds in human health and their effectiveness against various food borne and plant pathogens. A review. J Food Biosyst Eng 7:1–19
Levine DP (2006) Vancomycin: a history. Clin Infect Dis. 42 (Suppl 1):S5-S12
Lima M, de Sousa CP, Fernandez-Prada C, Harel J, Dubreuil J, De Souza E (2019) A review of the current evidence of fruit phenolic compounds as potential antimicrobials against pathogenic bacteria. Microb Pathog 130:259–270
Lin S-W, Lin C-J, Yang J-C (2017) Rifamycin SV MMX for the treatment of traveler's diarrhea. Expert Opin Pharmacother 18(12):1269–1277
Liu S, Liu X, Guo L, Che Y, Liu L (2013) 2H-Pyran-2-one and 2H-Furan-2-one derivatives from the plant endophytic fungus Pestalotiopsis fici. Chem Biodivers 10(11):2007–2013
Mahmood H, Jamshidi S, Mark Sutton J, Rahman K (2016) Current advances in developing inhibitors of bacterial multidrug efflux pumps. Curr Med Chem 23(10):1062–1081
Mahomoodally MF, Mollica A, Stefanucci A, Aumeeruddy MZ, Poorneeka R, Zengin G (2018) Volatile components, pharmacological profile, and computational studies of essential oil from Aegle marmelos (Bael) leaves: a functional approach. Ind Crop Prod 126:13–21
Maistro EL, Terrazzas PM, Perazzo FF, Gaivão IONDM, Sawaya ACHF, Rosa PCP (2019) Salix alba (white willow) medicinal plant presents genotoxic effects in human cultured leukocytes. J Toxic Environ Health A 82(23–24):1223–1234
Majdan M, Kiss AK, Hałasa R, Granica S, Osińska E, Czerwińska ME (2020) Inhibition of neutrophil functions and antibacterial effects of Tarragon (Artemisia dracunculus L) infusion—phytochemical characterization. Front Pharmacol:11
Majer J, Martin JR, Egan RS, Corcoran JW (1977) Antibiotic glycosides. 8. Erythromycin D, a new macrolide antibiotic. J Am Chem Soc 99(5):1620–1622
Makroo H, Naqash F, Fayaz S (2021) Antimicrobial compounds to improve foods' shelf life. Food Waste Recovery:461–482
Mandal SK, Maji AK, Mishra SK, Ishfaq PM, Devkota HP, Silva AS, Das N (2020) Goldenseal (Hydrastis canadensis L.) and its active constituents: a critical review of their efficacy and toxicological issues. Pharmacological research:105085
Mander L, Liu H-W (2010) Comprehensive natural products II: chemistry and biology, vol 1. Elsevier
Manzoor A, Dar IH, Bhat SA, Ahmad S (2020) Flavonoids: health benefits and their potential use in food systems. In: Functional food products and sustainable health. Springer, pp. 235–256
Martelli G, Giacomini D (2018) Antibacterial and antioxidant activities for natural and synthetic dual-active compounds. Eur J Med Chem 158:91–105
Marwa C, Fikri-Benbrahim K, Ou-Yahia D, Farah A (2017) African peppermint (Mentha piperita) from Morocco: chemical composition and antimicrobial properties of essential oil. J Adv Pharm Technol Res 8(3):86

Memar MY, Ghotaslou R, Samiei M, Adibkia K (2018) Antimicrobial use of reactive oxygen therapy: current insights. Infect Drug Resist 11:567

Mesic A, Mahmutović-Dizdarević I, Tahirović E, Durmišević I, Eminovic I, Jerković-Mujkić A, Bešta-Gajević R (2021) Evaluation of toxicological and antimicrobial activity of lavender and immortelle essential oils. Drug Chem Toxicol 44(2):190–197

Meylan S, Porter CB, Yang JH, Belenky P, Gutierrez A, Lobritz MA, Park J, Kim SH, Moskowitz SM, Collins JJ (2017) Carbon sources tune antibiotic susceptibility in Pseudomonas aeruginosa via tricarboxylic acid cycle control. Cell Chem Biol 24(2):195–206

Mima T, Kvitko BH, Rholl DA, Page MG, Desarbre E, Schweizer HP (2011) In vitro activity of BAL30072 against Burkholderia pseudomallei. Int J Antimicrob Agents 38(2):157–159

Mingeot-Leclercq M-P, Glupczynski Y, Tulkens PM (1999) Aminoglycosides: activity and resistance. Antimicrob Agents Chemother 43(4):727–737

Moloney MG (2016) Natural products as a source for novel antibiotics. Trends Pharmacol Sci 37(8):689–701

Mondol MAM, Farthouse J, Islam MT, Schüffler A, Laatsch H (2017) Metabolites from the endophytic fungus Curvularia sp. M12 act as motility inhibitors against Phytophthora capsici zoospores. J Nat Prod 80(2):347–355

Moumni M, Romanazzi G, Najar B, Pistelli L, Ben Amara H, Mezrioui K, Karous O, Chaieb I, Allagui MB (2021) Antifungal activity and chemical composition of seven essential oils to control the main seedborne fungi of cucurbits. Antibiotics 10(2):104

Mulani MS, Kamble EE, Kumkar SN, Tawre MS, Pardesi KR (2019) Emerging strategies to combat ESKAPE pathogens in the era of antimicrobial resistance: a review. Front Microbiol 10:539

Nelson ML, Levy SB (2011) The history of the tetracyclines. Ann N Y Acad Sci 1241(1):17–32

Nicol M, Mlouka MAB, Berthe T, Di Martino P, Jouenne T, Brunel J-M, Dé E (2019) Antipersister activity of squalamine against Acinetobacter baumannii. Int J Antimicrob Agents 53(3):337–342

de Oliveira EF, Tosati JV, Tikekar RV, Monteiro AR, Nitin N (2018) Antimicrobial activity of curcumin in combination with light against Escherichia coli O157: H7 and listeria innocua: applications for fresh produce sanitation. Postharvest Biol Technol 137:86–94

Orey C (2019) The healing powers of essential oils: a complete guide to Nature's Most magical medicine, vol 8. Citadel Press

Oyedemi BO, Kotsia E, Stapleton PD, Gibbons S (2019) Capsaicin and gingerol analogues inhibit the growth of efflux-multidrug resistant bacteria and R-plasmids conjugal transfer. J Ethnopharmacol 245:111871

Parai D, Banerjee M, Dey P, Chakraborty A, Islam E, Mukherjee SK (2018) Effect of reserpine on Pseudomonas aeruginosa quorum sensing mediated virulence factors and biofilm formation. Biofouling 34(3):320–334

Patkari M, Mehra S (2013) Transcriptomic study of ciprofloxacin resistance in Streptomyces coelicolor A3 (2). Mol BioSyst 9(12):3101–3116

Paukner S, Riedl R (2017) Pleuromutilins: potent drugs for resistant bugs—mode of action and resistance. Cold Spring Harb Perspect Med 7(1):a027110

Pedroso LA, Campos VP, Pedroso MP, Barros AF, Freire ES, Resende FM (2019) Volatile organic compounds produced by castor bean cake incorporated into the soil exhibit toxic activity against Meloidogyne incognita. Pest Manag Sci 75(2):476–483

Petchiappan A, Chatterji D (2017) Antibiotic resistance: current perspectives. Acs Omega 2(10):7400–7409

Petruzzi L, Corbo MR, Sinigaglia M, Bevilacqua A (2017) Microbial spoilage of foods: fundamentals. In: The microbiological quality of food. Elsevier, pp. 1–21

Pierzynowski SG, Sharma P, Sobczyk J, Garwacki S, Barej W, Weström B (1993) Comparative study of antibacterial activity of pancreatic juice in six mammalian species. Pancreas 8(5):546–550

Pitt JI, Hocking AD (2009) Fungi and food spoilage, vol 519. Springer, Cham

Rath SK, Singh S, Kumar S, Wani NA, Rai R, Koul S, Khan IA, Sangwan PL (2019) Synthesis of amides from (E)-3-(1-chloro-3, 4-dihydronaphthalen-2-yl) acrylic acid and substituted amino acid esters as NorA efflux pump inhibitors of Staphylococcus aureus. Bioorg Med Chem 27(2):343–353

Reiter J, Hübbers AM, Albrecht F, Leichert LIO, Slusarenko AJ (2020) Allicin, a natural antimicrobial defence substance from garlic, inhibits DNA gyrase activity in bacteria. Int J Med Microbiol 310(1):151359

Sabo VA, Knezevic P (2019) Antimicrobial activity of Eucalyptus camaldulensis Dehn. Plant extracts and essential oils: a review. Ind Crop Prod 132:413–429

Saeed F, Afzaal M, Tufail T, Ahmad A (2019) Use of natural antimicrobial agents: a safe preservation approach. Act Antimicrob food Packag:published online January 30

Sarkic A, Stappen I (2018) Essential oils and their single compounds in cosmetics—a critical review. Cosmetics 5(1):11

Shaheen A, Afridi W, Mahboob S, Sana M, Zeeshan N, Ismat F, Mirza O, Iqbal M, Rahman M (2019) Reserpine is the new addition into the repertoire of AcrB efflux pump inhibitors. Mol Biol 53(4):596–605

Sharma A, Magotra A, Rath SK, Wazir P, Nandi U, Koul S, Sangwan PL, Gupta AP, Singh G (2018) In-vitro and in-vivo pharmacokinetics of IS01957, p-coumaric acid derivative using a validated LC–ESI–MS/MS method in mice plasma. J Pharm Investig 48(5):565–574

Sharma A, Gour A, Bhatt S, Rath SK, Malik TA, Dogra A, Sangwan PL, Koul S, Abdullah ST, Singh G (2019) Effect of IS01957, a Para-coumaric acid derivative on pharmacokinetic modulation of diclofenac through oral route for augmented efficacy. Drug Dev Res 80(7):948–957

Shentu X, Zhan X, Ma Z, Yu X, Zhang C (2014) Antifungal activity of metabolites of the endophytic fungus Trichoderma brevicompactum from garlic. Braz J Microbiol 45(1):248–254

Silva GH, Zeraik ML, De Oliveira CM, Teles HL, Trevisan HC, Pfenning LH, Nicolli CP, Young MC, Mascarenhas YP, Abreu LM (2017) Lactone derivatives produced by a Phaeoacremonium sp., an endophytic fungus from Senna spectabilis. J Nat Prod 80(5):1674–1678

Silva-Hughes AF, Wedge DE, Cantrell CL, Carvalho CR, Pan Z, Moraes RM, Madoxx VL, Rosa LH (2015) Diversity and antifungal activity of the endophytic fungi associated with the native medicinal cactus Opuntia humifusa (Cactaceae) from the United States. Microbiol Res 175:67–77

Singh DD (2018) Assessment of antimicrobial activity of hundreds extract of twenty Indian medicinal plants

Singh R, Kumar M, Mittal A, Mehta PK (2017a) Microbial metabolites in nutrition, healthcare and agriculture. 3. Biotech 7(1):15

Singh S, Singh SK, Chowdhury I, Singh R (2017b) Understanding the mechanism of bacterial biofilms resistance to antimicrobial agents. Open Microb J 11:53

Siriwach R, Kinoshita H, Kitani S, Igarashi Y, Pansuksan K, Panbangred W, Nihira T (2014) Bipolamides a and B, triene amides isolated from the endophytic fungus Bipolaris sp. MU34. J Antibiot 67(2):167–170

Spížek J, Řezanka T (2004) Lincomycin, clindamycin and their applications. Appl Microbiol Biotechnol 64(4):455–464

Stavri M, Piddock LJ, Gibbons S (2007) Bacterial efflux pump inhibitors from natural sources. J Antimicrob Chemother 59(6):1247–1260

Suresh JI, Sona N (n.d.) Fungal endophytes, biodiversity and biopotential applications. In: Fungi bio-prospects in sustainable agriculture, environment and nano-technology. Elsevier, pp 107–115

Swain S, Rautray TR (2021) Estimation of trace elements, antioxidants, and antibacterial agents of regularly consumed Indian medicinal plants. Biol Trace Elem Res 199(3):1185–1193

Szponder T, Wessely-Szponder J, Sobczyńska-Rak A (2018) The neutrophil response to rabbit antimicrobial extract after implantation of biomaterial into a bone/cartilage defect. In Vivo 32(6):1345-1351

Takó M, Kerekes EB, Zambrano C, Kotogán A, Papp T, Krisch J, Vágvölgyi C (2020) Plant phenolics and phenolic-enriched extracts as antimicrobial agents against food-contaminating microorganisms. Antioxidants 9(2):165

Taware R, Abnave P, Patil D, Ramanpillai Rajamohananan P, Raja R, Soundararajan G, Chandra Kundu G, Kumar Dhondhiram Kharat M, Pai K, Ahmad A (2015) Trichothecin from endophytic fungus Trichothecium sp. and its anticancer effect on murine melanoma and breast cancer cell lines. Curr Biochem Eng 2(1):73–80

Tegos G, Haynes M, Strouse J, Khan M, Bologa C, Oprea T, Sklar L (2011) Microbial efflux pump inhibition: tactics and strategies. Curr Pharm Des 17(13):1291–1302

Timofeeva L, Kleshcheva N (2011) Antimicrobial polymers: mechanism of action, factors of activity, and applications. Appl Microbiol Biotechnol 89(3):475–492

Torres KA, Lima SMRR, Ueda SMY (2016) Activity of the aqueous extract of Schinus terebinthifolius Raddi on strains of the Candida genus. Rev Bras Ginecol Obstet 38(12):593–599

Trenin A (2013) Microbial models in screening of inhibitors of sterol biosynthesis. Antibiotiki i khimioterapiia= Antibiotics and chemoterapy [sic] 58(7–8):3–11

Tsuchiya H (2017) Anesthetic agents of plant origin: a review of phytochemicals with anesthetic activity. Molecules 22(8):1369

Tuyen CK, Le LT (2021) Plant extracts: antimicrobial properties, mechanisms of action and applications. In: Advanced antimicrobial materials and applications. Springer, pp. 257–283

Valente S, Cometto A, Piombo E, Meloni GR, Ballester A-R, González-Candelas L, Spadaro D (2020) Elaborated regulation of griseofulvin biosynthesis in Penicillium griseofulvum and its role on conidiation and virulence. Int J Food Microbiol 328:108687

Vasireddy L, Bingle LE, Davies MS (2018) Antimicrobial activity of essential oils against multidrug-resistant clinical isolates of the Burkholderia cepacia complex. PLoS One 13(8):e0201835

VasudhaUdupa A, Gowda B, Kumarswammy B, Shivanna M (2021) The antimicrobial and antioxidant property, GC–MS analysis of non-edible oil-seed cakes of neem, madhuca, and simarouba. Bull Nat Res Centre 45(1):1–14

Wang X, Radwan MM, Taráwneh AH, Gao J, Wedge DE, Rosa LH, Cutler HG, Cutler SJ (2013) Antifungal activity against plant pathogens of metabolites from the endophytic fungus Cladosporium cladosporioides. J Agric Food Chem 61(19):4551–4555

Wang J, Ma K, Ruan M, Wang Y, Li Y, Fu YV, Song Y, Sun H, Wang J (2018) A novel cecropin B-derived peptide with antibacterial and potential anti-inflammatory properties. PeerJ 6:e5369

Wei L, Zhao J, Meng Y, Guo Y, Luo C (2020) Antibacterial activity, safety and preservative effect of aminoethyl-phloretin on the quality parameters of salmon fillets. LWT 118:108874

Wink M (2015) Modes of action of herbal medicines and plant secondary metabolites. Medicines 2(3):251–286

Wright PM, Seiple IB, Myers AG (2014) The evolving role of chemical synthesis in antibacterial drug discovery. Angew Chem Int Ed 53(34):8840–8869

Wu Q, Patocka J, Nepovimova E, Kuca K (2019) Jatropha gossypiifolia L. and its biologically active metabolites: a mini review. J Ethnopharmacol 234:197–203

Xiao J, Zhang Q, Gao Y-Q, Shi X-W, Gao J-M (2014) Antifungal and antibacterial metabolites from an endophytic aspergillus sp. associated with Melia azedarach. Nat Prod Res 28(17):1388–1392

Xiao S, Cui P, Shi W, Zhang Y (2020) Identification of essential oils with activity against stationary phase Staphylococcus aureus. BMC Comple Med Ther 20(1):1–10

Xu J, Xu Z, Zheng W (2017) A review of the antiviral role of green tea catechins. Molecules 22(8):1337

Yan YQ, Fu YJ, Wu S, Qin HQ, Zhen X, Song BM, Weng YS, Wang PC, Chen XY, Jiang ZY (2018) Anti-influenza activity of berberine improves prognosis by reducing viral replication in mice. Phytother Res 32(12):2560–2567

Yim J, Smith JR, Rybak MJ (2017) Role of combination antimicrobial therapy for vancomycin-resistant enterococcus faecium infections: review of the current evidence. Pharmacother J Human Pharmacol Drug Therapy 37(5):579–592

Ying Y, Wang H, Xi X, Ma C, Liu Y, Zhou M, Du Q, Burrows JF, Wei M, Chen T (2019) Design of N-terminal derivatives from a novel dermaseptin exhibiting broad-spectrum antimicrobial activity against isolates from cystic fibrosis patients. Biomol Ther 9(11):646

Yuan M, Song X, Lv W, Xin Q, Wang L, Gao Q, Zhang G, Liao W, Lian S, Jing T (2019) Effect of anacardic acid against echinococcosis through inhibition of VEGF-induced angiogenesis. Vet Res 50(1):1–11

Chapter 10
Polyphenols as Emerging Antimicrobial Agents

Ashraf Ali, Antonio Parisi, and Giovanni Normanno

Abbreviations

AMR	Antimicrobial resistance
CHQA	3-*p*-trans-Coumaroyl-2-hydroxyquinic acid
EGCG	Epigallocatechin-3-gallate
GTE	Green tea extract
MIC	Minimum inhibitory concentration
MRSA	Methicillin-resistant *Staphylococcus aureus*
PACs	Proanthocyanidins
RMA	Resistance modification agent
STEC	Shiga toxin-producing *Escherichia coli*
UTI	Urinary tract infection
WHO	World Health Organization

10.1 Introduction

Polyphenols represent one of the largest groups of natural compounds obtained from the plant kingdom which have several medicinal properties (Di Lorenzo et al. 2021; Durazzo et al. 2019; Mehany et al. 2021). Until now, about 8000 phenolic compounds have been characterized starting from simplest bioactive phytochemicals containing single substituted phenyl rings to highly polymeric substances (Di Lorenzo et al. 2021; Cheynier 2005; Daglia 2012; Ciulu et al. 2018; Aliaño-González et al. 2020). They are important secondary metabolites in plants and are considered as some of the most beneficial natural molecules due to their bioactive properties (Durazzo et al. 2019). They are extracted from fruits, vegetables, seeds, nuts, stems, and floral parts of the plants. Polyphenols are found universally in a

A. Ali · G. Normanno (✉)
Department of Sciences of Agriculture, Food, Natural Resources and Engineering (DAFNE), University of Foggia, Foggia, Italy
e-mail: giovanni.normanno@unifg.it

A. Parisi
Zooprofilactic Institute of Apulia and Basilicata, Foggia, Italy

N. Akhtar et al. (eds.), *Emerging Modalities in Mitigation of Antimicrobial Resistance*,
https://doi.org/10.1007/978-3-030-84126-3_10

variety of foods, medicinal supplements, and products of cosmetic use, and have gained most attention in the last two decades due to growing indication of their helpful health outcome (Durazzo et al. 2019; Cheynier 2005; Crozier et al. 2009; Castro-Barquero et al. 2020). Increased consumption of polyphenol-rich plant products including whole grain cereals, vegetables, and fruits leads to a decreased risk of several severe health complications and pathologies associated with reactive oxygen species (oxidative stress), including cardiac complications, diabetes, severe inflammation, cancers, and several degenerative pathologies (Castro-Barquero et al. 2020; Tsao 2010; Vetrani et al. 2018; Giacco et al. 2020; Grosso et al. 2017; Kopustinskiene et al. 2020; Van de Velde et al. 2019; Csepregi et al. 2020; Tajner-Czopek et al. 2020). Oxidant properties of polyphenol can be classified as either pro-oxidant or antioxidant depending on the design of specific polyphenol and the cellular redox setting, which may enhance oxidant scavenging proteins or reduce the level of oxidized proteins and lipids (Tajner-Czopek et al. 2020; Chen et al. 2014; Yu et al. 2012; Julibert et al. 2019). Plant polyphenols are scavengers of metal chelators as well as free radicals. They are very effective in inhibiting the peroxidation of lipid molecules and display numerous physiological and biological actions as antioxidants. These molecules contribute to the defense mechanism in plants and take part in adaptation and pigmentation activities (Fischer et al. 2018; Ma et al. 2018a). In addition to health benefits against various diseases, several extracts rich in these polyphenols can serve as excellent inhibitors of several pathogenetic bacteria (Julibert et al. 2019; Gutiérrez-Del-Río et al. 2018; Redondo-Blanco et al. 2020; Makhuvele et al. 2020; Simonetti et al. 2020).

Many bacteria can form biofilms and can cause serious contamination and infection. Cell-to-cell chemical communication in bacteria is performed by quorum sensing, which plays decisive role in antibiotic resistance, formation of biofilm, proliferation, survival, and production of toxins (Abisado et al. 2018). Disruption of this signaling pathway can make way to control these pathogenic organisms and prevent them from causing infection and disease (Tsao 2010; Defoirdt 2018; Mion et al. 2018, 2019). These bioactive compounds can be obtained by various physical and chemical extraction methods. Some of the effective methods include enzyme-associated extraction techniques with application of carbohydrase active microorganisms or enzymes and solid-phase fermentation (Yu et al. 2012; Aherne and O'Brien 2002; Cowan 1999; Brglez Mojzer et al. 2016; Iglesias-Carres et al. 2019; Mesquita and Monteiro 2018; Heng et al. 2017; Capriotti et al. 2014). Overall, the effectiveness of the polyphenol and its derivatives depends on the target microorganism, their membrane and cell structure, the time and duration of treatment. The antimicrobial effect of these polyphenols also depends on the type of solvent used and which extraction method was applied and on the plant cultivars. The compound can also affect efflux pump expression and their functions besides causing down-regulation of genes linked with the virulence factor of microorganism.

The discovery of penicillin revolutionized the treatment of bacterial diseases but the overuse of antibiotics has created an opening for potential future pandemic leading to the emergence of antibiotic resistance pathogens (AMR), which is now the greatest threat to public health around the globe. The microbes display antibiotic resistance either through the expression of resistance phenotype or due to the inherent resistance genes inside them. The appearance of multidrug-resistant bacteria like methicillin-resistant *Staphylococcus aureus* (MRSA) is giving more nightmares to clinicians than resistance to a single antibiotic (Abreu et al. 2012). Studies on resistance modification agents (RMAs) which can counter the problem of AMR have been explored and polyphenolic compounds can provide one such alternative. The synergistic approach between resistance modification agents (RMAs) like polyphenols and antibiotics can be a more effective strategy to control the phenomenon of AMR.

An interesting area of future research is the interaction of polyphenol and gut microbiota. It is hypothesized that polyphenolic compounds favor the growth of beneficial bacteria while doing reverse with harmful bacteria (Marín et al. 2015). Some research indicates that there is positive correlation between blueberry drink and bifidobacteria (Vendrame et al. 2011), while extracts obtained from green tea modulate activities of bacteria such as *Salmonella typhimurium, Clostridium difficile*, and *Escherichia coli* (Lee et al. 2006; Pacheco-Ordaz et al. 2018). There is also an indication that phenolic extracts may stimulate the beneficial effects of probiotics (de Souza et al. 2019). The efforts are underway to exploit polyphenols as treatment option for COVID-19 and other viruses (Paraiso et al. 2020; Annunziata et al. 2020).

Polyphenols can be classified based on source of origin, the structure of the chemicals, and their biological function (Di Lorenzo et al. 2021; Durazzo et al. 2019; Singla et al. 2019) (Figs. 10.1, 10.2, 10.3, 10.4, 10.5, 10.6, and 10.7). Chemically, one aromatic ring and presence of one or more hydroxyl (–OH) groups are characteristic of phenolic compounds which are further divided into flavonoids and non-flavonoids (Figs. 10.1, 10.2, 10.3, 10.4, 10.5, 10.6, and 10.7). Flavonoids are defined as cluster of low molecular weight polyphenolic compounds obtained from various plant sources. Their synthesis is done by phenylalanine and acetic acid and combination of their derivatives. Flavonoid structure is composed of flavonoid nucleus, consisting of three phenolic rings designated as A, B & C (Tsao 2010;

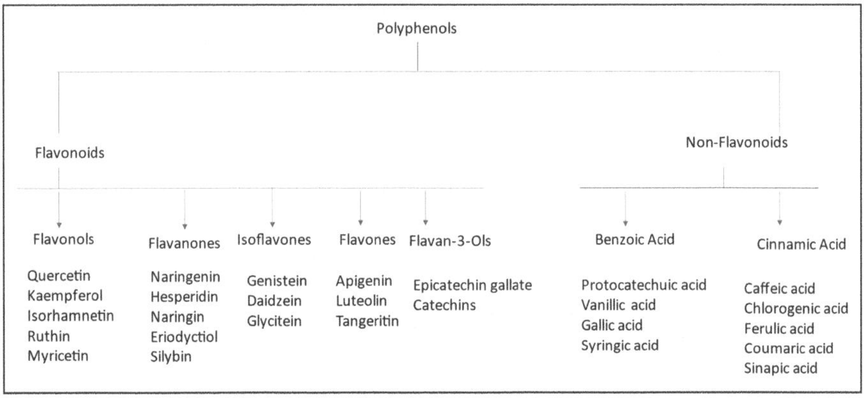

Fig. 10.1 Polyphenols classification with some examples

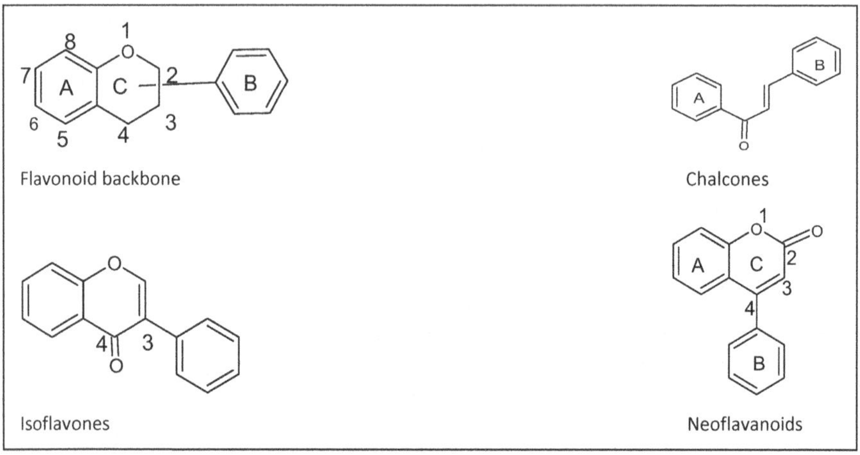

Fig. 10.2 Basic flavonoid structures

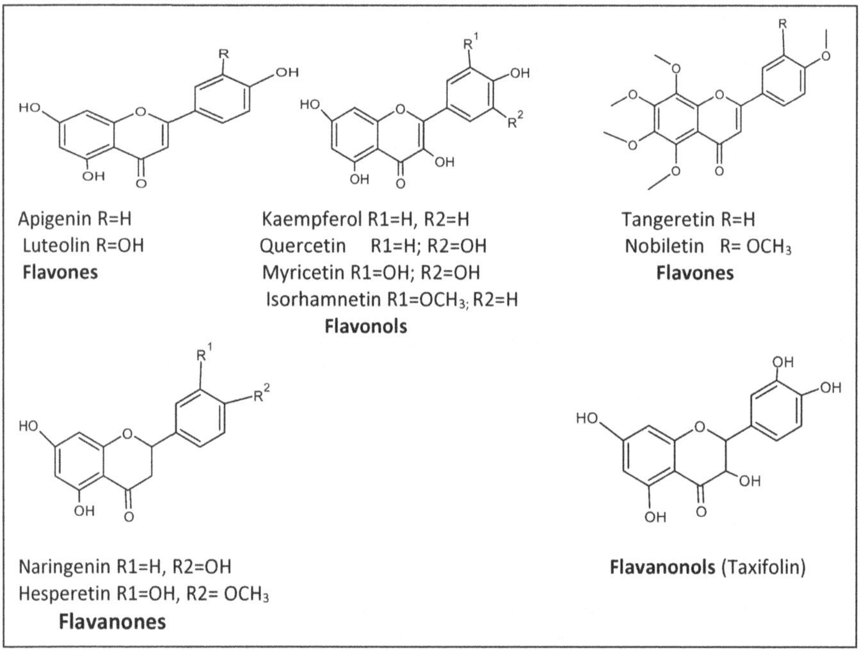

Fig. 10.3 Structure of flavones, flavonols, flavanones, and flavanonols

Aherne and O'Brien 2002) (Fig. 10.2). Flavonoids can be further differentiated into different groups based on the presence or absence of double bonds between carbon atom C2 and C3 and a hydroxyl group (–OH) at C3 of ring C. Based on these structural modifications, they are further subdivided into different subfamilies. The

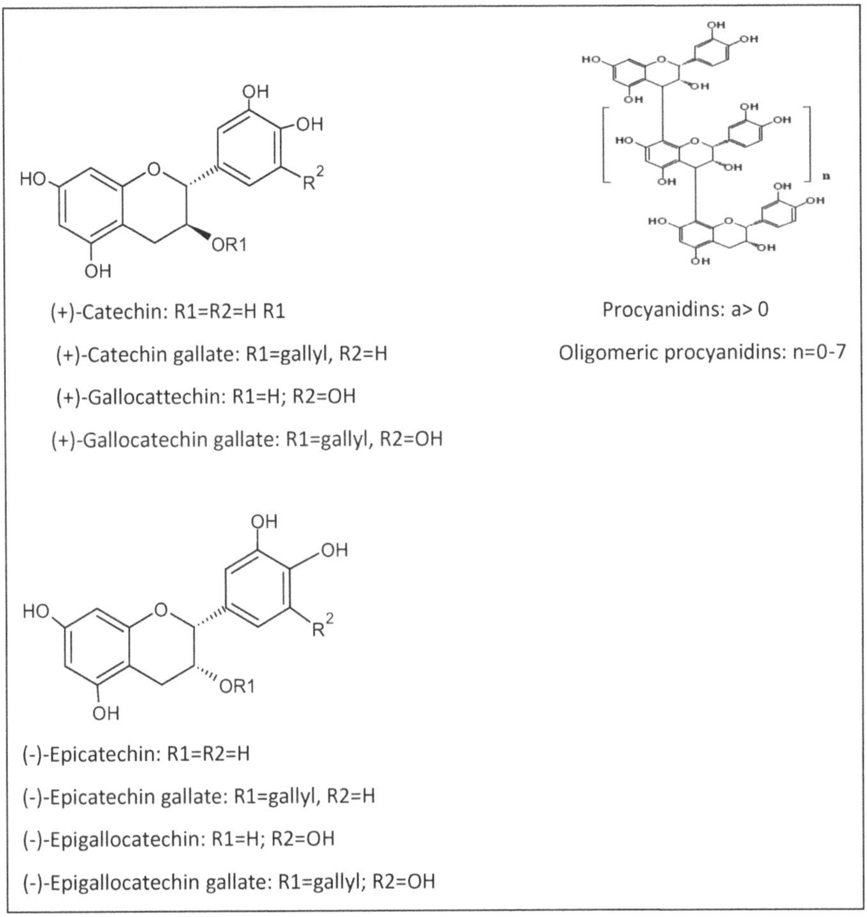

Fig. 10.4 Flavanols and procyanidins

subfamilies are classified as flavonols, flavones, isoflavones, flavanones, and flavan3-ols (Singla et al. 2019; Yu et al. 2020; Křížová et al. 2019) (Figs. 10.2, 10.3, 10.4, and 10.5). The position and number of hydroxyl groups on the phenol ring play a major role in their biological properties. It is hypothesized that these groups are responsible for their relative toxicity toward different pathogens, and there is a positive correlation between hydroxylation (no of –OH groups) and toxicity of the compounds (Cowan 1999).

Flavonols are classified as a cluster of polyphenolic compounds having two benzene rings attached by a linear three carbon chain (Figs. 10.3 and 10.4). Flavonols are derivatives of flavonoids with a ketone group. Flavonols are present in abundant quantities in several fruits and vegetables. Apples, grapes, berries (Blueberries), onions, kale, lettuce, and tomatoes are rich sources of flavonols. Due to their antioxidant activity, they possess several benefits to health, which are suggested to be

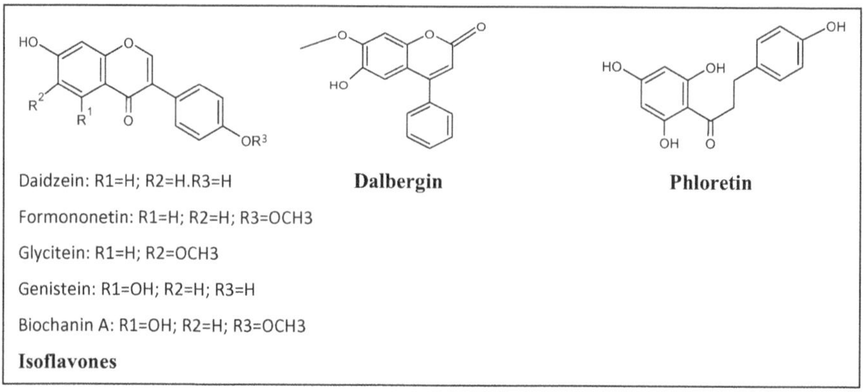

Fig. 10.5 Typical isoflavones, neoflavones found in plants

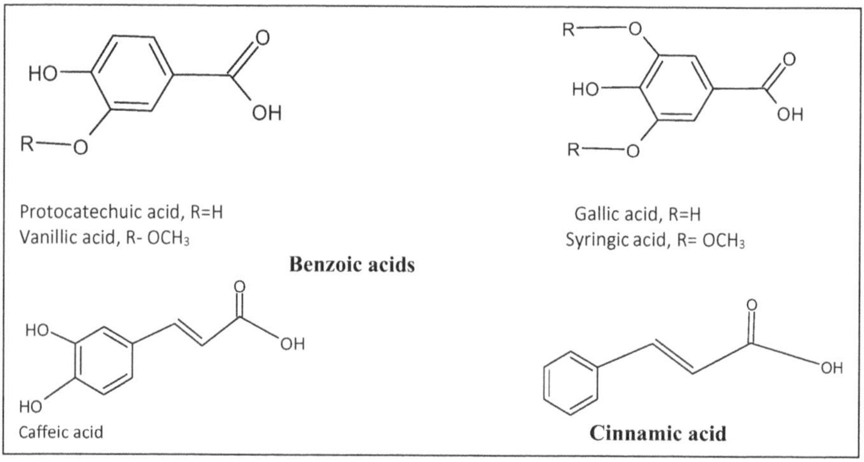

Fig. 10.6 Typical phenolic acids (non-flavonoids) in plants

due to the presence of phenolic hydroxyl functional group (OH) on the B ring (Kim et al. 2006).

Flavones are almost structurally like flavonols except that there is presence of oxygen at C3 (Figs. 10.2, 10.3, and 10.4) (Di Lorenzo et al. 2021; Daglia 2012; Das 2020). Isoflavones (Fig. 10.5) have the B ring attached at C3 rather than at the C2 position. They are mostly found in leguminous family of plants like soybean. Genistein and daidzein are two examples of isoflavones found in soy along with glycetein, biochanin A, and formonetin. Red clovers are also good source of this compound. In neoflavanoids Ring B is attached at C4 position of ring C. Neoflavanoids are rarely found in many plants, but Dalbergin is the most frequent and commonly circulated in the plant world. In flavanones, D2,3 double bond is absent and there is presence of a chiral center at C2. Flavan-3-ols, also called catechin, are a complex

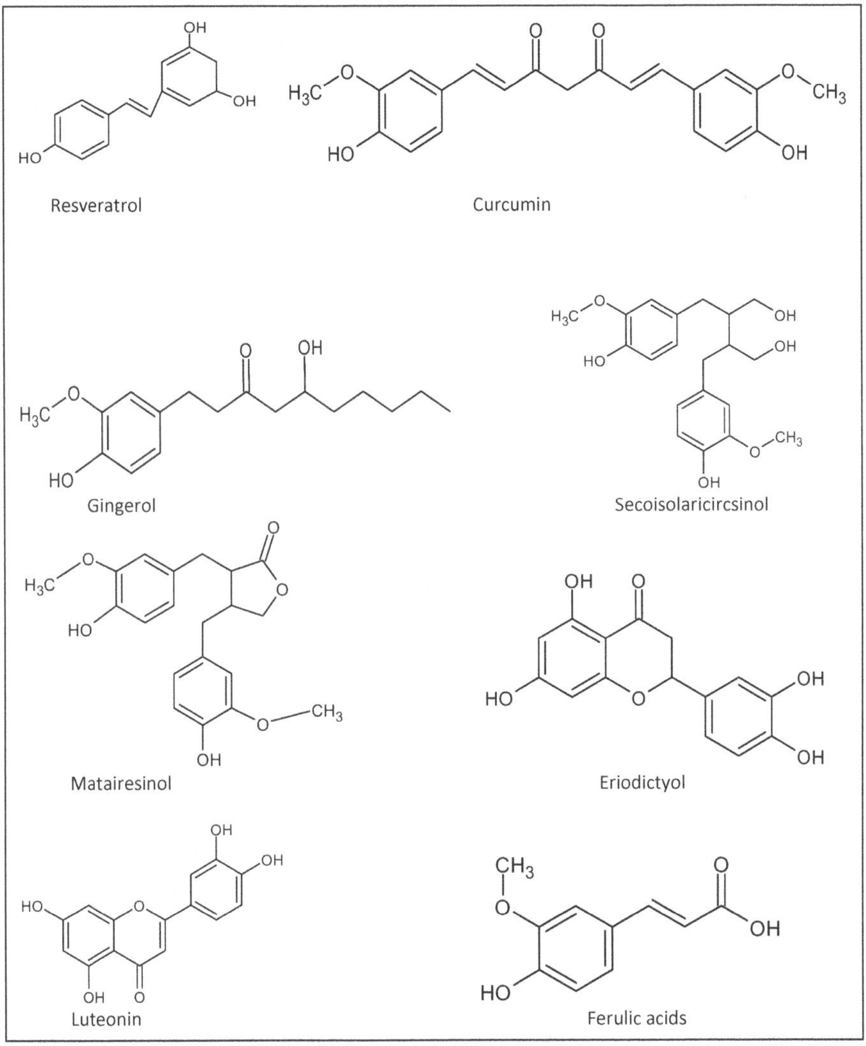

Fig. 10.7 Other important polyphenols

subclass of flavonoids. In their structure, the two chiral carbons at C2 and C3 of the monomeric flavan-3-ol produces four isomers for each level of B ring hydroxylation (Figs. 10.3 and 10.4). Isomer with *trans* configuration is named catechin while one with *cis* configuration is called epicatechin.

Proanthocyanidins are polymers of catechin and epicatechin because anthocyanidins are produced by acid-catalyzed cleavage of the polymeric chains. Although chalcones lack the heterocyclic ring at C they are still counted as part of the flavonoid family (Fig. 10.2). Grapes, apples, and blueberries are rich sources of flavanols. Monomeric flavanols like catechin and epicatechin and their byproducts (e.g., gallocatechin) are the major flavonoids found in tea leaves, cacao beans, and chocolates (Tsao 2010).

Tannins are another class of polyphenols with mol weight ranging from 500 to 3000 Da but some tannins have been found with mol weight ranging from 3000 to over 30,000 Da (Serrano et al. 2009; Fraga-Corral et al. 2021). They can be classified based on their solubility and extractability and on the basis of their chemical structure. Tannins can be further categorized into two major groups, either as proanthocyanidins or condensed tannins and hydrolyzable tannins depending upon the structure of monomer (Giacco et al. 2020).

The classical example of non-flavonoid polyphenolic compounds is phenolic acid. Phenolic acids are further sub-divided into two main groups, benzoic acid and cinnamic acid derivatives depending upon C1–C6 and C3–C6 backbones (Fig. 10.6) (Tsao 2010). Some polyphenolic compounds with their target microorganism are mentioned in Tables 10.1 and 10.2. Some other polyphenols are depicted in Fig. 10.7 (Tsao 2010).

10.2 Type of Flavonoids and Their Antimicrobial Activity

Flavonoids are one of the most widely researched polyphenols (a type of phytochemical) due to their numerous health benefits. They belong to a class of secondary metabolites with polyphenolic structure. They provide enormous health benefits, and they are part of various pharmaceuticals, nutraceutical, cosmetic, and medicinal products. They can modulate important cellular and enzymatic functions. They can inhibit activities of several enzymes like cyclo-oxygenase (COX), xanthine oxidase (XO), lipoxygenase, and phosphoinositide 3-kinase. They are available in abundant quantities in fruits, vegetables, certain beverages, and green tea extract (GTE). Since ancient times flavonoids from plant and herbal extracts were utilized as folk medicine across the world. They display several health benefits, for example, as protective agents for peptic ulcer, cancers related to stomach, and treatment of rheumatic pain (Di Lorenzo et al. 2021; Rupasinghe 2020; Kozłowska and Szostak-Wegierek 2014; Samec et al. 2021; Ullah et al. 2020). Flavonoids display antioxidant and anti-inflammatory properties besides their effectiveness as antibacterial agents. The antimicrobial potential of flavonoids has been increasingly explored and many studies across the world have reported identification of different chemicals having antibacterial characteristics (Marín et al. 2015; Sarbu et al. 2019).

There are different mechanisms through which flavonoids can target bacterial growth. Inhibition of nucleic acid synthesis and targeting the function of cytoplasmic membrane are the most widely used strategies adopted by flavonoids, especially by flavonoids with B ring hydroxylation (Sarbu et al. 2019; Farhadi et al. 2019). A natural phenolic compound, 3-*p*-trans-coumaroyl-2-hydroxyquin ic acid (CHQA) from needles of *Cedrus deodara*, has been exploited as strong antibacterial agent. Extraction of 3-*p*-trans-coumaroyl-2-hydroxyquinic acid

Table 10.1 Polyphenolic compounds and their target microorganisms

Compound	Target bacteria	References
Galangin	*K. pneumoniae*	Chen and Huang (2011)
Kaempferol	*K. pneumoniae* *S. aureus*	Shao et al. (2016), Randhawa et al. (2016), Wu et al. (2013)
Quercetin	*K. pneumoniae* *S. aureus* *S. haemolyticus* *E. coli* O157:H7 *S. pyogenes*	Chen and Huang (2011), Liu et al. (2010), Siriwong et al. (2015)
Myricetin	*K. pneumoniae* *S. aureus*	Chen and Huang (2011), Huang et al. (2015)
Luteolin	*S. pyogenes* *S. aureus* *E. coli*	Liu et al. (2010), Siriwong et al. (2015)
Chrysin	*P. lachrymans* *B. subtilis* *S. haemolyticus* *S. aureus* *E. coli* *A. tumefaciens* *B. subtilis* *S. haemolyticus*	Liu et al. (2010), Siriwong et al. (2015)
Camellia sinensis polyphenols	*S. aureus* *S. marcescens* *P. aeruginosa* *H. pylori* *E. coli*	Ankolekar et al. (2011), Thakur et al. (2016), Yi et al. (2014), Radji et al. (2013)
ECGC	*S. aureus* *S. mutans* *S. pyogenes* *E. faecalis* *E. coli* *S. pyogenes* *S. typhi* *P. gingivalis* *P. aeruginosa* *K. pneumoniae*	Cui et al. (2012), Cho et al. (2011), Lee et al. (2009a), Betts et al. (2015), Asahi et al. (2014)
Naringenin	*S. typhimurium* *P. aeruginosa*	Vandeputte et al. (2011), Vikram et al. (2011)
Hesperetin	*S. aureus* *A. hydrophila*	Abuelsaad et al. (2013), Bakar et al. (2012)
Malvidin	*K. pneumoniae*	Gopu et al. (2015)
Daidzein	*Vancomycin-resistant* *E. faecalis*	Chin et al. (2012)
Gallic acid	*E. coli* *S. mutans*	Shao et al. (2015, 2016), Dwivedi et al. (2016)

(continued)

Table 10.1 (continued)

Compound	Target bacteria	References
Resveratrol	*V. cholerae* *P. mirabilis*	Ma et al. (2018b), Klančnik et al. (2017), Ferreira et al. (2014)
Coumarins	*H. pylori*	Basile et al. (2009), Takeda et al. (2007)
Lacinartin	*P. gingivalis*	Marquis et al. (2012)
Furocoumarins	*E. coli O157:H7* *S. typhimurium*	Girennavar et al. (2008)
Blueberry proanthocyanidins	*L. monocytogenes* *H. pylori* *S. typhimurium* *E. coli*	Joshi et al. (2014), Lacombe et al. (2013), Park et al. (2011)
Proanthocyanidins	*E. coli* *P. aeruginosa* *S. epidermidis*	Zeng et al. (2020)

(CHQA) from *Cedrus deodara* has demonstrated effective antimicrobial activity towards 11 food-borne pathogens like *S. aureus*. The MIC values obtained against such pathogens are between 2.5 and 10 mg/mL. CHQA is responsible for damaging the cytoplasmic membrane of bacteria and induces seepage of intracellular components. It can act as a future natural antibacterial candidate against various food pathogens (Wu et al. 2016). Sophoraflavanone B is a prenylated flavonoid extracted from the roots of *Desmodium caudatum*. It can directly target the peptidoglycan layer and inhibit biosynthesis of cell wall inhibiting MRSA (MIC observed between 15.6 and 31.25 μg/mL) (Mun et al. 2014). Urease, a major virulence factor of *H. pylori*, is the target of many phenolic compounds such as 4-deoxy analogs of flavonoids which acts as a potent inhibitor of urease (Xiao et al. 2013). Curcumin isolated from turmeric has displayed beneficial activity by altering the cell membrane of *S. aureus* and *E. coli*. It inhibits sortase A from *S. aureus* (IC50 13.8 μg/mL) (Tyagi et al. 2015). Haloemodins are classified as semisynthetic natural anthraquinone derivative having antimicrobial properties. They target DNA gyrase in MRSA and vancomycin-resistant *Enterococcus faecium* showing effective inhibition (Duan et al. 2014). In a recently conducted study, the molecular characteristics exploring antibacterial activity of prenylated (iso)flavonoids against MRSA are evaluated (Kalli et al. 2021). Phloretin, a kind of flavonoid, obtained from apple was able to block biofilm formation in *E. coli O157:H7*. It was successful in limiting curli genes, i.e., csgA and csgB, responsible for the production of fimbriae (Lee et al. 2011).

Table 10.2 Polyphenolic compounds with their target microorganisms with MIC and source of extraction

Name of plant	Common name	Target pathogens	MIC (mg/mL)	Polyphenol derivatives	References
Eugenia caryophyllata	Cloves	*B. subtilis*	1.6	Flavonoids Terpenoids	Shehadi et al. (2014)
Mentha piperita	Mint	*B. subtilis*	0.5×10^2	Terpenoids Flavonoids	Shehadi et al. (2014), Othman et al. (2019)
Rosemarinus officinalis	Rosemary	*B. subtilis*	2.96	Phenolics Flavonoids Terpenoids	Shehadi et al. (2014), Othman et al. (2019)
Prunus avium	Cherry	*B. subtilis*	4	Phenolics	Othman et al. (2019)
Matricaria aurea L.	Golden chamomile	*B. subtilis* *S. pyogenes* *S. aureus* *K. pneumoniae*	0.4 0.4 50 0.4	Phenols Phenolic acids	Rizwana (2016)
Annona squamosa L.	Custard Apple	*S. aureus* *E. faecalis* *E. coli*	50	Quinones Flavonoids	Mohamad et al. (2017)
Premna resinosa	Mukarakara	*S. aureus* *B. subtilis* *E. coli* *S. typhimurium* *E. faecalis* *S. flexneri*	0.01–1	Quercetin Kaempferol	Albadawi et al. (2017)
Premna resinosa	Mukarakara	*MRSA* *M. tuberculosis*	0.031 0.006	Quercetin Kaempferol	Njeru et al. (2015)
Pheonix dactylifera L.	Date palm	*P. aeruginosa*	0.06	Flavonoid Glycosides Quercetin Apigenin Luteolin	Selim et al. (2012)
Thymus vulgaris *Boswellia carterii*	Thyme Myrrh	*S. aureus* *B. cereus* *A. flavus* *F. oxysporum*	2–4% (v/v)	Rosmarinus acid Caffeic acid Carnosol Flavonoids	Al-Juraifani (2009)
Olea europaea	Olive	*S. aureus* *B. cereus*	0.08 0.04	Flavonoids	Malik (2015)
Eruca sativa	Arugula	*S. aureus* *B. cereus*	0.06 0.02	Flavonoids	Malik (2015)

(continued)

Table 10.2 (continued)

Name of plant	Common name	Target pathogens	MIC (mg/mL)	Polyphenol derivatives	References
Azadirachta indica	Neem	*B. subtilis* *S. aureus* *M. roseus*	0.05–0.2	Flavonoids Saponins Anthocyanins	Othman et al. (2019)
Zingiber officinale	Ginger	*S. aureus*, *B. subtilis*	0.21–1.33	Tannins	Shohayeb et al. (2013)
Curcuma Longa	Turmeric	*P. gingivalis*	0.012	Tannins	Sha and Garib (2019)
Thymbra spicata L.	Mediterranean thyme	*MDR S. aureus* *K. pneumoniae*	0.006–0.012 0.012	Alkaloids	Haroun and Al-Kayali (2016)
Salvadora persica L.	Miswak	*S. aureus* *S. mutans* *S. faecalis* *S. pyogenic* *L. acidophilus* *P. aeruginosa* *C. albicans*	3.12 1.56 0.781 3.12 6.25 6.25 6.25	Saponins Alkaloids Terpenoids Flavonoids	Al-Bayati and Al-Mola (2008), Al-Bayati and Sulaiman (2008), Sher et al. (2011)
Origanum majorana	Oregano	*ESBL E. coli* *K. pneumoniae*	0.0025–0.08	Rosmarinic acid	Abdel-Massih et al. (2010), Daoud et al. (2011), Hossain et al. (2014)
Trigonella foenum-graecum	Fenugreek	*ESBL E. coli* *K. pneumoniae*	0.0025–0.08	Stilbenes	Abdel-Massih et al. (2010), Daoud et al. (2011)
Quercus infectoria	Aleppo oak	*B. cereus* *Y. enterocolitica*	0.00256 0.00512	Flavonoids	Shariatifar et al. (2014)
Galla rhois		*Salmonella* *E. coli*	0.004–0.125 0.03	Methyl gallate	Choi et al. (2014), Madikizela et al. (2013)
Boletus edulis	Wild polish mushroom	Gram-positive bacteria	0.156–5	Flavonoids	Nowacka et al. (2015)
Kaempferia pandurata Roxb	Temu kunci	*S. aureus*	0.06–2	Natural chalcone	Cushnie and Lamb (2011), Rukayadi et al. (2009)
Helichrysum italicum (Roth)	Curry plant	*S. aureus*	0.001–0.004	Phloroglucinol	Taglialatela-Scafati et al. (2013)

(continued)

Table 10.2 (continued)

Name of plant	Common name	Target pathogens	MIC (mg/mL)	Polyphenol derivatives	References
Hypericum beanie	St John's wort	*MRSA*	0.016–0.032	Acylphloroglucinols	Shiu and Gibbons (2006)
Cedrus deodara	Cedar	*S. aureus*	2.5–10	Natural phenolic compound	Wu et al. (2016)
Desmodium caudatum	Salparni	MRSA	0.016–0.031	Flavonoid	Mun et al. (2014)
Citrus bergamia Risso	Bergamot	*E. coli* *P. putida* *S. enterica*	0.25–0.08	Flavonoid	Mandalari et al. (2007)
Polymnia fruticose	Ming aralia	*H. pylori*	0.087	Flavanone	Zhang et al. (2008)
P. fruticose	*Potentilla* species	*P. aeruginosa*, and a fungus *Candida albicans*	0.78–6.25	Caffeic acid Ferulic acid	Wang et al. (2013a)
Vaccinium virgatum	Blueberry	*S. epidermidis* *P. aeruginosa*	4	Chlorogenic acid	Zimmer et al. (2014)

10.2.1 Flavonols and Their Antimicrobial Activity

Flavonols are derivatives of flavonoids having a ketonic group in their structure, and they are building blocks of proanthocyanins. The most common example of flavonols is kaempferol, quercetin, myricetin, and fisetin which are abundantly found in a variety of fruits and vegetables. The rich sources of flavonols are lettuce, tomatoes, onions, kale, apples, grapes, and berries besides tea and red wine. Flavonols are found in large varieties of plants, apart from algae and fungi. Some of the most frequently found flavanols are galangin and quercetin which possess good antimicrobial activity (Rauha et al. 2000; Basile et al. 2000; Arima et al. 2002; Rizzo et al. 2014). Flavonols like myricetin, morin, rhamnetin, and quercetin show inhibitory activity against the growth of *Chalamydophyla pneumoniae*, an obligate intracellular Gram-negative bacterium responsible for community and nosocomial pneumonia. Quercetin and apigenin, the most abundant flavonoids can block D-alanine by blocking D-alanine ligase, an important enzyme in the assembly of peptidoglycan precursors of the cell wall in *H. pylori* and *E. coli* (Wu et al. 2008).

Resveratrol, a natural polyphenol, is a promising candidate against different pathogenic bacteria like *Campylobacter jejuni, M. smegmatis*, and *Arcobacter cryaerophilus* (Ma et al. 2018b; Klančnik et al. 2017; Ferreira et al. 2014; Lechner et al. 2008). Resveratrol, at 40 μM concentration in combination with quercetin at 20 μM, was able to inhibit the growth of *C. pneumoniae.* In some reports, rhamnetin was more potent than quercetin, probably due to its hydrophobic nature (Alvesalo et al. 2006). Application of clarithromycin, a macrolide, and ofloxacin, a fluoroquinolone, was beneficial in the treatment of chlamydial infections although the therapy was not completely successful (Rizzo et al. 2014). A synergistic approach with quercetin and ofloxacin or clarithromycin may provide better alternative in treating this type of infection (Rizzo et al. 2014). Some important enzymes including beta-ketoacyl acyl carrier protein synthase (KAS) II and III are responsible for generation of precursor of bacterial cell membrane and they are targets of phenolic compounds. The phenolic derivatives also target enzymes involved in elongation cycle of fatty acids and biosynthesis such as FabG, FabI, and FabZ. 3,6-Dihydroxyflavone, a flavonol, can also bind effectively to KAS III and KAS I and can inhibit *E. coli* with MIC at 512 μg/mL(Farhadi et al. 2019).

Propolis, containing main components like flavonoids and phenolic acid esters, is a non-toxic natural extract having multiple pharmacological characteristics and is found in several parts of the world, including Europe, China, and North America (Patel 2016; Rivera-Yañez et al. 2020). The main polyphenolic components of propolis analyzed by HPLC are galangin, pinocembrin, apigenin, quercitrin, myricetin, kaempferol, caffeic acid, and its phenylethyl ester (Koru et al. 2007; Barrientos et al. 2013; Santos et al. 2020). They displayed various antimicrobial activities against several bacteria, parasites, viruses, and fungus (Bankova et al. 1995; Kujumgiev et al. 1999; Sforcin et al. 2000; Orsi et al. 2005; Bosio et al. 2000; Almuhayawi 2020). The main component participating in these antimicrobial activities are galangin and pinocembrin (Bosio et al. 2000; Hegazi et al. 2000). The antimicrobial property of propolis depends on the solvent used for extraction as ethanol and acetone extracts were observed to be more potent against the most microorganism and this action was most inhibitory in a slightly acidic (pH-6) environment (Ivančajić et al. 2010).

Propolis and its various polyphenolic extracts are very effective in inhibition of biofilms and caries development by *Streptococcus mutans*. Inhibitory activity of various propolis components was studied on several oral microorganisms like *Veillonellaparvula, Actinomycescesnaesundii, Prevotella oralis*, and *Lactobacillus acidophilus*. Gram-positive anaerobic were more susceptible toward these extracts compared to Gram-negative ones. Propolis isolates from Chile were found to inhibit *Candida* spp. The extracts are very potent against isolates from the oral cavity of removable dentures and from cariogenic bacteria *Lactobacillus fermentum* (Koru et al. 2007*;* Przybyłek and Karpiński 2019*;* Saavedra et al. 2011). The exact mechanism by which propolis displays its antibacterial activity is still not clear. Some researchers suggest that the polyphenolic compounds present in propolis such as flavonoids probably act at a site on the cytoplasmic membranes or cell wall of bacteria causing structural and functional changes affecting their pathogenic mechanism (Almuhayawi 2020; Przybyłek and Karpiński 2019).

Kaempferol (3,4′,5,7-tetrahydroxyflavone) is a natural flavanol and a regular flavonoid which can be extracted from different plants and plant-derived foods, including spinach, kale, beans, tea, and broccoli. It is characterized as a yellow strong glasslike structure that has a low dissolvability rate in water but is profoundly soluble in ethanol, ethers, and dimethyl sulfoxide. It has a high liquefying point (276–278 °C) and is utilized in different drug items for its medical advantages and in food items for nutritional benefits. Kaempferol is exceptionally utilized in dietary enhancements because of its micronutrients, which help in treating chronic diseases and type 1 diabetes. Also, it finds applications in the therapy of sicknesses and infectious diseases caused by viruses and bacteria. Effectiveness of kaempferol against different pathogenic microorganisms, including fluconazole-resistant *C. albicans* and MRSA, is already well demonstrated (Shao et al. 2016; Randhawa et al. 2016; Holler et al. 2012a). A natural glycoside product of kaempferol (kaempferol rhamnoside) isolated from *Persea lingue* has potential to increase the antimicrobial activity of ciprofloxacin in a NorA overexpressed *S. aureus* (Holler et al. 2012a).

Another rich source of flavonols are white and red onion (*Allium cepa*), which contain very high amounts of quercetin-4″-glucoside and quercetin-3,4″-*O*-diglucoside (Sidhu et al. 2019). Extracts of white onion displayed higher inhibitory activity against Gram-negative bacteria such as *E. coli* and moderate activity against *S. aureus*. Compared to white onion, red onion is less effective in inhibition of pathogens as it displayed no activity against *S. aureus* (Benmalek et al. 2013). Different range of antimicrobial activity was shown by isolated dry buds of *A. cepa*. as *S. aureus* was found to be most susceptible, while *Pseudomonas aeruginosa* and *Salmonella typhi* are the most resistant (Bakht et al. 2014). Quercetin is one of the most studied flavanols and a major bioflavonoid is also quite effective against *E. coli*. It probably inhibits DNA gyrase and probably binds to β subunit of *E. coli* DNA gyrase and blocks the ATPase activity (Plaper et al. 2003). Many factors are responsible for potential gyrase inhibition, such as altering the permeability and pharmacokinetics of the membrane. The inherent mechanisms of gyrase inhibitors lie in the activity against the target enzyme (Hilliard et al. 1995). The binding mechanism of some flavanols including ellagic acid, quercetin, and 4-quinolons were studied by Hillmond et al. (Hilliard et al. 1995) against *E. coli* DNA gyrase.

10.2.2 Flavones and Isoflavone and Their Antimicrobial Activity

Citrus fruits like orange and lemons are rich sources of flavones such as tangeritin, luteolin, and apigenin. In citrus fruits, these polyphenols are found as conjugated polymethyl flavones while in the diet they appear as O-glycosides and C-glycosides (Hollman and Arts 2000; Hostetler et al. 2017). They show antimicrobial activity against several bacteria and pathogens (Liu et al. 2013a; Stojanović et al. 2005). Soyabean and other leguminous plants are a rich source of isoflavones. Soyabean contains ample amount of isoflavones like daidzein and genistein (Osawa et al. 1992; Leuner et al. 2013). Antimicrobial activity of genistein and other isoflavones was tested against several bacteria like *Vibrio harveyi, E. coli*, and *Bacillus subtilis* (Ulanowska et al. 2006). Among them *Vibrio harveyi* was most sensitive while *E. coli* did not show any inhibition. Genistein was able to alter cell morphology at a final concentration of 0.1 mM. Baicalein, a flavone extracted from roots of plant *Scutellaria Lateriflora, Thymus vulgaris*, and *Scutellaria baicalensis*, has displayed a range of antimicrobial activities. Extract isolated from *Scutellaria baicalensis* has been tested against several bacterial strains (Lu et al. 2011).

Baicalein has the potential to reverse anti-AMR of β-lactam antibiotics against MRSA by inhibition of NorA efflux pump. Baicalein shows greater antimicrobial activity with a combination of tetracycline against *E. coli* (Fujita et al. 2005). Biochanin A, an isoflavone, was found to be quite effective against a range of pathogens like MRSA, chlamydia, and mycobacterium (Zou et al. 2014; Hanski et al. 2014; Cannalire et al. 2017). Two methoxylated flavones from *Artemisia annua*, Chrysosplenol-D and chrysoplenetin, have shown inhibitory activity against NorA efflux pump (Stermitz et al. 2003). Two other classes of phenolic compounds, isoflavonoids and flavonolignans, are able to inhibit Nor A efflux pump and can enhance the potency of norfloxacin and berberine (Morel et al. 2003). Orobol from *Lupinus argenteus* and silybin, a flavonolignan from the famous medicinal plant *Silybum marianum*, was effective against *S. aureus* (Morel et al. 2003; Stermitz et al. 2001). Morin (a type of flavonol), a compound isolated from the bark of *Rhus verniciflua*, was able to inhibit *S. aureus* by targeting sortase A and B (Kang et al. 2006).

10.2.3 Flavanones and Their Antimicrobial Activity

Citrus fruits like oranges, lemons, and grapes are the exclusive source of flavanones such as naringenin, eriodictyol, and hesperetin. Grapefruit contains a higher amount of naringenin than hesperetin (Testai and Calderone 2017; Barreca et al. 2017). Finnish plant rich in Naringenin displayed potent antibacterial activity mainly toward Gram-positive bacteria like *Micrococcus luteus, S. epidermidis, S. aureus*, and *B. subtilis*, while against Gram-negative bacteria like *E. coli* and *P. aeruginosa* there was moderate or little activity (Rauha et al. 2000; Radulescu et al. 2020). In one study, flavanones isolated from peel of bergamot (*Citrus bergamia Risso*) were tested for their antimicrobial activity against different bacteria and eriodictyol displayed the highest antimicrobial activity (Observed MIC 250–800 μg/mL) (Mandalari et al. 2007). Several in vitro studies showed antibacterial activity of hesperetin and naringenin against *Helicobacter pylori* (Bae et al. 1999; Fukai et al. 2002; Tombola et al. 2003; Burger et al. 2002). Naringenin displays its antibacterial effects by disruption of the cytoplasmic membrane and targeting DNA of *Staphylococcus aureus* (Wang et al. 2017, Wang et al. 2018; Duda-Madej et al. 2020). In a recent study, it was demonstrated that Pompia and lemon juices have inhibitory and antibiofilm activity against pathogenic bacteria like *Streptococcus aureus, Pseudomonas aeruginosa*, and *Enterococcus faecalis* (Barberis et al. 2020). A methoxy derivative of the flavanone naringenin extracted from *Polymnia fruticose* called Sakuranetin successfully blocked FabZ in *H. pylori* with MIC 87.3 μM (Zhang et al. 2008).

10.2.4 Flavan-3-ols as Antimicrobial Agent

Fresh leaves of green tea (*Camellia sinensis*) are a major source of polyphenols, particularly catechins, a monomeric flavanols. High number of flavan-3-ol monomers are obtained from green tea, the main component found is (−)-epigallocatechin-3-gallate (EGCG), containing around 60% of the total catechin content followed by (−)-epigallocatechin (EGC) containing around 20% catechin. The other components include (−)-epicatechin-3-gallate (ECG) with approximately 14% and (−)-epicatechin (EC) with approximately 6% catechin content (McKay and Blumberg 2002; Khan and Mukhtar 2018). Catechins can cause damage to the bacterial membrane and appear to have more potency against Gram-positive bacteria. The presence of lipopolysaccharide makes Gram-negative bacteria more resistant to these chemicals. Like other polyphenols, tea polyphenol also displays a wide range of health benefits due to its antioxidant property. It also shows anticancer properties as well as being beneficial in cardiovascular complications (Navarro-Martínez et al. 2006; Hemaiswarya et al. 2008). McNaught in mid-1990s first showed antimicrobial activity of tea when he observed that brewed black tea could kill *B. melitensis* and *Typhi* (McNaught 1906). The extract obtained from green tea displayed anti-adhesive properties as it can inhibit the adherence of several microorganisms on the host cell membrane (Lee et al. 2009b). The extract obtained from green tea is found to be promising against the inhibition of MRSA (Taylor 2020). Cho et al. explored the antibacterial effect of polyphenols obtained from green tea on 30 clinical isolates of *S. aureus,* including 13 MRSA isolates (Cho et al. 2008). He observed MIC in the range of 50–80 μg/mL for both the MRSA and the non-MRSA strains (Cho et al. 2008). It is also found that green tea extract was a potent inhibitor of *E. coli* in cases of urinary tract infections (UTI), EGC compounds which are oozed in urine prevented 99% of strains tested at MIC concentration of $\leq$4.0 mg/mL (Reygaert and Jusufi 2013; Yang et al. 1998). The antimicrobial potential of green tea extract was also studied against *P. aeruginosa* (Radji et al. 2013; Taylor 2020; Jazani et al. 2007; Yi et al. 2010). Green tea extract displayed significant inhibitory activity against clinical MDR isolates of *P. aeruginosa* (MIC and MBC 2.06 ± 1.76 and 2.54 ± 2.22 mg/mL, respectively). ECGC has shown inhibitory potential against 18 clinical isolates of nosocomial pathogen *Stenotrophomonas maltophilia* by inhibiting dihydrofolate reductase. The compound has the potential to replace trimethoprim in patients who cannot tolerate the side effects associated with this antibiotic (Mun et al. 2014).

In one interesting activity, the antimicrobial effect of Chinese green tea beverages was analyzed. The study was performed through different parts of the digestive system with events after duodenal, gastric, and gastroduodenal parts were analyzed. After passing through different digestive processes green tea has shown significant activity against *Staphylococcus* with different MIC concentrations at each step. The anti-*Staphylococcus* activity was performed by different chromatographic methods like RP HPLC and mass spectrometry. Catechin concentration of green tea was resistant to decomposition under gastric condition while it was reversed in the case of duodenal and gastroduodenal processes. Galloxyl catechin, not native catechin, is the component responsible for antibacterial behavior of green tea which is generated after decomposition of duodenal digestion (Marchese et al. 2014).

EGCG is considered the most potent among the various tea catechins due to its broad range and increased antimicrobial activities (Reygaert and Jusufi 2013). It shows effective antibacterial activity against *E. coli O157:H7, P. aeruginosa, S. aureus,* and *S. mutans* (Cui et al. 2012). Various morphological changes are induced by EGCG in *S. aureus*. EGCG binds with the peptidoglycan layer of bacterial membrane making the cell surface wrinkled and aggregated. The compound can also make structural changes in bacterial proteins and peptides by disruption of H-bonding. The effect is induced by the hydroxyl group or galloyl group of the compound (Oneda et al. 2003; Zhao et al. 2001; Yoda et al. 2004). EGCG also effects temporary change to the cell wall of *E. coli O157:H7,* a foodborne pathogen responsible for hemorrhagic colitis and uremic syndrome. Pore-like lesions were induced by EGCG in bacteria which also show concentration-dependent effect. Hydrogen peroxide production was mainly responsible for damage to the cell wall of Gram-negative bacteria (Ikigai et al. 1993). Transcription of the key virulence gene was also altered by this compound as it also inhibited biofilm formation and swarming motility to this food-borne pathogen. Extracellular release of Shiga toxin from *E. coli O157:H7* was also hindered by EGCG (Sugita-Konishi et al. 1999). This compound is also effective in inhibition of biofilm and blocking the swarming motility of other Gram-negative bacteria like *Burkholderia cepacian* and staphylococcal isolates (Huber et al. 2003). Several Gram-negative bacteria show resistance to carbapenems, a β-lactam, having broad-spectrum antibacterial activity. The resistance mechanism involves increased assembly of carbapenem hydrolyzing β-lactamases(Livermore 1992). *Klebsiella pneumoniae* is also resistant to imipenem and EGCG is very effectively used as inhibitor and stressor to these disease-causing bacteria with observed MIC between 300 and 650 μg/mL. EGCG also shows synergistic activity with imipenem against such bacteria (Cho et al. 2011). Green tea catechins, especially EGCG, is also very effective against different viruses (Xu et al. 2017; Yang et al. 2014; Carneiro et al. 2016; Raekiansyah et al. 2018).

10.3 Tannins and Its Antimicrobial Activity

Tannins are also a type of phenolic molecule found in several parts of the plant, barks, leaves, fruits, wood, and roots (Serrano et al. 2009; Chai et al. 2018). The oxygenation pattern and degree of polymerization determine the biological properties of tannins. Several researchers indicate that tannins are very effective against filamentous fungi, yeast, and other microorganisms (Guo et al. 2018). The antimicrobial effect of tannins is well-defined against several pathogens (Chung et al. 1993). They can act by several mechanisms such as by inhibition of extracellular enzymes and targeting oxidative phosphorylation in bacteria (Guo et al. 2018; Olchowik-Grabarek et al. 2020; Schofield et al. 2001). Chebulinic acid, a kind of tannin isolated from *Terminalia chebula*, has inhibitory potential against quinolone-resistant *M. tuberculosis* DNA gyrase (Patel et al. 2015). The importance of chebulinic acid as DNA gyrase inhibitor and antituberculosis agent can be explored further as it can be a promising candidate for AMR mitigation.

10.3.1 Proanthocyanidins

Proanthocyanidins (PACs) are polymers of flavan-3-ol, having a common C6–C3–C6 flavone skeleton (Esquivel-Alvarado et al. 2021). The source includes fruits, berries, leaves, and seeds (Zeng et al. 2020; Rauf et al. 2019). Most of the research about PACs is conducted on berries, which are rich sources of dietary polyphenols, particularly they contain a considerable number of PACs with structural and compositional diversity depending on the type of berry species. Besides PACs, berries are also rich source of many flavonols (especially blackcurrant, cranberry, lingonberry, and whortleberry), anthocyanidins (especially bilberry and black currant), hydroxycinnamic acids, and hydroxybenzoic acid (Puupponen-Pimiä et al. 2005a; Hellström et al. 2007; Kähkönen et al. 2001). Other sources of PACs include beer, wine, beverages, and some common fruit juices. Proanthocyanidins and flavan-3-ols are effectively used in the prevention and treatment of pathogens associated with periodontitis (Nawrot-Hadzik et al. 2021a, b).

Proanthocyanidins which comprise A-type linkage have been extracted from cranberry and these PACs display potent antibacterial and antiadhesive properties (Baranowska and Bartoszek 2016; Narwojsz et al. 2019; Luís et al. 2017). They are effective in inhibiting *E. coli* attachment to uroepithelium (Reid et al. 2001; Raguzzini et al. 2020). *E. coli* is responsible for more than 80% of all urinary tract infections acquired in the community and it is found that frequently taking cranberry juice can reduce the risk of urinary tract infections (Luís et al. 2017; Singh et al. 2016; Barbosa-Cesnik et al. 2011), also its exposure can also dislodge pre-attached *E. coli* (Singh et al. 2016; Howell et al. 2005; Guay 2009). In one meta-analysis, cranberry intake was recommended for the treatment of urinary tract infection, particularly in patients having recurrent episodes (Das 2020; Luís et al. 2017; Gbinigie et al. 2020). It is evident that plant tannins are effective toward *E. coli O157:*H7 and this potency relies on the resources of tannins (Wang et al. 2013b; Bumunang et al. 2020; Min et al. 2007), particularly purple prairie clover condensed tannins prevented *E. coli O157:H7* growth at concentrations as minimum as 20 μg/mL (Štumpf et al. 2020; Liu et al. 2013b).

Cranberry (*Vaccinium macrocarpon*), a rich source of proanthocyanidins, has been extensively explored as natural weapons against periodontal diseases (Feghali et al. 2012). In one interesting study, Orophenol (a mixture of polyphenols from different wild berries) displayed interesting anti-caries properties against *Streptococcus mutans* (Souissi et al. 2021).

Low bush wild blueberry is a rich source of polyphenols like PACs, anthocyanins, and monomeric phenolic acid. They are very effective in reducing the growth of *E. coli O157:H7,* depending upon the equivalent content of gallic acid used. Low bush wild blueberry significantly increased permeability of *E. coli* cell membrane making it susceptible for inhibition (Lacombe et al. 2013). Virulence of *S. mutans* was slowed down by PACs as they can inhibit bacterial adhesion to apatitic surface, and it also inhibited the growth of biofilm by oral bacteria in vitro (Souissi et al. 2021; Yamanaka et al. 2004; Singhal et al. 2020; Sánchez et al. 2020). Isolates obtained from proanthocyanidins were able to reduce the acidogenicity of *S. mutans* biofilms in in vitro conditions (Yamanaka et al. 2004; Gregoire et al. 2007). According to some reports, cranberry juice provides a promising therapeutic option for diseases related to biofilm formation (Souissi et al. 2021). Tannin extracts from plants like *Anadenanthera* and *Commiphora* in Brazil were able to inhibit the formation of biofilm as they disrupt bacterial membrane and inhibit matrix production; they are used as folk medicine in Brazil (Trentin et al. 2013). Tannin extract particularly proanthocyanidins have bacteriostatic property and they also exhibit anti-adhesive property against *P. aeruginosa* with observed MIC between 1 and 4 mg/mL (Trentin et al. 2013).

10.3.2 Hydrolyzable Tannins

Hydrolyzable tannins are made of complex monomeric or oligomeric polyester molecules of phenolic acid and α-glucose or related sugar molecule or polyol. They are synthesized by large varieties of plants and trees and are found in barks, wood, leaves, and galls of such plants. Esters with hexahydroxydiphenic acid are named as ellagitannins while the compound is named as gallotannin when the acidic part of Hydrolyzable tannins is gallic acid ester. These compounds display several useful biological properties such as anticancer, antimicrobial, and antioxidant activity (Yang and Liu 2014; Valverde Malaver et al. 2019; Smeriglio et al. 2017; Ekambaram et al. 2016). A promising inhibitory activity was displayed toward Shiga toxin-producing (STEC) *E. coli O157:H7* in murine model by ellagitannins extracted from *Quercus infectoria.* These compounds can provide alternative therapeutic option against infection caused by this enterohemorrhagic organism as these compounds can remove the colonization of *E. coli O157:H7* in the intestinal tract of mice (Voravuthikunchai et al. 2012).

Antibacterial and antiviral potency of various other berries and their phenolic extracts were also studied toward many pathogens (Puupponen-Pimiä et al. 2005a, b; Nohynek et al. 2006). Strawberry containing little amounts of ellagitannins displayed moderate activity while cloudberry and raspberry isolates were highly effective against *Staphylococcus* spp. and *Salmonella* spp. (Nohynek et al. 2006). Extracts from Finnish berry rich in ellagitannins were found to have potent antibacterial activity, the highest antimicrobial activity was observed from berries from genus Rubus (cloudberry and raspberry abundant in ellagitannins) (Rauha et al. 2000). Another rich source of hydrolyzable tannins is methanol extract of pomegranate containing high amount of ellagic acid and gallic acid, punicalins, and punicalagins (Reddy et al. 2007; Saeed et al. 2018). Oligomeric ellagitannins are most potent antimicrobial component extracted from pomegranate, although it also contains anthocyanins and flavanols showing good activity alone or in combination with other compounds (Reed et al. 2005; Fourati et al. 2020; Elshafie et al. 2021; Das et al. 2021). Compound extracted from pomegranate show activity against several enteric pathogenic bacteria including *E. coli* O157:H7 (Pagliarulo et al. 2016; Voravuthikunchai et al. 2004), *Listeria monocytogenes* (Lucas and Were 2009), *V. cholerae* (Guevara et al. 1994; Mathabe et al. 2006), *Salmonella* spp. (Wafa et al. 2017; Rani and Khullar 2004), and some viruses (Moradi et al. 2020). Additionally, methanolic peel extract of pomegranate also displayed excellent inhibitory activity against *S. mutans, L. acidophilus*, and *S. salivarius* (Fourati et al. 2020; Abdollahzadeh et al. 2011). Peel extract of *Punica Granatum* is also very effective against oral candidiasis (Madugula et al. 2017; Kumar et al. 2020; Paul et al. 2018).

10.4 Non-flavonoid Polyphenols and Their Antimicrobial Activity

Non-flavonoid polyphenolic compounds include phenolic acids. Many free phenolic acids are found in fruits and vegetables, although they mostly occur in the bound form in grains and seeds (Veloso et al. 2020; Chandrasekara and Shahidi 2010). Some polyphenolic compounds like (−) epicatechin and oleuropeinglucoside, caffeic acid, gallic acid, ferulic acid, phloridzin, etc., act as anti-quorum sensing molecules (Asfour 2018). Inhibitory potential of such molecules was confirmed by disk diffusion assay based on pigment inhibition in *Chromobacterium violaceum* (CV12472) by Borges et al. (2014). They observed that this polyphenolic compound can alter the quorum sensing mechanism of *C. violaceum* CviI/CviR homologs of LuxI/LuxR systems (Borges et al. 2014).

Caffeic acid displayed antibacterial activity against *S. aureus* and *Escherichia coli* having MIC equal to 1600 μg/mL (Lim et al. 2016). Gallic acid another kid of phenolic acid was found to be quite effective against *Enterococcus faecalis* (Aires et al. 2013), and it also showed good inhibitory activity against the bacteria *Campylobacter* with MIC equal to 125 μg/mL (Sarjit et al. 2015). Gallic acid also showed inhibitory activity against *P. aeruginosa* (MIC equal to 500 μg/mL), *S. aureus* (MIC equal to 1750 μg/mL), *E. coli* (MIC equal to 1500 μg/mL), and *Listeria monocytogenes* (MIC equal to 2000 μg/mL). Ferulic acid was also effective against bacteria *E. coli* and *P. aeruginosa* (MIC equal to 100 μg/mL), *L. monocytogenes* (MIC equal to 1250 μg/mL), and *S. aureus* (MIC equal to 1100 μg/mL). The target of gallic acid and ferulic acid was bacterial cell wall in *E. coli, S. aureus*, and *P. aeruginosa* generating local damage and leading to leakage of cellular material (Borges et al. 2013).

Many respiratory pathogenic bacteria were the target of phenolic compounds like gallic acid and ethyl gallate isolated from wine (Cueva et al. 2012; Friedman

2014; Lima et al. 2019). Several respiratory pathogens like *E. faecalis, S. aureus, Moraxella catarrhalis*, and *P. aeruginosa* are susceptible toward these compounds. *M. catarrhalis* is the most sensitive bacteria toward some non-flavonoid molecules. Also, these phenolic compounds were found to be more potent toward Gram-negative bacteria compared to Gram-positives ones (Cueva et al. 2012). Gallic and caffeic acid inhibited the synthesis of α-hemolysin displaying anti-staphylococcal activity. They worked with different mechanisms as gallic acid targeted the adhesive property of *S. aureus*, and caffeic acid made interference with the structure of cell membrane (Dos Santos et al. 2018; Luís et al. 2014). Some phenolic compounds like cinnamic acid and *p*-coumaric acid displayed anti-adhesive property against pathogenic bacteria *L. monocytogenes*. Other effective inhibitors against bacterial pathogens are ferulic acid and caffeic acid. These studies showed that there is a positive correlation between hydroxylation of the cinnamic acid with antibacterial activity (Pernin et al. 2019; Vasconcelos et al. 2018). Some phenolic compounds isolated from blueberries like chlorogenic acid and caffeic acid were able to inhibit biofilm formation in pathogenic bacteria like *S. epidermidis* and *P. aeruginosa*. Chlorogenic acid displayed the antibacterial activity at the concentration of 4 mg/mL, against *S. epidermidis* and *P. aeruginosa*. Caffeic acid was also able to inhibit biofilm formation in *S. epidermidis*, but it was not effective in inhibition of growth of this bacteria (Zimmer et al. 2014). In one study, it was observed that coffee can inhibit biofilm formation in *S. mutans* and chlorogenic acid; trigonelline and caffeine were the major parts showing highest anti-adhesive activity (Daglia et al. 1998). Antibacterial activities against bacteria causing periodontitis were shown by catechol (MIC 4.9–312.5 μg/mL) and pyrogallol (MIC 2.4–2500 μg/mL) (Shahzad et al. 2015). In one study antimicrobial activity of 22 polyphenolic compounds was observed against several Gram-positive and Gram-negative bacteria and they found pyrogallol-based compounds were more effective compared to catechol or resorcinol (Taguri et al. 2006).

Wild polish mushroom is found to be quite rich in polyphenols such as protocatechuic acid, vanillic acid, syringic acid, 4-hydroxybenzoic acid, caffeic acids, and *p*-coumaric and ferulic acids. These compounds displayed intermediate antibacterial activity with MIC ranging from 156 to 5000 μg/mL against a variety of bacteria showing stronger activity against Gram-positive bacteria (Nowacka et al. 2015). *Galla rhois*, a plant from Korea, is a rich source of methyl gallate that showed antibacterial activity against *Salmonella* with MIC between 3.9 and 125 μg/mL (Choi et al. 2014). It also demonstrated effective antimicrobial activity against *E. coli* and *Shigella flexneri* with MIC of 30 μg/mL (Madikizela et al. 2013). Another rich source of polyphenol is *Potentilla* species which is quite rich in compounds like caffeic acid, ferulic acid, rutin, hyperoside, (+)-catechin, and ellagic acid. *P. fruticose* displayed the highest effect against *P. aeruginosa*, and a fungus *Candida albicans* with MIC ranging from 0.78 to 6.25 mg/mL (Wang et al. 2013a). The synergistic activity of isoquercitrin (10 μg/mL) with gallic acid (10 μg/mL) was quite effective in inhibiting the growth of *S. aureus* as their combined MIC was found to be 10 times less compared to their individual activities (Soberón et al. 2014).

The ethanol extract of *Searsia chirindensis* containing myricetin-3-*O*-arabinopyranoside, myricetin-3-*O*-rhamnoside, methyl gallate, kaempferol-3-*O*-rhamnoside, and quercetin-3-*O*-arabinofuranoside was effective against many bacteria like *E. coli, S. flexneri, Campylobacter jejuni,* and *S. aureus* (Madikizela et al. 2013). The seed kernel of *Mangifera indica* L. was a rich source of phenolic acids like methyl gallate, gallic acid, and pentagalloylglucopyranose. The extracts displayed inhibitory activity against MRSA, and they also show better synergistic activity with penicillin G against MRSA (Jiamboonsri et al. 2011). A compound extracted from *Kaempferia pandurata* Roxb, called Panduratin A (a natural chalcone), displayed inhibiting activity against *S. aureus* with MIC ranging from 0.06 to 2 μg/mL (Cushnie and Lamb 2011; Rukayadi et al. 2009). Aerial part of the plant *Hypericum olympicum* L. contains acylphloroglucinols (Olympicins A–E) which showed antibacterial activity against MRSA and multidrug-resistant *S. aureus* (Shiu et al. 2012). Arzanol is a very effective strong anti-staphylococcal drug that is isolated from *Helichrysum talinum* sp. and is also very potent against different *S. aureus strains* (MIC equal to 1–4 μg/mL) (Taglialatela-Scafati et al. 2013).

Two potent acylphloroglucinols, 1,5-dihydroxy-2-(2′-methylpropionyl)-3-methoxy-6-methylbenzene, and 1,5-dihydroxy-2-(2′-methylbutanoyl)-3-methoxy-6-methylbenzene isolated from *Hypericum beanie* were found to be inhibitory against multidrug-resistant *S. aureus* with MIC between 16 and 32 μg/mL (Shiu and Gibbons 2006). β-Lactamase and penicillin-binding protein 2a (PBP2a) inhibitory activity was shown by epicatechin gallate, catechin gallate, and epigallocatechin gallate as they show antibacterial activity against *S. aureus* and MRSA (Yam et al. 1998; Hu et al. 2002; Sudano Roccaro et al. 2004). *Epigallocatechin gallate* displayed its activity by inhibiting Tet(K) pump and upgrading intracellular concentration of tetracycline and ultimately its activity. These compounds also reduced the MIC of norfloxacin four times against *S. aureus* with NorA (Gibbons et al. 2004). Bromelain is a protein-degrading enzyme and found in large amounts in fresh pineapples applied tenderizing meat. It has been found that that bromelain targets the outer membrane by disintegrating the surface membrane protein in Gram-negative bacteria, causing leakage, swelling, and damage to the cells (Zharfan et al. 2017).

10.5 Synergistic Antimicrobial Activities of Polyphenols and Antibiotics

Beta-lactam antibiotics having a β-lactam nucleus in their chemical structure is one of the most recommended drugs for the treatment of bacterial infections. But regular use of these antibiotics has resulted in bacteria developing resistance to these antibiotics. The drug-resistant bacteria are emerging as a global public health problem. The potential of modulating β-lactam resistance in *S. aureus* through polyphenolic compounds is quite promising (Shiota et al. 1999; Stapleton et al. 2004a; Hu et al. 2001). Several studies have documented the anti-MRSA activity of EGCG, a type of catechins. Different hypotheses have been proposed regarding synergistic activity between EGCG and β-lactam antibiotics. The target site of EGCG is the same as the target site for antibiotics as both targets the peptidoglycan layer (Zhao et al. 2001; Yam et al. 1998), and has been shown to inhibit penicillinase production (Zhao et al. 2001). According to some report, EGC is more potent than EGCG in modulating β-lactam resistance. The effect is more powerful when used in combination with oxacillin against *Staphylococcus*. The occurrence of two rather three hydroxyl groups on the B ring of catechin is a possible cause of this synergistic effect (Shiota et al. 1999; Stapleton et al. 2004b). The combination therapy of cefotaxime (a β-lactam antibiotic) and EGCG is found to be very effective toward extended-spectrum beta-lactamase (ESBL) *E. coli* as a topical application to skin. ECGC was very effective in reducing tetracycline resistance in staphylococci by preventing tetracycline efflux pump, Tet(K) (Sudano Roccaro et al. 2004).

Another synergistic activity of polyphenols was displayed against amoxicillin-resistant *E. coli* isolates by application of luteolin (80 μg/mL) in combination with amoxicillin (70 μg/mL). After 6 h of treatment the mixture of amoxicillin and luteolin reduced the bacterial cell count to 1×10^3 CFU mL^{-1}. Luteolin can work by reduction of β-lactamase activity, by disrupting protein and peptidoglycan synthesis, and it can alter permeability of outer and inner bacterial cell membranes (Eumkeb et al. 2012a, b). Ceftazidime in combination with flavones, apigenin, and naringenin displayed antimicrobial effects against ceftazidime-resistant *Enterobacter cloacae* in vitro. Apigenin was quite useful in reversing resistance activity by affecting the cell structure and the blockage of β-lactamase activity (Eumkeb and Chukrathok 2013). Other flavanols such as galangin and kaempferide also displayed similar activity and inhibited the hydrolysis of benzylpenicillin (Sudano Roccaro et al. 2004). Chalcones are another class of phenolic compound that show a synergistic effect in mitigating AMR. Chalcones extracted from *Dalea versicolor* inhibited NorA efflux pump and effectively decreased the MIC of erythromycin against *S. aureus* (Belofsky et al. 2004). Holler et al. studied library of 117 natural and synthetic chalcones for their antimicrobial activity. Among them the potency of two chalcones was equivalent to reserpine, an alkaloid with compelling NorA efflux pump-repressive activity (Holler et al. 2012b). Synergistic activity between various green and black tea extracts in combination with several antibiotics was studied toward some enteropathogens by Tiwari et al. who reported better antibacterial activity by green tea as compared to black tea and more effectiveness against enteropathogenic *Shigella dysenteriae*. Tea extract also displayed superior synergistic activity when applied in combination with chloramphenicol (Tiwari et al. 2005).

10.6 Conclusions

The discovery of penicillin as the first antibiotic was an outstanding scientific breakthrough in medicine which was responsible for mankind gaining tremendous advantage and an important tool to treat and control bacterial infections. It also spawned a whole new area of research for development of new and more effective antibiotics creating several new milestones in treating infectious diseases. However, the emergence of life-threatening drug-resistant pathogens has become a very important problem that can impact and impair the global public health in a serious way. The development of new effective antimicrobials with a different mechanism of action against infectious pathogens remains an urgent task. Scientists are looking to plant defense molecules as a new ray of hope and more focus is on the combination of plant extracts and antibiotics for their synergistic activity against such pathogens. Phenol and phenolic acid compounds are the potential plant origin molecules that can provide a novel preventive and therapeutic pathway in combination with antibiotics against such pathogens. Recently, research on bioactive compounds, particularly polyphenols, has gained momentum. Polyphenolic compounds like flavonols and tannins isolated from different plant sources have displayed antimicrobial, antiviral, and antifungal activities. Green tea, berry extracts, mushrooms, curcumin, and propolis are rich source of polyphenolic compounds and found to display potent activity against several bacteria. Pyrogallol-based extracts are showing better results than other polyphenol derivatives such as resorcinol, gallic acid, and ferulic acid and in destabilizing the cell wall of *S. aureus, E. coli, P. aeruginosa,* etc., causing leakage of cellular contents.

Plant polyphenols exert antibacterial activity by targeting the physiology of bacteria by various mechanisms. They can alter bacterial membrane affecting membrane proteins and permeability. They also affect various virulence factors associated with the pathogenicity of microorganisms. They can downregulate bacterial enzymes and toxins and affect the signaling receptor molecules. The other important mechanisms include the inhibition of biofilm. Polyphenols are more potent when used in combination with different antibiotics reducing the phenomenon of AMR.

Several methods are available to evaluate the antimicrobial activity of polyphenols and their derivatives. The assays include disk diffusion and agar dilution method and micro and macro dilution method. The size of the inoculum, the size of the well or proper disk as well as the incubation period are variables, which must be considered in various assays. Although polyphenols are promising targets for their antimicrobial properties and AMR, there are many discrepancies available, mainly in the application of different assays and solvents used. Many *in vitro* susceptibility tests and marks are needed as tentative breakpoints for the most encouraging polyphenolic compounds. More such studies are needed to confirm pathogen as susceptible or resistant to an environmental agent, and the authenticity of the susceptibility method is fundamental and important to confirm the in vitro and in vivo activities. In summary, polyphenols represent possible alternatives as antimicrobial agents that can be exploited either alone or in combination with antibiotics for treatment and management of infectious diseases. However, much research further is needed in this area to exploit this opportunity, especially, in an era of antibiotic-resistant bacteria which can be considered as a future looming public health crisis.

Acknowledgment The authors would like to thank Prof. Amin Beg, King Abdul Aziz University, Jeddah, Saudi Arabia for improving the language of the manuscript.

Conflict of Interest There is no conflict of interest among the authors.

References

Abdel-Massih R, Abdou E, Baydoun E, Daoud Z (2010) Antibacterial activity of the extracts obtained from *Rosmarinus officinalis*, *Origanum majorana*, and *Trigonella foenum-graecum* on highly drug-resistant gram negative bacilli. J Bot 2010:464087

Abdollahzadeh S, Mashouf R, Mortazavi H, Moghaddam M, Roozbahani N, Vahedi M (2011) Antibacterial and antifungal activities of punica granatum peel extracts against oral pathogens. J Dent (Tehran) 8(1):1–6

Abisado RG, Benomar S, Klaus JR, Dandekar AA, Chandler JR (2018) Bacterial quorum sensing and microbial community interactions. mBio 9(3):e02331–e02317

Abreu AC, McBain AJ, Simões M (2012) Plants as sources of new antimicrobials and resistance-modifying agents. Nat Prod Rep 29(9):1007–1021

Abuelsaad AS, Mohamed I, Allam G, Al-Solumani AA (2013) Antimicrobial and immunomodulating activities of hesperidin and ellagic acid against diarrheic Aeromonas hydrophila in a murine model. Life Sci 93(20):714–722

Aherne SA, O'Brien NM (2002) Dietary flavonols: chemistry, food content, and metabolism. Nutrition 18(1):75–81

Aires A, Marques E, Carvalho R, Rosa EA, Saavedra MJ (2013) Evaluation of biological value and appraisal of polyphenols and glucosinolates from organic baby-leaf salads as antioxidants and antimicrobials against important human pathogenic bacteria. Molecules 18(4):4651–4668

Albadawi DA, Mothana RA, Khaled JM, Ashour AE, Kumar A, Ahmad SF et al (2017) Antimicrobial, anticancer, and antioxidant compounds from Premna resinosa growing in Saudi Arabia. Pharm Biol 55(1):1759–1766

Al-Bayati FA, Al-Mola HF (2008) Antibacterial and antifungal activities of different parts of Tribulus terrestris L. growing in Iraq. J Zhejiang Univ Sci B 9(2):154–159

Al-Bayati FAY, Sulaiman K (2008) In vitro antimicrobial activity of Salvadora persica L. extracts against some isolated oral pathogens in Iraq. Turkish J Biol 32:57–62

Aliaño-González MJ, Jarillo JA, Carrera C, Ferreiro-González M, Álvarez J, Palma M et al (2020) Optimization of a novel method based on ultrasound-assisted extraction for the quantification of anthocyanins and total phenolic compounds in blueberry samples (Vaccinium corymbosum L.). Foods 9(12):1763

Al-Juraifani AA (2009) Antimicrobial activity of some medicinal plants used in Saudi Arabia. Can J Pure Appl Sci 5:1509–1512

Almuhayawi MS (2020) Propolis as a novel antibacterial agent. Saudi J Biol Sci 27(11):3079–3086

Alvesalo J, Vuorela H, Tammela P, Leinonen M, Saikku P, Vuorela P (2006) Inhibitory effect of dietary phenolic compounds on Chlamydia pneumoniae in cell cultures. Biochem Pharmacol 71(6):735–741

Ankolekar C, Johnson D, Pinto Mda S, Johnson K, Labbe R, Shetty K (2011) Inhibitory potential of tea polyphenolics and influence of extraction time against Helicobacter pylori and lack of inhibition of beneficial lactic acid bacteria. J Med Food 14(11):1321–1329
Annunziata G, Sanduzzi Zamparelli M, Santoro C, Ciampaglia R, Stornaiuolo M, Tenore GC et al (2020) May polyphenols have a role against coronavirus infection? An overview of in vitro evidence. Front Med 7:240
Arima H, Ashida H, Danno G (2002) Rutin-enhanced antibacterial activities of flavonoids against Bacillus cereus and Salmonella enteritidis. Biosci Biotechnol Biochem 66(5):1009–1014
Asahi Y, Noiri Y, Miura J, Maezono H, Yamaguchi M, Yamamoto R et al (2014) Effects of the tea catechin epigallocatechin gallate on Porphyromonas gingivalis biofilms. J Appl Microbiol 116(5):1164–1171
Asfour HZ (2018) Anti-quorum sensing natural compounds. J Microsc Ultrastruct 6(1):1–10
Bae EA, Han MJ, Kim DH (1999) In vitro anti-Helicobacter pylori activity of some flavonoids and their metabolites. Planta Medica 65(5):442–443
Bakar NS, Zin NM, Basri DF (2012) Synergy of flavone with vancomycin and oxacillin against vancomycin-intermediate Staphyloccus aureus. Pak J Pharm Sci 25(3):633–638
Bakht J, Khan S, Shafi M (2014) In vitro antimicrobial activity of Allium cepa (dry bulbs) against Gram positive and Gram negative bacteria and fungi. Pak J Pharm Sci 27(1):139–145
Bankova V, Christov R, Kujumgiev A, Marcucci MC, Popov S (1995) Chemical composition and antibacterial activity of Brazilian propolis. Z Naturforsch C J Biosci 50(3-4):167–172
Baranowska M, Bartoszek A (2016) Antioxidant and antimicrobial properties of bioactive phytochemicals from cranberry. Postepy higieny i medycyny doswiadczalnej (Online) 70:1460–1468
Barberis A, Deiana M, Spissu Y, Azara E, Fadda A, Serra PA et al (2020) Antioxidant, antimicrobial, and other biological properties of pompia juice. Molecules 25(14):3186
Barbosa-Cesnik C, Brown MB, Buxton M, Zhang L, DeBusscher J, Foxman B (2011) Cranberry juice fails to prevent recurrent urinary tract infection: results from a randomized placebo-controlled trial. Clin Infect Dis 52(1):23–30
Barreca D, Gattuso G, Bellocco E, Calderaro A, Trombetta D, Smeriglio A et al (2017) Flavanones: citrus phytochemical with health-promoting properties. BioFactors 43(4):495–506
Barrientos L, Herrera CL, Montenegro G, Ortega X, Veloz J, Alvear M et al (2013) Chemical and botanical characterization of Chilean propolis and biological activity on cariogenic bacteria Streptococcus mutans and Streptococcus sobrinus. Braz J Microbiol 44(2):577–585
Basile A, Sorbo S, Giordano S, Ricciardi L, Ferrara S, Montesano D et al (2000) Antibacterial and allelopathic activity of extract from Castanea sativa leaves. Fitoterapia 71(Suppl 1):S110–S116
Basile A, Sorbo S, Spadaro V, Bruno M, Maggio A, Faraone N et al (2009) Antimicrobial and antioxidant activities of coumarins from the roots of Ferulago campestris (Apiaceae). Molecules 14(3):939–952
Belofsky G, Percivill D, Lewis K, Tegos GP, Ekart J (2004) Phenolic metabolites of Dalea versicolor that enhance antibiotic activity against model pathogenic bacteria. J Nat Prod 67(3):481–484
Benmalek Y, Yahia OA, Belkebir A, Fardeau ML (2013) Anti-microbial and anti-oxidant activities of Illicium verum, Crataegus oxyacantha ssp monogyna and Allium cepa red and white varieties. Bioengineered 4(4):244–248
Betts JW, Sharili AS, Phee LM, Wareham DW (2015) In vitro activity of epigallocatechin gallate and quercetin alone and in combination versus clinical isolates of methicillin-resistant Staphylococcus aureus. J Nat Prod 78(8):2145–2148
Borges A, Ferreira C, Saavedra MJ, Simões M (2013) Antibacterial activity and mode of action of ferulic and gallic acids against pathogenic bacteria. Microb Drug Resist 19(4):256–265
Borges A, Serra S, Cristina Abreu A, Saavedra MJ, Salgado A, Simões M (2014) Evaluation of the effects of selected phytochemicals on quorum sensing inhibition and in vitro cytotoxicity. Biofouling 30(2):183–195
Bosio K, Avanzini C, D'Avolio A, Ozino O, Savoia D (2000) In vitro activity of propolis against Streptococcus pyogenes. Lett Appl Microbiol 31(2):174–177

Brglez Mojzer E, Knez Hrnčič M, Škerget M, Knez Ž, Bren U (2016) Polyphenols: extraction methods, antioxidative action, bioavailability and anticarcinogenic effects. Molecules 21(7):901
Bumunang EW, Ateba CN, Stanford K, Niu YD, Wang Y, McAllister TA (2020) Activity of bacteriophage and complex tannins against biofilm-forming shiga toxin-producing Escherichia coli from Canada and South Africa. Antibiotics (Basel) 9(5):257
Burger O, Weiss E, Sharon N, Tabak M, Neeman I, Ofek I (2002) Inhibition of Helicobacter pylori adhesion to human gastric mucus by a high-molecular-weight constituent of cranberry juice. Crit Rev Food Sci Nutr 42(3 Suppl):279–284
Cannalire R, Machado D, Felicetti T, Santos Costa S, Massari S, Manfroni G et al (2017) Natural isoflavone biochanin A as a template for the design of new and potent 3-phenylquinolone efflux inhibitors against Mycobacterium avium. Eur J Med Chem 140:321–330
Capriotti AL, Cavaliere C, Crescenzi C, Foglia P, Nescatelli R, Samperi R et al (2014) Comparison of extraction methods for the identification and quantification of polyphenols in virgin olive oil by ultra-HPLC-QToF mass spectrometry. Food Chem 158:392–400
Carneiro BM, Batista MN, Braga ACS, Nogueira ML, Rahal P (2016) The green tea molecule EGCG inhibits Zika virus entry. Virology 496:215–218
Castro-Barquero S, Tresserra-Rimbau A, Vitelli-Storelli F, Doménech M, Salas-Salvadó J, Martín-Sánchez V et al (2020) Dietary polyphenol intake is associated with HDL-cholesterol and a better profile of other components of the metabolic syndrome: a PREDIMED-plus sub-study. Nutrients 12(3):689
Chai WM, Huang Q, Lin MZ, Ou-Yang C, Huang WY, Wang YX et al (2018) Condensed tannins from longan bark as inhibitor of tyrosinase: structure, activity, and mechanism. J Agric Food Chem 66(4):908–917
Chandrasekara A, Shahidi F (2010) Content of insoluble bound phenolics in millets and their contribution to antioxidant capacity. J Agric Food Chem 58(11):6706–6714
Chen CC, Huang CY (2011) Inhibition of Klebsiella pneumoniae DnaB helicase by the flavonol galangin. Protein J 30(1):59–65
Chen L, Xin X, Yuan Q, Su D, Liu W (2014) Phytochemical properties and antioxidant capacities of various colored berries. J Sci Food Agric 94(2):180–188
Cheynier V (2005) Polyphenols in foods are more complex than often thought. Am J Clin Nutr 81(1 Suppl):223s–229s
Chin YP, Tsui KC, Chen MC, Wang CY, Yang CY, Lin YL (2012) Bactericidal activity of soymilk fermentation broth by in vitro and animal models. J Med Food 15(6):520–526
Cho YS, Schiller NL, Oh KH (2008) Antibacterial effects of green tea polyphenols on clinical isolates of methicillin-resistant Staphylococcus aureus. Curr Microbiol 57(6):542–546
Cho YS, Oh JJ, Oh KH (2011) Synergistic anti-bacterial and proteomic effects of epigallocatechin gallate on clinical isolates of imipenem-resistant Klebsiella pneumoniae. Phytomedicine 18(11):941–946
Choi JG, Mun SH, Chahar HS, Bharaj P, Kang OH, Kim SG et al (2014) Methyl gallate from Galla rhois successfully controls clinical isolates of Salmonella infection in both in vitro and in vivo systems. PLoS One 9(7):e102697
Chung K-T, Stevens SE Jr, Lin W-F, Wei CI (1993) Growth inhibition of selected food-borne bacteria by tannic acid, propyl gallate and related compounds. Lett Appl Microbiol 17(1):29–32
Ciulu M, Cádiz-Gurrea ML, Segura-Carretero A (2018) Extraction and analysis of phenolic compounds in rice: a review. Molecules 23(11):2890
Cowan MM (1999) Plant products as antimicrobial agents. Clin Microbiol Rev 12(4):564–582
Crozier A, Jaganath IB, Clifford MN (2009) Dietary phenolics: chemistry, bioavailability and effects on health. Nat Prod Rep 26(8):1001–1043
Csepregi R, Temesfői V, Das S, Alberti Á, Tóth CA, Herczeg R et al (2020) Cytotoxic, antimicrobial, antioxidant properties and effects on cell migration of phenolic compounds of selected transylvanian medicinal plants. Antioxidants 9(2):166

Cueva C, Mingo S, Muñoz-González I, Bustos I, Requena T, del Campo R et al (2012) Antibacterial activity of wine phenolic compounds and oenological extracts against potential respiratory pathogens. Lett Appl Microbiol 54(6):557–563
Cui Y, Oh YJ, Lim J, Youn M, Lee I, Pak HK et al (2012) AFM study of the differential inhibitory effects of the green tea polyphenol (-)-epigallocatechin-3-gallate (EGCG) against Gram-positive and Gram-negative bacteria. Food Microbiol 29(1):80–87
Cushnie TP, Lamb AJ (2011) Recent advances in understanding the antibacterial properties of flavonoids. Int J Antimicrob Agents 38(2):99–107
Daglia M (2012) Polyphenols as antimicrobial agents. Curr Opin Biotechnol 23(2):174–181
Daglia M, Papetti A, Dacarro C, Gazzani G (1998) Isolation of an antibacterial component from roasted coffee. J Pharm Biomed Anal 18(1-2):219–225
Daoud Z, Abdou E, Abdel-Massih R (2011) Antibacterial activity of Rheum rhaponticum, Olea europaea, and Viola odorata on ESBL producing clinical isolates of Escherichia coli and Klebsiella pneumoniae. Int J Pharm Sci Res 2:1669–1678
Das S (2020) Natural therapeutics for urinary tract infections—a review. Future J Pharm Sci 6(1):64
Das AK, Nanda PK, Chowdhury NR, Dandapat P, Gagaoua M, Chauhan P et al (2021) Application of pomegranate by-products in muscle foods: oxidative indices, colour stability, shelf life and health benefits. Molecules 26(2):467
Defoirdt T (2018) Quorum-sensing systems as targets for antivirulence therapy. Trends Microbiol 26(4):313–328
Di Lorenzo C, Colombo F, Biella S, Stockley C, Restani P (2021) Polyphenols and human health: the role of bioavailability. Nutrients 13(1):273
Dos Santos JFS, Tintino SR, de Freitas TS, Campina FF, de A Menezes IR, Siqueira-Júnior JP et al (2018) In vitro e in silico evaluation of the inhibition of Staphylococcus aureus efflux pumps by caffeic and gallic acid. Comp Immunol Microbiol Infect Dis 57:22–28
Duan F, Li X, Cai S, Xin G, Wang Y, Du D et al (2014) Haloemodin as novel antibacterial agent inhibiting DNA gyrase and bacterial topoisomerase I. J Med Chem 57(9):3707–3714
Duda-Madej A, Kozłowska J, Krzyżek P, Anioł M, Seniuk A, Jermakow K et al (2020) Antimicrobial O-alkyl derivatives of naringenin and their oximes against multidrug-resistant bacteria. Molecules 25(16):3642
Durazzo A, Lucarini M, Souto EB, Cicala C, Caiazzo E, Izzo AA et al (2019) Polyphenols: a concise overview on the chemistry, occurrence, and human health. Phytother Res 33(9):2221–2243
Dwivedi GR, Tiwari N, Singh A, Kumar A, Roy S, Negi AS et al (2016) Gallic acid-based indanone derivative interacts synergistically with tetracycline by inhibiting efflux pump in multidrug resistant E. coli. Appl Microbiol Biotechnol 100(5):2311–2325
Ekambaram SP, Perumal SS, Balakrishnan A (2016) Scope of hydrolysable tannins as possible antimicrobial agent. Phytother Res 30(7):1035–1045
Elshafie HS, Caputo L, De Martino L, Sakr SH, De Feo V, Camele I (2021) Study of bio-pharmaceutical and antimicrobial properties of pomegranate (Punica granatum L.) leathery exocarp extract. Plants (Basel) 10(1):153
Esquivel-Alvarado D, Alfaro-Viquez E, Krueger CG, Vestling MM, Reed JD (2021) Classification of proanthocyanidin profiles using matrix-assisted laser desorption/ionization time-of-flight mass spectrometry (MALDI-TOF MS) spectra data combined with multivariate analysis. Food Chem 336:127667
Eumkeb G, Chukrathok S (2013) Synergistic activity and mechanism of action of ceftazidime and apigenin combination against ceftazidime-resistant Enterobacter cloacae. Phytomedicine 20(3-4):262–269
Eumkeb G, Siriwong S, Phitaktim S, Rojtinnakorn N, Sakdarat S (2012a) Synergistic activity and mode of action of flavonoids isolated from smaller galangal and amoxicillin combinations against amoxicillin-resistant Escherichia coli. J Appl Microbiol 112(1):55–64
Eumkeb G, Siriwong S, Thumanu K (2012b) Synergistic activity of luteolin and amoxicillin combination against amoxicillin-resistant Escherichia coli and mode of action. J Photochem Photobiol B Biol 117:247–253

Farhadi F, Khameneh B, Iranshahi M, Iranshahy M (2019) Antibacterial activity of flavonoids and their structure-activity relationship: an update review. Phytother Res 33(1):13–40
Feghali K, Feldman M, La VD, Santos J, Grenier D (2012) Cranberry proanthocyanidins: natural weapons against periodontal diseases. J Agric Food Chem 60(23):5728–5735
Ferreira S, Silva F, Queiroz JA, Oleastro M, Domingues FC (2014) Resveratrol against Arcobacter butzleri and Arcobacter cryaerophilus: activity and effect on cellular functions. Int J Food Microbiol 180:62–68
Fischer N, Seo EJ, Efferth T (2018) Prevention from radiation damage by natural products. Phytomedicine 47:192–200
Fourati M, Smaoui S, Hlima HB, Elhadef K, Braïek OB, Ennouri K et al (2020) Bioactive compounds and pharmacological potential of pomegranate (Punica granatum) seeds—a review. Plant Foods Hum Nutr 75(4):477–486
Fraga-Corral M, Otero P, Echave J, Garcia-Oliveira P, Carpena M, Jarboui A et al (2021) By-products of agri-food industry as tannin-rich sources: a review of tannins' biological activities and their potential for valorization. Foods 10(1):137
Friedman M (2014) Antibacterial, antiviral, and antifungal properties of wines and winery byproducts in relation to their flavonoid content. J Agric Food Chem 62(26):6025–6042
Fujita M, Shiota S, Kuroda T, Hatano T, Yoshida T, Mizushima T et al (2005) Remarkable synergies between baicalein and tetracycline, and baicalein and beta-lactams against methicillin-resistant Staphylococcus aureus. Microbiol Immunol 49(4):391–396
Fukai T, Marumo A, Kaitou K, Kanda T, Terada S, Nomura T (2002) Anti-Helicobacter pylori flavonoids from licorice extract. Life Sci 71(12):1449–1463
Gbinigie OA, Spencer EA, Heneghan CJ, Lee JJ, Butler CC (2020) Cranberry extract for symptoms of acute, uncomplicated urinary tract infection: a systematic review. Antibiotics (Basel) 10(1):12
Giacco R, Costabile G, Fatati G, Frittitta L, Maiorino MI, Marelli G et al (2020) Effects of polyphenols on cardio-metabolic risk factors and risk of type 2 diabetes. A joint position statement of the Diabetes and Nutrition Study Group of the Italian Society of Diabetology (SID), the Italian Association of Dietetics and Clinical Nutrition (ADI) and the Italian Association of Medical Diabetologists (AMD). Nutr Metab Cardiovasc Dis 30(3):355–367
Gibbons S, Moser E, Kaatz GW (2004) Catechin gallates inhibit multidrug resistance (MDR) in Staphylococcus aureus. Planta Med 70(12):1240–1242
Girennavar B, Cepeda ML, Soni KA, Vikram A, Jesudhasan P, Jayaprakasha GK et al (2008) Grapefruit juice and its furocoumarins inhibits autoinducer signaling and biofilm formation in bacteria. Int J Food Microbiol 125(2):204–208
Gopu V, Kothandapani S, Shetty PH (2015) Quorum quenching activity of Syzygium cumini (L.) Skeels and its anthocyanin malvidin against Klebsiella pneumoniae. Microb Pathog 79:61–69
Gregoire S, Singh AP, Vorsa N, Koo H (2007) Influence of cranberry phenolics on glucan synthesis by glucosyltransferases and Streptococcus mutans acidogenicity. J Appl Microbiol 103(5):1960–1968
Grosso G, Stepaniak U, Micek A, Stefler D, Bobak M, Pająk A (2017) Dietary polyphenols are inversely associated with metabolic syndrome in Polish adults of the HAPIEE study. Eur J Nutr 56(4):1409–1420
Guay DR (2009) Cranberry and urinary tract infections. Drugs 69(7):775–807
Guevara JM, Chumpitaz J, Valencia E (1994) [The in vitro action of plants on Vibrio cholerae]. Revista de gastroenterologia del Peru 14(1):27–31
Guo J, Sun W, Kim JP, Lu X, Li Q, Lin M et al (2018) Development of tannin-inspired antimicrobial bioadhesives. Acta Biomater 72:35–44
Gutiérrez-Del-Río I, Fernández J, Lombó F (2018) Plant nutraceuticals as antimicrobial agents in food preservation: terpenoids, polyphenols and thiols. Int J Antimicrob Agents 52(3):309–315
Hanski L, Genina N, Uvell H, Malinovskaja K, Gylfe Å, Laaksonen T et al (2014) Inhibitory activity of the isoflavone biochanin A on intracellular bacteria of genus Chlamydia and initial development of a buccal formulation. PLoS One 9(12):e115115

Haroun MF, Al-Kayali RS (2016) Synergistic effect of Thymbra spicata L. extracts with antibiotics against multidrug-resistant Staphylococcus aureus and Klebsiella pneumoniae strains. Iranian J Basic Med Sci 19(11):1193–1200
Hegazi AG, Abd El Hady FK, Abd Allah FA (2000) Chemical composition and antimicrobial activity of European propolis. Z Naturforsch C J Biosci 55(1-2):70–75
Hellström J, Sinkkonen J, Karonen M, Mattila P (2007) Isolation and structure elucidation of procyanidin oligomers from Saskatoon berries (Amelanchier alnifolia). J Agric Food Chem 55(1):157–164
Hemaiswarya S, Kruthiventi AK, Doble M (2008) Synergism between natural products and antibiotics against infectious diseases. Phytomedicine 15(8):639–652
Heng MY, Katayama S, Mitani T, Ong ES, Nakamura S (2017) Solventless extraction methods for immature fruits: evaluation of their antioxidant and cytoprotective activities. Food Chem 221:1388–1393
Hilliard JJ, Krause HM, Bernstein JI, Fernandez JA, Nguyen V, Ohemeng KA et al (1995) A comparison of active site binding of 4-quinolones and novel flavone gyrase inhibitors to DNA gyrase. Adv Exp Med Biol 390:59–69
Holler JG, Christensen SB, Slotved HC, Rasmussen HB, Gúzman A, Olsen CE et al (2012a) Novel inhibitory activity of the Staphylococcus aureus NorA efflux pump by a kaempferol rhamnoside isolated from Persea lingue Nees. J Antimicrob Chemother 67(5):1138–1144
Holler JG, Slotved HC, Mølgaard P, Olsen CE, Christensen SB (2012b) Chalcone inhibitors of the NorA efflux pump in Staphylococcus aureus whole cells and enriched everted membrane vesicles. Bioorg Med Chem 20(14):4514–4521
Hollman PCH, Arts ICW (2000) Flavonols, flavones and flavanols—nature, occurrence and dietary burden. J Sci Food Agric 80(7):1081–1093
Hossain MB, Camphuis G, Aguiló-Aguayo I, Gangopadhyay N, Rai DK (2014) Antioxidant activity guided separation of major polyphenols of marjoram (Origanum majorana L.) using flash chromatography and their identification by liquid chromatography coupled with electrospray ionization tandem mass spectrometry. J Sep Sci 37(22):3205–3213
Hostetler GL, Ralston RA, Schwartz SJ (2017) Flavones: food sources, bioavailability, metabolism, and bioactivity. Adv Nutr 8(3):423–435
Howell AB, Reed JD, Krueger CG, Winterbottom R, Cunningham DG, Leahy M (2005) A-type cranberry proanthocyanidins and uropathogenic bacterial anti-adhesion activity. Phytochemistry 66(18):2281–2291
Hu ZQ, Zhao WH, Hara Y, Shimamura T (2001) Epigallocatechin gallate synergy with ampicillin/sulbactam against 28 clinical isolates of methicillin-resistant Staphylococcus aureus. J Antimicrob Chemother 48(3):361–364
Hu ZQ, Zhao WH, Asano N, Yoda Y, Hara Y, Shimamura T (2002) Epigallocatechin gallate synergistically enhances the activity of carbapenems against methicillin-resistant Staphylococcus aureus. Antimicrob Agents Chemother 46(2):558–560
Huang YH, Huang CC, Chen CC, Yang KJ, Huang CY (2015) Inhibition of Staphylococcus aureus PriA helicase by flavonol kaempferol. Protein J 34(3):169–172
Huber B, Eberl L, Feucht W, Polster J (2003) Influence of polyphenols on bacterial biofilm formation and quorum-sensing. Z Naturforsch C J Biosci 58(11-12):879–884
Iglesias-Carres L, Mas-Capdevila A, Bravo FI, Aragonès G, Muguerza B, Arola-Arnal A (2019) Optimization of a polyphenol extraction method for sweet orange pulp (Citrus sinensis L.) to identify phenolic compounds consumed from sweet oranges. PLoS One 14(1):e0211267
Ikigai H, Nakae T, Hara Y, Shimamura T (1993) Bactericidal catechins damage the lipid bilayer. Biochim Biophys Acta 1147(1):132–136
Ivančajić S, Mileusnić I, Milošević DC (2010) In vitro antibacterial activity of propolis extracts on 12 different bacteria in conditions of 3 various pH values. Arch Biol Sci 62(4):915–934
Jazani NH, Shahabi S, Ali AA, Zartoshti M (2007) Antibacterial effects of water soluble green tea extracts on multi-antibiotic resistant isolates of Acinetobacter sp. Pak J Biol Sci 10(9):1477–1480

Jiamboonsri P, Pithayanukul P, Bavovada R, Chomnawang MT (2011) The inhibitory potential of Thai mango seed kernel extract against methicillin-resistant Staphylococcus aureus. Molecules 16(8):6255–6270

Joshi SS, Howell AB, D'Souza DH (2014) Cronobacter sakazakii reduction by blueberry proanthocyanidins. Food Microbiol 39:127–131

Julibert A, Bibiloni MDM, Bouzas C, Martínez-González M, Salas-Salvadó J, Corella D et al (2019) Total and subtypes of dietary fat intake and its association with components of the metabolic syndrome in a mediterranean population at high cardiovascular risk. Nutrients 11(7):1493

Kähkönen MP, Hopia AI, Heinonen M (2001) Berry phenolics and their antioxidant activity. J Agric Food Chem 49(8):4076–4082

Kalli S, Araya-Cloutier C, Hageman J, Vincken J-P (2021) Insights into the molecular properties underlying antibacterial activity of prenylated (iso)flavonoids against MRSA. Sci Rep 11(1):14180

Kang SS, Kim JG, Lee TH, Oh KB (2006) Flavonols inhibit sortases and sortase-mediated Staphylococcus aureus clumping to fibrinogen. Biol Pharm Bull 29(8):1751–1755

Khan N, Mukhtar H (2018) Tea polyphenols in promotion of human health. Nutrients 11(1):39

Kim JD, Liu L, Guo W, Meydani M (2006) Chemical structure of flavonols in relation to modulation of angiogenesis and immune-endothelial cell adhesion. J Nutr Biochem 17(3):165–176

Klančnik A, Šikić Pogačar M, Trošt K, Tušek Žnidarič M, Mozetič Vodopivec B, Smole MS (2017) Anti-Campylobacter activity of resveratrol and an extract from waste Pinot noir grape skins and seeds, and resistance of Camp. jejuni planktonic and biofilm cells, mediated via the CmeABC efflux pump. J Appl Microbiol 122(1):65–77

Kopustinskiene DM, Jakstas V, Savickas A, Bernatoniene J (2020) Flavonoids as anticancer agents. Nutrients 12(2):457

Koru O, Toksoy F, Acikel CH, Tunca YM, Baysallar M, Uskudar Guclu A et al (2007) In vitro antimicrobial activity of propolis samples from different geographical origins against certain oral pathogens. Anaerobe 13(3-4):140–145

Kozłowska A, Szostak-Wegierek D (2014) Flavonoids—food sources and health benefits. Roczniki Panstwowego Zakladu Higieny 65(2):79–85

Křížová L, Dadáková K, Kašparovská J, Kašparovský T (2019) Isoflavones. Molecules 24(6):1076

Kujumgiev A, Tsvetkova I, Serkedjieva Y, Bankova V, Christov R, Popov S (1999) Antibacterial, antifungal and antiviral activity of propolis of different geographic origin. J Ethnopharmacol 64(3):235–240

Kumar KSP, Samlin SS, Siva B, Sudharshan R, Vignesswary A, Divya K (2020) Punica granatum as a salutiferous superfruit in the treatment of oral candidiasis—an in-vitro study. J Oral Maxillofac Pathol 24(1):188–189

Lacombe A, Tadepalli S, Hwang CA, Wu VC (2013) Phytochemicals in lowbush wild blueberry inactivate Escherichia coli O157:H7 by damaging its cell membrane. Foodborne Pathog Dis 10(11):944–950

Lechner D, Gibbons S, Bucar F (2008) Plant phenolic compounds as ethidium bromide efflux inhibitors in Mycobacterium smegmatis. J Antimicrob Chemother 62(2):345–348

Lee HC, Jenner AM, Low CS, Lee YK (2006) Effect of tea phenolics and their aromatic fecal bacterial metabolites on intestinal microbiota. Res Microbiol 157(9):876–884

Lee KM, Kim WS, Lim J, Nam S, Youn M, Nam SW et al (2009a) Antipathogenic properties of green tea polyphenol epigallocatechin gallate at concentrations below the MIC against enterohemorrhagic Escherichia coli O157:H7. J Food Protect 72(2):325–331

Lee JH, Shim JS, Chung MS, Lim ST, Kim KH (2009b) In vitro anti-adhesive activity of green tea extract against pathogen adhesion. Phytother Res 23(4):460–466

Lee J-H, Regmi Sushil C, Kim J-A, Cho Moo H, Yun H, Lee C-S et al (2011) Apple flavonoid phloretin inhibits Escherichia coli O157:H7 biofilm formation and ameliorates colon inflammation in rats. Infect Immun 79(12):4819–4827

Leuner O, Havlik J, Hummelova J, Prokudina E, Novy P, Kokoska L (2013) Distribution of isoflavones and coumestrol in neglected tropical and subtropical legumes. J Sci Food Agric 93(3):575–579
Lim A, Subhan N, Jazayeri JA, John G, Vanniasinkam T, Obied HK (2016) Plant phenols as antibiotic boosters: in vitro interaction of olive leaf phenols with ampicillin. Phytother Res 30(3):503–509
Lima MC, Paiva de Sousa C, Fernandez-Prada C, Harel J, Dubreuil JD, de Souza EL (2019) A review of the current evidence of fruit phenolic compounds as potential antimicrobials against pathogenic bacteria. Microb Pathog 130:259–270
Liu H, Mou Y, Zhao J, Wang J, Zhou L, Wang M et al (2010) Flavonoids from Halostachys caspica and their antimicrobial and antioxidant activities. Molecules 15(11):7933–7945
Liu R, Zhang H, Yuan M, Zhou J, Tu Q, Liu JJ et al (2013a) Synthesis and biological evaluation of apigenin derivatives as antibacterial and antiproliferative agents. Molecules 18(9):11496–11511
Liu XL, Hao YQ, Jin L, Xu ZJ, McAllister TA, Wang Y (2013b) Anti-Escherichia coli O157:H7 properties of purple prairie clover and sainfoin condensed tannins. Molecules 18(2):2183–2199
Livermore DM (1992) Carbapenemases. J Antimicrob Chemother 29(6):609–613
Lu Y, Joerger R, Wu C (2011) Study of the chemical composition and antimicrobial activities of ethanolic extracts from roots of Scutellaria baicalensis Georgi. J Agric Food Chem 59(20):10934–10942
Lucas DL, Were LM (2009) Anti-Listeria monocytogenes activity of heat-treated lyophilized pomegranate juice in media and in ground top round beef. J Food Protect 72(12):2508–2516
Luís Â, Silva F, Sousa S, Duarte AP, Domingues F (2014) Antistaphylococcal and biofilm inhibitory activities of gallic, caffeic, and chlorogenic acids. Biofouling 30(1):69–79
Luís Â, Domingues F, Pereira L (2017) Can cranberries contribute to reduce the incidence of urinary tract infections? A systematic review with meta-analysis and trial sequential analysis of clinical trials. J Urol 198(3):614–621
Ma H, Johnson SL, Liu W, DaSilva NA, Meschwitz S, Dain JA et al (2018a) Evaluation of polyphenol anthocyanin-enriched extracts of blackberry, black raspberry, blueberry, cranberry, red raspberry, and strawberry for free radical scavenging, reactive carbonyl species trapping, anti-glycation, anti-β-amyloid aggregation, and microglial neuroprotective effects. Int J Mol Sci 19(2):461
Ma DSL, Tan LT-H, Chan K-G, Yap WH, Pusparajah P, Chuah L-H et al (2018b) Resveratrol-potential antibacterial agent against foodborne pathogens. Front Pharmacol 9:102
Madikizela B, Aderogba MA, Van Staden J (2013) Isolation and characterization of antimicrobial constituents of Searsia chirindensis L. (Anacardiaceae) leaf extracts. J Ethnopharmacol 150(2):609–613
Madugula P, Reddy S, Koneru J, Rao AS, Sruthi R, Dalli DT (2017) "Rhetoric to reality"—efficacy of punica granatum peel extract on oral candidiasis: an in vitro study. J Clin Diagn Res 11(1):ZC114–ZC117
Makhuvele R, Naidu K, Gbashi S, Thipe VC, Adebo OA, Njobeh PB (2020) The use of plant extracts and their phytochemicals for control of toxigenic fungi and mycotoxins. Heliyon 6(10):e05291
Malik SN (2015) Antibacterial activity of olive (Olea europaea) leaves and arugula (Eruca sativa) seeds extract. Int J Pharmacogn Phytochem Res 7:307–310
Mandalari G, Bennett RN, Bisignano G, Trombetta D, Saija A, Faulds CB et al (2007) Antimicrobial activity of flavonoids extracted from bergamot (Citrus bergamia Risso) peel, a byproduct of the essential oil industry. J Appl Microbiol 103(6):2056–2064
Marchese A, Coppo E, Sobolev A, Rossi D, Mannina L, Daglia M (2014) Influence of in vitro simulated gastroduodenal digestion on the antibacterial activity, metabolic profiling and polyphenols content of green tea (Camellia sinensis). Food Res Int 63:182–191
Marín L, Miguélez EM, Villar CJ, Lombó F (2015) Bioavailability of dietary polyphenols and gut microbiota metabolism: antimicrobial properties. Biomed Res Int 2015:905215

Marquis A, Genovese S, Epifano F, Grenier D (2012) The plant coumarins auraptene and lacinartin as potential multifunctional therapeutic agents for treating periodontal disease. BMC Complement Alternat Med 12:80

Mathabe MC, Nikolova RV, Lall N, Nyazema NZ (2006) Antibacterial activities of medicinal plants used for the treatment of diarrhoea in Limpopo Province, South Africa. J Ethnopharmacol 105(1-2):286–293

McKay DL, Blumberg JB (2002) The role of tea in human health: an update. J Am Coll Nutr 21(1):1–13

McNaught JG (1906) On the action of cold or lukewarm tea on Bacillus typhosus. J Royal Army Med Corps 7:372–373

Mehany T, Khalifa I, Barakat H, Althwab SA, Alharbi YM, El-Sohaimy S (2021) Polyphenols as promising biologically active substances for preventing SARS-CoV-2: a review with research evidence and underlying mechanisms. Food Biosci 40:100891

Mesquita E, Monteiro M (2018) Simultaneous HPLC determination of flavonoids and phenolic acids profile in Pêra-Rio orange juice. Food Res Int 106:54–63

Min ER, Pinchak WE, Anderson RC, Callaway TR (2007) Effect of tannins on the in vitro growth of Escherichia coli O157:H7 and in vivo growth of generic Escherichia coli excreted from steers. J Food Protect 70(3):543–550

Mion S, Rémy B, Plener L, Chabrière E, Daudé D (2018) [Prevent bacteria from communicating: divide to cure]. Annales pharmaceutiques francaises 76(4):249–264

Mion S, Rémy B, Plener L, Chabrière É, Daudé D (2019) [Quorum sensing and quorum quenching: how to disrupt bacterial communication to inhibit virulence?]. Medecine sciences : M/S 35(1):31–38

Mohamad N, Majid EM, Falah AS, Layla C, Akram H, Ali C et al (2017) Antibacterial, antioxidant and antiproliferative activities of the hydroalcoholic extract of the Lebanese Annona Squamosa L. seeds. Int Res J Pharm 8:1–7

Moradi MT, Karimi A, Rafieian-Kopaei M, Rabiei-Faradonbeh M, Momtaz H (2020) Pomegranate peel extract inhibits internalization and replication of the influenza virus: an in vitro study. Avicenna J Phytomed 10(2):143–151

Morel C, Stermitz FR, Tegos G, Lewis K (2003) Isoflavones as potentiators of antibacterial activity. J Agric Food Chem 51(19):5677–5679

Mun SH, Joung DK, Kim SB, Park SJ, Seo YS, Gong R et al (2014) The mechanism of antimicrobial activity of sophoraflavanone B against methicillin-resistant Staphylococcus aureus. Foodborne Pathog Dis 11(3):234–239

Narwojsz A, Tańska M, Mazur B, Borowska EJ (2019) Fruit physical features, phenolic compounds profile and inhibition activities of cranberry cultivars (Vaccinium macrocarpon) compared to wild-grown cranberry (Vaccinium oxycoccus). Plant Foods Hum Nutr 74(3):300–306

Navarro-Martínez MD, García-Cánovas F, Rodríguez-López JN (2006) Tea polyphenol epigallocatechin-3-gallate inhibits ergosterol synthesis by disturbing folic acid metabolism in Candida albicans. J Antimicrob Chemother 57(6):1083–1092

Nawrot-Hadzik I, Matkowski A, Hadzik J, Dobrowolska-Czopor B, Olchowy C, Dominiak M et al (2021a) Proanthocyanidins and flavan-3-ols in the prevention and treatment of periodontitis-antibacterial effects. Nutrients 13(1):165

Nawrot-Hadzik I, Matkowski A, Kubasiewicz-Ross P, Hadzik J (2021b) Proanthocyanidins and flavan-3-ols in the prevention and treatment of periodontitis-immunomodulatory effects, animal and clinical studies. Nutrients 13(1):239

Njeru SN, Obonyo MA, Nyambati SO, Ngari SM (2015) Antimicrobial and cytotoxicity properties of the crude extracts and fractions of Premna resinosa (Hochst.) Schauer (Compositae): Kenyan traditional medicinal plant. BMC Complement Alternat Med 15:295

Nohynek LJ, Alakomi HL, Kähkönen MP, Heinonen M, Helander IM, Oksman-Caldentey KM et al (2006) Berry phenolics: antimicrobial properties and mechanisms of action against severe human pathogens. Nutr Cancer 54(1):18–32

Nowacka N, Nowak R, Drozd M, Olech M, Los R, Malm A (2015) Antibacterial, antiradical potential and phenolic compounds of thirty-one polish mushrooms. PLoS One 10(10):e0140355

Olchowik-Grabarek E, Sekowski S, Bitiucki M, Dobrzynska I, Shlyonsky V, Ionov M et al (2020) Inhibition of interaction between Staphylococcus aureus α-hemolysin and erythrocytes membrane by hydrolysable tannins: structure-related activity study. Sci Rep 10(1):11168

Oneda H, Shiihara M, Inouye K (2003) Inhibitory effects of green tea catechins on the activity of human matrix metalloproteinase 7 (matrilysin). J Biochem 133(5):571–576

Orsi RO, Sforcin JM, Funari SR, Bankova V (2005) Effects of Brazilian and Bulgarian propolis on bactericidal activity of macrophages against Salmonella Typhimurium. Int Immunopharmacol 5(2):359–368

Osawa K, Yasuda H, Maruyama T, Morita H, Takeya K, Itokawa H (1992) Isoflavanones from the heartwood of Swartzia polyphylla and their antibacterial activity against cariogenic bacteria. Chem Pharm Bull 40(11):2970–2974

Othman L, Sleiman A, Abdel-Massih RM (2019) Antimicrobial activity of polyphenols and alkaloids in middle eastern plants. Front Microbiol 10:911

Pacheco-Ordaz R, Wall-Medrano A, Goñi MG, Ramos-Clamont-Montfort G, Ayala-Zavala JF, González-Aguilar GA (2018) Effect of phenolic compounds on the growth of selected probiotic and pathogenic bacteria. Lett Appl Microbiol 66(1):25–31

Pagliarulo C, De Vito V, Picariello G, Colicchio R, Pastore G, Salvatore P et al (2016) Inhibitory effect of pomegranate (Punica granatum L.) polyphenol extracts on the bacterial growth and survival of clinical isolates of pathogenic Staphylococcus aureus and Escherichia coli. Food Chem 190:824–831

Paraiso IL, Revel JS, Stevens JF (2020) Potential use of polyphenols in the battle against COVID-19. Curr Opin Food Sci 32:149–155

Park YJ, Biswas R, Phillips RD, Chen J (2011) Antibacterial activities of blueberry and muscadine phenolic extracts. J Food Sci 76(2):M101–M105

Patel S (2016) Emerging adjuvant therapy for cancer: propolis and its constituents. J Diet Suppl 13(3):245–268

Patel K, Tyagi C, Goyal S, Jamal S, Wahi D, Jain R et al (2015) Identification of chebulinic acid as potent natural inhibitor of M. tuberculosis DNA gyrase and molecular insights into its binding mode of action. Comput Biol Chem 59(Pt A):37–47

Paul S, Mohanram K, Kannan I (2018) Antifungal activity, gas chromatographic-mass spectrometric analysis and in silico study of punica granatum peel extracts against fluconazole resistant strains of candida species. Curr Pharm Biotechnol 19(3):250–257

Pernin A, Guillier L, Dubois-Brissonnet F (2019) Inhibitory activity of phenolic acids against Listeria monocytogenes: deciphering the mechanisms of action using three different models. Food Microbiol 80:18–24

Plaper A, Golob M, Hafner I, Oblak M, Solmajer T, Jerala R (2003) Characterization of quercetin binding site on DNA gyrase. Biochem Biophys Res Commun 306(2):530–536

Przybyłek I, Karpiński TM (2019) Antibacterial properties of propolis. Molecules 24(11):2047

Puupponen-Pimiä R, Nohynek L, Alakomi HL, Oksman-Caldentey KM (2005a) The action of berry phenolics against human intestinal pathogens. Biofactors 23(4):243–251

Puupponen-Pimiä R, Nohynek L, Hartmann-Schmidlin S, Kähkönen M, Heinonen M, Määttä-Riihinen K et al (2005b) Berry phenolics selectively inhibit the growth of intestinal pathogens. J Appl Microbiol 98(4):991–1000

Radji M, Agustama RA, Elya B, Tjampakasari CR (2013) Antimicrobial activity of green tea extract against isolates of methicillin-resistant Staphylococcus aureus and multi-drug resistant Pseudomonas aeruginosa. Asian Pac J Trop Biomed 3(8):663–667; discussion 6

Radulescu C, Buruleanu LC, Nicolescu CM, Olteanu RL, Bumbac M, Holban GC et al (2020) Phytochemical profiles, antioxidant and antibacterial activities of grape (Vitis vinifera L.) seeds and skin from organic and conventional vineyards. Plants (Basel) 9(11):1470

Raekiansyah M, Buerano CC, Luz MAD, Morita K (2018) Inhibitory effect of the green tea molecule EGCG against dengue virus infection. Arch Virol 163(6):1649–1655

Raguzzini A, Toti E, Sciarra T, Fedullo AL, Peluso I (2020) Cranberry for bacteriuria in individuals with spinal cord injury: a systematic review and meta-analysis. Oxid Med Cell Longev 2020:9869851

Randhawa HK, Hundal KK, Ahirrao PN, Jachak SM, Nandanwar HS (2016) Efflux pump inhibitory activity of flavonoids isolated from Alpinia calcarata against methicillin-resistant Staphylococcus aureus. Biologia 71(5):484–493

Rani P, Khullar N (2004) Antimicrobial evaluation of some medicinal plants for their anti-enteric potential against multi-drug resistant Salmonella typhi. Phytother Res 18(8):670–673

Rauf A, Imran M, Abu-Izneid T, Iahtisham Ul H, Patel S, Pan X et al (2019) Proanthocyanidins: a comprehensive review. Biomed Pharmacother 116:108999

Rauha JP, Remes S, Heinonen M, Hopia A, Kähkönen M, Kujala T et al (2000) Antimicrobial effects of Finnish plant extracts containing flavonoids and other phenolic compounds. Int J Food Microbiol 56(1):3–12

Reddy MK, Gupta SK, Jacob MR, Khan SI, Ferreira D (2007) Antioxidant, antimalarial and antimicrobial activities of tannin-rich fractions, ellagitannins and phenolic acids from Punica granatum L. Planta Med 73(5):461–467

Redondo-Blanco S, Fernández J, López-Ibáñez S, Miguélez EM, Villar CJ, Lombó F (2020) Plant phytochemicals in food preservation: antifungal bioactivity: a review. J Food Protect 83(1):163–171

Reed JD, Krueger CG, Vestling MM (2005) MALDI-TOF mass spectrometry of oligomeric food polyphenols. Phytochemistry 66(18):2248–2263

Reid G, Hsiehl J, Potter P, Mighton J, Lam D, Warren D et al (2001) Cranberry juice consumption may reduce biofilms on uroepithelial cells: pilot study in spinal cord injured patients. Spinal Cord 39(1):26–30

Reygaert W, Jusufi I (2013) Green tea as an effective antimicrobial for urinary tract infections caused by Escherichia coli. Front Microbiol 4:162

Rivera-Yañez N, Rivera-Yañez CR, Pozo-Molina G, Méndez-Catalá CF, Méndez-Cruz AR, Nieto-Yañez O (2020) Biomedical properties of propolis on diverse chronic diseases and its potential applications and health benefits. Nutrients 13(1):78

Rizwana H (2016) MSAaDAS. Antimicrobial activity and chemical composition of flowers of Matricaria aurea a native herb of Saudi Arabia. Int J Pharmacol 12:576–586

Rizzo A, Carratelli CR, Losacco A, Iovene MR (2014) Antimicrobial effect of natural polyphenols with or without antibiotics on Chlamydia pneumoniae infection in vitro. Microb Drug Resist 20(1):1–10

Rukayadi Y, Lee K, Han S, Yong D, Hwang JK (2009) In vitro activities of panduratin A against clinical Staphylococcus strains. Antimicrob Agents Chemother 53(10):4529–4532

Rupasinghe HPV (2020) Special issue "flavonoids and their disease prevention and treatment potential": recent advances and future perspectives. Molecules 25(20):4746

Saavedra N, Barrientos L, Herrera C, Alvear M, Montenegro G, Salazar L (2011) Effect of Chilean propolis on cariogenic bacteria Lactobacillus fermentum. Ciencia e investigación agraria 38:117–125

Saeed M, Naveed M, BiBi J, Kamboh AA, Arain MA, Shah QA et al (2018) The promising pharmacological effects and therapeutic/medicinal applications of Punica Granatum L. (pomegranate) as a functional food in humans and animals. Recent Pat Inflamm Allergy drug Discov 12(1):24–38

Samec M, Liskova A, Koklesova L, Mersakova S, Strnadel J, Kajo K et al (2021) Flavonoids targeting HIF-1: implications on cancer metabolism. Cancers 13(1):130

Sánchez MC, Ribeiro-Vidal H, Bartolomé B, Figuero E, Moreno-Arribas MV, Sanz M et al (2020) New evidences of antibacterial effects of cranberry against periodontal pathogens. Foods 9(2):246

Santos LM, Fonseca MS, Sokolonski AR, Deegan KR, Araújo RP, Umsza-Guez MA et al (2020) Propolis: types, composition, biological activities, and veterinary product patent prospecting. J Sci Food Agric 100(4):1369–1382

Sarbu LG, Bahrin LG, Babii C, Stefan M, Birsa ML (2019) Synthetic flavonoids with antimicrobial activity: a review. J Appl Microbiol 127(5):1282–1290

Sarjit A, Wang Y, Dykes GA (2015) Antimicrobial activity of gallic acid against thermophilic Campylobacter is strain specific and associated with a loss of calcium ions. Food Microbiol 46:227–233

Schofield P, Mbugua DM, Pell AN (2001) Analysis of condensed tannins: a review. Anim Feed Sci Technol 91(1):21–40

Selim S, Alfy S, Alruwaili M, Abdo A, Al JS (2012) Susceptibility of imipenem-resistant Pseudomonas aeruginosa to flavonoid glycosides of date palm (Phoenix dactylifera L.) tamar growing in Al Madinah, Saudi Arabia. Afr J Biotechnol 11:416

Serrano J, Puupponen-Pimiä R, Dauer A, Aura AM, Saura-Calixto F (2009) Tannins: current knowledge of food sources, intake, bioavailability and biological effects. Mol Nutr Food Res 53(Suppl 2):S310–S329

Sforcin JM, Fernandes A Jr, Lopes CA, Bankova V, Funari SR (2000) Seasonal effect on Brazilian propolis antibacterial activity. J Ethnopharmacol 73(1-2):243–249

Sha AM, Garib BT (2019) Antibacterial effect of curcumin against clinically isolated porphyromonas gingivalis and connective tissue reactions to curcumin gel in the subcutaneous tissue of rats. BioMed Res Int 2019:6810936

Shahzad M, Millhouse E, Culshaw S, Edwards CA, Ramage G, Combet E (2015) Selected dietary (poly)phenols inhibit periodontal pathogen growth and biofilm formation. Food Funct 6(3):719–729

Shao D, Li J, Li J, Tang R, Liu L, Shi J et al (2015) Inhibition of gallic acid on the growth and biofilm formation of Escherichia coli and Streptococcus mutans. J Food Sci 80(6):M1299–M1305

Shao J, Zhang M, Wang T, Li Y, Wang C (2016) The roles of CDR1, CDR2, and MDR1 in kaempferol-induced suppression with fluconazole-resistant Candida albicans. Pharm Biol 54(6):984–992

Shariatifar N, Fathabad AE, Khaniki GJ, Nasrabadi HG (2014) Evaluation of the antibacterial activity of essential oil and aqueous and ethanolic extracts of Quercus infectoria leaves on food-borne pathogenic bacteria. Int J Pharm Sci Res (IJPSR) 5:709–713

Shehadi M, Awada F, Oleik R, Chokr A, Hamze K, Abou Hamdan H (2014) Comparative analysis of the anti-bacterial activity of four plant extracts. Int J Curr Res Acad Rev 2(6):83–94

Sher H, Al-Yemeni MN, Wijaya L (2011) Ethnobotanical and antibacterial potential of Salvadora persica L.: a well known medicinal plant in Arab and Unani system of medicine. J Med Plants Res 5:1224–1229

Shiota S, Shimizu M, Mizushima T, Ito H, Hatano T, Yoshida T et al (1999) Marked reduction in the minimum inhibitory concentration (MIC) of beta-lactams in methicillin-resistant Staphylococcus aureus produced by epicatechin gallate, an ingredient of green tea (Camellia sinensis). Biol Pharm Bull 22(12):1388–1390

Shiu WK, Gibbons S (2006) Anti-staphylococcal acylphloroglucinols from Hypericum beanii. Phytochemistry 67(23):2568–2572

Shiu WK, Rahman MM, Curry J, Stapleton P, Zloh M, Malkinson JP et al (2012) Antibacterial acylphloroglucinols from Hypericum olympicum. J Nat Prod 75(3):336–343

Shohayeb M, Abdel-Hameed E, Bazaid S (2013) Antimicrobial activity of tannins and extracts of different parts of Conocarpus erectus. Int J PharmBio Sci 3:544–553

Sidhu JS, Ali M, Al-Rashdan A, Ahmed N (2019) Onion (Allium cepa L.) is potentially a good source of important antioxidants. J Food Sci Technol 56(4):1811–1819

Simonetti G, Brasili E, Pasqua G (2020) Antifungal activity of phenolic and polyphenolic compounds from different matrices of Vitis vinifera L. against human pathogens. Molecules 25(16):3748

Singh I, Gautam LK, Kaur IR (2016) Effect of oral cranberry extract (standardized proanthocyanidin-A) in patients with recurrent UTI by pathogenic E. coli: a randomized placebo-controlled clinical research study. Int Urol Nephrol 48(9):1379–1386

Singhal R, Patil P, Siddibhavi M, Ankola AV, Sankeshwari R, Kumar V (2020) Antimicrobial and antibiofilm effect of cranberry extract on Streptococcus mutans and Lactobacillus acidophilus: an in vitro study. Int J Clin Pediatr Dent 13(1):11–15

Singla RK, Dubey AK, Garg A, Sharma RK, Fiorino M, Ameen SM et al (2019) Natural polyphenols: chemical classification, definition of classes, subcategories, and structures. J AOAC Int 102(5):1397–1400

Siriwong S, Thumanu K, Hengpratom T, Eumkeb G (2015) Synergy and mode of action of ceftazidime plus quercetin or luteolin on Streptococcus pyogenes. Evid Based Complement Alternat Med 2015:759459

Smeriglio A, Barreca D, Bellocco E, Trombetta D (2017) Proanthocyanidins and hydrolysable tannins: occurrence, dietary intake and pharmacological effects. Br J Pharmacol 174(11):1244–1262

Soberón JR, Sgariglia MA, Dip Maderuelo MR, Andina ML, Sampietro DA, Vattuone MA (2014) Antibacterial activities of Ligaria cuneifolia and Jodina rhombifolia leaf extracts against phytopathogenic and clinical bacteria. J Biosci Bioeng 118(5):599–605

Souissi M, Ben Lagha A, Chaieb K, Grenier D (2021) Effect of a berry polyphenolic fraction on biofilm formation, adherence properties and gene expression of Streptococcus mutans and its biocompatibility with oral epithelial cells. Antibiotics (Basel) 10(1):46

de Souza EL, de Albuquerque TMR, dos Santos AS, Massa NML, de Brito Alves JL (2019) Potential interactions among phenolic compounds and probiotics for mutual boosting of their health-promoting properties and food functionalities—a review. Crit Rev Food Sci Nutr 59(10):1645–1659

Stapleton PD, Shah S, Hamilton-Miller JM, Hara Y, Nagaoka Y, Kumagai A et al (2004a) Anti-Staphylococcus aureus activity and oxacillin resistance modulating capacity of 3-O-acyl-catechins. Int J Antimicrob Agents 24(4):374–380

Stapleton PD, Shah S, Anderson JC, Hara Y, Hamilton-Miller JM, Taylor PW (2004b) Modulation of beta-lactam resistance in Staphylococcus aureus by catechins and gallates. Int J Antimicrob Agents 23(5):462–467

Stermitz FR, Beeson TD, Mueller PJ, Hsiang J, Lewis K (2001) Staphylococcus aureus MDR efflux pump inhibitors from a Berberis and a Mahonia (sensu strictu) species. Biochem Syst Ecol 29(8):793–798

Stermitz FR, Cashman KK, Halligan KM, Morel C, Tegos GP, Lewis K (2003) Polyacylated neohesperidosides from Geranium caespitosum: bacterial multidrug resistance pump inhibitors. Bioorg Med Chem Lett 13(11):1915–1918

Stojanović G, Radulović N, Hashimoto T, Palić R (2005) In vitro antimicrobial activity of extracts of four Achillea species: the composition of Achillea clavennae L. (Asteraceae) extract. J Ethnopharmacol 101(1-3):185–190

Štumpf S, Hostnik G, Primožič M, Leitgeb M, Salminen JP, Bren U (2020) The effect of growth medium strength on minimum inhibitory concentrations of tannins and tannin extracts against E. coli. Molecules 25(12):2947

Sudano Roccaro A, Blanco AR, Giuliano F, Rusciano D, Enea V (2004) Epigallocatechin-gallate enhances the activity of tetracycline in staphylococci by inhibiting its efflux from bacterial cells. Antimicrob Agents Chemother 48(6):1968–1973

Sugita-Konishi Y, Hara-Kudo Y, Amano F, Okubo T, Aoi N, Iwaki M et al (1999) Epigallocatechin gallate and gallocatechin gallate in green tea catechins inhibit extracellular release of Vero toxin from enterohemorrhagic Escherichia coli O157:H7. Biochim Biophys Acta 1472(1-2):42–50

Taglialatela-Scafati O, Pollastro F, Chianese G, Minassi A, Gibbons S, Arunotayanun W et al (2013) Antimicrobial phenolics and unusual glycerides from Helichrysum italicum subsp. microphyllum. J Nat Prod 76(3):346–353

Taguri T, Tanaka T, Kouno I (2006) Antibacterial spectrum of plant polyphenols and extracts depending upon hydroxyphenyl structure. Biol Pharm Bull 29(11):2226–2235

Tajner-Czopek A, Gertchen M, Rytel E, Kita A, Kucharska AZ, Sokół-Łętowska A (2020) Study of antioxidant activity of some medicinal plants having high content of caffeic acid derivatives. Antioxidants (Basel) 9(5):412
Takeda K, Utsunomiya H, Kakiuchi S, Okuno Y, Oda K, Inada K et al (2007) Citrus auraptene reduces Helicobacter pylori colonization of glandular stomach lesions in Mongolian gerbils. J Oleo Sci 56(5):253–260
Taylor PW (2020) Interactions of tea-derived catechin gallates with bacterial pathogens. Molecules 25(8):1986
Testai L, Calderone V (2017) Nutraceutical value of citrus flavanones and their implications in cardiovascular disease. Nutrients 9(5):502
Thakur P, Chawla R, Chakotiya AS, Tanwar A, Goel R, Narula A et al (2016) Camellia sinensis ameliorates the efficacy of last line antibiotics against carbapenem resistant Escherichia coli. Phytother Res 30(2):314–322
Tiwari TP, Bharti SK, Kaur HD, Dikshit RP, Hoondal GS (2005) Synergistic antimicrobial activity of tea & antibiotics. Indian J Med Res 122(1):80–84
Tombola F, Campello S, De Luca L, Ruggiero P, Del Giudice G, Papini E et al (2003) Plant polyphenols inhibit VacA, a toxin secreted by the gastric pathogen Helicobacter pylori. FEBS Lett 543(1-3):184–189
Trentin DS, Silva DB, Amaral MW, Zimmer KR, Silva MV, Lopes NP et al (2013) Tannins possessing bacteriostatic effect impair Pseudomonas aeruginosa adhesion and biofilm formation. PLoS One 8(6):e66257
Tsao R (2010) Chemistry and biochemistry of dietary polyphenols. Nutrients 2(12):1231–1246
Tyagi P, Singh M, Kumari H, Kumari A, Mukhopadhyay K (2015) Bactericidal activity of curcumin I is associated with damaging of bacterial membrane. PLoS One 10(3):e0121313
Ulanowska K, Tkaczyk A, Konopa G, Wegrzyn G (2006) Differential antibacterial activity of genistein arising from global inhibition of DNA, RNA and protein synthesis in some bacterial strains. Arch Microbiol 184(5):271–278
Ullah A, Munir S, Badshah SL, Khan N, Ghani L, Poulson BG et al (2020) Important flavonoids and their role as a therapeutic agent. Molecules 25(22):5243
Valverde Malaver CL, Colmenares Dulcey AJ, Rial C, Varela RM, Molinillo JMG, Macías FA et al (2019) Hydrolysable tannins and biological activities of Meriania hernandoi and Meriania nobilis (Melastomataceae). Molecules 24(4):746
Van de Velde F, Esposito D, Grace MH, Pirovani ME, Lila MA (2019) Anti-inflammatory and wound healing properties of polyphenolic extracts from strawberry and blackberry fruits. Food Res Int 121:453–462
Vandeputte OM, Kiendrebeogo M, Rasamiravaka T, Stévigny C, Duez P, Rajaonson S et al (2011) The flavanone naringenin reduces the production of quorum sensing-controlled virulence factors in Pseudomonas aeruginosa PAO1. Microbiology 157(Pt 7):2120–2132
Vasconcelos NG, Croda J, Simionatto S (2018) Antibacterial mechanisms of cinnamon and its constituents: a review. Microb Pathog 120:198–203
Veloso DJ, Abrão F, Martins CHG, Bronzato JD, Gomes B, Higino JS et al (2020) Potential antibacterial and anti-halitosis activity of medicinal plants against oral bacteria. Arch Oral Biol 110:104585
Vendrame S, Guglielmetti S, Riso P, Arioli S, Klimis-Zacas D, Porrini M (2011) Six-week consumption of a wild blueberry powder drink increases bifidobacteria in the human gut. J Agric Food Chem 59(24):12815–12820
Vetrani C, Vitale M, Bozzetto L, Della Pepa G, Cocozza S, Costabile G et al (2018) Association between different dietary polyphenol subclasses and the improvement in cardiometabolic risk factors: evidence from a randomized controlled clinical trial. Acta Diabetol 55(2):149–153
Vikram A, Jesudhasan PR, Jayaprakasha GK, Pillai SD, Jayaraman A, Patil BS (2011) Citrus flavonoid represses Salmonella pathogenicity island 1 and motility in S. Typhimurium LT2. Int J Food Microbiol 145(1):28–36

Voravuthikunchai S, Lortheeranuwat A, Jeeju W, Sririrak T, Phongpaichit S, Supawita T (2004) Effective medicinal plants against enterohaemorrhagic Escherichia coli O157:H7. J Ethnopharmacol 94(1):49–54

Voravuthikunchai SP, Suwalak S, Mitranan W (2012) Ellagitannin from Quercus infectoria eradicates intestinal colonization and prevents renal injuries in mice infected with Escherichia coli O157:H7. J Med Microbiol 61(Pt 10):1366–1372

Wafa BA, Makni M, Ammar S, Khannous L, Hassana AB, Bouaziz M et al (2017) Antimicrobial effect of the Tunisian Nana variety Punica granatum L. extracts against Salmonella enterica (serovars Kentucky and Enteritidis) isolated from chicken meat and phenolic composition of its peel extract. Int J Food Microbiol 241:123–131

Wang SS, Wang DM, Pu WJ, Li DW (2013a) Phytochemical profiles, antioxidant and antimicrobial activities of three Potentilla species. BMC Complement Alternat Med 13:321

Wang Y, Jin L, Ominski KH, He M, Xu Z, Krause DO et al (2013b) Screening of condensed tannins from Canadian prairie forages for anti-Escherichia coli O157:H7 with an emphasis on purple prairie clover (Dalea purpurea Vent). J Food Protect 76(4):560–567

Wang LH, Wang MS, Zeng XA, Xu XM, Brennan CS (2017) Membrane and genomic DNA dual-targeting of citrus flavonoid naringenin against Staphylococcus aureus. Integr Biol 9(10):820–829

Wang LH, Zeng XA, Wang MS, Brennan CS, Gong D (2018) Modification of membrane properties and fatty acids biosynthesis-related genes in Escherichia coli and Staphylococcus aureus: implications for the antibacterial mechanism of naringenin. Biochim Biophys Acta Biomembr 1860(2):481–490

Wu D, Kong Y, Han C, Chen J, Hu L, Jiang H et al (2008) D-Alanine:D-alanine ligase as a new target for the flavonoids quercetin and apigenin. Int J Antimicrob Agents 32(5):421–426

Wu T, He M, Zang X, Zhou Y, Qiu T, Pan S et al (2013) A structure-activity relationship study of flavonoids as inhibitors of E. coli by membrane interaction effect. Biochim Biophys Acta 1828(11):2751–2756

Wu Y, Bai J, Zhong K, Huang Y, Qi H, Jiang Y et al (2016) Antibacterial activity and membrane-disruptive mechanism of 3-p-trans-coumaroyl-2-hydroxyquinic acid, a novel phenolic compound from pine needles of Cedrus deodara, against Staphylococcus aureus. Molecules 21(8):1084

Xiao ZP, Peng ZY, Dong JJ, He J, Ouyang H, Feng YT et al (2013) Synthesis, structure-activity relationship analysis and kinetics study of reductive derivatives of flavonoids as Helicobacter pylori urease inhibitors. Eur J Med Chem 63:685–695

Xu J, Xu Z, Zheng W (2017) A review of the antiviral role of green tea catechins. Molecules 22(8):1337

Yam TS, Hamilton-Miller JM, Shah S (1998) The effect of a component of tea (Camellia sinensis) on methicillin resistance, PBP2′ synthesis, and beta-lactamase production in Staphylococcus aureus. J Antimicrob Chemother 42(2):211–216

Yamanaka A, Kimizuka R, Kato T, Okuda K (2004) Inhibitory effects of cranberry juice on attachment of oral streptococci and biofilm formation. Oral Microbiol Immunol 19(3):150–154

Yang B, Liu P (2014) Composition and biological activities of hydrolyzable tannins of fruits of Phyllanthus emblica. J Agric Food Chem 62(3):529–541

Yang CS, Chen L, Lee MJ, Balentine D, Kuo MC, Schantz SP (1998) Blood and urine levels of tea catechins after ingestion of different amounts of green tea by human volunteers. Cancer Epidemiol Biomarkers Prev 7(4):351–354

Yang ZF, Bai LP, Huang WB, Li XZ, Zhao SS, Zhong NS et al (2014) Comparison of in vitro antiviral activity of tea polyphenols against influenza A and B viruses and structure-activity relationship analysis. Fitoterapia 93:47–53

Yi SM, Zhu JL, Fu LL, Li JR (2010) Tea polyphenols inhibit Pseudomonas aeruginosa through damage to the cell membrane. Int J Food Microbiol 144(1):111–117

Yi S, Wang W, Bai F, Zhu J, Li J, Li X et al (2014) Antimicrobial effect and membrane-active mechanism of tea polyphenols against Serratia marcescens. World J Microbiol Biotechnol 30(2):451–460

Yoda Y, Hu ZQ, Zhao WH, Shimamura T (2004) Different susceptibilities of Staphylococcus and Gram-negative rods to epigallocatechin gallate. J Infect Chemother 10(1):55–58

Yu W, Fu YC, Wang W (2012) Cellular and molecular effects of resveratrol in health and disease. J Cell Biochem 113(3):752–759

Yu Y, Pauli GF, Huang L, Gan LS, van Breemen RB, Li D et al (2020) Classification of flavonoid metabolomes via data mining and quantification of hydroxyl NMR signals. Anal Chem 92(7):4954–4962

Zeng YX, Wang S, Wei L, Cui YY, Chen YH (2020) Proanthocyanidins: components, pharmacokinetics and biomedical properties. Am J Chin Med 48(4):813–869

Zhang L, Kong Y, Wu D, Zhang H, Wu J, Chen J et al (2008) Three flavonoids targeting the beta-hydroxyacyl-acyl carrier protein dehydratase from Helicobacter pylori: crystal structure characterization with enzymatic inhibition assay. Protein Sci 17(11):1971–1978

Zhao WH, Hu ZQ, Okubo S, Hara Y, Shimamura T (2001) Mechanism of synergy between epigallocatechin gallate and beta-lactams against methicillin-resistant Staphylococcus aureus. Antimicrob Agents Chemother 45(6):1737–1742

Zharfan RS, Purwono PB, Mustika A (2017) Antimicrobial activity of pineapple (Ananas comosus L. Merr) extract against multidrug-resistant of Pseudomonas aeruginosa: an in vitro study. Indonesian J Trop Infect Dis 2017:6

Zimmer KR, Blum-Silva CH, Souza AL, Wulffschuch M, Reginatto FH, Pereira CM et al (2014) The antibiofilm effect of blueberry fruit cultivars against Staphylococcus epidermidis and Pseudomonas aeruginosa. J Med Food 17(3):324–331

Zou D, Xie K, Wang H, Chen Y, Xie M (2014) [Inhibitory effects of biochanin A on the efflux pump of methicillin-resistant Staphylococcus aureus (MRSA)]. Wei sheng wu xue bao = Acta microbiologica Sinica 54(10):1204–1211

Chapter 11
Antimicrobial Peptides and Small Molecules as Antibiotics Substitute

Nidhi Sethi

11.1 Introduction

With the incessant emergence of resistance, there has been an ever-rising demand for antibiotics. However, the development of new antibiotic molecules has reached an impasse, thereby posing an increased threat to public health. This crisis could have been easily averted or at the least be delayed by prudent use of the existing antibiotics, i.e., by limiting the usage of antibiotics only when required and with appropriate dosing for treatment. A large fraction of antibiotics are being administered to livestock these days. Although pharmaceutical giants consider investing in development of new antibiotics a non-profitable enterprise, scientists around the world have been exploring different perspectives toward their endeavor to find alternatives to the current overexploited antibiotic regime. Keeping in mind the urgency to find new antibiotics, it becomes equally important to strategize upon preserving the existing ones (by correct adjuvant selection) for their prolonged use. Numerous substitutes to classic antibiotics treating specific diseases have been identified. These include antimicrobial peptides, bacteriocins, SMAMPs (synthetic mimics of antimicrobial peptides), IDR peptides (innate defense regulatory peptides), antibacterial oligonucleotides foldamers, antibacterial nucleic acids, and immune stimulation by P4 peptide each one offering its own advantages and disadvantages.

Antimicrobial peptides (AMPs) are endogenously produced small molecules accounting for a large group of diverse peptides. AMPs, alternatively known as "host defense peptides", "alarmins", or "defensins" have not only gained significance since their discovery in the latter half of the twentieth century but also have researchers reconsider the way they have been reflecting upon the immune defense mechanisms and physiology of human diseases. These peptides have been isolated

N. Sethi (✉)
Bio-organic and Photochemistry Laboratory, Department of Pharmaceutical Sciences, Guru Nanak Dev University, Amritsar, India

N. Akhtar et al. (eds.), *Emerging Modalities in Mitigation of Antimicrobial Resistance*,
https://doi.org/10.1007/978-3-030-84126-3_11

from a variety of multicellular and unicellular organisms such as animals (vertebrates and invertebrates), plants, fungi, and bacteria. Blood cells (macrophages and neutrophils), epithelial cells, reproductive tract, hemocytes, body fat, etc., have been the warehouses for AMPs. Hundreds of AMPs have been identified, mostly containing amino acid residues ranging between 10 and 50 units. AMPs have shown a broad spectrum of activity against a number of Gram-positive and Gram-negative bacteria, fungi, protozoa, and viruses; also exhibiting cytotoxic effects toward human sperm and tumor cells. Most AMPs are cationic in nature, having an amphipathic structure which facilitates their binding to the membrane interface and likely to be their mechanism of action against various microbes. AMPs have been of therapeutic importance having been indicated for their use as topical antiseptics, anti-inflammatory, oral mucositis, candidiasis, impetigo, chronic respiratory infection, cystic fibrosis, acute bacterial skin infections, and inflammatory bowel disease; also encompassing to overcome the challenge of resistance.

11.2 AMPs from Diverse Natural Origin

Sundry antimicrobial peptides have been identified and isolated from almost all types of organisms such as insects (ants, flies, spiders), plants, animals (including humans), and even bacteria themselves. AMPs are composed of different types and sizes of amino acid residues forming the primary and secondary structure. The unifying factor in these diverse peptides is the N-terminal domain and a cationic C-terminal domain exhibiting antimicrobial activity. Their occurrence in human skin has provided ample revelation on the vitality of AMPs to mammalian defense system and has unlocked a new era for the admiration of these peptides in human health and disease. In this section, several examples of antimicrobial peptides occurring naturally from various sources have been discussed.

11.2.1 AMPs Isolated from Insects

Since the unearthing of AMPs from cecropia silk moth, *Hyalophora cecropia*, a number of peptides have been recognized in various insects. Named as cecropins, these molecules are linear, amphipathic in nature, and exert activity against parasites, bacteria, and fungi. A breakthrough in the discovery of insect AMPs can be attributed to the septic injury in fruit fly, *Drosophila melanogaster*, resulting in an immune response by producing AMPs such as drosomycin, drosocin, cecropins, attacin, diptericin, and metchnikowin possessing antifungal and antibacterial properties (Reddy et al. 2004). Venom of ponerine ant, *Pachycondyla goeldii*, has been found to be rich in 15 types of peptides displaying insecticidal and antibacterial properties. As the name suggests, these peptides are called ponericidins and have properties similar to cecropins and dermaseptins (Peravali et al. 2013). Other examples of insect AMPs include defensins (from flesh fly, *Sarcophaga peregrina* and from larvae of black-brown fly, *Phormia terranovae*), thanatin (from hemipteran insect, *Podisus maculiventris*), acaloleptins (from the hemolymph of beetle, *Acalolepta luxuriosa*), lycotoxins and cupiennin-1 (from the venom of spiders, *Cupiennius salei* and *Lycosa carolinensis*), etc. (Gabay and Almeida 1993; Kuhn-Nentwig et al. 2002; Reddy et al. 2004; Vetterli et al. 2018). These peptides demonstrate good insecticidal, antimicrobial, and mild hemolytic activity.

11.2.2 AMPs Isolated from Invertebrates

In addition to insects, vertebrates and invertebrates also contribute to a large number of AMPs. For instance, tachyplesins (from immune cells of Japanese horseshoe crab, *Tachypleus tridentatus*) and polyphemusin (from hemocytes of horseshoe crab, *Limulus polyphemus*), both are small in size, cysteine-rich, have intramolecular disulfide bridging patterns and exhibit carboxy-terminal amidation (Edwards et al. 2017; Powers et al. 2004). Other peptide, tachycitin, a 73-amino acid peptide also isolated from *T. Tridentatus*, has shown specific binding affinity for chitin (essential constituent of fungal cell wall); defensin, has also been purified from the hemocytic granules of *T. tridentatus*, displaying activity against both Gram-positive and Gram-negative bacteria; penaeidins isolated from the hemolymph of shrimp, *Penaeus vannamei*, show activity against filamentous fungi only; mytilin and mytimycin obtained from mussel, *Mytilus edulis*, are cysteine-rich with activity against Gram-positive bacteria and filamentous fungi respectively; androctonin, a 25-residue molecule, obtained from scorpion, *Androctonus australis*, forming two intramolecular disulfide linkages is active against fungi, Gram-positive and Gram-negative bacteria (Mandard et al. 1999; Sperstad et al. 2011).

11.2.3 AMPs Isolated from Vertebrates

Vertebrates have also served as abundant sources of AMPs. These include defensins, cathelicidins, bactenecins, histatins, protegrins, etc. α-defensins, are small peptides varying from 29 to 35 amino acids with three intramolecular disulfide bonds. So far, six β-defensins have been identified in humans, four from granulocytes and lymphocytes, also called human neutrophil peptides (HNP 1–4) and two expressed by the paneth cells in intestine (Human defensin, HD-5 and 6). β-defensins contain approximately 45 amino acid residues and their synthesis can be induced by cytokines (Schneider et al. 2005). Cathelicidins (CAP-18 and LL-37) are composed of highly conserved cathelin domain, stored in neutrophils as propeptides and are released upon proteolytic stimulation (Dijksteel et al. 2021; Kang et al. 2017). Bactenecins (Bac-5 and Bac-7) are cathelicidin-derived peptides isolated from bovine cattle, exhibit antimicrobial activity (Kościuczuk et al. 2012). Histatins are a family of histidine-rich proteins which have been isolated from human saliva and exhibit moderate activity against *Candida albicans* (Reddy et al. 2004). Protegrins are antimicrobial peptides, having 16–18 residues and are isolated from neutrophils of pig (Kosikowska and Lesner 2016).

11.3 AMPs of Synthetic Origin

Advances in the field of organic chemistry, computational chemistry, and pharmacology have streamlined the process of unwinding the amino acid sequence, size, primary, secondary, and 3D (three-dimensional) structure along with finding the mode of action and therapeutic potential of the naturally occurring AMPs and design strategies to synthesize their derivatives. The designing of a synthetic AMP postulated to be used therapeutically must consider length of amino acid chain (minimum 7–8 residue), net charge, helicity, hydrophobicity, amphipathicity, solubility as well as ease of synthesis, involving least steps, producing desired effectiveness against pathogen of interest with high therapeutic index and minimum or no adverse effects as important physicochemical parameters before synthesis (Bahar and Ren 2013; Park and Hahm 2005). Various methodologies have been developed for the synthesis of new antimicrobial peptides of known or foreseen sequences. These include alteration of known AMP sequences or biophysical modeling comprehending peptidic activity, e.g., WR12 is a novel 12-residue peptide composed of arginine and tryptophan has been found to exhibit activity against multidrug-resistant staphylococci, MRSA (methicillin-resistant *Staphylococcus aureus*), linezolid-resistant *S. aureus*, vancomycin-resistant *S. aureus*, and methicillin-resistant *Staphylococcus epidermidis* infections (Mohamed et al. 2016). D-IK8 is another 8-residue peptide having a β-sheet structure and has an activity profile similar to WR12 (Mohamed et al. 2016). MSI-78 or pexiganan is a 22-amino acid derivative synthesized from magainin II and has been found to be more potent than parent molecule magainin II against both Gram-positive and Gram-negative bacteria (Dijksteel et al. 2021). CP-11, a derivative of indolicidin, contains an additional disulfide bond and a trp-trp (tryptophan-tryptophan) cross-linking than indolicin and has higher protease stability without any change in antimicrobial action (Kang et al. 2014). P60.4 is another synthetic AMP, a derivative of LL-37, synthesized by removal of asparagine and glutamine amino acids. Addition of two cationic arginine units into the parent sequence displayed diminished cytotoxic effects on eukaryotic cells but has been successfully used in nasal formulations acting against MRSA (Bahar and Ren 2013). AMP72, AMP126, and AMP2041 are three peptides synthesized from arenicin, protegrin, and thanatin respectively possessing low cytotoxicity compared to the parent AMPs and showed dose-dependent antimicrobial potency against many Gram-negative bacteria (Bahar and Ren 2013). Tet-20 is a synthetic peptide exhibiting broad spectrum of antimicrobial activity and inhibits biofilm formation without any toxic effects (Gao et al. 2011). A synthetic histatin derivative dhvar4 has been investigated against oral flora and it decreased the number of viable biofilm cells (Ricci et al. 2020). Other examples include MUC17 12-mer-L and MUC17 20-mer (derived from MUC7, a humans saliva AMP) which inhibit biofilm formation against *Streptococcus mutans* (Bahar and Ren 2013); NRC-16 is a pleurocidin-like cationic AMP which inhibits *Pseudomonas aeruginosa* biofilm formation (Gopal et al. 2013); $(RW)_4$-NH_2 (a cationic peptide) inhibits formation of biofilms and kill *Escherichia coli* HM22 persister cells in planktonic cultures (Bahar and Ren 2013).

11.4 Classification of AMPs

Most AMPs are amphipathic in nature having clusters of positively charged and hydrophobic chains of amino acids. Due to the amphipathic nature, AMPs interact with the phospholipids (negative charge) and fatty acids (hydrophobic) embedded in the microbial plasma membranes. These act by forming a pore on the membrane and thus leading to lysis and release of components of the cytosol (Wimley 2010). Isolation and characterization of these peptides through chromatography and NMR (nuclear magnetic resonance) spectroscopy respectively has helped understand their 3D structure and function. Based upon the type of amino acid present, conformational and sequence analysis, these peptides have been divided into several categories as tabulated under Table 11.1 (Reddy et al. 2004).

11.4.1 α-Helical AMPs

These are highly amphipathic in nature, having helical structure with hydrophobic and cationic charged surfaces. Cecropins was the first α-helical AMPs isolated from cecropia moth, *Hyalophora cecropia* and have the propensity to attain helical form in various organic co-solvents. Another group of α-helical AMPs includes magainins. Magainins (I and II) have been isolated from skin glands of African clawed frog, *Xenopus laevis,* are 21–27 residues long. These have been documented to be broad-spectrum antimicrobial agents, potent against a number of pathogenic microorganisms and cancer cells (murine and human cancer cells), exhibiting cidal activity. Magainins are easy to synthesize, selective in action, are less prone to bacterial resistance and hence, have been developed as therapeutic agents (Jacob and Zasloff 1994; Reddy et al. 2004).

11.4.2 Cysteine-Rich AMPs

Defensins were the first cysteine-rich AMPs to be isolated from human granules. These are the human neutrophil peptides, HNP-1, HNP-2, and HNP-3, composed of 30 amino acids rich in cysteine residues and are found in numerous organisms such

Table 11.1 Classification of AMPs based upon structural features (amino acid type, conformation, and sequence)

S.no.	Type	Characteristic features	Example	Activity potential
1.	α-Helical	Highly amphipathic, helix with hydrophobic and cationic charged surfaces	Cecropins	Antimicrobial, anticancer
			Magainins	Antimicrobial, anticancer, wound healing
2.	Cysteine-rich	Rich in cysteine residues, pairs of cysteine residues forming intramolecular disulfide linkages	Defensins (HNP-1, HNP-2, HNP-3)	Antimicrobial, immune stimulation
			Drosomycin	Antimicrobial
3.	β-sheeted	Form a single β-hairpin structure, approximately 20 residues long, one or two disulfide linkages	Tachyplesins	Antimicrobial, antitumor
			Polyphemusin II	Antimicrobial
			Thanatin	Antimicrobial
			Lactoferricin B	Antimicrobial
4.	AMPs rich in regular amino acids	High content of regular amino acids	Histatin	Antimicrobial
			Cathelicidins	Antimicrobial, immunomodulation, wound healing, angiogenesis
			Indolicidins	Antimicrobial, antiprotozoal, hemolytic
			Tritripticin	Antimicrobial, hemolytic
			Bactenecins	Antimicrobial
5.	AMPs with rare modified amino acids	Composed of amino acids such as lanthionine, 3-methyllanthionine, dehydroalanine, dehydrobutyrine, DH-amino acids	Leucocin A	Antimicrobial
			Gramicidins	Antimicrobial

as plants, insects, and mammals. Apart from neutrophils, human defensins are also present in intestinal paneth cells, keratinocytes, genitourinary, pulmonary, and reproductive epithelia. Usually defensins contain six cysteine residues linked to each other through three disulfide bonds between C1-C4, C2-C5, and C3-C6 position. Defensins have a broad spectrum of activity and possess cytotoxic potential against parasites, bacteria, virus, fungi, and host cells. In addition to antimicrobial activity, defensins also tend to increase TNF-α (tumor necrosis factor-α) and IL-1 (interleukin-6) levels, promotes PG-D2 (prostaglandin-D2) production, stimulate histamine release, and exhibit antiviral potency against adenoviruses. Drosomycin, another cysteine-rich AMP, has been isolated from drosophila and possesses four disulfide linkages made up of three β strands which are antiparallel and form an α helix between the first two strands (Koczulla and Bals 2003; Reddy et al. 2004).

11.4.3 β-Sheet AMPs

β-Sheet AMPs contain approximately 20 amino acid residues with one or more disulfide bonds, acquiring a stable single β-hairpin-like structure. Some of the examples include tachyplesins and polyphemusin II found in horseshoe crab, thanatin, lactoferricin B, etc. Both tachyplesins and polyphemusin II form β-sheets which are antiparallel to each other and are connected to a β-turn bridged through two disulfide linkages. Thanatin has been isolated from the spined soldier bug, *Podisus maculiventris* and has an antiparallel β-sheet retained by one disulfide linkage (Andreu and Rivas 1998; Reddy et al. 2004). Lactoferricin B is a lactoferrin (milk protein) derivative possessing 25 amino acid residues adopting a β-sheet structure maintained by a single disulfide linkage. It is also present in tears and saliva. Studies have suggested that it also exerts antiviral effect against HIV (human immunodeficiency virus) and neutralizes endotoxic activity by bacteria in addition to antimicrobial activity (Hwang et al. 1998; Reddy et al. 2004).

11.4.4 AMPs Rich in Regular Amino Acids

Certain AMPs are majorly composed of regular amino acids such as histidine, arginine, and proline in their structure. These peptides orient themselves into α-helical or β-sheet conformations. This class of AMPs includes histatin, cathelicidins (CAP-18, protegrins, prophenins, PR-39, and porcine myeloid AMPs), indolicidins, tritripticin, bactenecins (Bac-5 and Bac-7), etc. (Koczulla and Bals 2003). Histatin has first been isolated from human saliva and is abundant in histidine amino acid residues. It has been documented to be active against *Candida albicans* (Andreu and Rivas 1998). Cathelicidins unlike histatins are rich in proline residues; indolicidins and tritripticin in tryptophan residues while PR-39 in arginine residues. Cathelicidins embody a large family of peptides which are structurally diverse. These have been found in the skin from wound fluid and were found to be components of machinery responsible for wound repair inside our body. Later, these have been isolated from rat, mouse, goat, pig, rabbit, guinea pig, monkey, and sheep as well. Cathelicidins enclose a cathelin protein precursor domain (conserved) and a C-terminal antimicrobial domain (variable). Apart from being potent antimicrobials, cathelicidins have been reported to be efficacious in angiogenesis, wound healing, and certain immune mechanisms in human body (Izadpanah and Gallo 2005; Reddy et al. 2004).

Indolicidin is a tridecapeptide isolated from bovine neutrophils and has been documented to exert antimicrobial activity against a number of bacteria, HIV-1 virus, fungi, and protozoa through membrane permeabilization (Vergis et al. 2019). Tritripticin is a 13 amino acid long AMP member of the cathelicidin family residing in neutrophils. Apart from being a good antimicrobial agent acting against various bacteria and fungi, it also exhibits hemolytic activity (Sharma et al. 2013). Bactenecins (Bac) are a group of small molecule cationic antimicrobial peptides (23 amino acid residues) found in ruminant cattle, containing arginine, cysteine, and hydrophobic residues. These have a folded structure because of the disulfide linkage formation. Bactenecins have been reported to show antimicrobial activity against Gram-negative bacteria (Munshi et al. 2020; Radermacher et al. 1993; Sun et al. 2019; Wu and Hancock 1999).

11.4.5 AMPs with Rare Modified Amino Acids

Certain AMPs are composed of atypical or uncommon amino acids unlike the ones discussed in previous four classes. These are composed of rare modified amino acids such as 3-methyllanthionine, lanthionine, dehydrobutyrine, dehydroalanine, and DH-amino acids. Examples include leucocin A and gramicidins (Reddy et al. 2004). Leucocin A is a non-lanthionine peptide isolated from lactic acid bacterium, *Leuconostoc gelidum* and is composed of 37 amino acid residues. It is a ribosomally synthesized antimicrobial peptide exhibiting activity against Gram-positive bacterial strains from different genus such as *Carnobacterium*, *Enterococcus*, *Lactobacillus*, *Pediococcus*, *Leuconostoc*, *Listeria*, and *Clostridium* (Etayash et al. 2014). Gramicidins are linear polypeptides composed of 15 alternating D,L-amino acids obtained from *Bacillus brevis* (aerobic soil bacterium). Gramicidins are active against a wide variety of bacterial species. They are composed of DH-amino acids that help them to acquire an unusual cyclic β-hairpin (Killian 1992).

11.5 Structure of AMPs

Due to diversity and complexities in the structure of proteins, there is insufficient data available on the three-dimensional structure of the currently used AMPs. Hence, it becomes cumbersome to generalize a single model deciphering the structure of AMPs. Usually, antimicrobial peptides contain basic amino acids such as arginine and lysine which impart a net positive charge on them and certain hydrophobic amino acids. Three-dimensionally, these peptides are stabilized by disulfide linkages or fold themselves into α-helices or β-sheets of amphiphilic molecules. The amphiphilic arrangement of peptides having a segregated hydrophilic and hydrophobic portions interact electrostatically at the interface where the negatively charged bacterial membrane engages with the positively charged peptides. The hydrophobic groups however, assist in their penetration into the lipophilic section of the bacterial membrane. Structure-activity relationship studies on β-sheet AMPs have shown that the disulfide bridges do not contribute to the antimicrobial activity but are requisite for membrane translocation (Baltzer and Brown 2011; Kang et al. 2017).

11.6 Mechanism of Action for Antimicrobial Peptides

11.6.1 Most Validated Mechanisms of Action for AMPs

Continuous efforts have been made by the researchers around the world to deduce a precise mechanism of action for antimicrobial peptides based upon their size, sequence, 3D structure, cationic nature, and hydrophobicity. Although there is not much clarity on the exact mechanism, a few key features necessary for the antimicrobial activity of these peptides have been figured out. These include the amphipathic conformation of peptides, positive charge due to basic amino acids present, disulfide bonds for membrane translocation, and negatively charged phospholipid head groups on bacterial cell membrane play a vital role in the peptide-membrane interactions. However, there has been complete agreement that the antimicrobial action proceeds through cell membranes disruption. Shai-Matsuzaki-Huang (SMH) model was the first of its kind which could efficiently explain the interactions between the bacterial cell membrane and antimicrobial peptides, elucidating their antimicrobial mechanism (Baltzer and Brown 2011). The model suggests that there occur electrostatic interactions between the anionic heads of the bacterial membrane and the cationic residues of AMPs. Then the AMPs fold into amphiphilic entities which tend to acquire a 3D structure upon interaction with the bacterial membrane. This leads to displacement of lipids, followed by thinning of bacterial membrane outer leaflet, peptide aggregation, and channel formation which facilitate peptide transport from surface to the interior of the target cell. However, scientists have not reached a consensus on the steps involved in pore formation and translocation of the peptides. Because of the diversity in the structure and sequencing of AMPs, no single model is adequate to fit all peptides. So far, four plausible mechanisms mediating through membrane permeabilization and explaining the method of channel formation have been proposed (Fig. 11.1). In the barrel-stave mechanism, there is a voltage-gated transmembrane channel formation wherein the non-polar elements of the peptide are at interface with the bacterial membrane lipids; thereby leading to formation of hydrophilic pore/channel traversing through the membrane. Like barrel stave model, the peptide in toroidal model has a transmembrane topology followed by formation of toroidal pores driven by electrostatic interactions. These pores are covered by head groups of the membrane phospholipids leading to changes in membrane curvature. In the sinking-raft model, AMPs rest on the membrane surface and initiate increase in membrane curvature. The peptides self-aggregate on the membrane, leading to sinking of AMPs into it and creating transient pores. As the name suggests, the last mechanism involves formation of a carpet of AMPs around the target cell membrane, thereby affecting the membrane integration. The lipid bilayer gets bent backward, leading to formation of micelles and final collapse of the membrane (Dijksteel et al. 2021; Mahlapuu et al. 2016). Operating through these mechanisms, AMPs successfully tamper the membrane permeability, transport of ions and metabolites and loss of membrane potential, hence, facilitating loss of membrane function. It has also been reported that the slight differences in the

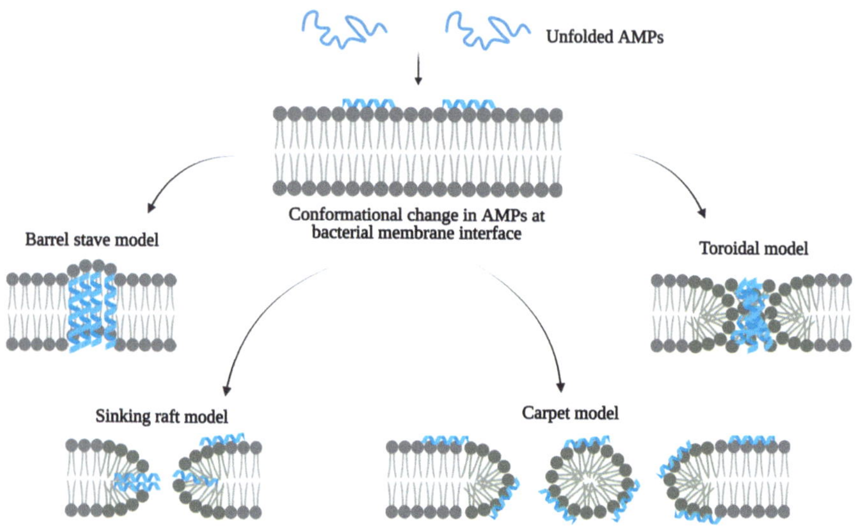

Fig. 11.1 Mechanism of action for antimicrobial peptides (AMPs) mediated through barrel stave model, toroidal model, sinking raft model, and carpet model

composition of the microbial cell membranes and the structure and sequence of the individual peptide are responsible for the selectivity of AMPs (Andreu and Rivas 1998; Dijksteel et al. 2021; Kang et al. 2014).

11.6.2 Other Reported Mechanisms for Antimicrobial Peptides

There is sizeable amount of data showing that antimicrobial peptides act not just through membrane permeabilization but there are various other mechanisms responsible for the antimicrobial activity. For instance, their interference to host protein production (e.g., PR-39); inhibition of specific membrane protein synthesis (e.g., attacins or gloverin); inhibition of protein and RNA synthesis (e.g., Bac-5 and Bac-7); interference with DNA without inhibiting its synthesis (buforins); breakdown of single strand of DNA (defensins); halting of DNA synthesis (PR-39), and production of hydrogen peroxide or synthesis of stress proteins (Andreu and Rivas 1998).

11.7 Multi-functional Roles of AMPs

With the aim of improving the selectivity and efficacy, a lot of researchers in recent years have devoted their resources in the development of new antimicrobial peptides and exploring the multifaceted roles of the existing ones. In addition to antimicrobial potential, AMPs are involved in a myriad of other functions such as histamine release, induction of syndecan synthesis, inhibition of steroid synthesis, inflammatory mediators having impact on various cellular processes like wound repair, cytokine release, cell proliferation, chemotaxis, immune induction, and protease-antiprotease balance. The currently known AMPs have been found to exhibit broad range of antimicrobial activity against all types of pathogenic microorganisms, e.g., human lactoferrin B on oral administration, effectively help reduce *E. coli* caused inflammation and infection in the urinary tract; cecropins and indolicidin have been found to be active against *P. aeruginosa* and have been documented to be co-administered with antibiotics like clarithromycin and polymyxin E for clinical use; ranalexin and buforins-1 have greater efficacy against methicillin-resistant strains of *S. epidermidis* than rifampicin; cathelicidins display potential against oral *Fusobacterium* species and *Candida* species; defensins and protegrins help reduce colonization of *Mycobacterium tuberculosis*; cecropins have shown to exert fungicidal action against fungi like *Aspergillus* species and *Fusarium* species and dermaseptins show activity against *Cryptococcus neoformans* fungi (Andreu and Rivas 1998; Koo and Seo 2019). Some of the wider applications of AMPs have been discussed in this section.

11.7.1 Wound Repair

Among numerous applications of AMPs, healing of wound is one of the earliest known mechanisms contributing to their discovery in mammalian skin. Studies suggest that AMPs have been found to be expressed in a variety of skin disorders. For example, cathelicidin (PR-39) stimulates fibroblast present in dermis which promotes the synthesis of proteoglycans syndecan-1 and syndecan-4. Syndecans are a small family of transmembrane domain proteins acting as co-receptors and interacting with a variety of ligands. These are composed of a proteinaceous core and 3–5 glycosaminoglycan side chains essential for action of numerous growth factors. Similarly, cathelicidin (LL-37) is also involved in wound repair mediated through keratinocyte proliferation and angiogenesis. Mechanistically, the process of wound repair by AMPs involves killing of bacteria and stimulation of host repair phenomena. AMPs tend to curb the proliferation of pathogens in the skin. This can be proven by inactivation of cathelicidins in pig lesions or by deletion of cathelicidin gene in mice both leading to increased and long-lasting bacterial growth (Izadpanah and Gallo 2005).

11.7.2 Skin Infections

Granulysin, a T-cell AMP, has been found to be significantly mediating the immune response (through association with macrophages) against leprosy pathogen, *Mycobacterium leprae* as it is evident from the analysis of tissue samples of leprotic patients. It has also been reported that they have six times higher expression in the skin (predominantly in CD-4/T cells) of patients suffering from tuberculoid leprosy than those with lepromatous type. The expression of granulysin in leprosy is further confirmed by the non-involvement of perforin (helps pore formation), proving it to be an effective indicator for cell-mediated immunity. AMPs have also been documented to play a vital role in another skin condition such as atopic dermatitis (AD) or eczema. AD is generally associated with periodic reoccurrence of skin infections. The patient is prone to attack by all types of pathogenic organisms such as bacteria, virus, and fungi. Evidences have suggested that production of AMP (LL-37 and HBD-2 in particular) was reduced during inflammation. However, on comparison between AD and psoriasis, the production of AMPs (cathelicidin and HBD-2) is decreased in AD while in case of psoriasis, AMPs have elevated levels during inflammation. This indicates that although both AD and psoriasis involve inflammation of skin and it is expected to have greater levels of AMPs, these behave distinctively in different conditions. The decrease in levels could be attributed to the elevated levels of Th-2 cytokines which suppress the expression of AMPs. Furthermore, AMPs have shown very good antimicrobial potential against *S. aureus*, herpes simplex virus (HSV), and vaccinia virus (VV), all three implicated to be affecting AD patients, which is evident from inflammation of keratinocytes. Thus, explaining the lack or deficiency of AMPs being associated with the susceptibility of AD patients to pathogens (Izadpanah and Gallo 2005).

11.7.3 Antiviral Activity

Evidences suggest that AMPs have protective effects against certain viral strains. For instance, cathelicidins induction is observed during human papilloma virus (HPV) infections such as verruca vulgaris and condyloma acuminatum development. Cathelicidins are also reported to have been expressed during the cutaneous proliferation of vaccinia virus (VV) infection. However, there is no indication that cathelicidins prevent the spread of virus or are involved in the process of resistance. Hence, AMPs can further be explored for their exploitation in skin infections involving viral intervention (Andreu and Rivas 1998; Izadpanah and Gallo 2005).

11.7.4 Antitumoral Activity

Several studies have shown that tumor cells unlike non-specialized cells offer greater liability toward antibiotic peptides. For instance, cecropins and magainins have displayed activity toward various human cancer cell lines (Andreu and Rivas 1998).

11.7.5 AMPs Against Sexually Transmitted Infections

Investigations on AMPs have shown them to be active against sexually transmitted pathogens. For example, defensins prevent entry and spread of syphilis bacterium *Treponema pallidum*, HIV, HSV-2 (herpes simplex virus type-2) viruses; protegrins act against *Neisseria gonorrhoeae, Chlamydia trachomatis*; cathelicidins have also shown activity against syphilis bacterium; magainins exhibit inhibitory effect against a number of pathogens responsible for sexually transmitted infections (Andreu and Rivas 1998).

11.7.6 Anti-malarial Activity

Dermaseptin-derived peptides have been found to show activity against malaria parasite *Plasmodium falciparum* in human erythrocytes, suggesting the likelihood of their use in intracellular infections. Possibly, these act by lysis of infected host cell, thereby destroying the intraerythrocytic malarial parasites. Aminoheptanoyl and isobutyryl dermaseptin derivatives have been documented to display better selectivity toward the parasite than the host cell (with reduced hemolysis). Plant antimicrobial peptides thionins have also been found to exhibit in vitro activity against *Leishmania donovani* (Mirski et al. 2018).

11.7.7 Spermicidal Agents

Among the broader spectrum of magainins' activity, they have been documented to display effects on spermatozoids by bringing morphological and functional changes in them. Studies show that dermaseptins affect motility of sperm in a time and dose-dependent manner by absolute stoppage of motility. Thus, there is sufficient evidence suggesting that AMPs can be potential candidates for use as contraceptive agents (Andreu and Rivas 1998; Zairi et al. 2009).

11.7.8 Buco-Dental Infections

A few examples of AMPs active against buccal cavity infections have been discussed. Histatins present in human saliva act as natural defenses for oral cavity with cidal action against *Candida* species and *Bacteroides gingivalis*. Similarly, human defensins are active against a number of oral Gram-negative bacteria and protegrins kill bacteria such as *Actinobacillus actinomycetemcomitans* and *Eikenella corrodens* (Andreu and Rivas 1998).

11.7.9 Ocular Infections

Antimicrobial peptides such as magainins, defensins, cecropin, and its derivatives have been proposed for their use in preserving contact lens disinfectants, cornea storage medium, and ocular antiseptics. Cecropin A has been found to be as efficacious as gentamycin toward *P. aeruginosa* infection clearance (Andreu and Rivas 1998).

11.7.10 Kostmann's Syndrome

Kostmann's syndrome is an autosomal recessive disorder characterized by a severe congenital neutropenia. Children born with this condition are deficient in neutrophils (a type of white blood cells helping in fighting infections) and are always vulnerable to recurrent infections and periodontal disease. Investigation into the neutrophils of the affected patients has revealed that there was a decrease in α-defensin levels and miniscule amount of cathelicidins with a normal oxidative burst; thereby implicating the role of functional AMPs in Kostmann's syndrome (Izadpanah and Gallo 2005).

11.7.11 Food Preservatives

In today's modern world, when packaged food has reached every nook and corner of our households in one form or another, retention of the freshness, taste, and shelf life of food product gains utmost significance. Food industry uses a number of natural and synthetic chemicals for preservation. The pre-requisite for any chemical to be used as preservative is its safety and stability profiling. The current regime of preservatives faces the biggest challenge of being vulnerable to microorganisms. Therefore, there has always been a never-ending quest to synthesize or isolate newer preservatives which are not susceptible to microbial attack. A number of AMPs have been recognized which have shown potential for their use in food industry and preventing food spoilage. Presently, pediocin has been approved by USFDA (United States Food and Drug Administration) for its commercial use. Certain AMPs have been tested and have displayed good activity at varying pH and temperature with no adverse effects. Hence, AMPs can be designated as promising candidates for their use as additives/food preservatives and development of newer peptides with enhanced activity and safety (Mor 2000; Piper et al. 1998; van Kan et al. 1998).

11.7.12 Miscellaneous Uses

Exploiting the AMPs to be fighting against an array of microorganisms, these peptides can also be used to prevent contamination/infections in therapeutic medical equipments such as catheters and orthopedic devices. Literature reports have shown that some cationic antimicrobial peptides tend to bind to bacterial lipopolysaccharides (endotoxins) when bound to a polymer/solid support resin and help the removal of these endotoxins (Mor 2000).

11.8 AMPs as Targets for Therapy

As discussed in the previous section, although numerous peptides, natural or synthetic have been identified to be efficacious against all types of pathogenic microbes, there has always been a constant quest to explore the existing ones for their potential and find or synthesize new derivatives which are small in size, easy to synthesize, have broad range of activity, fast acting, lesser adverse effects, lesser risk of developing resistance and having wider applications (Bahar and Ren 2013). Studies suggest that AMPs show promising results in vitro, indicative of their in vivo efficiency. Prior for use of any substance or chemical compound in therapy, it should be capable of overcoming the barriers of stability, bioavailability, route of administration, enzymatic breakdown, undesirable toxicity, and lower production cost. Advances in spectroscopic techniques have helped deduction of secondary peptide structure, thereby making it easier to understand their mode of action. Few AMPs and their synthetic analogues have been approved by USFDA for their commercial use. These include smaller peptides like pediocin, magainins, and its derivatives (in wound healing), gramicidin S and polymyxin B (topical application), defensins (synergistic effect), and many others (Dijksteel et al. 2021). These peptides are not only potent when used alone but also show synergy in combination with other antibiotics. Since their discovery, although a number of peptides have been successfully developed, a substantial amount of effort is still required to explore their therapeutic potential.

11.9 Challenges Faced by Antimicrobial Peptides

Since the discovery of first antibiotic, Penicillin by Alexander Fleming, antibiotic resistance has emerged to be a serious hazard for antibiotics in therapy. However, this has left researchers and pharmaceutical industry a little disenchanted toward antibiotics but the emphasis still lies on the development of antibiotics capable of enduring resistance. Several mechanisms unraveling antibiotic resistance in microbes have evolved. Interactions between bacterial membrane and peptides play key role in the process of resistance. Studies have shown that changes in bacterial membranes are responsible for resistance. These modifications include reduction in negative charge at the interface of interaction, thereby preventing AMP entry into cell mediated through inhibition of peptide aggregation. One such example is the bacterium *S. aureus* which operates through reduction of anionic charge on the bacterial membrane, leading to the repulsion and development of resistance. Another mechanism contributing to resistance is the modification of lipopolysaccharide composition through integration of fatty acids, phosphoethanolamine, and aminoarabinose followed by reduced membrane permeability in bacterium, *Salmonella enterica* (Fjell et al. 2012). In addition to these protective modifications, bacteria can also develop resistance through alteration in their membrane fluidity and hydrophobicity. Other mechanisms to resistance can be membrane-bound efflux pumps (in case of *Neisseria gonorrhoeae*) which expel AMPs from the cell and attack by proteases in the outer membrane with subsequent cleavage (in case of *S. enterica*; Baltzer and Brown 2011).

Similar to conventional antibiotics, AMPs also have a tendency to develop resistance among bacteria when used therapeutically. The frequency of exposure of a therapeutic peptide decides its progression toward resistance. Bacteria use their capability to alter cell membrane and develop resistance. Several lineages of bacteria such as *Pseudomonas fluorescens* and *E. coli* have evolved the mechanisms of resistance (Baltzer and Brown 2011). It has been hypothesized that the diverse existence of quite a number of AMPs in almost all types of organisms can be an indication that these may have co-evolved with bacteria as a response to their attack, thereby explaining their high specificity. Hence, limited exposure or judicious use of these peptides can be an effective way of combating acquired resistance.

Cross-resistance is a consequence of bacteria developing/evolving a mechanism that is efficient not just against one type of antibiotic but also maintains effectiveness against a variety of other antimicrobial molecules. Similar to currently used antibiotics, cross-resistance raises grave concern over the therapeutic or clinical use of AMPs as an alternative to the present ones. Cross-resistance poses a serious threat to these peptides as they are vital part of the innate defense system of plants and animals. Certain studies have displayed emergence of cross-resistance to both human and non-human AMPs but at very low levels. Hence, one should not over speculate about the concept of cross-resistance in AMPs, rather more research in this area can be enlightening (Baltzer and Brown 2011).

11.10 AMPs and Analogues Under Clinical Trials and in Clinical Use

Several approaches have been employed to assess AMPs and their derivatives for their clinical expediency. A few have already entered various phases of clinical trials and some of these have been successfully approved for their therapeutic usage. For instance, Pexiganan (magainin derivative) is in phase III clinical trials for treatment of mild infections associated with diabetic foot ulcers (Koo and Seo 2019); PAC-113 (histatin-5 derivative) is a human salivary peptide, undergoing phase II trials for treatment of oral candidiasis in immunocompromised HIV-infected patients (Cheng et al. 2018); hLF1–11 (human lactoferrin derivative) is undergoing phase I/II testing for treatment of infections in hematopoietic stem cell transplantation (HSCT) recipients (Costa et al. 2019); iseganan (protegrin-1 derivative) is undergoing phase III clinical trial for use in ventilator-associated pneumonia caused by *Klebsiella pneumoniae* and *P. aeruginosa* infections (Dijksteel et al. 2021); rBPI-21 (human neutrophil peptide, bactericidal permeability increasing protein, BPI derivative) is undergoing phase III trials to treat Crohn's disease and severe pediatric meningococcaemia (Dijksteel et al. 2021); omiganan (indolicidin derivative) is undergoing phase III clinical testing for the treatment of topical skin antisepsis and rosacea and phase II trial for vulvar intraepithelial neoplasia, AD and acne (Koo and Seo 2019); LL-37 (human cathelicidin) is presently in the phase IIb trial conducted for the treatment of chronic leg ulcers (Koo and Seo 2019); OP-145 (LL-37 derivative) is currently undergoing phase II clinical trial to treat chronic middle ear infection (Koo and Seo 2019); LTX-109 (synthetic tripeptide) is a phase I/II clinical trial for treatment of MRSA impetigo (Dijksteel et al. 2021).

At present, a great deal work is ongoing to improve AMP development and production (natural and synthetic), aiming to be cost-effective with therapeutical potential against all types of bacterial or viral infections. As of now, the application of AMPs has been limited to the treatment of local infections but scientists are hopeful of achieving promising results with wider therapeutic application from some of the peptides. Some of these peptides under investigation and in different stages of clinical trials have been discussed (Table 11.2) while those approved for clinical use have been enlisted in Table 11.3.

11.11 Other Small Molecule Alternatives to Conventional Antibiotics

The current state of affairs for antibiotics provides overwhelming evidence toward their diminishing effectiveness rendered by bacterial resistance. Therefore, the focus has to be shifted toward devising alternative therapies to combat pathogenic attack and use them clinically. With advances in synthetic organic chemistry, biotechnology, and genetic engineering, scholars have been successful in finding new areas that can be used in substitution to antibiotics. These promising strategies include use of small molecules such as antimicrobial peptides, bacteriocins, SMAMPs, IDR peptides, antibacterial oligonucleotides, foldamers, antibacterial nucleic acids, and immune stimulation by P4 peptide (Table 11.4). As already discussed, broad spectrum, small molecule antimicrobial peptides have drawn significant attention. A few of them are in clinical use and many in the clinical trial

Table 11.2 AMPs and their derivatives under investigation and in clinical trials (Costa et al. 2019; Dijksteel et al. 2021; Koo and Seo 2019)

AMP	Source	Postulated clinical use	Trial phase	Company/collaborator
Pexiganan	Magainin analogue	Diabetic foot ulcers infection	III	Dipexium Pharmaceuticals
PAC-113	Histatin analogue	Oral candidiasis in HIV-seropositive patients	II	Pacgen Biopharmaceuticals
Lactoferrin	Histatin analogue	HIV infection and nosocomial infection in critically ill patients	II	Jason Baker/Ventri Bioscience/ Minneapolis Medical Research Foundation Queen's University
Bovine lactoferrin	Histatin analogue	Neonatal sepsis Necrotizing enterocolitis	III III	Aga Khan University/ University of Sydney/ United States Agency for International Development
hLF1-11	Human lactoferrin derivative	Complications among HSCT recipient infections Bacteremia caused due to *Staphylococcus epidermidis*	I/II I/II (failed)	AM-Pharma
Talactoferrin	Human lactoferrin derivative	Oral solution for nosocomial infection in preterm infants	I/II	Agennix/NIH
PXL01	Lactoferrin analogue	Postsurgical adhesions	II	ProMore Pharma
Iseganan	Protegrin-1 derivative	Oral mucositis (patients undergoing radiation therapy for cancer of head and neck) and ventilator-associated pneumonia	III (failed)	IntraBiotics Pharmaceuticals
POL7080 (murepavadin)	Iseganan derivative	Ventilator-associated pneumonia and non-cystic fibrosis-associated bronchiectasis	II	Polyphor, Lda
Neuprex (rBPI-21)	Human neutrophil peptide, BPI derivative	Crohn's disease and severe pediatric meningococcemia	III	Xoma Ltd.

(continued)

Table 11.2 (continued)

AMP	Source	Postulated clinical use	Trial phase	Company/collaborator
Omiganan (MBI 226 or MX 226)	Indolicidin derivative	Rosacea Severe inflammatory acne vulgaris Central venous catheter infection	III II III	Cutanea Life Sciences Cutanea Life Sciences Mallinacckrodt
LL-37	hCAP18 derivative	Hard to heal venous leg ulcers	II	Lipopeptide AB
OP-145	LL-37 derivative	Chronic suppurative otitis media	II	OctoPlus BV
AP-214	α-MSH derivative	Post-surgical organ failure	II	Action Pharma A/S
SGX942 (Dusquetide)	IDR-1 analogue	Oral mucositis	III	Soligenix
Delmitide (RDP58)	HLA derivative	Inflammatory bowel disease	II	Genzyme
LTX-109	Synthetic tripeptide	MRSA/impetigo	I/II	Lytix Biopharma
EA-230	Oligopeptide	Renal failure and sepsis protection	II	Exponential Biotherapies
Surotomycin (CB-315)	Cyclic lipopeptide	*Clostridium difficile* infections	III	Cubist Pharmaceuticals
NVB-302	Lantibiotic	*Clostridium difficile* infections	III	Novacta
Ramoplanin (NTI-851)	Glycolipodepsipeptide	*Clostridium difficile* infections, VRE	III	Nano-therapeutics
LFF571	Semisynthetic thiopeptide	*Clostridium difficile* infections	II	Novartis Pharmaceuticals Co.
Friulimicin	Cyclic lipopeptide	MRSA/pneumonia	I	MerLion Pharmaceuticals
DPK-060	Kininogen derivative	Acute external otitis	II	ProMore Pharma
XOMA-629 (XMP-629)	Derivative of BPI	Impetigo/acne rosacea	III	Xoma Ltd.
PMX-30063 (Brilacidin)	Defensin mimetic	Acute bacterial skin infections	II	PolyMedix
XF-73 (Exeporfinium chloride)	Porphyrin derivative	*Staphylococcal* infection	II	Destiny Pharma Ltd.
CZEN-002	α-MSH derivative	Antifungal activity	II	Zengen
Ghrelin	Endogenous peptide	Chronic respiratory infection	II	University of Miyazaki; Papworth Hospital
Wap-8294A2 (Lotilibcin)	Produced by *Lysobacter* sp.	Gram-positive bacteria	I/II	aRigen
Novexatin (NP213)	Cyclic cationic peptide	Fungal nail infection	II	Novabiotics

(continued)

Table 11.2 (continued)

AMP	Source	Postulated clinical use	Trial phase	Company/collaborator
p2TA (AB103)	Synthetic peptide	Necrotic tissue infection	III	Atox Bio Ltd.
C16G2	Synthetic peptide	*Streptococcus* mutans	II	C3 Jian Inc.
D2A21	Synthetic peptide	Burn wound infections	III	Demegen
GSK1322322	Synthetic hydrazide	Bacterial skin infections	II	GlaxoSmithKline
PL-5	Synthetic peptide	Skin infections	I	China Food and Drug Administration (CFDA)

HIV human immunodeficiency virus, *HSCT* hematopoietic stem cell transplantation, *MSH* melanocyte stimulating hormone, *IDR* innate defense regulator, *HLA* human leukocyte antigen, *MRSA* methicillin-resistant *Staphylococcus aureus*, *VRE* vancomycin-resistant *Enterococci*, *BPI* bactericidal permeability increasing protein

pipeline. Several other approaches are also being applied to find appropriate antibiotic.

11.11.1 Bacteriocins

The first among these are bacteriocins, a subcategory of AMPs offering advantage over AMPs by being non-toxic to mammalian cells and selective against particular bacterial strains (narrowing the spectrum for bacteriocins). Bacteriocins are diverse, ribosomally synthesized small peptides produced by bacteria for their survival (kill other bacterial population), active against multidrug-resistant pathogenic microbes. They have mechanism similar to AMPs, acting through membrane pore formation. Another mechanism of action for bacteriocins involves inhibition of peptidoglycan biosynthesis, e.g., Nisin is a lantibiotic produced by Gram-positive bacteria, *Lactococcus lactis*, containing rare amino acids such as lanthionine, 3-methyllanthionine, dehydrobutyrine, and dehydroalanine. It is active against Gram-positive bacteria and acquires β-turn conformation on association to dodecylphosphocholine (Koczulla and Bals 2003). Nisin has been approved for commercial use in food industry and acts by inhibiting peptidoglycan biosynthesis. Lactococcin A is another bacteriocin, acting through pore formation in the membrane; thiopeptides and bottromycin target translation; Microcin B17 (MccB17), Microcin J25 (MccJ25), and Microcin C7-C51 (MccC7-C51) bacteriocins acting by traversing through the membrane of Gram-negative bacteria, thereby affecting DNA, RNA, and proteins metabolism; thuricin is selective toward *Clostridium difficile* and does not affect commensal bacteria. An added advantage to bacteriocins is slow development of resistance. For instance, nisin has been a perfect case of

Table 11.3 FDA-approved AMPs in clinical use (Costa et al. 2019; Usmani et al. 2017)

AMP	Characteristic features	Clinical use	Route of administration
Bacitracin	Mixture of cyclic peptides obtained from *Bacillus subtilis*; the side chain thiol of Cys residue undergoes intramolecular condensation with the foregoing peptide bond, resulting in thiazolidine ring formation	Skin and eye infections; use restricted to infants with *S. aureus* pneumonia and purulent pleuritis	Topical application Intramuscular
Polymyxin B	Mixture of cyclic lipopeptides obtained from *Bacillus polymyxa*	Last line treatment for MDR Gram-negative bacterial infections	Parenteral administration (intramuscular, intravenous, intrathecal, ophthalmic)
Tyrothricin	Mixture of cyclic polypeptides obtained from *Bacillus brevis*	Local treatment; skin and oropharyngeal mucous membrane infections; active against Gram-positive bacteria	Topical application only; parenterally (highly toxic)
Polymyxin E (colistins)	Mixture of cyclic lipopeptides from *Bacillus colistinus*	GIT infections caused by *E. coli* and *Salmonella* spp.; MDR gram-negative bacteria such as *P. aeruginosa, K. pneumoniae*, *Acinetobacter* spp. and not responding to other antibiotics	Parenteral administration (intramuscular, intravenous)
Gramicidin D or gramicidin	Mixture of linear pentadecapeptides obtained from *Bacillus brevis* with formylated N-terminal residue and peptide bond with ethanolamine at C-terminal tryptophan residue	Skin lesions, surface wounds, and eye infections	External use only
Gramicidin S	Cyclic peptide biosynthesized from gramicidin in *Bacillus brevis*; contain two identical pentapeptides coupled head to tail	Effective against gram-negative and positive bacteria and fungi; treat genital ulcers caused by STD	Topical application only
Daptomycin	Cyclic lipopeptide obtained from *S. roseosporus*, acylated with decanoic acid at N-terminal and contains unusual residue kynurenine at C-terminal	Complex skin infections caused by susceptible strains of Gram-positive bacteria	Intravenous injection

S. aureus: Staphylococcus aureus, *GIT:* gastrointestinal, *E. coli: Escherichia coli*, *MDR:* multidrug-resistant, *P. aeruginosa: Pseudomonas aeruginosa*, *K. pneumoniae: Klebsiella pneumoniae*, *STD:* sexually transmitted disease, *S. roseosporus: Streptomyces roseosporus* antimicrobial remaining unaffected by resistance after decades of use (Allen et al. 2014; Ghosh et al. 2019). Bacteriocins have also been reported to be resilient toward extreme environmental conditions like temperature and UV radiations (van der Helm 2014). These advantages of bacteriocins establish them as potential alternatives to classical antibiotics.

11.11.2 Innate Defense Regulatory (IDR) Peptides

It is well known fact that AMPs modulate the innate immune system for defense against pathogenic bacteria. Certain peptides lacking in direct antibacterial activity but exhibit immunomodulatory and antiendotoxin activities against pathogens are called "innate defense regulatory" (IDR) peptides. Investigations have shown that these peptides do not have any direct antimicrobial activity in mice but activate cellular machinery and suppress harmful inflammation and protect from succumbing to severe bacterial and malarial infections. SGX 942 is one of such peptides undergoing phase I clinical trials, indicated in the treatment of bacterial infection (Ghosh et al. 2019). IDR peptides can hence be projected as promising candidates to be used over conventional antibiotics.

11.11.3 Synthetically Designed Strategies Synthetic Mimics of Antimicrobial Peptides (SMAMPs)

An attempt has been made to simplify the complexities in the structure of AMPs and preserve the ensemble of pharmacophoric elements in their chemical structure and retain the biological properties of AMPs. This has led to the design of polymeric mimics of AMPs, novel biomimetic/peptidomimetic oligomers, SMAMPs, or small molecules which mimic the essential properties of antimicrobial peptides. The objective is to develop approaches that can trounce over toxicity, enzymatic degradation (protease lability), and high cost of manufacture of AMPs. Alternative molecular designing and synthetic strategies have helped develop small molecules mimicking the properties of AMPs. In this approach, facial amphiphilicity is integrated into small molecules by using H-bonding motifs. For instance, Brilacidin is a defensin mimetic undergoing phase II clinical trials for use in acute bacterial skin infections. In this global COVID-19 pandemic, brilacidin like many other compounds has also been evaluated for its antiviral properties against SARS-CoV-2 and it has been found that brilacidin strongly decreases the viral load when the virus is pre-incubated with brilacidin. It demonstrates early viral inhibition in vitro against Washington and Italian strain (Bakovic et al. 2021). LTX-109 is another synthetic tripeptide, currently in phase I/II clinical trial for treatment of MRSA impetigo (Dijksteel et al. 2021). Ceragenins are cholic acid derivatives possessing

Table 11.4 Small molecule alternatives to conventional antibiotics (Czaplewski et al. 2016; Ghosh et al. 2019; van der Helm 2014)

Strategy	Characteristics	Advantages over conventional antibiotics	Possible disadvantages
Antimicrobial peptides	Endogenously produced large group of diverse peptides made up of 10–50 units Proceed through cell membranes disruption	Resilient to resistance development Broad-spectrum activity	Expensive large-scale production Susceptible to proteolytic degradation Toxicity
Bacteriocins	Ribosomally synthesized Selective to bacterial strains Mechanism similar to AMPs	Specificity toward infective bacteria Resistant to heat and UV	Expensive large-scale production Susceptible to proteolytic degradation
IDR peptides	Act through immunomodulation and antiendotoxin activity	No resistance development as no direct antimicrobial activity	Expensive large-scale production Susceptible to proteolytic degradation
SMAMPs	Small molecules mimicking essential properties of antimicrobial peptides	Ease of synthesis Good tissue distribution Resilient to resistance development Broad-spectrum activity	Toxicity Route of administration
Antibacterial oligonucleotides	Oligomeric backbone modified with morpholino or phosphorodiamidate moieties Utilize antisense technology Work by inhibiting expression of resistance causing genes	No resistance development as no direct antimicrobial activity	Relatively new approach, require extensive research
Foldamer	Short sequence-specific oligomers based on arylamide and β-amino acid backbones Fold into well-defined secondary structures	Structural simplicity and relative ease of synthesis Less toxicity Low cost of manufacture	Relatively new approach, require extensive research
Antibacterial nucleic acids	Utilize nucleic acids to kill bacteria directly	Ease of synthesis Resilient to resistance development	Expensive method

(continued)

Table 11.4 (continued)

Strategy	Characteristics	Advantages over conventional antibiotics	Possible disadvantages
Immune stimulation by P4 peptide	28 amino acid residue synthetic peptide Enhance phagocytic killing of bacteria	No resistance development as no direct antimicrobial activity	Relatively new approach, require extensive research

AMPs antimicrobial peptides, *IDR* innate defense regulatory peptides, *SMAMPS* synthetic mimics of antimicrobial peptides

antimicrobial properties which are in preclinical stage of investigation. Binaphthyl-based dicationic peptides and xanthone derivatives have been synthesized by addition of amino acids or short peptides to the hydrophobic groups and show significant antimicrobial activity (Ghosh et al. 2019). Ghosh et al. have synthesized several aryl-alkyl-lysine compounds which have shown bactericidal properties, activity against fungi, malaria parasite, and Ebola virus. Synthetic simple lipidated and biphenyl-based lysines have shown potent activity against intracellular bacteria in vivo. A library of norspermidine-based compounds have been synthesized, documented to exhibit potential against Gram-positive and Gram-negative bacteria (Ghosh et al. 2019). Further development and extensive clinical testing of these small molecules can pave way for their therapeutic use as alternative antibiotics.

11.11.4 Antibacterial Oligonucleotides

Antisense technology has proven to be a useful tool in tackling the problem of antimicrobial resistance. A number of oligonucleotides and their analogues have been generated by this technique which involves inhibition of the expression of mutant genes or in this case the resistance-causing genes. It therefore becomes important to identify and select the undesirable gene before silencing it. Diverse antisense oligonucleotides are hybridized with mRNA, thereby inhibiting the expression of genes responsible for resistance, mediating through different mechanisms (Jani et al. 2021). The oligomeric backbone is first modified with morpholino and phosphorodiamidate groups also known as phosphorodiamidate morpholino oligomers (PMOs) to prevent resistance by nuclease enzymatic degradation and then conjugated with cell-penetrating peptides (PPMOs) to facilitate transport of oligomers across the bacterial membrane. The PPMOs have displayed in vitro and in vivo antibacterial activity against *E. coli, S. enterica*, and *Acinetobacter baumannii*. Another oligonucleotide, phosphorothioate oligodeoxynucleotides (PSODNs) have been found to activate RNase H activity leading to mRNA degradation followed by antibacterial potential. The PPMOs exhibit bacterial growth inhibition when used along with colistin and polymyxin B. On combining with β-lactams and tobramycin, PPMOs disrupted biofilms and inhibited *P. aeruginosa* (Ghosh et al. 2019).

Hence, oligonucleotides embrace a diverse group of molecules capable of fighting against a number of pathogens (bacteria and virus), cancer, and genetic disorders.

11.11.5 Foldamers

Foldamers are short, sequence-specific oligomers that fold into well-defined secondary structures, postulated to act as antimicrobial agents. Similar to AMPs, foldamers typify the positively charged amphiphilic structures responsible for the activity of AMPs. In the past decade, several novel small molecule foldamers with specificity for bacterial membranes and greater activity have been synthesized and have shown excellent potency toward multidrug-resistant bacterial infections in animal models. A variety of phenylene-ethynylene, arylamide, acrylate, and hydrocarbon-based foldamers have been developed by various researchers and investigations on deducing their activity potential and mechanism of action are still in process. Guichard et al. have synthesized urea-based enantiomerically pure antimicrobial foldamers (Violette et al. 2006). While several others are undergoing human clinical trials for use as antibiotic in the treatment of multi drug-resistant *S. aureus* (Tew et al. 2010).

11.11.6 Antibacterial Nucleic Acids

Studies are being conducted by researchers to investigate certain nucleic acids that can kill bacteria directly. However, it is too early to be sure about nucleic acids exhibiting antibacterial action as studies are still at preliminary stages. Advances in microbial genetics techniques can help develop greater number of nucleic acid antibiotics (Czaplewski et al. 2016).

11.11.7 Immune Stimulation by P4 Peptide

P4 peptide is a synthetic 28 amino acid residue peptide derived from the functional site at the surface of bacterium, *Streptococcus pneumoniae* on exposure to virulence factor PsaA (pneumococcal surface adhesion) and has been found to stimulate opsonophagocytic uptake and killing in mice invasive disease experimental models for *S. pneumoniae* infection. This peptide enhances phagocytic killing of bacteria. When P4 (intranasal) and IgG (immunoglobulin G, intraperitoneal) are administered in mouse, there was 100% survival and significant reduction in bacterial burden (Czaplewski et al. 2016). Hence, P4 peptide and many more such peptides acting through immunostimulation can provide an alternative to antimicrobial therapy as it is devoid of resistance.

11.12 Conclusions

Immense focus lays on the development and building of libraries of new novel classes of small molecule antimicrobials competent of surmounting the growing problem of microbial resistance. In order to combat antimicrobial resistance and fight infections caused by pathogenic microorganisms, numerous alternative classes of antibiotics have been developed. Some of the approaches have advanced to clinical trial phases while others are still nascent at the laboratory stage. With the advent of large number of antimicrobials from our microbiome and their diverse actions, peptide-based antimicrobials have drawn considerable attention from the scientific community as well as from the pharma sector to further explore their potential applications in various fields. Studies have shown encouraging results for natural and synthetic peptides for their use in bacterial, mycobacterial, sexually transmitted, buco-dental, ocular and viral infections, as anticancer agent, wound repair, Kostmann's syndrome, as food preservatives, contraceptive, etc. AMPs have also shown desirable efficacy when used alone or in synergy with conventional antibiotics. Cumulative data from clinical studies conducted on several AMP molecules entering different phases further consolidate the claims to their therapeutic utility. Despite all the optimism toward AMPs, there are certain challenges associated with them *viz a viz* unwanted toxicity, lack of selectivity, stability of the peptide or its formulation and cost-effectiveness for large-scale production. Therefore, the focal point remains to be the optimization of strategies for candidate identification, having appropriate pharmacodynamic and pharmacokinetic properties (short half-life, high binding affinity, and high therapeutic index) and analyzing their biological functions. Peptide or non-peptide small scaffolds that can efficiently integrate these principles can be strongly considered for the future development of new and improved generations of peptides; making their employment safer for end user and a lucrative enterprise for the manufacturer.

References

Allen HK, Trachsel J, Looft T, Casey TA (2014) Finding alternatives to antibiotics. Ann N Y Acad Sci 1323(1):91–100

Andreu D, Rivas L (1998) Animal antimicrobial peptides: an overview. Pept Sci 47(6):415–433

Bahar AA, Ren D (2013) Antimicrobial peptides. Pharmaceuticals 6(12):1543–1575

Bakovic A, Risner K, Bhalla N, Alem F, Chang TL, Weston WK, Harness JA, Narayanan A (2021) Brilacidin demonstrates inhibition of SARS-CoV-2 in cell culture. Viruses 13(2):271

Baltzer SA, Brown MH (2011) Antimicrobial peptides—promising alternatives to conventional antibiotics. J Mol Microbiol Biotechnol 20(4):228–235

Cheng K-T, Wu C-L, Yip B-S, Yu H-Y, Cheng H-T, Chih Y-H, Cheng J-W (2018) High level expression and purification of the clinically active antimicrobial peptide P-113 in Escherichia coli. Molecules 23(4):800

Costa F, Teixeira C, Gomes P, Martins MCL (2019) Clinical application of AMPs. Antimicrob Peptides 1117:281–298

Czaplewski L, Bax R, Clokie M, Dawson M, Fairhead H, Fischetti VA, Foster S, Gilmore BF, Hancock RE, Harper D (2016) Alternatives to antibiotics—a pipeline portfolio review. Lancet Infect Dis 16(2):239–251

Dijksteel GS, Ulrich MM, Middelkoop E, Boekema BK (2021) Lessons learned from clinical trials using antimicrobial peptides (AMPs). Front Microbiol 12:287

Edwards IA, Elliott AG, Kavanagh AM, Blaskovich MA, Cooper MA (2017) Structure–activity and −toxicity relationships of the antimicrobial peptide tachyplesin-1. ACS Infect Dis 3(12):917–926

Etayash H, Norman L, Thundat T, Stiles M, Kaur K (2014) Surface-conjugated antimicrobial peptide leucocin a displays high binding to pathogenic gram-positive bacteria. ACS Appl Mater Interfaces 6(2):1131–1138

Fjell CD, Hiss JA, Hancock RE, Schneider G (2012) Designing antimicrobial peptides: form follows function. Nat Rev Drug Discov 11(1):37–51

Gabay JE, Almeida RP (1993) Antibiotic peptides and serine protease homologs in human polymorphonuclear leukocytes: defensins and azurocidin. Curr Opin Immunol 5(1):97–102

Gao G, Lange D, Hilpert K, Kindrachuk J, Zou Y, Cheng JT, Kazemzadeh-Narbat M, Yu K, Wang R, Straus SK (2011) The biocompatibility and biofilm resistance of implant coatings based on hydrophilic polymer brushes conjugated with antimicrobial peptides. Biomaterials 32(16):3899–3909

Ghosh C, Sarkar P, Issa R, Haldar J (2019) Alternatives to conventional antibiotics in the era of antimicrobial resistance. Trends Microbiol 27(4):323–338

Gopal R, Lee JH, Kim YG, Kim M-S, Seo CH, Park Y (2013) Anti-microbial, anti-biofilm activities and cell selectivity of the NRC-16 peptide derived from witch flounder, Glyptocephalus cynoglossus. Mar Drugs 11(6):1836–1852

van der Helm A (2014) An overview of alternatives for conventional antibiotics

Hwang PM, Zhou N, Shan X, Arrowsmith CH, Vogel HJ (1998) Three-dimensional solution structure of lactoferricin B, an antimicrobial peptide derived from bovine lactoferrin. Biochemistry 37(12):4288–4298

Izadpanah A, Gallo RL (2005) Antimicrobial peptides. J Am Acad Dermatol 52(3):381–390

Jacob L, Zasloff M (1994) Potential therapeutic applications of magainins and other antimicrobial agents of animal origin. Antimicrob Pept 186:197–223

Jani S, Ramirez MS, Tolmasky ME (2021) Silencing antibiotic resistance with antisense oligonucleotides. Biomedicine 9(4):416

van Kan E, van der Bent A, van Amerongen A (1998) Antimicrobial peptides as food preservatives. In: SON-study section for protein research, 7-8 December, Lunteren, The Netherlands

Kang S-J, Park SJ, Mishig-Ochir T, Lee B-J (2014) Antimicrobial peptides: therapeutic potentials. Expert Rev Anti-Infect Ther 12(12):1477–1486

Kang H-K, Kim C, Seo CH, Park Y (2017) The therapeutic applications of antimicrobial peptides (AMPs): a patent review. J Microbiol 55(1):1–12

Killian JA (1992) Gramicidin and gramicidin-lipid interactions. Biochim Biophys Acta 1113(3–4):391–425

Koczulla AR, Bals R (2003) Antimicrobial peptides. Drugs 63(4):389–406

Koo HB, Seo J (2019) Antimicrobial peptides under clinical investigation. Pept Sci 111(5):e24122

Kościuczuk EM, Lisowski P, Jarczak J, Strzałkowska N, Jóźwik A, Horbańczuk J, Krzyżewski J, Zwierzchowski L, Bagnicka E (2012) Cathelicidins: family of antimicrobial peptides. A review. Mol Biol Rep 39(12):10957–10970

Kosikowska P, Lesner A (2016) Antimicrobial peptides (AMPs) as drug candidates: a patent review (2003–2015). Expert Opin Ther Pat 26(6):689–702

Kuhn-Nentwig L, Müller J, Schaller J, Walz A, Dathe M, Nentwig W (2002) Cupiennin 1, a new family of highly basic antimicrobial peptides in the venom of the spider Cupiennius salei (Ctenidae). J Biol Chem 277(13):11208–11216

Mahlapuu M, Håkansson J, Ringstad L, Björn C (2016) Antimicrobial peptides: an emerging category of therapeutic agents. Front Cell Infect Microbiol 6:194

Mandard N, Sy D, Maufrais C, Bonmatin J-M, Bulet P, Hetru C, Vovelle F (1999) Androctonin, a novel antimicrobial peptide from scorpion Androctonus australis: solution structure and molecular dynamics simulations in the presence of a lipid monolayer. J Biomol Struct Dyn 17(2):367–380

Mirski T, Niemcewicz M, Bartoszcze M, Gryko R, Michalski A (2018) Utilisation of peptides against microbial infections—a review. Ann Agric Environ Med 25:205–210

Mohamed MF, Abdelkhalek A, Seleem MN (2016) Evaluation of short synthetic antimicrobial peptides for treatment of drug-resistant and intracellular Staphylococcus aureus. Sci Rep 6(1):1–14

Mor A (2000) Antimicrobial peptides. In: Kirk-Othmer encyclopedia of chemical technology. Wiley, New York

Munshi T, Sparrow A, Wren BW, Reljic R, Willcocks SJ (2020) The antimicrobial peptide, Bactenecin 5, supports cell-mediated but not humoral immunity in the context of a mycobacterial antigen vaccine model. Antibiotics 9(12):926

Park Y-K, Hahm K-S (2005) Antimicrobial peptides (AMPs): peptide structure and mode of action. BMB Rep 38(5):507–516

Peravali J, Kotra S, Sobha K, Nelson R, Rajesh K, Pulicherla K (2013) Antimicrobial peptides: an effective alternative for antibiotic therapy. Mintage J Pharm Med Sci 2(2):1–7

Piper P, Mahé Y, Thompson S, Pandjaitan R, Holyoak C, Egner R, Mühlbauer M, Coote P, Kuchler K (1998) The Pdr12 ABC transporter is required for the development of weak organic acid resistance in yeast. EMBO J 17(15):4257–4265

Powers J-PS, Rozek A, Hancock RE (2004) Structure–activity relationships for the β-hairpin cationic antimicrobial peptide polyphemusin I. Biochim Biophys Acta 1698(2):239–250

Radermacher S, Schoop V, Schluesener H (1993) Bactenecin, a leukocytic antimicrobial peptide, is cytotoxic to neuronal and glial cells. J Neurosci Res 36(6):657–662

Reddy K, Yedery R, Aranha C (2004) Antimicrobial peptides: premises and promises. Int J Antimicrob Agents 24(6):536–547

Ricci M, Horváti K, Juhász T, Szigyártó I, Török G, Sebák F, Bodor A, Homolya L, Henczkó J, Pályi B (2020) Anionic food color tartrazine enhances antibacterial efficacy of histatin-derived peptide DHVAR4 by fine-tuning its membrane activity. Q Rev Biophys 53:e5

Schneider JJ, Unholzer A, Schaller M, Schäfer-Korting M, Korting HC (2005) Human defensins. J Mol Med 83(8):587–595

Sharma R, Lomash S, Salunke DM (2013) Putative bioactive motif of tritrpticin revealed by an antibody with biological receptor-like properties. PLoS One 8(9):e75582

Sperstad SV, Haug T, Blencke H-M, Styrvold OB, Li C, Stensvåg K (2011) Antimicrobial peptides from marine invertebrates: challenges and perspectives in marine antimicrobial peptide discovery. Biotechnol Adv 29(5):519–530

Sun C, Gu L, Hussain MA, Chen L, Lin L, Wang H, Pang S, Jiang C, Jiang Z, Hou J (2019) Characterization of the bioactivity and mechanism of bactenecin derivatives against food-pathogens. Front Microbiol 10:2593

Tew GN, Scott RW, Klein ML, DeGrado WF (2010) De novo design of antimicrobial polymers, foldamers, and small molecules: from discovery to practical applications. Acc Chem Res 43(1):30–39

Usmani SS, Bedi G, Samuel JS, Singh S, Kalra S, Kumar P, Ahuja AA, Sharma M, Gautam A, Raghava GP (2017) THPdb: database of FDA-approved peptide and protein therapeutics. PLoS One 12(7):e0181748

Vergis J, Malik SS, Pathak R, Kumar M, Ramanjaneya S, Kurkure NV, Barbuddhe SB, Rawool DB (2019) Antimicrobial efficacy of indolicidin against multi-drug resistant enteroaggregative Escherichia coli in a Galleria mellonella model. Front Microbiol 10:2723

Vetterli SU, Zerbe K, Müller M, Urfer M, Mondal M, Wang S-Y, Moehle K, Zerbe O, Vitale A, Pessi G (2018) Thanatin targets the intermembrane protein complex required for lipopolysaccharide transport in Escherichia coli. Sci Adv 4(11):eaau2634

Violette A, Fournel S, Lamour K, Chaloin O, Frisch B, Briand J-P, Monteil H, Guichard G (2006) Mimicking helical antibacterial peptides with nonpeptidic folding oligomers. Chem Biol 13(5):531–538
Wimley WC (2010) Describing the mechanism of antimicrobial Peptide action with the interfacial activity model. ACS Chem Biol 5(10):905–17
Wu M, Hancock RE (1999) Improved derivatives of bactenecin, a cyclic dodecameric antimicrobial cationic peptide. Antimicrob Agents Chemother 43(5):1274–1276
Zairi A, Tangy F, Bouassida K, Hani K (2009) Dermaseptins and magainins: antimicrobial peptides from frogs' skin—new sources for a promising spermicides microbicides—a mini review. J Biomed Biotechnol 2009:452567

Chapter 12
Molecular Mechanisms of Antimicrobial Resistance in *Staphylococcus aureus* Biofilms

Vanessa Silva, José L. Capelo, Gilberto Igrejas, and Patrícia Poeta

12.1 Biofilms: Introduction

Biofilms are multicellular complex systems which can attach to biotic and abiotic surfaces. Bacterial biofilms were first described by Antony van Leeuwenhoek in 1684. Leeuwenhoek scraped plaque biofilm off his own teeth and studied the sample (Valen and Scheie 2018). Biofilms are a common phenomenon in nature. Arthur Henrici in 1933 reported that the majority of aquatic

V. Silva
Microbiology and Antibiotic Resistance Team (MicroART), Department of Veterinary Sciences, University of Trás-os-Montes and Alto Douro (UTAD), Vila Real, Portugal

Department of Genetics and Biotechnology, University of Trás-os-Montes and Alto Douro (UTAD), Vila Real, Portugal

Functional Genomics and Proteomics Unit, University of Trás-os-Montes and Alto Douro (UTAD), Vila Real, Portugal

University NOVA of Lisboa, Lisboa, Caparica, Portugal

J. L. Capelo
BIOSCOPE Group, LAQV@REQUIMTE, Chemistry Department, Faculty of Science and Technology, NOVA University of Lisbon, Almada, Portugal

Proteomass Scientific Society, Costa de Caparica, Portugal

G. Igrejas
Department of Genetics and Biotechnology, University of Trás-os-Montes and Alto Douro (UTAD), Vila Real, Portugal

Functional Genomics and Proteomics Unit, University of Trás-os-Montes and Alto Douro (UTAD), Vila Real, Portugal

University NOVA of Lisboa, Lisboa, Caparica, Portugal

N. Akhtar et al. (eds.), *Emerging Modalities in Mitigation of Antimicrobial Resistance*,
https://doi.org/10.1007/978-3-030-84126-3_12

microorganisms were not found in the form of individual cells swimming freely but clustered on solid submerged surfaces (Henrici 1933). Nevertheless, it was not only until 1970s that scientists investigate the role of bacterial biofilms. Costerton and Høiby reported that bacteria should not only be considered as planktonic cells but also as sessile cells (Høiby 2014). The term biofilm was introduced in 1978 by Costerton referring to heterogenous bacterial aggregates surrounded by a matrix that allows their attachment to inorganic or organic surfaces (Costerton et al. 1978). Bacteria can grow as free-living planktonic cells or as sessile cells that attach to surface within biofilms. Biofilm formation is a developmental process which consists in the aggregation of planktonic cells and their transition to a surface-associated lifestyle (Chua et al. 2014). In nature, bacteria rarely grow as planktonic cultures. Instead, they predominantly exist as communities of sessile cells (Hobley et al. 2015; Berlanga and Guerrero 2016). The presence of biofilms can affect negatively several different areas, such as the food, industry, environment, and biomedical sectors. For instance, one of the most documented effects of microbial biofilms in the industrial processes is the biofouling which has a negative impact since it leads to corrosion (Mehrabi et al. 2020). Biofouling can affect several different fields such as drinking water systems, marine aquaculture, agricultural industry, food, beverage, and milk industries, among others (Flemming 2020). In the medical setting, biofilms (including both bacterial and fungal biofilms) are responsible for around 80% of all human infections (Fox and Nobile 2012; Dunn et al. 2020). Biofilms are usually associated with medical devices such as catheters, mechanical heart valves, joint prostheses, and orthopedic devices but can also be related with other infections such as endocarditis and osteomyelitis (Kwiecinski et al. 2019; Balaure and Grumezescu 2020). Nosocomial infections, either biofilm-associated or not, have a tremendous social and economic impact. In the United States, it is estimated that 1.96 million cases of biofilm-related infections occur, leading to 268,000 deaths, with more than $US18 billion spent in treatment costs (Rumbaugh and Sauer 2020). In the European Union and European Economic Area, it is estimated that 8.9 million health-care-associated infections occur each year (Carl et al. 2018). Thus, biofilms formed by pathogenic and multidrug-resistant organisms create a major public health concern. In fact, biofilm-related infections are extremely difficult to eradicate since biofilms have high cell densities and the antimicrobial resistance of biofilms is up to 1000-fold higher than that of planktonic cells (Olsen 2015). Biofilm-forming bacteria display characteristics, activities, patterns, and behavior that are distinct from those of planktonic cells (Corning 2002).

P. Poeta (✉)
Microbiology and Antibiotic Resistance Team (MicroART), Department of Veterinary Sciences, University of Trás-os-Montes and Alto Douro (UTAD), Vila Real, Portugal

University NOVA of Lisboa, Lisboa, Caparica, Portugal
e-mail: ppoeta@utad.pt

The bacterial community can respond non-homogeneously and adaptively to external physicochemical conditions which may vary depending on the requirements of the bacterial community (Stubbendieck et al. 2016). Within a biofilm there are microenvironments that favor the survival of these communities that can be modified according to their metabolic needs. For instance, anaerobic bacteria may live within biofilms in a modified microenvironment despite the presence of dissolved oxygen in the medium (Kato et al. 1997). Biofilms also express novel phenotypic characteristics due to genetic variations that may occur through mutations or recombination (Penesyan et al. 2019). Although most studies on biofilms have been performed with a single strain, in nature, biofilms are usually formed by different species of bacteria and/or fungi that can form a single community in symbiosis. This synergism seen among the bacteria living in biofilm brings some advantages, such as metabolic cooperation, passive resistance, quorum-sensing system, and a larger pool of genes that allow them a high adaptation and survival (Wolcott et al. 2013; Madsen et al. 2016). Consequently, this cooperation generates more severe infections such as those reported in biofilm-related infections caused by *Staphylococcus aureus* together with *Pseudomonas aeruginosa* (DeLeon et al. 2014; Hotterbeekx et al. 2017). Nevertheless, neutral or competitive interactions may also exist between some species which leads to significant effects of interaction or a decrease in productivity, respectively (Ballén et al. 2020). Another important feature of the biofilm-living mode is the capacity that all bacteria have to communicate among themselves which is known as Quorum-sensing (QS). QS is a type of cell-cell communication that regulates the expression of genes in a medium with high bacterial density allowing them to carry out a wide range of complex social activities (Müller et al. 2006). Bacteria that bind to the biofilm produce a signal molecule (self-inducer) that sends the message to the other bacteria nearby and the close proximity between cells of a biofilm facilitates the communication by QS. QS is necessary for the biofilm maturation stage and for the expression of virulence factors (Kong et al. 2006). Finally, one of the most important biofilm characteristics is the bacterial resistance conferred by this community mode of living. Biofilms have an organizational structure that shields them from host defense mechanisms and external antibiotics (De la Fuente-Núñez et al. 2013). They have a physical barrier, mainly constituted by the extracellular matrix, that prevents the action of the host defense mechanisms (De la Fuente-Núñez et al. 2013). The bacterial resistance of a biofilm may be due to low metabolic activity of bacteria in a biofilm, existence of "anaerobic niches" in the depth of a biofilm and slow or incomplete penetration of the antibiotic. All of these aspects are covered in more depth in the next topics.

As mentioned before, the majority of biofilm development studies are conducted using single-species biofilms. Several genus and species of bacteria are known to be strong biofilm producers. Among them we can highlight *S. aureus*, *S. epidermidis*, *P. aeruginosa*, and *Klebsiella* pneumoniae (Mohammed et al. 2018). In this chapter, we will focus on *S. aureus* biofilms.

12.2 *Staphylococcus aureus*

S. aureus is a Gram-positive, catalase, and coagulase-positive organism that colonizes the nostrils and mouth of several mammals including humans (Silva et al. 2020). However, when there is a breakdown of this usual balance of the organism, *S. aureus* is able to cause a wide range of infections from mild skin and soft tissue infections to life-threatening ones such as endocarditis and osteomyelitis (Silva et al. 2020). *S. aureus* belongs to the ESKAPE family of antimicrobial resistance pathogens that includes other bacteria that are responsible for the majority of nosocomial infections. Due to the fact that antimicrobial resistance is spreading rapidly it is estimated that it may cause ten million deaths per year by 2050 (O'Neill 2016). *S. aureus* has the ability to easily acquire antibiotic resistance determinants which, together with the wide range of virulence factors they produce, make its infections even harder to treat (Silva et al. 2020). *S. aureus* is a notable example of a human pathogen that is able to cause biofilm-related infections which are recalcitrant to clearance by antimicrobials and the host immune system (Suresh et al. 2019). These bacteria are a leading cause of biofilm-related infections associated with catheters and implanted medical devices, such as heart valves, pacemakers, and prosthetic joints (Hogan et al. 2016; Zhang et al. 2020). In most cases, the biofilm-related infection becomes chronic and the treatment requires a second surgery to remove and replace the implant (Hogan et al. 2016). Biofilm formation impairs the action of the host immune system and antibiotics due to *S. aureus* phenotypic characteristics and the biofilm matrix (Silva et al. 2020).

12.3 *S. aureus* Stages of Biofilm Development

Biofilm formation is a dynamic and cyclic process. Although many studies have reported different biofilm architectures, it is known that, regardless of the bacterial species, biofilm formation comprises the same developmental stages and general features (Rumbaugh and Sauer 2020). Based on in vitro research on single species biofilms, biofilm formation was classically described in 3 phenotypically distinct stages: attachment, maturation, and dispersion (Fig. 12.1), although recently a few authors consider two more stages: multiplication and exodus (Miao et al. 2017; Moormeier and Bayles 2017). The development of biofilm formation is triggered by unfavorable conditions such as external attack, physical conditions, or nutrient limitations (Liu et al. 2015). For the process of biofilm formation, there are a variety of physical and metabolic interactions necessary for adhesion, growth, and survival, in addition to increasing the resistance of these groups to hostile environments for their development.

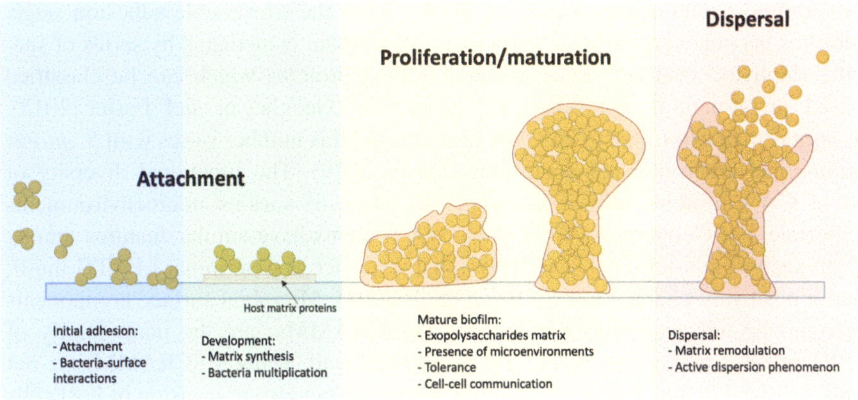

Fig. 12.1 Schematic representation of a biofilm formation. Attachment: reversible attachment of the planktonic cells followed by the adhesion to the surface. Proliferation and maturation: maturation and accumulation of bacteria in multilayered clusters surrounded by extracellular polymeric substances. Dispersal: dissemination of bacterial cells from the aggregate biofilm and reactivation of the planktonic state through dispersal

12.3.1 Attachment

Biofilm formation of *S. aureus* starts with the initial attachment of planktonic bacterial cells to biotic or abiotic surfaces, followed by the formation of bacterial aggregates (Schilcher and Horswill 2020). This step depends on physical factors such as surface charge, roughness, topography, and temperature; and on chemical factors such as pH, substrate, dissolved oxygen, and medium composition. The initial attachment may be reversible or irreversible. The reversible adhesion to the surfaces or subtracts is characterized by a weak interaction which depends on environmental factors and the genetic regulation of bacteria (Armbruster and Parsek 2018). Moreover, in this stage, the bacterial film is more susceptible to the action of antimicrobials than in the next stages. Reversible cell adhesion to surfaces is accomplished by van der Waals forces, steric interactions, electrostatic interaction, and hydrophobic forces (Van Loosdrecht et al. 1987; Gilbert et al. 1991; Carpentier and Cerf 1993). These forces are not strong enough to hold the cells together, since the bonds between cells and substrate are weak. Therefore, for cell adhesion to become permanent, the extracellular components of bacterial cells are needed (Liu et al. 2004). Cell adhesion becomes irreversible due to the presence of exopolysaccharides (EPS) on the outer surface of cells which act as adhesives, producing strong bonds between bacteria or between bacteria and the substrate. EPS includes proteins, phospholipids, nucleic acids, and polysaccharides which bond to bacteria and surfaces by stronger forces such as covalent, hydrophobic, hydrogen bonds, and dipole-dipole interactions (Liu et al. 2004). After the irreversible adhesion stage, biofilms become very difficult to remove. Attachment is mediated by series of specific staphylococcal cell wall-anchored (CWA) proteins which can be classified based on structural and functional properties (Geoghegan and Foster 2015). *S. aureus* may exhibit up to 24 CWA proteins and this number varies with *S. aureus* strains and environmental conditions (Foster 2019). The functional diversity of these CWA proteins allows *S. aureus* to adapt to various microenvironments (Speziale and Geoghegan 2015). All CWA proteins have similar features among them such as a N-terminal secretory signal sequence, a C-terminal LPXTG motif, and a positively charged tail (O'Brien et al. 2002). Microbial surface components recognizing adhesive matrix molecules (MSCRAMMs) are the main family of CWA proteins that are anchored in *S. aureus* cell wall. Many MSCRAMMs are not only involved in attachment and biofilm formation but also in invasion of host cells and tissues, evasion of immune responses, and virulence (Kang et al. 2013). MSCRAMMs are responsible for the initial attachment to native tissues, being able to bind to one or several different host extracellular matrix proteins such as fibronectin, fibrogen, laminin, and elastin; to indwelling medical devices and plasma-coated biological surfaces (Seo et al. 2008). The MSCRAMMs family includes many proteins such as fibronectin binding proteins, clumping factor proteins, biofilm-associated protein, surface-anchored proteins, elastin binding protein, fibrinogen binding protein, laminin binding protein, and collagen binding protein. *S. aureus* has two main fibronectin binding proteins, FnBA and FnBB, which are encoded by the *fnbA* and *fnbB* genes, respectively. *S aureus* may have only one or both *fnb* genes (Peacock et al. 2000). Both proteins are involved in bacterial invasion of the endothelial cells (Peacock et al. 2002). The N-terminal A region of FnB proteins has the ability to bind to fibrinogen and elastin (Murai et al. 2016). Fibrinogen/fibrin is one of the major plasma proteins accumulated on implanted medical devices. The ability of *S. aureus* to adhere to fibrinogen results in instantaneous clumping of bacterial cells. Two receptors for fibrinogen, ClfA and ClfB, promote binding of *S. aureus* to fibrinogen (Ní Eidhin et al. 1998; Hall et al. 2003). ClfA and ClfB have all the traditional features of the CWA proteins and are also members of the serine-aspartate repeat (Sdr) protein subfamily (Josefsson et al. 1998). Sdr proteins are cell wall-anchored proteins with large serine-aspartate repeat domains and are also involved in adhesion (Josefsson et al. 1998). In *S. aureus*, SdrC, SdrD, and SdrE are the known proteins belonging to the Sdr protein subfamily (Pi et al. 2020). All three share a conserved structural organization; however, unlike SdrD, and SdrE, SdrC is essential for biofilm growth via homophilic interaction (Barbu et al. 2014). The biofilm-associated protein (Bap) was first described as a large, multi-domain, cell surface-anchored protein that promotes biofilm formation. This protein has never been found in humans and was described as a cattle-specific pathogenic factor of biofilm formation being highly associated with bovine mastitis (Shukla and Rao 2017; Naushad et al. 2020). Apart from this, the prevalence of Bap among staphylococci is very low and its occurrence is significantly higher in coagulase-negative strains (Shukla and Rao 2017). The elastin binding protein is encoded by the *ebps* gene. This protein is responsible for attachment to host cells via binding to elastin promoting colonization of mammalian tissues by *S. aureus* (Kot et al. 2018). However, the presence of *ebps* does not seem to be essential for biofilm formation since the inactivation of *ebps* has minimal effect on binding of *S. aureus* to elastin (Roche et al. 2004). The *S. aureus* ability to bind to fibrinogen is granted by the fibrinogen binding protein (encoded by the *fib* gene) which is an important factor in promoting wound infection, foreign body infection, and endocarditis (Hartford et al. 2001). *S. aureus* strains may also harbor the *eno* gene that encodes for the α-enolase. α-Enolase is able to bind laminin and can also mediate the binding of *S. aureus* to laminin which may contribute to the dissemination of *S. aureus* cells by blood leading to tissue invasion (Kot et al. 2018). Finally, the collagen-binding protein, encoded by the *cna* gene, is a mediator of the staphylococci adhesion to collagenous substrates and tissues. The presence of this protein is also necessary for the adhesion of *S. aureus* to cartilaginous tissues (Valotteau et al. 2017). Another important CWA protein family is the near-iron transporter (NEAT) motif family. These proteins are essential for *S. aureus* survival against the host immune system and are responsible for promoting biofilm formation under iron starvation (Hammer and Skaar 2011). One of the most recognized CWA protein is the staphylococcal protein A (SpA) which is also used for *S. aureus* molecular typing. Spa protein is essential for the pathogenesis of staphylococcal infections and significantly contributes to biofilm-associated infections development (Merino et al. 2009). It is known that *S. aureus* has the ability to attach to abiotic surfaces, such as plastic surfaces. The particular attachment can be mediated by teichoic acids and/or the surface protein SasC which are not involved in *S. aureus* attachment on biotic surfaces (Gross et al. 2001; Reffuveille 2017). Many of these proteins and factors play a role both in attachment and accumulation. After initial attachment to either biotic or abiotic surfaces, mediated by all these proteins and factors, the adhesion becomes irreversible. This step is the is the first to the maturation of a future biofilm.

12.3.2 Maturation/Proliferation

After the attachment, *S. aureus* maturates and accumulates in multilayered clusters surrounded by an extracellular matrix (ECM). The initial biofilm comprises a thin layer that thickens until it forms a mushroom or tower-shaped structure in the mature biofilm (Balaure and Grumezescu 2020). These structures, and also channels, form to facilitate nutrient delivery to deeper layers of the biofilm. Moreover, the architecture of biofilms derives from a division of labor between cells within a biofilm since several factors, such as morphology, phenotype, gene expression, and differentiation stage, may vary between biofilm cells (McDougald et al. 2012). The biofilm structural integrity depends upon the ECM which is produced by the organisms that make up the biofilm. In general, ECM in *S. aureus* biofilms is a complex mixture of polysaccharides, teichoic acids, extracellular DNA (eDNA), lipids, proteins, and other compounds; and the individual components present within the biofilms depends on the *S. aureus* strains and external conditions (Schwartz et al. 2016;

Sugimoto et al. 2018). Generally, ECM is responsible for protecting bacteria against desiccation, oxidation, action of antibiotics, metals, and the action of elements of the immune system (Kostakioti et al. 2013). Also, the ECM allows physical and social interactions which enhanced the rate of gene exchange and an increased tolerance to antimicrobials (Kostakioti et al. 2013). Microorganisms in biofilms are metabolically and functionally integrated consortia that, due to the three-dimensional structure of the biofilm, present differences in the metabolic and gene expression patterns based on the position of the cells in the ECM (Berlanga and Guerrero 2016). The incorporation and exchange of genetic material through cell-cell interaction is also facilitated within biofilms due to the presence of DNA strands trapped in the ECM and also to the three-dimensional structure that allows the proximity between cells. The ECM is also the first line of defense since, due to its physical and biochemical properties, prevents the diffusion of antimicrobial agents (Kumaran et al. 2018). Polysaccharides play a fundamental role in the biofilm matrix since they provide architectural support and protection. In staphylococcal biofilms, several polysaccharides have been described; however, the first and main determinant of the attachment and multiplication phases is the polysaccharide intracellular adhesion (PIA) which consists of polymeric *N*-acetyl-glucosamine (PNAG) (Cue et al. 2012). PIA production is dependent upon four genes, *icaADBC*, encoded by the *ica* operon. The operon i*caADBC* encodes IcaA and IcaB proteins (N-acteylglucosamine transferases), IcaC (predicted exporter) and IcaD (deacetylase) (Miao et al. 2019). IcaA together with IcaD have been shown to produce an N-acetylglucosamine oligomer. IcaC is responsible for chain growth and for the transport of IcaAD-synthesized oligomers across the cell membrane. IcaB is responsible for the deacetylation of PIA which results in a positively charged PIA promoting the interaction of PIA with cell surface (Fig. 12.2) (Gerke et al. 1998; Vuong et al. 2004; Cue et al. 2012). Furthermore, an atomic force microscopy study reported that PIA charged positively promotes cell-cell adhesion between *S. aureus* and polyanionic teichoic and lipo-teichoic acids (Formosa-Dague et al. 2016). *icaR* gene is a member of the TetR family of regulatory proteins and is located upstream of the *ica* operon. IcaR is one of the most important factors involved in *icaADBC* regulation since by binding to the *icaA* promoter region it represses the *ica* operon expression (You et al. 2014). Therefore, a significant increase in *icaADBC* expression occurs when *icaR* is inactivated leading to a higher PIA production (Conlon et al. 2002). Yu et al. (2017) reported a new repressor gene that recognizes and binds to the 5-nucleotide motif, located within the *icaR-icaA* intergenic region, decreasing biofilm PIA productions (Yu et al. 2017). Recently, several studies have reported that environmental conditions including temperature, low oxygen, osmotic pressure, glucose, and sub-inhibitory concentrations of some antibiotics influence the expression and production of PIA (Mhatre et al. 2014; Weidenmaier and Lee 2015; Miao et al. 2019). Initially, it was thought that PIA was essential for staphylococcal biofilm formation; however, in recent years, an increased number of studies have shown that some strains of *Staphylococcus* produced biofilm independently of the presence and expression of *ica* operon responsible for PIA synthesis (Hennig et al. 2007; O'Gara 2007; Lauderdale et al. 2009). These studies have demonstrated that PIA-independent

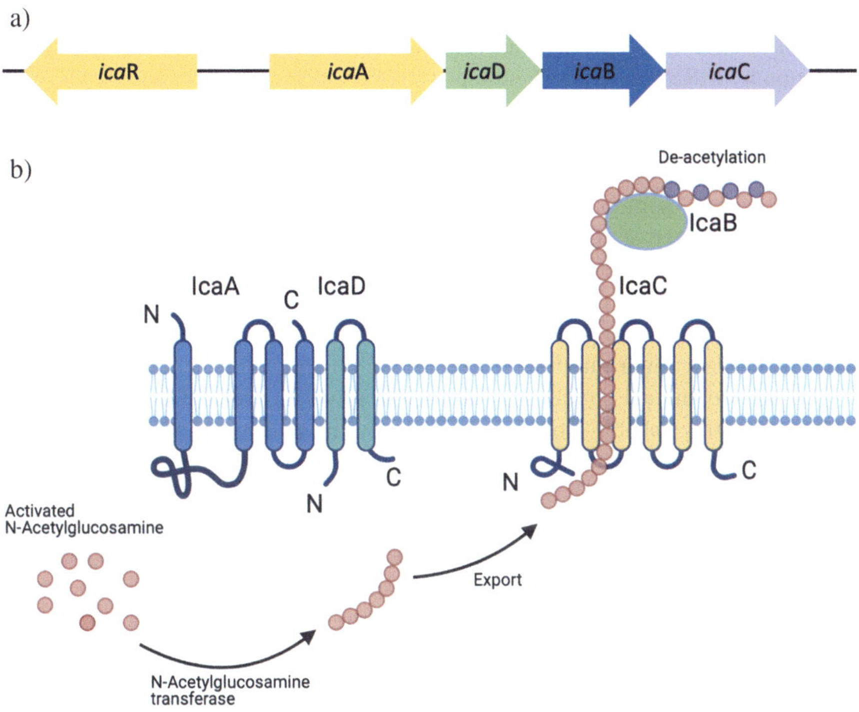

Fig. 12.2 PIA production is dependent upon four genes, *ica*ADBC, encoded by the *ica* operon. (**a**) The *ica* operon (**b**) The production of exopolysaccharide poly-N-acetylglucosamine (PNAG; also known as PIA) (adapted from Otto 2009)

biofilm formation is promoted by several cell wall-attached proteins discussed in the previous section showing that these proteins can replace the PIA function in *ica*-deficient *S. aureus* strains during biofilm formation and development. The Bap protein in bovine mastitis, FnBPs in human methicillin-resistant *S. aureus* (MRSA), and accumulation-associated protein (Aap) in *S. epidermidis* were among the biofilm-related proteins in the first studies reporting PIA-independent biofilm formation (Cucarella et al. 2001; Rohde et al. 2005; O'Neill et al. 2008). Mempel et al. reported a relationship between β-lactam susceptibility and the biofilm phenotype. In this study, it was shown that PIA production varied with the different levels of β-lactam susceptibility in several *S. epidermidis* isolates (Mempel et al. 1994; Mempel et al. 1995). *ica*-dependent biofilm formation was associated with methicillin susceptibility in *S. aureus* with PIA production found to be essential for biofilm formation in methicillin-susceptible *S. aureus* (MSSA) but not MRSA (O'Neill et al. 2007). In fact, the capacity of biofilm formation was suppressed by the deletion of the *ica* locus (Valle et al. 2003; O'Neill et al. 2007). It has been shown that *ica* operon expression is activated by the presence of NaCl in media. Unlike MSSA strains, and since *ica* locus was found not to influence MRSA biofilm formation, NaCl does not induce biofilm formation of MRSA strains, yet, the addition of glucose to growth media seems to enhance biofilm formation of MRSA strains (O'Neill et al. 2007). Indeed, glucose and NaCl were shown to promote biofilm formation by MRSA and MSSA strains through acidification of culture media and induction of PIA production, respectively (O'Neill et al. 2007; Boles and Horswill 2008). Although PIA is a major component in the biofilm ECM, studies have been showing that some *S. aureus* strains were able to form biofilm PIA-independent being composed by proteins and eDNA (Sugimoto et al. 2018). Sugimoto et al. (2018) have shown that eDNA has different impacts on the development and structural integrity of both PIA-dependent and PIA-independent biofilms. Furthermore, the authors reported that the quantity of eDNA does not seem to influence the development of robust biofilms (Sugimoto et al. 2018). In fact, studies have shown that eDNA may play a more important role in biofilm formation than CWA proteins since *S. aureus* mutant strains deficient in the production of MSCRAMMs grew biofilms indistinguishable from those grown by the wild type (Moormeier et al. 2014; Kavanaugh et al. 2019). In contrast, in a *S. aureus* mutant strain lacking the murine hydrolase AtlA, the biofilm formation was reduced, which is consistent with the fact that AtlA can also function as an autolysin that is responsible for the release of eDNA that incorporates into the matrix (Bose et al. 2012). Therefore, it seems that eDNA plays a crucial role in biofilm maturation. Other proteins, such as extracellular adherence protein (Eap) and beta toxin (Hlb), that are secreted proteins, also play a role in biofilm maturation. Nevertheless, the level of importance of these proteins vary largely between strains (Lister and Horswill 2014). The accessory gene regulator (*agr*) system is also implicated in biofilm formation. The deletion of *agr* enhances biofilm formation by MRSA strains while its deletion does not seem to influence biofilm formation in MSSA strains (O'Neill et al. 2007). Accordantly, *agr* deletion does not significantly alter *ica*ADBC expression or PIA production (Vuong et al. 2003). Furthermore, activation of the *agr* system was shown to initiate biofilm dispersal (Boles and Horswill 2008).

12.3.3 Dispersal

The dynamic equilibrium of a biofilm is reached when its outermost layers begin to generate planktonic cells that are metabolically active and capable of dividing. Following biofilm maturation/proliferation, cells within the biofilm can reactivate to a planktonic state through dispersal. Nevertheless, studies have shown that dispersal cells have a unique phenotype, different from sessile or planktonic cells (Guilhen et al. 2017). Furthermore, biofilm-dispersed cells seem to overexpress common virulence factors when compared with planktonic or sessile cells (Hay and Zhu 2015; MacKenzie et al. 2015). Biofilm dispersal can comprise three stages: detachment of cells from the biofilm, movement of bacterial cells to a new site, and attachment of the cells to a surface/substrate in the new site (Kaplan 2010). Biofilm dispersal represents a critical step in bacterial lifestyle involving coordination of several environmental signals, signaling messengers, and molecular effectors

(Guilhen et al. 2017). Biofilm cell dispersal can be promoted by several cues, such as changes in oxygen concentration and cell density, lack of nutrients, and intense competition, resulting in total or partial biofilm dispersal (Balaure and Grumezescu 2020). Cue-specific sensory proteins detect these changes and stimulate a response that leads to production of effectors within the biofilm that are responsible for the bacterial escape from the biofilm structure. Studies have reported a wide range of signals regulatory pathways and effectors (Guilhen et al. 2017). The dispersal of cells from a biofilm can lead to new and subsequent infections in the host since cells can disseminate from the original site of infection into the host and recolonize other available host sites. Indeed, biofilm cells dispersal plays an important role in chronic infections (Speziale and Geoghegan 2015). Biofilm dispersal may also be an important approach to treat persistent *S. aureus* infections since intentional dispersal of a biofilm together with antibiotic treatment can lead to the exposure and death of metabolically active bacterial cells (Lister and Horswill 2014). No single mechanism of biofilm dispersal is used by all bacteria. The main biofilm dispersal mechanism used by *S. aureus* is the production of several exo-enzymes and surfactants to degrade the ECM. The effectiveness of the different biofilm dispersal mechanisms is dependent on the *S. aureus* strain and the matrix composition (Kiedrowski et al. 2011; Lister and Horswill 2014). In fact, in PIA-dependent biofilms, the use of proteins to induce enzymatic self-destruction, is less effective in dispersing biofilm cells and, contrarily, mechanisms specifically targeting PIA have no effect on the dispersal of PIA-independent biofilms (Lister and Horswill 2014). The different stages of *S. aureus* biofilm formation can be controlled by several different mechanisms, such as phenol-soluble polypeptides, several surface proteins, pH change, protease, DNase, and quorum sensing, among others (Boles and Horswill 2011). Notably, the activation of the *agr* quorum-sensing in *S. aureus* inhibits its own biofilm formation and dispersed the established biofilms. In fact, Boles and Horswill (2008) showed that the *agr* system triggers a dispersal pathway, detaching cells from the biofilm and reverting them to a planktonic state and consequently to a more antibiotic susceptible state (Boles and Horswill 2008).

12.4 Impact of Quorum Sensing on *S. aureus* Gene Expression and Biofilm Formation

Studies suggest that certain bacteria use the production, release, exchange, and detection of signaling molecules to measure population density and control their behavior in response to variations in the number of cells. However, for a long period of time, the phenomenon of chemical signaling between cells was considered to be limited to a few bacterial species. Fuqua et al. (1994) described the phenomenon that bacteria know their concentration in a given environment and decide when to start the expression of a certain set of genes to develop a specific response simultaneously in response to eventual changing conditions (Fuqua et al. 1994). Currently, it is known that intercellular communication is not exclusive to certain species but that is common in the bacterial ecosystem and this process known as quorum sensing (QS) is fundamental in the formation of biofilms (Schauder and Bassler 2001). The occurrence of QS has been demonstrated in Gram-negative and -positive bacteria and it involves intra- and intercellular signals that generate benefits for groups of microorganisms. Nevertheless, cell-cell interactions are facilitated by the close proximity between bacteria within the biofilm (Flemming and Wingender 2010). QS plays an important role in biofilm development. Each bacterium that binds to a surface generates a signal molecule, and the more bacteria that bind the more the signal increases until it reaches a threshold leading to the bacterial differentiation in the biofilm. Through QS, bacteria regulate or synchronize changes in their genetic expression based on their population density, such as the production of antibiotics, toxins, pigments, exopolysaccharides (formation of biofilms), motility, and other virulence factors. Bacterial cells lacking these signals produce biofilms with a weak architecture (Lazar 2011; Khan et al. 2019). QS is based on the diffusion and subsequent detection of molecules that act as a signal called "autoinducers". Bacteria also possess a receptor that can specifically detect the respective auto-inducer. When it binds to the receptor, it activates the transcription of certain genes, including those for inducer synthesis (Zhang and Li 2016). Gram-negative bacteria diffuse low molecular weight compounds called acyl-homoserine-lactones (AHLs) into the intercellular space whereas in Gram-positive bacteria, such as *S. aureus*, the auto-inductive-modified oligopeptides produced by these bacteria act as a signal for other cells (Abraham 2006). Two QS systems have been identified in *S. aureus:* the *agr* system that is common in Gram-positive bacteria and the *LuxS*/AI-2 system that is expressed in both Gram-positive and -negative bacteria (Kong et al. 2006). The *agr* system is one of the most important factors and it has been characterized as an operon in *S. aureus*. Agr system has great relevance in the control and regulation of cell density and the expression of virulence genes. The promoters P2 and P3 are included in *agr* operon and regulate the transcription of RNAII and RNAIII, respectively (Morfeldt et al. 1995). The *agr* locus consists of four genes: *agr*B, *agr*D, *agr*C, and *agr*A. *agr*B encodes a membrane protein responsible for translocation and modification of AgrD. The *agr*D gene encodes the precursor of autoinducing peptide (AIP). *agr*C and *agr*A genes are located in the RNAII encoding region. *agr*C encodes a membrane receptor protein of the AIP signal. AIP binds to *agr*C which activates the phosphorylation of *agr*A which, in turn, binds to P2 and P3 inducing the expression of RNAII and RNAIII (Queck et al. 2008). The increasing cell density leads to the accumulation of AIP which activates the *agr* system. RNAIII is the key effector of *agr* since its concentration increases with the increase in cell density. RNAIII has been associated with the increasing in virulence factors and surface proteins, such as SpA, FnBPA, FnBPB, α-hemolysin, and Panton-Valentine leukocidin (Queck et al. 2008; Cheung et al. 2011). Due to the increase in the prevalence of antimicrobial resistance among clinically relevant bacterial strains, their high capacity to form biofilms and the failure of vaccines to prevent infection, new alternatives to antibiotics are required. Therefore, much attention has been given to chemical inhibition of QS, particularly the *agr* system in *S. aureus* (Sully et al.

2014). In fact, it has been shown that *agr*-negative strains display intermediate resistance to glycopeptides such as vancomycin (Howden et al. 2010). Due to the increase in the prevalence of antimicrobial resistance among clinically relevant bacterial strains, their high capacity to form biofilms and the failure of vaccines to prevent infection, new alternatives to antibiotics are required. Despite the fact that chemical inhibitors of *agr* have been detected and identified, it seems that none of them have demonstrated efficacy in vivo (Sully et al. 2014).

12.5 Biofilm Resistance, Tolerance, and Persistence

Biofilm cells can be between 10 and 1000 times more resistant than planktonic cells corresponding to a large number of broad-spectrum antibiotics and biocides oxidants such as chlorine, iodine, or ozone (Govan and Deretic 1996). In biofilm, the mechanisms of antibiotic resistance are altered and bacteria showing susceptibility to a given antibiotic become quiescent, increasing their tolerance to that compound (Singh et al. 2017). Nevertheless, resistance to antibiotics and disinfectants in biofilms is also dependent on the species and strain of microorganisms, on the physiological characteristics of the biofilm, on the phenotypic changes in the cells themselves, and on the age of the biofilm. Biofilm eradication is a very difficult process, since biofilm inactivation and removal are distinct processes, that is, biofilms may be inactive but remain attached to the surface (Wolfmeier et al. 2018). Furthermore, antibiotics may diminish the number of bacterial cells within biofilms, but cannot completely eradicate the pathogen which leads to refractory/relapsing infections (Dufour et al. 2010). Antimicrobial resistance is defined as the ability of bacteria to grow in the presence of high concentrations of antibiotics or an increase in the minimum inhibitory concentration (MIC) to the antimicrobial agent, necessary to stop cell growth. Biofilms, on the other hand, present various resistance mechanisms that enable the growth under antimicrobial therapy. Currently, it is thought that the high antimicrobial resistance in biofilms is the combined result of different mechanisms, such as presence of persister cells, starvation-induced slow growth, reduced metabolic rate, and the restricted penetration and sequestration of antimicrobials through the matrix. Furthermore, it has been shown that biofilm cells express genes that encode various xenobiotic efflux and export systems (Rumbaugh and Sauer 2020). Studies have shown that there are several mechanisms associated with this phenomenon of resistance (Brown et al. 1988; Costerton et al. 1999; Anderl et al. 2000; Brooun et al. 2000; Gilbert et al. 2002; Walters et al. 2003; Lewis 2007): delayed penetration of the antimicrobial agent due to the exopolysaccharides in the matrix; lower growth rate inducing physiological changes; phenotypic changes within biofilms; quorum-sensing; the expression of efflux pumps; and the development of dormant persister cells (Dufour et al. 2010). Although the primary function of the ECM is protection, it has been shown that the ECM can act as an impermeable barrier to limit antimicrobial penetration. The delaying of the antimicrobials diffusion is due to the exopolysaccharides decreasing the rate of transport of the molecule into the biofilm and changing the reaction of the antimicrobial agent with the matrix (Sutherland 2001; Donlan 2002). The ECM protection can also be due to the direct binding of the antibiotics by the EPS matrix. After antibiotic treatment, the cells located in the outer layer of the biofilm die due to their closer exposure whereas cells embedded deep inside the biofilm are able to survive (Dufour et al. 2010). The ECM can also be considered a chemically active barrier. Diffusion obstruction is probably effective against small antimicrobial peptides, toxic cationic heavy metals, and positively charged antibiotics, such as aminoglycosides, due to the negative charge of exopolysaccharides (Nichols et al. 1988). On the other hand, the ECM has no effect on the diffusion of uncharged antibiotics such as β-lactams (Zheng and Stewart 2002). In fact, although the ECM protects biofilm cells against penetration of antimicrobials, it may not be a general protection mechanism since many antibiotics are able to penetrate into the biofilm. Studies have shown that ECM of *S. aureus* decreases the activity of oxacillin, cefotaxime, vancomycin, teicoplanin, moderately reduce the activity of linezolid and quinupristin/dalfopristin but susceptibility to rifampicin, amikacin, ranbezolid, and ciprofloxacin remains unaffected (Mathur et al. 2005; Kostenko et al. 2007; Singh et al. 2010). Other studies showed that phenotypic changes within biofilms, such as adoption of a distinct and protected phenotype, may occur contributing to the antimicrobial resistance (Dufour et al. 2010; Corona and Martinez 2013). The fact that the biofilm contains a great diversity of environmental niches, presenting different nutrients and pH in different zones, contributes for such resistance. However, this phenotypic heterogeneity is mainly due to different concentration gradients experienced within biofilms. Cells located deep inside the biofilm have a lower growth rate, due to limited nutrients and oxygen, which is similar to the stationary phase in planktonic cells. Consequently, the decreased metabolic activity of bottom layer cells contributes to antibiotic tolerance. Contrarily, cells located in the external layers have a growth rate similar to the exponentially growing planktonic cells (Stewart and Franklin 2008). Nevertheless, in general, organisms organized in biofilms have a lower growth rate than planktonic cells, thus taking on antimicrobial agents more slowly. Indeed, all antibiotics are more effective in growing cells. For instance, penicillin does not have an effect on non-growing cells and the effect rate is proportional to the growth rate. However, other antibiotics such as cephalosporins, aminoglycosides, and fluoroquinolones have an effect on cells in stationary phase (Lode et al. 1998; Rosenberg et al. 2020). Oxygen concentration and pH gradient established within the biofilms, although indirectly, also contribute to decreased growth rate.

Bacteria may show antimicrobial resistance in the planktonic state which is also responsible for the biofilm antimicrobial tolerance. However, the antimicrobial resistance by itself is not sufficient to explain the higher antimicrobial tolerance of the biofilms nor the recurrence of biofilm infections. Some studies showed that environmental changes, such as temperature, low oxygen availability, and DNA damage, among others, trigger gene expression and the transfer of genes between bacteria leading to resistance and survival in the presence of certain antibiotics (Beloin and Ghigo 2005; Resch et al. 2005). Other mechanisms have been described, such as the action of reactive agents located in the superficial layers of the biofilm which deactivate antibiotics. Another mechanism responsible for antibiotic tolerance of biofilm is the presence of drug efflux pumps. Studies have shown that sub-inhibitory concentrations of antibiotics induce membrane-bound drug efflux pumps both in Gram-positive and -negative bacteria (Lynch and Robertson 2008). These drug efflux pumps can have several functions and can have specificity for a single antibacterial agent and a broad specificity for structurally unrelated antimicrobials (known as multidrug efflux pumps) (Dufour et al. 2010). Five families of efflux pumps associated with antimicrobial resistance are known. Four of those families use proton motive force as an energy source: multidrug and toxin extrusion, small multidrug resistance, major facilitator superfamily, and resistance-nodulation-division; while ATP-binding cassette uses ATP hydrolysis as an energy source (Pao et al. 1998; Jack et al. 2001; Lubelski et al. 2007; Kuroda and Tsuchiya 2009; Nikaido and Takatsuka 2009). The resistance-nodulation-division efflux pumps have only been found in Gram-negative bacteria (Alav et al. 2018). The major facilitator superfamily of efflux pumps are the most known among Gram-positive bacteria (Alav et al. 2018). Several studies have been conducted to better understand the role of efflux pumps on *S. aureus* biofilm tolerance. An early study conducted by Beenken et al. (2004) reported a difference in the expression of several efflux and transporter genes in biofilm cells when compared to the planktonic cells (Beenken et al. 2004). The lipoprotein streptococcal adhesion PsaA homolog (SA0587) gene and multidrug resistance protein B (SA2142) gene encode for the ATP-binding cassette and major facilitator superfamily, respectively. In a study by Resch et al. (2005), the expression of genes during biofilm formation was analyzed. After just 8 h of growth, SA0589 and SA2142 genes were expressed at least five-fold more in biofilm than in planktonic cells (Resch et al. 2005). Another study reported the presence of multidrug resistant gene cluster including the macrolide efflux pump *msr*A in *S. aureus* (Weigel et al. 2007). The expression of NorB and NorC, which are multidrug-resistant pumps conferring resistance to quinolones, tetraphenylphosphonium, cetrimide, and other antimicrobial agents, was detected during *S. aureus* biofilm growth (He and Ahn 2011). Furthermore, *norB* is expressed in *S. aureus* in response to acid environments and reduced oxygen which are conditions that are found in deep layers of *S. aureus* biofilms since cells switch to anaerobic respiration and the environment has a low pH (Costa et al. 2013). QS and intracellular signaling molecules are also implicated in biofilm tolerance. The *agr* system in *S. aureus* is related to the QS system as reported above. The Agr expression in *S. aureus* biofilms has been associated with antimicrobial resistance. The RNAIII production of *agr*-positive staphylococci is induced by sub-inhibitory levels of ciprofloxacin, mupirocin, and rifampin (Yarwood et al. 2004). Hamamelitannin has been shown to block QS through the TraP QS system and to decrease *S. aureus* biofilm tolerance to vancomycin (Kiran et al. 2008; Brackman et al. 2013). Another mechanism of antibiotic tolerance in biofilms is the presence of persister cells. These cells enter into a state of dormancy which allows their survival in stress conditions, namely during antimicrobial treatment, since antibiotics target cell growth (Jayaraman 2008; Lewis 2010). This resistance mechanism is very successful since, in general, resistance mechanisms prevent the binding of the antibiotic to the target and, in contrast, in persister cells the targets are shut down (Dufour et al. 2010).

12.6 Conclusions

S. aureus is a relevant pathogen for both human and animal health. *S. aureus* has the capacity to easily acquire antimicrobial resistance which, together with its wide range of virulence factors that supports its evasion of the host immune system, turn it into a worrisome pathogen. Furthermore, *S. aureus* strains have a high ability to form biofilms which leads to recurrent and chronic infections. The relation between antibiotic resistance and biofilm phenotypes is of much interest since both contribute to the success of *S. aureus* infections. Many transcriptomic and proteomic studies have been conducted in order to better understand the mechanism underlying *S. aureus* biofilm formation, dispersal, and tolerance to antimicrobial agents. Given the importance of improving biofilm-related infections treatment, the eradication of biofilms, and eventually the development of a new vaccine, additional studies are necessary to fill specific knowledge gaps.

Acknowledgments This work was funded by the R&D Project CAREBIO2: Comparative assessment of antimicrobial resistance in environmental biofilms through proteomics—towards innovative theranostic biomarkers, with reference NORTE-01-0145-FEDER-030101 and PTDC/SAU-INF/30101/2017, financed by the European Regional Development Fund (ERDF) through the Northern Regional Operational Program (NORTE 2020) and the Foundation for Science and Technology (FCT). This work was supported by the Associate Laboratory for Green Chemistry-LAQV, which is financed by national funds from FCT/MCTES (UIDB/50006/2020 and UIDP/50006/2020). Vanessa Silva is grateful to FCT (Fundação para a Ciência e a Tecnologia) for financial support through the PhD grant SFRH/BD/137947/2018.

Bibliography

Abraham W-R (2006) Controlling biofilms of gram-positive pathogenic bacteria. Curr Med Chem 13:1509–1524

Alav I, Sutton JM, Rahman KM (2018) Role of bacterial efflux pumps in biofilm formation. J Antimicrob Chemother 73:2003–2020. https://doi.org/10.1093/jac/dky042

Anderl JN, Franklin MJ, Stewart PS (2000) Role of antibiotic penetration limitation in *Klebsiella pneumoniae* biofilm resistance to ampicillin and ciprofloxacin. Antimicrob Agents Chemother 44:1818–1824

Armbruster CR, Parsek MR (2018) New insight into the early stages of biofilm formation. Proc Natl Acad Sci 115:4317–4319

Balaure PC, Grumezescu AM (2020) Recent advances in surface Nanoengineering for biofilm prevention and control. Part I: molecular basis of biofilm recalcitrance. Passive anti-biofouling Nanocoatings. Nanomaterials 10:1230

Ballén V, Ratia C, Cepas V, Soto SM (2020) *Enterococcus faecalis* inhibits *Klebsiella pneumoniae* growth in polymicrobial biofilms in a glucose-enriched medium. Biofouling 36:846–861. https://doi.org/10.1080/08927014.2020.1824272

Barbu EM, Mackenzie C, Foster TJ, Höök M (2014) SdrC induces staphylococcal biofilm formation through a homophilic interaction. Mol Microbiol 94:172–185. https://doi.org/10.1111/mmi.12750

Beenken KE, Dunman PM, McAleese F, Macapagal D, Murphy E, Projan SJ, Blevins JS, Smeltzer MS (2004) Global gene expression in *Staphylococcus aureus* biofilms. J Bacteriol 186:4665–4684. https://doi.org/10.1128/JB.186.14.4665-4684.2004

Beloin C, Ghigo J-M (2005) Finding gene-expression patterns in bacterial biofilms. Trends Microbiol 13:16–19

Berlanga M, Guerrero R (2016) Living together in biofilms: the microbial cell factory and its biotechnological implications. Microb Cell Factories 15:165. https://doi.org/10.1186/s12934-016-0569-5

Boles BR, Horswill AR (2008) Agr-mediated dispersal of *Staphylococcus aureus* biofilms. PLoS Pathog 4:e1000052

Boles BR, Horswill AR (2011) Staphylococcal biofilm disassembly. Trends Microbiol 19:449–455

Bose JL, Lehman MK, Fey PD, Bayles KW (2012) Contribution of the *Staphylococcus aureus* Atl AM and GL murein hydrolase activities in cell division, autolysis, and biofilm formation. PLoS One 7:e42244

Brackman G, De Meyer L, Nelis HJ, Coenye T (2013) Biofilm inhibitory and eradicating activity of wound care products against *Staphylococcus aureus* and *Staphylococcus epidermidis* biofilms in an in vitro chronic wound model. J Appl Microbiol 114:1833–1842. https://doi.org/10.1111/jam.12191

Brooun A, Liu S, Lewis K (2000) A dose-response study of antibiotic resistance in *Pseudomonas aeruginosa* biofilms. Antimicrob Agents Chemother 44:640–646

Brown MRW, Allison DG, Gilbert P (1988) Resistance of bacterial biofilms to antibiotics a growth-rate related effect? J Antimicrob Chemother 22:777–780

Carl S, Katrien L, Tommi K, Enrico R, Pete K, Luisa MM, Béatrice J, Susan H, Sonja H, Outi L, Jacqui R, Aleksander D, Walter Z, Diamantis P, Monnet DL, Group. the H-AIPS (2018) Prevalence of healthcare-associated infections, estimated incidence and composite antimicrobial resistance index in acute care hospitals and long-term care facilities: results from two European point prevalence surveys, 2016 to 2017. Eur Secur 23:1800516

Carpentier B, Cerf O (1993) Biofilms and their consequences, with particular reference to hygiene in the food industry. J Appl Bacteriol 75:499–511

Cheung GYC, Wang R, Khan BA, Sturdevant DE, Otto M (2011) Role of the accessory gene regulator agr in community-associated methicillin-resistant *Staphylococcus aureus* pathogenesis. Infect Immun 79:1927–1935

Chua SL, Liu Y, Yam JKH, Chen Y, Vejborg RM, Tan BGC, Kjelleberg S, Tolker-Nielsen T, Givskov M, Yang L (2014) Dispersed cells represent a distinct stage in the transition from bacterial biofilm to planktonic lifestyles. Nat Commun 5:4462. https://doi.org/10.1038/ncomms5462

Conlon KM, Humphreys H, O'Gara JP (2002) *ica*R encodes a transcriptional repressor involved in environmental regulation of *Ica* operon expression and biofilm formation in *Staphylococcus epidermidis*. J Bacteriol 184:4400–4408

Corning PA (2002) The re-emergence of "emergence": a venerable concept in search of a theory. Complexity 7:18–30

Corona F, Martinez JL (2013) Phenotypic resistance to antibiotics. Antibiot 2(2):237–255

Costa SS, Viveiros M, Amaral L, Couto I (2013) Multidrug efflux pumps in *Staphylococcus aureus*: an update. Open Microbiol J 7:59

Costerton JW, Geesey GG, Cheng K-J (1978) How bacteria stick. Sci Am 238:86–95

Costerton JW, Stewart PS, Greenberg EP (1999) Bacterial biofilms: a common cause of persistent infections. Science (80) 284:1318–1322

Cucarella C, Solano C, Valle J, Amorena B, Lasa Í, Penadés JR (2001) Bap, a *Staphylococcus aureus* surface protein involved in biofilm formation. J Bacteriol 183:2888–2896

Cue D, Lei M, Lee C (2012) Genetic regulation of the intercellular adhesion locus in staphylococci. Front Cell Infect Microbiol 2:38

De la Fuente-Núñez C, Reffuveille F, Fernández L, Hancock REW (2013) Bacterial biofilm development as a multicellular adaptation: antibiotic resistance and new therapeutic strategies. Curr Opin Microbiol 16:580–589

DeLeon S, Clinton A, Fowler H, Everett J, Horswill AR, Rumbaugh KP (2014) Synergistic interactions of *Pseudomonas aeruginosa*. Infect Immun 82:4718–4728. https://doi.org/10.1128/IAI.02198-14

Donlan RM (2002) Biofilms: microbial life on surfaces. Emerg Infect Dis 8:881

Dufour D, Leung V, Lévesque CM (2010) Bacterial biofilm: structure, function, and antimicrobial resistance. Endod Top 22:2–16. https://doi.org/10.1111/j.1601-1546.2012.00277.x

Dunn MJ, Fillinger RJ, Anderson LM, Anderson MZ (2020) Automated quantification of *Candida albicans* biofilm-related phenotypes reveals additive contributions to biofilm production. Npj Biofilms Microb 6:36. https://doi.org/10.1038/s41522-020-00149-5

Flemming H-C (2020) Biofouling and me: my Stockholm syndrome with biofilms. Water Res 173:115576. https://doi.org/10.1016/j.watres.2020.115576

Flemming H-C, Wingender J (2010) The biofilm matrix. Nat Rev Microbiol 8:623–633. https://doi.org/10.1038/nrmicro2415

Formosa-Dague C, Feuillie C, Beaussart A, Derclaye S, Kucharíková S, Lasa I, Van Dijck P, Dufrêne YF (2016) Sticky matrix: adhesion mechanism of the staphylococcal polysaccharide intercellular Adhesin. ACS Nano 10:3443–3452. https://doi.org/10.1021/acsnano.5b07515

Foster TJ (2019) Surface proteins of *Staphylococcus aureus*. Gram-Positive Pathog:599–617

Fox EP, Nobile CJ (2012) A sticky situation. Transcription 3:315–322. https://doi.org/10.4161/trns.22281

Fuqua WC, Winans SC, Greenberg EP (1994) Quorum sensing in bacteria: the LuxR-LuxI family of cell density-responsive transcriptional regulators. J Bacteriol 176:269

Geoghegan JA, Foster TJ (2015) Cell wall-anchored surface proteins of *Staphylococcus aureus*: many proteins, multiple functions. *Staphylococcus aureus* 95–120

Gerke C, Kraft A, Süßmuth R, Schweitzer O, Götz F (1998) Characterization of the N-Acetylglucosaminyl transferase activity involved in the biosynthesis of the *Staphylococcus epidermidis* polysaccharide intercellular Adhesin. J Biol Chem 273:18586–18593

Gilbert P, Evans DJ, Evans E, Duguid IG, Brown MRW (1991) Surface characteristics and adhesion of *Escherichia coli* and *Staphylococcus epidermidis*. J Appl Bacteriol 71:72–77

Gilbert P, Allison DG, McBain AJ (2002) Biofilms *in vitro* and *in vivo*: do singular mechanisms imply cross-resistance? J Appl Microbiol 92:98S–110S

Govan JR, Deretic V (1996) Microbial pathogenesis in cystic fibrosis: mucoid *Pseudomonas aeruginosa* and *Burkholderia cepacia*. Microbiol Rev 60:539–574

Gross M, Cramton SE, Götz F, Peschel A (2001) Key role of teichoic acid net charge in *Staphylococcus aureus* colonization of artificial surfaces. Infect Immun 69:3423–3426. https://doi.org/10.1128/IAI.69.5.3423-3426.2001

Guilhen C, Forestier C, Balestrino D (2017) Biofilm dispersal: multiple elaborate strategies for dissemination of bacteria with unique properties. Mol Microbiol 105:188–210. https://doi.org/10.1111/mmi.13698

Hall AE, Domanski PJ, Patel PR, Vernachio JH, Syribeys PJ, Gorovits EL, Johnson MA, Ross JM, Hutchins JT, Patti JM (2003) Characterization of a protective monoclonal AntibodyRecognizing *Staphylococcus aureus* MSCRAMM protein clumping FactorA. Infect Immun 71:6864–6870. https://doi.org/10.1128/IAI.71.12.6864-6870.2003

Hammer ND, Skaar EP (2011) Molecular mechanisms of *Staphylococcus aureus* iron acquisition. Annu Rev Microbiol 65:129–147

Hartford OM, Wann ER, Höök M, Foster TJ (2001) Identification of residues in the *Staphylococcus aureus* fibrinogen-binding MSCRAMM clumping factor a (ClfA) that are important for ligand binding. J Biol Chem 276:2466–2473

Hay AJ, Zhu J (2015) Host intestinal signal-promoted biofilm dispersal induces *vibrio cholerae* colonization. Infect Immun 83:317–323

He X, Ahn J (2011) Differential gene expression in planktonic and biofilm cells of multiple antibiotic-resistant *salmonella typhimurium* and *Staphylococcus aureus*. FEMS Microbiol Lett 325:180–188

Hennig S, Wai SN, Ziebuhr W (2007) Spontaneous switch to PIA-independent biofilm formation in an Ica-positive *Staphylococcus epidermidis* isolate. Int J Med Microbiol 297:117–122

Henrici AT (1933) Studies of freshwater bacteria: I. A direct microscopic technique. J Bacteriol 25:277–287. https://doi.org/10.1128/JB.25.3.277-287.1933

Hobley L, Harkins C, MacPhee CE, Stanley-Wall NR (2015) Giving structure to the biofilm matrix: an overview of individual strategies and emerging common themes. FEMS Microbiol Rev 39:649–669. https://doi.org/10.1093/femsre/fuv015

Hogan S, Zapotoczna M, Stevens NT, Humphreys H, Gara JP, Neill E (2016) Eradication of *Staphylococcus aureus* catheter-related biofilm infections using ML:8 and Citrox. Antimicrob Agents Chemother 60:5968–5975. https://doi.org/10.1128/AAC.00910-16

Høiby N (2014) A personal history of research on microbial biofilms and biofilm infections. Pathog Dis 70:205–211. https://doi.org/10.1111/2049-632X.12165

Hotterbeekx A, Kumar-Singh S, Goossens H, Malhotra-Kumar S (2017) *In vivo* and *in vitro* Interactions between P*seudomonas aeruginosa* and *Staphylococcus* spp. Front Cell Infect Microbiol. 7:106

Howden BP, Davies JK, Johnson PDR, Stinear TP, Grayson ML (2010) Reduced vancomycin susceptibility in *Staphylococcus aureus*, including vancomycin-intermediate and heterogeneous vancomycin-intermediate strains: resistance mechanisms, laboratory detection, and clinical implications. Clin Microbiol Rev 23:99–139. https://doi.org/10.1128/CMR.00042-09

Jack DL, Yang NM, Saier MH (2001) The drug/metabolite transporter superfamily. Eur J Biochem 268:3620–3639

Jayaraman R (2008) Bacterial persistence: some new insights into an old phenomenon. J Biosci 33:795–805

Josefsson E, McCrea KW, Eidhin DN, O'Connell D, Cox J, Hook M, Foster TJ (1998) Three new members of the serine-aspartate repeat protein multigene family of *Staphylococcus aureus*. Microbiology 144:3387–3395

Kang M, Ko Y-P, Liang X, Ross CL, Liu Q, Murray BE, Höök M (2013) Collagen-binding microbial surface components recognizing adhesive matrix molecule (MSCRAMM) of gram-positive bacteria inhibit complement activation via the classical pathway. J Biol Chem 288:20520–20531

Kaplan JB (2010) Biofilm dispersal: mechanisms, clinical implications, and potential therapeutic uses. J Dent Res 89:205–218. https://doi.org/10.1177/0022034509359403

Kato MT, Field JA, Lettinga G (1997) Anaerobe tolerance to oxygen and the potentials of anaerobic and aerobic cocultures for wastewater treatment. Brazilian J Chem Eng 14(4)

Kavanaugh JS, Flack CE, Lister J, Ricker EB, Ibberson CB, Jenul C, Moormeier DE, Delmain EA, Bayles KW, Horswill AR (2019) Identification of extracellular DNA-binding proteins in the biofilm matrix. MBio 10:e01137–e01119. https://doi.org/10.1128/mBio.01137-19

Khan F, Javaid A, Kim Y-M (2019) Functional diversity of quorum sensing receptors in pathogenic bacteria: interspecies, intraspecies and interkingdom level. Curr Drug Targets 20:655–667

Kiedrowski MR, Kavanaugh JS, Malone CL, Mootz JM, Voyich JM, Smeltzer MS, Bayles KW, Horswill AR (2011) Nuclease modulates biofilm formation in community-associated methicillin-resistant *Staphylococcus aureus*. PLoS One 6:e26714

Kiran MD, Adikesavan NV, Cirioni O, Giacometti A, Silvestri C, Scalise G, Ghiselli R, Saba V, Orlando F, Shoham M (2008) Discovery of a quorum-sensing inhibitor of drug-resistant staphylococcal infections by structure-based virtual screening. Mol Pharmacol 73:1578–1586

Kong K-F, Vuong C, Otto M (2006) *Staphylococcus quorum* sensing in biofilm formation and infection. Int J Med Microbiol 296:133–139

Kostakioti M, Hadjifrangiskou M, Hultgren SJ (2013) Bacterial biofilms: development, dispersal, and therapeutic strategies in the dawn of the postantibiotic era. Cold Spring Harb Perspect Med 3:a010306

Kostenko V, Ceri H, Martinuzzi RJ (2007) Increased tolerance of *Staphylococcus aureus* to vancomycin in viscous media. FEMS Immunol Med Microbiol 51:277–288
Kot B, Sytykiewicz H, Sprawka I (2018) Expression of the biofilm-associated genes in methicillin-resistant *Staphylococcus aureus* in biofilm and planktonic conditions. Int J Mol Sci 19
Kumaran D, Taha M, Yi Q, Ramirez-Arcos S, Diallo J-S, Carli A, Abdelbary H (2018) Does treatment order matter? Investigating the ability of bacteriophage to augment antibiotic activity against *Staphylococcus aureus* biofilms. Front Microbiol 9:127
Kuroda T, Tsuchiya T (2009) Multidrug efflux transporters in the MATE family. Biochim Biophys Acta (BBA)-proteins. Proteomics 1794:763–768
Kwiecinski JM, Jacobsson G, Horswill AR, Josefsson E, Jin T (2019) Biofilm formation by *Staphylococcus aureus* clinical isolates correlates with the infection type. Infect Dis (Auckl) 51:446–451. https://doi.org/10.1080/23744235.2019.1593499
Lauderdale KJ, Malone CL, Boles BR, Morcuende J, Horswill AR (2009) Biofilm dispersal of community-associated methicillin-resistant *Staphylococcus aureus* on orthopedic implant material. J Orthop Res 28:55–61. https://doi.org/10.1002/jor.20943
Lazar V (2011) Quorum sensing in biofilms–how to destroy the bacterial citadels or their cohesion/power? Anaerobe 17:280–285
Lewis K (2007) Persister cells, dormancy and infectious disease. Nat Rev Microbiol 5:48–56
Lewis K (2010) Persister cells. Annu Rev Microbiol 64:357–372
Lister JL, Horswill AR (2014) *Staphylococcus aureus* biofilms: recent developments in biofilm dispersal. Front Cell Infect Microbiol 4:178. https://doi.org/10.3389/fcimb.2014.00178
Liu Y, Yang S-F, Li Y, Xu H, Qin L, Tay J-H (2004) The influence of cell and substratum surface hydrophobicities on microbial attachment. J Biotechnol 110:251–256. https://doi.org/10.1016/j.jbiotec.2004.02.012
Liu J, Prindle A, Humphries J, Gabalda-Sagarra M, Asally M, Dong-yeon DL, Ly S, Garcia-Ojalvo J, Süel GM (2015) Metabolic co-dependence gives rise to collective oscillations within biofilms. Nature 523:550–554
Lode H, Borner K, Koeppe P (1998) Pharmacodynamics of fluoroquinolones. Clin Infect Dis 27:33–39
Lubelski J, Konings WN, Driessen AJM (2007) Distribution and physiology of ABC-type transporters contributing to multidrug resistance in bacteria. Microbiol Mol Biol Rev 71:463–476
Lynch AS, Robertson GT (2008) Bacterial and fungal biofilm infections. Annu Rev Med 59:415–428
MacKenzie KD, Wang Y, Shivak DJ, Wong CS, Hoffman LJL, Lam S, Kröger C, Cameron ADS, Townsend HGG, Köster W (2015) Bistable expression of CsgD in *salmonella enterica* serovar *typhimurium* connects virulence to persistence. Infect Immun 83:2312–2326
Madsen JS, Røder HL, Russel J, Sørensen H, Burmølle M, Sørensen SJ (2016) Coexistence facilitates interspecific biofilm formation in complex microbial communities. Environ Microbiol 18:2565–2574. https://doi.org/10.1111/1462-2920.13335
Mathur T, Singhal S, Khan S, Upadhyay D, Fatma T, Rattan A (2005) Adverse effect of staphylococci slime on in vitro activity of glycopeptides. Jpn J Infect Dis 58:353
McDougald D, Rice SA, Barraud N, Steinberg PD, Kjelleberg S (2012) Should we stay or should we go: mechanisms and ecological consequences for biofilm dispersal. Nat Rev Microbiol 10:39–50. https://doi.org/10.1038/nrmicro2695
Mehrabi Z, Taheri-Kafrani A, Asadnia M, Razmjou A (2020) Bienzymatic modification of polymeric membranes to mitigate biofouling. Sep Purif Technol 237:116464. https://doi.org/10.1016/j.seppur.2019.116464
Mempel M, Feucht H, Ziebuhr W, Endres M, Laufs R, Grüter L (1994) Lack of *mec*A transcription in slime-negative phase variants of methicillin-resistant *Staphylococcus epidermidis*. Antimicrob Agents Chemother 38:1251–1255
Mempel M, Müller E, Hoffmann R, Feucht H, Laufs R, Grüter L (1995) Variable degree of slime production is linked to different levels of beta-lactam susceptibility in *Staphylococcus epidermidis* phase variants. Med Microbiol Immunol 184:109–113

Merino N, Toledo-Arana A, Vergara-Irigaray M, Valle J, Solano C, Calvo E, Lopez JA, Foster TJ, Penadés JR, Lasa I (2009) Protein A-mediated multicellular behavior in *Staphylococcus aureus*. J Bacteriol 191:832–843

Mhatre E, Monterrosa RG, Kovács ÁT (2014) From environmental signals to regulators: modulation of biofilm development in gram-positive bacteria. J Basic Microbiol 54:616–632

Miao J, Liang Y, Chen L, Wang W, Wang J, Li B, Li L, Chen D, Xu Z (2017) Formation and development of *staphylococcus* biofilm: with focus on food safety. J Food Saf 37:e12358

Miao J, Lin S, Soteyome T, Peters BM, Li Y, Chen H, Su J, Li L, Li B, Xu Z, Shirtliff ME, harro JM (2019) Biofilm formation of *Staphylococcus aureus* under food heat processing conditions: first report on CML production within biofilm. Sci Rep 9:1312. doi: https://doi.org/10.1038/s41598-018-35558-2

Mohammed YHE, Manukumar HM, Rakesh KP, Karthik CS, Mallu P, Qin H-L (2018) Vision for medicine: *Staphylococcus aureus* biofilm war and unlocking key's for anti-biofilm drug development. Microb Pathog 123:339–347. https://doi.org/10.1016/j.micpath.2018.07.002

Moormeier DE, Bayles KW (2017) *Staphylococcus aureus* biofilm: a complex developmental organism. Mol Microbiol 104:365–376. https://doi.org/10.1111/mmi.13634

Moormeier DE, Bose JL, Horswill AR, Bayles KW (2014) Temporal and stochastic control of *Staphylococcus aureus* biofilm development. MBio 14;5(5):e01341-14

Morfeldt E, Taylor DV, Von Gabain A, Arvidson S (1995) Activation of alpha-toxin translation in *Staphylococcus aureus* by the trans-encoded antisense RNA, RNAIII. EMBO J 14:4569–4577

Müller J, Kuttler C, Hense BA, Rothballer M, Hartmann A (2006) Cell–cell communication by quorum sensing and dimension-reduction. J Math Biol 53:672–702

Murai M, Moriyama H, Hata E, Takeuchi F, Amemura-Maekawa J (2016) Variation and association of fibronectin-binding protein genes *fnb*A and *fnb*B in *Staphylococcus aureus* Japanese isolates. Microbiol Immunol 60:312–325. https://doi.org/10.1111/1348-0421.12377

Naushad S, Nobrega DB, Naqvi SA, Barkema HW, De Buck J (2020) Genomic analysis of bovine *Staphylococcus aureus* isolates from Milk to elucidate diversity and determine the distributions of antimicrobial and virulence. mSystems 5:e00063-20. https://doi.org/10.1128/mSystems.00063-20

Ní Eidhin D, Perkins S, Francois P, Vaudaux P, Höök M, Foster TJ (1998) Clumping factor B (ClfB), a new surface-located fibrinogen-binding adhesin of *Staphylococcus aureus*. Mol Microbiol 30:245–257. https://doi.org/10.1046/j.1365-2958.1998.01050.x

Nichols WW, Dorrington SM, Slack MP, Walmsley HL (1988) Inhibition of tobramycin diffusion by binding to alginate. Antimicrob Agents Chemother 32:518–523

Nikaido H, Takatsuka Y (2009) Mechanisms of RND multidrug efflux pumps. Biochim Biophys Acta (BBA)-proteins. Proteomics 1794:769–781

O'Brien L, Kerrigan SW, Kaw G, Hogan M, Penadés J, Litt D, Fitzgerald DJ, Foster TJ, Cox D (2002) Multiple mechanisms for the activation of human platelet aggregation by *Staphylococcus aureus*: roles for the clumping factors ClfA and ClfB, the serine–aspartate repeat protein SdrE and protein A. Mol Microbiol 44:1033–1044. doi: https://doi.org/10.1046/j.1365-2958.2002.02935.x

O'Gara JP (2007) *Ica* and beyond: biofilm mechanisms and regulation in *Staphylococcus epidermidis* and *staphylococcus aureus*. FEMS Microbiol Lett 270:179–188

O'Neill J (2016) Tackling drug-resistant infections globally: final report and recommendations

O'Neill E, Pozzi C, Houston P, Smyth D, Humphreys H, Robinson DA, O'Gara JP (2007) Association between methicillin susceptibility and biofilm regulation in *Staphylococcus aureus* isolates from device-related infections. J Clin Microbiol 45:1379–1388

O'Neill E, Pozzi C, Houston P, Humphreys H, Robinson DA, Loughman A, Foster TJ, O'Gara JP (2008) A novel *Staphylococcus aureus* biofilm phenotype mediated by the fibronectin-binding proteins, FnBPA and FnBPB. J Bacteriol 190:3835–3850

Olsen I (2015) Biofilm-specific antibiotic tolerance and resistance. Eur J Clin Microbiol Infect Dis 34:877–886

Otto M (2009) *Staphylococcus epidermidis* - the "accidental" pathogen. Nat Rev Microbiol 7:555–567. https://doi.org/10.1038/nrmicro2182
Pao SS, Paulsen IT, Saier MH (1998) Major facilitator superfamily. Microbiol Mol Biol Rev 62:1–34
Peacock SJ, Day NPJ, Thomas MG, Berendt AR, Foster TJ (2000) Clinical isolates of *Staphylococcus aureus* exhibit diversity in fnb genes and adhesion to human fibronectin. J Infect 41:23–31
Peacock SJ, Moore CE, Justice A, Kantzanou M, Story L, Mackie K, O'Neill G, Day NPJ (2002) Virulent combinations of adhesin and toxin genes in natural populations of *Staphylococcus aureus*. Infect Immun 70:4987–4996
Penesyan A, Nagy SS, Kjelleberg S, Gillings MR, Paulsen IT (2019) Rapid microevolution of biofilm cells in response to antibiotics. Npj Biofilms Microb 5:34. https://doi.org/10.1038/s41522-019-0108-3
Pi Y, Chen W, Ji Q (2020) Structural basis of *Staphylococcus aureus S*urface protein SdrC. Biochemistry 59:1465–1469. https://doi.org/10.1021/acs.biochem.0c00124
Queck SY, Jameson-Lee M, Villaruz AE, Bach T-HL, Khan BA, Sturdevant DE, Ricklefs SM, Li M, Otto M (2008) RNAIII-independent target gene control by the *agr* quorum-sensing system: insight into the evolution of virulence regulation in *Staphylococcus aureus*. Mol Cell 32:150–158
Reffuveille F (2017) *Staphylococcus aureus* biofilms and their impact on the medical Field. In: Josse J (ed). IntechOpen, Rijeka, p Ch. 11
Resch A, Rosenstein R, Nerz C, Götz F (2005) Differential gene expression profiling of *Staphylococcus aureus* cultivated under biofilm and planktonic conditions. Appl Environ Microbiol 71:2663–2676
Roche FM, Downer R, Keane F, Speziale P, Park PW, Foster TJ (2004) The N-terminal a domain of fibronectin-binding proteins a and B promotes adhesion of *Staphylococcus aureus* to elastin. J Biol Chem 279:38433–38440
Rohde H, Burdelski C, Bartscht K, Hussain M, Buck F, Horstkotte MA, Knobloch JK, Heilmann C, Herrmann M, Mack D (2005) Induction of *Staphylococcus epidermidis* biofilm formation via proteolytic processing of the accumulation-associated protein by staphylococcal and host proteases. Mol Microbiol 55:1883–1895
Rosenberg CR, Fang X, Allison KR (2020) Potentiating aminoglycoside antibiotics to reduce their toxic side effects. PLoS One 15:e0237948
Rumbaugh KP, Sauer K (2020) Biofilm dispersion. Nat Rev Microbiol 18:571–586. https://doi.org/10.1038/s41579-020-0385-0
Schauder S, Bassler BL (2001) The languages of bacteria. Genes Dev 15:1468–1480
Schilcher K, Horswill AR (2020) Staphylococcal biofilm development: structure, regulation, and treatment strategies. Microbiol Mol Biol Rev 84:e00026–e00019. https://doi.org/10.1128/MMBR.00026-19
Schwartz K, Ganesan M, Payne DE, Solomon MJ, Boles BR (2016) Extracellular DNA facilitates the formation of functional amyloids in *Staphylococcus aureus* biofilms. Mol Microbiol 99:123–134
Seo Y-S, Lee DY, Rayamahji N, Kang ML, Yoo HS (2008) Biofilm-forming associated genotypic and phenotypic characteristics of *staphylococcus* spp. isolated from animals and air. Res Vet Sci 85:433–438
Shukla SK, Rao TS (2017) *Staphylococcus aureus* biofilm removal by targeting biofilm-associated extracellular proteins. Indian J Med Res 146:S1–S8. https://doi.org/10.4103/ijmr.IJMR_410_15
Silva V, Capelo JL, Igrejas G, Poeta P (2020) Molecular epidemiology of *Staphylococcus aureus* lineages in wild animals in Europe: a review. Antibiotics 9:122. https://doi.org/10.3390/antibiotics9030122
Singh R, Ray P, Das A, Sharma M (2010) Penetration of antibiotics through *Staphylococcus aureus* and *Staphylococcus epidermidis* biofilms. J Antimicrob Chemother 65:1955–1958. https://doi.org/10.1093/jac/dkq257

Singh S, Singh SK, Chowdhury I, Singh R (2017) Understanding the mechanism of bacterial biofilms resistance to antimicrobial agents. Open Microbiol J 11:53–62. https://doi.org/10.2174/1874285801711010053

Speziale P, Geoghegan JA (2015) Biofilm formation by staphylococci and streptococci: structural, functional, and regulatory aspects and implications for pathogenesis. Front Cell Infect Microbiol 5:31

Stewart PS, Franklin MJ (2008) Physiological heterogeneity in biofilms. Nat Rev Microbiol 6:199–210

Stubbendieck RM, Vargas-Bautista C, Straight PD (2016) Bacterial communities: interactions to scale. Front Microbiol 7:1234. https://doi.org/10.3389/fmicb.2016.01234

Sugimoto S, Sato F, Miyakawa R, Chiba A, Onodera S, Hori S, Mizunoe Y (2018) Broad impact of extracellular DNA on biofilm formation by clinically isolated methicillin-resistant and-sensitive strains of *Staphylococcus aureus*. Sci Rep 8:1–11

Sully EK, Malachowa N, Elmore BO, Alexander SM, Femling JK, Gray BM, DeLeo FR, Otto M, Cheung AL, Edwards BS, Sklar LA, Horswill AR, Hall PR, Gresham HD (2014) Selective chemical inhibition of agr quorum sensing in *Staphylococcus aureus* promotes host defense with minimal impact on resistance. PLoS Pathog 10:e1004174

Suresh MK, Biswas R, Biswas L (2019) An update on recent developments in the prevention and treatment of *Staphylococcus aureus* biofilms. Int J med Microbiol 309:1–12. https://doi.org/10.1016/j.ijmm.2018.11.002

Sutherland IW (2001) Biofilm exopolysaccharides: a strong and sticky framework. Microbiology 147:3–9

Valen H, Scheie AA (2018) Biofilms and their properties. Eur J Oral Sci 126:13–18. https://doi.org/10.1111/eos.12425

Valle J, Toledo-Arana A, Berasain C, Ghigo J, Amorena B, Penadés JR, Lasa I (2003) SarA and not σB is essential for biofilm development by *Staphylococcus aureus*. Mol Microbiol 48:1075–1087

Valotteau C, Prystopiuk V, Pietrocola G, Rindi S, Peterle D, De Filippis V, Foster TJ, Speziale P, Dufrêne YF (2017) Single-cell and single-molecule analysis unravels the multifunctionality of the *Staphylococcus aureus* collagen-binding protein Cna. ACS Nano 11:2160–2170. https://doi.org/10.1021/acsnano.6b08404

Van Loosdrecht MC, Lyklema J, Norde W, Schraa G, Zehnder AJ (1987) Electrophoretic mobility and hydrophobicity as a measured to predict the initial steps of bacterial adhesion. Appl Environ Microbiol 53:1898–1901

Vuong C, Gerke C, Somerville GA, Fischer ER, Otto M (2003) Quorum-sensing control of biofilm factors in *Staphylococcus epidermidis*. J Infect Dis 188:706–718

Vuong C, Kocianova S, Voyich JM, Yao Y, Fischer ER, DeLeo FR, Otto M (2004) A crucial role for exopolysaccharide modification in bacterial biofilm formation, immune evasion, and virulence. J Biol Chem 279:54881–54886

Walters MC, Roe F, Bugnicourt A, Franklin MJ, Stewart PS (2003) Contributions of antibiotic penetration, oxygen limitation, and low metabolic activity to tolerance of *Pseudomonas aeruginosa* biofilms to ciprofloxacin and tobramycin. Antimicrob Agents Chemother 47:317–323

Weidenmaier C, Lee JC (2015) Structure and function of surface polysaccharides of *Staphylococcus aureus*. *Staphylococcus aureus* 57–93

Weigel LM, Donlan RM, Shin DH, Jensen B, Clark NC, McDougal LK, Zhu W, Musser KA, Thompson J, Kohlerschmidt D (2007) High-level vancomycin-resistant *Staphylococcus aureus* isolates associated with a polymicrobial biofilm. Antimicrob Agents Chemother 51:231–238

Wolcott R, Costerton JW, Raoult D, Cutler SJ (2013) The polymicrobial nature of biofilm infection. Clin Microbiol infect 19:107–112. Doi. https://doi.org/10.1111/j.1469-0691.2012.04001.x

Wolfmeier H, Pletzer D, Mansour SC, Hancock REW (2018) New perspectives in biofilm eradication. ACS Infect Dis 4:93–106

Yarwood JM, Bartels DJ, Volper EM, Greenberg EP (2004) Quorum sensing in *Staphylococcus aureus* biofilms. J Bacteriol 186:1838–1850

You Y, Xue T, Cao L, Zhao L, Sun H, Sun B (2014) *Staphylococcus aureus* glucose-induced biofilm accessory proteins, GbaAB, influence biofilm formation in a PIA-dependent manner. Int J med Microbiol 304:603–612. https://doi.org/10.1016/j.ijmm.2014.04.003

Yu L, Hisatsune J, Hayashi I, Tatsukawa N, Sato'o Y, Mizumachi E, Kato F, Hirakawa H, Pier GB, Sugai M (2017) A novel repressor of the locus discovered in clinically isolated super-biofilm-elaborating *Staphylococcus aureus*. MBio 8:e02282–e02216. https://doi.org/10.1128/mBio.02282-16

Zhang W, Li C (2016) Exploiting quorum sensing interfering strategies in gram-negative bacteria for the enhancement of environmental applications. Front Microbiol 6:1535

Zhang K, Li X, Yu C, Wang Y (2020) Promising therapeutic strategies against microbial biofilm challenges. Front Cell Infect Microbiol 10:359

Zheng Z, Stewart PS (2002) Penetration of rifampin through *Staphylococcus epidermidis* biofilms. Antimicrob Agents Chemother 46:900–903

Chapter 13
Antimicrobial Resistance: Meaning and Developing Realization

Atul N. Chandu, Samsher Singh, and Santosh Kumar Rath

13.1 Introduction

Antimicrobial resistance (AMR) is the capacity of a microorganism to resist the growth-inhibitory or killing activity of an antimicrobial to which the specific bacterial species is normally susceptible. The availability of antibiotics to treat infectious diseases has significantly improved the health and life prospects of humans, as well as the health and welfare of animals. However, use of antibiotics results in selection for antimicrobial resistance in bacteria (Verraes et al. 2013).

AMR is a fast-growing public health issue overshadowing modern medicine worldwide. The development of pathogens with a variety of resistance mechanisms is making infection control and treatment strategies increasingly ineffective. However, it is still difficult to accurately estimate the extent of AMR (Mattar et al. 2020). Development of antimicrobial resistance has added significantly to the impacts of various infectious diseases, as well as overall health care costs. Although we have a large number of antimicrobial agents, this is paralleled by continuously increasing evidence of antimicrobial resistance in the present era (Beceiro et al. 2013; Aslam et al. 2018). This continuous increase in antimicrobial resistance is a

A. N. Chandu
Department of Pharmaceutical Microbiology, Danteswari College of Pharmacy, Jagdalpur, Chhattisgarh, India

S. Singh (✉)
Lee Kong Chian School of Medicine, Nanyang Technological University Singapore, 59 Nanyang Drive, Singapore
e-mail: samsher.singh@ntu.edu.sg

S. K. Rath (✉)
Department of Pharmaceutical Chemistry, Danteswari College of Pharmacy, Jagdalpur, Chhattisgarh, India
e-mail: santosh@pharmadanteswari.org

N. Akhtar et al. (eds.), *Emerging Modalities in Mitigation of Antimicrobial Resistance*,
https://doi.org/10.1007/978-3-030-84126-3_13

global health issue, accounting for about 700,000 deaths annually (World Health Organization 2019). Diseases caused by drug-resistant microbes—the majority of which are respiratory infections, sexually transmitted infections and urinary tract infections (UTIs)—are very hard to treat and are increasing all the time. So-called wonder drugs are becoming ineffective; the effect of this is currently being witnessed by the world and will become even more serious in the future. In an attempt to help control them, the World Health Organization (WHO) has classified resistant infections as one of the top ten global health threats (Prestinaci et al. 2015).

AMR spreads as a result of antimicrobial selection pressure (ASP) and clonal dissemination. Dissemination of resistant microbes can occur following selection pressure and is the main cause of the spread of AMR in settings of close personal contact, such as hospitals and day care centres (Davey et al. 2010).

Inappropriate and extensive use of antimicrobials has led to the present scenario, which is posing serious threats for the future. It is evident that human, animal, food and environmental factors play significant roles in microbial evolution, and all of these are interconnected, demonstrating the need for a coordinated, multisectoral 'one health' approach (Wang and Schaffner 2011; Newell et al. 2010).

A return to the pre-antibiotic era is fast becoming a reality in many parts of the world (Aminov 2010); it is clear that we are entering an era in which the impact of antibiotics will be minimal. It is therefore highly relevant to know the details of AMR mechanisms and the routes of resistance so that therapies, practices, surveillance and research can be adjusted to minimize the increase of resistance in the future (Larsson et al. 2018; Brown and Wright 2016).

Principally—and contrary to the dogma promulgated a few years ago—the increase and spread of resistance are not caused by or restricted to hospitals and excessive use of antibiotics. The presence of resistant bacterial strains in sewage, in water, in food, in soil and elsewhere demonstrates the participation of environmental pressures on these strains, and by associating these data with the presence of antibiotics in the environment and their use in veterinary therapy and agriculture, researchers from all over the world are expanding their studies beyond hospitals. This reflects the concept that the involvement of environmental factors is somehow partly responsible for antibiotic resistance (Lupo et al. 2012; Balsalobre et al. 2014). It is recommended that countries prioritize national action plans (NAPs) to scale-up financing and capacity-building efforts; put in place stronger regulatory systems and support awareness programmes for responsible and prudent usage of antimicrobials by professionals in human, animal and plant health; and invest in ambitious research and development of new technologies to combat antimicrobial resistance (Chua et al. 2021; Bungau et al. 2021).

Because of the continuous increase in drug-resistant infections, there is an urgent need for new antimicrobials. AMR is defined as inefficiency of an antibiotic for killing the same type of pathogenic bacteria that were previously killed by the same antibiotic, which may be due to genetic inheritable changes followed by clonal propagation or phenotypic adaptations to survive under antibiotic pressure (Vincent and Uphoff 2020).

The continued increase in antimicrobial resistance has led to fewer treatment options being available for patients, and it is believed to increase morbidity and mortality. In the present scenario where common infections are becoming resistant, we are facing more severe infections needing more extensive treatment, longer courses of illness often requiring extended periods of hospitalization, and huge increases in treatment costs (Reygaert 2018; Michael et al. 2014).

To prevent AMR, it is necessary first to understand what causes it. In simple terms, random mutations can lead to microorganisms becoming immune to a certain type of drug, while uncontrolled or unnecessary ambient use of these drugs creates transformative pressure making resistant microorganisms proliferate (Rodríguez-González et al. 2019; Liu et al. 2021).

13.2 Overview of Current Antimicrobial Resistance

Common infections were previously cured with frequently used antibiotics globally, but now resistance is common among UTIs, sepsis, sexually transmitted infections and some forms of diarrhoea (Micoli et al. 2021; Ahmad 2021). According to the Global Antimicrobial Resistance and Use Surveillance System (GLASS), the rate of resistance to ciprofloxacin for treatment of UTIs is 8.4–92.9% in cases caused by *Escherichia coli* and 4.1–79.4% in those caused by *Klebsiella pneumoniae* (Cheong et al. 2021; Veeraraghavan et al. 2018). Worldwide, there have been reports of resistance to carbapenems, an antibiotic class that includes drugs of last resort for curing intestinal infections caused by *K. pneumoniae* bacteria, which are now considered life threatening and are among the most common types of hospital-acquired infections (Chen et al. 2021). As a result of this resistance scenario, carbapenems are unable to cure more than 50% of such cases in some countries (Veeraraghavan et al. 2018). A similar scenario is now seen with fluoroquinolones, which are losing their effectiveness to treat UTIs caused by *E. coli*, although they were previously the drugs of choice in such cases (Nazir et al. 2021; Langner et al. 2021). To treat *E. coli*–related UTIs and carbapenem-resistant Enterobacteriaceae, colistin is now considered the drug of last resort, but resistance reports from several countries indicate that these infections are becoming harder to cure, and, at present, we have no alternative drugs that are effective in such cases (Ara et al. 2021; Kelly et al. 2017).

Another major concern is resistance of *Staphylococcus aureus* to multiple classes of antibiotics (Rath et al. 2019). These bacteria are among those that live commensally on the skin and are responsible for a plethora of infections in the community, as well as hospital-acquired infections (Foster and Geoghegan 2015). Methicillin (methicillin) was previously effective against them, but increasing resistance has resulted in evolution of methicillin-resistant *S. aureus* (MRSA), which has now spread worldwide and is associated with a 64% greater risk of mortality than drug-sensitive infections (Romandini et al. 2021; Lee et al. 2018). Over time, MRSA has evolved resistance to every antibiotic, and this has consequently made such infections very hard to treat. Selection pressure from increasing antibiotic use has provided the pathogen with opportunities to develop complex survival mechanisms (Craft et al. 2019). Furthermore, the problem is become more complex because of decreased interest in antibiotic development by pharmaceutical industries, given the high levels of investment that are required and the risk that increasing resistance will rapidly make the results of such huge efforts redundant anyway (Årdal et al. 2020).

Resistance in some variants of *Neisseria gonorrhoeae* makes them hard to eradicate with sulfonamides (sulphonamides), penicillins, tetracycline, macrolides, fluoroquinolones and some old cephalosporins (Unemo and Shafer 2014; Bodoev and Il'Ina 2015). The only currently viable option for treatment of *N. gonorrhoeae* infections is monotherapy with the injectable extended-spectrum cephalosporin (ESC) ceftriaxone (Kularatne et al. 2018; Unemo and Shafer 2014).

Tuberculosis is a leading cause of mortality among infectious diseases. Antibiotic-resistant strains are hindering control of the tuberculosis epidemic around the globe (Bendre et al. 2021). Rifampicin is considered a surrogate marker of multidrug resistant tuberculosis and there were around half million (range 400000–535000) cases having rifampicin/multidrug resistant tuberculosis in 2019 (WHO, 2020; Neimz et al. 2012; Smith et al. 2012; Denkinger et al. 2014). Treatment for tuberculosis is very lengthy (6–18 months long for MDR or extensively drug-resistant (XDR) tuberculosis) and less effective than treatment of susceptible tuberculosis (Mesfin et al. 2018; Cheng et al. 2020). It has been estimated that fewer than 60% of MDR/rifampicin-resistant tuberculosis cases are treated successfully. Resistance to drugs used for treatment of MDR/XDR tuberculosis is continuing to rise and poses a major threat (Shivekar et al. 2020).

There are increasing reports of morbidity and mortality caused by variant strains of nosocomial Gram-negative pathogens such as carbapenem-resistant *Acinetobacter baumanii* (CRAB) (Isler et al. 2018). The current treatment for infections caused by CRAB relies on cefiderocol, a cephalosporin that is active against many MDR Gram-negative pathogens and was recently approved by the US Food and Drug Administration (FDA) for clinical use against CRAB with a variety of resistance. However, there have been reports of resistance to cefiderocol too, which poses a risk of increasing in the future as clonal propagation and selection will encourage these resistant variants to spread (Isler et al. 2018).

13.3 Status of Antimicrobial Drug Discovery and Development, and Evolution of Antimicrobial Resistance

The earliest uses of antimicrobials had a dramatic impact in reducing mortality from serious and life-threatening bacterial diseases in comparison with the absence of such therapy in the pre-antibiotic era. Subcutaneous use of sulfanilamide (sulphanilamide) lowers the mortality rate associated with acute meningococcal meningitis by 70–90% (Heselpoth 2014). With the approval of an increasing number of agents within the same scaffold classes and the lack of new drug classes, it is perhaps not surprising that development of antibacterial agents would reach a saturation point (Shlaes and Bradford 2018; Heselpoth 2014).

Today, antimicrobials are the third most profitable class of drugs for pharmaceutical companies, surpassed only by central nervous system drugs and cardiovascular drugs. A recent study showed that the top five disease categories (including heart diseases, pulmonary conditions, mental disorders, cancers and hypertension) accounted for 31% of the increase in health care expenditure between 1987 and 2000 (De Oliveira et al. 2020; Ng 2015).

The development of newer antibiotics and further evolution of pathogens resistant to the same antibiotics have had significant impacts on the discovery of modern antimicrobial agents. Between about 1930 and 2005, various new antibiotics were developed (Fig. 13.1). However, from 1940 onwards, pathogens started to be

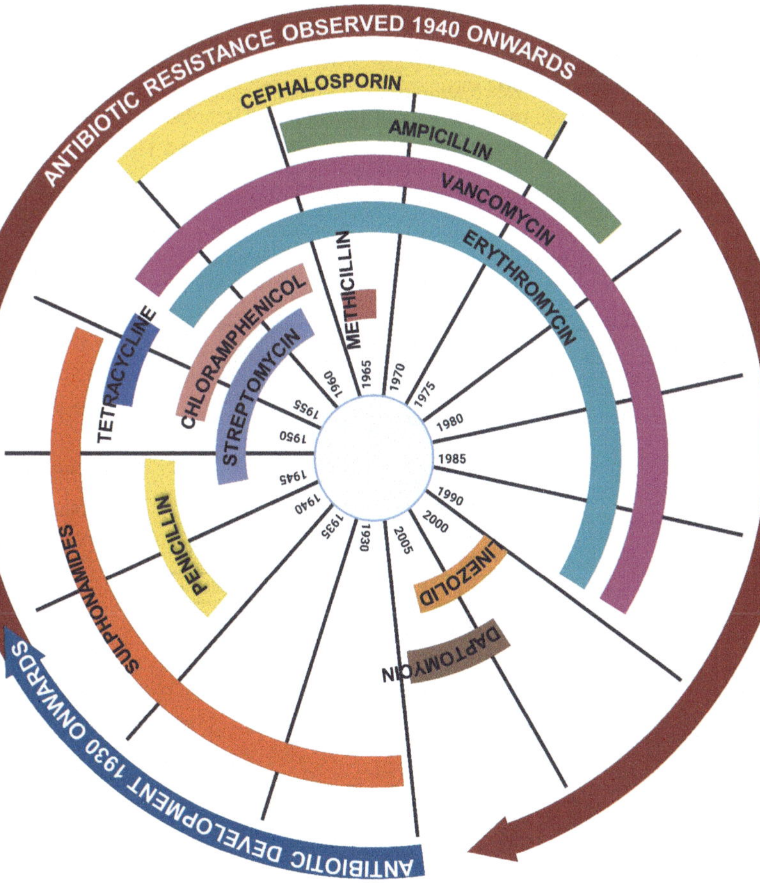

Fig. 13.1 History of antibiotic discovery and development of antibiotic resistance from 1930 to 2005

identified that were resistant to antibiotics that had previously been effective against those species. Federal agencies are aware of the needs to address the present challenges related to antimicrobial resistance and to simultaneously give more focus to stimulating antimicrobial drug development. These agencies have been using their combined resources to attempt to address these needs.

A strategy of modification of existing antimicrobials was initiated and successfully implemented during a period when the rate of discovery of novel drug classes suddenly slowed in the 1970s, and growing resistance problems forced researchers to look into possible modifications of the existing arsenal that could confer improved antimicrobial activity, improved sensitivity towards resistant organisms and less toxicity or adverse effect towards host. Possible approaches to tap into novel antimicrobial diversity include exploring ecological niches other than soils (such as the marine environment (Hughes and Fenical 2010)), borrowing antimicrobial peptides and compounds from animals and plants (Hancock and Sahl 2006), mimicking natural lipopeptides of bacteria and fungi (Aminov 2010; Makovitzki et al. 2006), accessing uncultivated portions of microbiota through a metagenomic approach (Gupta et al. 2018) and, finally, using the complete synthetic route pioneered during the early years of the antibiotic era.

The generally observed mechanisms of bacterial resistance to various antibiotics mainly include the following (Rao et al. 2018) (Figs. 13.2 and 13.3):

1. Enzymatic degradation leads to inactivation of antibiotics.
2. Enzymatic modification can alter the cellular targets of antibiotics, making them unable to bind effectively to the target site.

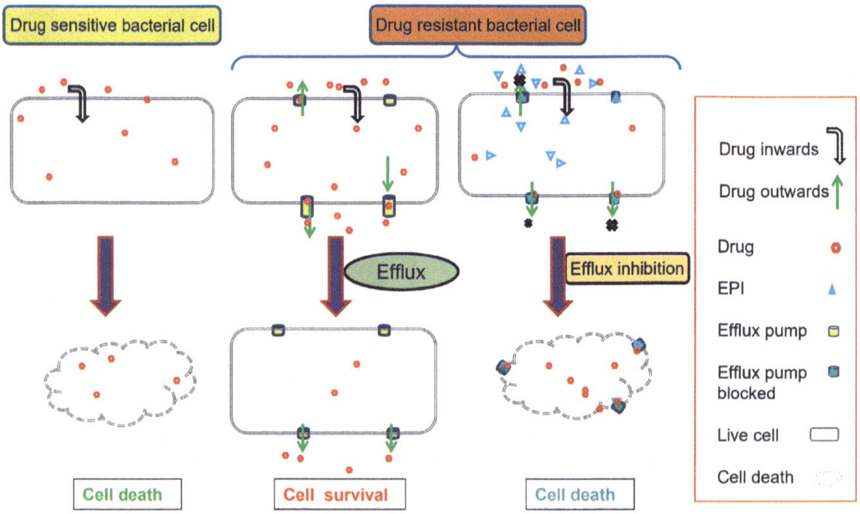

Fig. 13.2 Evolution of antibiotic resistance in bacterial cells and efflux inhibition. *EPI* efflux pump inhibitor

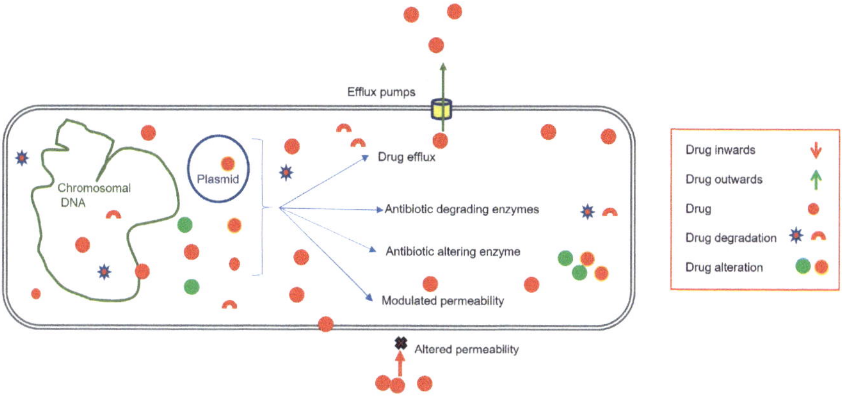

Fig. 13.3 Mechanisms behind antibiotic resistance

3. Alteration of cell membranes leads to abridged permeability of the antibiotic into the bacterial cytoplasm.
4. Active extrusion of antibiotics by membrane efflux pumps reduces cellular concentrations of antibiotics (Rao et al. 2018; Rath et al. 2019; Blair et al. 2015).

Cloning, polymerase chain reaction and gene expression techniques have been applied to detect natural genes in random recombinant clones derived from bacterial DNA libraries from soils and sediments (Mukherjee and Reddy 2020). A potential problem is that identification of functional resistance requires gene expression (transcription and translation) of the cloned genes in a heterologous host. To date, only *E. coli* has been used (Olivenza et al. 2019). Many questions remain; the roles of these environmental reservoirs in clinical resistance development are still hypothetical, and the primary metabolic functions of microbial populations are, as yet, unknown (Davies and Davies 2010).

Most bacteria were initially assumed to be commensal organisms and were considered harmless to the human system (Méthot and Alizon 2014). This is contrary to the current scientific evidence in which disease-causing bacteria have been identified alongside use of drugs in treating disease conditions. Despite the challenges emanating from clinical use of antibiotics, there is growing use of these molecules in bacterial infection treatment (Gupta and Datta 2019). However, various resistant strains have evolved and threatened the effectiveness of most antibiotics in clinical management of disease conditions. Molecular understanding of bacterial existence and its activities is key in preventing bacterial resistance (Ghai and Ghai 2018). In recent times, bioinformatics has played a vital role in drug discovery, gene sequencing, gene alignment and genera proteomics study. Adequate knowledge of the processes of bacterial resistance can precipitate the bioinformatics approach needed to unravel such resistance cases (Kaushik et al. 2018). Recently, new approaches to research involving bacterial resistance to antibiotics have appeared; these involve combinations of molecular understanding of bacterial systems with knowledge of bioinformatics. Consequently, many molecules have been developed to curb resistance associated with different bacterial infections. Because of increased emphasis on the clinical relevance of antibiotics, the synergy between in silico study and in vivo study is well cemented, and this facilitates the discovery of potent antibiotics (Ndagi et al. 2020). Transfer of genetic traits through DNA is known to be the most common method of bacterial resistance; it involves acquisition of foreign DNA material through horizontal gene transfer (Munita and Arias 2016; Ndagi et al. 2020).

13.4 The Drug Discovery Pipeline and the Future

In respect of innovation in the development of new drugs and therapeutic biological products, the US FDA Center for Drug Evaluation and Research (CDER) reinforces the activities of the pharmaceutical industry at every step of the work. The closeness of new drugs and new biological products frequently offers new treatment options. Advances in health care achieved through previous historical innovations and approvals of antimicrobial drugs by the US FDA between 2012 and 2020 are listed in Table 13.1.

13.5 Drug-Resistant Microbes

In 1994, a global drug resistance surveillance programme was started in order to monitor trends in drug resistance. Its first report was published in 1997 and included data from 35 geographical settings for the period of 1994–1996 (Pablos-Méndez and Raviglione 1998).

After the first antituberculosis drugs—streptomycin, *para*-aminosalicylic acid and isoniazid—were introduced, resistance to particular drugs was noted in clinical isolates of *Mycobacterium tuberculosis* (Crofton and Mitchison 1948). This brought about a need to measure resistance accurately and easily. To control the drug resistance epidemic, it is necessary to gain insight into how *M. tuberculosis* develops drug resistance. This knowledge will help us to grasp how to avert the threat of drug resistance in addition to identifying genes associated with resistance to new drugs. *M. tuberculosis*, an aetiological agent of life-threatening disease, remains a major health problem, causing 1.5 million human deaths in 2019. Because *M. tuberculosis* has an ability to reside and replicate inside host tissue, it requires months-long treatment, which is difficult and often requires drug combination therapies, but these have a low success rate. Pervasive use of antibiotics has resulted in MDR, XDR and totally drug-resistant (TDR) infections, in addition to latent or persistent *M. tuberculosis* infections, for which antibiotic treatment is ineffective.

Acinetobacter species are non-fermentative, Gram-negative coccobacilli. There are at least 21 different *Acinetobacter* genospecies, 9 of which have been given formal species names (Lim et al. 2007). Risk factors for colonization or infection

Table 13.1 Antimicrobial drugs approved by the US Food and Drug Administration from 2012 to 2020 (Andrei et al. 2018; Bhutani et al. 2021; Ebied et al. 2020)

Generic drug names	Approval date	Brand name	Targeted microorganisms	Source	Purposes
Cefiderocol	November 14, 2019	Fetroja	Gram-negative bacteria	Siderophore cephalosporins	To treat complicated urinary tract infections in cases with only limited or no alternative treatment options
Imipenem, cilastatin and relebactam	July 16, 2019	Recarbrio	Gram-negative bacteria	β-Lactam antibiotic	To treat complicated urinary tract and complicated intra-abdominal infections
Rifamycin	November 16, 2018	Aemcolo	Mycobacteria	*Amycolatopsis rifamycinica*	To treat complicated diarrhoea
Omadacycline	October 2, 2018	Nuzyra	Gram-positive and Gram-negative bacteria	Semisynthetic compound derived from tetracycline	To treat community-acquired bacterial pneumonia and acute bacterial skin and skin structure infections
Plazomicin	June 25, 2018	Zemdri	*Escherichia coli*, *Klebsiella pneumoniae*, *Proteus mirabilis* and *Enterobacter cloacae*	Derived from sisomicin	To treat complicated urinary tract infections in adults
Secnidazole	September 15, 2017	Solosec	*Atopobium vaginae*	Nitroimidazole	To treat bacterial vaginosis
Meropenem and vaborbactam	August 29, 2017	Vabomere	*Escherichia coli*, *Klebsiella pneumoniae* and *Enterobacter cloacae* species		To treat complicated urinary tract infections in adults

(continued)

Table 13.1 (continued)

Generic drug names	Approval date	Brand name	Targeted microorganisms	Source	Purposes
Delafloxacin	June 19, 2017	Baxdela	Gram-negative and gram-positive bacteria	Fluoroquinolone antibiotic	To treat acute bacterial skin infections
Bezlotoxumab	October 21, 2016	Zinplava	*Clostridium difficile*		To reduce recurrent *Clostridium difficile* infections in patients aged ≥18 years
Isavuconazonium sulfate	March 6, 2015	Cresemba	*Aspergillus* and Mucorales	Prodrug of isavuconazole	To treat invasive aspergillosis and invasive mucormycosis
Luliconazole	2014	Luzu	*Trichophyton rubrum*, *Microsporum gypseum* and *Epidermophyton floccosum*	Imidazole antifungal	To treat athlete's foot, jock itch and ringworm caused by dermatophytes
Pneumococcal 13-valent conjugate vaccine	2013	Prevnar-13	*Pneumococcus* sp.	*Streptococcus pneumoniae*	For vaccination against pneumococcus
Tobramycin	2013	Tobrex	*Pseudomonas* sp.	Aminoglycoside antibiotic derived from streptomycin	To treat exacerbations of chronic infection with *Pseudomonas aeruginosa* in patients with cystic fibrosis

with MDR *Acinetobacter* species include prolonged hospitalization, intensive care unit admission, cessation of mechanical ventilation, colonization pressure, exposure to antimicrobial agents, recent surgery, invasive measures and fluctuations in illness severity. AMR among *Acinetobacter* species has increased substantially in the past decade. MDR *Acinetobacter baumannii* is a opportunistic pathogen in the health care setting, where it causes infections such as bacteraemia, pneumonia, meningitis, UTIs and wound infections. This organism's ability to endure a wide range of environmental conditions and to survive on surfaces for prolonged periods of time make it a frequent cause of sudden infections and an endemic health care–associated pathogen. The resistance mechanisms used by *Acinetobacter* species are similar to those used by *Pseudomonas* species, but *Acinetobacter* species have not been studied as extensively. The mechanisms of resistance can generally be classified into four categories: (1) activity of antimicrobial-inactivating enzymes, (2) reduced access to bacterial targets, (3) mutations that change targets or cellular functions, and (4) reduced drug uptake or drug efflux.

Pseudomonas aeruginosa is a leading cause of opportunistic infections. It can grow on a wide variety of substrates and can change its properties in response to changes in the environment. The identification of the entire genome sequence of *P. aeruginosa* and application of powerful DNA array techniques to reveal microbial gene expression under in vivo conditions should provide us with a clearer insight into the mechanisms that are involved (Eliopoulos et al. 2008; Lambert 2002). Inappropriate empirical use of antibiotic therapy for MDR/XDR *P. aeruginosa* infections has been associated with increased mortality and increased hospital costs. Antibiotic combinations are often used for treatment of these infections, although the value of such combination therapy in comparison with that of monotherapy is debatable. Use of amikacin and colistin in combination with antipseudomonal antibiotics offers the greatest coverage against MDR/XDR *P. aeruginosa,* but they are both associated with side effects and toxicity. In this scenario, new antibiotics with activity against MDR/XDR *P. aeruginosa* have been developed, and they represent a viable alternative option for treatment of infections caused by this organism (Asempa et al. 2019).

N. gonorrhoeae is an intracellular Gram-negative diplococcus. It exclusively infects human mucosa and can spread secondarily to other tissues. New strains of *N. gonorrhoeae* have now appeared that are resistant to the first-line treatment, ceftriaxone. The wide variety of gonococcal surface antigens and other factors in the immune response against *N. gonorrhoeae* weaken the likelihood of further development of this therapeutic resource (Salmerón et al. 2020).

The development and spread of resistance in *N. gonorrhoeae* have occurred mainly by addition of new DNA via conjugation and transformation, and determinants of resistance may be located on the chromosome or on extrachromosomal elements (plasmids). In contrast to Plasmid-mediated resistance can spread or transmit inter as well intraspecies levels, while chromosome-mediated resistance remains confined to the same species or descendants. Both forms of resistance can occur in one organism, and resistance to multiple antibiotics is common (Salmerón et al. 2020; Keese 2008).

Antifungal resistance can be innate (known as primary resistance) or acquired (known as secondary resistance). Many such resistance mechanisms have been described, such as biofilm formation (especially in *Candida albicans*), failure of intracellular drug accumulation or drug target alterations. The genome plasticity of human fungal pathogens is strongly associated with their ability to acquire resistance to antifungals. Some researchers have recommended use of other pharmacological classes and repurposing of old drugs used either as single antifungal agents or in combination with other known antifungal drugs (Gulati and Nobile 2016; Nobile and Johnson 2015).

The resistance of *Enterobacter* species to third-generation cephalosporins—such as cefizox (ceftizoxime), claforan (cefotaxime), tazimide (ceftazidime) and vantin (cefpodoxime)—is greater and is mostly due to widespread use of AmpC β-lactamases. Treatment with these antibiotics may cause selection of mutants for AmpC overproduction. Modulation of the phenotype by host bacteria makes gene transmission less obvious, and this may, in part, explain why tracking and control of carbapenem resistance in Enterobacteriaceae has been especially difficult (Iredell et al. 2016; Blair et al. 2015). The evolution and spread of resistance in Enterobacteriaceae complicate the treatment of serious infections. Approximately 20% of health care–associated infections by *K. pneumoniae* and 31% of infections by *Enterobacter* species are due to strains that are resistant to third-generation cephalosporins. Resistance of *K. pneumoniae* to third-generation cephalosporins is typically caused by acquisition of plasmids containing genes encoding broad-spectrum β-lactamases, and these plasmids often also carry other resistance genes (Chong et al. 2011).

13.6 Global Approaches for Control of Drug Resistance

A global action plan (GAP) provides a framework for controlling the spread of antimicrobial resistance. At a meeting in 2015, the WHO World Health Assembly (WHA) adopted a GAP with five objectives to control the spread of antimicrobial resistance: (1) to improve awareness and understanding of antimicrobial resistance, (2) to strengthen knowledge through surveillance and research, (3) to reduce the incidence of infection, (4) to optimize use of antimicrobial agents and (5) to ensure sustainable investment in countering antimicrobial resistance (World Health Organization 2015).

To reap effective outcomes from the GAP, a multisectoral coordination action involves a 'one health' approach. This requires coordinated action among numerous international sectors and national agencies, including human and veterinary medicine; the agriculture, finance and environmental sectors; and well-informed consumers. The GAP recognizes the various efforts that nations need to make in order to combat antimicrobial resistance. It also discusses the economic factors that nations should keep in mind to encourage the development of alternative novel products by pharmaceutical companies (World Health Organization 2015).

13.7 The Indian Scenario in Brief

Following the WHA meeting in 2015, many countries agreed on and implemented their NAPs. In April 2017, India published its NAP for 2017–2021, documenting its priorities and implementation strategies for curbing AMR in India. The report was submitted to the 70th meeting of the WHA, held in Switzerland in May 2017 (Government of India 2017; Ranjalkar and Chandy 2019). The objectives of the NAP include improved awareness, enhancement of surveillance measures, strengthened infection prevention and control, research and development, promotion of investment and collaborative activities to control AMR. On the basis of the NAP, various states of India have taken initiatives to make their own state action plans.

13.8 Conclusion

Antimicrobial resistance is one of the major causes of the current decline in public health and is increasing the challenges posed by disease in human society. This problem has also been accelerated by inappropriate prescribing and usage of antimicrobials; overconsumption of medicines; prolonged hospitalization; treatment failures; food spoilage; antimicrobial resistance among stakeholders; and improper control of, and regulations for, antimicrobial use in animals and food production, with consequent environmental exposure—as well as lack of newer drug development. However, proper functioning of public health care systems, promotion of good pharmacy practices, timely identification of diseases and adherence to standard treatment guidelines, rational use of antibiotics and immunization can be considered successful examples of implementation of appropriate action plans to help avoid antimicrobial resistance problems in the world. A global, collaborative and healthy research environment for new drug discovery; public commitment; and appropriate awareness about use of the available antibiotics are crucial factors in minimization or avoidance of bacterial resistance–related problems. We are still far from reaching that goal, but continued efforts, bit by bit, may lead to success in the coming decades.

References

Ahmad SI (2021) Phage therapy of human bacterial infections: a systematic review. In: Ahmad SI (ed) Human viruses: diseases, treatments and vaccines. Springer, Cham, pp 663–692

Aminov RI (2010) A brief history of the antibiotic era: lessons learned and challenges for the future. Front Microbiol. 1:134

Andrei S, Valeanu L, Chirvasuta R, Stefan M-G (2018) New FDA approved antibacterial drugs: 2015–2017. Discoveries (Craiova) 6(1):e81. https://doi.org/10.15190/d.2018.1

Ara B, Urmi UL, Haque TA, Nahar S, Rumnaz A, Ali T, Alam MS, Mosaddek ASM, Rahman NAA, Haque M (2021) Detection of mobile colistin-resistance gene variants (*mcr-1* and *mcr-2*) in urinary tract pathogens in Bangladesh: the last resort of infectious disease management colistin efficacy is under threat. Expert Rev Clin Pharmacol 14(4):513–522. https://doi.org/10.1080/17512433.2021.1901577

Årdal C, Balasegaram M, Laxminarayan R, McAdams D, Outterson K, Rex JH, Sumpradit N (2020) Antibiotic development—economic, regulatory and societal challenges. Nat Rev Microbiol 18(5):267–274. https://doi.org/10.1038/s41579-019-0293-3

Asempa TE, Nicolau DP, Kuti JL (2019) In vitro activity of imipenem–relebactam alone or in combination with amikacin or colistin against *Pseudomonas aeruginosa*. Antimicrob Agents Chemother 63(9):e00997–e00919. https://doi.org/10.1128/AAC.00997-19

Aslam B, Wang W, Arshad MI, Khurshid M, Muzammil S, Rasool MH, Nisar MA, Alvi RF, Aslam MA, Qamar MU (2018) Antibiotic resistance: a rundown of a global crisis. Infect Drug Resist 11:1645

Balsalobre LC, Dropa M, Matté MH (2014) An overview of antimicrobial resistance and its public health significance. Braz J Microbiol 45(1):1–6. https://doi.org/10.1590/S1517-83822014005000033

Beceiro A, Tomás M, Bou G (2013) Antimicrobial resistance and virulence: a successful or deleterious association in the bacterial world? Clin Microbiol Rev 26(2):185–230

Bendre AD, Peters PJ, Kumar J (2021) Tuberculosis: past, present and future of the treatment and drug discovery research. Curr Res Pharmacol Drug Discov 2:100037. https://doi.org/10.1016/j.crphar.2021.100037

Bhutani P, Joshi G, Raja N, Bachhav N, Rajanna PK, Bhutani H, Paul AT, Kumar R (2021) US FDA approved drugs from 2015–June 2020: a perspective. J Med Chem 64(5):2339–2381

Blair JM, Webber MA, Baylay AJ, Ogbolu DO, Piddock LJ (2015) Molecular mechanisms of antibiotic resistance. Nat Rev Microbiol 13(1):42–51

Bodoev I, Il'Ina E (2015) Molecular mechanisms of formation of drug resistance in *Neisseria gonorrhoeae*: history and prospects. Mol Genet Microbiol Virol 30(3):132–140

Brown ED, Wright GD (2016) Antibacterial drug discovery in the resistance era. Nature 529(7586):336–343

Bungau S, Tit DM, Behl T, Aleya L, Zaha DC (2021) Aspects of excessive antibiotic consumption and environmental influences correlated with the occurrence of resistance to antimicrobial agents. Curr Opin Environ Sci Health 19:100224. https://doi.org/10.1016/j.coesh.2020.10.012

Chen D, Liying X, Hong D, Zhao Y, Hu X, Shi S, Chen F (2021) Epidemiology of resistance of carbapenemase-producing *Klebsiella pneumoniae* to ceftazidime–avibactam in a Chinese hospital. J Appl Microbiol. https://doi.org/10.1111/jam.15166

Cheng Q, Zhao G, Wang X, Wang L, Lu M, Li Q, Wu Y, Huang Y, Jia Q, Xie L (2020) Nomogram for individualized prediction of incident multidrug-resistant tuberculosis after completing pulmonary tuberculosis treatment. Sci Rep 10(1):13730. https://doi.org/10.1038/s41598-020-70,748-x

Cheong T, Ahn J, Kim YS, Pai H, Kim B (2021) Quantitative evaluation of the economic impact of antimicrobial resistance on the treatment of community-acquired acute pyelonephritis in Korea.

Chong Y, Ito Y, Kamimura T (2011) Genetic evolution and clinical impact in extended-spectrum β-lactamase-producing *Escherichia coli* and *Klebsiella pneumoniae*. Infecti Genet Evol 11(7):1499–1504

Chua AQ, Verma M, Hsu LY, Legido-Quigley H (2021) An analysis of national action plans on antimicrobial resistance in Southeast Asia using a governance framework approach. Lancet Reg HealthWest Pac 7:100084. https://doi.org/10.1016/j.lanwpc.2020.100084

Craft KM, Nguyen JM, Berg LJ, Townsend SD (2019) Methicillin-resistant *Staphylococcus aureus* (MRSA): antibiotic-resistance and the biofilm phenotype. MedChemComm 10(8):1231–1241

Crofton J, Mitchison D (1948) Streptomycin resistance in pulmonary tuberculosis. Br Med J 2(4588):1009. https://doi.org/10.1136/bmj.2.4588.1009

Davey P, Sneddon J, Nathwani D (2010) Overview of strategies for overcoming the challenge of antimicrobial resistance. Expert Rev Clin Pharmacol 3(5):667–686

Davies J, Davies D (2010) Origins and evolution of antibiotic resistance. Microbiol Mol Biol Rev 74(3):417–433

De Oliveira DM, Forde BM, Kidd TJ, Harris PN, Schembri MA, Beatson SA, Paterson DL, Walker MJ (2020) Antimicrobial resistance in ESKAPE pathogens. Clin Microbiol Rev 33(3):e00181–e00119

Denkinger CM, Pai M, Dowdy DW. Do we need to detect isoniazid resistance in addition to rifampicin resistance in diagnostic tests for tuberculosis? PLoS One 2014; 9:e84197

Ebied AM, Patel KH, Cooper-DeHoff RM (2020) New drugs approved in 2019. Am J Med 133(6):675–678

Eliopoulos GM, Maragakis LL, Perl TM (2008) *Acinetobacter baumannii*: epidemiology, antimicrobial resistance, and treatment options. Clin Infect Dis 46 8):1254–1263
Foster TJ, Geoghegan JA (2015) Chapter 37—*Staphylococcus aureus*. In: Tang Y-W, Sussman M, Liu D, Poxton I, Schwartzman J (eds) Molecular medical microbiology, vol 2. Academic, London, pp 655–674. https://doi.org/10.1016/B978-0-12-397,169-2.00037-8
Ghai I, Ghai S (2018) Understanding antibiotic resistance via outer membrane permeability. Infect Drug Resist 11:523
Global tuberculosis report (2020) Geneva: World Health Organization; 2020
Government of India (2017) National action plan on antimicrobial resistance (NAP-AMR) 2017–2021. Ministry of Health & Family Welfare, New Delhi
Gulati M, Nobile CJ (2016) *Candida albicans* biofilms: development, regulation, and molecular mechanisms. Microbes Infect 18(5):310–321
Gupta V, Datta P (2019) Next-generation strategy for treating drug resistant bacteria: antibiotic hybrids. Indian J Med Res 149(2):97
Gupta N, Vats S, Bhargava P (2018) Sustainable agriculture: role of metagenomics and metabolomics in exploring the soil microbiota. In: Choudhary DK, Kumar M, Prasad R, Kumar V (eds) In silico approach for sustainable agriculture. Springer, Singapore, pp 183–199
Hancock RE, Sahl H-G (2006) Antimicrobial and host-defense peptides as new anti-infective therapeutic strategies. Nat Biotechnol 24(12):1551–1557
Heselpoth RD (2014) Engineering enhanced structural stability to the streptococcal bacteriophage endolysin PlyC. Dissertation, University of Maryland
Hughes CC, Fenical W (2010) Antibacterials from the sea. Chemistry 16(42):12512–12525. https://doi.org/10.1002/chem.201001279
Iredell J, Brown J, Tagg K (2016) Antibiotic resistance in Enterobacteriaceae: mechanisms and clinical implications. BMJ. 352:h6420. https://doi.org/10.1136/bmj.h6420
Isler B, Doi Y, Bonomo RA, Paterson DL (2018) New treatment options against carbapenem-resistant *Acinetobacter baumannii* infections. Antimicrob Agents Chemother 63(1):e01110–e01118
Kaushik AC, Kumar A, Bharadwaj S, Chaudhary R, Sahi S (2018) Bioinformatics techniques for drug discovery: applications for complex diseases. Springer, Cham
Keese P (2008) Risks from GMOs due to horizontal gene transfer. Environ Biosaf Res 7(3):123–149. https://doi.org/10.1051/ebr:2008014
Kelly AM, Mathema B, Larson EL (2017) Carbapenem-resistant Enterobacteriaceae in the community: a scoping review. Int J Antimicrob Agents 50(2):127–134
Kularatne R, Maseko V, Gumede L, Kufa T (2018) Trends in *Neisseria gonorrhoeae* antimicrobial resistance over a ten-year surveillance period, Johannesburg, South Africa, 2008–2017. Antibiotics 7(3):58
Lambert P (2002) Mechanisms of antibiotic resistance in *Pseudomonas aeruginosa*. J R Soc Med 95(Suppl 41):22
Langner JL, Chiang KF, Stafford RS (2021) Current prescribing practices and guideline concordance for the treatment of uncomplicated urinary tract infections in women. Am J Obstet Gynecol 225(3):272.e1–272.e11. https://doi.org/10.1016/j.ajog.2021.04.218
Larsson DJ, Andremont A, Bengtsson-Palme J, Brandt KK, de Roda Husman AM, Fagerstedt P, Fick J, Flach C-F, Gaze WH, Kuroda M (2018) Critical knowledge gaps and research needs related to the environmental dimensions of antibiotic resistance. Environ Int. 117:132–138
Lee AS, de Lencastre H, Garau J, Kluytmans J, Malhotra-Kumar S, Peschel A, Harbarth S (2018) Methicillin-resistant *Staphylococcus aureus*. Nat Rev Dis Primers 4:18033. https://doi.org/10.1038/nrdp.2018.33
Lim YM, Shin KS, Kim J (2007) Distinct antimicrobial resistance patterns and antimicrobial resistance–harboring genes according to genomic species of *Acinetobacter* isolates. J Clin Microbiol 45(3):902–905. https://doi.org/10.1128/JCM.01573-06
Liu Y, Tong Z, Shi J, Li R, Upton M, Wang Z (2021) Drug repurposing for next-generation combination therapies against multidrug-resistant bacteria. Theranostics 11(10):4910

Lupo A, Coyne S, Berendonk TU (2012) Origin and evolution of antibiotic resistance: the common mechanisms of emergence and spread in water bodies. Front Microbiol 3:18. https://doi.org/10.3389/fmicb.2012.00018

Makovitzki A, Avrahami D, Shai Y (2006) Ultrashort antibacterial and antifungal lipopeptides. Proc Natl Acad Sci 103(43):15997–16,002

Mattar C, Edwards S, Baraldi E, Hood J (2020) An overview of the global antimicrobial resistance research and development hub and the current landscape. Curr Opin Microbiol 57:56–61

Mesfin EA, Beyene D, Tesfaye A, Admasu A, Addise D, Amare M, Dagne B, Yaregal Z, Tesfaye E, Tessema B (2018) Drug-resistance patterns of *Mycobacterium tuberculosis* strains and associated risk factors among multi drug–resistant tuberculosis suspected patients from Ethiopia. PLoS One 13(6):e0197737

Méthot P-O, Alizon S (2014) What is a pathogen? Toward a process view of host–parasite interactions. Virulence 5(8):775–785

Michael CA, Dominey-Howes D, Labbate M (2014) The antimicrobial resistance crisis: causes, consequences, and management. Front Public Health 2:145

Micoli F, Bagnoli F, Rappuoli R, Serruto D (2021) The role of vaccines in combatting antimicrobial resistance. Nat Rev Microbiol 19:287–302. https://doi.org/10.1038/s41579-020-00506-3

Mukherjee A, Reddy MS (2020) Metatranscriptomics: an approach for retrieving novel eukaryotic genes from polluted and related environments. 3 Biotech 10:71. https://doi.org/10.1007/s13205-020-2057-1

Munita JM, Arias CA (2016) Mechanisms of antibiotic resistance. In: Kudva IT, Cornick NA, Plummer PJ, Zhang Q, Nicholson TL, Bannantine JP, Bellaire BH (eds) Virulence mechanisms of bacterial pathogens, 5th edn. ASM Press, Washington, DC, pp 481–511. https://doi.org/10.1128/9781555819286.ch17

Nazir H, Aziz M, Mirani ZA, Sheikh AS, Saeed MQ, Khan AA, Ruby T, Rauf N (2021) Correlation between antibiotic resistance and phylogenetic types among multidrug-resistant *Escherichia coli* isolated from urinary tract infections. Iranian J Basic Med Sci 24(3):400

Ndagi U, Falaki AA, Abdullahi M, Lawal MM, Soliman ME (2020) Antibiotic resistance: bioinformatics-based understanding as a functional strategy for drug design. RSC Adv 10(31):18451–18,468

Newell DG, Koopmans M, Verhoef L, Duizer E, Aidara-Kane A, Sprong H, Opsteegh M, Langelaar M, Threfall J, Scheutz F (2010) Food-borne diseases—the challenges of 20 years ago still persist while new ones continue to emerge. Int J Food Microbiol 139:S3–S15

Ng R (2015) Drugs: from discovery to approval, 3rd edn. Wiley, Hoboken

Niemz A, Boyle DS (2012) Nucleic acid testing for tuberculosis at the point-of-care in high-burden countries. Expert Rev Mol Diagn 12:687–701

Nobile CJ, Johnson AD (2015) *Candida albicans* biofilms and human disease. Annu Rev Microbiol 69:71–92

Olivenza DR, Nicoloff H, Sánchez-Romero MA, Cota I, Andersson DI, Casadesús J (2019) A portable epigenetic switch for bistable gene expression in bacteria. Sci Rep 9:11261. https://doi.org/10.1038/s41598-019-47,650-2

Pablos-Méndez A, Raviglione MC, Laszlo A, Binkin N, Rieder HL, Bustreo F, Cohn DL, Lambregts-van Weezenbeek CS, Kim SJ, Chaulet P, Nunn P (1998) Global surveillance for antituberculosis-drug resistance, 1994–1997. World Health Organization–International Union against Tuberculosis and Lung Disease Working Group on Anti-tuberculosis Drug Resistance Surveillance. New England Journal of Medicine. 338 (23): 1641–1649. https://doi.org/10.1056/NEJM199806043382301

Prestinaci F, Pezzotti P, Pantosti A (2015) Antimicrobial resistance: a global multifaceted phenomenon. Pathog Glob Health 109(7):309–318

Ranjalkar J, Chandy SJ (2019) India's national action plan for antimicrobial resistance—an overview of the context, status, and way ahead. J Fam Med Primary Care 8(6):1828

Rao M, Padyana S, Dipin K, Kumar S, Nayak B, Varela M (2018) Antimicrobial compounds of plant origin as efflux pump inhibitors: new avenues for controlling multidrug resistant pathogens. J Antimicrob Agents 4(1):159. https://doi.org/10.4172/2472-1212.1000159

Rath SK, Singh S, Kumar S, Wani NA, Rai R, Koul S, Khan IA, Sangwan PL (2019) Synthesis of amides from (E)-3-(1-chloro-3,4-dihydronaphthalen-2-yl)acrylic acid and substituted amino acid esters as NorA efflux pump inhibitors of *Staphylococcus aureus*. Bioorg Med Chem 27(2):343–353. https://doi.org/10.1016/j.bmc.2018.12.008

Reygaert WC (2018) An overview of the antimicrobial resistance mechanisms of bacteria. AIMS Microbiol 4(3):482

Rodríguez-González A, Zanin M, Menasalvas-Ruiz E (2019) Public health and epidemiology informatics: can artificial intelligence help future global challenges? An overview of antimicrobial resistance and impact of climate change in disease epidemiology. Yearbook Med Inform 28(1):224

Romandini A, Pani A, Schenardi PA, Pattarino GAC, De Giacomo C, Scaglione F (2021) Antibiotic resistance in pediatric infections: global emerging threats, predicting the near future. Antibiotics 10(4):393

Salmerón P, Viñado B, El Ouazzani R, Hernández M, Barbera MJ, Alberny M, Jané M, Larrosa N, Pumarola T, Hoyos-Mallecot Y (2020) Antimicrobial susceptibility of *Neisseria gonorrhoeae* in Barcelona during a five-year period, 2013 to 2017. Eurosurveillance 25(42):1900576. https://doi.org/10.2807/1560-7917.ES.2020.25.42.1900576

Shivekar SS, Kaliaperumal V, Brammacharry U, Sakkaravarthy A, Raj CV, Alagappan C, Muthaiah M (2020) Prevalence and factors associated with multidrug-resistant tuberculosis in South India. Sci Rep 10:17552. https://doi.org/10.1038/s41598-020-74,432-y

Shlaes DM, Bradford PA (2018) Antibiotics—from there to where? How the antibiotic miracle is threatened by resistance and a broken market and what we can do about it. Pathog Immunity 3(1):19

Smith SE, Kurbatova EV, Cavanaugh JS, Cegielski JP. Global isoniazid resistance patterns in rifampinresistant and rifampin-susceptible tuberculosis. Int J Tuberc Lung Dis 2012; 16:203–5

Unemo M, Shafer WM (2014) Antimicrobial resistance in *Neisseria gonorrhoeae* in the 21st century: past, evolution, and future. Clin Microbiol Rev 27(3):587–613

Veeraraghavan B, Jesudason MR, Prakasah JAJ, Anandan S, Sahni RD, Pragasam AK, Bakthavatchalam YD, Selvakumar RJ, Dhole T, Rodrigues C (2018) Antimicrobial susceptibility profiles of Gram-negative bacteria causing infections collected across India during 2014–2016: study for monitoring antimicrobial resistance trend report. Indian J Med Microbiol 36(1):32–36

Verraes C, Van Boxstael S, Van Meervenne E, Van Coillie E, Butaye P, Catry B, De Schaetzen M-A, Van Huffel X, Imberechts H, Dierick K (2013) Antimicrobial resistance in the food chain: a review. Int J Environ Res Public Health 10(7):2643–2669

Vincent MS, Uphoff S (2020) Bacterial phenotypic heterogeneity in DNA repair and mutagenesis. Biochem Soc Trans 48(2):451–462

Wang HH, Schaffner DW (2011) Antibiotic resistance: how much do we know and where do we go from here? Appl Environ Microbiol 77(20):7093–7095

WHO. (2019, April 29). New report calls for urgent action to avert antimicrobial resistance crisis. Retrieved October 30, 2021, from https://www.who.int/news-room/detail/29-04-2019-new-report-calls-for-urgent-actionto-avertantimicrobial-resistance-crisis

World Health Organization (2015) WHO global action plan to minimize poliovirus facility-associated risk after type-specific eradication of wild polioviruses and sequential cessation of oral polio vaccine use: GAPIII. World Health Organization, Geneva

Chapter 14
Chemical Diversity in Fungi: Strategies to Mitigate Antimicrobial Resistance

Neha Kapoor, Richa Bhardawaj, and Lokesh Gambhir

14.1 Introduction

The emergence of antibiotic resistance is a swiftly evolving phenomenon in microorganisms owing to the excess utilization of antibiotics. In the current scenario, antimicrobial resistance (AMR) is considered as the prime threat to public health worldwide because of the increasing emergence of resistance observed in pathogenic bacteria. Acquiring AMR at a fastidious pace by many pathogenic microorganisms, which are no longer susceptible to the available antibiotics, requires an urgent attention for the development of effective preventive and treatment regimes. There exists an imperative need to cater to the threat of growing AMR emerging as a global crisis in healthcare system. The World Health Organization (WHO) has long been documented the need for an improved and synchronized comprehensive approach to combat the AMR. *Staphylococcus aureus* is a major pathogen contributing to the escalation in AMR incidence with the emergence of multidrug-resistant strains such as methicillin-resistant (MRSA) and vancomycin-resistant (VRSA) *S. aureus* capable of causing various infections viz. bacteremia, pneumonia, skin, soft tissue, endocarditis, and osteoarticular infections (Kumar and Schweizer 2005; Tong et al. 2015). In the context of hospital-acquired infection, *Enterococcus faecalis*, MRSA, *Klebsiella pneumonia, Pseudomonas aeruginosa*, and extended-spectrum β-lactamase-producing *Escherichia coli* (ESBL-producing *E. coli*) are reported to secrete a diverse array of resistance and virulence factors.

A significant rise in antibiotic-resistant Gram-negative bacteria, *Pseudomonas aeruginosa*, isolated from clinical setup (Fridkin and Gaynes 1999) has been well

N. Kapoor · L. Gambhir (✉)
School of Applied Sciences, Suresh Gyan Vihar University, Jaipur, Rajasthan, India

R. Bhardawaj
Department of Botany, IIS University, Jaipur, Rajasthan, India

N. Akhtar et al. (eds.), *Emerging Modalities in Mitigation of Antimicrobial Resistance*,
https://doi.org/10.1007/978-3-030-84126-3_14

documented for being responsible for 10% of total hospital-acquired infections (Nagaveni et al. 2011). There are many routes via which microorganisms confer resistance against various antimicrobial agents. The main cornerstone treatment method entails the utilization of β-lactam class of antibiotics against a range of infectious diseases. Many pathogenic counterparts such as *Enterobacterales* (Paterson 2008*) are exhibiting resistance to* β-lactam antibiotics which can be channelized through efflux pumps, distorted antibiotic binding sites, and enzyme β-lactamases activity (Peirano and Pitout 2019). Cephalosporins and carbapenems are inactivated by many pathogenic microorganisms due to the cleavage of β-lactamase-mediated β-lactam ring (Pfeifer 2010). AMR has also found to be acquired through increased expression of certain genes and novel mutational events in few microbes (Barlow and Hall 2002).

According to the report of the Center for Disease Control and Prevention (CDC), a sharp increase in the mortality rate of approximately 23,000 deaths per year on an average was observed due to MDR-based infections. The growing burden of AMR has prompt the implementation of comprehensive, coordinated national and international action plans. The foremost financial report deciphering the impact of AMR gave a frightening picture depicting the exponential rise in AMR-related deaths from 700,000 to 10 million annually by 2050, if the issue remains unaddressed (O'Neill 2014, De Kraker et al. 2016). It would cost trillions of USD in the healthcare industry. The World Bank estimates that an additional 28 million people could be forced into extreme poverty by 2050, through shortfalls in economic output, unless resistance is contained. It is hypothesized that drug-resistant diseases could kill more people than cancer. As per the population modeling analysis to ascertain the burden of infections caused by antibiotic-resistant bacteria among EU and European Economic Area (EEA) in 2015, the overt effects of infections with MDR bacteria predominantly crop up in hospitals (63.5% infection cases) and other healthcare settings claiming approximately 72.4% of attributable deaths and 74.9% of DALYs per 100,000 population (Cassini et al. 2019). The inappropriate exploitation of antimicrobial agents for purposes other than treatment of infections has contributed in the selection for AMR in food production environments.

The inadequacy of appropriate strategies to mitigate growing AMR prevalence is posing adverse healthcare effects among patients. Hence, conventional treatment remedies of MDR infections are facing enormous challenge and are a matter of serious concern among healthcare facilities. Keeping in consideration the challenges being faced by existing antibiotics against MDR, the healthcare setup is in dire need of the development of novel, broad-spectrum, and effective scaffolds for their probable exploitation as antibiotics or innovative strategies to surmount antibiotic resistance.

14.2 Strategies for Antimicrobial Resistance Management

The rapid emergence of antibiotic-resistant bacteria in the last few decades seeks immediate, innovative research and development strategies. After the serendipitous discovery of penicillin by Sir Alexander Fleming, the golden age for the development of many more antibiotics against deadly pathogens was marked. However, all of these amazing discoveries can impose adverse effects because microbial pathogens gradually evolve new ways to dodge the antibacterial drugs. AMR bacteria were categorized into three classes viz. critical, high, and medium based on their extent of resistance, mortality rate, and treatability as per the list proposed by WHO 2017. An alarming and frightening condition persists in infections caused by the Gram-negative bacteria (ESKAPE) viz. *Enterococcus faecium, Staphylococcus aureus, Klebsiella pneumoniae, Acinetobacter baumanii, Pseudomonas aeruginosa*, and *Enterobacter.* These bacteria display resistance against numerous antibiotic drugs of the current pipeline including carbapenems and third-generation cephalosporins, being considered as last-resort antibiotics. *Mycobacterium tuberculosis*, the causative agent of tuberculosis, is also included in the category of MDR bacteria and is being considered a top global health priority. In the last two decades, antibiotics therapy is facing major setbacks because of the absence of any antimicrobials in the pipeline. It is imperative to conserve the efficacy of antibiotic therapy and to refuel the current pipeline of antibacterial drugs. Further, the development of new drugs is very slow due to a lack of investment. Researchers are continuously investigating ways to reduce the incidence of antimicrobial resistance. Detailed insight into the exact mechanism by which bacteria confer resistance can provide a foundation for the development of novel treatment strategies. The researchers are in persistent search to reinforce the existing antimicrobial arsenal to conquer MDR bacterial defenses either by advocating fresh approaches/sources or by exploration of novel activity profiling of old drugs. However, the development of newer drugs seems to face both economic and scientific obstacles due to reduced profit and being derivative of previous drugs faces resistance from bacteria respectively.

Based on the above facts, it has become vital to hunt for novel and more effective antimicrobial drugs with potent scaffold and mechanism of action. Across the globe, scientific community is investing in huge efforts to unravel the new resources for the probable discovery of new moieties to combat antimicrobial resistance. In order to address the problem of AMR, World Health Organization (WHO) in agreement with members of Food and Agriculture Organization (FAO) and World Organization for Animal Health endorsed a joint Global Action Plan on Antimicrobial Resistance in year 2015. Furthermore, Natural products still remain the most vital reserve for the discovery of new and potential drug molecules (Newman and Cragg 2020, Strobel and Daisy 2003). For many decades, microorganisms are reported to manufacture diverse antibiotics that function in an antagonistic capacity in nature and most of the antibacterial agents in clinical or preclinical trials are either microbial products or analogs (Huang and Lin 2017, Pham et al. 2019). Henceforth, exploration of antimicrobial compounds from microorganisms is a promising way to convene the mounting risk of drug-resistant strains of human and plant pathogens.

14.3 Fungi as a Goldmine of Novel Chemistries for Curbing Antimicrobial Resistance

For more than a decade, numerous antimicrobial compounds have been discovered from fungi, with many more antimicrobials still being put ahead for clinical investigation every year (Saxena et al. 2019). Fungi are an enormously diverse group of organisms, with about 230,000 species distributed widely essentially in every ecosystem. Fungi are considered to be a persuasive reservoir of novel bioactive compounds with antimicrobial potency against currently drug-resistant strains of pathogenic microorganisms (Table 14.1). Five decades ago, a fungal metabolite, Aspergillomarasmine A was discovered which was more recently repurposed based on its capacity to combat the antibiotic resistance associated with metallo-β-lactamase7. Two Azaphilone compounds viz. penicilazaphilone B and C, purified from *P. sclerotiorum* M-22 inhabiting rotted leaf was reported to exhibit potent antibacterial activity ESBL *K. pneumoniae* ATCC 700603 with MIC values of 500 and 15.63 μg/ml respectively (Zhou et al. 2016). Another study highlighted the antimicrobial action of xylitol and oxysporone recovered from *Pestalotia* spp., associated with the mangrove plant, *Heritiera fomes* against MRSA with MIC value of 32–128 μg/ml (Nurunnabi et al. 2018).

Three new cerebrosides, alternarosides A–C, and new diketopiperazine alkaloid alternarosin A, purified from ethyl acetate extract of halotolerant *Alternaria raphani* exhibited antibacterial activity against *Escherichia coli, Bacillus subtilis*, and *Candida albicans*, with MIC values ranging from 70 to 400 μM (Wang et al. 2009). As per the study conducted by Nogueira et al. 2006, ethyl acetate fraction of phytopathogenic fungi *Diplodia maydis* and *Sclerotium rolfsii* recovered from the culture of mays and soya respectively exhibited potent antibacterial activity on the multidrug-resistant bacteria *Acinetobacter baumannii, Enterococcus cloacae, Klebsiella pneumoniae, Proteus mirabilis*, and *Staphylococcus aureus.* Of both the pathogenic fungi, *S. rolfsii* was observed to exert much potent activity against MDR *Acinetobacter baumannii, Klebsiella pneumonia,* followed by *Shigella* spp. and *Staphylococcus aureus.*

In the study documented by Arivudainambi et al. (2014a, b), ethyl acetate extract of endophytic fungus *Pestalotiopsis virgatula* VN2 and *Alternaria alternata VN3* isolated from the medicinal plant *Vitex negundo* L exert promising antibacterial activity against multidrug-resistant strains of *Staphylococcus aureus* and *Pseudomonas aeruginosa.* A partially purified fraction E from ethyl acetate extract of *P. virgatula* was found to inhibit *S. aureus* strain 4 with a zone size of 19.8 mm. Similarly, ethyl acetate extracts of three endophytes, including *Guignardia*, and *Cladosporium* recovered from the leaves of mangrove plant *Aegiceras corniculatum* displayed promising inhibitory activity against MDR bacteria under in vitro conditions. *Colletotrichum* spp. and *Guignardia* spp. inhibited MDR, *K. pneumoniae* with MIC of 4 and 8 μg/ml respectively. Further, *Colletotrichum* also exhibited promising inhibitory activity against *A. baumanii* with MIC of 0.5 μg/ml (Bin et al. 2014).

Table 14.1 Brief listing of natural potent bioactive compounds potent in mitigating the antimicrobial resistance isolated from diverse fungi

S. No	Name of compound	Source fungi	Target organism	References
1.	Sphaeropsidin A	*D. corticola*	• Methicillin-resistant *S. aureus* (MRSA) • *P. aeruginosa*	Roscetto et al. (2020)
2.	Fonsecinone A	*Aspergillus* spp. *Z120*	• ESBL-producing *E. coli* • *P. aeruginosa* • *E. faecalis*	He et al. (2016)
3.	Gliotoxin	*Aspergillus fumigatus*	• Methicillin-resistant *S. aureus* (MRSA) 2.08 • ESBL-producing *E. coli* 4.09 • Vancomycin-resistant *Enterococcus faecalis* VRE van A V4932	Stefan Svahn et al. (2012)
4.	• Diaporthin • Orthosporin	*Diaporthe terebinthifolii* LGMF907	Methicillin-sensitive *Staphylococcus aureus*, and methicillin-resistant *S. aureus*	de Medeiros et al. (2018)
5.	• Siccayne • Deacetoxyyanuthone A	*Penicillium* spp.	• Gram-positive bacteria • MRSA • Multidrug resistant	Li et al. (2005)
6.	Naphtho-γ-pyrones	*Penicillium* spp. HK1-22	MRSA strains	Zheng et al. (2019)
7.	Chlorohydroaspyrones A and B	*Exophiala* spp.	*S. aureus*, methicillin-resistant *S. aureus*, and multidrug-resistant *S. aureus*.	Zhang et al. (2008)
8.	Penicillstressol, isopenicillstressol, and 0Z-isocitreoviridinol (0.5–1 μg/ml)	*Penicillium* spp.	MRSA	Auckloo et al. 2017
9.	Chaetoglobosin A Chaetoglobosin C	*Chaetomium globosum* *Aspergillus fumigatus* isolate AF3-093A	MRSA ATCC 33591 *S. aureus* ATCC 43300 *Staphylococcus aureus*, methicillin-resistant *S. aureus*, and *Mycobacterium tuberculosis* H37R	(Dissanayake et al. 2016; Flewelling et al. 2015)
10.	((3S)-3,8-dihydroxy-6,7-dimethyl-a-tetralone (3))	*Cladosporium* spp. JJM22	MRSA CMCC(B) 63303	Wu et al. (2018)
11.	(22E, 24R)5, 8-Epidioxy-5a, 8a-ergosta-6,22E-dien-3ß-ol	*Cytospora* spp.	MRSA GIM1.771	Deng et al. (2018)

(continued)

Table 14.1 (continued)

S. No	Name of compound	Source fungi	Target organism	References
12.	Penicipyrrodiether A Pyrrospirone J	*Penicillium* spp. ZZ380	MRSA ATCC 43300 MRSA	Song et al. (2019)
13.	(*Z*)-Octadec-9-enamide (oleamide)	*Penicillium* spp. ArCSPf	MRSA	Farha and Hatha (2019)
14.	Versicolorin B Averufin	*A. versicolor* MF180151	MRSA	Hu et al. (2019)
15.	-cytochalasa-5(6),13-diene-1,21-dione-7,18-dihydroxy-16,18-dimethyl-10-phenyl-(7S*, 13E, 16S*, 18R*)	*Daldinia eschscholtzii* HJ001	MRSA ATCC 33591	Yang et al. (2018)

Two novel compounds Xanalteric acids I and II along with 11 known secondary metabolites purified from an endophytic fungus *Alternaria* spp. from the mangrove plant *Sonneratia alba* were evaluated for antimicrobial efficacy. Xanalteric acids I and II were observed to exert weak antibacterial activity against multidrug-resistant *Staphylococcus aureus*, whereas alternusin displayed broad-spectrum antimicrobial activity against several additional multidrug-resistant bacterial and fungal strains (Kjer et al. 2009). A bioactive compound namely, altechromone, was isolated from endophytic fungi *Alternaira brassicicola* ML-P08, which displayed antimicrobial activity against *B. subtilis, E. coli, Pseudomonas fluorescens*, and *C. albicans*, with MIC values of 3.9, 3.9, 1.8, and 3.9 μg/ml, respectively (Gu 2009). An endophytic fungus *Chaetomium cupreum* inhabiting ornamental plant *Mussaenda luteola* was reported to produce resorcinol type of lipid with potent antibacterial activity against *Mycobacterium, E. coli* (ATCC 25922), and *S. aureus* (ATCC 25923) with MIC of 6.3 μg/ml (Shylaja et al. 2018).

There exist numerous reports documenting the immense potential of various fungal endophytes against multidrug-resistant bacteria: *Aspergillus* neobridgeri (Sadrati et al. 2020) and *Aspergillus tubingensis* (Yadav et al. 2016), *Diaporthe terebinthifolii* LGMF907 (de Medeiros et al. 2018), *Phomopsis prunorum* (Qu et al. 2020), *P. griseofulvum* TPL25 (Luo et al. 2015), *Alternaria* GFAV15 (Yadav et al. 2020), *Alternaria tenuissima* OE7, *Pestalotia* spp. (Nurunnabi et al. 2018), Penicillium spp. CPCC400817 (Qi et al. 2019) and *Trichoderma koningiopsis* QA-3 (Shi et al. 2020). Fungal endophytes namely *Aspergillus terreus* and *Trichoderma virens* were found to secrete butyrolactone I and 9-epiviridol respectively. Both of the compounds displayed antibacterial activity against methicillin-resistant *S. aureus* (MRSA) with MIC values in the range of 128–256 μg/ml (Ratnaweera et al. 2018).

Mushroom, a fruiting body of any fungus, is being given considerate importance as food commodity and due to its medicinal attributes worldwide (Hobbs 1996). There are many reports documenting the antibacterial potency of various mushrooms against MDR bacteria. Methanolic extracts of frozen fungus *Amanita virosa* (Fr) Bertill and *Cortinarius praestans* Cordier collected from Slovenia exhibited the strongest antibacterial activity against *Pseudomonas aeruginosa* and *Staphylococcus aureus* (Janes et al. 2007) The methanolic extract of fruiting bodies and mycelial biomass of *Coprinopsis cinerea* (C2), coprophilous basidiomycetous mushroom, recovered from horse dung showed significant antibacterial activity against methicillin-resistant *Staphylococcus aureus* (MRSA) and carbapenemase-producing *Klebsiella pneumoniae* with a zone of inhibition ranging between 10 and 14 mm in diameter (MohanKumar and Savitha 2019). Two farnesyl hydroquinones viz. ganomycins A and B from *Ganoderma pfeifferi* were also reported to exhibit potent antibacterial action toward *S. aureus* and *MRSA* (Mothana et al. 2000). Methanolic extract of the fruiting bodies of *Tapinella atrotomentosa* yielded osmundalatone, 5-hydroxy-hex-2-en-4-olide and spiromentin C with strong inhibiting activity against MRSA (SZMC 6270) with MIC value of 250 μg/ml (Beni et al. 2018). Similarly, fruiting body extracts of *Pleurotus sajor-caju* (Fr.) Singer also displayed antistaphylococcal activity against MSSA and MRSA with a MIC value of 10 mg/ml.

Furthermore, the potential of fungal nanoparticles against multidrug-resistant microorganism has been embroiled in the literature. In the study reported by Neethu et al. 2018, silver nanoparticles of algicolous endophytic fungus, *Penicillium polonicum*, isolated from the marine green alga *Chetomorpha antennina* was found to exert strong bactericidal potential against multidrug-resistant, biofilm-forming *Acinetobacter baumanii* with MIC as well as MBC of 15.62 μg/ml and 31.24 μg/ml respectively. In a similar kind of study, silver nanoparticles of *Penicillium italicum* isolated from wasp nest soil were reported to exhibit antibacterial and antifungal activity against multidrug bacterial pathogens like *Staphylococcus aureus*, *Vibrio parahaemolyticus*, *E. coli*, and *Shewanella putrefaciens* (Nayak et al. 2018). A recent study reported the biocidal activity of silver, copper, and zinc nanoparticles of *Fusarium solani* KJ 623702 against multidrug-resistant (MDR) *P. aeruginosa* and *S. aureus* as well as the mycotoxigenic *Aspergillus awamori, A. fumigatus,* and *F. oxysporum* based on the diameter of inhibition zone. The potent antibacterial effect was exerted by spherical AgNPs (13.70 nm) against *P. aeruginosa* with MIC of 21.33 μg/ml, whereas zinc oxide nanoparticles were found to be highly effective against *F. oxysporum* with MIC value of 24.7 μg/ml (El Sayed and El-Sayed 2020).

14.4 Strategies Adopted to Explore the Chemical Diversity of Fungi

14.4.1 Genome Mining

Genome mining refers to identifying the gene arrays responsible for producing secondary metabolites as natural products in the genomes of sequenced microorganism (Lee et al. 2020). Genome mining is entirely dependent on bioinformatics tools and analysis of the enzyme coding gene clusters with experimental verification of the expression in terms of product (Chavali and Rhee 2018). Initially, the concept of genome mining was validated in *Streptomyces coelicolor* and *Streptomyces avermitilis*, where a lot of unexplored hidden potential was explored which led to the extension of the genome mining-based chemical prospecting to other microbes (Gross 2009). Various approaches of genome mining were put forward to accomplish this task (Fig. 14.1); novel technologies including single-cell genomics and metagenomics generated enormous data to be analysed. Multiple tactics are being used at present for big-data mining in microbial communities like:

Classical genome mining: It is the explicit exploration for genes that code for enzymes that play a vital role in the biosynthesis of secondary metabolites in (meta) genomic sequences. Aided by the analysis using high-end softwares, classical genome mining revolutionized the discovery of natural products in the past decade.

Amplicon sequencing which details the composition of the targeted microbial community uses specific marker genes including bacterial 16S rRNA and fungal internal transcribed spacer (ITS). Amplicon sequencing assigns and analyses the presence of different strains or types residing in uncultured microbial community. Its disadvantage lies in its disability in functional annotation. Therefore, metagenomics sequencing is used for resolving phylogeny and functions of strains present in the microbial biome (de Fátima Alves et al. 2018). Metagenomics sequencing can still have certain disadvantages like losing the genomes of low abundant microbes and closely related microbes can be read as repeats. Solution to such issues can be answered by metagenomics synergistically with single-cell sequencing. The approach may narrate the DNA compartmentalization into cells and may connect functions to their matching species (Embree et al. 2014).

Single-cell sequencing advantageously can differentiate and authenticate the individual functions in the community. Connecting the authenticated functions to specific species also helps in generating high-quality genome for species with low abundance. It has proven pivotal in identifying bacteria with alternative genetic codes. Due to all above said advantages single-cell genomics proved to be substantial to allow meaningful data mining and its interpretation of single-cell RNA sequence. Single-cell protein analyses are now a routine in analyzing transcriptomics of the cell (Hwang et al. 2018). The relationship between transcriptomic variation and genomic variation among a population of single cells was

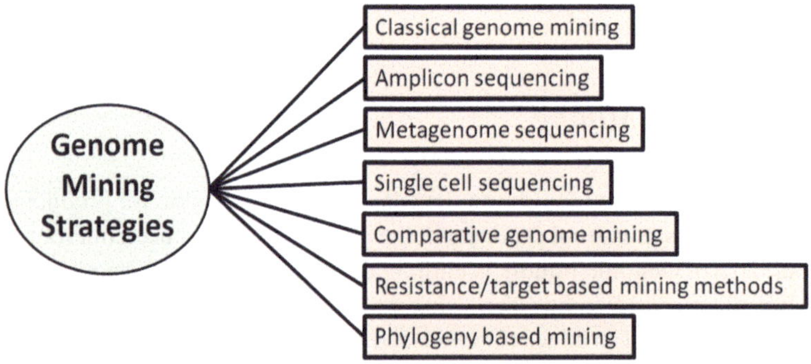

Fig. 14.1 The scheme representing the strategies adopted across the globe in genome mining-based prospecting of microbial flora for medicinal chemistries against AMR

demonstrated by multiomics single-cell sequencing—DR-seq and G&T-seq. A positive correlation was shown between gene expression and variation in the DNA copy number in individual cells. On the other hand, analysis of G&T-seq data corroborated the association among (sub) chromosomal copy number and expression level of genes located in variable regions in single cells. Both DR-seq and G&T-seq are capable in deciphering the single nucleotide variants (SNVs) in matched single-cell genomes and transcriptomes (Macaulay et al. 2017).

Metagenome sequencing: To uncover the wealth of novel genes so as to reveal the answers and understanding the processes of various biosynthetic pathways. Microbial-host metagenomics mining has proven to be significant to generate the hypothesis that can be validated experimentally. Metagenomic data mining can be used to uncover the correlations between the presence of functional genes and chemical diversity in various biosynthetic pathways (Delmont et al. 2011). The approach can also be used to retrieve the enzymes for biocatalysis and industrial use. By building an in silico metagenomic library from high-throughput sequencing data a resource is created that can be swiftly mined for enzymes (Jeffries et al. 2016; Guazzaroni et al. 2010).

Comparative genome mining: Microbial chemical diversity of high therapeutic gain needs no introduction. Genome comparison can highlight the genera which are diverse and profound producers of therapeutically and industrially important compounds. Gene networking, genome comparison can give a significant insight into the abundance of unique (unique to a single strain) and "extended families" (present in multiple strains) sequences (Shi et al. 2020). It can also identify new biosynthetic gene clusters as producers of natural compounds with potent antimicrobial activity. Comparative genomic approaches can be employed to recognize the signature genes present in the microorganism-producing antibiotic. Various algorithms have been developed to identify the gene clusters for several fungal antibiotics including searching for secondary metabolite genes of interest, like anti-SMASH, BAGEL, NP searcher, and PRISM. Mining for duplicated or altered housekeeping enzymes has also been used as strategies to predict novel biosynthetic clusters (Williams et al. 2020; Tran et al. 2019). Genome mining in combination with tools of molecular biology has paved the way for multiple novel proteins, including lantipeptides, cyanobactins, microviridins, and lasso peptides. Violaceae (violet), Rubiaceae (coffee), Cucurbitaceae, and Apocynaceae have also been mined for cyclopeptides which are head-to-tail cyclized peptides with a knotted arrangement of disulfide bonds (Hetrick and van der Donk 2017). Their structural complexity grants them bioactivities such as antibiotic, anti-cancer, and anti-inflammatory properties (Vassaux et al. 2019). Horizontal and vertical gene transfer is the underlining mechanism in the attainment and maintenance of precious secondary metabolites as confirmed by comparative genome mining which distinguished core and hypervariable regions in *Amycolatopsis* genomes (Adamek et al. 2018).

Resistance/target-based mining methods: Several potent bioactive compounds are toxic in nature that in turn results in impairment of the producing microorganism. Therefore, a self-defense mechanism is essential for the survivability of the organism. To detect secondary metabolite gene clusters based on the self-resistant mechanisms of an antibiotic, producing organism forms the basis of resistance-based genome mining. A classical example of self-resistance mechanism is the duplication of the target gene that leads to resistance to the compound. This second resistant duplicated gene is most often recognized as the culprit producer of toxic compound. FRIGG (fungal resistance gene-directed genome mining) algorithm uses homology patterns for identifying biosynthetic gene cluster (Kjærbølling et al. 2019). Resistance mechanism may vary which include efflux pumps, removal of toxic compounds by enzyme degradation, and mitigation of the binding between the antibiotic and its target (D'Costa et al. 2011; Alvarez-Ortega et al. 2011). Resistance-based genome mining approach has also been successful in cancer and antibiotic therapy (Panter et al. 2018).

Phylogeny-based mining methods: Continuum of evolution in biosynthetic gene clusters led to structural diversities in natural products. Molecular phylogeny is a frequently used technique to demarcate the evolutionary history of a gene sequence. A major advantage of the phylogenetic approach is its fastidious analysis. A single gene is used by phylogeny-based mining to prospect a plethora of microbial metagenomes to identify bioactive compounds producing biosynthetic gene clusters. In the phylogeny-guided mining approach, marker biosynthetic genes based phylogenetic trees provide a road map to prioritize biosynthetic gene clusters and subsequent molecule expression studies (Kang 2017). Two of the most common enzyme families associated with natural product biosynthesis are PKSs and NRPSs. These genes are accountable for the biological synthesis of the several bioactive metabolites of microbial origin. Polyketide and nonribosomal peptide biosynthetic pathways are multienzyme complexes that sequentially construct natural products in an assembly line process from carboxylic acid and amino acid building blocks (Miyanaga et al. 2018). The types of bacteriocins and their distribution among various phyla viz. *Actinobacteria, BV4, Firmicutes, Glidobacteria*, and *Proteobacteria* are analyzed using phylogenetic genome mining. Phylogenetic trees from these alignments are used to study evolutionary relationships between bacteriocins and producer species (Joynt and Seipke 2018).

Ribosomal RNAs and mitochondrial cytochrome oxidase genes have been exploited for establishing phylogenetic relationship between leaf beetles, longhorn beetles, and weevils (Pistone et al. 2016, Aoki et al. 2018). 16s rRNA sequences are commonly used to represent the evolution of entire microbial genomes and its speciation (Clarridge 2004). Indolocarbazoles are a family of natural precursor to multiple anticancer drugs. Its biosynthesis involves the dimerization of two oxygenated tryptophans. Due to their clinical significance, the phylogeny-guided approach was employed to discover novel indolocarbazoles from soil metagenomes (Chang et al. 2015). An improved understanding of these phylogenetic patterns can direct bioprospecting efforts in combination with the metagenomic sequencing. A web tool called Natural Product Domain Seeker (NaPDoS) provides an automated solution to measure the microbial diversity and secondary metabolite biosynthetic gene diversity. NaPDoS analysis is based on the phylogenetic relationships of sequence tags derived from polyketide synthase (PKS) and non-ribosomal peptide synthetase (NRPS) genes, respectively (Ziemert et al. 2012). Other softwares include MEGA and Seaview which produce efficient sequence alignments and phylogenetic trees (Hall 2013; Gouy et al. 2010).

14.5 Genome Mining as Tool of Novel Chemistries

Advances in genomics have resulted in the accumulation of large quantity of DNA sequence data in public databases. This availability of enormous publicly accessible data has made it possible for bioinformatics intervention to discover the novel chemistries of natural origin. This has greatly facilitated the identification and analysis of gene clusters likely to encode natural product biosynthetic pathways in sequenced genomes. However, the advancement in genomics has stimulated the development of multiple approaches and tools which were not available earlier to discover the natural bioactive chemistries (Zerikly and Challis 2009).

Biosynthetic gene cluster (BGC) analysis tools have been developed to efficiently predict the structure based on functional and architectural domains (Fig. 14.2). These tools have been used for broad profiling for modular assembly-line complexes, such as polyketide synthases (PKSs) and non-ribosomal peptide synthetases (NRPSs) (Tietz et al. 2017).

Biosynthetic gene cluster (BGC) analysis tools:

1. **anti-SMASH (antibiotics and secondary metabolite analysis shell):** Identification and annotation of secondary metabolite gene clusters. It combines automated identification of gene clusters with compound-specific analysis algorithms. It functions by analyzing the single genome (and integrates the

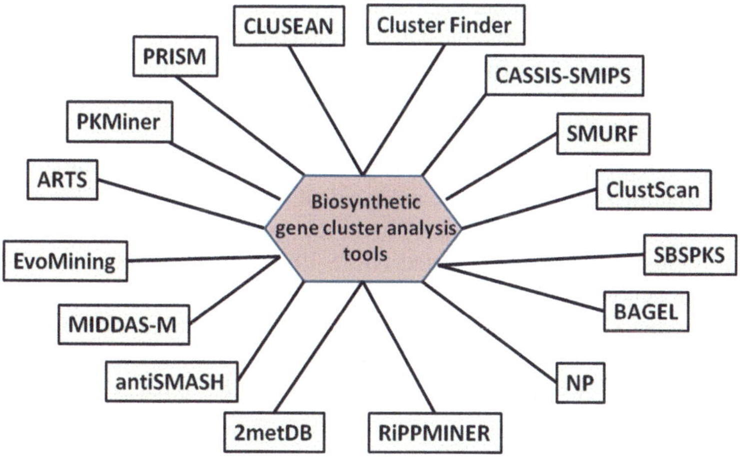

Fig. 14.2 A schematic representation of the web-based tools of biosynthetic gene cluster analysis available for identifying the secondary metabolite producing gene clusters in microorganisms based on the detection of signature gene clusters

RiPP output with analysis of other BGC classes). Along with bacterial genome mining, anti-SMASH is also used for fungal (fungiSMASH) and plant (plantiSMASH) genomes. It works by comparing encoded gene products with a manually curated library of hidden Markov models, describing a range of natural products producing biosynthetic genes (Blin et al. 2013).

2. **PRISM** (PRediction Informatics for Secondary Metabolism): Identification of nonribosomal peptides, type I and II polyketides, and RiPPs (Skinnider et al. 2017). It is used for structure prediction of nonribosomal peptides and polyketides. It executes structure prediction by recognizing biosynthetic genes attained by searching the query sequence with hidden Markov models library of secondary metabolism. It depends on the homology between experimentally characterized biosynthetic enzyme and gene clusters. PRISM is used for the analysis of prokaryotic genome and has limitation in eukaryotic setting
3. **BAGEL** (BActeriocin GEnome mining tool): It is one of the earliest tools developed for the identification of bacteriocin BGCs. BAGEL searches for RiPPs (also defined as class I bacteriocins by BAGEL), class II and III bacteriocins. BAGEL recognizes potent RiPP and bacteriocin open reading frames by employing the combination of genetic context and knowledge-based motif databases. It takes into account the processing, modification, transport, and regulation of RiPPs and bacteriocins. BAGEL2 focused on using profile hidden Markov models for bacteriocin identification led to updating of manually curated databases of known bacteriocins. BAGEL3 was aimed at discovering cyanobactins, sactipeptides, and linaridins producing biosynthetic genes (Jong et al. 2006).
4. **CLUSEAN** (CLUster SEquence ANalyzer): It is a BioPerl-based software pipeline for the annotation of gene clusters responsible for biosynthesis of antibiotics or compounds with novel bioactivities. CLUSEAN contains components for automated homology search, protein domain identification, and substrate prediction for the biosynthetic enzymes. CLUSEAN integrates standard analysis tools, like BLAST and HMMer, with specific tools for the identification of the functional domains and motifs in NRPS/PKS (Weber et al. 2009; Weber and Blin 2013).
5. **Cluster finder**: It is a probability algorithm based on hidden Markow model for predicting signature enzyme independent biosynthetic metabolic gene clusters in bacterial genomes. It is used in the recognition of gene clusters including NRPS clusters, PKS clusters, and NRPS-PKS hybrid clusters from different bacterial genomes (Cimermancic et al. 2014; Naughton et al. 2017).
6. **NP** searcher (Natural Product searcher) is an open-source software program used for prospecting genomes of microorganisms to recognize NRPS, PKS, or hybrid NRPS-PKS gene clusters with their putative bioactive product (Ren et al. 2020).
7. **ClustScan**: It is similar to NP searcher and appropriate for DNA sequences. Cluster Scanner is a commercial software package for gene cluster identification intended for NRPS, PKS, and hybrid NRPS-PKS enzymes. It also guides

the prediction of protein structure attained by the identified gene clusters (Cullum et al. 2011).

8. **SMURF** (Secondary Metabolite Unknown Regions Finder) is dedicated for the analyses of fungal genomes. It dictates cluster boundaries by employing combination of physical genomic distance and enzymes (Khaldi et al. 2010).
9. **CASSIS-SMIPS** (Cluster Assignment by Island of Sites) is also dedicated for fungal genome mining. It exploits the principle of high-density promoter regions in clustered genes compared to non-clustered genes. SMIPS (Secondary Metabolites by InterProScan) is used to predict anchor genes based on domains encoding biosynthetic enzymes in eukaryotic genomic sequences (Wolf et al. 2016).
10. **EvoMining**: It is based on the analyses of evolutionary theory and phylogenomics for predicting secondary metabolite-producing gene clusters in bacterial genomes (Sélem-Mojica et al. 2019).
11. **MIDDAS-M** (Motif-Independent De novo Detection Algorithm for Secondary Metabolite) gene clusters is another platform aimed for the detection of metabolite gene clusters in fungal genome in absence of signature enzymes (Chavali and Rhee 2018).
12. **ARTS** (Antibiotic Resistant Target Seeker) provides a platform for swift recognition of potent biosynthetic gene clusters. It is based on the cumulated analysis of target-directed genome mining and antibiotic gene cluster predictions. It depicts housekeeping genes shoeing resistance owing to gene duplication, colocalization with gene clusters, and horizontal gene transfer. Further advanced versions of ARTS allow the analysis of bacterial kingdom and its comparison with other genomes (Direnc et al. 2020; Alanjary et al. 2017).
13. **2metDB**: Genome mining for polyketides and nonribosomal peptides. 2metDB is a unique tool that offers the mining of PKS and NRPS gene clusters in whole-genome protein FASTA files. It combines the biochemical pattern and gene sequence knowledge to produce translated gene sequence data. Thus, it can be readily used to decipher the structural metabolites coded by the gene clusters (Bachmann and Ravel 2019).
14. **PKMiner**: Genome mining for type II polyketide synthases. PKMiner identifies novel BCGs of PKSs and polyketides based on aromatase and cyclise domains. It is used extensively in exploring novel polyketides that are synthesized by type II polyketide synthases and exhibits high pharmacological significance (Kim and Yi 2012).
15. **SBSPKS**: Sequence analysis of polyketide synthases. SBSPKS is used for multiple chemical analyses of experimentally characterized BGCs encoding PKS/NRPS. Its updated versions provide a platform for structural comparison of a secondary metabolite with intermediate and final products of BGCs. It employs profile-based searches based on HMM. It provides tools of identifying condensation, epimerization, and cyclization domains of NRPS (Khater et al. 2017).
16. **RiPPMINER**: It is used for genome mining and deciphering chemical structures of RiPPs, and BGCs. RiPPMiner predicts chemical structures based on predecessor peptides of lanthipeptides, lasso peptides, cyanobactins, and thio-

peptides. It contains two major modules based on peptides and genomes. RiPPMiner-peptide and genome takes sequence data as input and identifies class, structure, crosslinks, and cleavage sites for RiPP families and BGCs respectively (Agrawal et al. 2017).

14.6 Some Important Compound Analyzed and Identified Using Genome Mining Tools

14.6.1 Lasso Peptides

Lasso peptides are a lariat knot-shaped class of natural products present in bacteria that undergoes post-translational modification. Lasso peptides have a spectrum of bioactivities. It acts as receptor antagonist, inhibits enzyme and bacterial growth. The BGC of lasso peptides includes a total of three genes out of which one codes for a precursor protein and two for enzymes. Various genome mining approaches have been used to identify diverse lasso peptides in bacterial and archaeal phylum. Expression of gene cluster in *Escherichia coli* led to the production of an astexin-1 lasso peptide (Maksimov et al. 2012). The discovery of capistruin produced by *Burkholderia thailandensis* in 2008 marks the initiation point of genome mining era of lasso peptides. The precursor-centric approach and genome mining based on mass spectrometry have been the successful strategies adopted for lasso peptide prospecting (Maksimov and Link 2014). Ortholog neighborhood-based prospecting of mcjB-like genes predicted hypothetical lasso peptides producing multiple gene clusters in *Bacillus, Burkholderia, Caulobacter*, and *Sphingopyxis* (Duquesne et al. 2007).

14.6.2 Bacteriocin

Bacteriocins are group of antimicrobial peptides produced by bacteria, these are ribosomally synthesized peptides. Bacteriocin is a non-growth-associated product. However, it is widely abundant in bacteria owing to its genetic determinants. Phylogeny-based genome mining methods have been used to analyze and identify bacteriocins in various microbial biomes (Kjos et al. 2011). Microbial genome mining resulted in the discovery of 145 bacteriocin gene clusters coding for 290 potent bacteriocin precursors. Seven groups have been identified in the mentioned gene clusters based on comparative genomics (Wang et al. 2011). BAGEL3 uses an identification approach that combines direct mining and context genes-based indirect mining for lanthipeptides, sactipeptides, and glycocins (Jong et al. 2010; van Heel et al. 2013; Javan et al. 2018; Egan et al. 2018).

14.6.3 Cyanobactins

Cyanobactins are a cyclic ribosomal peptides produced by cyanobacteria. Genome mining predicted a horizontal transfer and precursor gene diversification dictated production of complex cyanobactins. Cyanobactin gene clusters are present in the *Oscillatoria, Arthrospira*, and *Microcystis* genera whereas absent from other generas including *Prochlorococcus* and *Synechococcus*. Phylogeny-based genome mining elucidated that cyanobactin biosynthetic gene clusters are grouped together into a single well-resolved clade, i.e., 31 cyanobactin gene clusters in a set of 126 cyanobacterial genomes (Leikoski et al. 2013).

14.6.4 Methanobactins

Methanobactins (Mbns) are a family of copper-binding bioactive compound in methanotrophic bacteria. Mbns are secreted under copper starvation conditions and then re-internalized as a copper source for the enzyme particulate methane monooxygenase. Genome mining studies have led to the identification and classification of operons with a number of putative transports, regulatory and biosynthetic proteins (Kenney and Rosenzweig 2013). Genome mining revealed the presence of Mbn-like precursor peptides encoding operons in 16 new species of both methanotrophs and non-methanotrophs. Aminotransferases, sulfotransferases, and flavin adenine dinucleotide (FAD)-dependent oxidoreductases were some other BGCs identified in these families (Laura et al. 2017).

14.6.5 Polyketide

Polyketides are produced by frequent decarboxylation and condensation cycles on polyketide synthases enzymes. Polyketides have been distributed across multiple and diverse bacterial genomes. Polyketides are diverse pharmacologically relevant secondary metabolites with exciting ecological functions. Polyketides include macrolides, enediynes, polyphenols, polyenes, and numerous other structural scaffolds. Many clinically relevant bioactive agents, such as antibiotics, immunosuppressants, cytotoxins, and cholesterol-lowering substances, are derived from polyketides. Bacillus and related genera in the Bacillales are remarkable producers of highly diverse polyketides (PKs). Bacillaene, Difficidin, Macrolactin, Paenimacrolidin, and Paenilamicin are some PKS isolated and identified from bacterial genomes (Boddy 2014; Helfrich et al. 2014; Aleti et al. 2015; Crawford et al. 2008; Chooi et al. 2013).

14.7 Genome Mining in Biosynthetic Gene Cluster-Based Families to Map the Pathway

Biosynthetic gene clusters (BGCs) are prearranged groups of genes concerned with the production of secondary specialized metabolites. A single BGC can produce one or multiple (up to over 100) comparable bioactive compounds with varying strength and/or specificity. Genome mining is the often prime move toward examining the potency of a microorganism to produce novel bioactive metabolites. BGCs can be associated with "superclusters" that function to produce two or more similar molecules. Access to whole-genome sequence data has highlighted the general occurrence of the cryptic BGCs, thereby renewing their attention for natural product discovery (Martinet et al. 2019). Different structural classes of BGCs exist including non-ribosomal peptide synthetases (NRPS), polyketide synthases (PKS), terpenes, and bacteriocins. NRPS and PKS are popular targets for natural product discovery as they are known to synthesize a diversity of antibiotics and immunosuppressants with enormous pharmaceutical potential (Chen et al. 2020).

Transcriptomics in combination with metabolite analysis of different plant tissues, developmental stages, and/or elicitor-treated material is also being used to great effect to identify candidate genes for the synthesis of natural products. To identify groups of BGCs that are functionally closely related and encode the production of the same or very similar molecules, approaches have been developed to group BGCs into gene cluster families (GCFs). By examining shared absence/presence patterns of GCFs and compound families (derived via molecular networking of MS/MS spectra across different microbial strains), one can connect BGCs to their expressed products (Kautsar et al. 2021). BiG-SLiCE is a tool that makes it possible to perform GCF reconstruction at very large scales in a computationally feasible way. BiG-FAM can be used to explore the biosynthetic repertoires of specific taxa, investigate the taxonomic and architectural diversity of BGCs of known function, and obtain insights into the novelty of newly sequenced BGCs. Minimum Information about a Biosynthetic Gene cluster (MIBiG) repository provides rich reference data to connect these GCFs to known products based on gene cluster homology, even with clusters from different fungal genera (Satria et al. 2021).

Many effective in silico-guided strategies have been designed using genetics, substrate labeling, and screening for predicted physicochemical properties to characterize novel medicinal chemistries from cryptic gene clusters (Kerstena et al. 2013). One of the comprehensively studied biosynthetic gene clusters from the microbiome is the colibactin pathway. The colibactin pathway has been implicated in colorectal tumorigenesis in mice models. The colibactin gene cluster is a ~55 kb BGC cluster that produces a family of polyketide-non-ribosomal peptide. This gene cluster is found among the *Enterobacteriaceae*, including *Escherichia coli, Citrobacter koseri, Klebsiella pneumoniae*, and *Enterobacter aerogenes*. The "pathway-targeted" molecular networking is used to more finely map expressed secondary metabolic pathways within complex metabolomes to aid in secondary metabolite identification and characterization (Vizcaino and Crawford 2016).

14.8 Conclusions

Resistance to antibiotic is ever evolving and escalating problem due to population burden, overuse of antibiotics in clinical settings, genetic selection pressure, and poor solid waste. Researchers across the globe are diligently working to find the solution or alternate strategy to curb the mounted resistance. Among the prime strategies adopted is to isolate a novel natural compound from the diverse chemical signatures of plants and microbes. Among this, the relevance of fungi as a repository of novel medicinal chemistries has been persistently substantiated in the literature. Multiple compounds have been isolated and characterized from fungi showcasing the promising potency to curb the antimicrobial resistance. The presented chapter provides an in-depth insight on the myriad of compounds put forth from fungal repository along with its potency to mitigate AMR. Another aspect of the chapter highlights the importance of the advanced strategy, i.e. genome mining adopted to prospect the microbial flora for the ability to produce antibiotics based on the detection of signature secondary metabolite genes. Along with the identification of the gene clusters, multiple comparisons with the curated cluster and compound database are required by using web-based tools. Plethora of advanced tools are available that are used for gene cluster identification, screening for patterns in gene regulation or content. A biosynthetic gene cluster analysis tool that provides an analysis of structure based on functional and architectural domain has been used for broad-scale profiling in deciphering the secondary metabolite gene clusters. The extensive work in the field of genome mining has led to the deposition of a large quantity of DNA sequence data in databases. This has led to the foundation of building tools that helped in potentiating the discovery of natural products.

Antimicrobial resistance in the microbial pathogens is a major challenge for scientific community that poses a threat of morbidity and mortality. Antimicrobial resistance is a major challenge in developing countries compared to developed countries owing to the inappropriate use of easily available antibiotics. Sincere efforts are direly needed to mount a mitigating response against this microbial resistome. The importance of novel chemistries in the present chapter is a promising feat in highlighting genome mining as one of the strategies of prospecting microorganisms for developing novel therapeutic chemistries.

Conflict of Interest There is no actual or potential conflict of interest among authors.

Author's Contribution LG conceived and designed the context of the chapter. RB and NK contributed in summarizing the genome mining and fungal chemical diversity respectively. NK, RB, and LG wrote the manuscript chapter.

References

Adamek M, Alanjary M, Sales-Ortells H, Goodfellow M, Bull AT, Winkler A, Wibberg D, Kalinowski J, Ziemert N (2018) Comparative genomics reveals phylogenetic distribution patterns of secondary metabolites in Amycolatopsis species. BMC Genomics 19:426

Agrawal P, Khater S, Gupta M, Sain N, Mohanty D (2017) RiPPMiner: a bioinformatics resource for deciphering chemical structures of RiPPs based on prediction of cleavage and cross-links. Nucleic Acids Res 45(W1):W80–W88

Alanjary M, Kronmiller B, Adamek M, Blin K, Weber T, Huson D, Philmus B, Ziemert N (2017) The Antibiotic Resistant Target Seeker (ARTS), an exploration engine for antibiotic cluster prioritization and novel drug target discovery. Nucleic Acids Res 45(W1):W42–W48

Aleti G, Sessitsch A, Brader G (2015) Genome mining: prediction of lipopeptides and polyketides from Bacillus and related Firmicutes. Comput Struct Biotechnol J 13:192–203

Alvarez-Ortega C, Wiegand I, Olivares J, Hancock REW, Martínez JL (2011) The intrinsic resistome of Pseudomonas aeruginosa to b-lactams. Virulence 2(2):144–146

Aoki T, Kasson MT, Berger MC, Freeman S, Geiser DM, O'Donnell K (2018) Fusarium oligoseptatum sp. nov., a mycosymbiont of the ambrosia beetle Euwallacea validus in the Eastern U.S. and typification of F. ambrosium. Fungal Syst Evol 1:23–39

Arivudainambi USE, Kanugula KA, Kotamraju S, Karunakaran C, Rajendran A (2014a) Cytotoxic and antibacterial activities of secondary metabolites from endophytic fungus *Pestalotiopsis virgatula* VN2. Curr Res Environ Appl Mycol 4(1):107–115

Arivudainambi USE, Kanugula AK, Kotamraju S, Karunakaran C, Rajendran A (2014b) Antibacterial effect of an extract of the endophytic fungus *Alternaria alternata* and its cytotoxic activity on MCF-7 and MDA MB-231 tumour cell lines. Biol Lett 51(1):7–17

Auckloo BN, Pan C, Akhter N, Wu B, Wu X, He S (2017) Stress-driven discovery of novel cryptic antibiotics from a marine fungus *Penicillium* sp. BB1122. Front Microbiol 8:1450. https://doi.org/10.3389/fmicb.2017.01450

Bachmann BO, Ravel J (2019) Methods for in silico prediction of microbial polyketide and nonribosomal peptide biosynthetic pathways from DNA sequence data. Methods Enzymol 181–217

Barlow M, Hall BG (2002) Origin and evolution of the AmpC β-lactamases of *Citrobacter freundii*. Antimicrob Agents Chemother 46:1190–1198. https://doi.org/10.1128/AAC.46.5.1190-1198.2002

Beni Z, Dekany M, Kovacs B, Csupor-Loffler B, Zomborszki ZP, Kerekes E, Vanyolos A (2018) Bioactivity-guided isolation of antimicrobial and antioxidant metabolites from the mushroom *Tapinella atrotomentosa*. Molecules 23:1082

Bin G, Yanping C, Hong Z, Zheng X, Yanqiu Z, Huaiyi F, Qiupin Z, Chenxiao Z (2014) Isolation, characterization and anti-multiple drug resistant (MDR) bacterial activity of endophytic fungi isolated from the mangrove plant, Aegiceras corniculatum. Trop J Pharm Res 13(4):593

Blin K, Medema MH, Kazempour D, Fischbach MA, Breitling R, Takano E, Weber T (2013) antiSMASH 2.0—a versatile platform for genome mining of secondary metabolite producers. Nucleic Acids Res 41(W1):W204–W212

Boddy CN (2014) Bioinformatics tools for genome mining of polyketide and non-ribosomal peptides. J Ind Microbiol Biotechnol 41:443–450

Cassini A, Högberg LD, Plachouras D, Quattrocchi A, Hoxha A, Simonsen GS, Colomb-Cotinat M, Kretzschmar ME, Devleesschauwer B, Cecchini M, Ouakrim DA, Oliveira TC, Struelens MJ, Suetens C, Monnet DL, Burden of AMR Collaborative Group (2019) Attributable deaths and disability-adjusted life-years caused by infections with antibiotic-resistant bacteria in the EU and the European Economic Area in 2015: a population-level modelling analysis. Lancet Infect Dis 19(1):56–66. https://doi.org/10.1016/S1473-3099(18)30605-4. Epub 2018 Nov 5. PMID: 30409683; PMCID: PMC6300481

Chang FY, Ternei MA, Calle PY, Brady SF (2015) Targeted metagenomics: finding rare tryptophan dimer natural products in the environment. J Am Chem Soc 137(18):6044–6052

Chavali AK, Rhee SY (2018) Bioinformatics tools for the identification of gene clusters that biosynthesize specialized metabolites. Brief Bioinform 19(5):1022–1034
Chen R, Wong HL, Kindler GS, MacLeod FL, Benaud N, Ferrari BC, Burns BP (2020) Discovery of an abundance of biosynthetic gene clusters in Shark Bay microbial mats. Front Microbiol 11:1950
Cheung-Lee WL, Link AJ (2019) Genome mining for lasso peptides: past, present, and future. J Ind Microbiol Biotechnol 46(9–10):1371–1379
Chooi YH, Fang J, Liu H, Filler SG, Wang P, Tang Y (2013) Genome mining of a prenylated and immunosuppressive polyketide from pathogenic fungi. Org Lett 15(4):780–783
Cimermancic P, Medema MH, Claesen J (2014) Insights into secondary metabolism from a global analysis of prokaryotic biosynthetic gene clusters. Cell 158(2):412–421
Clarridge JE III (2004) Impact of 16S rRNA gene sequence analysis for identification of bacteria on clinical microbiology and infectious diseases. Clin Microbiol Rev 17(4):840–862
Crawford JM, Vagstada AL, Ehrlich KC, Townsend CA (2008) Starter unit specificity directs genome mining of polyketide synthase pathways in fungi. Bioorg Chem 36(1):16–22
Cullum J, Starcevic A, Diminic J, Zucko J, Long PF, Hranueli D (2011) *ClustScan*: an integrated program package for the detection and semiautomatic annotation of secondary metabolite clusters in genomic and metagenomic DNA datasets. In: de Bruijn FJ (ed) Handbook of molecular microbial ecology I: metagenomics and complementary approaches, pp 423–432
D'Costa VM, King CE, Kalan L, Morar M, Sung WWL, Schwarz C, Froese D, Zazula G, Calmels F, Debruyne R, Golding GB, Poinar HN, Wright GD (2011) Antibiotic resistance is ancient. Nature 477(7365):457–461
De Kraker MEA, Stewardson AJ, Harbarth S (2016) Will 10 million people die a year due to antimicrobial resistance by 2050? PLoS Med 13(11):e1002184. https://doi.org/10.1371/journal.pmed.1002184
Delmont TO, Malandain C, Prestat E, Larose C, Monier JM, Simonet P, Vogel TM (2011) Metagenomic mining for microbiologists. ISME J 5:1837–1843
Deng Q, Li G, Sun M, Yang X, Xu J (2018) A new antimicrobial sesquiterpene isolated from endophytic fungus *Cytospora* sp. from the Chinese mangrove plant Ceriops tagal. Nat Prod Res 34:1404–1408. https://doi.org/10.1080/14786419.2018.1512993
Direnc M, Mungan AM, Blin K, Weber T, Medema MH, Ziemert N (2020) ARTS 2.0: feature updates and expansion of the antibiotic resistant target seeker for comparative genome mining. Nucleic Acids Res 48:W546–W552
Dissanayake RK, Ratnaweera PB, Williams DE, Dilrukshi Wijayarathne C, Wijesundera RLC, Andersen RJ, de Silva ED (2016) Antimicrobial activities of endophytic fungi of the Sri Lankan aquatic plant Nymphaea nouchali and chaetoglobosin A and C, produced by the endophytic fungus Chaetomium globosum. Mycology 7(1):1–8. https://doi.org/10.1080/21501203.2015.1136708
Duquesne S, Destoumieux-Garzon D, Zirah S, Goulard C, Peduzzi J, Rebuffat S (2007) Two enzymes catalyze the maturation of a lasso peptide in Escherichia coli. Chem Biol 14:793–803
Egan K, Field D, Ross RP, Cotter PD, Hill C (2018) In silico prediction and exploration of potential bacteriocin gene clusters within the bacterial genus Geobacillus. Front Microbiol 9:2116
El Sayed MT, El-Sayed ASA (2020) Biocidal activity of metal nanoparticles synthesized by Fusarium solani against multidrug-resistant bacteria and Mycotoxigenic fungi. J Microbiol Biotechnol 30(2):226–236
Embree M, Nagarajan H, Movahedi N, Chitsaz H, Zengler K (2014) Single-cell genome and metatranscriptome sequencing reveal metabolic interactions of an alkane-degrading methanogenic community. ISME J 8:757–767
Farha AK, Hatha AM (2019) Bioprospecting potential and secondary metabolite profile of a novel sediment-derived fungus Penicillium sp. ArCSPf from continental slope of eastern Arabian Sea. Mycology. 2019. Jun 10(2):109–117
de Fátima Alves L, Westmann CA, Lovate GL, de Siqueira GMV, Borelli TC, Guazzaroni M-E (2018) Metagenomic approaches for understanding new concepts in microbial science. Int J Genomics 2018:2312987

Flewelling AJ, Bishop AI, Johnson JA, Graya CA (2015) Polyketides from an endophytic *Aspergillus fumigates* isolate inhibit the growth of Mycobacterium tuberculosis and MRSA. Nat Prod Commun 10(10):2015

Fridkin SK, Gaynes RP (1999) Antimicrobial resistance in intensive care units. Clin Chest Med 20(2):303–16, viii. https://doi.org/10.1016/s0272-5231(05)70143-x. PMID: 10386258

Gouy M, Guindon S, Gascuel O (2010) SeaView version 4: a multiplatform graphical user interface for sequence alignment and phylogenetic tree building. Mol Biol Evol 27(2):221–224

Gross H (2009) Genomic mining--a concept for the discovery of new bioactive natural products. Curr Opin Drug Discov Devel 12(2):207–219

Gu W (2009) Bioactive metabolites from *Alternaria brassicicola* ML-P08, an endophytic fungus residing in *Malus halliana*. World J Microbiol Biotechnol 25:1677. https://doi.org/10.1007/s11274-009-0062-y

Guazzaroni ME, Vieites YBM, Al-ramahi N, Cortés AL, Ghazi P, Ferrer GN (2010) Metagenomic mining of enzyme diversity. In: Timmis KN (ed) Handbook of hydrocarbon and lipid microbiology. Springer, Berlin, pp 2911–2927

Hall BG (2013) Building phylogenetic trees from molecular data with MEGA. Mol Biol Evol 30(5):1229–1235

He Y, Tian J, Chen X, Sun W, Zhu H, Li Q, Lei L, Yao G, Xue Y, Wang J, Li H, Zhang Y (2016) Fungal naphtho-γ-pyrones: Potent antibiotics for drug-resistant microbial pathogens. Scientific Reports 6, 24291. https://doi.org/10.1038/srep24291

van Heel AJ, Jong AD, Montalbán-López M, Kok J, Kuipers OP (2013) BAGEL3: automated identification of genes encoding bacteriocins and (non-)bactericidal posttranslationally modified peptides. Nucleic Acids Res 41(W1):W448–W453

Helfrich EJN, Reiter S, Piel J (2014) Recent advances in genome-based polyketide discovery. Curr Opin Biotechnol 29:107–115

Hetrick KJ, van der Donk WA (2017) Ribosomally synthesized and post-translationally modified peptide natural product discovery in the genomic era. Curr Opin Chem Biol 38:36–44

Hobbs C (1996) Introduction. In: Medicinal mushrooms; an exploration of tradition, healing & culture, 3rd edn. Interweave Press, Loveland, pp 1–2

Hu J, Li Z, Gao J, He H, Dai H, Xia X, Liu C, Zhang L, Song F (2019) New diketopiperazines from a marine-derived fungus strain Aspergillus versicolor MF180151. Mar Drugs 17:262. https://doi.org/10.3390/md17050262

Huang T, Lin S (2017) Microbial natural products: a promising source for drug discovery. J Appl Microbiol Biochem 1(2):5

Hwang B, Lee JH, Bang D (2018) Single-cell RNA sequencing technologies and bioinformatics pipelines. Exp Mol Med 50:1–14

Janeš D, Kreft S, Jurc M, Seme K, Štrukelj B (2007) Antibacterial activity in higher fungi (mushrooms) and endophytic fungi from Slovenia. Pharm Biol 45(9):700–706. https://doi.org/10.1080/13880200701575189

Javan RR, van Tonder AJ, King JP, Harrold CL, Brueggemann AB (2018) Genome sequencing reveals a large and diverse repertoire of antimicrobial peptides. Front Microbiol 9:2012

Jeffries JWE, Dawson N, Orengo C, Moody TS, Quinn DJ, Hailes HC, Ward JM (2016) Metagenome mining: a sequence directed strategy for the retrieval of enzymes for biocatalysis. Chemistry Select 1:2217–2220

Jong AD, van Hijum SAFT, Bijlsma JJE, Kok J, Kuipers OP (2006) BAGEL: a web-based bacteriocin genome mining tool. Nucleic Acids Res 34:W273–W279

Jong AD, van Heel AJ, Kok J, Kuipers OP (2010) BAGEL2: mining for bacteriocins in genomic data. Nucleic Acids Res 38:W647–W651

Joynt R, Seipke RF (2018) A phylogenetic and evolutionary analysis of antimycin biosynthesis. Microbiology 164:28–39

Kang HS (2017) Phylogeny-guided (meta)genome mining approach for the targeted discovery of new microbial natural products. J Ind Microbiol Biotechnol 44(2):285–293

Kautsar SA, Blin K, Shaw S, Weber T, Medema MH (2021) BiG-FAM: the biosynthetic gene cluster families database. Nucleic Acids Res 49(D1):D490–D497

Kenney GE, Rosenzweig AC (2013) Genome mining for methanobactins. BMC Biol 11:17

Kerstena RD, Ziemerta N, Gonzalez DJ, Dugganc BM, Nizet V, Dorrestein PC, Moore BS (2013) Glycogenomics as a mass spectrometry-guided genome-mining method for microbial glycosylated molecules. PNAS 110:E4415

Khaldi N, Seifuddin FT, Turner G et al (2010) SMURF: genomic mapping of fungal secondary metabolite clusters. Fungal Genet Biol 47(9):736–741. https://doi.org/10.1016/j.fgb.2010.06.003

Khater S, Gupta M, Agrawal P, Sain N, Prava J, Gupta P, Grover M, Kumar N, Mohanty D (2017) SBSPKSv2: structure-based sequence analysis of polyketide synthases and non-ribosomal peptide synthetases. Nucleic Acids Res 45(W1):W72–W79

Kim J, Yi GS (2012) PKMiner: a database for exploring type II polyketide synthases. BMC Microbiol 8(12):169

Kjærbølling I, Tammi Vesth T, Andersen MR (2019) Resistance gene-directed genome mining of 50 Aspergillus species. mSystems 4(4):e00085–e00019

Kjer J, Wray V, Edrada-Ebel R, Ebel R, Pretsch A, Lin W, Proksch P (2009) Xanalteric acids I and II and related phenolic compounds from an endophytic *Alternaria* sp. isolated from the mangrove plant *Sonneratia alba*. J Nat Prod 72(11):2053–2057

Kjos M, Borrero J, Opsata M, Birri DJ, Holo H, Cintas LM, Snipen L, Hernández PE, Nes IF, Diep DB (2011) Target recognition, resistance, immunity and genome mining of class II bacteriocins from gram-positive bacteria. Microbiology 157:3256–3267

Kumar A, Schweizer HP (2005) Bacterial resistance to antibiotics: active efflux and reduced uptake. Adv Drug Deliv Rev 57(10):1486–1513. https://doi.org/10.1016/j.addr.2005.04.004

Laura MKD, Kenney GE, Rosenzweig AC (2017) Methanobactins: from genome to function. Metallomics 9:7–20

Lee N, Hwang S, Kim J, Cho S, Palsson B, Cho BK (2020) Genome mining approaches for the identification of secondary metabolite biosynthetic gene clusters in Streptomyces. Comput Struct Biotechnol J 18:1548–1556

Leikoski N, Liu L, Jokela J, Wahlsten M, Gugger M, Calteau A, Permi P, Kerfeld CA, Sivonen K, Fewer DP (2013) Genome mining expands the chemical diversity of the Cyanobactin family to include highly modified linear peptides. Chem Biol 20:1033–1043

Li X, Kim MK, Lee U, Kim SK, Kang JS, Choi HD, Son BW (2005) Myrothenones A and B, cyclopentenone derivatives with tyrosinase inhibitory activity from the marine-derived fungus Myrothecium sp. Chem Pharm Bull (Tokyo). 53(4):453–455. https://doi.org/10.1248/cpb.53.453

Luo ZP, Lin HY, Ding WB, He HL, Li YZ (2015) Phylogenetic diversity and antifungal activity of endophytic fungi associated with Tephrosia purpurea. Mycobiology. 43:435–43

Macaulay LC, Ponting CP, Voet T (2017) Single-cell multiomics: multiple measurements from single cells. Trends Genet 33(2):155–168

Maksimov MO, Link AJ (2014) Prospecting genomes for lasso peptides. J Ind Microbiol Biotechnol 41(2):333–344

Maksimov MO, Pelczer I, Link AJ (2012) Precursor-centric genome-mining approach for lasso peptide discovery. PNAS 109(38):15223–15228

Martinet L, Naômé A, Deflandre B, Maciejewska M, Tellatin D, Tenconi E, Smargiasso N, de Pauw E, van Wezel GP, Rigalia S (2019) A single biosynthetic gene cluster is responsible for the production of Bagremycin antibiotics and ferroverdin iron chelators. MBio 10(4):e01230–e01219

de Medeiros AG, Savi DC, Mitra P, Shaaban KA, Jha AK, Thorson JS, Rohr J, Glienke C (2018) Bioprospecting of Diaporthe terebinthifolii LGMF907 for antimicrobial compounds. Folia Microbiol 63(4):499–505. https://doi.org/10.1007/s12223-018-0587-2

Miyanaga A, Kudo F, Eguchi T (2018) Protein-protein interactions in polyketide synthase-nonribosomal peptide synthetase hybrid assembly lines. Nat Prod Rep 35(11):1185–1209. https://doi.org/10.1039/c8np00022k. PMID: 30074030

MohanKumar S, Savitha J (2019) Antibacterial activity of extracts of *Coprinopsis Cinerea*, a coprophilous fungus against multidrug resistant nosocomial pathogens. Clin Microbiol Res 2(1):1–8. https://doi.org/10.31487/j.CMR.2019.01.01

Mothana RA, Jansen R, Julich WD, Lindequist U (2000) Ganomycins A and B, new antimicrobial farnesyl hydroquinones from the basidiomycete *Ganoderma pfeifferi*. J Nat Prod 63:416–418

Mudalungu CM, Richter C, Wittstein K, Abdalla MA, Matasyoh JC, Stadler M, Suussmuth RD (2016) Laxitextines A and B, cyathane xylosides from the tropical fungus *Laxitextum incrustatum*. J Nat Prod 79:894

Nagaveni S, Rajeshwari H, Oli AK, Patil SA, Chandrakanth RK (2011) Widespread emergence of multidrug resistant Pseudomonas aeruginosa isolated from CSF samples. Indian J Microbiol 51(1):2–7. https://doi.org/10.1007/s12088-011-0062-0

Naughton LM, Romano S, O'Gara F, Dobson ADW (2017) Identification of secondary metabolite gene clusters in the Pseudovibrio genus reveals encouraging biosynthetic potential toward the production of novel bioactive compounds. Front Microbiol 8:1494

Nayak BK, Nanda A, Prabhakar V (2018) Biogenic synthesis of silver nanoparticle from wasp nest soil fungus, Penicillium italicum and its analysis against multi drug resistance pathogens. Biocatal Agric Biotechnol 16:412–418

Neethu S, Midhun SJ, Radhakrishnan EK, Jyothis M (2018) Green synthesized silver nanoparticles by marine endophytic fungus *Penicillium polonicum* and its antibacterial efficacy against biofilm forming, multidrug-resistant *Acinetobacter baumanii*. Microb Pathog 116:263–272. https://doi.org/10.1016/j.micpath.2018.01.033

Newman DJ, Cragg GM (2020) Natural products as sources of new drugs over the nearly four decades from 01/1981 to 09/2019. J Nat Prod 83(3):770–803. https://doi.org/10.1021/acs.jnatprod.9b01285

Nogueira MA, Diaz G, Andrioli W, Falconi FA, Stangarlin JR (2006) Secondary metabolites from Diplodia maydis and Sclerotium rolfsii with antibiotic activity. Braz J Microbiol 37(1):14–16. https://doi.org/10.1590/S1517-83822006000100002

Nurunnabi TR, Nahar L, Al-Majmaie S, Rahman SMM, Sohrab MH, Billah MM, Ismail FMD, Rahman MM, Sharples GP, Sarker SD (2018) Anti-MRSA activity of oxysporone and xylitol from the endophytic fungus *Pestalotia* sp. growing on the Sundarbans mangrove plant Heritiera fomes. Phyther Res 32:348–354

O'Neill J (2014) Antimicrobial Resistance: Tackling a crisis for the health and wealth of nations. London: Review on Antimicrobial Resistance. https://wellcomecollection.org/works/rdpck35v

Panter F, Krug D, Baumanna S, Müller R (2018) Self-resistance guided genome mining uncovers new topoisomerase inhibitors from myxobacteria. Chem Sci 9:4898–4908

Paterson DL (2008) Impact of antibiotic resistance in gram-negative bacilli on empirical and definitive antibiotic therapy. Clin Infect Dis 47(suppl 1):S14–S20. https://doi.org/10.1086/590062

Peirano G, Pitout JDD (2019) Extended-spectrum β-lactamase-producing *Enterobacteriaceae*: update on molecular epidemiology and treatment options. Drugs 79:1529–1541. https://doi.org/10.1007/s40265-019-01180-3

Pfeifer Y (2010) Resistance to cephalosporins and carbapenems in Gram-negative bacterial pathogens. Int J Med Microbiol 300:371–379. https://doi.org/10.1016/j.ijmm.2010.04.005

Pham JV, Yilma MA, Adriana F, Majid MT, Nicholas M, Walker JR, Eunji K, Je CH, Reynolds JM, Chong SM, Ryeol PS, Joon YY (2019) A review of the microbial production of bioactive natural products and biologics. Front Microbiol 10:1404

Pistone D, Mugu S, Jordal BH (2016) Genomic mining of phylogenetically informative nuclear markers in bark and Ambrosia beetles. PLoS One 11(9):e0163529

Qi X, Li X, Zhao J, He N, Li Y, Zhang T, Wang S, Yu L, Xie Y (2019) GKK1032C, a new alkaloid compound from the endophytic fungus Penicillium sp. CPCC 400817 with activity against methicillin-resistant S. aureus. J Antibiot (Tokyo) 72:237–240

Qu HR, Yang WW, Zhang XQ, Lu ZH, Deng ZS, Guo ZY, CaoF, Zou K, Proksch P (2020) Antibacterial bisabolane sesquiterpenoids and isocoumarin derivatives from theendophytic fungus Phomopsis prunorum Phytochem Lett 37:1–4

Ratnaweera PB, Chandula WR, Jayasundera KU, Herath SD, Abira S, Williams DE, Andersen RJ, Dilip de Silva E (2018) Antibacterial activities of endophytic fungi isolated from six Sri Lankan plants of the family Cyperaceae. Bangladesh J Pharmacol 13(3):264–272

Ren H, Shi C, Zhao H (2020) Computational tools for discovering and engineering natural product biosynthetic pathways. iScience 23(1):100795. https://doi.org/10.1016/j.isci.2019.100795

Roscetto E, Masi M, Esposito M, Di Lecce R, Delicato A, Maddau L, Calabrò V, Evidente A, Catania MR (2020) Anti-Biofilm Activity of the Fungal Phytotoxin Sphaeropsidin A Against Clinical Isolates of Antibiotic-Resistant Bacteria. Toxins (Basel). 12(7):444 https://doi.org/10.3390/toxins12070444

Sadrati N, Zerroug A, Demirel R, Bakli S, Harzallah D (2020) Antimicrobial activity of secondary metabolites produced by *Aspergillus neobridgeri* isolated from *Pistacia lentiscus* against multi-drug resistant bacteria. Bangladesh J Pharmacol 15:82–95

Satria AK, Blin K, Shaw S, Weber T, Medema MH (2021) BiG-FAM: the biosynthetic gene cluster families database. Nucleic Acids Res 49(D1):D490–D497

Saxena S, Chhibber M, Singh IP (2019) Fungal bioactive compounds in pharmaceutical research and development. Curr Bioactive Compounds 15:211

Sélem-Mojica N, Aguilar C, Gutiérrez-García K, Martínez-Guerrero CE, Barona-Gómez F (2019) EvoMining reveals the origin and fate of natural product biosynthetic enzymes. Microb Genom 5(12):e000260. https://doi.org/10.1099/mgen.0.000260

Shi XS, Li HL, Li XM, Wang DJ, Li X, Meng LH, Zhou XW, Wang BG (2020) Highly oxygenated polyketides produced by Trichoderma koningiopsis QA-3, an endophytic fungus obtained from the fresh roots of the medicinal plant Artemisia argyi. Bioorg Chem 94:103448

Skinnider MA, Merwin NJ, Johnston CW, Magarvey NA (2017) PRISM 3: expanded prediction of natural product chemical structures from microbial genomes. Nucleic Acids Res 45(W1):W49–W54

Song T, Tang M, Ge H, Chen M, Lian X, Zhang Z (2019) Novel bioactive penicipyrroether A and pyrrospirone J from the marine-derived Penicillium sp. ZZ380. Mar Drugs 17:292

Stefan Svahn K, Göransson U, El-Seedi H, Lars B, Joakim Larsson DG, Olsen B, Chryssanthou E (2012) Antimicrobial activity of filamentous fungi isolated from highly antibiotic-contaminated river sediment. Infect Ecol Epidemiol 2(1):11591. https://doi.org/10.3402/iee.v2i0.11591

Strobel G, Daisy B (2003) Bioprospecting for Microbial Endophytes and Their Natural Products. Microbiologyand Molecular Biology Reviews, 67:491–502. https://doi.org/10.1128/MMBR.67.4.491-502.

Shylaja G, Sasikumar K, Sathiavelu A (2018) Antimycobacterial potential of resorcinol type lipid isolated from Chaetomium cupreum, an endophytic fungus from Mussaenda luteola. Bangladesh J Pharmacol 13:114–119

Tietz JI, Schwalen CJ, Patel PS, Maxson T, Blair PM, Tai H-C, Zakai UI, Mitchell DA (2017) A new genome-mining tool redefines the lasso peptide biosynthetic landscape. Nat Chem Biol 13:470–485

Tong SYC, Davis JS, Eichenberger E, Holland TL, Fowler Jr VG (2015) *Staphylococcus aureus* infections: epidemiology, pathophysiology, clinical manifestations, and management. Clin Microbiol Rev 28(3):603–661. https://doi.org/10.1128/CMR.00134-14

Tran PN, Yen M-R, Chiang C-Y, Lin H-C, Chen P-Y (2019) Detecting and prioritizing biosynthetic gene clusters for bioactive compounds in bacteria and fungi. Appl Microbiol Biotechnol 103:3277–3287

Vassaux A, Meunier L, Vandenbol M, Baurain D, Fickers P, Jacques P, Leclère V (2019) Nonribosomal peptides in fungal cell factories: from genome mining to optimized heterologous production. Biotechnol Adv 37(8):107449

Vizcaino MI, Crawford JM (2016) Secondary metabolic pathway-targeted metabolomics. Methods Mol Biol 1401:175–195

Walsh CJ, Guinane CM, Hill C, Ross RP, O'Toole PW, Cotter PD (2015) In silico identification of bacteriocin gene clusters in the gastrointestinal tract, based on the Human Microbiome Project's reference genome database. BMC Microbiol 15:183

Wang W, Wang Y, Tao H, Peng X, Liu P, Zhu W (2009) Cerebrosides of the halotolerant fungus *Alternaria raphani* isolated from a sea salt field. J Nat Prod 72(9):1695–1698

Wang H, Fewer DP, Sivonen K (2011) Genome mining demonstrates the widespread occurrence of gene clusters encoding bacteriocins in Cyanobacteria. PLoS One 6(7):e22384

Weber T, Blin K (2013) CLUSEAN, overview. In: Nelson K (ed) Encyclopedia of metagenomics. Springer, New York. https://doi.org/10.1007/978-1-4614-6418-1_706-4

Weber T, Rausch C, Lopeza P, Hoof I, Gaykovab V, Huson DH, Wohlleben W (2009) CLUSEAN: a computer-based framework for the automated analysis of bacterial secondary metabolite biosynthetic gene clusters. J Biotechnol 140:13–17

Williams AN, Sorout N, Cameron AJ, Stavrinides J (2020) The integration of genome mining, comparative genomics, and functional genetics for biosynthetic gene cluster identification. Front Genet 11:600116

Wolf T, Shelest V, Nath N, Shelest E (2016) CASSIS and SMIPS: promoter-based prediction of secondary metabolite gene clusters in eukaryotic genomes. Bioinformatics 32(8):1138–1143

Wu JT, Zheng CJ, Zhang B, Zhou XM, Zhou Q, Chen GY, Zeng ZE, Xie JL, Han CR, Lyu JX (2018) Two new secondary metabolites from a mangrovederived fungus Cladosporium sp. JJM22. Nat Prod Res 33:34–40

Yadav M, Yadav A, Kumar S, Yadav JP (2016) Spatial and seasonal influences on culturable endophytic mycobiota associated with different tissues of *Eugenia jambolana* Lam. and their antibacterial activity against MDR strains. BMC Microbiol 16:44

Yadav V, Singh A, Mathur N, Yadav R (2020) Isolation and characterization of Alternaria GFAV15, an endophytic fungus from green fruit of Tinospora cordifolia (Willd) Miers from semi-arid region, South African Journal of Botany, 134:343–348. https://doi.org/10.1016/j.sajb.2020.04.003

Yang LJ, Liao HX, Bai M, Huang GL, Luo YP, Niu YY, Zheng CJ, Wang CY (2018) One new cytochalasin metabolite isolated from a mangrove-derived fungus Daldinia eschscholtzii HJ001. Nat Prod Res 32:208–213. https://doi.org/10.1080/14786419.2017.1346641

Zerikly M, Challis GL (2009) Strategies for the discovery of new natural products by genome mining. Chem Biol Chem 10:625–633

Zhang D, Yang X, Kang JS, Choi HD, Son BW (2008) Chlorohydroaspyrones A and B, antibacterial aspyrone derivatives from the marine-derived fungus *Exophiala* sp. J Nat Prod 71:1458–1460

Zhang X, Wang TT, Xu QL, Xiong Y, Zhang L, Han H, Xu K, Guo WJ, Xu Q, Tan RX, Ge HM (2018) Genome mining and comparative biosynthesis of meroterpenoids from two phylogenetically distinct fungi. Angew Chem 130(27):8316–8320

Zheng Y, Liang Z, Shen N, Liu W, Zhou X-J, Fu X-M, Chen M, Wang C-Y (2019) New naphtho-γ-pyrones isolated from marine-derived fungus Penicillium sp. HK1-22 and their antimicrobial activities. Mar Drugs 17:322. https://doi.org/10.3390/md17060322

Zhou S, Wang M, Zhao H, Huang Y, Lin Y, Tan G, Chen S (2016) Penicilazaphilone C, a new antineoplastic and antibacterial azaphilone from the marine fungus Penicillium sclerotiorum. Arch Pharm Res 39:1621–1627

Ziemert N, Podell S, Penn K, Badger JH, Allen E, Jensen PR (2012) The natural product domain seeker NaPDoS: a phylogeny based bioinformatic tool to classify secondary metabolite gene diversity. PLoS One 7(3):e34064

Chapter 15
Phage Therapy as an Alternative Treatment in the Fight Against AMR: Real-World Problems and Possible Futures

Rajni Kaur and Nidhi Sethi

15.1 Introduction

The emergence of resistance amongst antimicrobials has raised grave concerns within the scientific community to further develop antibiotics to fight pathogens. The worldwide rise and spread of new resistant strains of microbes have undermined our capacity to treat common infections, which has resulted in delayed recovery, disability and death (https://www.who.int/news-room/fact-sheets/detail/antibiotic-resistance). The development of antimicrobial resistance (AMR) is a continuous process that evolves over time. After a certain period of time, the antibiotic drug loses its effectiveness towards the same pathogenic microbe (such as bacteria, fungi and parasites) it was previously effective. Antimicrobial resistance has led to a lack of availability of efficacious antibiotics and consequently increasing the risk of infectious spreads, severe illnesses and deaths. The year 2019 report released by the United Nations Ad hoc Interagency Coordinating Group (UN IACG) predicts that antimicrobial resistance could cause ten million deaths every year by 2050. Economically, antimicrobial resistance could constrain up to 24 million individuals to extreme poverty by 2030. Currently, at least 700,000 deaths occur every year globally due to drug-resistant diseases. Data show that only in the United States (US), 223,900 cases of antibiotic resistance-driven *Clostridioides difficile* had been reported in 2017, of which there were 12,800 casualties (https://www.cdc.gov/drugresistance/biggest-threats). Antibiotic resistance also adds to various complications in vulnerable patients going through chemotherapy, surgery, joint replacement

R. Kaur
Biomedical Department, University of Windsor, Ontario, Canada
e-mail: rkaur_msc17@thapar.edu

N. Sethi (✉)
Bio-organic and Photochemistry Laboratory, Department of Pharmaceutical Sciences, Guru Nanak Dev University, Amritsar, India

N. Akhtar et al. (eds.), *Emerging Modalities in Mitigation of Antimicrobial Resistance*,
https://doi.org/10.1007/978-3-030-84126-3_15

and dialysis, and is considered a huge threat to the public health sector of developing countries and throughout the world. Global data from the year 2017 reflect that around ten million cases of tuberculosis (TB), an infectious disease caused by *Mycobacterium tuberculosis*, had been reported which included 558,000 cases of rifampicin-resistant patients, of which 82% were multi-drug resistant (MDR). There has always been greater number of TB cases (100 per 100,000 or higher) in underdeveloped and developing countries like sub-Saharan Africa, India, and Southeast Asia due to poor understanding of prescription or treatment, inappropriate choices and insufficient dosing of antibiotics, which attributes to their misuse and further adds to the increase in antibiotic resistance (Park et al. 2019). Extreme use owes to easy accessibility of antimicrobial medications as the same can easily be purchased without the prescription of a qualified health professional. It has been clinically proven that prolonged and unscrupulous use of antimicrobials is probably the main reason behind the development and spread of antimicrobial-resistant infections. Other factors include exceptionally vulnerable immunosuppressed patients (cancer patients, transplant receipt and AIDS patients), delicate elderly patients, or long stay in hospitals (Prestinaci et al. 2015). The inability of antibiotics to treat infectious diseases showcases the uncertainty towards the future of the healthcare system. Hence, there lies a constant need to design and develop new antibiotics that could guard us against widespread microbes causing infections and find alternative strategies to reduce the burden on the current overexploited antibiotic therapy. Several therapies have been developed in the fight against antimicrobial resistance. These include probiotics, prebiotics, synbiotics, bacteriocins, SMAMPs (synthetic mimics of antimicrobial peptides), antimicrobial peptides, IDR (innate defence regulator) peptides, peptidomimetics, bacteriophages, vaccines and immunoglobulins, antibacterial oligonucleotides foldamers, antibacterial nucleic acids and immune stimulation by P4 peptide, etc., each one offering its own benefits and limitations. Amongst the mentioned alternatives, this chapter focuses on bacteriophages and how they can be efficiently used clinically to cure bacterial diseases in the form of phage therapy, covering all the aspects of this alternative treatment.

15.2 Antimicrobial Resistance (AMR): The Crisis

Antimicrobial resistance is a natural phenomenon that occurs when microbes are constantly exposed to antibiotic drugs. Whenever an antibiotic is used against disease-causing microbes, the most susceptible microbes are either inhibited or killed, while some are resilient and have a natural resistance to the attack, as expressed in the species. Significant factors contributing to AMR include the unscrupulous use of antimicrobials, patient incompetence to follow the prescribed therapy properly, and unavailability of new drugs within a particular class of antimicrobials that could replace the ineffective ones. Broadly, resistance can be of two types: intrinsic resistance and acquired resistance. In the case of intrinsic resistance, microbes do not possess the target for the antibiotic, while in the case of acquired resistance, bacteria device a mechanism that allows them to escape the action of the antibacterial drug molecule. Acquired resistance arises through the acquisition of host genetic material or mutations. These two forms of resistance are the most prominent ones and impart greater chances for the survival and multiplication of pathogen (Annunziato 2019; Prestinaci et al. 2015). Antibiotics have consistently been considered the most significant discovery over the last 70 years. Antibiotics are agents used to prevent and treat microbial infections in humans. Between 1950 and 1970, the use of antibiotics formally began with the so-called "golden age" of antimicrobial drugs. Unfortunately, the beginning of the antibiotics era tragically corresponds to the rise of the phenomenon of antimicrobial resistance (Fig. 15.1).

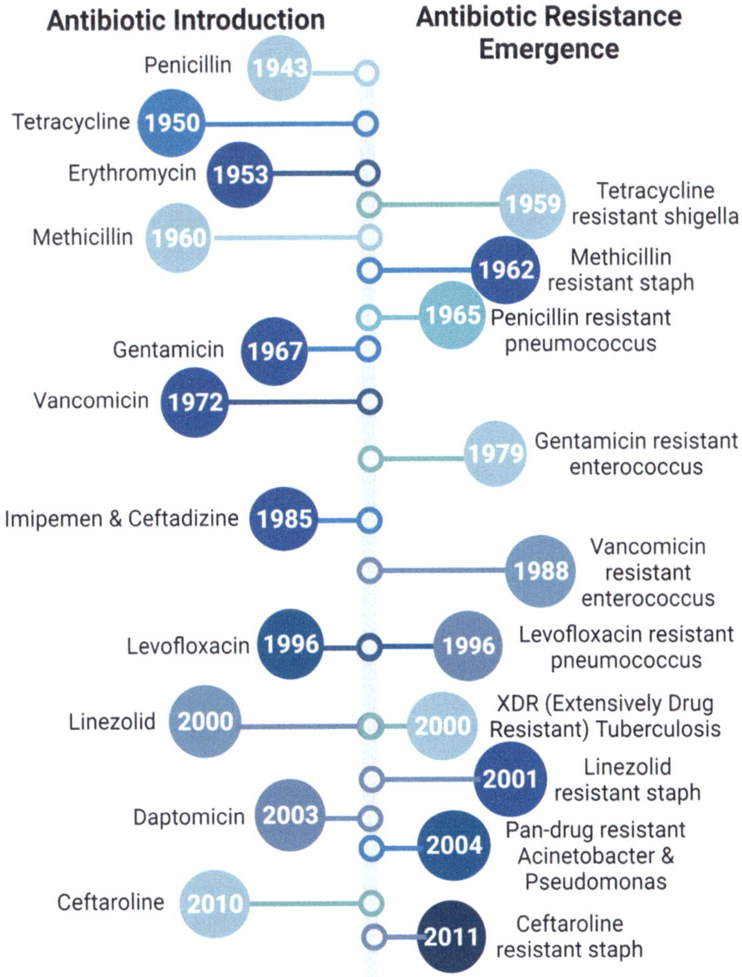

Fig. 15.1 Diagram showing major milestones in the journey of antibiotics and evolution of antimicrobial resistance (Annunziato 2019)

Since the discovery of penicillin, the first antibiotic, Alexander Flemming had expressed concerns and cautioned the scientific community for the higher demand of antibiotics which later could lead to their abuse. This has been widely acknowledged now that, while the utilization of antimicrobials has prompted the control and even elimination of infectious disease; their abuse and/or overuse has led to the development of resistant strains. After a couple of years of the golden age of antimicrobials, an alarming sign of developing resistance has been observed (Annunziato 2019).

15.3 Phage Biology Basics

Bacteriophages or simply phages are viruses that infect bacteria instead of humans, small in size (20–200 nm), diverse and the most abundant biological entities estimated to be around 10^{31}, amounting to ten folds the bacterial population. These are composed of protein or proteolipidic polyhedral capsid containing nucleic acid (mainly DNA and RNA) fragments and complex appendages. The capsid is generally joined to a tail made up of a helical protein structure necessary for the adsorption of the virion (viral particle) to the bacterial cell (Brives and Pourraz 2020). Characterized as natural intracellular bacterial parasites, phages lack the ability to reproduce independently and completely depend on the host bacterial cell for its survival. Phages bind to specific receptors on the bacterium's surface and inject the genetic material into the host cell. After entry into the host, phages may follow a lysogenic or temperate and lytic or virulent path. In the first case, these may amalgamate their genetic material into the host's genome and reproduce to form daughter cells. In the latter case, phages take control over the bacterial replication machinery and produce the next generation of phages followed by consequent lysis of the host cell (Lin et al. 2017). Figure 15.2 demonstrates the lytic and lysogenic life cycle of a bacteriophage. As illustrated, phage attaches to the host cell and rapidly kills the infected cell at the end of the growth cycle. However, in the lysogenic phase, the phage introduces its genetic material into the host genome or stays undetected as plasmids inside their host cells by embedding into the bacterial chromosome (the prophage state) (Inal 2003).

Despite the advances in molecular biology techniques, it still is a cumbersome task to classify the vast varieties of phages. The International Committee on Taxonomy of Viruses (ICTV), an international organization, emphasizes the classification of phages based upon taxonomical properties such as the nature of genetic material, morphology of the virion, physico-chemical characteristics and genome sequencing data. The largest order of bacteriophages, i.e. *Caudovirales*, constitutes more than 96% population of all the bacteriophages, includes caudate viruses having double-stranded DNA and virions having an icosahedral capsid with the DNA and a tail (enabling connection between host and phage to transport its genetic material into the bacterial cell). The order *Caudovirales* further constitute three families: *Myoviridae* (capsid size ~150 nm and a contractile tail); *Siphoviridae*

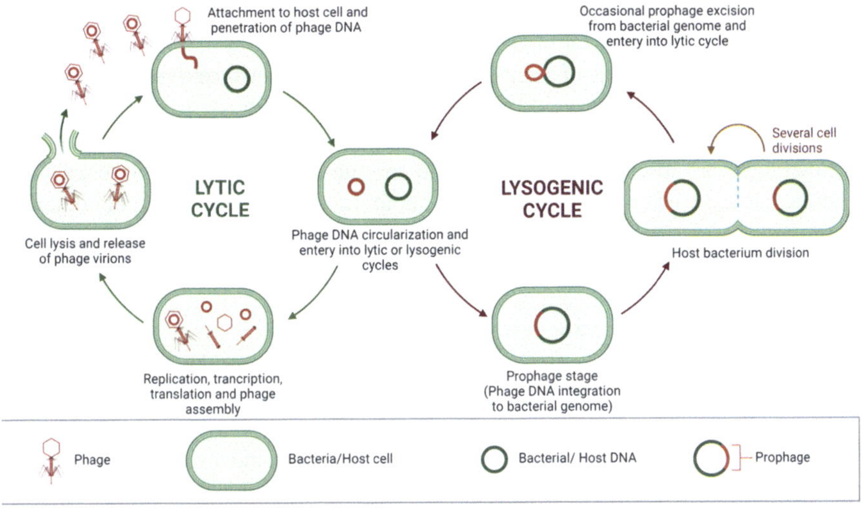

Fig. 15.2 Diagram showing schematic representation of lytic and lysogenic replication cycles of bacteriophage (Brives and Pourraz 2020)

(capsid size ~50–60 nm, with a long, flexible and non-contractile tail) and *Podoviridae* (capsid size ~50–60 nm and a short tail). The remaining 4% of the phage population demonstrate greater morphological diversity (representing as filamentous, cubic or pleomorphic viruses) and hence are studied on an individual basis (Royer et al. 2021).

15.4 Discovery and Early History of Bacteriophages

Earliest reports for bacteriophages date back to 1896 by Ernst Hankin who reported the antibacterial activity of waters from the rivers Ganges and Yamuna in India against *Vibrio cholera* infection (Hankin 1896). Two years later, a Russian bacteriologist also reported similar phenomenon while working with *Bacillus subtilis* bacteria (Samsygina and Boni 1984). Later it was in the early twentieth century that phages were concurrently discovered by British microbiologist Frederick Twort (1915) and French-Canadian microbiologist Felix d'Herelle (1917). It was Frederick Twort who first hypothesized the possibility of a virus to be exhibiting activity against bacteria. However, Twort did not pursue phage research further but it was d'Herelle who methodically explored the nature of phages and investigated their capacity to work as therapeutic agents (Summers 2016). While working at the Pasteur Institute in Paris, d'Herelle utilized the newly discovered phage therapy during World War I, being in-charge of the bacteriological examination of the cases. During this period, he investigated extreme and atypical cases of bacillary dysentery amongst French soldiers stationed at Maisons-Laffitte commune on the north-western suburbs of Paris. While working on the faecal samples, d'Herelle found out that the not so familiar bacterial culture tend to lyse and clear in some time. When small portions of the cleared cultures were transferred to other samples of bacterial cultures, these would clear them as well. The mysterious antimicrobial activity is retained even after the removal of bacteria from the culture solutions. Following these observations, he went on to coin the term "Bacteriophage"with reference to assumed microbes that bring about the lysis and the lysis phenomenon itself (Ackermann 1917). d'Herelle expanded his research to United States, France, and Soviet Georgia by setting up phage therapy centres at several places (Carlton 1999; Myelnikov 2018).

Kharkov Mechnikov Institute was established in Kharkov, Soviet Ukraine in the year 1886 to carry out bacteriological studies investigated the Donbass region affected by frequent outbreaks of typhoid, scarlet fever and dysentery between years 1929 and 1935. Several therapies such as scarlet fever serum, typhoid vaccine and bacteriophage against dysentery were tried. The scientists and technicians were successful in making phage preparations against Shiga strain of dysentery bacillus (*Shigella dysenteriae*). Later Mel'nyk also isolated bacteriophages from areas near those water sources and used phage preparations against Shiga sub-strains (cultured from the faecal matter of patients). Similar to d'Herelle's findings, the phages were found to lyse the infective microbes, as depicted by clear cultures. The consequent clear cultures were first mixed and then filtered through the bacterial filter to trap bacteria and debris; permitting the passage of phage for further use. This area-specific use of phages to fight localized strains helped deducing the specificity of bacteriophages (Myelnikov 2018). However, with the advent of antibiotics, a decline in further development and use of phage therapy was observed. Initially, two audits on phage therapy were published in the Journal of the American Medical Association (JAMA) in 1934 and 1941, explaining challenges in the utilization of phages and issues concerning their efficacy. During the same period, several derivatives of sulphonamides or sulfa drugs (potent antibacterial compounds) were introduced in Germany and there was enormous production and utilization of antibiotics in the United States in the 1940s. Hence, these sulfa drugs were considered one of the noteworthy accomplishments in the history of therapeutic agents also contributing to the sharp decline of phage therapy in the western world (Brives and Pourraz 2020). However, to overcome the earnest problem of antibiotic resistance arising due to multi-drug-resistant bacteria, interest in phage therapy has revived.

In the year 2017, WHO (World Health Organization) released a list of some disease-causing microbes consisting of antibiotic-resistant bacteria that pose the greatest risk to global well-being. Some susceptible microbes included carbapenem-resistant *Acinetobacter*, *Candida auris*, *Clostridioides difficile*, carbapenem-resistant *Enterobacteriaceae*, drug-resistant *Neisseria gonorrhoeae*, drug-resistant *Campylobacter*, drug-resistant *Candida* vancomycin-resistant *Enterococci* (VRE), erythromycin-resistant Group A *Streptococcus* and clindamycin-resistant group B *Streptococcus*. Therefore, the need to develop phage therapy for the treatment of common bacterial infections like sepsis, some forms of diarrhoea and sexually transmitted infections increases substantially as the world is running out of effective

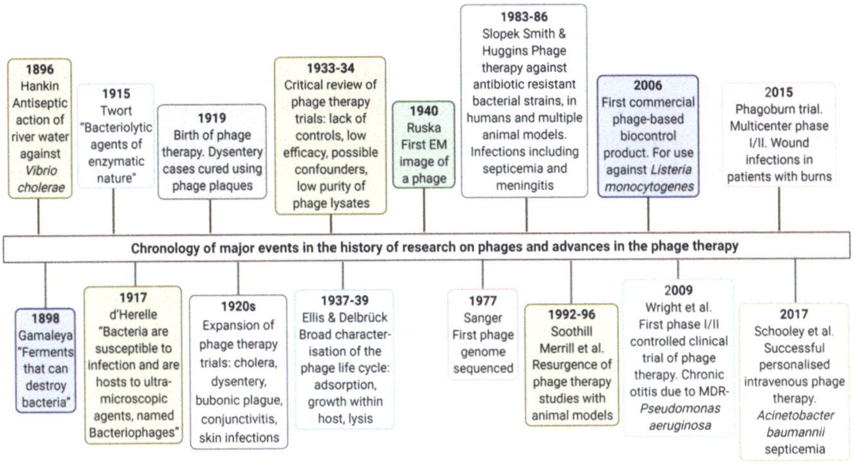

Fig. 15.3 Chronology of major events in the history of research on phages and advances in the phage therapy (Gordillo Altamirano and Barr 2019)

antibiotics (https://www.who.int/news/item/29-04-2019-new-report-calls-for-urgent-action-to-avert-antimicrobial-resistance-crisis). A brief timeline of events occurring in the development of bacteriophages and phage therapy has been summarized in Fig. 15.3.

15.5 Mechanism of Action of Bacteriophages

Based upon replication cycle, bacteriophages can proceed through lytic and lysogenic mechanisms (Fig. 15.2). Bacterial lysis occurs when the infected cells tend to release the lytic phage progeny. Scientific investigations on lytic mechanism of phages provide insights into the phage therapy. Some bacteriophages utilize amurins (single proteins) which inhibit the synthesis of bacterial peptidoglycan whereas others use the holin-lysin systems. In the year 1992, a model was hypothesized which stated that lysis by phage involved two proteins: holins and endolysins. The model was principally based on molecular and genetic studies carried out on phage lambda. During the morphogenesis of the infectious cycle, the holin encoded by lambda S gets accumulated in the membrane, forming large pores in the cytoplasmic membrane and assisting the endolysins to escape and attack the peptidoglycan and consequently affecting the physical integrity of cell wall of bacteria (Moghadam et al. 2020; Young 2014). Holins are an extremely different class of small phage-encoded membrane proteins. There are three classes of holins based on the differences in amino acids sequence. Class I incorporates proteins that have more than 95 amino acids residues and are represented by *Escherichia coli* phage λ S105 protein and *Staphylococcus aureus* bacteriophage p68 hol15 protein. Class II holins have 65–95 amino acid residues and are characterized by *Clostridium perfringens* bacteriophage Φ3626 hol3626 protein and Lambdoid phage 21 S protein. Class III holins are represented by phage ΦCP26F holin (Dewey et al. 2010). Genetic and biochemical methodologies on S105 suggest that it is the most extensively studied holin. It is a 105 amino acid residue polypeptide with three transmembrane domains (TMDs) encoded by the S gene of phage. S105 produces lethal lesions (~340 nm in diameter) in the bacterial lipid bilayer.

Endolysins, also known as phage lysins, are enzymatic proteins acting through hydrolysing peptidoglycan and subsequent cell wall degradation. They possess muralytic action on the bacterial cell wall mediated through amidase, endopeptidase or glycosidase action leading to bacterial cell destruction. At the end of their phage replication cycle, penetration into the peptidoglycan layer leads to osmotic lysis which eventually causes bacterial cell death and promotes virion progeny release (Cisek et al. 2017). The structural differences in the cell wall of gram-positive and gram-negative bacteria reflect upon the differences between their enzyme targets (Schmelcher et al. 2012). The impact of endolysins on bacterial cell walls is scheduled by certain holin genes which are specified in the dual start model. In this model, the holin gene is an open reading frame that encodes two proteins, holins and antiholins. The proportion of holin to antiholin (holin antagonist) decides the time of release of endolysins. For instance, Class I holin gene, the S gene of bacteriophage lambda not only encodes the effector holin, S105 but also an inhibitor, S107 with a Met1-Lys2-Met3 extension at the terminus (Shi et al. 2012).

15.6 Benefits and Limitations of Phage Therapy

Phage therapy has numerous advantages that make it an attractive option in contrast to antibiotics. Bacteriophages have high specificity for bacterium, unlike antibiotics, which are more active against a wide spectrum of microorganisms and probably cause secondary infections, e.g. with yeast, dysbiosis and other side effects (Romero-Calle et al. 2019). In addition, phages are dominant in nature and producing new phages takes fewer time rather than production of antibiotics. Thus, the development of phage therapy is inexpensive as compared to the production of antibiotics. One of the reasons why antibiotics are rendered ineffective is their metabolism and excretion by the body without reaching the site of infection. However, phages on administration replicate at the site of infection where the host organisms are located and then spread throughout the body. A comparative between phage therapy with conventional antibiotics has been summed up in Table 15.1 (El-Shibiny and El-Sahhar 2017; Romero-Calle et al. 2019).

Generally, phages target the surface receptors of the bacterial cell through virulence factors which later kill the bacteria by rupturing cell wall and cell membrane and eventually leading to the death of bacteria. Also, it is difficult for bacteria to develop resistance against phages but there are a few reports citing incidents of resistance within bacterial species attributed mainly to alteration and attenuation of

Table 15.1 Comparison between antibiotic and phage therapy

Characteristic feature	Antibiotic therapy	Phage therapy
Specificity	Low	High
Side effects	Moderate to high	Usually low
Spread of bacterial resistance	Broad spectrum	Narrow spectrum with some exceptions, phages tend not to cross-species limit
Delivery target	Moderate	Moderate to good. Can enter the blood-brain barrier
Formulation	Fixed	Mostly fixed, sometimes variable
Regulation	Well established	Under process
Kinetics	Single hit	Single hit or self-replicating
Development costs	High	Low to moderate
Immunogenicity	Not fixed	Likely low, but not well established
Clinical validation	Numerous trial studies	Relatively few trials studies
Fate of the drug	Metabolic degradation of the molecule, as it functions	Exponential growth in numbers, so that the drug replicates at the site of diseases, where it is required

phage virulence (El-Shibiny and El-Sahhar 2017). On comparison of phage therapy to traditional antibiotic treatments for use in the cure of bacterial infections, phage therapy offers extraordinary advantages over them. However, the phage therapy like other therapies is not devoid of its limitations. Despite being the largest group of organic entities on earth and offering numerous advantages, application of phage therapy is accompanied by various limitations as outlined in Table 15.2 (Moghadam et al. 2020).

15.7 Phage Therapy against Bacterial Infections in Humans

Since their discovery in early 1900s, bacteriophages have been widely used clinically to treat numerous bacterial infections accompanied by cases of conflicting outcomes in phage trials in the 1930s. Issues concerning the safety and efficacy of phage therapy were also raised in this period because of lack and inappropriate characterization, problems in production and purification of phage preparation (Gordillo Altamirano and Barr 2019). To overcome these obstacles, several clinical trials have been carried out and a few have been completed in recent years (Furfaro et al. 2018). Studies have shown that humans are constantly exposed to bacteriophages everyday and they help in their well-being. However, a number of factors still need to be explored while performing the clinical trials (Parracho et al. 2012). Bacteriophages have been utilized to treat bacterial diseases occurring in diverse body locales with different arrangements. Eastern Europe has a long history of using bacteriophages against bacterial infections (Summers 1999). Various mixed phage cocktails are sold in Russian drug stores as registered items. The

Table 15.2 Advantages and disadvantages of phage therapy in the treatment of bacterial infections

Advantages of phage therapy	Disadvantages of phage therapy
Active against both gram-positive and -negative bacteria	Bacteria capable of developing resistance against phages
Faster isolation and lower cost of development	Do not replicate in absence of target organism
Relatively lower side effects	Phages may act as carriers of bacterial virulence factors or antibiotic-resistance genes
Wider application in food preservation	Immune system perceives phages as invaders and hence, can be removed rapidly
Disruption of bacterial biofilms, XDR and MDR	Lack of regulatory guidelines
Can affect immune system by diminishing C-reactive protein mean values and leukocyte count	Release of endotoxins, superantigens and induction of inflammatory cascade after phage lysis of bacteria that may lead to multiple organ failure
Help reducing damage to normal microbiome	Insufficient data on the function of genes obtained from known phage genomes
Prevent overgrowth of secondary pathogen	Extrapolation of *in vitro* phage growth data to *in vivo* response is difficult
Rapid distribution throughout the body	Needs to identify phage specificity for exact host bacterium
No cross-resistance development to antibiotics	Exclusivity while using lytic phages is a concern
Cocktail of phages, more advantageous, has greater impact on target bacteria	Time-consuming task to diagnose a pathogen in clinical microbiology laboratory and then use a specific bacteriophage for treatment
Recognize different receptors on the surface of cell	Not yet recognized as pharmaceutical drugs
	No health insurance cover provided for phage treatment

XDR: extensively drug-resistant, *MDR:* multidrug-resistant

bacteriophage therapy has been best examined for topical application on bacterial infections occurring in human skin (Sulakvelidze et al. 2001). The first human phage therapy has been accounted for the treatment of skin infections caused by *Staphylococcus aureus*. During the 1980s, phage preparations at the Eliava Institute, Georgia were given to human volunteers with no reported adverse effects. They were mostly effective on immunocompromised patients, new born children and in cases of pelvic inflammatory illnesses. However, it has also been observed that phage preparations have high efficacy in early phases of diseases (Abedon et al. 2011). It is a known fact that burn surfaces quickly get colonized by microorganisms which are fit for producing biofilms and are regularly impervious to multiple antibiotics. Phage treatment might have been utilized to treat burns and prevent sepsis during and after world war II in Eastern Europe (Morozova et al. 2018).

The largest clinical trial on phage therapy carried out in Europe was the Phago Burn trial, in the year 2013. This clinical trial fulfilled the criteria for both Good Manufacturing Practices (GMP) and Good Clinical Practices (GCP). In this randomized but controlled phase I/II clinical trial, 27 patients experiencing burn wounds infections were enrolled from hospitals situated in France and Belgium. Patients were randomly treated with phage therapy with a cocktail of 12 lytic phages and a standard preparation (1% sulfadiazine silver emulsion cream) to compare efficacy and tolerability of both therapies in patients infected by *Pseudomonas aeruginosa*. Both the formulations were applied topically for 7 days with a 14-day check. In general, bacterial load in burn wounds was less in treated groups but the process was slower than in the control group. No adverse effects were observed in the phage treated group. The restricted adequacy of the phage cocktail was accounted for an essential drop of phage titre. This resulted in lower concentration of phages in the patients than the ones assessed first. Also, in these participants, phage treatment failed as bacteria were found to be resistant to low doses of phage (Jault et al. 2019).

A combination of eight bacteriophages targeted against *P. aeruginosa*, *S. aureus* and *Escherichia coli* were applied on venous leg ulcers in around 39 patients and no side effects were observed relating to the treatment (Rhoads et al. 2009). Another randomized clinical trial was conducted on a group of 24 patients with a persistent otitis externa condition using a topical phage preparation against *P. aeruginosa* and the patients were successfully treated without any adverse effects. Besides this, a combination of *P. aeruginosa* and *S. aureus* phages was developed for application on the skin of infected patients (Wright et al. 2009). In Eastern Europe, oral phage delivery has also been utilized against intestinal diseases nowadays. Nonetheless, before studying the efficacy of phage therapy on numerous people in clinical trials, it is fundamental to first evaluate their safety in humans to guarantee that they do not cause adverse reactions when given orally (Sulakvelidze et al. 2001). While most clinical trials have failed to provide promising evidence of the efficacy of the phage therapy, some of the few case analyses show phage therapy as a successful method to treat life-threatening infections (Pires et al. 2020). For example, a novel method has been used to prepare personalized therapeutic bacteriophages cocktails to protect patients from life-threatening multidrug resistance caused by *Acinetobacter baumanni* infection. This case further consolidated the utilization of phage therapy in treating patients experiencing MDR bacterial infections with limited therapeutic options (Schooley et al. 2017). It is important that the western world should acknowledge the phage therapy and ramp up the clinical trials for their commercial use. There are numerous observational studies that have been performed but have encountered restrictions such as sample sizes and inadequacy in control groups. Every clinical case requires robust clinical trial data to be submitted to the regulators for the improvement of existing clinical guidelines (Payne and Jansen 2003).

15.8 Phage Therapy: Problems and Alternative Solutions

As an antibacterial therapeutic, phage therapy has not been successful in completely replacing antibiotics as alternative. In order to establish phages as therapeutics, lot more investigations need to be done to uncover every aspect of bacteriophages. Due to reluctance of using phages, they have not been tested for their effects systemically in animal models and hence, they have a rather simplified pharmacokinetic profile. Merrils and coworkers in 1973 first reported infusion of high titre of lambda phage into non-immune germ-free mice. They concluded that phages were quickly cleared by the spleen, other filtering organs and liver (Geier et al. 1973). It has been observed that selected phage strains can remain in circulation for over a period of many days and continue to be accessible for interaction with possible systemic bacterial infectious agents by genetic selection method (Vitiello et al. 2005). This capacity of phages to stay in the circulation for a longer period and clear the toxins and antibiotic resistance genes may be helpful in the administration of phages. This property can further help design and develop phages which are effective with improved activity for specific bacterial strain (Adhya et al. 2014). Another limitation of phages is their moderately narrow spectrum of activity. Apparently, it was thought that the fundamental restricting factor for their clinical use is the number of phages accessible to doctors, which appears to be irrational after recognition of phages to be the most abundant organic entity on earth. The uniqueness of phages for bacterial strains infers the need to have a wide variety of phages with strong lytic action for fighting pathogenic bacteria in the given patient.

After the First International Congress on Microbes and Viruses in 2010 at the Pasteur Institute in Paris, an article on bacteriophages was published in 2011 in which the authors proposed two approaches known as 'prêt-à porter' and 'sur-mesure'. In the 'prêt-à-porter' approach, fixed cocktails of several phages (around 10) are brought together to ensure that at least one the phage will be effective on the bacterial strain carried by the patient. This approach helps standardization of phage cocktail for commercial use whereas in the 'sur-mesure' approach, the patients will be administered only few phages that are specifically active on the infecting strain (Brives and Pourraz 2020). Phage preparations are often contaminated by bacterial debris, especially, macromolecules derived from the host bacteria and culture media with significant amount of endotoxins (Raetz et al. 2007). Injection of even modest quantities can result in cell injury, toxic shock and can be fatal to patients. However, various endotoxin removal methodologies and commercial kits have been developed which help in purification of phage preparations (Merabishvili et al. 2009).

Other key problem associated with phages is the inability to establish scientific proof of phage efficacy. Several articles in defence of phage therapy have been published in Europe in the recent couple of years, stating that the deficiency of clinical preliminaries exhibits the efficacy of phage due to their specificity whereas in infectious diseases, the antibiotics developed in the course of recent years are themselves quite specific, which has not kept them from demonstrating their adequacy (Morrison and Ulevitch 1978).

15.9 Future Challenges and Role of Bacteriophages

Unlike other antimicrobials such as antibiotics, phages show greater diversity in mechanism of action and can be much safer in many instances (Chan et al. 2013). The challenges for phage therapy include: how to put forward the positive attributes of phage therapy in presence of existing regulatory practices and how phages would find a way into the current economic models that support the distribution and utilization of antibacterial agents (Fig. 15.4) (Międzybrodzki et al. 2012). Phage therapy appears to flourish especially in regions where the regulatory authorities are moderate or with lesser restrictions (for example, in Poland and some countries of undivided Soviet Union). Scientists and researchers believe that phage therapy shall show expansion in years to come as promising candidates for use as antibacterial agents to overcome infections in western medicine system. Thus, it is anticipated that in the following 5–10 years, phage therapy would ascertain its way into clinical practice to cure life-threatening, ongoing chronic bacterial infections that cannot be treated using available antibacterial drugs (Sommer and Dantas 2011). To scale up the production of phages, there are significant issues that have not been satisfactorily addressed. These include removal of endotoxins and pyrogens which are delivered from the ruptured cells during phage lysis after administration of phage preparation. However, it is quite possible to optimize such outcomes. Recently, an endotoxin removal kit has been introduced which purifies the phage preparation before use in clinical trials (Merabishvili et al. 2009). Literature reports show that preliminary clinical studies have been attempted on cocktail of phages and have cleared phase I, showing no adverse reactions related to the utilization of phage cocktails, thus paving way to phase II clinical trials (e.g. APS Biocontrol Limited, Dundee, United Kingdom). Clearance of phage therapy in phase I and II encourages positive model for subsequent investigation in phase III (Rhoads et al. 2009). Another major barrier in bacteriophage delivery is the removal of phages by the immune system. Kim et al. studied the immune response by human body after phage administration by conjugating phage with polyethylene glycol. Results revealed that there was a decrease in the levels of helper T cells and an increase in the blood circulation time in contrast to unmodified phages (Kim et al. 2008).

Currently, phage therapy has been investigated only on fast dividing bacterial species and not on slow dividing bacterial species, where it needs to be developed

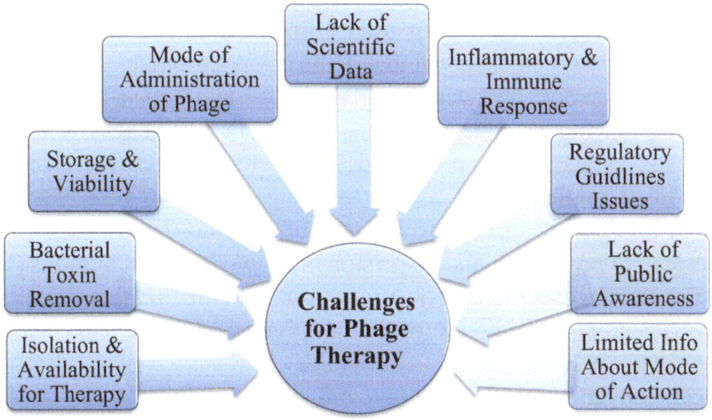

Fig. 15.4 Major challenges faced by phage therapy

and explored. Phage therapy can provide a promising solution for extremely slow dividing *Mycobacterium africanum* and *Mycobacterium leprae* (Dąbrowska 2019; Van Belleghem et al. 2019). Many challenges lie ahead for the therapeutic application of phages such as elimination of endotoxins in the final product, delivery system for the host, inadequacy in identification of quick and immediate therapy, activity potential of phages against specific host bacteria, public awareness and the absence of regulatory guidelines (Manohar et al. 2019). A huge amount of effort is required from government, organizations and academic researchers to translate phage therapy from clinical trials to markets.

15.10 Emerging Approaches in Phage Therapy

Phage therapy has been an area of interest that offers endless advantages over other antimicrobial therapies already threatened by antibiotic-resistant pathogens. Combining antibiotic and phage therapy, or the use of different phages as cocktails might be the most reassuring strategies for the treatment of bacterial infections (Manohar et al. 2019). For example, effectiveness of bacteriophages alone or in combination of amoxicillin, for eradication of biofilm formed by *Klebsiella pneumoniae* B5055 and planktonic cells has been assessed (Bedi et al. 2009). It has also been observed that individual treatments had restricted success rates. However, repeated phage treatments have shown to increase the biovolume of biofilms, as reported in *P. aeruginosa* PA01 (Henriksen et al. 2019). In contrast, the combination of phage and antibiotics leads to biofilm eradication (Díaz-Pascual et al. 2019). Consequently, when an antibiotic is administered in combination with specific bacteriophage, significant destruction of biofilm has been observed. Hence, the phages could be used effectively in conjugation to antibiotic therapy (Bedi et al. 2009). Mechanistically, antibiotics have substantial effects on biofilm structures and enhance the intrusion of biofilm by bacteriophages. An additional benefit of this combination therapy would be its ability to control emergence of resistant mutants that, in any case, progress effectively after utilizing antibiotic and phage therapies separately. It has also been observed that while using antibiotics and phages simultaneously, synergism does not occur for all the phage-antibiotic combinations. It has been reported that higher concentration of antibiotics to phages in combination therapy can also antagonize the proliferation of bacteriophages (Jansen et al. 2018).

Increase in the antimicrobial activity of phages has also been reported when administered along with enzymes. The use of enzyme depolymerase alongside phages has lead to prevention and dispersion of *Staphylococcus* biofilms (Gutiérrez et al. 2015). Other agents that can be effective in combination therapy with phages include hydrogen peroxide, honey, probiotics, xylitol, chlorine and cobalt (II) sulphate (Pires et al. 2020).

Advancement in synthetic biology and engineering methods has enabled improvement in phages by producing novel genomes. Engineered phages have enhanced antimicrobial activity along with the limitations of phage therapies (narrow host range and host immunity) (Kim et al. 2019). Engineered phages based upon lytic mechanism of phages have been developed. One or more temperate phages are genetically engineered with a lytic phage for use as therapeutic. The most common approach comprises of turning the phage into lytic one by genetic engineering. An illustration of this approach is the utilization of a cocktail made from one natural lytic phage and two engineered temperate phages to effectively treat a 15-year-old patient with cystic fibrosis with a disseminated *Mycobacterium abscessus* infection (Dorscht et al. 2009).

A well-known approach to overcome the problem of narrow range of activity is to use a combination of phages with different spectra in a single cocktail. However, the use of cocktails requires greater optimization to improve their performance. In addition, mixing up a diverse group of phages can lead to more challenges for approval and manufacturing. Subsequently, it would be ideal to create engineered phages with broad-spectrum and enhanced antimicrobial activity instead of simply combining multiple phages. For example, use of a recombinant phage (T3/7) produced by combination of a T3-based hybrid phage whose tail fibre gene I7 is a recombinant between those of T3 and T7. The hybrid phage T3/7 has been found to show better adsorption, effectiveness and a wider host range than both T3 and T7 phages used independently. To enhance the phage antimicrobial activity against biofilms, some enzymes such as lactonase and dispersin B have also been engineered into phage T7 (Lin et al. 2012). Researchers have proposed the use of a restriction enzyme on a non-specific target of a host genome that can be fundamentally extended to 'CRISPR antibacterial' technique which focuses on a specific gene vital for bacterial growth and survival and even more significantly, antibacterial species with their accessible phages (Kim et al. 2019). Hence, engineering methodologies can potentially improve phages' antimicrobial properties, and such phages should be considered therapeutic choices. Additionally, phages that are engineered have simpler patentability and hence are of more commercial interest.

15.11 Conclusions

Undoubtedly, bacteriophages offer vast diversity and have huge potential for development as antimicrobial therapy in a 'post-antibiotic era' world which is short of medications essential for our fight against surging microbial infections. To ensure maximum clinical efficacy of phage therapy, fast-track research needs to be carried out on pathogenic bacteria affecting patients, followed by isolation, identification and screening against individual phage strains or available approved phage cocktails. So far, antimicrobial resistance continues to be one of the preeminent threats to global health, which requires significant intervention at different strata of the society. Considering the limited armamentarium of antibiotics available and an insufficiency in production of newer ones, and despite all its limitations, phages are a nature's gift to humanity which is safe and effective strategy for combating bacterial infections, with the added advantage of controlling and preventing multidrug-resistant organisms.

To overcome individual limitations of a therapy, a combination therapy approach, wherein two or more therapies can be concurrently used, holds potential in extending the effectiveness of the currently available antimicrobials. The prospects of phage preparations to be used in combination with antibiotics, probiotics and vaccines against chronic infections and resistant pathogenic bacteria can help reduce human illnesses significantly. Measures can be taken in this direction to encourage pharmaceutical giants to invest in development of phage therapy and promote international collaborations amongst academia, doctors and pharma sector, resulting in ensuring advancements in phage therapy. Hence, comprehensive research is the requisite to achieve approval of the regulatory agencies for their inclusion to our repertoire of treatments as well as commercial use.

In conclusion, the looming crisis of antimicrobial resistance has left us with no other choice but to device multidimensional strategies to deal with widespread bacterial infections. The phage therapy offers enormous possibilities and can be instrumental in redefining modern biology by assisting in understanding the biological processes at the molecular level. Therefore, it is time to deliberate the benefits of phage therapy in its entirety with its targeted compassionate use on patients left with no other alternative treatment.

References

Abedon ST, Kuhl SJ, Blasdel BG, Kutter EM (2011) Phage treatment of human infections. Bacteriophage 1(2):66–85

Ackermann H-W (1917) On an invisible microbe antagonistic to dysentery bacilli. C R Acad Sci 165:373–375

Adhya S, Merril CR, Biswas B (2014) Therapeutic and prophylactic applications of bacteriophage components in modern medicine. Cold Spring Harb Perspect Med 4(1):a012518

Annunziato G (2019) Strategies to overcome antimicrobial resistance (AMR) making use of non-essential target inhibitors: a review. Int J Mol Sci 20(23):5844

Bedi MS, Verma V, Chhibber S (2009) Amoxicillin and specific bacteriophage can be used together for eradication of biofilm of Klebsiella pneumoniae B5055. World J Microbiol Biotechnol 25(7):1145–1151

Brives C, Pourraz J (2020) Phage therapy as a potential solution in the fight against AMR: obstacles and possible futures. Palgrave Commun 6(1):1–11

Carlton RM (1999) Phage therapy: past history and future prospects. Arch Immunol Ther Exp 47:267–274

Chan BK, Abedon ST, Loc-Carrillo C (2013) Phage cocktails and the future of phage therapy. Future Microbiol 8(6):769–783

Cisek AA, Dąbrowska I, Gregorczyk KP, Wyżewski Z (2017) Phage therapy in bacterial infections treatment: one hundred years after the discovery of bacteriophages. Curr Microbiol 74(2):277–283

Dąbrowska K (2019) Phage therapy: what factors shape phage pharmacokinetics and bioavailability? Systematic and critical review. Med Res Rev 39(5):2000–2025

Dewey JS, Savva CG, White RL, Vitha S, Holzenburg A, Young R (2010) Micron-scale holes terminate the phage infection cycle. Proc Natl Acad Sci 107(5):2219–2223

Díaz-Pascual F, Hartmann R, Lempp M, Vidakovic L, Song B, Jeckel H, Thormann KM, Yildiz FH, Dunkel J, Link H (2019) Breakdown of Vibrio cholerae biofilm architecture induced by antibiotics disrupts community barrier function. Nat Microbiol 4(12):2136–2145
Dorscht J, Klumpp J, Bielmann R, Schmelcher M, Born Y, Zimmer M, Calendar R, Loessner MJ (2009) Comparative genome analysis of listeria bacteriophages reveals extensive mosaicism, programmed translational frameshifting, and a novel prophage insertion site. J Bacteriol 191(23):7206–7215
El-Shibiny A, El-Sahhar S (2017) Bacteriophages: the possible solution to treat infections caused by pathogenic bacteria. Can J Microbiol 63(11):865–879
Furfaro LL, Payne MS, Chang BJ (2018) Bacteriophage therapy: clinical trials and regulatory hurdles. Front Cell Infect Microbiol 8:376
Geier MR, Trigg ME, Merril CR (1973) Fate of bacteriophage lambda in non-immune germ-free mice. Nature 246(5430):221–223
Gordillo Altamirano FL, Barr JJ (2019) Phage therapy in the postantibiotic era. Clin Microbiol Rev 32(2):e00066–e00018
Gutiérrez D, Briers Y, Rodríguez-Rubio L, Martínez B, Rodríguez A, Lavigne R, García P (2015) Role of the pre-neck appendage protein (Dpo7) from phage vB_SepiS-phiIPLA7 as an anti-biofilm agent in staphylococcal species. Front Microbiol 6:1315
Hankin E (1896) An OUTBREAK of CHOLERA in an OFFICERS' MESS. Br Med J 2(1878):1817
Henriksen K, Rørbo N, Rybtke ML, Martinet MG, Tolker-Nielsen T, Høiby N, Middelboe M, Ciofu O (2019) P. aeruginosa flow-cell biofilms are enhanced by repeated phage treatments but can be eradicated by phage–ciprofloxacin combination: —monitoring the phage–P. aeruginosa biofilms interactions. Pathog Dis 77(2):ftz011
Inal JM (2003) Phage therapy: a reappraisal of bacteriophages as antibiotics. Arch Immunol Ther Exp 51(4):237–244
Jansen M, Wahida A, Latz S, Krüttgen A, Häfner H, Buhl EM, Ritter K, Horz H-P (2018) Enhanced antibacterial effect of the novel T4-like bacteriophage KARL-1 in combination with antibiotics against multi-drug resistant Acinetobacter baumannii. Sci Rep 8(1):1–12
Jault P, Leclerc T, Jennes S, Pirnay JP, Que Y-A, Resch G, Rousseau AF, Ravat F, Carsin H, Le Floch R (2019) Efficacy and tolerability of a cocktail of bacteriophages to treat burn wounds infected by Pseudomonas aeruginosa (PhagoBurn): a randomised, controlled, double-blind phase 1/2 trial. Lancet Infect Dis 19(1):35–45
Kim KP, Cha JD, Jang EH, Klumpp J, Hagens S, Hardt WD, Lee KY, Loessner MJ (2008) PEGylation of bacteriophages increases blood circulation time and reduces T-helper type 1 immune response. Microb Biotechnol 1(3):247–257
Kim B-O, Kim ES, Yoo Y-J, Bae H-W, Chung I-Y, Cho Y-H (2019) Phage-derived antibacterials: harnessing the simplicity, plasticity, and diversity of phages. Viruses 11(3):268
Lin T-Y, Lo Y-H, Tseng P-W, Chang S-F, Lin Y-T, Chen T-S (2012) A T3 and T7 recombinant phage acquires efficient adsorption and a broader host range. PLoS One 7(2):e30954
Lin DM, Koskella B, Lin HC (2017) Phage therapy: an alternative to antibiotics in the age of multi-drug resistance. World J Gastrointest Pharmacol Ther 8(3):162
Manohar P, Tamhankar AJ, Leptihn S, Ramesh N (2019) Pharmacological and immunological aspects of phage therapy. Infect Microbes Dis 1(2):34–42
Merabishvili M, Pirnay J-P, Verbeken G, Chanishvili N, Tediashvili M, Lashkhi N, Glonti T, Krylov V, Mast J, Van Parys L (2009) Quality-controlled small-scale production of a well-defined bacteriophage cocktail for use in human clinical trials. PLoS One 4(3):e4944
Międzybrodzki R, Borysowski J, Weber-Dąbrowska B, Fortuna W, Letkiewicz S, Szufnarowski K, Pawełczyk Z, Rogóż P, Kłak M, Wojtasik E (2012) Clinical aspects of phage therapy. Adv Virus Res 83:73–121
Moghadam MT, Amirmozafari N, Shariati A, Hallajzadeh M, Mirkalantari S, Khoshbayan A, Jazi FM (2020) How phages overcome the challenges of drug resistant bacteria in clinical infections. Infect Drug Resist 13:45
Morozova VV, Kozlova YN, Ganichev DA, Tikunova NV (2018) Bacteriophage treatment of infected diabetic foot ulcers. In: Bacteriophage therapy. Springer, Berlin, pp 151–158

Morrison D, Ulevitch R (1978) The effects of bacterial endotoxins on host mediation systems. A review. Am J Pathol 93(2):526
Myelnikov D (2018) An alternative cure: the adoption and survival of bacteriophage therapy in the USSR, 1922–1955. J Hist Med Allied Sci 73(4):385–411
Park M, Satta G, Kon OM (2019) An update on multidrug-resistant tuberculosis. Clin Med 19(2):135
Parracho HM, Burrowes BH, Enright MC, McConville ML, Harper DR (2012) The role of regulated clinical trials in the development of bacteriophage therapeutics. J Mol Genet Med 6:279
Payne RJ, Jansen VA (2003) Pharmacokinetic principles of bacteriophage therapy. Clin Pharmacokinet 42(4):315–325
Pires DP, Costa AR, Pinto G, Meneses L, Azeredo J (2020) Current challenges and future opportunities of phage therapy. FEMS Microbiol Rev 44(6):684–700
Prestinaci F, Pezzotti P, Pantosti A (2015) Antimicrobial resistance: a global multifaceted phenomenon. Pathog Glob Health 109(7):309–318
Raetz CR, Reynolds CM, Trent MS, Bishop RE (2007) Lipid a modification systems in gram-negative bacteria. Annu Rev Biochem 76:295–329
Rhoads D, Wolcott R, Kuskowski M, Wolcott B, Ward L, Sulakvelidze A (2009) Bacteriophage therapy of venous leg ulcers in humans: results of a phase I safety trial. J Wound Care 18(6):237–243
Romero-Calle D, Guimarães Benevides R, Góes-Neto A, Billington C (2019) Bacteriophages as alternatives to antibiotics in clinical care. Antibiotics 8(3):138
Royer S, Morais AP, da Fonseca Batistão DW (2021) Phage therapy as strategy to face post-antibiotic era: a guide to beginners and experts. Arch Microbiol 203:1271–1279
Samsygina G, Boni E (1984) Bacteriophages and phage therapy in pediatric practice. Pediatriia (4):67–70
Schmelcher M, Donovan DM, Loessner MJ (2012) Bacteriophage endolysins as novel antimicrobials. Future Microbiol 7(10):1147–1171
Schooley RT, Biswas B, Gill JJ, Hernandez-Morales A, Lancaster J, Lessor L, Barr JJ, Reed SL, Rohwer F, Benler S (2017) Development and use of personalized bacteriophage-based therapeutic cocktails to treat a patient with a disseminated resistant Acinetobacter baumannii infection. Antimicrob Agents Chemother 61(10):e00954–e00917
Shi Y, Yan Y, Ji W, Du B, Meng X, Wang H, Sun J (2012) Characterization and determination of holin protein of Streptococcus suis bacteriophage SMP in heterologous host. Virol J 9(1):1–11
Sommer MO, Dantas G (2011) Antibiotics and the resistant microbiome. Curr Opin Microbiol 14(5):556–563
Sulakvelidze A, Alavidze Z, Morris JG Jr (2001) Bacteriophage therapy. Antimicrob Agents Chemother 45(3):649–659
Summers WC (1999) Felix dHerelle and the origins of molecular biology. Yale University Press, Yale
Summers WC (2016) Félix Hubert d'Herelle (1873–1949): history of a scientific mind. Bacteriophage 6(4):e1270090
Van Belleghem JD, Dąbrowska K, Vaneechoutte M, Barr JJ, Bollyky PL (2019) Interactions between bacteriophage, bacteria, and the mammalian immune system. Viruses 11(1):10
Vitiello CL, Merril CR, Adhya S (2005) An amino acid substitution in a capsid protein enhances phage survival in mouse circulatory system more than a 1000-fold. Virus Res 114(1–2):101–103
Wright A, Hawkins C, Änggård E, Harper D (2009) A controlled clinical trial of a therapeutic bacteriophage preparation in chronic otitis due to antibiotic-resistant Pseudomonas aeruginosa; a preliminary report of efficacy. Clin Otolaryngol 34(4):349–357
Young R (2014) Phage lysis: three steps, three choices, one outcome. J Microbiol 52(3):243–258

Part III
Novel Strategies to Combat AMR

Chapter 16
Omics and In Silico Approaches in the Surveillance and Monitoring of Antimicrobial Resistance

Ajith M. Thomas, Lija L. Raju, and S. Sudheer Khan

16.1 Introduction

The discovery and industrial production of antibiotics in the early twentieth century were one of the most significant accomplishments in medical history. These drugs increased the quality and expectancy of life (Mohr 2016). Antibiotic abuse has unfortunately resulted in creating multidrug-resistant pathogens. Deaths and infections linked with multidrug-resistant pathogens have surged in recent years (Chan 2016). Drug-resistant diseases cause at least 700,000 deaths globally a year, including 230,000 deaths from multidrug-resistant tuberculosis, and these numbers could increase to ten million deaths globally per year by 2050 under the most alarming scenario if no action is taken. Moreover, it is predicted that around 2.4 million people can lose their lives in high-income countries between 2015 and 2050 without continued effort to manage AMR (Interagency Coordination Group on Antimicrobial Resistance 2019). The diversity of resistant species, the regions affected by drug resistance, and single-organism resistance are enormous and growing (Levy and Marshall 2004). Microbe gains antimicrobial resistance via a variety of mechanisms (Alekshun and Levy 2007) like producing enzymes that catabolize the antibiotic, expressing efflux pumps that remove the antimicrobials, changing the antibiotic's target to prevent binding, triggering an alternate pathway that impedes the action of

A. M. Thomas
Department of Botany and Biotechnology, St. Xavier's College, Thumba, Thiruvananthapuram, India

L. L. Raju
Department of Zoology, Mar Ivanios College, Thiruvananthapuram, India

S. S. Khan (✉)
Nanobiotechnology Laboratory, Department of Biotechnology, Bannari Amman Institute of Technology, Sathyamangalam, Tamil Nadu, India
e-mail: sudheerkhan@bitsathy.ac.in

N. Akhtar et al. (eds.), *Emerging Modalities in Mitigation of Antimicrobial Resistance*,
https://doi.org/10.1007/978-3-030-84126-3_16

the drugs, and eliminating the antibiotic by downregulating transmembrane porins through which drugs enter the cell (Vila et al. 2007). Drug resistance is mobile, and resistance genes can be passed from one taxonomic community to another through bacteriophages, plasmids, transposons, and other means. Many genes carrying a drug resistance trait can amass in the same organism, and these genes are typically directed toward a family of antibiotics. In the absence of transposons and plasmids (which typically mediate high-level resistance), bacteria undergo development from low-level to high-level resistance through chromosome mutations (Levy and Marshall 2004).

The word "antibiotic resistome" denotes the set of all antibiotic resistance genes found in microorganisms. This term refers to the idea of an antibiotic resistance gene reservoir in environmental samples. Pathogen-derived ARGs make up a small portion of the resistome. Antibiotic resistance genes from antibiotic producers and cryptic resistance genes are examples of resistance genes from nonpathogenic bacteria. The resistome also includes several genes that code proteins with reasonable resistance or antibiotic binding functions that may develop into resistance factors because of selection pressure during evolution. The ultimate external cause of antibiotic resistance is resistance precursor genes (Wright 2007). The widespread use of high-throughput sequencing (HTS) technology has made it easier to identify microbial communities from numerous samples. The growing scale of datasets has created new challenges for exploring and analyzing the information. The identification and classification of antimicrobial resistance genes (ARGs) in large datasets are aided by state-of-the-art databases and analysis pipelines (Yin et al. 2018). However, since bioinformatic platforms can vary significantly, it is critical to choose the resources that are best suited to the user's requirements. This chapter aims to provide detailed information on omic approaches and the most important bioinformatic platforms for identifying and characterizing ARGs.

16.2 Omics of Antimicrobial Resistance

The advancement of many "omic" technologies now allows the high-throughput tracking of the relative abundance of several biomolecules. Genomics aids in the understanding of the whole genome during various biological states. Transcriptomics, which analyzes the whole transcriptome; proteomics, which includes whole proteome studies; metabolomics, which determines various parameters of metabolomes; interactomics, which deals with molecular interactions; and fluxomics, which deals with dynamic alterations of molecules inside a cell, are some common "omic" platforms that have aided microbiology. Yet, no single "omic" analysis can entirely decipher the challenges of microbiology; hence, a unification of many omic approaches, i.e., a "multi-omic" approach, is necessary to evaluate antimicrobial resistance (Zhang et al. 2010). This multi-omic approach in microbes has revealed the involvement of several pathways in determining AMR (Fig. 16.1).

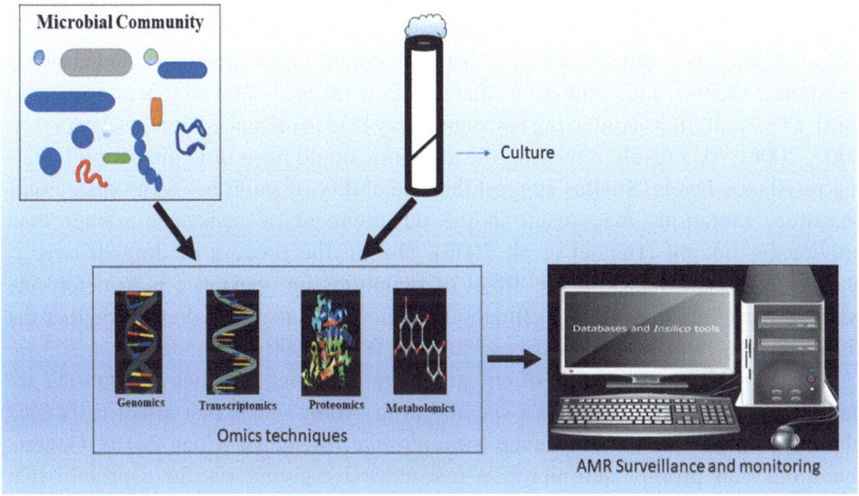

Fig. 16.1 Multi-omic approaches in antimicrobial resistance

16.3 Genomics in Antimicrobial Resistance

Genome sequencing has shown that all bacteria have a significant number of resistance genes. Bacteria have many genes that code for efflux pumps. Microbes like *P. aeruginosa*, an opportunistic pathogen found in a variety of habitats including freshwater and soil, have a remarkable array of efflux pumps (Amsalu et al. 2020). These pumps give microbes the most versatility possible, allowing them to thrive in a variety of environments and increase pathogenicity and biofilm formation (Piddock 2006). Enzymes may contribute to microbial tolerance, one of which is β-lactamases. β-Lactamases have a long history dating back millions of years (Bush 2018). β-Lactamases are the most common enzyme in Gram-negative bacteria that confer resistance to β-lactam antibiotics. Bacteria are thought to be protected from naturally occurring β-lactams by a group of enzymes called β-lactamases, which currently number about 2800 unique proteins (Bush 2018). These β-lactamases are tightly regulated and can be induced shortly by antibiotic exposure. β-Lactamases encoded by genes in resistance plasmids and transposons found in clinical isolates are constitutively expressed because the host organism prefers to live in conditions where antibiotics are abundant: In such cases, regulated expression offers a few disadvantages (Wright 2007). Antibiotic resistance mutations have a negative impact on microorganisms. If the antibiotic is withdrawn, species that are sensitive to the antibiotic outcompete antibiotic-resistant mutants with lower health, weakening antibiotic tolerance in the microbes. As a result, antibiotic cycling has been proposed as a strategy for combating antibiotic resistance. Cycling or rotation consists of the sequential or periodic use of two or more antibiotics not sharing a common mechanism of resistance. Cycling has been proposed as a putative tool for decreasing current resistance rates or deterring the emergence of resistance (Martínez and Rojo 2011). In most cases, however, it has been demonstrated that, first, compensatory mutations may reinstate fit condition while preserving antibiotic resistance; second, a few mutations that confer resistance have no fitness costs; and third, a few mutations conferring resistance may lead to enhanced fitness (Andersson 2003, 2006). As a result, removing the antibiotic could have little impact on lowering resistance levels. Studies suggest that the ability of microbes to produce compensatory mutations may result in the development of resistance sooner than previously thought (Handel et al. 2006). During the process of drug discovery, another important factor of the effect of mutations on resistance is to determine whether these mutants have less fitness in the host organism and decide whether the compensatory mutations are easily identified (Wright 2007).

Antimicrobial resistance is often caused by mutation, but many mutations are needed for robust resistance to a specific drug. Many genes that affect resistance phenotype indirectly also influence antimicrobial resistance (Chan 2016). Genetic mutations can prompt antimicrobial resistance by giving rise to a protein that impairs the antimicrobial agent, producing a different enzyme that is not suppressed by the antimicrobials, resulting in target alterations, posttranscriptional and posttranslational changes in the target (Levy and Marshall 2004), lessened uptake of the antimicrobial agent, and altering actions of efflux pumps and overproduction of the target. It is perplexing how antimicrobial resistance and genetic events are connected in such a complex way (Chan 2016). Antimicrobial resistance is associated with mutations not only in target protein-coding genes but also in a wide range of other genes whose products participate in cellular functions such as DNA repair, replication, transmembrane transport, energy production, and virulence. Comparative genomic study of strains with varying antimicrobial sensitivity is critical for understanding complex antimicrobial resistance mechanisms. The original and antimicrobial genomes can be compared after selecting the corresponding resistant mutants for locating possible target molecules for the drugs. Mutations occur in genes that code for proteins with unknown functions. Antimicrobial resistance phenotype may be significantly influenced by concomitant mutations (Chernov et al. 2019). Today, scientists can disclose the metabolic and genetic potential of new antimicrobials and antimicrobial resistance genes using genomes or genomic fragments assembled from metagenomic studies (Medvedeva et al. 2016). The role of membrane vesicles in antimicrobial resistance has been probed through advanced genomic research. Extracellular vesicles, which channel molecules such as proteins, DNA, RNA, and lipids, can promote intercellular interactions and microorganism adaptation to their surroundings. Antimicrobial resistance determinants are horizontally transferred across these structures (Caruana and Walper 2020). Extracellular vesicles, which help in secretion and intercellular interactions, are uniformly conserved. Although the exact mechanisms of their creation remain unknown, efforts to find vesicular mutants have failed, suggesting that these structures may play an important role in cell biology. Extracellular vesicles may thus be an intriguing target for antimicrobial drug production, both to suppress pathogens and to prevent antimicrobial resistance from spreading through extracellular vesicles (Yang et al. 2018).

16.4 Transcriptomics in Antimicrobial Resistance

Transcriptomics, also known as global gene expression analysis or genome-wide expression profiling, is a technique used for analyzing the entire collection of mRNA molecules (transcripts). Low-throughput methods for calculating single transcripts, such as northern blots and quantitative polymerase chain reaction (qPCR), were employed in the early studies of gene expression. Methods for quantification of gene expression via transcriptomics have advanced over the last few decades (Schena et al. 1995). Moreover, transcriptomics has made it possible to quantitate the expression of mRNA between different states at the genome scale, while sequencing and comparative genomics, which focus on DNA, are undynamic information and normally do not change significantly in response to short-term environmental changes (Zhang et al. 2010). The initial transcriptomic studies were carried out with hybridization-based microarrays, which offer an economical high-throughput choice (Schena et al. 1995). These approaches, however, have many drawbacks, like the need for a prior understanding of the sequences being investigated, cross-hybridization artifacts in the study of highly related sequences, and the inability to reliably measure low and high gene expression (Casneuf et al. 2007). Sequence-based approaches, in comparison with hybridization-based methods, have been designed to elucidate the transcriptome by specifically identifying the transcript sequence.

Next-generation sequencing (NGS) has transformed transcriptomics by allowing RNA analysis via cDNA sequencing. This method, known as RNA sequencing (RNA-Seq), outperforms earlier methods and has redefined the study of the transcriptome complexity and dynamism. Alternative splicing, gene expression, and allele-specific expression can all be studied more thoroughly and quantitatively with RNA-Seq. Deep profiling of the transcriptome and recent developments in the RNA-Seq workflow have facilitated the ability to explain various physiological and pathological conditions and have lightened the labor involved from sample preparation to sequencing platforms to bioinformatic data analysis (Kukurba and Montgomery 2015). Small RNAs (sRNAs) regulate resistance by base-pairing with mRNAs that are necessary for conferring resistance. Antibiotics cause bacteria to drastically alter their sRNA pool. An antibiotic produces a distinct sRNA profile, which may have effects that aid in overcoming the antibiotic challenge. As a result, regulatory RNAs, such as sRNAs, and their protein interaction partners, such as Hfq, may function as antimicrobial chemotherapy targets (Dersch et al. 2017). A more reliable RT-PCR technique can be used to verify transcriptome analysis through microarray-dependent hybridization. Transposon insertion sequencing (TIS) techniques like *transposon sequencing (Tn-Seq)* (van Opijnen and Levin 2020), *transposon-directed insertion site sequencing* (TraDIS) (Langridge et al. 2009), *insertion sequencing* (INseq) (Goodman et al. 2009), or *high-throughput insertion tracking by deep sequencing* (HITS) (Gawronski et al. 2009) may be used to further validate candidate genes or gene products known as possible targets through genomic and transcriptomic techniques, with the expected suppressive impacts of such mutations. TIS can provide details of the relative impact that disrupts each genomic feature on antibiotic sensitivity and can provide a better perception of how resistance appears, besides guiding the development of new strategies to target resistant bacteria. These studies have shown that antibiotics may have specific targets (e.g., in cell wall synthesis, DNA replication, or protein synthesis) and the genome encodes the bacterial response to antibiotics. Drugs like fluoroquinolones target topoisomerase IV and DNA gyrase, which are enzymes involved in DNA replication. Many TIS experiments cannot assay these targets directly owing to their inherent essentiality. TIS profiles generated under fluoroquinolone exposure implicate the involvement of other genes in DNA replication and repair, like *recN* and *xseA (*van Opijnen and Camilli 2012*;* Geisinger et al. 2018*)*. Antibiotic TIS profiles for each antibiotic tested, and each organism screened, show a role for genes beyond those related to the primary target, indicating the significance of genes with varied functions, like amino acid and carbohydrate metabolism, energy generation, and transport. This shows that although we have some knowledge of the mechanism involved in the inhibition of bacteria by an antibiotic, TIS can unravel this multifactorial process. Yet, conclusions can only be reached after a protein-level validation of antimicrobial targets (Chernov et al. 2019).

16.5 Proteomics in Antimicrobial Resistance

Proteomics is an important method for the study of the emergence and spread of antimicrobial resistance. They play an important role in understanding the molecular mechanisms of bacterial pathogenesis and determinants of diseases. Proteomics' physical associations contribute to methods for developing pathogen-specific treatment strategies that reduce antibiotic resistance. Proteomics is now commonly used to study microbial gene expression because of the recent rapid advances in whole-genome and protein sequencing techniques. Therefore, proteomics has become a dependable tool to study microbes (Khodadadi et al. 2020). Proteomics was first used to footprint the proteomes or sub-proteomes of antibiotic-sensitive and antibiotic-resistant microbes, as well as to classify the differences in protein abundance, a process known as "discovery proteomics." This form of study was later questioned because not all variations in abundant proteins lead to antibiotic resistance. To validate the proposed functions of the proteins of interest, functional validations such as gene deletion and overexpression, as well as enzyme activity, must be performed (Peng et al. 2019). The proteome is a complex system that reacts quickly to changes in the environment. Modification in the pattern of gene expression because of antimicrobials can be used to identify components of the cell that are impaired besides identifying their functions. The unique proteomic responses that appear against antimicrobials allow for the fast identification of mechanisms, compounds involved, and their cellular targets (Wenzel and Bandow 2011). In general, proteomics of AMR can be divided into two groups: the one that involves comparing resistant and susceptible bacteria, and responses of microbes to antibiotics. Generally, in the first method the least abundant proteins are related to secretion and metabolism, specifically *OmpW* (bacteriocin receptor). The commonly translated proteins are those involved in cell wall biogenesis, known resistance mechanisms, polysaccharide metabolism, and transport. The most cited of these is *Tolc*, an outer membrane channel. In the second case, studies on bacterial response to antibiotic challenge showed that proteins commonly involved are chaperones and those associated with stress response, amino acid, and energy metabolism. Interestingly, a few proteins involved in amino acid and energy metabolism are upregulated, while others are downregulated, indicating the complexity of antibiotic response. Researchers have also reported that some unique mechanisms or groups of proteins are produced at low or indiscernible levels in the unchallenged resistant organisms, and are regulated on antibiotic exposure pointing to them acting as potential targets in novel therapies (Pérez-Llarena and Bou 2016). The design of molecular networks and pathways is for classifying broad datasets obtained using omic technologies. Genes, mRNAs, proteins, metabolites, epigenetic factors, and posttranslational regulators found in the network's "nodes" could be used to produce new antimicrobials. Studying metabolites (the final substrates of the biochemical pathways) can help with the construction of such connections based on data from "omic" approach (Chernov et al. 2019).

16.6 Metabolomics in Antimicrobial Resistance

Metabolomics is rapidly evolving as one of the platforms of "omics." Metabolomics is the systematic identification and quantitation of all metabolites in a given organism or biological sample (Idle and Gonzalez 2007). Metabolomics offers high-throughput metabolite analysis and a forum for comparing biological samples at the same time, revealing their critical roles in metabolism, interorgan, and intraorgan communication. Because of the diverse chemical structure of metabolites, metabolome analysis is particularly difficult (Chandra Mohana et al. 2018). Modification in microbial metabolism, like those linked to biofilm formation or dormancy, regulates antibiotic susceptibility, hinting that bacterial metabolism and antibiotic resistance are correlated. Bacterial pathogen resistome analysis also shows that the production of intrinsic resistance necessitates the coordinated activity of several components, many of them may function in bacterial metabolism (Martínez and Rojo 2011). There is an exometabolome besides the cellular metabolome, which contains metabolites released into the exterior of the cell and is extremely important in the formation of AMR (Birkenstock et al. 2012). Profiling more than 500 intracellular and extracellular putative metabolites in 190 evolved populations revealed that carbon and energy metabolism strongly constrained the evolutionary trajectories, in terms of both speed and mode of resistance acquisition. This analysis, together with genome sequencing of resistant populations, identified condition-dependent compensatory mechanisms of antibiotic resistance, such as the shift from respiratory to fermentative metabolism of glucose upon overexpression of efflux pumps (Zampieri et al. 2017). Changes in the levels of metabolites that occur during the advent of AMR can aid in the identification of previously unknown antimicrobial therapy targets. Inhibiting these targets with other pharmaceuticals may act as a selection pressure, reducing the rate with which AMR mutations are fixed. As a result, this method may be employed in the development of integrated and multitargeted antimicrobial therapies, as well as new treatments that develop selection pressure on resistant phenotypes to return them to a susceptible state (Chernov et al. 2019).

16.7 Meta-omic Approaches in Antimicrobial Resistance

Meta-omic approaches have allowed working on uncultured species found in the environment. The widespread existence of ARGs in several presumably sterile habitats, despite no documented exposure to synthetic antimicrobials, is one of the most intriguing findings from functional metagenomic studies (Chernov et al. 2019). Random sequencing or PCR amplification of target genes is commonly used to analyze metagenomic clones. Alternatively, functional metagenomics can be used to find genes whose function is not known based on their sequence (Torres-Cortés et al. 2011). State-of-the-art high-throughput functional metagenomics helps scientists to multiplex up to 400 functional metagenomic selections on a single Illumina sequencing lane. A dedicated customized tool like PARFuMS (parallel annotation and reassembly of functional metagenomic selections) aids scientists to perform demultiplexing, quality filtering, trimming, assembly of the reads into full-length metagenomic fragments, and annotation in a single automated step, significantly lowering the cost of the experiment (Boolchandani et al. 2017). Metagenomic sequencing has aided the finding of new antimicrobial-resistant genes and their interactions by providing a potent way of obtaining information about ARGs, or "resistomes," that are the hallmark of different ecosystems. Current metagenomic approaches rely heavily on sequence similarity computation to predict antibiotic resistance attributes, which have substantial drawbacks. To allow consistent and accurate ARG recognition, such computations of similarity require a quality and cutting-edge antimicrobial resistance gene reference and annotation database. Furthermore, the absence of a systematic target gene for alignment limits the reach of such studies to previously identified ARGs. However, through using structural (3D) analogs, computational attempts have been made to annotate new antimicrobial-resistant genes in order to extend the sequence repertoire (Arango-Argoty et al. 2020).

16.8 Bioinformatics in Antimicrobial Resistance

Antimicrobial resistance gene prediction necessitates assurance of the data quality of genome, data trimming, genome assembly, and annotation. All of these procedures can now be performed in a short period using bioinformatic analysis, depending on factors such as computer processing capacity, database efficiency, and Internet speed (Chan 2016). The fast amassing of data on ARG necessitates the development of sophisticated databases for organizing and effortless accessibility to data. These databases not only store data, but also include bioinformatic resources for analyzing data for genes encoding possible antimicrobial biosynthetic pathways and AMR in microorganisms isolated from various sources, including microbes having unknown culture conditions (Chernov et al. 2019). Commercial or open-access platforms may be used to detect ARGs, and both require varying degrees of user skills. Freely available services are typically available online at public genome data centers or as an independent program for installation and use on a computer. Most bioinformatic software can process assembled sequences or raw reads, but to avoid ambiguous/false-positive results, users must check the validity of sequence that is submitted (Lal Gupta et al. 2020). As of now, there are close to 47 publicly available bioinformatic tools for detecting AMR in DNA or amino acid sequence data (Hendriksen et al. 2019). In this chapter, we discuss a few AMR detection methods and databases that are currently available.

16.9 Comprehensive Antibiotic Resistance Database (CARD)

CARD is a curated database offering reference sequences of protein and DNA, detection models, and the molecular biology of bacterial AMR. The antibiotic resistance ontology (ARO), developed to merge with software design for resistome analysis, such as CARD's Resistance Gene Identifier (RGI), aims at offering reference data of high quality (Alcock et al. 2020). The RGI uses BLAST to compare genome or assemblies to CARD and makes an all-around report with projected genes and targeted drug groups concerned with antibiotic resistance. This encompasses antibiotic resistance caused by mutations in their targets or by antibiotic resistance gene products. The information about resistance genes in CARD is connected to entries in Web resources like GenBank, PubChem, PubMed, and Protein Data Bank, allowing a connection between literature and databases (McArthur and Wright 2015). CARD has evolved significantly since 2017, with comprehensive reference sequence curation, updating the ontological framework, curation of over 500 novel antimicrobial resistance detection models, and development of a new classification approach and analytical methods. As of now, a new resistome and variant module, for example, offers statistical analysis and description of computer-assisted predicted resistance variants from 82 pathogens and over 100,000 genomes. A researcher can summarize predicted resistance using CARD, key out patterns in AMR mobility, and identify novel resistance variants (Alcock et al. 2020).

16.10 ResFinder

ResFinder is a widely cited tool that recognizes both short reads and assembled genome or contigs for tracking and locating gained ARGs in complete or partial sequences of bacterial isolates using KMA- and BLAST-based approaches. This tool includes a built-in reference ARG database that users can download and use for custom pipelines (Zankari et al. 2012). ResFinder is centered on a database of over 2000 ARGs that is searched using BLAST. A 50% identity over at least 3/5 of the length of the gene was laid as the criterion for notifying a match between a gene in the ResFinder and the input genome (Kleinheinz et al. 2014). ResFinder is a freely available Web-based tool. To make a prediction, the following four steps need to be taken: (1) Select the file containing the genome sequence applying the "Browse" button. (2) Under "Pick Antimicrobial configuration," the researcher can select the antimicrobial agents against which resistance genes are to be identified. ResFinder looks for resistance genes for all 12 forms of antimicrobial agents by default. (3) The "Select threshold for percent ID" dropdown menu is used to set the percent identity threshold among genes in the ResFinder database and genes in the input sequence. By default, the ResFinder database reports only genes in the input genome that are 98% similar to genes in the ResFinder. (4) The input genome's format has to be defined using the "Select sort of your reads" dropdown menu. The input sequence is assumed to be a draft or full genome in FASTA format by design. The time taken to study a genome is decided by a variety of factors. It is usually less than 10 min (Kleinheinz et al. 2014). ResFinder includes only acquired ARGs, making it ineffective for resistance detection caused by chromosomal mutations (Lal Gupta et al. 2020). The latest version of ResFinder was improved by rewriting the program and developing and broadening the database of PointFinder and ResFinder. It uses the same Web interface as old ResFinder versions. It also produces *in silico* antibiograms other than detecting AMR genes and chromosomal mutations. Antimicrobial resistance genes (ResFinder), gene mutations specific to chromosomes facilitating antimicrobial resistance (PointFinder), genotype-to-phenotype translation, and species-specific modules for *in silico* antibiograms are all included in the new version of ResFinder (Bortolaia et al. 2020).

16.11 PointFinder

A new Internet-aided tool for whole-genome sequence-based tracking of AMR in bacterial pathogens is linked to chromosomal point mutations. PointFinder is made up of a chromosomal gene database with reference sequences in fasta format and a database for chromosomal mutation with positions of codons and substitutions. Only hits with an identity of less than 80% are further examined by PointFinder, which uses BLASTN to determine the similarity for each gene in the chromosomal gene database. The software compares each position in the query with the corresponding position in the database sequence for each alignment. The chromosomal mutation database is used to save and compare all mismatches (Zankari et al. 2017). In their databases, both CARD and ARG-ANNOT have attempted to include chromosomal point mutations. Since both ARG-ANNOT and CARD do not take into account the microbial species, both methods produce mutations and sequences linked to mutational resistance, which are unrelated to the bacteria in question. To use these tools, the user must have prior knowledge of the mutational genes and mutations that they are searching. To fix some of these concerns, PointFinder was built to make the identification of chromosomal point mutations linked to resistance more user-friendly. The Web tool's functionality is user-friendly and easy to understand, describing observed nucleotide and amino acid codon shifts, mutations, and expected resistance (Zankari et al. 2017).

16.12 MEGARes

MEGARes is designed for high-throughput sequencing of already reported antimicrobial resistance genes in metagenomic results. It aims to offer an extensive database that includes all available reference sequences for genetic determinants of antimicrobial resistance. It also employs an acyclic hierarchical annotation system that aids high-throughput statistical and classification analysis. MEGARes is not an alternative to source databases like ResFinder and CARD, which offer comprehensive accession details, but instead provides an easier, hierarchical ontological framework that is specifically modeled for high-throughput analysis of metagenomic resistome data. MEGARes also emphasizes reported sequences than concentrating on discovering new variants. Metagenomic data are a source for discovering new AMR genes, which necessitate the use of specialized bioinformatic tools (e.g., fAR-Gene) (Berglund et al. 2019), Meta-MARC (Lakin et al. 2019) and PCM (Ruppé et al. 2019). MEGARes, on the other hand, aims to make bioinformatic detection of already reported sequences related to resistance to antimicrobial drugs, metals, and biocides that are more reproducible and statistically relevant (Doster et al. 2020).

16.13 PRAP (Pan Resistome Analysis Pipeline)

PRAP (Pan Resistome Analysis Pipeline) is a freely available pipeline for detecting antibiotic resistance genes (ARGs), characterizing pan-resistomes using annotations, and predicting the impact of ARG on resistance phenotypes using machine learning. PRAP helps to predict AMR phenotypes straight from whole-genome sequences, allowing researchers to dig deeper into possible ARG features. PRAP's workflow is divided into three sections: input file preprocessing, ARG recognition, and pan-resistome characterization. PRAP supports a number of sequence file formats for input data preprocessing, including raw reads files (fastq), GenBank annotation files (gb), fasta nucleic acid files (fna), and fasta amino acid files (faa). PRAP retrieves protein-coding sequences (CDSs) from GenBank annotation files and generates both fna and faa files. In a nutshell, PRAP allows the rapid recognition of ARGs from multiple genome files besides helping in the discovery of possible ARG transmission and distribution "laws" within the population (He et al. 2020).

16.14 Other Databases

Antibiotic Resistance Genes Database (ARDB) was an effort to make preliminary ontology and gene catalog for ARG. ARDB is a database that integrates virtually all the freely accessible ARG and offers an authentic annotation tool to scientists looking into the molecular biology of bacterial resistance (Liu and Pop 2009). Since ARDB was last updated in 2009, it does not have any newly explored ARGs (Arango-Argoty et al. 2020). ARG-ANNOT was the maiden database to locate point mutations in target genes in chromosomes linked to AMR. ARG-ANNOT maintains a repertoire containing sequences of genes linked to mutational resistance and particulars on the gene's location and mutation. Since ARG-ANNOT does not trace these mutations automatically, the user must manually search along the alignment to find them (Zankari et al. 2017). AMRFinderPlus is a tool developed by NCBI that uses protein annotations and assembled nucleotide sequences to identify AMR genes and resistance-associated point mutations. The pathogen detection pipeline utilizes AMRFinderPlus, and the results are shown in NCBI's Isolate Browser. AMRFinderPlus relies on curated collections of hidden Markov models and NCBI's curated Reference Gene Database (Feldgarden et al. 2019). Another Web-based tool, the Resistome Tracker, was created to make resistance data more available. It was created for the National Antimicrobial Resistance Monitoring System to help researchers study the contagious property of resistance by providing vast volumes of resistome data available to a wide range of users. Resistome Tracker helps in the discovery of new resistance factors, variations in the occurrence of resistance genes, and the spread of resistance genes over time. Resistome Tracker is also updated on a regular basis, allowing users to identify early resistance risks. Here, data dashboards are solely dedicated to the study and visualization of AMR genes derived from the NCBI's full genomes. User orchestrated searches for the data are possible with Resistome Tracker. It allows users to start a search with a particular resistance allele and finish with an evolutionary study of similar strains since it is directly linked to the NCBI pathogen database. It is currently based on foodborne bacteria, but it can be tweaked to look for AMR genes in any genome (Hendriksen et al. 2019). The structured ARG database (SARG) was developed by incorporating sequence information from ARDB and CARD in a hierarchical structure. SARG allows ARG-like sequences found through similarity searches to be automatically classified into various types and subtypes, saving time and effort. Using SARG, an online research pipeline (ARGs-OAP) is also open. ARGs-OAP was created to make analyzing metagenomic sequences for ARG data easier using the Galaxy Web server and SARG, which has a user-friendly interface. Users may use this pipeline to assess ARG profiles from many samples with one step. The pipeline aims to better understand the distribution and propagation of ARGs in various environments by using the power of metagenomic research (Yang et al. 2016). PathoFact is a pipeline for identifying virulence factors, toxins from bacteria, and ARGs in metagenomic data that is simple to use, modular, and reproducible. In addition, our tool blends pathogenicity factor estimation with the detection of mobile genetic elements. This adds complexity to the study by taking into account the genetic background of the relevant genes. PathoFact's modules for toxins, virulence factors, and ARGs may also be used separately, making it a versatile and portable method (de Nies et al. 2021).

AMR reference databases are divided into those for detecting resistance to specific antimicrobials in specific bacteria and solutions for detecting any AMR influencing sequences in any nucleotide or amino acid sequence. AMR reference databases have significant variations that users must know in order to choose the best database for their needs. First, the criteria for inclusion in AMR reference databases vary. For example, for CARD, submissions have to be published in peer-reviewed journals, but publication is not a prerequisite in ResFinder. A GenBank number is required for each gene, and GenBank entries need to be analyzed by an expert. In addition, the kinds of submissions vary between databases, with most containing antimicrobial resistance genes and some concerned with mutations in chromosomal genes facilitating antimicrobial resistance. Ultimately, the databases for AMR vary in terms of entry format (fasta, json, etc.), download capability, and curation capacity and frequency. To lessen ill-suited outputs due to differences in algorithms, synchronization of the databases is needed (Hendriksen et al. 2019). In theory, *in silico* AMR detection works by querying input data like nucleotide or protein sequence for the existence of a predetermined set of AMR determinants stored in reference databases using a search algorithm. Sequence data can be processed as reads or assemblies by various bioinformatic applications (Boolchandani et al. 2019). When utilizing assembly-based approaches, differences between assemblers can compromise the outcome's comparability. Following assembly, BLAST and hidden Markov model searches and so on are used for the comparison of the input data with the antimicrobial resistance reference databases. BLAST-based tools can produce a variety of results depending on the default settings for gene length and similarity percentage. If the settings are too high or too low, this can have a negative impact on specificity. Furthermore, approaches based on assembly are very intensive computationally. Despite these drawbacks, assembly-based tools can have more worth in AMR monitoring by allowing genetic analysis of the causative AMR genes, such as their mobilizing potential. To match reads to AMR databases, read-based methods may use a variety of tools, such as Bowtie2, BWA, and KMA (Boolchandani et al. 2019). The KMA (k-mer alignment) was recently developed to directly map raw reads against redundant AMR databases. In comparison with other methods of mapping such as BWA, which were designed for broad reference genomes such as the human genome and then applied scientifically to microbiology, the KMA tool was designed particularly for speedy and precise bacterial genome analyses. KMA uses the Needleman–Wunsch algorithm to correctly align extensions from k-mer seeds, as well as k-mer seeding to speed up mapping. To ensure a precise choice of templates, multi-mapping reads are handled by adopting a new sorting strategy (ConClave scheme) (Clausen et al. 2018). Read-based techniques enable the detection of antimicrobial resistance genes that are present in low abundance and genes that might miss out during incomplete assemblies (Boolchandani et al. 2019). The complexity of the analyzed dataset, the study's objectives, and the explicit questions posed influence the choice of suitable bioinformatic methods and ARG tools.

16.15 Global Collaborations for the Surveillance of AMR

AMR is a global problem in developing countries, where the burden is high and the process of replacing older antibiotics with new ones is daunting due to budget constraints (Ganguly et al. 2011). For addressing the factors influencing AMR, one must consider the healthcare system, the efficiency of regulatory authorities, the socioeconomic status, and geographical properties of a particular country (Winters and Gelband 2011). Many regulatory authorities, collaborations, and bodies are set up internationally for the management of AMR. In this session, we will briefly describe a few important ones. The Global Antibiotic Resistance Partnership (GARP) was started in 2009 for developing action plans and proposals relevant to low- and middle-income countries (Ganguly et al. 2011). GARP aims to define policy solutions by investigating the particular contexts of target countries. National working groups, with support from the Center for Disease Dynamics, Economics, and Policy (CDDEP), developed strategies tailored to local conditions in the GARP countries, based on the data collected and analyzed in this study. The strategies include reducing the antibiotic demand by cramping the occurrence of infections in hospitals and by controlling the use of antibiotics in humans and livestock (Winters and Gelband 2011). In India, a network of microbiology laboratories (Indian Network for Surveillance of Antimicrobial Resistance—INSAR) at reputed hospitals was formed with support from WHO. The network monitors AMR and assesses the extent of complications associated with it (Joshi et al. 2013). The FDA Centre for Veterinary Medicine (CVM) launched the National Antimicrobial Resistance Monitoring System (NARMS) in 1996 as a means to monitor the influence of antimicrobial use in agriculture moreover studying the evolution of AMR in clinical isolates (Gilbert et al. 2007). NARMS monitors AMR in microbes like *Campylobacter*, *Escherichia coli* O157:H7, non-Typhi *Salmonella*, *Enterococcus*, *Salmonella* Typhi, and *Shigella* (Holmes and Chiller 2004).

A monitoring system concentrating on European countries is the European Surveillance of Antimicrobial Consumption (ESAC-EU) project that was launched in 2001. ESAC is a network of monitoring systems that collect reliable data on the use of antibiotics in European countries (Coenen et al. 2007). ESAC-Net collects and analyzes data on antimicrobial consumption from both communities and hospitals. The data are used to give timely information on indicators of antimicrobial consumption (European Centre for Disease Prevention and Control 2021a). Another globally accepted initiative is the South African Society for Clinical Microbiology (SASCM). The objectives of SASCM include the monitoring of AMR patterns in the public and private medical sector in South Africa like the publication of AMR patterns in selected publications. SASCM handles the oversight of antimicrobial surveillance activities, and it is incumbent upon the society to educate and raise awareness around antimicrobial susceptibility testing issues that have implications for treatment and surveillance data (Singh-Moodley et al. 2018). Following the announcement of the World Health Organization's (WHO) Global Action Plan on Antimicrobial Resistance in 2015, AMR has begun to receive international political publicity. The member states of the Pan American Health Organization (PAHO) supported this in a resolution adopted in September 2015, emphasizing the political allegiance of all associated countries to develop AMR action plans. The Global Action Plan lays out five strategies for containing AMR: education and awareness, surveillance and study, infection prevention and control, effective antimicrobial use and research, and the establishment of the economic case for long-term investment. A multisectoral One Health response is key to implementing these action plans. Countries have made progress in policies, measures, and strategies since the adoption of the Global Action Plan to ensure the successful design and execution of their action plans in line with the regional plan of action and the One Health strategy for containing AMR (Pan American Health Organization 2021).

The European Antimicrobial Resistance Surveillance Network (EARS-Net) is Europe's largest publicly funded surveillance system for AMR. Data from EARS-Net are crucial in raising awareness among public health officials, scientists, and the general public. The European Antimicrobial Resistance Surveillance System (EARSS), established in 1998, is the predecessor of EARS-Net. The objectives of EARS-Net are to collect comparable and accurate AMR data, analyze spatiotemporal tendencies of AMR in Europe, and encourage the implementation and improvement of national AMR surveillance programs (European Centre for Disease Prevention and Control 2021b). GLASS (National Antimicrobial Resistance Surveillance System) is a WHO-supported system that was introduced in October 2015 to promote a systematic approach for the collection, review, and sharing of data related to AMR at the global and regional levels. GLASS aims to promote national surveillance frameworks and harmonized global standards, estimate the global scope and burden of AMR using selected metrics, analyze and monitor global AMR data regularly, identify emerging resistance and its international spread, inform the implementation of targeted prevention and control initiatives, and evaluate the impact of interventions (Basheera 2020). All these agencies and collaborations result in an effective exchange of information, and this can certainly help in the effective monitoring of AMR.

16.16 Conclusions

The NGS platform, combined with bioinformatic tools, is nowadays becoming the tool for the study and prediction of AMR. This is primarily due to reducing expense and reduced overall time spent on the experiment. Today, for typical bacteria, the entire process of analyzing AMR from raw sequence reads needs only 3 min on a laptop using Mykrobe predictor software (Bradley et al. 2015). The co-carriage of particular genes having various multidrug-resistant patterns, their genetic background including the possibility for horizontal transfer, allelic trends over time, and their distribution can be revealed using bioinformatic analysis. Furthermore, the presence of co-resistances not identified by typical drug workflows, such as those conferred by disinfectant and heavy metals, is revealed. This degree of "deep surveillance" will reveal additional potential modulators of AMR perseverance and phylogeny as well as the possibility of a more polished microbial risk assessment based on traits that confer resistance. The kind of approved input data, the existence of tools for searching within an AMR determinant database that can be unique to a tool or cloned from other resources, and the search method used (based on mapping or alignment) are some factors that differentiate bioinformatic resources. As a result, each method has potency and constraints in terms of sensitivity and specificity of AMR determinant detection, as well as in terms of implementation, which have been highlighted in benchmarking exercises and research papers for some methods. Today, the number of tools and DNA sequence data available is increasing, allowing for worldwide pathogen and AMR monitoring based on genomic data. Calibration of databases and pipelines, and prediction of phenotype based on data are, however, needed for extensive workflows. A dependable genomic workflow to assay AMR gene necessitates well-curated, harmonized, and constantly updated reference databases to ensure comparable results. Even though there are challenges, the bioinformatic approach is certainly a high-throughput method to do monitoring and surveillance of AMR globally.

References

Alcock BP, Raphenya AR, Lau TTY, Tsang KK, Bouchard M, Edalatmand A et al (2020) CARD 2020: antibiotic resistome surveillance with the comprehensive antibiotic resistance database. Nucleic Acids Res 48(D1):D517–D525

Alekshun MN, Levy SB (2007) Molecular mechanisms of antibacterial multidrug resistance. Cell 128(6):1037–1050

Amsalu A, Sapula SA, De Barros LM, Hart BJ, Nguyen AH, Drigo B et al (2020) Efflux pump-driven antibiotic and biocide cross-resistance in Pseudomonas aeruginosa isolated from different ecological niches: a case study in the development of multidrug resistance in environmental hotspots. Microorganisms 8(11):1647

Andersson DI (2003) Persistence of antibiotic resistant bacteria. Curr Opin Microbiol 6(5):452–456

Andersson DI (2006) The biological cost of mutational antibiotic resistance: any practical conclusions? Curr Opin Microbiol 9(5):461–465

Arango-Argoty GA, Guron GKP, Garner E, Riquelme MV, Heath LS, Pruden A et al (2020) ARGminer: a web platform for the crowdsourcing-based curation of antibiotic resistance genes. Bioinformatics 36(9):2966–2973

Basheera V (2020) Global antimicrobial resistance; a peek in to the GLASS data. Asian J Pharm Health Sci [Internet] [cited 2021 Apr 24];10(1). http://ajphs.com/article/2020/10/1/2197-2207

Berglund F, Österlund T, Boulund F, Marathe NP, Larsson DGJ, Kristiansson E (2019) Identification and reconstruction of novel antibiotic resistance genes from metagenomes. Microbiome 7(1):52

Birkenstock T, Liebeke M, Winstel V, Krismer B, Gekeler C, Niemiec MJ et al (2012) Exometabolome analysis identifies pyruvate dehydrogenase as a target for the antibiotic triphenylbismuthdichloride in multiresistant bacterial pathogens. J Biol Chem 287(4):2887–2895

Boolchandani M, Patel S, Dantas G (2017) Functional metagenomics to study antibiotic resistance. Methods Mol Biol 1520:307–329

Boolchandani M, D'Souza AW, Dantas G (2019) Sequencing-based methods and resources to study antimicrobial resistance. Nat Rev Genet 20(6):356–370

Bortolaia V, Kaas RS, Ruppe E, Roberts MC, Schwarz S, Cattoir V et al (2020) ResFinder 4.0 for predictions of phenotypes from genotypes. J Antimicrob Chemother 75(12):3491–3500

Bradley P, Gordon NC, Walker TM, Dunn L, Heys S, Huang B et al (2015) Rapid antibiotic-resistance predictions from genome sequence data for Staphylococcus aureus and Mycobacterium tuberculosis. Nat Commun 6(1):10063

Bush K (2018) Past and present perspectives on β-lactamases. Antimicrob Agents Chemother [Internet] [cited 2021 Mar 18];62(10). https://aac.asm.org/content/62/10/e01076-18

Caruana JC, Walper SA (2020) Bacterial membrane vesicles as mediators of microbe – microbe and microbe – host community interactions. Front Microbiol [Internet] [cited 2021 Mar 19];11. https://www.frontiersin.org/articles/10.3389/fmicb.2020.00432/full

Casneuf T, Van de Peer Y, Huber W (2007) In situ analysis of cross-hybridisation on microarrays and the inference of expression correlation. BMC Bioinformatics 8(1):461

Chan K-G (2016) Whole-genome sequencing in the prediction of antimicrobial resistance. Expert Rev Anti-Infect Ther 14(7):617–619

Chandra Mohana N, Yashavantha Rao HC, Rakshith D, Mithun PR, Nuthan BR, Satish S (2018) Omics based approach for biodiscovery of microbial natural products in antibiotic resistance era. J Genet Eng Biotechnol 16(1):1–8

Chernov VM, Chernova OA, Mouzykantov AA, Lopukhov LL, Aminov RI (2019) Omics of antimicrobials and antimicrobial resistance. Expert Opin Drug Discov 14(5):455–468

Clausen PTLC, Aarestrup FM, Lund O (2018) Rapid and precise alignment of raw reads against redundant databases with KMA. BMC Bioinformatics 19(1):307

Coenen S, Ferech M, Haaijer-Ruskamp FM, Butler CC, Stichele RHV, Verheij TJM et al (2007) European Surveillance of Antimicrobial Consumption (ESAC): quality indicators for outpatient antibiotic use in Europe. Qual Saf Health Care 16(6):440–445

Dersch P, Khan MA, Mühlen S, Görke B (2017) Roles of regulatory RNAs for antibiotic resistance in bacteria and their potential value as novel drug targets. Front Microbiol [Internet] [cited 2021 Mar 19];8. https://www.ncbi.nlm.nih.gov/pmc/articles/PMC5418344/

Doster E, Lakin SM, Dean CJ, Wolfe C, Young JG, Boucher C et al (2020) MEGARes 2.0: a database for classification of antimicrobial drug, biocide and metal resistance determinants in metagenomic sequence data. Nucleic Acids Res 48(D1):D561–D569

European Centre for Disease Prevention and Control (2021a) About the network [Internet] [cited 2021 Apr 24]. https://www.ecdc.europa.eu/en/about-us/partnerships-and-networks/disease-and-laboratory-networks/esac-net-about

European Centre for Disease Prevention and Control (2021b) European Antimicrobial Resistance Surveillance Network (EARS-Net) [Internet] [cited 2021 Apr 24]. https://www.ecdc.europa.eu/en/about-us/partnerships-and-networks/disease-and-laboratory-networks/ears-net

Feldgarden M, Brover V, Haft DH, Prasad AB, Slotta DJ, Tolstoy I et al (2019) Validating the AMRFinder tool and resistance gene database by using antimicrobial resistance genotype-phenotype correlations in a collection of isolates. Antimicrob Agents Chemother 63(11):e00483

Ganguly NK, Arora NK, Chandy SJ, Fairoze MN, Gill JP, Gupta U et al (2011) Rationalizing antibiotic use to limit antibiotic resistance in India. Indian J Med Res 134:281–294
Gawronski JD, Wong SMS, Giannoukos G, Ward DV, Akerley BJ (2009) Tracking insertion mutants within libraries by deep sequencing and a genome-wide screen for Haemophilus genes required in the lung. Proc Natl Acad Sci U S A 106(38):16422–16427
Geisinger E, Mortman NJ, Vargas-Cuebas G, Tai AK, Isberg RR (2018) A global regulatory system links virulence and antibiotic resistance to envelope homeostasis in Acinetobacter baumannii. PLoS Pathog 14(5):e1007030
Gilbert JM, White DG, McDermott PF (2007) The US national antimicrobial resistance monitoring system. Future Microbiol 2(5):493–500
Goodman AL, McNulty NP, Zhao Y, Leip D, Mitra RD, Lozupone CA et al (2009) Identifying genetic determinants needed to establish a human gut symbiont in its habitat. Cell Host Microbe 6(3):279–289
Handel A, Regoes RR, Antia R (2006) The role of compensatory mutations in the emergence of drug resistance. PLoS Comput Biol 2(10):e137
He Y, Zhou X, Chen Z, Deng X, Gehring A, Ou H et al (2020) PRAP: Pan Resistome analysis pipeline. BMC Bioinformatics 21(1):20
Hendriksen RS, Bortolaia V, Tate H, Tyson GH, Aarestrup FM, McDermott PF (2019) Using genomics to track global antimicrobial resistance. Front Public Health [Internet] [cited 2021 Mar 6];7. https://www.frontiersin.org/articles/10.3389/fpubh.2019.00242/full
Holmes CN, Chiller TM (2004) National Antibiotic Resistance Monitoring System for enteric bacteria. Emerg Infect Dis 10(11):2061
Idle JR, Gonzalez FJ (2007) Metabolomics. Cell Metab 6(5):348–351
Interagency Coordination Group on Antimicrobial Resistance (2019) No time to wait: securing the future from drug-resistant infections. World Health Organization [cited 2021 Apr 25]. https://www.who.int/antimicrobial-resistance/interagency-coordinationgroup/IACG_final_report_EN.pdf?ua=1
Joshi S, Ray P, Manchanda V, Bajaj J, Chitnis DS, Gautam V et al (2013) Methicillin resistant Staphylococcus aureus (MRSA) in India: prevalence & susceptibility pattern. Indian J Med Res 137(2):363–369
Khodadadi E, Zeinalzadeh E, Taghizadeh S, Mehramouz B, Kamounah FS, Khodadadi E et al (2020) Proteomic applications in antimicrobial resistance and clinical microbiology studies. Infect Drug Resist 13:1785–1806
Kleinheinz KA, Joensen KG, Larsen MV (2014) Applying the ResFinder and VirulenceFinder web-services for easy identification of acquired antibiotic resistance and E. coli virulence genes in bacteriophage and prophage nucleotide sequences. Bacteriophage 4(2):e27943
Kukurba KR, Montgomery SB (2015) RNA sequencing and analysis. Cold Spring Harb Protoc 2015(11):951–969
Lakin SM, Kuhnle A, Alipanahi B, Noyes NR, Dean C, Muggli M et al (2019) Hierarchical Hidden Markov models enable accurate and diverse detection of antimicrobial resistance sequences. Commun Biol [Internet] [cited 2021 Mar 21];2. https://www.ncbi.nlm.nih.gov/pmc/articles/PMC6684577/
Lal Gupta C, Kumar Tiwari R, Cytryn E (2020) Platforms for elucidating antibiotic resistance in single genomes and complex metagenomes. Environ Int 138:105667
Langridge GC, Phan M-D, Turner DJ, Perkins TT, Parts L, Haase J et al (2009) Simultaneous assay of every salmonella Typhi gene using one million transposon mutants. Genome Res 19(12):2308–2316
Levy SB, Marshall B (2004) Antibacterial resistance worldwide: causes, challenges and responses. Nat Med 10(12):S122–S129
Liu B, Pop M (2009) ARDB—Antibiotic Resistance Genes Database. Nucleic Acids Res 37(suppl_1):D443–D447
Martínez JL, Rojo F (2011) Metabolic regulation of antibiotic resistance. FEMS Microbiol Rev 35(5):768–789

McArthur AG, Wright GD (2015) Bioinformatics of antimicrobial resistance in the age of molecular epidemiology. Antimicrob Microb Syst Biol 27:45–50
Medvedeva ES, Davydova MN, Mouzykantov AA, Baranova NB, Grigoreva TY, Siniagina MN et al (2016) Genomic and proteomic profiles of Acholeplasma laidlawii strains differing in sensitivity to ciprofloxacin. Dokl Biochem Biophys 466(1):23–27
Mohr KI (2016) History of antibiotics research. Curr Top Microbiol Immunol 398:237–272
de Nies L, Lopes S, Busi SB, Galata V, Heintz-Buschart A, Laczny CC et al (2021) PathoFact: a pipeline for the prediction of virulence factors and antimicrobial resistance genes in metagenomic data. Microbiome 9(1):49
van Opijnen T, Camilli A (2012) A fine scale phenotype–genotype virulence map of a bacterial pathogen. Genome Res 22(12):2541–2551
van Opijnen T, Levin HL (2020) Transposon insertion sequencing, a global measure of gene function. Annu Rev Genet 54(1):337–365
Pan American Journal of Public Health. Special issue on antimicrobial resistance, vol 30, no. 6. December 2011 - PAHO/WHO | Pan American Health Organization [Internet] [cited 2021 Apr 24]. https://www.paho.org/en/documents/pan-american-journal-public-health-special-issue-antimicrobial-resistance-vol-30-no-6-0
Peng B, Li H, Peng X (2019) Proteomics approach to understand bacterial antibiotic resistance strategies. Expert Rev Proteomics 16(10):829–839
Pérez-Llarena FJ, Bou G (2016) Proteomics as a tool for studying bacterial virulence and antimicrobial resistance. Front Microbiol [Internet] [cited 2021 Apr 25];7. https://www.frontiersin.org/articles/10.3389/fmicb.2016.00410/full
Piddock LJV (2006) Multidrug-resistance efflux pumps - not just for resistance. Nat Rev Microbiol 4(8):629–636
Ruppé E, Ghozlane A, Tap J, Pons N, Alvarez A-S, Maziers N et al (2019) Prediction of the intestinal resistome by a three-dimensional structure-based method. Nat Microbiol 4(1):112–123
Schena M, Shalon D, Davis RW, Brown PO (1995) Quantitative monitoring of gene expression patterns with a complementary DNA microarray. Science 270(5235):467–470
Singh-Moodley A, Ismail H, Perovic O (2018) An overview of antimicrobial resistance surveillance among healthcare-associated pathogens in South Africa. Afr J Lab Med 7(2):1–6
Torres-Cortés G, Millán V, Ramírez-Saad HC, Nisa-Martínez R, Toro N, Martínez-Abarca F (2011) Characterization of novel antibiotic resistance genes identified by functional metagenomics on soil samples. Environ Microbiol 13(4):1101–1114
Vila J, Martí S, Sánchez-Céspedes J (2007) Porins, efflux pumps and multidrug resistance in Acinetobacter baumannii. J Antimicrob Chemother 59(6):1210–1215
Wenzel M, Bandow JE (2011) Proteomic signatures in antibiotic research. Proteomics 11(15):3256–3268
Winters C, Gelband H (2011) Part I. The global antibiotic resistance partnership (GARP). S Afr Med J 101(8 pt 2):556–557
Wright GD (2007) The antibiotic resistome: the nexus of chemical and genetic diversity. Nat Rev Microbiol 5(3):175–186
Yang Y, Jiang X, Chai B, Ma L, Li B, Zhang A et al (2016) ARGs-OAP: online analysis pipeline for antibiotic resistance genes detection from metagenomic data using an integrated structured ARG-database. Bioinformatics 32(15):2346–2351
Yang J, Kim EK, McDowell A, Kim Y-K (2018) Microbe-derived extracellular vesicles as a smart drug delivery system. Transl Clin Pharmacol 26(3):103
Yin X, Jiang X-T, Chai B, Li L, Yang Y, Cole JR et al (2018) ARGs-OAP v2.0 with an expanded SARG database and Hidden Markov models for enhancement characterization and quantification of antibiotic resistance genes in environmental metagenomes. Bioinformatics 34(13):2263–2270
Zampieri M, Enke T, Chubukov V, Ricci V, Piddock L, Sauer U (2017) Metabolic constraints on the evolution of antibiotic resistance. Mol Syst Biol [Internet] [cited 2021 Apr 25];13(3). https://www.ncbi.nlm.nih.gov/pmc/articles/PMC5371735/

Zankari E, Hasman H, Cosentino S, Vestergaard M, Rasmussen S, Lund O et al (2012) Identification of acquired antimicrobial resistance genes. J Antimicrob Chemother 67(11):2640–2644
Zankari E, Allesøe R, Joensen KG, Cavaco LM, Lund O, Aarestrup FM (2017) PointFinder: a novel web tool for WGS-based detection of antimicrobial resistance associated with chromosomal point mutations in bacterial pathogens. J Antimicrob Chemother 72(10):2764–2768
Zhang W, Li F, Nie L (2010) Integrating multiple "omics" analysis for microbial biology: application and methodologies. Microbiol Read Engl 156(Pt 2):287–301

Chapter 17
Antimicrobial Activity of Bioactive Compounds (Thymoquinone and Eugenol) and Its Nanoformulation Therapeutic Potential

Jasmeet Kaur, Mohamad Taleuzzaman, and Iqra Rahat

17.1 Introduction

Essential oils, generally representing the essence of the all-natural plant and common feature of all plant extracts, have been widely used since ancient times due to their wide range of pharmacological effects. In fact, essential oils are important bioactive constituents of medicines from herbal plants for the development of good health with very few side effects. Further, based on recent scientific publications, eugenol and thymoquinone claim to be interesting bioactive compounds exerting beneficial effects on humans and possessing broad-spectrum antimicrobial activity.

17.1.1 Chemistry and Sources

These compounds comprise isoprene units (C_6H_8) and are being used to control, cure, or prevent microbial infections. Eugenol, belonging to the class of phenylpropanoids, is amphipathic hydroxyphenyl propene with the chemical formula $C_{10}H_{12}O_2$. This aromatic yellowish oily liquid has a pleasant odor and taste. It is chemically expressed as 4-allyl-2-methoxy phenol or 1-allyl-4-hydroxy-3-methoxybenzene (Mishra et al. 2013).

J. Kaur
Department of Pharmacognosy, Faculty of Pharmacy, School of Pharmaceutical Education and Research, Jamia Hamdard, New Delhi, India

M. Taleuzzaman (✉)
Department of Pharmaceutical Chemistry, Faculty of Pharmacy, Maulana Azad University, Village Bujhawar, Tehsil Luni, Jodhpur, Rajasthan, India

I. Rahat
Department of Pharmaceutics, Glocal School of Pharmacy, Glocal University, Saharanpur, Uttar Pradesh, India

N. Akhtar et al. (eds.), *Emerging Modalities in Mitigation of Antimicrobial Resistance*,
https://doi.org/10.1007/978-3-030-84126-3_17

17.1.1.1 Eugenol

Eugenol was first isolated as a volatile compound in the year 1929 from *Eugenia caryophyllata* and was thus named after this genus. Eugenol is present in a variety of herbal plants including *Melissa officinalis* L. (lemon balm) (Jaganathan and Supriyanto 2012), *Abutilon indicum* L. (country mallow) (Kumar et al. 2013), *Aegle marmelos* (Indian bael), *Illicium anisatum* (Japanese star anise) (Jaganathan and Supriyanto 2012), *Myristica fragrans* (nutmeg), *Pogostemon cablin* (patchouli) (Chen et al. 2013), *Piper betel* (betel) (Gundala and Aneja 2014), *Acorus calamus* (sweet flag) (Mythili Avadhani et al. 2013), and *Aphanamixis polystachya* (pithraj) (Shaikh et al. 2012). Also, an important source of eugenol includes *Syzygium aromaticum* (clove buds and leaves), various *Cinnamon* species such as *Cinnamomum zeylanicum*, *C. verum*, and *C. cassia*, and *Ocimum* species such as *Ocimum basilicum* L, *O. sanctum* L, *O. americanum* L, *O. gratissimum* L, *O. micranthum* Wild, *O. minimum* L., and *O. adscendens* Wild (Marchese et al. 2017). Furthermore, eugenol can be commercially produced via biotransformation method involving various microorganisms such as *Corynebacterium* sp., *Bacillus cereus*, *Pseudomonas* sp., and *Streptomyces* sp. and chemically synthesized by using allyl chloride for allylation of guaiacol (Mishra et al. 2013; Molina et al. 2013).

Eugenol is a highly volatile compound with limited solubility and low chemical stability. Due to volatility and limited stability, it is highly sensitive to oxidation and chemical interactions (Majeed et al. 2016). Upon administration, eugenol is rapidly absorbed and metabolized in the liver with an excretion rate of 95% within 24 h. Thus, it is necessary to encapsulate eugenol to improve its efficacy and solubility while preventing its early absorption (Majeed et al. 2016; Joris et al. 2008). A wide range of biological activities has been reported to be associated with eugenol such as analgesic, antiviral, antispasmodic, anticancer, anti-inflammatory, antiseptic, and antioxidant. It has been reported to be used for the treatment of many diseases like acne, arthritis, warts, asthma, diarrhea, skin diseases, and various allergies (Arana-Sánchez et al. 2010; Shah et al. 2013).

17.1.1.2 Thymoquinone

On the other hand, thymoquinone is a naturally occurring phytochemical compound that belongs to the chemical class of benzoquinones. It is a main bioactive constituent present in the seeds of the plant *Nigella sativa* (black seed), belonging to the Ranunculaceae family. The chemical formula of thymoquinone is $C_{10}H_{12}O_2$ with a molecular weight of 162.4 g/mol and is chemically expressed as 2-isopropyl-5-methylbenzo-1, 4-quinone. The substance was first isolated in the year 1963 and appears to be a potential candidate against many diseases (Abu Khader and Khan 2017). Besides, *Nigella sativa* being the main source of thymoquinone, the substance is present in a large number of plants such as *Nigella sativa* L. seeds (black bread weed) (Havlik et al. 2006), *Satureja* (winter savory) (Kubátová et al. 2001), *Origanum* (oregano) (Lukas et al. 2009) (Economakis et al. 2002; Ipek et al. 2005),

Agastache (giant hyssops), *Monarda* (bee balm), *Cupressus*, *Coridothymus* (wild thyme), *Thymus* (garden thyme) (Fleisher and Fleisher 2000), and *Juniperus* (Mastelić et al. 2000). The solubility of the substance ranges from 0.549 to 0.740 mg/ml for a period of 24–72 h and is thus categorized as a fat-soluble molecule. Therefore, the drug is highly unstable in an aqueous solution. Upon oral administration, the drug undergoes chemical and enzymatic alterations in the gastrointestinal tract and follows hepatic first-pass metabolism (Abu Khader and Khan 2017). The use of nanocarriers for the enhancement of bioavailability of thymoquinone seems to be a plausible alternative. Interestingly, *Nigella sativa* seeds have been traditionally reported to be used for the treatment of various ailments such as asthma, cough, headache, bronchitis, fever, influenza, and rheumatism. Moreover, many kinds of literature have reported the beneficial effects of *Nigella sativa* seeds as antidiabetic, immunomodulatory, anxiolytic, antioxidative, anti-inflammatory, antidepressants, antipsychotic, anticonvulsant, gastroprotective, antibacterial, hepatoprotective, cardioprotective, anticancer, and memory impairments (El-Far 2015; Darakhshan et al. 2015; El-Far et al. 2014).

17.1.1.3 Significance of Natural Products

Herbal drugs being natural products are valued for their use in various health ailments since the prehistoric period. Moreover, as these drugs are plant extracts, the side effects and other complications associated with these drugs are very few. Herbs are a rich source of ingredients that plays a critical role in drug synthesis and development. Many herbs are used for a wide range of biomedical applications such as gene therapy, tissue engineering, the medicinal industry, and skin regeneration. Many herbs are used as blood cleansers, which are responsible for eliminating and altering metabolic toxins. Hence, these drugs can be used to improve immunity and are used in fevers, as Chirata. Herbs such as aloe, ginseng, cinnamon, black pepper, and sandalwood are used in case of boils, wounds, and sores. Many herbs also possess antibiotic properties such as turmeric. It works by inhibiting the growth of harmful microbes, germs, and bacteria. Moreover, it is widely used since ancient times for healing cuts and wounds. Some herbs such as marshmallow roots act as antacids by neutralizing the gastric acid. Traditional Indian medicine practitioners recommend the use of Indian sage for snake bites. Many herbs such as ginger, eucalyptus, cardamom, and cloves possess the expectorant property and are used in cough syrups. These herbs are responsible for the elimination of mucus from the lungs. Apart from these, herbs are widely used as an ingredient in the food, dye industry, pest control, and perfume industry. Despite these advantages, herbal drugs possess low solubility, an increased absorption, higher instability, and rapid metabolism and excretion. Therefore, the administration of herbal drugs requires modification for achieving enhanced efficacy and to achieve sustained release. To achieve a rational dosage regime of herbal drugs, it is important to understand the bioavailability and pharmacokinetic profile of these drugs. These parameters require the usage of new technologies such as nanotechnology for the formulation of herbal drugs to achieve site-specific delivery at a predetermined rate. Nanoformulation of herbal drugs includes many benefits such as targeted drug delivery, enhanced component solubility, reduced dose, increased absorption, reduced metabolism and elimination of bioactive ingredients, and enhanced bioavailability (Huang and Chang 2009). It helps in improving stability and achieving precision targeting, thereby improving the effectiveness of phytoconstituent. This can be achieved by incorporation or encapsulation of herbal drugs in a suitable carrier system such as liposomes, nanoparticles, and nanoemulsions, which can convert a poorly available herbal drug into a favorably bioavailable drug candidate.

17.2 Pharmacology

17.2.1 Eugenol

The antimicrobial activity of eugenol is due to the presence of free radical scavengers in the molecules. These free radicals are responsible for blocking the production of reactive oxygen species (ROS). Several pieces of kinds of literature have been reported for antimicrobial activity of eugenol. In the year 2004, Burt along with his coworkers hypothesized that eugenol prevents enzyme action by binding to the proteins (Burt 2004). Later in 2006, Gill and Holley demonstrated that eugenol affects membrane nonspecific permeability and alters ion and ATP transport by disrupting the cytoplasmic membrane (Gill and Holley 2006). This result was further validated by Filgueiras and Vanetti and concluded that eugenol increases phosphate ions permeability upon addition of bacterial concentration (Filgueiras et al. 2006). In particular, eugenol has been tested against a broad spectrum of microorganisms that are responsible for causing infectious diseases in humans. For example, microorganisms responsible for causing respiratory infections such as *Stenotrophomonas maltophilia* and *S. pneumonia* were found to be most susceptible to eugenol with inhibition zone (IZ): 28 mm, followed by an inhibition zone of 27 mm for *Haemophilus influenzae*. The researchers reported *S. aureus* and *Klebsiella pneumonia* to be the most resistant strains with IZ of 18 and 10 mm (Fabio et al. 2007). In the same year, other researchers reported the antibacterial activity of eugenol against various Gram-negative bacteria including *E. coli*, *S. typhimurium*, *S. typhi*, *Shigella flexneri*, and *P. aeruginosa* (Dabur et al. 2007). In 2009, Pei et al. demonstrated the antibacterial activity of eugenol against *E. coli* and reported its minimum bacterial concentration of 1600 mg/l using broth macrodilution assay (Pei et al. 2009). Devi along with his coworkers demonstrated that eugenol alters membrane permeability leading to extensive loss of ions and cellular content followed by cell death. The authors evaluated this mechanism of action for antibacterial activity against *Salmonella typhi* (Devi et al. 2010). In the later years, Das along with his coworkers demonstrated the antibacterial activity of eugenol against *S. aureus*. The authors evaluated that eugenol was able to produce intracellular reactive oxygen species, which results in inhibition of cell growth, cell membrane disruption, and DNA damage followed by decomposition of cell and ultimately leading to cell death (Das et al. 2016). Another investigation shows the activity of eugenol against bacterial enzymes such as amylase, protease, ATPase, and histidine carboxylase (Hyldgaard et al. 2012). The major mechanism of action of eugenol is described in Fig. 17.1.

Besides being highly effective against Gram-negative and Gram-positive bacteria, some researchers have evaluated its activity against fungi. Eugenol being lipophilic in nature, the authors evaluated that this may enter in the lipid bilayer membrane resulting in altering its fatty acyl chains, which ultimately disturbs the membrane fluidity and permeability. Ahmad along with his coworkers evaluated the activity of eugenol against *Candida* spp. The researchers found that eugenol inhibited the H^+-ATPase and glucose-stimulated H^+-extrusion of *Candida* spp. at a minimum inhibitory concentration of 500 µg/ml (Ahmad et al. 2010). Moreover, other researchers showed that the treatment of *Candida albicans* with eugenol and methyl eugenol resulted in induction of oxidative stress by increasing superoxide dismutase levels and elevating ROS, which causes peroxidation of lipids present in cytoplasmic membrane leading to cell death (Khan et al. 2011). In fact, the first study for antifungal activity of eugenol was conducted in 1982 against 31 strains of *Candida albicans* and 33 strains of *Cryptococcus* reforms (Boonchird and Flegel 1982). Moreover, the alteration in the structure of eugenol results in more potent derivatives of eugenol. For example, the addition of morpholinyl group in the eugenol results in much higher antibacterial activity as compared to eugenol (Abrão et al.

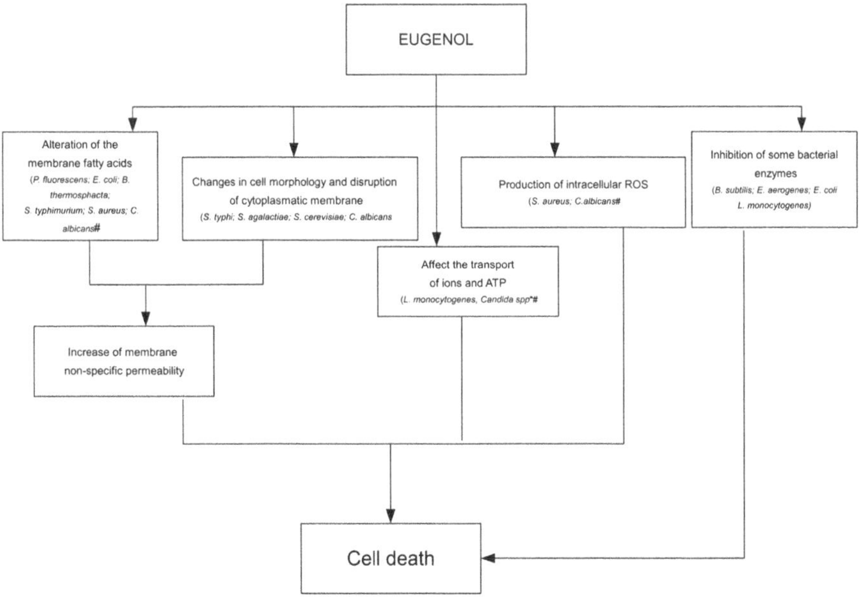

Fig. 17.1 Eugenol mechanism of action against bacteria and fungi

2015). The activity of per acetyl glucoside derivative of eugenol showed better potency against *Candida glabrata* and was less cytotoxic than eugenol (Hipólito et al. 2018) (Fig. 17.1).

To exhibit the synergistic effect of eugenol, it is necessary to combine the essential oil or standard antibiotics with the eugenol. This synergism results in various interactions such as interfering of substance with the cell wall, which results in an increased sensitivity toward the other substance and inhibitory effects of both substances in the same biochemical pathways. For example, eugenol and thymol result in synergism where thymol allows eugenol to reach the cytoplasm and interact with enzymes by disintegrating and disrupting the outer cell membrane of Gram-negative bacteria (Bassolé and Juliani 2012). Another example involves the additive and synergistic effect of Streptomycin and eugenol against *L. monocytogenes* and *S. typhimurium* (Liu et al. 2015).

17.2.2 Thymoquinone

Interestingly, the tiny seeds of *Nigella sativa* are a rich source of thymoquinone. Many scientific articles have been published, which refer to the antimicrobial activity of thymoquinone and black seed. This bioactive constituent shows potent activity against a broad spectrum of Gram-negative and Gram-positive bacteria such as *Listeria monocytogenes, Micrococcus luteus, Bacillus aureus, Enterococcus faecalis, Staphylococcus aureus, Staphylococcus epidermidis, Escherichia coli, Pseudomonas aeruginosa, Typhimurium, Vibrio* spp., and Salmonella enteric by preventing the formation of bacterial biofilm (Chaieb et al. 2011). The diverse therapeutic action of thymoquinone is demonstrated in Fig. 17.2.

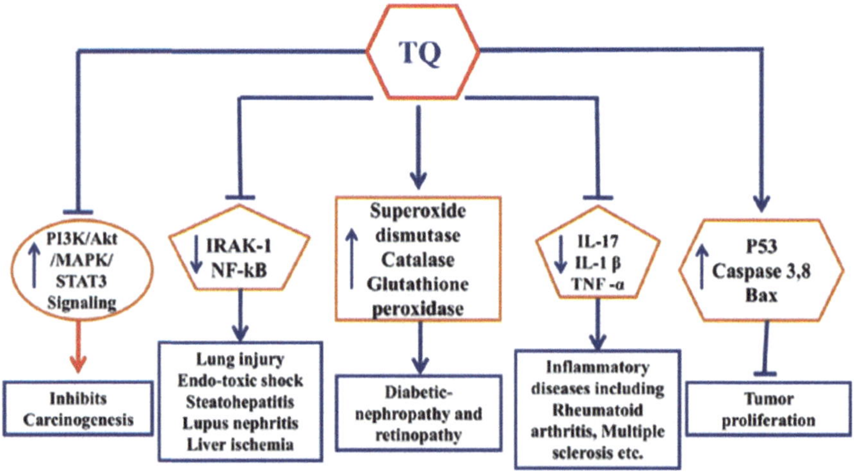

Fig. 17.2 Mechanism of action of thymoquinone

Interestingly, the antibacterial effect of oil of *Nigella sativa* has been reported to be effective against multidrug-resistant bacteria such as *Staphylococcus aureus* (Ugur et al. 2016). Moreover, the food industry uses black seed oil as an antimicrobial agent during the production of food to prevent food spoilage. Arici along with his coworkers reported the ability of black seed oil to inhibit the growth of spoilage, and pathogenic and lactic acid bacteria (Arici et al. 2005). Hasan et al. reported the antibacterial effect of methanolic and water extracts of black seed against Gram-positive bacteria such as *Streptococcus pyrogen, Proteus vulgaris, Klebsiella pneumonia*, and *Pseudomonas aeruginosa* (Hasan et al. 2013). The literature revealed the antifungal property of thymoquinone against *Fusarium solani* and *Aspergillus niger*. Moreover, the researchers were able to compare the antifungal property of thymoquinone with amphotericin B, an antifungal drug (Aljabre et al. 2015). Interestingly, the in vivo antifungal property of aqueous extract of black seed against candidiasis resulted in a high inhibitory effect with many-fold decrease in kidney, liver, and spleen (Khan et al. 2003). Other researchers evaluated the methanolic and ethanolic extracts of black seed and reported higher potency of black seed extract than that of standard amphotericin B against *Candida albicans, C. tropicalis, Cryptococcus laurentii, Cryptococcus albidus, Aspergillus fumigates*, and *Aspergillus flavus*. Also, literature revealed the antiviral activity of *Nigella sativa* oil. For example, Salem and Hussain reported the protective effect of black seed oil against murine cytomegalovirus. They found that the oil was able to completely inhibit the virus in the liver and spleen (Hussain and Hussain 2016; Salem et al. 2011). Despite these, the black seed oil was found to be effective against the influenza virus, laryngotracheitis virus, and hepatitis C virus. Further, the administration of black seed leads to improvement in clinical conditions and oxidative stress and results in decreasing viral load in the patients (Barakat et al. 2013). Therefore, while referring to the amount of literature present, eugenol and thymoquinone prove to be effective antimicrobial agents against various bacteria, fungi, and viruses. Moreover, the drugs are relatively safe with long-term traditional use. Thus, it is important to develop novel antimicrobial drugs using natural products and this can be achieved by applying nanotechnology and by completely understanding the mechanism of action of both drugs on the microbial cell.

17.3 Nanoformulation of Active Compounds

As discussed above, herbal medicine is an ideal candidate for antimicrobial activity against various Gram-negative and Gram-positive bacteria. They act by disrupting bacterial cell membranes or by damaging intracellular biomolecules as well as DNA and proteins. Moreover, chemical antimicrobial agents possess high toxicity and side effects. Therefore, the focus has been shifted toward formulating an antimicrobial drug incorporating natural sources. Since natural products are poorly soluble, highly unstable, and undergo chemical and enzymatic alterations, it is necessary to incorporate nanotechnology for the development and designing of new drugs for the treatment of microbial infection. The introduction of nanotechnology plays a promising role in natural medicine by improving the pharmacokinetic profile of the drug candidate. Nanotechnology involves encapsulation or coating of a drug candidate in an appropriate polymer, which results in increasing the surface area, thereby increasing the solubility, permeability, and delivery of the drug to the target cells. The enhancement of drug delivery can be achieved by using chitosan, carbon, and ceramic nanoparticles. Various types of nanoformulations include nanoemulsions, nanocrystals, self-nanoemulsifying drug delivery systems, polymeric nanoparticles, polymeric micelles, lipid nanoparticles, and carbon nanotubes (Fig. 17.3). Among these, nanoemulsions, nanocrystals, and polymeric nanoparticles are widely used approaches. These methods are effective in reducing the particle size to the nanoscale. For example, nanoparticles of paclitaxel have been reported to cause a many-fold increase in bioavailability causing higher therapeutic activity. Nanocrystal and nanosuspension formation results in increased dissolution, saturation solubility, and relatively higher bioavailability (Zhang et al. 2011; Thadkala et al. 2014). Besides these, dendrimers, polymeric micelles, and self-nanoemulsifying drug delivery systems are gaining much attention. Nanomicelles have been reported to improve the stability of formulation and inducing a strong inhibitory effect in tumors. Dendrimers have a high affinity for both hydrophobic and hydrophilic drugs and have the ability to increase the solubility, transfect the cells, sustain the release of drugs, and enhance penetration and bioavailability of drugs (Pathak and Raghuvanshi 2015). Despite various advantages, the toxic effects of nanomedicines are related to their small sizes and intrinsic toxicity to the surface. This can be overcome by the development of herbal nanomedicines. Various literatures have been reported regarding different nanoformulations of thymoquinone and eugenol, which demonstrate enhanced physicochemical properties of the drugs (Fig. 17.3).

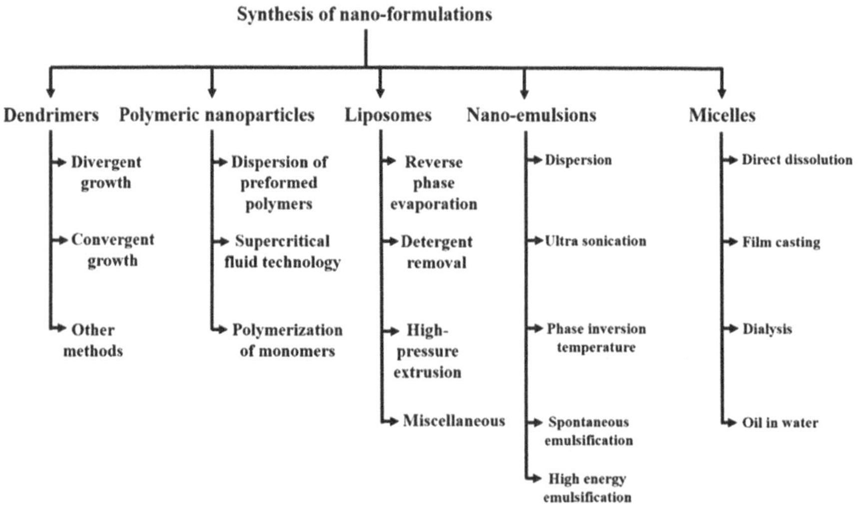

Fig. 17.3 Types of nanoformulations

17.3.1 Eugenol

Eugenol exhibits limited solubility and chemical stability, and due to its rapid absorption and metabolism, the rate of elimination is high. It is highly sensitive to chemical interaction and undergoes oxidation. This factor leads to poor bioavailability of eugenol. Therefore, encapsulation is necessary to improve the aqueous solubility and to prevent early absorption of eugenol resulting in improved efficiency and bioavailability (Jahangir et al. 2020). Nanoemulsion of eugenol with other carrier oil results in enhancement of transdermal permeation and stability of the drug. Encapsulation of eugenol with β-cyclodextrin shows promising antimicrobial activity of eugenol alone; similarly, the same formulation was developed by the molecular inclusion method. After freeze-drying, the developed formulation was smaller in size and showed negative zeta potential along with higher efficiency of encapsulated eugenol. Chitosan-based nanoparticles result in increased thermal stability of eugenol (Taleuzzaman et al. 2020). The synergistic effect of eugenol and carvacrol using chitosan for encapsulation shows effective antimicrobial activity against *S. aureus* and *E. coli*. Zhao along with his coworkers evaluated the ulcerogenic activity of oral microemulsion of ibuprofen and eugenol ester (Zhao et al. 2005). The solubility of optimized formulation was many folds higher than that in water. Further, according to Garg and his coworkers, solid lipid nanoparticles resulted in a sixfold increase in the activity of eugenol as compared to the free form of eugenol (Garg and Singh 2014). In another study, eugenol nanoemulsion gel was prepared. The optimized nanoemulsion was converted into Carbopol gel for the treatment of periodontitis. The formulation showed antibacterial activity and was effective in the treatment of periodontitis by acting as an anesthetic and analgesic. Another similar study involves the preparation and evaluation of controlled release eugenol mucoadhesive tablet for periodontitis. The composition includes Carbopol and hydroxyl methylcellulose in the ratio of 1:1, 1:2, and 2:1 along with 10 mg eugenol. Controlled drug release of the eugenol formulation was observed for a period of 8 hours. The Carbopol results in prolonged release of eugenol. The eugenol showed promising results as antibacterial, anesthetic, and analgesic (Razzaq et al. 2018).

Said and his coworkers formulated a water-based microemulsion of eugenol. The authors used the oil titration method for the incorporation of oil and eugenol in the Tween 20 solution. The formulation was further evaluated for antioxidant and antimicrobial activity. The antioxidant activity was measured by DPPH-free radical scavenging assay, while antimicrobial activity was measured by the agar disk diffusion method. The results showed promising antimicrobial activity of eugenol microemulsion in comparison with the crude eugenol (Hamed et al. 2012). In another study, novel eugenol–chitosan nanoemulsion was prepared by ultrasonication method. The nanoemulsion was prepared using chitosan and eugenol at a ratio of 1: 1 and Tween 20 at 1% w/v at 450 W for 6 min. The resulting formulation was spherical in shape and showed promising antioxidant and antimicrobial activity. Moreover, the formulation was stable at a greater extent and showed prolong storage (Shao et al. 2018). The prepared eugenol-loaded solid lipid nanoparticles were evaluated for antifungal activity. The formulation was used in immunosuppressed rats for the treatment of oral candidiasis. The formulation showed faster drug release, higher efficiency, and greater antifungal activity. Poly(DL-lactide-co-glycolide) (PLGA) nanoparticles of eugenol and trans-cinnamaldehyde were developed for the evaluation of an antimicrobial activity. The formulation had greater efficiency in inhibition of Gram-positive bacteria (*Listeria* spp.) and Gram-negative bacteria (*Salmonella* spp.). In another study, eugenol-loaded nanoemulsion was prepared for the antiproliferative action of eugenol against liver and cancer cells. The apoptosis was measured by flow cytometry as a result of reactive oxygen species generation, and cell viability was measured by MTT assay. Several other works of literature have also been reported for the development and evaluation of eugenol-loaded nanoformulations for anti-inflammatory, anticancer, antibacterial, antifungal, and analgesic activity with greater therapeutic efficiency of eugenol after encapsulation. Gaysinsky et al. formulated eugenol microemulsion using carvacrol as micelle for encapsulation. The authors evaluated that an efficient antimicrobial activity could be the result of surfactant. They concluded that the surfactant was able to increase the permeability of the bacterial cell wall. Further, the addition of a hydrophilic surfactant may result in increasing the solubility of the drug. Also, other scientists concluded that improvement in the efficiency of a drug could be due to the adsorption of carvacrol to the bacterial cell surface, which further increases the concentration of eugenol in the target organ. According to the literature, antimicrobial microemulsion interacts with bacterial cell walls by the addition of surfactant. This interaction results in loss of cellular components leading to loss of viability of bacteria. Literature reveals that microemulsion formation results in distortion of a phospholipid bilayer, which further disrupts the plasma membrane leading to cell death (Gaysinsky et al. 2008) (Table 17.1).

17.3.2 Thymoquinone

Literature review reveals the use of thymoquinone nanoparticles as anticancer, anti-inflammatory, antimicrobial, antioxidant, and immunomodulatory. Bakal along with his coworkers evaluated the antibiotic property of thymoquinone against multidrug-resistant bacteria (Bakal et al. 2017). Also, other researchers compared the antifungal activity of different nanoparticles such as ketoconazole, amphotericin B, and thymoquinone against *Candida albicans* yeast and biofilm. They evaluated that antifungal activity of thymoquinone nanoparticles (160 μg/ml) was two- to fourfold higher than that of amphotericin B nanoparticles (0.31 μg/ml) and ketoconazole nanoparticles (0.62 μg/ml) (Randhawa et al. 2015). Nihei et al. developed a thymoquinone nanoparticle using a cold wet milling system and evaluated its pharmacokinetic profile. The researchers found that the synthesized

Table 17.1 Eugenol Nanoformulations

S. No.	Nanoformulation	Facts	References
1.	Microemulsion	Prepared microemulsion of eugenol and clove bud essential oil (CEO) investigates the antimicrobial activity by DPPH-free radical scavenging assay and agar disk dilution method. Outcome of these methods had shown that the microemulsion of eugenol was more effective than CEO	Hamed et al. (2012)
2.	Eugenol oil nanoemulsion.	Seed cotton is infected by fusarium wilt, and it can be controlled by eugenol oil nanoemulsion. The developed formulation-optimized result indicated that the Z-average diameter was 80 nm, and its shape is spherical. Three different concentrations of this formulation were evaluated by experiments in vitro and in vivo. Outcome of this study reveals antimicrobial effect of nanoemulsion	Abd-Elsalam and Khokhlov (2015)
3.	Eugenol–chitosan nanoparticles	Antibacterial and antioxidant activity of eugenol–chitosan nanoparticles is examined. Antimicrobial activity was assayed with gram-positive bacterium (*Staphylococcus aureus*) and gram-negative bacterium (*Escherichia coli*). A better or equivalent activity is that of the unmodified chitosan nanoparticles	Chen et al. (2009)
4.	Eugenol-entrapped ethosome nanoparticles	The formulation released drug in sustained manner and it used against fruit anthracnose, antibacterial activity (>93%) against fruit pathogens had founded greater efficacy that free eugenol. The advantage of sustained drug release was prolonged antibacterial action	Jin et al. (2019)
5.	Clove oil nanoemulsion	Nanoemulsions (NEs) of clove oil prepared for the treatment of infection are caused by microorganisms *Candida albicans* and *Candida glabrata*. It is suitable for topical application, use co and pluronic F-127 in different concentrations for the preparation of NEs by ultrasonic processor. Benefits of this formulation are less irritation on skin, and low dose had shown therapeutic effects	de Oliveira de Siqueira et al. (2019)
6.	Eugenol gel for periodontal delivery.	Prepared a eugenol gel formulation incorporated in NLC for sustained delivery. Many advantages of this formulation such as high bioavailability, low-dose high therapeutic value, and effectively delivered	Pokharkar et al. (2011)

(continued)

Table 17.1 (continued)

S. No.	Nanoformulation	Facts	References
7.	Nonionic nanoemulsion.	Antifungal activity investigated of oil-in-water nanoemulsion, prepared through low-energy method using nonionic surfactants. Optimized this formulation against *Fusarium oxysporum* f.sp. lycopersici (FOL)	Sharma et al. (2018)
8.	Nanoformulation of eugenol	Formulation contains eugenol (oil phase), tween 80 (surfactant), and PEG (co-surfactant), a novel nanoemulsion—Carbopol®934 gel. Evaluate this formulation on the basis of their pH, drug content, syringe ability, and mucoadhesion on the animal goat buccal mucosa. This formulation shows significant antibacterial effects in the treatment of periodontal disease	Ahmad et al. (2019)
9.	Eugenol with nanoparticles.	Periodontal diseases cured by the eugenol, but lipophilic nature of the molecule create challenge for the formulation development. With the application of nanotechnology, the therapeutic efficacy of the drug enhanced. Nanoparticles used for the delivery of drug have potential for the treatment of periodontal disease. Nanoparticles help to modulate the drug kinetics and help to deliver the drug in the right site at right time	Garg et al. (2018)
10.	Quantification of eugenol in *Myristica fragrans* seed extract by HPTLC	Developed a HPTLC method for the estimation of eugenol in *Myristica fragrans* seed, this method was robust and can be used for the optimization of nanoformulation of eugenol	Taleuzzamana et al. (2017)

thymoquinone nanoparticle was more photostable than that of free thymoquinone. Moreover, there was a sixfold increase in the oral bioavailability of nanoformulation thymoquinone as compared to the free form of thymoquinone solution (Nihei et al. 2016). Further, it is necessary to develop a proper strategy to overcome the problems that occurred during the nanoformulation of thymoquinone. As a large amount of excipient is required in the preparation of nanothymoquinone, therefore excipient should be stable, biocompatible, and biodegradable and should be free from side effects. Also, a prodrug can be designed using in silico modeling where the ADME profile of thymoquinone can be acquired for developing drug-likeness of thymoquinone by modifying its structure during the encapsulation of thymoquinone within nanoformulation (van de Waterbeemd and Gifford 2003). Literature review reveals that the first thymoquinone-loaded nanoparticle was developed in the year 2010 by Shah et al. where preparation, characterization, and in vitro analysis of nanothymoquinone were evaluated. Nanoformulation is relatively advantageous to other methods such as facilitate drug release by modulation of physicochemical properties, thereby enhancing its biological behavior. Moreover, the drug can be protected from degradation in the gastrointestinal tract by encapsulation technique, which on the other hand enhances the stability of the drug and lower the dose. Besides these advantages, nanoformulations can be easily taken up lymphatic tissues resulting in effective and direct delivery of a drug in the blood and target organs. Various studies have been conducted where polymeric nanoparticles are used for intranasal delivery of thymoquinone to the brain. They concluded that upon encapsulation, nanoparticles were able to protect the core drug molecule from biological and chemical degradation. Further, being small in size, the nanoparticles were able to transport across the blood–brain barrier for efficient treatment of Parkinson's disease. Therefore, an increased bioavailability of drugs can be achieved in the brain. Thus, polymeric nanoparticles prove to be an attractive approach for in situ nasal gel formulation (El-Far et al. 2018).

Ganea et al. evaluated copolymeric nanoparticles of thymoquinone against the MDA-MB 231 cancer cell line and reported it to be much effective than that of free thymoquinone (Ganea et al. 2010). In the year 2016, Abdel mottled reported the anticolorectal properties of polycaprolactone-encapsulated thymoquinone nanoparticles using 6-week-old male BALB/c mice (Abdel-Mottaleb 2016). Sustained effects and enhanced proliferative activity of thymoquinone nanoparticles have been demonstrated against MCF 7 cell line by Shaarani et al. (2017). Anti-inflammatory effects of thymoquinone lipospheres have been reported by Jain et al. for psoriasis. The authors conducted in vitro and in vivo analyses. The in vitro study was conducted against RAW 264.7, a murine macrophage cell line, and decreased levels of interleukin 2, interleukin 6, and tumor necrosis factor-alpha were reported. On the other hand, the drug was able to reduce interleukin-17 and tumor necrosis factor-alpha in psoriatic skin of BALB/c mice (Jain et al. 2017). However, studies regarding safety concern for long-term use of thymoquinone are still lacking and need immediate attention (Table 17.2).

From the above findings, it can be concluded that nanotechnology is a rapidly emerging field for efficient drug delivery. Various efforts are being done to enhance bioavailability and patient compliance. The nanotechnology-based system appears to be a promising, effective, and safe drug delivery system for achieving site-specific drug targeting and overcoming the limitations of a potential drug candidate with a narrow therapeutic window and poor bioavailability. Therefore, nanotechnology has been widely used for the formulation of phytoconstituents. Despite the above advantages, it is important to understand detailed concepts of phytoconstituents and modern scientific methods to overcome the toxicity of bioactive constituents.

17.4 Toxicity of Bioactive Constituents

Since phytoconstituents are considered safe for administration, but due to broad range of application in the treatment of diseases, concern regarding the toxicological profile of natural products has also increased in recent years. Several

Table 17.2 Thymoquinone Nanoformulation

S. No.	Nanoformulation	Facts	References
1.	Thymoquinone-loaded PLGA nanoparticles	Formulation of TQ prepared, it encapsulated in nanoparticle PLGA (poly(DL-lactide-co-glycolide)) and examined its antimicrobial properties against *E. coli, Staphylococcus aureus*, and *Salmonella typhi* strains. DPPH radical scavenging activity was founded	Nallamuthu et al. (2013)
2.	Biofilm formation inhibited by TQ	Gram-positive bacteria and Gram-negative bacteria growth inhibited and also anti-biofilm activity. Studied with minimum inhibitory concentration in the range of 1.56–100 μg/ml. To understand the mechanism studied DiSC3, NPN, and ROS assay	Goel and Mishra (2018)
3.	Formulation of L-ascorbic acid and thymoquinone encapsulated in chitosan nanoparticles	Use nanoparticles to solve the challenge of combination hydrophilic and hydrophobic molecule delivery. Due to high surface area chitosan, NPs (CNPs), the L-ascorbic acid and thymoquinone, are more efficiently delivered and uptake by cells, prepared CNPs-LAA-TQ via ionic gelation routes because the preparation is nontoxic. Encapsulated together hydrophilic (LAA) and hydrophobic (TQ) in CNPs increase their therapeutic efficacy	Othman et al. (2018)
4.	Assessment of antibacterial activity of synthesized silver nanoparticles using *Nigella sativa*	Biosynthesized silver nanoparticles (Ag-NPs) from *Nigella sativa* (NS) seed extract using ethanol and chloroform. Studied antibacterial activity against some drug-resistant bacteria and it has been used for the treatment and prevention of drug-resistant bacteria. Microtiter broth dilution process using ciprofloxacin as standard for assessment of antibacterial activity of Ag-NPs–NS seed extract	Shehensha and Jyothi (2020)
5.	Nanoformulation of amphotericin-B, ketoconazole, and TQ	Amphotericin-B, ketoconazole, and TQ drugs encapsulated in nanoparticle for its better efficacy using the ball milling technique. Comparative study of nanoformulation and its conventional forms against *Candida albicans*, yeasts, and *Candida* biofilm have been done. Founded two to four times more effective disinfectant efficacy	Randhawa et al. (2015)
6.	Thymoquinone liposomes	Diabetic mice with infection of *C. albicans* treated with Liposomal formulation of TQ (Lip-TQ). Used dose of 10 mg/kg of Lip-TQ result founded that in blood glucose level reduced and systemic infection of *C. albicans* also. The survival rate of mice was 70% when treated with Lip-TQ, and in the case of free TQ, the survival rate was 20%	Khan et al. (2018)

(continued)

Table 17.2 (continued)

S. No.	Nanoformulation	Facts	References
7.	Thymoquinone niosomes	Quantification of TQ has been done by HPLC and the formulated TMQ-loaded niosomes (TMQNIOS). Studied various parameters such as permeation study, in vitro drug release, surface morphology and confocal laser scanning microscopic (CLSM), enhancement of permeation founded in TMQNIOS. Efficacy of this formulation is high	Gilani et al. (2019)
8.	Nanocarrier PEGylated LMW chitosan used for TQ	This formulation was stable and had better antimicrobial activity. Optimize the formulation-founded stable and constant particle size along with polydispersity index. Performed in vitro studied, results indicate that the percentage of micellization of TQ compared to mixed micelles	Vignesh Kumar et al. (2020)
9.	Thymoquinone-loaded coconut oil nanostructured lipid carriers (NLCs)	NLC was used for the development of TQ formulation, using also Compritol® 888 and coconut oil. Melt emulsification and sonification method used for the preparation of formulation using Poloxamer 188 as a surfactant. In in vitro study, a biphasic release pattern was observed, and burst release of drug in first 2 h followed by a gradual release	Fahmy et al. (2020)
10.	Encapsulation of thymoquinone using solid lipid nanoparticle	Formulated, thymoquinone-encapsulated solid lipid nanoparticle (TQ-SLNs) studied the thermal behavior result found that a melting enthalpy value of $\Delta H = 131.7$ J/g and attained peak at 69.5 °C indicate greater stability and reduced lipid crystal matrices; additionally, stability analysis of TQ-SLNs showed a stable encapsulation enhancement effect and almost increased five times bioavailability after oral administration	Surekha and Sumathi (2016)

pieces of literature have been reported for toxicological effects of thymoquinone and eugenol.

Though no chronic toxicity of thymoquinone was reported in the literature, the authors revealed various side effects along with acute and subacute toxicity in vitro and in vivo studies. In the year 2012, Harzallah et al. evaluated the oxidative damage effect of thymoquinone in the liver and kidney. The study involves the administration of intraperitoneal injection of thymoquinone at various doses (10, 20, 40, and 80 mg/kg) to BALB/c mice. An induction in oxidative damage activity was reported in higher doses of thymoquinone, i.e., 40 mg/kg and 80 mg/kg. The oxidative effect was recorded due to an increase in malondialdehyde levels and catalase activity. The authors further evaluated the genotoxic activity of thymoquinone at 80 mg/kg and found a significant increase in chromosomal aberration with a high DNA damage index in the organs. In another study, thymoquinone was administered intraperitoneally to albino rats for a different time period (1, 4, 7, 10, and

14 days) at different doses (0.5, 1, 2, 4, 6, and 8 mg/kg). The authors evaluated thymoquinone effects on blood lipids in the rat. The drug was able to reduce triglyceride levels, high-density lipoproteins (HDLs), and low-density lipoproteins (LDLs). The animals were safe at all doses except at 8 mg where animals died within 1 week of administration and showed signs of peritonitis (Mashayekhi-Sardoo et al. 2020).

Badary et al. evaluated the cytoprotective activity of thymoquinone in Swiss albino mice. Thymoquinone was added to drinking water at 100, 200, and 300 ppm concentration. The water was given to male Swiss albino mice for 90 days. Microscopic examination of tissues such as kidneys, heart, and liver showed no damage. Other serum concentrations such as urea, triglycerides, creatinine, and LDH were normal. Since no mortality occurred, therefore, no toxicity was recorded. However, a significant decrease in fasting blood sugar levels was observed. The results concluded that it was safe and possess cytoprotective activity (Badary et al. 1998). Ong et al. formulated thymoquinone nanoparticles by encapsulation of crude drug in lipid carrier. The resulting thymoquinone nanostructured lipid carrier (TQNLC) possesses high drug loading capacity and high encapsulation efficiency. The authors further conducted in vivo studies using female BALB/c mice. Oral administration of thymoquinone and TQNLC was done at three different doses such as 5, 50, and 300 mg/kg body weight. Death was seen in mice receiving 300 mg/kg within 24 h of treatment. Delayed death was seen after 48 h of treatment in the case of mice receiving TQNLC. No mortality was observed in mice receiving 5 and 50 mg/kg dose of thymoquinone and TQNLC. The LD50 of both the formulation was found to be 50–300 mg/kg and 300–2000 mg/kg. The authors concluded that encapsulation of thymoquinone results in reduced toxicity of thymoquinone. Therefore, encapsulation could be a promising approach for the delivery of thymoquinone. Later, the authors administered 1, 10, and 100 mg/kg of both formulations in both sexes of BALB/c mice for a period of 28 days. During the experimental period, no significant changes in biochemical and hematological profiles were seen. In the liver of both sexes, cell degeneration was observed following oral administration of 100 mg/kg dose of both formulations. However, no mortality was observed in either sex (Haron et al. 2018).

The results of literature research were consistent, and it can be concluded that it is completely safe to use thymoquinone for clinical applications as no chronic toxicity was observed in animal models following thymoquinone administration.

Though eugenol is considered safe in the food industry, toxicity studies of eugenol are of greater concern due to its wide range of applications in the pharmaceutical industry. Various adverse effects have been reported after the use of dental products containing eugenol including ulcers, tissue necrosis, irritation to the skin, and allergic dermatitis. Researchers found a similar result between eugenol and paracetamol poisoning. The authors reported a case study involving ingestion of 5–10 ml of clove oil by a 2-year-old boy. The patient was in a deep coma after 3 h of administration. The patient suffered from generalized seizures for 8 h, and within

24 h of administration of clove oil, the patient suffered from liver failure along with disseminated intravascular coagulopathy. The patient recovered and gained consciousness after 1 week of symptomatic treatment (Hartnoll et al. 1993). Gaulet and his coworkers studied the toxic effects of eugenol at anesthetic doses in clawed frogs. The authors demonstrated renal tubular apoptosis where distal tubules were affected in the medulla. They reported some morphological changes such as hepatic necrosis, adipose tissue hemorrhages, and the presence of hyaline membranes in the lung. In another study, the authors evaluated the cytotoxicity of eugenol using human osteoblastic cells. The toxic effects were evaluated by performing cytotoxicity and cell proliferation assay. They demonstrated significant periapical toxicity of eugenol. Eugenol inhibited cell proliferation by decreasing the activity of dehydrogenase enzymes. The IC50 was reported to be 0.75 mmol/l. However, the addition of *N*-acetyl-L-cysteine was able to protect the cells from the inhibitory action of eugenol (Ho et al. 2006).

The cytotoxicity of eugenol was determined in HL-60 human promyelocytic leukemia cells. Eugenol-treated HL-60 cells were evaluated for cytotoxicity and apoptosis. The results demonstrated apoptosis by a significant increase in fragmented DNA. Moreover, 50% of cell growth was inhibited by eugenol at a concentration of 23.7 μM. Further, the authors evaluated the putative pathway for the action of eugenol. They concluded that apoptosis occurred as a result of ROS generation, which induces mitochondrial permeability, the release of cytochrome c in the cytosol, and reduced levels of anti-apoptotic proteins (bcl-2) levels ultimately leading to cell death (Yoo et al. 2005). In another study, the authors demonstrated the effects of eugenol and cinnamaldehyde in isolated rat liver mitochondria. Due to the lipophilic character of both components, they were able to penetrate in the intracellular organs. Both compounds were able to decrease the membrane potential, inhibit Na^+/K^+ ATPase activity in the kidney and intestine, and stimulate ATPase at a concentration equal to or greater than 0.3 μM, ultimately disturbing mitochondrial functions. Toxicity study of aspirin eugenol ester was evaluated for a period of 15 days at a dose of 50, 1000, and 2000 mg/kg. The authors evaluated a significant increase in total bilirubin levels, alanine aminotransferase, and aspartate aminotransferase. However, no adverse effects were observed at a daily dose of 50 mg/kg. Another case study reported unexpected hypersensitivity reactions of zinc oxide eugenol in an 8-year-old boy during dental treatment. It was found that eugenol causes dermatitis with evident redness in the body parts including the face, neck, lower and upper limbs, and behind the ears. The composition consisted of 10% zinc oxide, 1% aqueous formaldehyde solution, and sodium chloride along with eugenol. The skin prick test revealed a negative response for all the compounds except for eugenol (Tammannavar et al. 2013).

Since extensive research has been done on the biological activities of eugenol, research related to the toxicity of eugenol is still limited. A long-term in vitro and in vivo human research is still undiscovered related to the chronic toxicity of eugenol.

17.5 Challenges for the Development of Herbal Nanoformulation

Herbal drugs have achieved great concern in today's era as people seek natural remedies. They have been used since ancient times for the treatment of diseases and maintenance of health. They serve as a common ingredient in homeopathic, ayurvedic, Unani, and another system of medicines and are considered safe. They are widely used in the treatment of cirrhosis, gall stones, menopause, asthma, Alzheimer's, acne, migraine, and impotence and as an antimicrobial, anti-anxiety, anti-inflammatory, antidiabetic, antispasmodic, hepatoprotective, and analgesic. Herbal drugs possess many advantages including fewer side effects, enhanced tolerance, and low cost, and are efficient. The main disadvantage related to herbal medicine involves complexity in the standardization of drugs, stability, and solubility of drugs. Due to poor solubility, phytochemicals suffer from poor absorption and poor bioavailability. Moreover, rapid clearance and high protein binding of phytochemicals pose a major problem in the design and development of the drug. Further challenges in the development of herbal medicine include toxicity issues of the plants, low yield of a bioactive constituent, limited literature for possible side effects, shortage of medicinal plants, and limited documentation for drug–phytochemical interactions. This can be overcome by using nanotechnology for the enhancement of efficacy, solubility, and stability of drugs.

Taxol being a potent anticancer agent isolated from the barks of *Taxus brevifolia* is capable of inhibiting tumor cells during mitosis. In the year 1992, the drug got the approval of the FDA against ovarian cancer, and in the year 1994, FDA approved its use for metastatic breast cancer. Due to low yield and difficulty in isolation of taxol, it accounts for a high cost. To improve the solubility of taxol, it was encapsulated in cremophor, but later researchers found that camphor was toxic for endothelial, epithelial, and living cells. However, further research for the treatment of cancer lead to the development of nanoformulation of paclitaxel bounded to albumin. The protein-bound paclitaxel reported higher solubility and higher tissue distribution. The drug was effective with other cytotoxic agents and prolonged the survival of patients. The formulation was further approved by FDA (Wani and Horwitz 2014). The targeted therapy has been widely applied in cancer to further enhance the efficacy of herbal medicines. This involves passive targeting, active targeting, and target-activated system. The passive targeting includes the incorporation of the drug in a nanoparticle that can passively reach the tumors. The active targeting involves the delivery of the herbal drug by attachment of ligands to the polymer. Target-activated systems are activated by pH or certain enzymes that respond to the target tissue. The major challenge in the development of nanoformulation includes its stability. Some literature reveals that nanoparticles are not able to hold the drug due to leakage of contents upon interaction with blood components. Another challenge is the residence time of nanoparticles in the metabolism system of the body. Stealth coating of nanoparticles with polyethylene glycol (PEG) was able to overcome this issue. However, the disadvantages of PEG are of great concern. Another issue related to nanoformulation involves changes in the structure of drug components, which can adversely affect the components of the formulation with negative effects. This can further alter the safety, efficacy, and stability of the herbal product. The acceptance of nanoformulation for clinical use has to undergo various standards and desirable pharmaceutical and physicochemical characterization including surface morphology, size, distribution, surface charge, chemistry, and pharmacokinetic profile of the nanoparticles. While these components seem to be similar to that are faced by any drug, the multicomponent nature, complexity, and a large number of additional variables related to nanoformulations make it difficult to control the processes and predict the biological behavior of the formulation in the body (Pathak and Raghuvanshi 2015).

Aseptic manufacturing of nanoparticles is another major challenge. Nanoformulation of the sterile product might get damaged by the techniques of sterilization such as autoclaving and gamma irradiation. In the case of flexible particles such as liposomal preparation, sterile filters can be used as an alternative. However, on the other hand, rigid particles such as polymers, silica particles, and metals may possess difficult infiltration due to the pore size of the particles and filtration membranes. The nanoformulation of dry materials is another challenge as they are distributed as aerosols. Such particles get deposited in the lung resulting in pulmonary toxicities such as pulmonary fibrosis, pleural effusion, and granuloma. Therefore, current challenges involve the design and development of methods to control the problems associated with interactions of nanoformulations with biological systems (Desai 2012).

New challenges related to nanoformulation involve a meeting of international standards for targeting efficiency, biocompatibility, and toxicity profile of the drug. Also, the ease of scale-up processes and the requirement of a multifunctional system for fulfilling biological and therapeutic demands are other challenges that need much concern in the future. The smaller size of nanoparticles captures more surface area per unit mass, making it more reactive to the cellular environment. This leads to the enhancement of the intrinsic toxicity of the particles. The major component of these particles is carbon, which may cause cardiovascular effects. Moreover, particles that are fine and ultrafine in nature can easily be inhaled resulting in an assessment of vasculature leading to cardiovascular dysfunction. Clearly, the design and development of herbal nanoformulation come up with a lot of obstacles.

The future of nanoformulation involves the design and development of personalized medicine for the clinical application of a particular drug in a subgroup getting benefit from the therapy. This approach can further reduce the size of clinical trials leading to immediate approval. Additionally, tissues such as tumors act as a natural biological barrier for the delivery of the drug to the organ. This can be controlled by a better understanding of the active transport mechanism and delivery of nanomedicines. The health authorities need to respond to such techniques that could provide steps for the development of quality, safe, and efficient drugs in the market and can lead to advances in drug development.

17.6 Conclusions

The literature reveals that the use of nanotechnology for the delivery of herbal medicines has been the most promising way to achieve the efficient biological activity. Thymoquinone and eugenol, active natural constituents derived from natural sources, are widely used in the treatment of a wide range of diseases as antimicrobial, antidiabetic, analgesic, anti-inflammatory, anesthetic, immunomodulatory, hepatoprotective, and antioxidant. These natural sources come up with several side effects and a lack of information regarding potential toxicity. Therefore, the incorporation of herbal drugs in novel drug delivery systems leads to enhanced bioavailability, sustained release action, and reduced toxicity. Extensive research has been done for the antimicrobial activity of thymoquinone and eugenol against a broad spectrum of Gram-positive and Gram-negative bacteria. The attempt to use bioactive constituent-loaded nanoparticles has its own advantages and could come up with an effective treatment strategy to be seen shortly.

Acknowledgement The authors are very thankful to the Faculty of Pharmacy, Maulana Azad University, Jodhpur, Rajasthan, India, for valuable support.

References

Abdel-Mottaleb MMA (2016) Biodegradable thymoquinone nanoparticles for higher therapeutic efficiency in murine colorectal cancer. Int J Pharm Pharm Res 7:436–450

Abd-Elsalam KA, Khokhlov AR (2015) Eugenol oil nanoemulsion: antifungal activity against Fusarium oxysporum f. sp. vasinfectum and phytotoxicity on cottonseeds. Appl Nanosci 5:255–265

Abrão PH, Pizi RB, de Souza TB, Silva NC, Fregnan AM et al (2015) Synthesis and biological evaluation of new eugenol Mannich bases as promising antifungal agents. Chem Biol Drug Des 86(4):459–465

Abu Khader MM, Khan SA (2017) Thymoquinone and nanoparticles: a promising approach for the clinical trials. J Bionanosci 11(4):258–265

Ahmad A, Khan A, Manzoor N, Khan LA (2010) Evolution of ergosterol biosynthesis inhibitors as fungicidal against Candida. Microb Pathog 48(1):35–41

Ahmad N, Ahmad FJ, Bedi S, Sharma S, Umar S, Ansari MA (2019) A novel nanoformulation development of eugenol and their treatment in inflammation and periodontitis. Saudi Pharm J 27(6):778–790

Aljabre SHM, Alakloby OM, Randhawa MA (2015) Dermatological effects of Nigella sativa. J Dermatol Dermatol Surg 19(2):92–98

Arana-Sánchez A, Estarrón-Espinosa M, Obledo-Vázquez EN, Padilla-Camberos E, Silva-Vázquez R, Lugo-Cervantes E (2010) Antimicrobial and antioxidant activities of Mexican oregano essential oils (Lippia graveolens H. B. K.) with different composition when microencapsulated in beta-cyclodextrin. Lett Appl Microbiol 50(6):585–590

Arici M, Sagdic O, Gecgel U (2005) Antibacterial effect of Turkish black cumin (Nigella sativa L.) oils. Grasas y Aceites 56:259–262

Badary OA, Al-Shabanah OA, Nagi MN, Al-Bekairi AM, Elmazar MMA (1998) Acute and subchronic toxicity of thymoquinone in mice. Drug Dev Res 44(2–3):56–61

Bakal SN, Bereswill S, Heimesaat MM (2017) Finding novel antibiotic substances from medicinal plants—antimicrobial properties of Nigella sativa directed against multidrug-resistant bacteria. Eur J Microbiol Immunol 7(1):92–98
Barakat EM, El Wakeel LM, Hagag RS (2013) Effects of Nigella sativa on outcome of hepatitis C in Egypt. World J Gastroenterol 19(16):2529–2536
Bassolé IH, Juliani HR (2012) Essential oils in combination and their antimicrobial properties. Molecules 17(4):3989–4006
Boonchird C, Flegel TW (1982) In vitro antifungal activity of eugenol and vanillin against Candida albicans and Cryptococcus neoformans. Can J Microbiol 28(11):1235–1241
Burt S (2004) Essential oils: their antibacterial properties and potential applications in foods—a review. Int J Food Microbiol 94(3):223–253
Chaieb K, Kouidhi B, Jrah H, Mahdouani K, Bakhrouf A (2011) Antibacterial activity of thymoquinone, an active principle of Nigella sativa and its potency to prevent bacterial biofilm formation. BMC Complement Altern Med 11:29
Chen F, Shi Z, Neoh KG, Kang ET (2009) Antioxidant and antibacterial activities of eugenol and carvacrol-grafted chitosan nanoparticles. Biotechnol Bioeng 104(1):30–39
Chen M, Zhang J, Lai Y, Wang S, Li P, Xiao J, Fu C, Hu H, Wang Y (2013) Analysis of Pogostemon cablin from pharmaceutical research to market performances. Expert Opin Investig Drugs 22(2):245–257
Dabur R, Gupta A, Mandal TK, Singh DD, Bajpai V, Gurav AM, Lavekar GS (2007) Antimicrobial activity of some Indian medicinal plants. Afr J Tradit Complement Altern Med 4(3):313–318
Darakhshan S, Bidmeshki Pour A, Hosseinzadeh Colagar A, Sisakhtnezhad S (2015) Thymoquinone and its therapeutic potentials. Pharmacol Res 95-96:138–158
Das B, Mandal D, Dash SK, Chattopadhyay S, Tripathy S, Dolai DP, Dey SK, Roy S (2016) Eugenol provokes ROS-mediated membrane damage-associated antibacterial activity against clinically isolated multidrug-resistant Staphylococcus aureus strains. Infect Dis (Auckl) 9:11–19
de Oliveira de Siqueira LB, dos Santos Matos AP, da Silva Cardoso V, Villanova JCO, Guimarães BCLR et al (2019) Clove oil nanoemulsion showed potent inhibitory effect against Candida spp. Nanotechnology 30:425101
Desai N (2012) Challenges in development of nanoparticle-based therapeutics. AAPS J 14(2):282–295
Devi KP, Nisha SA, Sakthivel R, Pandian SK (2010) Eugenol (an essential oil of clove) acts as an antibacterial agent against Salmonella typhi by disrupting the cellular membrane. J Ethnopharmacol 130(1):107–115
Economakis C, Skaltsa H, Demetzos C, Soković M, Thanos CA (2002) Effect of phosphorus concentration of the nutrient solution on the volatile constituents of leaves and bracts of Origanum dictamnus. J Agric Food Chem 50(22):6276–6280
El-Far AH (2015) Thymoquinone anticancer discovery: possible mechanisms. Curr Drug Discov Technol 12(2):80–89
El-Far AH, Bazh EK, Moharam MS (2014) Antioxidant and antinematodal effects of Nigella sativa and Zingiber officinale supplementations in ewes. Int J Pharm Sci Rev Res 26(1):222–227
El-Far AH, Al Jaouni SK, Li W, Mousa SA (2018) Protective roles of thymoquinone nanoformulations: potential nanonutraceuticals in human diseases. Nutrients 10(10):1369
Fabio A, Cermelli C, Fabio G, Nicoletti P, Quaglio P (2007) Screening of the antibacterial effects of a variety of essential oils on microorganisms responsible for respiratory infections. Phytother Res 21(4):374–377
Fahmy UA, Alaofi AL, Awan ZA et al (2020) Optimization of thymoquinone-loaded coconut oil nanostructured lipid carriers for the management of ethanol-induced ulcer. AAPS Pharm Sci Tech 21:137
Filgueiras TC, Vanetti, Dantas MC (2006) Effect of eugenol on growth and listeriolysin o production by Listeria monocytogenes. Braz Arch Biol Technol 49(3):405–409
Fleisher Z, Fleisher A (2000) Volatiles of Coridothymus capitatus chemotypes growing in Israel: aromatic plants of the holy land and the Sinai. Part XV. J Essent Oil Res 14(2):105–106

Ganea GM, Fakayode SO, Losso JN, van Nostrum CF, Sabliov CM, Warner IM (2010) Delivery of phytochemical thymoquinone using molecular micelle modified poly (D, L lactide-co-glycolide) (PLGA) nanoparticles. Nanotechnology 21(28):285104

Garg A, Singh S (2014) Targeting of eugenol-loaded solid lipid nanoparticles to the epidermal layer of human skin. Nanomedicine (Lond) 9(8):1223–1238

Garg V, Chawla K, Pawar SK (2018) Nanotechnology controlled local drug delivery system for the treatment of periodontitis. J Adv Med Med Res 26(6):1–17

Gaysinsky S, Davidson PM, McClements DJ et al (2008) Formulation and characterization of phytophenol-carrying antimicrobial microemulsions. Food Biophys 3:54–65

Gilani SJ, Imam SS, Ahmed A, Chauhan S, Mirza MA, Taleuzzaman M (2019) Formulation and evaluation of thymoquinone niosomes: application of developed and validated RP-HPLC method in the delivery system. Drug Dev Ind Pharm 45(11):1799–1806

Gill AO, Holley RA (2006) Inhibition of membrane bound ATPases of Escherichia coli and Listeria monocytogenes by plant oil aromatics. Int J Food Microbiol 111(2):170–174

Goel S, Mishra P (2018) Thymoquinone inhibits biofilm formation and has selective antibacterial activity due to ROS generation. Appl Microbiol Biotechnol 102:1955–1967

Gundala SR, Aneja R (2014) Piper betel leaf: a reservoir of potential xenohormetic nutraceuticals with cancer-fighting properties. Cancer Prev Res (Phila) 7(5):477–486

Hamed SF, Sadek Z, Edris A (2012) Antioxidant and antimicrobial activities of clove bud essential oil and eugenol nanoparticles in alcohol-free microemulsion. J Oleo Sci 61(11):641–648

Haron AS, Syed Alwi SS, Saiful Yazan L, Abd Razak R, Ong YS et al (2018) Cytotoxic effect of thymoquinone-loaded nanostructured lipid carrier (TQ-NLC) on liver cancer cell integrated with hepatitis B genome, Hep3B. Evid Based Complement Alternat Med 2018:1549805

Hartnoll G, Moore D, Douek D (1993) Near fatal ingestion of oil of cloves. Arch Dis Child 69(3):392–393

Hasan NA, Nawahwi MZ, Malek HA (2013) Antimicrobial activity of nigella sativa seed extract. Sains Malaysiana

Havlik J, Kokoska L, Vasickova S, Valterova I (2006) Chemical composition of essential oil from the seeds of Nigella arvensis L. and assessment of its actimicrobial activity. Flavour Fragr J 21(4):713–717

Hipólito TMM, Bastos GTL, Barbosa TWL, de Souza TB, Coelho LFL et al (2018) Synthesis, activity, and docking studies of eugenol-based glucosides as new agents against Candida sp. Chem Biol Drug Des 92(2):1514–1524

Ho YC, Huang FM, Chang YC (2006) Mechanisms of cytotoxicity of eugenol in human osteoblastic cells in vitro. Int Endod J 39(5):389–393

Huang S, Chang WH (2009) Advantages of nanotechnology-based Chinese herb drugs on biological activities. Curr Drug Metab 10(8):905–913

Hussain DA, Hussain MM (2016) Nigella sativa (black seed) is an effective herbal remedy for every disease except death—a Prophetic statement which modern scientists confirm unanimously: a review. Landline 4(2):27–57

Hyldgaard M, Mygind T, Meyer RL (2012) Essential oils in food preservation: mode of action, synergies, and interactions with food matrix components. Front Microbiol 3:12

Ipek E, Zeytinoglu H, Okay S et al (2005) Genotoxicity and antigenotoxicity of Origanum oil and carvacrol evaluated by Ames Salmonella/microsomal test. Food Chem 93(3):551–556

Jaganathan SK, Supriyanto E (2012) Antiproliferative and molecular mechanism of eugenol-induced apoptosis in cancer cells. Molecules 17(6):6290–6304

Jahangir MA, Taleuzzaman M, Beg S, Verma S, Gilani SJ, Alam P (2020) A review of eugenol-based nanomedicine: recent advancements. Curr Bioact Comp 16:1

Jain A, Pooladanda V, Bulbake U, Doppalapudi S, Rafeeqi TA, Godugu C, Khan W (2017) Liposphere mediated topical delivery of thymoquinone in the treatment of psoriasis. Nanomedicine 13(7):2251–2262

Jin P, Yao R, Qin D, Chen Q, Du Q (2019) Enhancement in antibacterial activities of eugenol-entrapped ethosome nanoparticles via strengthening its permeability and sustained release. J Agric Food Chem 67:1371–1380

Joris M, Joris M, Noël D et al (2008) In vitro degradation and in vivo passage kinetics of carvacrol, thymol, eugenol and trans-cinnamaldehyde along the gastrointestinal tract of piglets. J Sci Food Agric 88(13):2371–2381

Khan MA, Ashfaq MK, Zuberi HS, Mahmood MS, Gilani AH (2003) The in vivo antifungal activity of the aqueous extract from Nigella sativa seeds. Phytother Res 17(2):183–186

Khan A, Ahmad A, Akhtar F, Yousuf S, Xess I, Khan LA, Manzoor N (2011) Induction of oxidative stress as a possible mechanism of the antifungal action of three phenylpropanoids. FEMS Yeast Res 11(1):114–122

Khan MA, Aldebasi YH, Alsuhaibani SA, AlSahli MA, Alzohairy MA, Khan A et al (2018) Therapeutic potential of thymoquinone liposomes against the systemic infection of Candida albicans in diabetic mice. PLoS One 13(12):e0208951

Kubátová A, Lagadec AJM, Miller DJ, Hawthorne SB (2001) Selective extraction of oxygenates from savory and peppermint using subcritical water, vol 16. Flavour Fragr J, pp 64–73

Kumar SS, Marella SS, Vipin A, Sharmistha M (2013) Evaluation of analgesic and anti-inflammatory activity of Abutilon indicum. Int J Drug Dev Res 5(1):99–102

Liu Q, Niu H, Zhang W, Mu H, Sun C, Duan J (2015) Synergy among thymol, eugenol, berberine, cinnamaldehyde and streptomycin against planktonic and biofilm-associated food-borne pathogens. Lett Appl Microbiol 60(5):421–430

Lukas B, Schmiderer C, Franz C, Novak J (2009) Composition of essential oil compounds from different Syrian populations of Origanum syriacum L. (Lamiaceae). J Agric Food Chem 57(4):1362–1365

Majeed H, Antoniou J, Hategekimana J, Sharif HR, Haider J, Liu F, Ali B, Liang Rong L, Ma J, Zhong F (2016) Influence of carrier oil type, particle size on in vitro lipid digestion and eugenol release in emulsion and nanoemulsions. Food Hydrocoll 52:415–422

Marchese A, Barbieri R, Coppo E, Orhan IE, Daglia M, Nabavi SF, Izadi M, Abdollahi M, Nabavi SM, Ajami M (2017) Antimicrobial activity of eugenol and essential oils containing eugenol: a mechanistic viewpoint. Crit Rev Microbiol 43(6):668–689

Mashayekhi-Sardoo H, Rezaee R, Karimi G (2020) An overview of in vivo toxicological profile of thymoquinone. Toxin Rev 39(2):115–122

Mastelić J, Miloš M, Kuštrak D, Radonić A (2000) Essential oil and glycosidically bound volatile compounds from the needles of common Juniper (Juniperus communis L.). Croat Chem Acta 73:585–593

Mishra S, Sachan A, Sachan SG (2013) Production of natural value-added compounds: an insight into the eugenol biotransformation pathway. J Ind Microbiol Biotechnol 40(6):545–550

Molina G, Pimentel MR, Pastore GM (2013) Pseudomonas: a promising biocatalyst for the bioconversion of terpenes. Appl Microbiol Biotechnol 97(5):1851–1864

Mythili Avadhani MN, Immanuel Selvaraj C, Rajasekharan PE, Tharachand C (2013) The sweet and bitterness of sweet flag (Acorus calamus L.)—a review. Res J Pharm Biol Chem Sci 4(2):598–610

Nallamuthu IA, Parthasarathi, Khanum F (2013) Thymoquinone-loaded PLGA nanoparticles: antioxidant and anti-microbial properties. Int Curr Pharm J 2(12):202–207

Nihei T, Suzuki H, Aoki A, Yuminoki K, Hashimoto N, Sato H, Seto Y, Onoue S (2016) Development of a novel nanoparticle formulation of thymoquinone with a cold wet-milling system and its pharmacokinetic analysis. Int J Pharm 511(1):455–461

Othman N, Masarudin MJ, Kuen CY, Dasuan NA, Abdullah LC, Md. Jamil SNA (2018) Synthesis and optimization of chitosan nanoparticles loaded with l-ascorbic acid and thymoquinone. Nanomaterials 8(11):920

Pathak K, Raghuvanshi S (2015) Oral bioavailability: issues and solutions via nanoformulations. Clin Pharmacokinet 54(4):325–357

Pei RS, Zhou F, Ji BP, Xu J (2009) Evaluation of combined antibacterial effects of eugenol, cinnamaldehyde, thymol, and carvacrol against E. coli with an improved method. J Food Sci 74(7):M379–M383
Pokharkar VB, Shekhawat PB, Dhapte VV, Mandpe LP (2011) Development and optimization of eugenol loaded nanostructured lipid carriers for periodontal delivery. Int J Pharm Pharm Sci 3(4):138–143
Randhawa MA, Gondal MA, Al-Zahrani AH, Rashid SG, Ali A (2015) Synthesis, morphology and antifungal activity of nano-particulated amphotericin-B, ketoconazole and thymoquinone against Candida albicans yeasts and Candida biofilm. J Environ Sci Health A Tox Hazard Subst Environ Eng 50(2):119–124
Razzaq S, Hanif S, Syed MA, Iqbal J, Hassan SS, Raza SA, Riaz H, Abid F (2018) Development and evaluation of mucoadhesive buccal tablet containing metronidazole for the treatment of periodontitis and gingivitis. Pak J Pharm Sci 31(5):1903–1910
Salem ML, Alenzi FQ, Attia WY (2011) Thymoquinone, the active ingredient of Nigella sativa seeds, enhances survival and activity of antigen-specific CD8-positive T cells in vitro. Br J Biomed Sci 68(3):131–137
Shaarani S, Hamid SS, Mohd Kaus NH (2017) The influence of pluronic F68 and F127 nanocarrier on physicochemical properties, in vitro release, and antiproliferative activity of thymoquinone drug. Pharm Res 9(1):12–20
Shah B, Davidson PM, Zhong Q (2013) Nanodispersed eugenol has improved antimicrobial activity against Escherichia coli O157:H7 and Listeria monocytogenes in bovine milk. Int J Food Microbiol 161(1):53–59
Shaikh S, Dubey R, Dhande S, Joshi YM, Kadam VJ (2012) Phytochemical and pharmacological profile of Aphanamixis polystachya: an overview. Res J Pharm Tech 5(10):1260–1263
Shao Y, Wu C, Wu T, Li Y, Chen S, Yuan C, Hu Y (2018) Eugenol-chitosan nanoemulsions by ultrasound-mediated emulsification: formulation, characterization and antimicrobial activity. Carbohydr Polym 193:144–152
Sharma A, Sharma NK, Srivastava A, Kataria A, Dubey S, Sharma S, Kundu B (2018) Clove and lemongrass oil based non-ionic nanoemulsion for suppressing the growth of plant pathogenic Fusarium oxysporum f.sp. lycopersici. Ind Crop Prod 123(1):353–362
Shehensha S, Jyothi MV (2020) Synthesis of silver nanoparticles using Nigella sativa seed extracts and assessment of their antibacterial activity. Int J Res Pharm Sci 11(2):2525–2532
Surekha R, Sumathi T (2016) An efficient encapsulation of thymoquinone using solid lipid nanoparticle for brain targeted drug delivery: physicochemical characterization, pharmacokinetics, and bio-distribution studies. Int J Pharm Clin Res 8(12):1616–1624
Taleuzzaman M, Imam SS, Gilani SJ (2020) Clove oil/eugenol as the nanotechnological perspective for healthcare applications. In: Rahman M, Beg S, Kumar V, Ahmad F (eds) Nanomedicine for bioactives. Springer, Singapore. https://doi.org/10.1007/978-981-15-1664-1_16
Taleuzzamana M, Jahangir A, Gilania SJ (2017) Quantification and identification of bioactive eugenol in Myristica fragrans seed using validated high performance thin layer chromatography technique. Pharm Anal Acta 8:563
Tammannavar P, Pushpalatha C, Jain S, Sowmya SV (2013) An unexpected positive hypersensitive reaction to eugenol. BMJ Case Rep 2013:bcr2013009464
Thadkala K, Nanam PK, Rambabu B, Sailu C, Aukunuru J (2014) Preparation and characterization of amorphous ezetimibe nanosuspensions intended for enhancement of oral bioavailability. Int J Pharm Investig 4(3):131–137
Ugur AR, Dagi HT, Ozturk B, Tekin G, Findik D (2016) Assessment of in vitro antibacterial activity and cytotoxicity effect of Nigella sativa oil. Pharmacogn Mag 12(Suppl 4):S471–S474
van de Waterbeemd H, Gifford E (2003) ADMET in silico modelling: towards prediction paradise? Nat Rev Drug Discov 2(3):192–204
Vignesh Kumar S, Renuka Devi P, Hemananthan E (2020) In vitro studies to analyze the stability and bioavailability of thymoquinone encapsulated in the developed nanocarrier. J Dispers Sci Technol 41(2):243–256

Wani MC, Horwitz SB (2014) Nature as a remarkable chemist: a personal story of the discovery and development of Taxol. Anti-Cancer Drugs 25(5):482–487

Yoo CB, Han KT, Cho KS, Ha J, Park HJ et al (2005) Eugenol isolated from the essential oil of Eugenia caryophyllata induces a reactive oxygen species-mediated apoptosis in HL-60 human promyelocytic leukemia cells. Cancer Lett 225(1):41–52

Zhang J, Lv H, Jiang K, Gao Y (2011) Enhanced bioavailability after oral and pulmonary administration of baicalein nanocrystal. Int J Pharm 420(1):180–188

Zhao X, Chen D, Gao P, Ding P, Li K (2005) Synthesis of ibuprofen eugenol ester and its microemulsion formulation for parenteral delivery. Chem Pharm Bull (Tokyo) 53(10):1246–1250

Chapter 18
Nanoparticles as a Future Alternative Against Multiple Drug Resistance

Md. Meraj Ansari, Kaushik Kuche, Rohan Ghadi, Tushar Date, Dasharath Chaudhari, Faiyaz Ahmad Ansari, Rehan Khan, Akshay Vyawahare, and Sanyog Jain

18.1 Introduction

Microbial infections are a leading cause of chronic infections and mortality (Gargano and Hughes 2014). Antibiotics have been used for treatment against microbial infections due to their cost-effectiveness and potent outcome. However, several studies have reported that the widespread use of antibiotics has become a major cause of the development of multiple drug-resistant bacterial strains. Bacterial multiple drug resistance against antibiotics has become a major challenge for human health for the treatment of bacterial pathogenic disease (Desselberger 2000). The statistical data of research reports against infection caused by multiple drug-resistant bacteria to antibiotics have increased many folds day by day. These reports have shown that resistance developed by bacteria against multiple antimicrobial agents becomes a major threat to human medication in Europe. In Europe, the mortality rate from multiple drug-resistant bacterial infections is enhanced and the amount of money spent by the government becomes €1.5 billion per year of the economy (Rémy et al. 2015). Nowadays, about two million people in the USA are suffering from antibiotic-resistant microorganisms per year with a rate of mortality of about

M. M. Ansari · K. Kuche · R. Ghadi · T. Date · D. Chaudhari · S. Jain (✉)
Centre for Pharmaceutical Nanotechnology, Department of Pharmaceutics, National Institute of Pharmaceutical Education and Research, Mohali, Punjab, India
e-mail: sanyogjain@niper.ac.in

F. A. Ansari
Department of Neurology, Burdwan Medical College and Hospital, Burdwan, West Bengal, India

R. Khan · A. Vyawahare
Department of Nano-Therapeutics, Institute of Nano Science and Technology, Habitual Centre,
Mohali, Punjab, India

N. Akhtar et al. (eds.), *Emerging Modalities in Mitigation of Antimicrobial Resistance*,
https://doi.org/10.1007/978-3-030-84126-3_18

23,000 as a direct result. The development of new antibiotics or antimicrobial agent is not an easy task, and this gap in new antibiotics or antimicrobial agent has further become worse in the circumstances of antimicrobial resistance (Li and Webster 2018).

Nanotechnology is a multidisciplinary promising area of biological science in the advancement of nanoscale research (Wong et al. 2013). Interacting with biological molecules at the nanoscale opens up a broad field of research and application. There are different methods for the preparation of nanoparticles and classified into conventional and biological methods. A conventional method deals with the fabrication of nanoparticles by using chemicals by adopting the top-down approach and a bottom-up approach. However, by biological methods, nanoparticles are fabricated by using living things such as plants leaves, stems, seeds, flowers, latex, and microbes (Shamaila et al. 2016a). The main constituents of plant extracts are bioactive polyphenols, sugar, alkaloids, terpenoids, phenolic acids, etc., which may have a very important responsibility in reducing the metallic ions and their stabilization (Castro et al. 2011).

Nanotechnology is providing a platform to modify important properties of metals and develop nanoparticles with potential applications in drug delivery, diagnostics, cell labeling, biomarkers, and novel antimicrobial for the treatment of various diseases (Singh and Nalwa 2011). Hence, researchers are shifting toward future alternatives of multiple drug resistance and silver nanoparticles as well as gold nanoparticles may be alternative. Silver is precious heavy metal that has been used as antibacterial agent since ancient times. In the current scenario, due to the beginning of antibiotics progress, the uses of silver in medical applications against bacterial infections were day by day declined (Chen and Schluesener 2008). The manipulation in size of silver by using nanotechnology technique resulted in their enhanced antimicrobial effects. Due to changes in physicochemical properties, AgNPs have served as antibacterial agents because of their high ratio of surface area to volume and their exclusive chemical and physical characteristics (Kim et al. 2007). It has been reported that the AgNPs with size range of 10–100 nm have potent antimicrobial properties against Gram-negative and Gram-positive microorganisms (Morones et al. 2005). The antibacterial or antimicrobial activity of AgNPs are extensively studied by many researchers against the multiple drug-resistant strains as well as pathogenic bacteria. It was found that AgNPs act as potent weapons against the multiple drug resistant organisms such as vancomycin-resistant *Staphylococcus aureus*, ampicillin-resistant *Escherichia coli*, and erythromycin-resistant *Streptococcus pyogenes*.

Gold is another precious heavy metal. Gold nanoparticles (AuNPs) are commonly used in many biological applications such as medical, gene therapy, and biosensor for disease diagnosis. AuNPs are easily prepared by co-precipitation methods and have lower toxicity as compared to other metallic NPs such as AgNPs (Shamaila et al. 2016a, b). Gold nanoparticles and ionic form have been studied for antibacterial activities. Recent research reports showed that AuNPs have antibacterial properties (Zhou et al. 2012).

Bacteria became resistant to multiple antibiotics via various pathways; thus, the therapeutic efficacy of antibiotics against infectious microorganisms declines. Most of the bacterial antibiotic-resistant mechanisms are not relevant for nanoparticles (NPs) of AgNPs and AuNPs because these nanoparticles directly act on cell wall of bacteria after contract. AgNPs and AuNPs exert their antibacterial properties through differing metabolic and biochemical pathways, thus acts effectively against a variety of multiple drug-resistant microorganism stains. It is very difficult for bacteria to develop resistance against these NPs. This chapter covers a general layout of nanoparticle fabrication, the mechanism involved in the development of antibiotic resistance, and the mechanism of action of silver nanoparticles (AgNPs) and gold nanoparticles (AuNPs) in bacterial cells by which we may consider these NPs as a future alternative against MDR.

18.2 Antibiotic Resistance Mechanisms

Generally, the current running antibiotics or antimicrobials that are most frequently being used against harmful bacteria or microorganism during infection have their mainly three action sites such as the bacterial cell wall, replication of bacterial DNA, and translational of proteins. Unfortunately, bacteria become resistant against antimicrobial agent by induction of mutation in bacterial chromosomes or by intrinsic factor contribution. After the development of resistance against antimicrobial agent, the normal type of microorganism may be terminated by using of drug. However, the microorganism that became mutant can survive and spread the resistance via vertical or horizontal transfer of genetic material (Arzanlou et al. 2017). There are the following mechanisms by which bacterial cell may be able to develop resistance against antibiotics or antimicrobial agent and nonantibiotics or non-antimicrobial agent (Fig. 18.1).

1. **Drug influx**: The porin channel is responsible for drug influx in bacterial cell. It has been found that very poor drug influx occurs via porin channel due to modification in bacterial cell wall permeability by mutation (Gupta and Birdi 2017). For example, the resistant development in *Pseudomonas aeruginosa* against imipenem.
2. **Drug efflux**: It has been observed that majority of the antimicrobial drugs that enter the bacterial cells undergo rapid drug efflux from the target bacterial cell. For instance, conversion of *Pseudomonas aeruginosa* into a multiple drug-resistant pathogen due to mutation in the major regulatory protein responsible for rapid and active drug efflux. The rapid and active drug efflux and poor influx alter drug effective concentration at action site that may cause sublethal drug concentration at the site of drug action and leading cause of target-based drug resistance. Some of the microbial strains have developed antimicrobial resistance against some drugs such as sulfonamides, aminoglycosides, streptogramins, tetracyclines, and chloramphenicol (Munita and Arias 2016).
3. **Inactivation of drug molecules**: For the development of drug resistance, molecule inactivation via covalent binding is also a common mechanism. For exam-

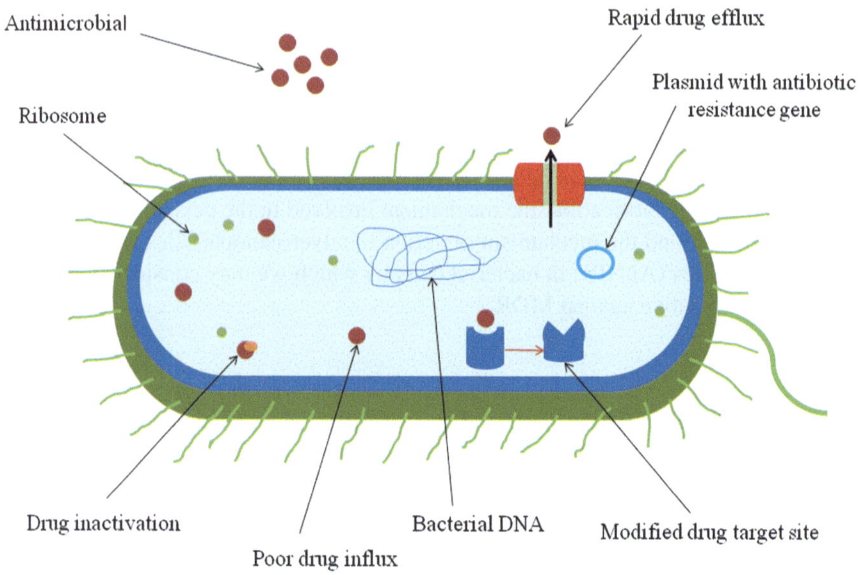

Fig. 18.1 General mechanism of bacterial resistance against antibacterial resistance

ple, β-lactam antibiotics (amoxicillin, ceftazidime, imipenem, penicillin, ampicillin, piperacillin, etc.) are inactivated by the cleavage of β-lactam ring in *H. influenza, S. aureus*, and *N. gonorrhea* (Munita and Arias 2016).

4. **Resistance plasmids**: Resistance plasmid is also called R plasmid, which has multiple resistance genes and is also responsible for transmission resistance. Bacteria have mobile genetic elements known as transposons, and these transposons facilitate horizontal genetic material transfer in condition of bulk metal ions and antibiotics (Baker-Austin et al. 2006).
5. **Modification/protection of target site**: Here the resistance occurs due to modification or alteration of microbial genes which induces conformational changes in the drug's target site making it inactive or therapeutically less effective. As for example, the rpoB gene of *Mycobacterium tuberculosis* after mutation become resistance to rifampin (Ao et al. 2012), and penicillin resistance in *Streptococcus pneumoniae* by the modification in penicillin-binding proteins (Kocaoglu et al. 2015).
6. **Biofilm formation**: Surface-associated microorganism generates three-dimensional biofilm made up of extracellular polysaccharide matrix that acts as protective cover for microorganisms against antibiotics or antimicrobial agents. The biofilm-forming bacterial strains are susceptible for gene mutation. Quorum sensing is another mechanism by which microbial gene expression may be responsible for secretion and synthesis of extracellular polymeric matrix. These extracellular polymeric matrixes make a protective cover around bacterial cell and prevent from high concentration of antimicrobial or antibiotics that is leading cause of some chronic bacterial infections like endocarditis, periodontitis,

cystic fibrosis, and prostatitis. Bioavailability of antibiotics reduces biofilm matrix due to slowing down diffusion of drugs. There are many scientific reports based on 3D bacterial biofilms as an inducer of resistance in several microorganisms (Liaqat et al. 2019).

These resistance development mechanisms of microorganism are leading cause of hindering the antibacterial drugs/antibiotic effectiveness against harmful microorganism.

18.3 Nanoparticle Fabrication Approaches

Nanotechnology is the branch of science, which deals with the fabrication of nanoparticles and their application in biological systems. There are different methods for the fabrication of nanoparticles; they are classified into conventional and biological methods (Fig. 18.2). By conventional methods, nanoparticles are fabricated by using chemicals by two basic approaches, the top-down and bottom-up approaches. The metallic substances are generally available in bulk materials, and the preparation of metallic nanoparticles from bulk materials into fine nanoparticle top-down approach will be applicable. But this approach has its limitations such as high energy consumption, formation of heterogeneous nanoparticles, very complex process of synthesis, and not eco-friendly, whereas bottom-up is a process in which we can easily prepare clusters of atoms and molecules by its self-assembly.

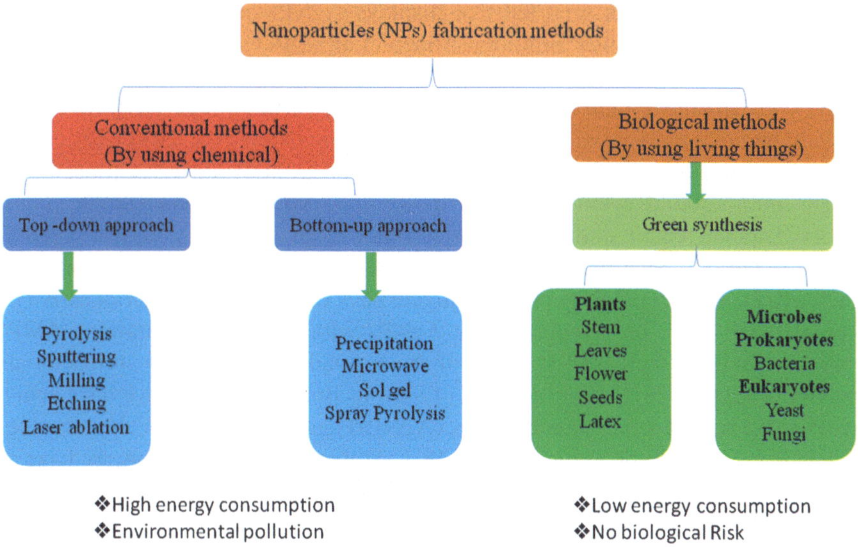

Fig. 18.2 Schematic representation of different approaches for nanoparticle fabrication

The bottom-up approach produces desired homogeneous metallic nanoparticles. By biological methods, we fabricate nanoparticles by using living things such as plants leaves, stems, seeds, flowers, latex, and microbes. Biological methods are less energy-consuming and eco-friendly.

18.3.1 Silver Nanoparticles (AgNPs)

Silver is a precious metal having antibacterial efficacy and has been recognized from ancient times (Reidy et al. 2013). Recently, various researchers have been reported that AgNPs as promising antimicrobial efficacy can be used as antimicrobial agent for the treatment against microbial infections (Natan and Banin 2017). The method of chemical reduction or green chemistry is a very traditional method for AgNP synthesis by using plant extract/microbial (Ribeiro et al. 2018). Many researchers have reported that the antibacterial activity of AgNP can be attributed to the cell death caused due to disruption of bacterial cell wall (Bondarenko et al. 2013), inactivation of respiratory chain enzyme, ROS production, and cellular components decomposition (Dakal et al. 2016). The permeability of bacterial cell membrane increased after the incorporation and adsorption of AgNPs due to depolarization of the bacterial cell wall. This ultimately makes the bacterial cell wall more permeable.

When AgNPs enter into the bacterial cell they induce ROS generation that is responsible for the inhibition of ATP production and DNA replication (Ramalingam et al. 2016). At the same time, AgNPs can release Ag^+ that is responsible for antibacterial activity by interacting with thiol-containing proteins (Durán et al. 2010). Another study has been reported that displays AgNPs as potent antimicrobial agent as compared to other antibiotics (Saeb et al. 2014) such as vancomycin or gentamicin against bacteria *P. aeruginosa*. One more study has reported that AgNPs are capable to inhibit *E. coli* and *S. aureus* growth through upregulation of ATPase pumps and expression of an antioxidant gene (Nagy et al. 2011). Recently, some researchers have been reported that AgNPs from P. amarus extract have potent antimicrobial agent against *P. aeruginosa* MDR bacteria (Singh et al. 2014). Another recent study has reported that rod-shaped and spherical AgNPs have a potent antibacterial agent for Gram-positive and Gram-negative bacteria (Acharya et al. 2018). AgNPs with conjugate have excellent antibacterial activity. Recently, Pal et al. have reported that AgNPs with a cysteine-containing AMP conjugate exhibit increased stability of Ag-S bond and enhanced antibacterial activity (Pal et al. 2016). On the basis of various previous reports, we have summarized the role of silver nanoparticles (AgNPs) against antimicrobial resistance and their mechanism of action in Table 18.1. At the end of this section, we have a clear view about AgNPs as possible candidature as antimicrobial agent against drug-resistant bacteria and still more extensive research work is required to explore more possibilities.

Table 18.1 Silver nanoparticles (AgNPs) against antimicrobial resistance and their mechanism of action

Targeted microorganism	Antibiotic resistance type	Mechanism of action	References
S. pneumoniae	Teicoplanin-resistant type	By generation of reactive oxygen species, silver ion uptake by cell membrane, intracellular cascades	Thapa et al. (2017)
P. aeruginosa	Ofloxacin-resistant type	By evading efflux pumps of multiple drugs	Ding et al. (2018)
S. mutans, Enterobacter cloacae	Biofilm formation	By reactive oxygen species-mediated disruption of cell membrane	Kulshrestha et al. (2017)
E. coli	MDR	By generation of reactive oxygen species	Siddiqi et al. (2018)
B. subtilis, S. aureus, K. pneumoniae	MDR	By the penetration of bacterial cell	Acharya et al. (2018)
A. baumannii	MDR	By structural change in cell wall permeability	Chang et al. (2017)
S. epidermidis, Mycobacterium smegmatis *Vibrio fluvialis, P. aeruginosa*	Formation biofilm/MDR	By adenosine monophosphate conjugation	Lambadi et al. (2015), Jaiswal et al. (2015)

18.3.2 Gold Nanoparticles (AuNPs)

Gold is precious heavy metal since ancient times, and it is inert and nontoxic in nature. From bulk gold materials, gold nanoparticles (AuNPs) can be prepared by gold salt through traditional chemical reduction method (Shah et al. 2014). The AuNPs can be synthesized most commonly by chloroauric acid reduction by citrate through chemical synthesis method (Fernandes and Baptista 2017).

According to some research reports, AuNPs have potential antimicrobial agents, but based on some researchers, controversy still remains (Shamaila et al. 2016a, b). AuNP concentrations have also played important role in bactericidal properties. According to some research reports, AuNPs generally do not serve as antimicrobial agent at low concentration and serve as weakly antibacterial agent at very high concentration (Shareena Dasari et al. 2015; Zhang et al. 2015). However, some research reports have described that the antimicrobial effect of AuNPs is associated with hindering ATPase activity, collapse in membrane potential, hindering the binding to tRNA of ribosomal subunit (Cui et al. 2012), and attachment to the respiratory chain of microorganism (Shamaila et al. 2016a, b).

Gold NPs in cationic and hydrophobic functionalized form have been reported as antibacterial agent for the Gram-positive and Gram-negative uropathogen. According to the research report, it has been reported that AuNPs are very less toxic in nature for mammalian cell and its chance to become resistant is very low (Li et al. 2014). Also, gold nanoparticles with ultra-thin graphite or graphite-like carbon nitride have served as potent antibacterial displaying potential against MDR Gram-positive as well as Gram-negative bacteria and high effectiveness against biofilms (Wang et al. 2017). Other studies have described that AuNPs to the conjugate of antibiotics enhance their inherent antibacterial potential against strain of multiple drug resistance (Payne et al. 2016). On the basis of previous reports, we have summarized the role of AuNPs against antimicrobial resistance and their mechanism of action in Table 18.2. In the end, keeping all research reports and researchers' views in our mind we can say that AuNPs have proved their possible candidature as antimicrobial agents against drug-resistant bacteria but still need more research work to further explore more possibilities in this field.

Table 18.2 Gold nanoparticles (AuNPs) against antimicrobial resistance and their mechanism of action

Targeted microorganism	Antibiotic resistance type	Mechanism of action	References
K. pneumoniae, *E. coli*	Cefotaxime-resistant type	By damage of bacterial genetic material, through bacterial cell wall disruption	Shaikh et al. (2017)
A. baumannii, *Proteus mirabilis*	Carbapenem-resistant type	By osmotic balance disturbance and disruption of bacterial cell wall integrity	Shaker and Shaaban (2017)
Streptococcus bovis, *E. Aerogenes*	Kanamycin-resistant type	By disruption of cell wall	Payne et al. (2016)
S. aureus	Methicillin-resistant type	By use of photothermal therapy with reactive oxygen species generation	Millenbaugh et al. (2015), Mocan et al. (2016), Hu et al. (2017), Ocsoy et al. (2017)
P. aeruginosa, *S. aureus*	Formation biofilm	By laser excitation for efficient photothermal response and efficient killing of biofilm	Yu et al. (2016), Pallavicini et al. (2014)
E. coli, *P. aeruginosa*, *B. Subtilis*	Formation of biofilm	By generation of reactive oxygen species	Wang et al. (2017)
S. epidermidis, *S. haemolyticus*	Formation of biofilm	By combination with antimicrobial agent	Roshmi et al. (2015)
S. aureus	MDR	By photoacoustic detection and photothermal therapy	Galanzha et al. (2012)
K. pneumoniae, *E. cloacae*	MDR	By use of photodynamic therapy	Khan et al. (2017)
E. coli	MDR	By generation of ROS	Zhang et al. (2013)
P. aeruginosa	Formation of biofilm/MDR	By conjugation with adenosine monophosphate.	Casciaro et al. (2017)
Salmonella typhimurium	Formation of biofilm/MDR	By adenosine monophosphate conjugates.	Yeom et al. (2016)

18.4 Mechanism of Action of Nanoparticles in Bacterial Cell

The antimicrobial or antibacterial potential of metallic nanoparticles against drug-resistant microbe and three-dimensional polymeric biofilms depend on a variety of factors. These factors are large surface area of nanoparticles in the contact with microorganism through intermolecular forces as van der Waals forces for polarization of particles, electrostatic attraction for long-range interaction, the particle size, and their stability collectively with drug concentration (Li et al. 2015). For the possible mechanism of nanoparticles, candidature as antimicrobial agent is inhibition of enzyme action, generation of reactive oxygen species for oxidative damage of microbes, change in gene expression, and protein deactivation.

Nowadays, metallic particles have been extensively used as potential antimicrobial agent and possible mechanism for action of nanoparticles to serve as antimicrobial agent is release of metal ion in bacterial cell, generation of reactive oxygen species for oxidative stress-mediated damage of microbes, and nonoxidative-mediated damage of microbes (Zaidi et al. 2017). The most common two metallic nanoparticles such as AgNPs and AuNPs have been included in this chapter, and their mechanism of action against microbes as antibacterial candidature is mentioned in Fig. 18.3.

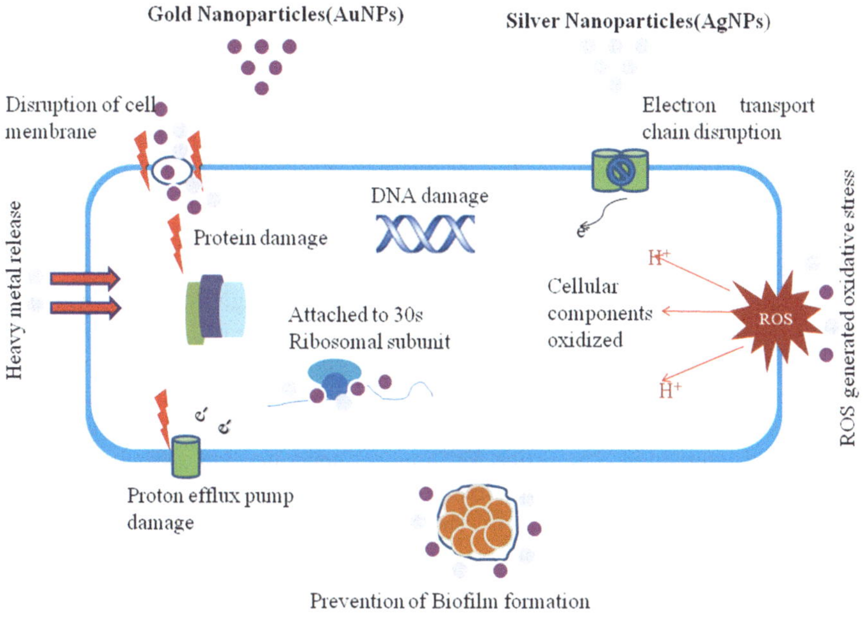

Fig. 18.3 Mechanism of action of AgNPs and AuNPs in bacterial cell

18.4.1 Mechanism of Action of AgNPs

In this section, we will explore the mechanism of action of AgNPs to prove their antibacterial potential against microbes. The mechanism of action has been classified into mainly two types, oxidative and nonoxidative. In oxidative damage, AgNPs generate reactive oxygen species that are responsible for oxidative-mediated death of microorganism. Therefore, AgNPs induced oxidative stress play important role in their antibacterial activity (Rudramurthy et al. 2016). Cellular oxidative metabolism generates abundant reactive oxygen species as by-product, and these reactive oxygen species are involved in cell survival and death, and meantime, these reactive oxygen species molecules play a very significant role in cell signaling and cell differentiation. During aerobic respiration in bacterial cell, reactive oxygen species are produced. These reactive oxygen species are neutralized by cellular intrinsic defense mechanism, but due to massive insult reactive oxygen species may cause oxidative damage of biomolecules and cell components (Li et al. 2012). The generation of massive reactive oxygen species causes cellular redox potential that is responsible for oxidative stress-mediated damage, alteration or modification in protein structure, deoxyribonucleic acid structure, and membrane lipid damage (Dwivedi et al. 2014). Production reactive oxygen species depends upon the chemical nature of nanoparticles. Silver had antimicrobial activity from ancient times, and metallic nanoparticle application against bacterial infection has become more popular in recent. A recent research report showed that AgNPs were generated superoxide radicals that lead to cellular oxidative damage and served as antibacterial efficacy (Zhang et al. 2013). Researchers have reported antibacterial effect of AgNPs against drug-resistant bacteria through silver ion (Ag^{+}) uptake across cell membrane. Further, the AgNPs can impart antibacterial effect through inhibition of drug efflux pump, intracellular cascades, production of reactive oxygen species, cell modifications affecting the membrane permeability, membrane disruption, and conjugation with adenosine monophosphate. These all are well summarized in Table 18.1.

The nonoxidative way includes the penetration of the AgNPs into the cell and disruption of the membrane permeability, alteration into the cellular components, and inhibition of the protein synthesis by incorporating into the purines and pyrimidines of the DNA and RNA. The AgNPs have a strong positive charge and the cell wall has a negative charge, and the AgNPs attach to the cell wall via electrochemical interaction. The active Ag^{+} ions are generated via dissociation after attachment to the cell wall; furthermore, Ag^{+} ions are generated by the Trojan horse mechanism (Singh et al. 2016). The Ag^{+} ion interaction causes disruption of the membrane integrity of lipid bilayer, hence the cellular leakage due to increased permeability (Agnihotri et al. 2014). This leads to cell death of bacteria due to loss of primary barrier for the defense.

18.4.2 Mechanism of Action of AuNPs

The AuNPs have been being extensively used as antibacterial agent. There are major concerns of researchers to explore the mechanism of action of AuNPs as antimicrobial agent. According to a research report, researchers have reported that reactive oxygen species generation and release of metal ion potentiate the antimicrobial properties of AuNPs via without coating with gold nanoparticles under ultraviolet electromagnetic radiation at the wavelength of 365 nm (Zhang et al. 2013). Another research report showed that the antimicrobial activity of AuNPs against *Klebsiella pneumoniae*, *Pseudomonas aeruginosa*, *E. coli*, and *Salmonella Typhi* was due to intracellular ROS-mediated oxidative stress (Umamaheswari et al. 2014). There are so many studies that have reported antimicrobial properties of AuNPs against drug-resistant microbes through reactive oxygen species-generated oxidative damage and well summarized in Table 18.2.

Another mechanism of action of NPs is a nonoxidative. In this mechanism, nanoparticles interact closely with bacterial cell wall and hamper them. As we all know that cell wall of microorganism is an outer protective layer and served as a defensive barrier that protects from environmental insults, hampering this protective layer definitely hampers the growth of microorganisms. The peptidoglycan polymer is the main constituent of Gram-negative bacteria. The bacterial outer membrane is made up of lipopolysaccharide and protein that are highly specialized in selective entry of macromolecules (Zaidi et al. 2017). Gram-positive bacteria cell wall is also composed of peptidoglycans and has several pores that allow the penetration by external or foreign particles that may able to interfere with bacterial cell proper functioning (Sarwar et al. 2015). So, these are possible ways by which gold nanoparticles serve as potent antibacterial agents and maybe probable candidates against microbial infection.

18.5 Mechanism of Action of NPs on Biofilm Formation

Biofilms are the three-dimensional film of polysaccharide matrix generated by microorganism. The biofilm has a great impact on public health. The microorganisms associated with generation of polymeric matrix are very less susceptible to antimicrobial action. Recently, researchers have reported that the formation of polymeric matrix with surface in the form of biofilm has become another way by which microorganisms protect themselves from antibiotics (Khameneh et al. 2016). By the formation of biofilm, cells adhere to the extracellular polymeric substance that acts as a protective barrier from antibiotics that is responsible for antibiotic resistance, inhibition of the most important process of phagocytosis during innate immune response by phagocytes, and the leading cause of serious health threat globally (Bjarnsholt 2013). However, the formation of polymeric three-dimensional matrix by microorganism has become a challenge for antimicrobial agent and meantime researchers have reported that nanoparticles can disrupt cell wall of microorganisms and generation of polymeric matrix that are the major causes to reduce microbial survival (Peulen and Wilkinson 2011; Chen et al. 2014; Miao et al. 2016). Therefore, the properties of AgNPs and AuNPs against biofilm formation give an extra advantage of these NPs to fight against microbial infection and a better optional candidate as antimicrobial agents.

18.6 Conclusions

Antimicrobial resistance has become global health threat. It has a great impact on global socioeconomic condition. Due to the development of drug resistance in bacteria, it becomes very crucial in medication against infectious disease in health sector. Microorganism becomes resistance to antimicrobial agents or antibiotics through alteration in their cell wall to cheek drug permeability from cell membrane, modification in target protein, inactivation of antimicrobial agent to bind bacterial ribosome, and development of active efflux of drugs. Nanomedicine is a budding branch of medicine that uses advancement in nanotechnology for the prevention and treatment of infectious disease. The advancement in nanotechnology has become an effective approach for the fabrication of metallic bulk materials in nanosized particles for the treatment against drug-resistant microbes. There are many metallic nanoparticles that have efficient antibacterial properties. Nowadays, there is no doubt on the potentiality of silver and gold nanoparticles against drug-resistant bacteria. AgNPs and AuNPs are well-established antimicrobial agents against both non-MDR and MDR bacterial strains. These NPs have developed antibacterial potential against multiple drug-resistant microorganisms via adhesion to the bacterial cell membrane and cell penetration, causing various structural disruptions and dysfunction through ROS-generated oxidative stress and inhibition of the formation of polymeric matrix from substrate in the form of three-dimensional biofilm by microorganism. The antibacterial activity of AgNPs and AuNPs may be the new alternative against multiple drug-resistant microbes. Furthermore, extensive studies are required on these nanoparticles to explore their potential as antibacterial agents. Also, there is a need for extensive research on these nanoparticles with an impact on gene level for wider application of these particles against MDR in healthcare system.

Acknowledgement This work was supported by the Department of Health Research (DHR), Ministry of Health and Family Welfare, Government of India. The author Md. Meraj Ansari is thankful to DHR for Young Scientist Award under DHR Young Scientist Scheme (YSS/2019/000052/PRCYSS).

Conflict of Interest There is no conflict of interest among the authors.

References

Acharya D, Singha KM, Pandey P, Mohanta B, Rajkumari J, Singha LP (2018) Shape dependent physical mutilation and lethal effects of silver nanoparticles on bacteria. Sci Rep 8:201. https://doi.org/10.1038/s41598-017-18590-6

Agnihotri S, Mukherji S, Mukherji S (2014) Size-controlled silver nanoparticles synthesized over the range 5–100 nm using the same protocol and their antibacterial efficacy. RSC Adv 4(8):3974–3983. https://doi.org/10.1039/C3RA44507K

Ao W, Aldous S, Woodruff E, Hicke B, Rea L, Kreiswirth B, Jenison R (2012) Rapid detection of rpoB gene mutations conferring rifampin resistance in Mycobacterium tuberculosis. J Clin Microbiol 50(7):2433–2440. https://doi.org/10.1128/JCM.00208-12

Arzanlou M, Chai WC, Venter H (2017) Intrinsic, adaptive and acquired antimicrobial resistance in Gram-negative bacteria. Essays Biochem 61(1):49–59. https://doi.org/10.1042/EBC20160063

Baker-Austin C, Wright MS, Stepanauskas R, McArthur JV (2006) Co-selection of antibiotic and metal resistance. Trends Microbiol 14(4):176–182. https://doi.org/10.1016/j.tim.2006.02.006

Bjarnsholt T (2013) The role of bacterial biofilms in chronic infections. APMIS. Supplementum (136):1–51. https://doi.org/10.1111/apm.12099

Bondarenko O, Ivask A, Käkinen A, Kurvet I, Kahru A (2013) Particle-cell contact enhances antibacterial activity of silver nanoparticles. PLoS One 8(5):e64060. https://doi.org/10.1371/journal.pone.0064060

Casciaro B, Moros M, Rivera-Fernández S, Bellelli A, de la Fuente JM, Mangoni ML (2017) Gold-nanoparticles coated with the antimicrobial peptide esculentin-1a(1-21)NH2 as a reliable strategy for antipseudomonal drugs. Acta Biomater 47:170–181. https://doi.org/10.1016/j.actbio.2016.09.041

Castro L, Blazquez ML, Munoz JA, Gonzalez F, Garcıa-Balboa C, Ballester A (2011) Biosynthesis of gold nanowires using sugar beet pulp. Process Biochem 46:1076–1082. https://doi.org/10.1016/j.procbio.2011.01.025

Chang TY, Chen CC, Cheng KM, Chin CY, Chen YH, Chen XA, Sun JR, Young JJ, Chiueh TS (2017) Trimethyl chitosan-capped silver nanoparticles with positive surface charge: their catalytic activity and antibacterial spectrum including multidrug-resistant strains of Acinetobacter baumannii. Colloids Surf B Biointerfaces 155:61–70. https://doi.org/10.1016/j.colsurfb.2017.03.054

Chen X, Schluesener HJ (2008) Nanosilver: a nanoproduct in medical application. Toxicol Lett 176(1):1–12. https://doi.org/10.1016/j.toxlet.2007.10.004

Chen CW, Hsu CY, Lai SM, Syu WJ, Wang TY, Lai PS (2014) Metal nanobullets for multidrug resistant bacteria and biofilms. Adv Drug Deliv Rev 78:88–104. https://doi.org/10.1016/j.addr.2014.08.004

Cui Y, Zhao Y, Tian Y, Zhang W, Lü X, Jiang X (2012) The molecular mechanism of action of bactericidal gold nanoparticles on Escherichia coli. Biomaterials 33(7):2327–2333. https://doi.org/10.1016/j.biomaterials.2011.11.057

Dakal TC, Kumar A, Majumdar RS, Yadav V (2016) Mechanistic basis of antimicrobial actions of silver nanoparticles. Front Microbiol 7:1831. https://doi.org/10.3389/fmicb.2016.01831

Desselberger U (2000) Emerging and re-emerging infectious diseases. J Infect 40(1):3–15. https://doi.org/10.1053/jinf.1999.0624

Ding F, Songkiatisak P, Cherukuri PK, Huang T, Xu XN (2018) Size-dependent inhibitory effects of antibiotic drug nanocarriers against Pseudomonas aeruginosa. ACS Omega 3:1231–1243. https://doi.org/10.1021/acsomega.7b01956

Durán N, Marcato PD, De Conti R, Alves OL, Costa FTM, Brocchi M (2010) Potential use of silver nanoparticles on pathogenic bacteria, their toxicity and possible mechanisms of action. J Braz Chem Soc 21:949–959. https://doi.org/10.1590/s0103-50532010000600002

Dwivedi S, Wahab R, Khan F, Mishra YK, Musarrat J, Al-Khedhairy AA (2014) Reactive oxygen species mediated bacterial biofilm inhibition via zinc oxide nanoparticles and their statistical determination. PLoS One 9(11):e111289. https://doi.org/10.1371/journal.pone.0111289

Fernandes AR, Baptista PV (2017) Gene silencing using multifunctionalized gold nanoparticles for cancer therapy. Methods Mol Biol 1530:319–336. https://doi.org/10.1007/978-1-4939-6646-2_19

Galanzha EI, Shashkov E, Sarimollaoglu M, Beenken KE, Basnakian AG, Shirtliff ME, Kim JW, Smeltzer MS, Zharov VP (2012) In vivo magnetic enrichment, photoacoustic diagnosis, and photothermal purging of infected blood using multifunctional gold and magnetic nanoparticles. PLoS One 7(9):e45557. https://doi.org/10.1371/journal.pone.0045557

Gargano LM, Hughes JM (2014) Microbial origins of chronic diseases. Annu Rev Public Health 35:65–82. https://doi.org/10.1146/annurev-publhealth-032013-182426

Gupta PD, Birdi TJ (2017) Development of botanicals to combat antibiotic resistance. J Ayurveda Integr Med 8(4):266–275. https://doi.org/10.1016/j.jaim.2017.05.004

Hu D, Li H, Wang B, Ye Z, Lei W, Jia F, Jin Q, Ren KF, Ji J (2017) Surface-adaptive gold nanoparticles with effective adherence and enhanced photothermal ablation of methicillin-resistant Staphylococcus aureus biofilm. ACS Nano 11(9):9330–9339. https://doi.org/10.1021/acsnano.7b04731

Jaiswal S, Bhattacharya K, McHale P, Duffy B (2015) Dual effects of β-cyclodextrin-stabilised silver nanoparticles: enhanced biofilm inhibition and reduced cytotoxicity. J Mater Sci Mater Med 26(1):5367. https://doi.org/10.1007/s10856-014-5367-1

Khameneh B, Diab R, Ghazvini K, Fazly Bazzaz BS (2016) Breakthroughs in bacterial resistance mechanisms and the potential ways to combat them. Microb Pathog 95:32–42. https://doi.org/10.1016/j.micpath.2016.02.009

Khan S, Khan SN, Meena R, Dar AM, Pal R, Khan AU (2017) Photoinactivation of multidrug resistant bacteria by monomeric methylene blue conjugated gold nanoparticles. J Photochem Photobiol B 174:150–161. https://doi.org/10.1016/j.jphotobiol.2017.07.011

Kim JS, Kuk E, Yu KN, Kim JH, Park SJ, Lee HJ, Kim SH, Park YK, Park YH, Hwang CY, Kim YK, Lee YS, Jeong DH, Cho MH (2007) Antimicrobial effects of silver nanoparticles. Nanomed Nanotechnol Biol Med 3(1):95–101. https://doi.org/10.1016/j.nano.2006.12.001

Kocaoglu O, Tsui HC, Winkler ME, Carlson EE (2015) Profiling of β-lactam selectivity for penicillin-binding proteins in Streptococcus pneumoniae D39. Antimicrob Agents Chemother 59(6):3548–3555. https://doi.org/10.1128/AAC.05142-14

Kulshrestha S, Qayyum S, Khan AU (2017) Antibiofilm efficacy of green synthesized graphene oxide-silver nanocomposite using Lagerstroemia speciosa floral extract: a comparative study on inhibition of gram-positive and gram-negative biofilms. Microb Pathog 103:167–177. https://doi.org/10.1016/j.micpath.2016.12.022

Lambadi PR, Sharma TK, Kumar P, Vasnani P, Thalluri SM, Bisht N, Pathania R, Navani NK (2015) Facile biofunctionalization of silver nanoparticles for enhanced antibacterial properties, endotoxin removal, and biofilm control. Int J Nanomedicine 10:2155–2171. https://doi.org/10.2147/IJN.S72923

Li B, Webster TJ (2018) Bacteria antibiotic resistance: new challenges and opportunities for implant-associated orthopedic infections. J Orthop Res 36(1):22–32. https://doi.org/10.1002/jor.23656

Li Y, Zhang W, Niu J, Chen Y (2012) Mechanism of photogenerated reactive oxygen species and correlation with the antibacterial properties of engineered metal-oxide nanoparticles. ACS Nano 6(6):5164–5173. https://doi.org/10.1021/nn300934k

Li X, Robinson SM, Gupta A, Saha K, Jiang Z, Moyano DF, Sahar A, Riley MA, Rotello VM (2014) Functional gold nanoparticles as potent antimicrobial agents against multi-drug-resistant bacteria. ACS Nano 8(10):10682–10686. https://doi.org/10.1021/nn5042625

Li H, Chen Q, Zhao J, Urmila K (2015) Enhancing the antimicrobial activity of natural extraction using the synthetic ultrasmall metal nanoparticles. Sci Rep 5:11033. https://doi.org/10.1038/srep11033

Liaqat I, Liaqat M, Ali S, Ali NM, Haneef U, Mirza SA, Tahir HM (2019) Biofilm formation, maturation and prevention: a review. J Bacteriol Mycol 6:1092

Miao L, Wang C, Hou J, Wang P, Ao Y, Li Y, Geng N, Yao Y, Lv B, Yang Y, You G, Xu Y (2016) Aggregation and removal of copper oxide (CuO) nanoparticles in wastewater environment and their effects on the microbial activities of wastewater biofilms. Bioresour Technol 216:537–544. https://doi.org/10.1016/j.biortech.2016.05.082

Millenbaugh NJ, Baskin JB, DeSilva MN, Elliott WR, Glickman RD (2015) Photothermal killing of Staphylococcus aureus using antibody-targeted gold nanoparticles. Int J Nanomedicine 10:1953–1960. https://doi.org/10.2147/IJN.S76150

Mocan L, Matea C, Tabaran FA, Mosteanu O, Pop T, Puia C, Agoston-Coldea L, Gonciar D, Kalman E, Zaharie G, Iancu C, Mocan T (2016) Selective in vitro photothermal nano-therapy of MRSA infections mediated by IgG conjugated gold nanoparticles. Sci Rep 6:39466. https://doi.org/10.1038/srep39466

Morones JR, Elechiguerra JL, Camacho A, Holt K, Kouri JB, Ramírez JT, Yacaman MJ (2005) The bactericidal effect of silver nanoparticles. Nanotechnology 16(10):2346–2353. https://doi.org/10.1088/0957-4484/16/10/059

Munita JM, Arias CA (2016) Mechanisms of antibiotic resistance. Microbiol Spectr 4(2). https://doi.org/10.1128/microbiolspec.VMBF-0016-2015

Nagy A, Harrison A, Sabbani S, Munson RS Jr, Dutta PK, Waldman WJ (2011) Silver nanoparticles embedded in zeolite membranes: release of silver ions and mechanism of antibacterial action. Int J Nanomedicine 6:1833–1852. https://doi.org/10.2147/IJN.S24019

Natan M, Banin E (2017) From nano to micro: using nanotechnology to combat microorganisms and their multidrug resistance. FEMS Microbiol Rev 41(3):302–322. https://doi.org/10.1093/femsre/fux003

Ocsoy I, Yusufbeyoglu S, Yılmaz V, McLamore ES, Ildız N, Ülgen A (2017) DNA aptamer functionalized gold nanostructures for molecular recognition and photothermal inactivation of methicillin-resistant Staphylococcus aureus. Colloids Surf B Biointerfaces 159:16–22. https://doi.org/10.1016/j.colsurfb.2017.07.056

Pal I, Brahmkhatri VP, Bera S, Bhattacharyya D, Quirishi Y, Bhunia A, Atreya HS (2016) Enhanced stability and activity of an antimicrobial peptide in conjugation with silver nanoparticle. J Colloid Interface Sci 483:385–393. https://doi.org/10.1016/j.jcis.2016.08.043

Pallavicini P, Donà A, Taglietti A, Minzioni P, Patrini M, Dacarro G, Chirico G, Sironi L, Bloise N, Visai L, Scarabelli L (2014) Self-assembled monolayers of gold nanostars: a convenient tool for near-IR photothermal biofilm eradication. Chem Commun (Camb) 50(16):1969–1971. https://doi.org/10.1039/c3cc48667b

Payne JN, Waghwani HK, Connor MG, Hamilton W, Tockstein S, Moolani H, Chavda F, Badwaik V, Lawrenz MB, Dakshinamurthy R (2016) Novel synthesis of kanamycin conjugated gold nanoparticles with potent antibacterial activity. Front Microbiol 7:607. https://doi.org/10.3389/fmicb.2016.00607

Peulen TO, Wilkinson KJ (2011) Diffusion of nanoparticles in a biofilm. Environ Sci Technol 45(8):3367–3373. https://doi.org/10.1021/es103450g

Ramalingam B, Parandhaman T, Das SK (2016) Antibacterial effects of biosynthesized silver nanoparticles on surface ultrastructure and nanomechanical properties of gram-negative bacteria viz. Escherichia coli and Pseudomonas aeruginosa. ACS Appl Mater Interfaces 8(7):4963–4976. https://doi.org/10.1021/acsami.6b00161

Reidy B, Haase A, Luch A, Dawson KA, Lynch I (2013) Mechanisms of silver nanoparticle release, transformation and toxicity: a critical review of current knowledge and recommendations for future studies and applications. Materials 6(6):2295–2350. https://doi.org/10.3390/ma6062295

Rémy V, Largeron N, Quilici S, Carroll S (2015) The economic value of vaccination: why prevention is wealth. J Market Access Health Policy 3. https://doi.org/10.3402/jmahp.v3.29284

Ribeiro A, Anbu S, Alegria E, Fernandes AR, Baptista PV, Mendes R, Matias AS, Mendes M, Guedes da Silva M, Pombeiro A (2018) Evaluation of cell toxicity and DNA and protein binding of green synthesized silver nanoparticles. Biomed Pharmacother 101:137–144. https://doi.org/10.1016/j.biopha.2018.02.069

Roshmi T, Soumya KR, Jyothis M, Radhakrishnan EK (2015) Effect of biofabricated gold nanoparticle-based antibiotic conjugates on minimum inhibitory concentration of bacterial isolates of clinical origin. Gold Bull 48:63–71. https://doi.org/10.1007/s13404-015-0162-4

Rudramurthy GR, Swamy MK, Sinniah UR, Ghasemzadeh A (2016) Nanoparticles: alternatives against drug-resistant pathogenic microbes. Molecules 21(7):836. https://doi.org/10.3390/molecules21070836

Saeb AT, Alshammari AS, Al-Brahim H, Al-Rubeaan KA (2014) Production of silver nanoparticles with strong and stable antimicrobial activity against highly pathogenic and multidrug resistant bacteria. ScientificWorldJournal 2014:704708. https://doi.org/10.1155/2014/704708

Sarwar A, Katas H, Samsudin SN, Zin NM (2015) Regioselective sequential modification of chitosan via azide-alkyne click reaction: synthesis, characterization, and antimicrobial activity of chitosan derivatives and nanoparticles. PLoS One 10(4):e0123084. https://doi.org/10.1371/journal.pone.0123084

Shah M, Badwaik V, Kherde Y, Waghwani HK, Modi T, Aguilar ZP, Rodgers H, Hamilton W, Marutharaj T, Webb C, Lawrenz MB, Dakshinamurthy R (2014) Gold nanoparticles: various methods of synthesis and antibacterial applications. Front Biosci 19:1320–1344. https://doi.org/10.2741/4284

Shaikh S, Rizvi S, Shakil S, Hussain T, Alshammari TM, Ahmad W, Tabrez S, Al-Qahtani MH, Abuzenadah AM (2017) Synthesis and characterization of cefotaxime conjugated gold nanoparticles and their use to target drug-resistant CTX-M-producing bacterial pathogens. J Cell Biochem 118(9):2802–2808. https://doi.org/10.1002/jcb.25929

Shaker MA, Shaaban MI (2017) Formulation of carbapenems loaded gold nanoparticles to combat multi-antibiotic bacterial resistance: in vitro antibacterial study. Int J Pharm 525(1):71–84. https://doi.org/10.1016/j.ijpharm.2017.04.019

Shamaila S, Sajjad AKL, Ryma N-u-A, Farooqi SA, Jabeen N, Majeed S, Farooq I (2016a) Advancements in nanoparticle fabrication by hazard free eco-friendly green routes. Appl Mater Today 5:150–199. https://doi.org/10.1016/j.apmt.2016.09.009

Shamaila S, Zafar N, Riaz S, Sharif R, Nazir J, Naseem S (2016b) Gold nanoparticles: an efficient antimicrobial agent against enteric bacterial human pathogen. Nanomaterials 6(4):71. https://doi.org/10.3390/nano6040071

Shareena Dasari TP, Zhang Y, Yu H (2015) Antibacterial activity and cytotoxicity of gold (I) and (III) ions and gold nanoparticles. Biochem Pharmacol Open Access 4(6):199. https://doi.org/10.4172/2167-0501.1000199

Siddiqi KS, Husen A, Rao R (2018) A review on biosynthesis of silver nanoparticles and their biocidal properties. J Nanobiotechnol 16(1):14. https://doi.org/10.1186/s12951-018-0334-5

Singh R, Nalwa HS (2011) Medical applications of nanoparticles in biological imaging, cell labeling, antimicrobial agents, and anticancer nanodrugs. J Biomed Nanotechnol 7(4):489–503. https://doi.org/10.1166/jbn.2011.1324

Singh K, Panghal M, Kadyan S, Chaudhary U, Yadav JP, arkas (2014) Green silver nanoparticles of Phyllanthus amarus: as an antibacterial agent against multi drug resistant clinical isolates of Pseudomonas aeruginosa. J Nanobiotechnol 12:40. https://doi.org/10.1186/s12951-014-0040-x

Singh J, Kaur G, Kaur P, Bajaj R, Rawat M (2016) A review on green synthesis and characterization of silver nanoparticles and their applications: a green nanoworld. World J Pharm Pharm Sci 7:730–762

Thapa R, Bhagat C, Shrestha P, Awal S, Dudhagara P (2017) Enzyme-mediated formulation of stable elliptical silver nanoparticles tested against clinical pathogens and MDR bacteria and development of antimicrobial surgical thread. Ann Clin Microbiol Antimicrob 16(1):39. https://doi.org/10.1186/s12941-017-0216-y

Umamaheswari K, Baskar R, Chandru K, Rajendiran N, Chandirasekar S (2014) Antibacterial activity of gold nanoparticles and their toxicity assessment. BMC Infect Dis 14:P64. https://doi.org/10.1186/1471-2334-14-S3-P64

Wang Z, Dong K, Liu Z, Zhang Y, Chen Z, Sun H, Ren J, Qu X (2017) Activation of biologically relevant levels of reactive oxygen species by Au/g-C3N4 hybrid nanozyme for bacteria killing and wound disinfection. Biomaterials 113:145–157. https://doi.org/10.1016/j.biomaterials.2016.10.041

Wong IY, Bhatia SN, Toner M (2013) Nanotechnology: emerging tools for biology and medicine. Genes Dev 27(22):2397–2408. https://doi.org/10.1101/gad.226837.113

Yeom JH, Lee B, Kim D, Lee JK, Kim S, Bae J, Park Y, Lee K (2016) Gold nanoparticle-DNA aptamer conjugate-assisted delivery of antimicrobial peptide effectively eliminates intracellular Salmonella enterica serovar Typhimurium. Biomaterials 104:43–51. https://doi.org/10.1016/j.biomaterials.2016.07.009

Yu Q, Li J, Zhang Y, Wang Y, Liu L, Li M (2016) Inhibition of gold nanoparticles (AuNPs) on pathogenic biofilm formation and invasion to host cells. Sci Rep 6:26667. https://doi.org/10.1038/srep26667

Zaidi S, Misba L, Khan AU (2017) Nano-therapeutics: a revolution in infection control in post antibiotic era. Nanomed Nanotechnol Biol Med 13(7):2281–2301. https://doi.org/10.1016/j.nano.2017.06.015

Zhang W, Li Y, Niu J, Chen Y (2013) Photogeneration of reactive oxygen species on uncoated silver, gold, nickel, and silicon nanoparticles and their antibacterial effects. Langmuir 29(15):4647–4651. https://doi.org/10.1021/la400500t

Zhang Y, Shareena Dasari TP, Deng H, Yu H (2015) Antimicrobial activity of gold nanoparticles and ionic gold. J Environ Sci Health C Environ Carcinog Ecotoxicol Rev 33(3):286–327. https://doi.org/10.1080/10590501.2015.1055161

Zhou Y, Kong Y, Kundu S, Cirillo JD, Liang H (2012) Antibacterial activities of gold and silver nanoparticles against Escherichia coli and bacillus Calmette-Guérin. J Nanobiotechnol 10:19. https://doi.org/10.1186/1477-3155-10-19

Chapter 19
Nano-Cargo Boarded Defensins to Combat Multidrug Resistance

Saurabh Bhorkade, Pawan Kumar Pandey, Md. Meraj Ansari, and Sanyog Jain

19.1 Introduction

Antibiotics are compounds that inhibit the growth of microorganisms through their cytostatic or cytotoxic action (Pizzolato-Cezar et al. 2019). The discovery of antibiotics was a providential event in the history of medicine as they have been used exhaustively over the decades to save numerous lives and decrease overall mortality vis-à-vis infectious disease (Davies and Davies 2010). Several classes of antibiotics have been developed over the past years and used for the set of infectious diseases. But lately, these antibiotics have started failing to provide the expected results and have become ineffective in eradicating infections due to the development of resistance in pathogenic microorganisms. An era of antibiotic resistance followed after an interminable failed attempt by the researchers to discover novel antibiotics, immoderate use by clinicians, and antibiotic abuse by humankind in medicine and agriculture (Wright and Sutherland 2007). The resistance gene can transfer vertically or horizontally to the microbes present in their surroundings. With exposure to other antibiotics for a more extended period, they might develop resistance toward more than one antibiotic. The condition is known as multidrug resistance, and this has been a significant challenge in medicine. The emergence of multidrug-resistant strains has led to serious concerns for clinicians as it creates a burden on existing resources. With many Gram-negative bacteria acquiring pan-resistant, it now appears as a severe threat to humankind. Novel approaches to combat multidrug

Saurabh Bhorkade and Pawan Kumar Pandey contributed equally to this work.

S. Bhorkade · P. K. Pandey · M. M. Ansari · S. Jain (✉)
Centre for Pharmaceutical Nanotechnology, Department of Pharmaceutics, National Institute of Pharmaceutical Education and Research, Mohali, Punjab, India
e-mail: sanyogjain@niper.ac.in

N. Akhtar et al. (eds.), *Emerging Modalities in Mitigation of Antimicrobial Resistance*,
https://doi.org/10.1007/978-3-030-84126-3_19

resistance of microbes are being investigated, and in this regard, the beacon of hope can be seen in novel microbicidal compounds like antimicrobial peptides (Magiorakos et al. 2012).

Scientists have indefatigable efforts to decipher new targets to intervene in the different mechanisms of resistance development and find novel compounds to combat antibiotic resistance (Lima et al. 2019). In recent studies, many approaches have been explored to combat the problems mentioned above. Preventing SOS response, blocking intracellular communication in bacteria to reduce the horizontal transfer of resistant genes, and inhibiting efflux transporters have been targeted to combat multidrug resistance in microorganisms (Wright and Sutherland 2007). Besides these, an appreciable amount of work has been done to discover novel antimicrobial compounds. Recent studies' most exciting outcome is the emergence of natural peptides as antimicrobial agents known as antimicrobial peptides (AMPs). Antimicrobial peptides are oligopeptides having a chain length of 5–100 amino acids equipped with antimicrobial activity. They have broad-spectrum and potent activity against numerous pathogenic microorganisms. First discovered in the 1980s, they have shown appealing results in conquering multidrug resistance. Moreover, they are ubiquitous in nature and found in plants, animals, bacteria, fungi, and viruses (Lei et al. 2019). AMPs are synthesized by almost all living organisms and form an indispensable component of the host defense mechanism. They may have specific or nonspecific interaction with the cell membrane of invading pathogen. Their mechanism of action has been found to be different based on their structure. Due to their cationic or amphiphilic nature, they interact with negatively charged components of the cell membrane to interfere with the cell wall synthesis. While conventional antibiotics intervene in cell membrane synthesis, the AMPs interact with precursor molecules essential for cell wall synthesis. These antimicrobial peptides can also evoke the immune system, resulting in increased cytotoxic and cytostatic activity against pathogenic microorganisms.

These peptides have been isolated from different organisms and have shown similar activity against microbial infection. The first-ever AMP was isolated from moths, i.e., cecropins A and B, which showed selective toxicity against bacteria, whereas magainin was the first AMP to be extracted from vertebrates. Later, Lehere et al. identified a class of AMP isolated from humans known as defensins. Defensins were first isolated from human neutrophils and were known as human neutrophilic peptides (Yamaguchi and Ouchi 2012). Other human antimicrobial peptides include cathelicidins and histatins. Defensins are nonglycosylated cationic AMPs with a molecular weight of below 5 kDa with six cysteine residues cross-linked via disulfide linkage. They have a vital role in the innate and adaptive immunity of the host organism. Recently, several studies were conducted on defensins' antimicrobial activity, but due to poor pharmacokinetics and stability-related problems, they have not been commercialized. Lately, the development of nanoformulation loaded with antimicrobial peptides has been explored to bypass the stability and pharmacokinetic-related limitations.

Nanotechnology and the development of nanoparticle-based drug delivery system are a well-established and promising field in therapeutics and diagnostics, which have been extensively studied to overcome the limitations of poor pharmacokinetic profile, limited solubility, poor absorption, probable adverse effects, and poor stability of numerous drug molecules. Different nanomaterials with unique properties have been designed to improve the efficacy and overall performance of the drug molecule with respect to pharmacokinetic profile (Teixeira et al. 2020). Nanoparticles with sizes in the range of 10–1000 nm have been helpful in various clinical setups due to their size and flexibility by fabricating them with different materials. Their surface properties allow them to act as cargos to deliver compounds ranging from simple molecules to complex ones like peptides and proteins. Several nanocarriers like metallic, inorganic, organic, and polymeric nanocarriers including dendrimers, micelles, and liposomes have played a pivotal role in delivering the perverse molecules to the targeted site (Mirza and Siddiqui 2014). These nanocargos can act as Trojan horse for drugs, which cannot be given through conventional formulation (Gulyaev et al. 1999). In the same line, defensins can also be loaded into these nanocarriers and thus overcoming the limitations of stability and poor pharmacokinetics. Defensins have shown promise in combating multidrug resistance, whereas some nanomaterials have their own antimicrobial activity; by combining these, there is possibility of getting an improved efficacy and overall clinical outcome.

19.2 Defensins

These are cationic (+2 to +11 net charge) in nature with molecular weight of 3.5–4.5 kDa. The term defensin was coined by Lehrer to describe the cationic peptides stabilized by disulfide linkage and having broad antimicrobial activities against fungi, bacteria, and viruses (Ganz et al. 1985; Selsted et al. 1985). The defensins represent a class of antimicrobial peptides that play important role in innate and adaptive immunity. Defensins have antimicrobial activity through innate immunity against bacteria, virus, and fungi (Fig. 19.1).

19.2.1 Classification of Defensins

Defensins, based on the structure and disulfide topology, are classified into three major subfamilies: α-defensins, β-defensins, and θ-defensins. Only α- and β-defensins are found in humans, whereas unique, circular, ladder-shaped θ defensins stabilized by parallel disulfide bonds are present in rhesus macaques leukocytes (Gabay et al. 1989). Apparently, humans contain RNA transcript of rhesus θ-defensins, but its subsequent translation is abolished by the presence of stop codon in upstream signal sequence (Lehrer et al. 2012). Both α- and β-defensins have triple straned β sheets with different lengths of protein segments between six cysteine residues.

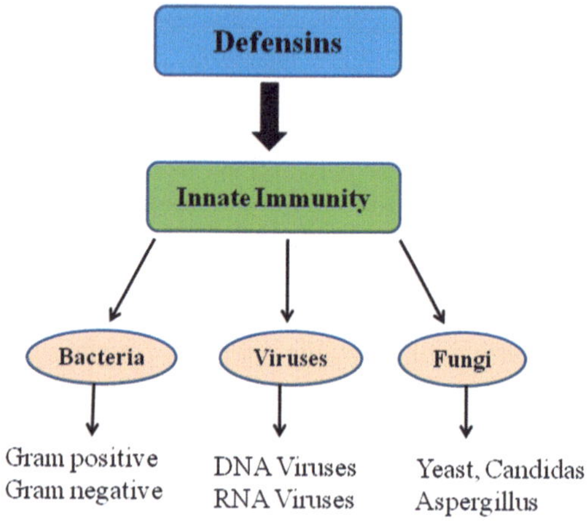

Fig. 19.1 Defensins as antimicrobial through innate immunity

19.2.1.1 α-Defensins

α-Defensins are present in neutrophils of humans, rats, guinea pigs, and rabbit macrophages. α-Defensins, based on expression and gene structure, are further classified into two groups, four human neutrophil peptides (HNP 1–4), also known as myeloid defensins, present in the azurophilic granules of neutrophils, and two human defensins (HD 5–6), also known as enteric defensins, secreted by colon and small intestine (Lehrer and Lu 2012; Nikitina et al. 2012). The three most abundant defensins present in the human neutrophils are HNP 1, HNP 2, and HNP 3. HNP 4 is also present in human neutrophils at concentration of <2% of HNP 3 levels. HNP 2 contains an extra amino acid (30 amino acids) compared to HNP 1 and HNP 3 (29 amino acids), whereas HNP 1 and HNP 3 differ in their amino acid residue at N-terminal. HNP 1 contains alanine, and HNP 3 has aspartate at its N-terminal. This replacement of single nucleotide residue at N-terminal by acidic aspartate amino acid makes HNP 3 less effective against *Staphylococcus aureus*, *Pseudomonas aeruginosa*, and *Escherichia coli*. HNP 4 contains 33 amino acids and has shown antimicrobial activity against *E. coli*, *Streptococcus faecalis*, and *Candida albicans* (Wang 2014). Azurophilic granules containing HNPs undergo restricted secretion and are fused with phagolysosomes releasing the content inside at high concentration and killing phagocytosed microbes.

Enteric defensins (HD 5 and HD 6) were isolated for the first time from the Paneth cells of the small intestinal crypt. These defensins play a crucial role in regulating intestinal microbes and protecting intestinal stem cells (Bevins et al. 1999; Bevins and Salzman 2011). HD 5 is also found in vaginal fluids (1–50 μg/ml) and is induced in the reproductive organs on infection of sexually transmitted diseases (Klotman et al. 2008; Porter et al. 2005). HD 5 have shown concentration-dependent antimicrobial activity against *Candida albicans, E. coli, Salmonella typhimurium*, and *Listeria monocytogenes*. Moreover, the antimicrobial efficacy of HD 5 against multidrug-resistant *Acinetobacter baumannii* is also confirmed by Wang and coworkers (Wang et al. 2018).

19.2.1.2 β-Defensins

β defensins differ from α-defensins in their disulfide linkage and have a short N-terminal α-helix. These were first time isolated from the bovine tracheal epithelium, where they were induced as a result of infection. Unlike α-defensins, the expression of human β-defensins (hBD) is transcriptionally regulated by keratinocytes of epithelial cells of the respiratory, gastrointestinal, and genitourinary tract. Scientists have reported more than 30 hBDs out of which only four (hBD 1 to hBD 4) have been studied in detail to date (Zasloff 2007). hBD 1 is expressed constitutively by the epithelial cells, whereas hBD 2 and hBD 3 are expressed in response to inflammation and infection (Ghosh et al. 2007). These hBDs also differ in their expression among the different tissues leading to tissue-specific upregulation. hBD 2 is abundantly expressed in the lungs, and hBD 3 in skin and tonsils, whereas hBD 4 is expressed highly in testicles and stomach (Tecle et al. 2010). β-defensins are also expressed in the genital tracts of males and females highlighting their protective roles in sexually transmitted diseases.

hBD 1 has shown strong microbicidal action against *P. aeruginosa* and *E. coli* and has proven effective in cystic fibrosis patients. On reduction of disulfide bond, hBD 1 acquires potent antimicrobial activity against *C. albicans*. On the other hand, hBD 2 has proven effective against Gram-negative bacteria like *E. coli* and *P. aeruginosa* along with *C. albicans, A. baumannii,* and S. *marcescens*. However, only bacteriostatic activity is observed against Gram-positive bacteria like *S. aureus*. Thus, hBD 2 can be employed in the treatment of gastric infection by *E. coli* and has also shown antitubercular properties.

The bactericidal activity of hBD 3 against Gram-negative bacteria like *P. aeruginosa, E. coli,* and *C. albicans* and Gram-positive bacteria like *S. aureus and Streptococcus pyogenes* is well established. Further, hBD 3 is also found effective against multidrug-resistant *S. aureus, P. aeruginosa, Enterococcus faecium, Acinetobacter baumannii, Stenotrophomonas maltophilia,* and vancomycin-resistant *Enterococcus faecium* (Dhople et al. 2006; Wang 2014). hBD 2 and hBD 3 have also shown anti-HIV activity by blocking viral replication (Pace et al. 2017). hBD 4 is found effective against bacterial infection *with P. aeruginosa*. Newly discovered β-defensins, hBD 5 and hBD 6, have shown antibacterial activity against *E. coli* and are widely expressed in the epididymis. Out of the most recently discovered defensins, hBD 19, hBD 23, hBD 27, and hBD 29, hBD 23 is found most effective against Gram-negative and Gram-positive bacteria like *S. aureus* and *E. coli* (Chow et al. 2012).

19.2.1.3 θ-Defensins

θ-Defensins are cyclic peptides with 18 amino acids connected with three disulfide linkages between cysteines. These were first time discovered in rhesus macaque leukocytes. Genomic analysis has shown the presence of θ defensins genes in genomes of macaques, colobus monkeys, siamangs, and orangutans. These defensins are resistant to proteases, have higher antibacterial efficacy, and are more stable due to their cyclic nature. Unlike acyclic peptides, the antimicrobial activity of these defensins is not dependent upon disulfide bonds and is more stable at physiological salt concentration giving them three times more activity against *E. coli* and *S. aureus* compared to acyclic defensins at osmotic conditions of body (Tang et al. 1999). One of the θ-defensins, rhesus θ-defensin 1 (RTD), has shown antimicrobial activity against multidrug-resistant *S. aureus* (methicillin-resistant *S. aureus*) and *P. aeruginosa* (Tai et al. 2015). The absence of θ-defensins increases the susceptibility of HIV infection highlighting its anti-HIV activity where the probable mechanism is blockade of viral entry (Conibear and Craik 2014; Wang 2014). However, anti-herpes and anti-influenza activities of these defensins are also observed (Doss et al. 2009; Yasin et al. 2004). θ-Defensins bind and neutralize bacterial toxins as well. For example, retrocyclin 1 and analogs of retrocyclin 1 have activity against anthrax and neutralize anthrax toxin (Wang et al. 2006). However, acyclic analogs of retrocyclin were unable to bind and neutralize bacterial toxins highlighting the role of sulfide bind and the cyclic backbone. Several attempts have been made to synthesize the cyclic analogs of these defensins having greater stability and broad antimicrobial activity.

19.3 Mechanism of Action of Defensins

AMPs like defensins are cationic in nature due to positively charged amino acid, which electrostatically interacts with the negatively charged cell membranes of bacteria. This interaction between defensins and cell surface leads to disruption of cell membrane, ultimately death of organism (Guilhelmelli et al. 2013). In contrast to human cell membrane, enriched with neutral phospholipids like phosphatidylethanolamine (PE), sphingomyelin (SM), and phosphatidylcholine (PC), bacterial cell wall is enriched with negatively charged phospholipids like phosphatidylserine (PS), phosphatidylglycerol (PG), and cardiolipin (CL). Additionally, Gram-positive bacteria have lipopolysaccharide-rich outer layer that imparts negative charge to the membrane. This contrasting nature of cell membranes is responsible for the selectivity of defensins toward bacteria. Further, the presence of cholesterol in human cell membrane imparts additional stability to human cell membrane making the interaction with AMPs less likely (Jin and Weinberg 2019).

The formation of pores inside cellular membrane is widely believed as the mechanism for antibacterial activity of AMPs resulting in the uncontrolled flow of ions across the membrane. Initially, interaction between the cationic amino acids of defensins and negatively charged phospholipids of cell membrane takes place leading to the accumulation of defensins on cell membrane (Epand et al. 2016). After a certain threshold, these AMPs self-assemble forming a pore inside the membrane. This process of pore formation is explained by three models: (1) barrel stave model, according to which there is perpendicular insertion of these AMPs inside the cellular membrane leading to the formation of pore, (2) carpet mechanism, which proposes parallel absorption of peptides on the membrane reaching a particular threshold, and this results in detergent-like effect causing membrane disintegration, and (3) toroidal pore mechanism, which also proposes perpendicular insertion of peptides but forming a curvature in the membrane with pores lined partly by phospholipid heads and partly by peptides (Yeaman and Yount 2003; Nguyen et al. 2011) (Fig. 19.2).

Variability among these mechanisms of action is also observed depending upon the type of defensin and its target. For example, antibacterial activity of HD 5 is by increasing the permeability of cellular membrane and by binding to DNA, thereby inhibiting cell replication. Similarly, HNP 1 acts against *E. coli* by destroying cell membrane, whereas against *S. aureus* by restricting wall precursor lipid II. In case of β-defensins, reduction in disulfide bridges is reported to increase the antibacterial activity. According to recently discovered mechanism, hBD 1 forms a Web-like structure with neutrophils and traps the bacteria. hBD 2 acts by binding to the membrane phospholipid leading to cell death, whereas hBd 3 is reported to act by two mechanisms, by binding with cellular lipid II precursor and by membrane destruction (Pachón-Ibáñez et al. 2017). Along with this generalized mechanism of action, i.e., pore formation and membrane permeabilization, other mechanisms are also

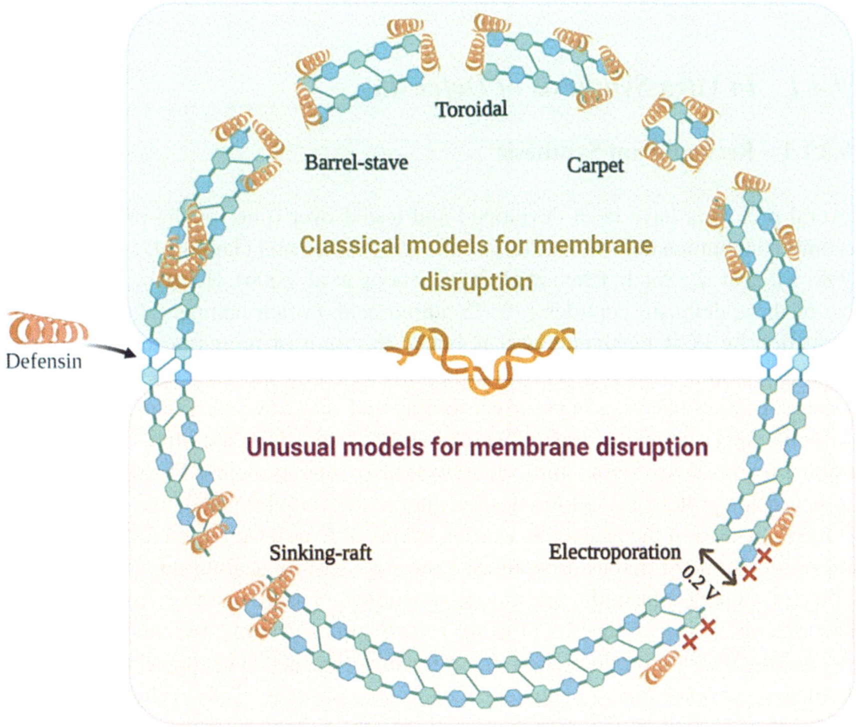

Fig. 19.2 Mechanism of action of defensin

depicted by the researchers, for example, inhibiting intracellular functions and bipolar synthesis (Yeaman and Yount 2003).

19.4 Defensins as Therapeutic Candidates

Multiple activities of the human defensins have drawn considerable attention of the scientist over the period, making them highly potential therapeutic candidates. Several companies have started investing or have future programs for engineering these peptide molecules into consumable therapeutic entity. The rapid development of antimicrobial resistance has further boosted the need for these novel antimicrobial agents as these defensins are less susceptible to resistance development when compared to other antimicrobial agents, and scientists across the globe have responded to it. Human defensins have ability to kill bacterial strains and enveloped and non-enveloped viruses. Defensins can also neutralize the bacterial toxin thus are ideally suited as broad anti-infective therapeutic agents. For example, recent study by Teles et al. has shown that the induction of endogenous gene regulation is effective against treatment of leprosy and other diseases (Teles et al. 2013).

19.4.1 In Vitro Synthesis of Defensins

19.4.1.1 Recombinant Synthesis

Several protocols have been developed and tested over time for the production of recombinant human defensins (Piers et al. 1993; Valore and Ganz 1997; Zhang et al. 1998; Xu et al. 2005a, b; Chen et al. 2006; Zhong et al. 2006). However, the smaller size of these defensin peptides (30–35 amino acids) often hampers the successful production by PCR technique as it is below the limit of protein size required for stable expression. This problem is further worsened by the presence of basic and cysteine residues resulting in incorrect folding and very low levels of expression. To overcome the issue related to smaller size of defensins, these are often expressed as fusion proteins with higher molecular weight partner proteins. Properly selected partner fusion protein also allows desired alteration of solubility suitable to particular protocol. In most of the protocols, various strains of *E. coli* were used for the production of recombinant defensin peptides. Low cost, ease in scaling up, ease in availability of suitable plasmids, and quick generation of large number of peptides in short time are some of benefits of using *E. coli* strains. Mainly, two main strategies are described for the production of recombinant defensins, (1) expression of insoluble inclusion bodies and (2) generation of soluble proteins. Both of these strategies involve the production of fusion proteins using the suitable carrier protein. Several other expression systems like yeast cells, tobacco plant cells, and infected insect cells were also employed for the production of recombinant defensins.

19.4.1.2 Chemical Synthesis

Due to smaller size of these peptides, chemical synthesis is often used as an alternative to recombinant technology for their production. Introduction of various non-natural modifications into these peptides is much easy using chemical synthesis when compared to recombinant technology. Further, ease in the purification of these peptides in the environment, which is simple and better defined compared to one in recombinant expression technology, is add-on benefit of this technology. However, synthetic peptides are generated in the nonfolded form and incorporation of folding, which involves multiple disulfide bonds, may be difficult in chemical synthesis. Low solubility of these peptides in basic solution due to the presence of cationic charge causes their precipitation during the process of synthesis. One of the protocols developed by Lu and coworkers involves amino acid sequence of mature defensins subjected to oxidative folding in the presence of 2 M urea and 25% DMF to minimize the aggregation and to avoid denaturation of proteins. Up to 80% of yield was achieved using these protocols, and four neutrophil defensins were folded and oxidized successfully (Wu et al. 2004).

19.4.2 Limitations of Defensins as Therapeutic Candidates

Although the defensins can be potential therapeutic agents, their progress as pharmaceuticals is majorly delayed due to some limitations. Very few of these novel peptides have made it up to the clinics, and most of them are natural cyclic peptides. One of the major factors limiting their development is high cost involved in the extraction of purified form of these AMPs, which largely affect the clinical trials. Until recently, biological studies of these defensins were limited by the lack of sufficient quantities of purified peptides. Efficient and economical production of pure form of these peptides is key for the clinical development of these therapeutic agents. Defensins are also reported to have nonspecific hemolytic activity and osmotic sensitivity to human cells. Low stability of these peptides and rapid elimination from the body is also a major hurdle as most of the gastrointestinal enzymes like carboxypeptidase, exopeptidases, and aminopeptidases rapidly digest these amino acid backbones. Thus, improvement is needed in terms of stability, targeted delivery, controlled release, transportation, and immunogenicity of these peptides before they can be used as therapeutic agents. The major challenge lies in the development of suitable formulation for these defensins to overcome above limitations.

19.4.3 Approaches to Overcome Limitations of Defensins

Notorious molecules like peptides have always been difficult to deliver, and there has been a continuous effort to improve delivery aspects of such components. Several researchers have applied different approaches to modify the properties of peptides like defensins. These include chemical modifications to increase the stability and protect it from enzymatic degradation, increase its intestinal absorption and half-life in plasma in human body, and increase its solubility and bioavailability. Many such modifications like cyclization, lipidization, and PEGylation are some of the methods that can be considered for modifying the properties of antimicrobial peptides. Fabrication of hybrid peptide is another method to improve the possibility of using antimicrobial peptides as therapeutics in near future.

Cyclization process can augment the stability of antimicrobial peptides through head-to-tail, head-to-side chain, tail-to-side chain, or side chain-to-side chain method. Nowadays, numerous methods exist to achieve cyclization of peptide molecules. Bacitracin is an example of cyclic peptide; after its discovery, the concept of cyclization has been a lucrative option for delivering peptides with improved stability. In a study conducted by Rozek et al., a linear peptide indolicidin was cyclized to achieve cell selectivity and at the same time stability in the presence of proteases up to 4.5-fold. Moreover, cyclization reduced hemolytic toxicity of the indolicidin, hence encouraging the possibility of oral delivery of the antimicrobial peptide (Rozek et al. 2003). However, this approach is not universal and some of the peptides have shown increased toxicity and decreased stability (Unger et al. 2001).

19.5 Nano-Cargos for Delivery of Antimicrobial Peptides

Nanomaterials have existed since the development of material science, but in last couple of decades they have gained significant attention. They were used in engineering and material sciences, but in last couple of decades they have attracted significant attention in therapeutic and diagnostic field (Pitkethly 2004). Many studies explored the potential of nanomaterials in drug delivery, which showed promising outcomes. Numbers of nanocarriers have been approved in last two decades with improved therapeutic efficacy and reduced toxicity. Many of these materials when reduced to nano-size have shown to have their own antimicrobial activity. They accentuate intracellular delivery of cargos, decrease frequency of administration, and attenuate the development of drug resistance strains by providing alternate mechanism for antimicrobial activity (Sánchez-López et al. 2020). Thus, loading of these antimicrobial peptides with nanocarriers can have ameliorating effect on microbicidal activity and it can further circumscribe the chances of development of resistance. Further are discussed some antimicrobial peptide-loaded nano-cargos for combating multidrug resistance.

19.5.1 Inorganic/Metallic Nanoparticles

In recent past, metallic nanoparticles fabricated with silver, gold, nickel, and silicon have been studied for their antimicrobial activity (Zhang et al. 2013). Metal nanoparticles have been shown to improve antimicrobial activity of drugs against resistant bacteria. They are easy to fabricate and provide uniform shape and size. Generation of reactive oxygen species is a principal mechanism of these metallic nanoparticles with respect to their antimicrobial activity. Studies have been conducted extensively with silver and gold NPs to explore their potential in providing better efficacy against resistant microbes (Abdeen et al. 2013) .

Tachyplesin-1, an antimicrobial peptide, has β-hairpin-like structure and has shown exemplary membranolytic activity, but it suffers from limitations in terms of stability and poor permeability. Shangshang et al. fabricated a multifunctional peptide-coated AgNPs to overcome the limitations of tachyplesin-1. The multifunctional peptide was equipped with a matrix metalloproteinase (MMP) cleavable sequence, a target peptide, and tachypresin-1 as an antimicrobial peptide, wherein the target peptide was pegylated in order to increase the circulation time in vivo, which was further assisted by MMP cleavable sequence. When this MFP reached the site of infection, the components get separated and by MMP which digests the MMP cleavable sequence releasing the tachyplesin-1 and restraining the spread of infection. It was observed that amalgamation of cationic MFP and anionic AgNPs led to improved bioactivity against MDR-AB, which opens new realm in curbing multidrug resistance (Li et al. 2020).

Gold nanoparticles have been investigated for their antimicrobial activity on both Gram-positive and Gram-negative bacteria. AuNPs have also shown ability to inhibit the growth of infection in MDR strains with lesser toxicity compared to other metallic NPs. Loading these AuNPs with antimicrobial peptides can further improve the antimicrobial therapy and is a preferential candidate for the development of nano-based delivery system of peptides for circumventing MDR (Li et al. 2014). Casciaro et al. demonstrated that covalently conjugated AMP esculentin-1a on AuNPs using poly(ethylene glycol) linker produced AuNP@Esc(1–21), which increased the microbicidal activity of esculentin-1a up to 15 times (Table 19.1). AuNPs formed were found to be less toxic to keratinocytes. Moreover, the nanoparticle system provided extended protection against proteolytic degradation and require lower concentration to break down membrane structure (Casciaro et al. 2017). Rai et al. illustrated the increased stability and antimicrobial activity of cecropin–melittin (CM)-conjugated gold nanoparticles. The CM-conjugated AuNPs with controlled size and polydispersity demonstrated higher stability in presence of proteolytic enzymes with significant increase in bactericidal activity. Moreover, the CM-AuNPs showed considerable reduction in side effects against human cells. The improved profile of the AMP indicated that the use of nanoparticle-based system for delivery of antimicrobial peptide can have multifarious merits in terms of efficacy and shows better therapeutic result against multidrug-resistant bacteria (Rai et al. 2016).

19.5.2 Polymeric Nanoparticles

Polymeric nanoparticles have gained significant attention in drug delivery with wide range of polymers to select from. Extensive numbers of publications are available on polymeric NPs. Most popular among polymeric nanoparticles are PLGA (poly-lactic-co-glycolic acid) owing to its biodegradability, biocompatibility, and multifaceted degradation kinetics (Sandreschi et al. 2016) (Table 19.1).

Chereddy et al. studied the wound-healing property of PLGA nanoparticle-loaded LL37 peptide, which is a component of host defense mechanism to counter the pathogenic microbes. PLGA nanoparticles have the merit in wound-healing formulation as they can supply lactate, which is required for faster recovery of epithelial membrane. The LL37-loaded PLGA nanoparticles fastened up the process of wound healing and increased the rate of angiogenesis. Furthermore, it demonstrated antimicrobial effect against *E. coli.* The authors concluded that LL37-PLGA NPs moderated inflammatory immune response and effectively restrained the infection owing to enhanced antimicrobial activity of LL37 peptide (Chereddy et al. 2014). Similarly, Chunhua et al. fabricated nisin/γ-PGA/chitosan nanoparticle, which demonstrated better MIC value compared to free drug. It was observed that drug release from Nisin/γ-PGA/chitosan NPs was pH-dependent manner. The Nisin-loaded NPs were efficiently able to restrain the growth of *E. coli* and *Listeria monocytogenes* at

Table 19.1 Nanoparticle-based delivery systems for AMPs

Nanocarrier	AMP	System	Reference
AuNPs	Magainin-1	Development of a one-step synthetic route form functionalization of AuNPs with AMPs	Rai et al. (2016)
AgNPs	MBP-1	Study of the synergistic antibacterial effect of plant peptide MBP-1 and AgNPs on infected wounds caused by *S. aureus*	Salouti et al. (2016)
Anionic mesoporous SiNPs	LL-37	Study of membrane interaction with AMPs in different types of mesoporous particles	Braun et al. (2016)
PLGA nanoparticles	Colistin	Development of a system of nano-embedded microparticles (NEM) for sustained delivery of CAMPs in the lung	d'Angelo et al. (2015)
PLGA nanoparticles	Plectasin	Intracellular antibacterial activity against *S. aureus* in epithelial cells	Jorrit Jeroen Water et al. (2015)
Hyaluronic acid nanogel	Novadin	Novel nanogel-based novadin delivery system	Jorrit Jeroen Water et al. (2015)
Pectin nanoparticles	Nisin	Study of a safe suitable antimicrobial system to be used in food industry. Influence of pectin degree of acetylation on NP properties	Krivorotova et al. (2016)
Chitosan/PGA nanoparticles	Nisin	Influence of chitosan coating on colloidal stability, loading capacity, and encapsulation efficiency	Wu et al. (2016)

lower concentrations compared to free drug and chitosan boarded Nisin nanoparticles. Hence, it implies that nanocarrier boarded antimicrobial peptides exhibits lower MIC values and thus are less susceptible to the development of resistance (Wu et al. 2016).

19.6 Nano-Cargo Boarded Defensins in Combating Multidrug Resistance

Defensins are a class of antimicrobial peptides found in humans, also known as human defensins primarily located in Paneth cells in intestinal crypts. Their discovery is a boon to combat infectious diseases and an alternative to combat MDR. Their commercialization is limited due to their poor stability and pharmacokinetics. Hence, some researchers have modified human defensins to make it suitable for in vitro conditions. The modified defensins have shown comparable antimicrobial activity with more circulation time. Furthermore, attaching these modified human defensins to nanoparticles significantly improves the therapeutic efficacy and overall outcome of treatment.

Zhao et al. modified human defensin by introducing couple of Arg moiety near C-terminal region of human defensin-5 (HD-5). The modified HD, i.e., T7E21R-HD5, was having an atypical dimer with stability against enzymatic hydrolysis. In addition, it showed potent microbicidal activity in saline solution making it an ideal candidate for in vitro bactericidal activity eliminating resistant bacteria and thus turns out to be a next-generation antibiotic. Mesoporous silica nanoparticles (MSNs), which provide controlled size, surface area, and biocompatibility, have been explored widely for better drug absorption, controlled release, and targeted delivery of drugs, and were prepared using CTAB as precursor for silica and TEOS as pore-forming agent, which was then loaded with T7E21R-HD5. Degradation of peptide in the stomach is a major limitation of oral peptide delivery. Hence, the MSN formed was further coated with succinylated–casein (SCN) layer, which prevents the release of peptide in acidic environment and helps in delivering and releases the peptide at the site of infection, i.e., intestine. The controlled release was observed with MSNs@T7E21R-HD5@SCN in the presence of trypsin indicating an improved drug release profile and successfully halted MDR *E. coli* in vivo. Further, it demonstrated subsided adverse effects in the host and inhibited the production of inflammatory factors, thereby reducing the intestinal distress. Loading of T7E21R-HD5 in SCN-coated MSNs enhanced the potency of modified defensins by 2.1-fold against MDR *E. coli* (Li and Shi 2014; Zhao et al. 2019).

The results obtained from the abovementioned study unveil the role of nanocargos in delivering peptide to overcome the limitation and improve their activity against multidrug-resistant bacteria.

PLGA nanoparticles form excellent polymeric nanocarriers for various drug molecules with great biocompatibility and easy manufacturing. They have been explored widely in drug delivery system; hence, they are always a suitable option for construction of nano-cargos. In another study by Liang et al., PLGA (polylactide-co-glycolic-acid) nanoparticles were loaded with recombinant bovine neutrophil defensin-5 (B5) and its activity was investigated against *Mycobacterium bovis*, which belongs to *Mycobacterium tuberculosis* family. Anti-TB medicines are now being less effective due to the development of resistance in *M. tuberculosis*, and it necessitates the development of novel strategies to counter the minacious disease. The authors developed PLGA nanoparticle encapsulating B5 peptide using solvent evaporation method. The surface of resulting NPs was negatively charged, and sustained release was observed under physiological environment. B5-PLGA NPs demonstrated significant reduction in microbial burden in lungs and spleen within 1 month. Loading of B5 in PLGA NPs was associated with improved immune response in terms of lung infection where secretion of inflammatory mediators was elevated (Liang et al. 2020).

Hangxiang et al. developed a noncovalently self-assembling human α-defensin-5 nanobiotic via formation of amide bond, which demonstrated broad spectrum of antimicrobial activity. The C-terminal myristoylated HD5 revealed intensified activity against *E. coli* and methicillin-resistant *Staphylococcus aureus* (MRSA) owing to its membranolytic activity. Moreover, it exhibited protection from MRSA and *E. coli*-induced sepsis, which can be lethal and limited the systemic infection and organ damage in rescued mice. The self-assembling human α-defensin-5 nanobiotic also revealed insignificant hemolytic activity and reduced toxicity in animals. The findings in abovementioned study revealed that modification of defensin to make self-assembling nanobiotics can be a crucial step to combat multidrug resistance. The formation of myristoylated-human α-defensin-5 was simple procedure, which not only improved the stability but also provided broad-spectrum activity to the antimicrobial peptide (Lei et al. 2018).

Defensin-loaded NPs have improved the therapeutic outcome in several cases and have established the role of nanocarriers in various fields including multidrug resistance. The encapsulation of naive or modified form of defensin has demonstrated improved toxicity and pharmacokinetic profile along with reduced ill effects related to host tissues. Nano-cargo boarded defensins were observed to have improved potency against MDR strains of bacteria.

19.7 Conclusions

Defensins, a major family of AMPs, play vital role in human innate immunity by protecting it from various strains of Gram-positive and Gram-negative bacteria, viruses, and fungi. Growing recent evidences have highlighted their role in the treatment of multidrug-resistant strains. Thus, defensins have great potential to be used as therapeutic agent for antimicrobial treatment. However, their applications are limited due to certain difficulties like low stability, higher toxicities, and high production cost. Some corrections are needed to be done for successful clinical translation of such peptides. Novel nanotechnological approaches offer a way out of such problems. Formulating these peptides into novel drug delivery systems like polymeric micelles, metallic nanoparticles, and lipidic nanoparticles has shown great promises by improving stability, half-life, selectivity, and even therapeutic potential. Herein, we envisage and optimize further development of novel nano-cargos loaded with defensins and other AMPs for the combat against multidrug resistance.

Conflict of Interest There is no conflict of interest among the authors.

References

Abdeen S, Isaac R, Geo S, Sornalekshmi S, Rose A, Praseetha PK (2013) Evaluation of antimicrobial activity of biosynthesized iron and silver nanoparticles using the fungi Fusarium oxysporum and Actinomycetes sp. on human pathogens. Nano Biomed Eng 5. https://doi.org/10.5101/nbe.v5i1.p39-45

Bevins CL, Salzman NH (2011) Paneth cells, antimicrobial peptides and maintenance of intestinal homeostasis. Nat Rev Microbiol 9(5):356–368. https://doi.org/10.1038/nrmicro2546

Bevins CL, Martin-Porter E, Ganz T (1999) Defensins and innate host defence of the gastrointestinal tract. Gut 45(6):911–915. https://doi.org/10.1136/gut.45.6.911

Braun K, Pochert A, Lindén M, Davoudi M, Schmidtchen A, Nordström R, Malmsten M, Membrane interactions of mesoporous silica nanoparticles as carriers of antimicrobial peptides, J. Colloid Interface Sci. 475 (2016) 161–170. https://doi.org/10.1016/J.JCIS.2016.05.002

Casciaro B, Moros M, Rivera-Fernández S, Bellelli A, de la Fuente JM, Mangoni ML (2017) Gold-nanoparticles coated with the antimicrobial peptide esculentin-1a(1-21)NH2 as a reliable strategy for antipseudomonal drugs. Acta Biomater 47:170–181. https://doi.org/10.1016/j.actbio.2016.09.041

Chen H, Xu Z, Cen P (2006) High-level expression of human beta-defensin-2 gene with rare codons in E. coli cell-free system. Protein Pept Lett 13(2):155–162. https://doi.org/10.2174/092986606775101724

Chereddy KK, Her C-H, Comune M, Moia C, Lopes A, Porporato PE, Vanacker J, Lam MC, Steinstraesser L, Sonveaux P, Zhu H, Ferreira LS, Vandermeulen G, Préat V (2014) PLGA nanoparticles loaded with host defense peptide LL37 promote wound healing. J Control Release 194:138–147. https://doi.org/10.1016/j.jconrel.2014.08.016

Chow BT, Soto M, Lo BL, Crosby DC, Camerini D (2012) Antibacterial activity of four human beta-defensins: HBD-19, HBD-23, HBD-27, and HBD-29. Polymers 4(1):747–758. https://doi.org/10.3390/polym4010747

Conibear AC, Craik DJ (2014) The chemistry and biology of theta defensins. Angew Chem Int Ed 53(40):10612–10623. https://doi.org/10.1002/anie.201402167

d'Angelo I, Casciaro B, Miro A, Quaglia F, Mangoni ML, Ungaro F, Overcoming barriers in Pseudomonas aeruginosa lung infections: Engineered nanoparticles for local delivery of a cationic antimicrobial peptide, Colloids Surfaces B Biointerfaces. 135(2015) 717–725. https://doi.org/10.1016/J.COLSURFB.2015.08.027

Davies J, Davies D (2010) Origins and evolution of antibiotic resistance. Microbiol Mol Biol Rev 74(3):417–433. https://doi.org/10.1128/MMBR.00016-10

Dhople V, Krukemeyer A, Ramamoorthy A (2006) The human beta-defensin-3, an antibacterial peptide with multiple biological functions. Biochim Biophys Acta 1758(9):1499–1512. https://doi.org/10.1016/j.bbamem.2006.07.007

Doss M, White MR, Tecle T, Gantz D, Crouch EC, Jung G, Ruchala P, Waring AJ, Lehrer RI, Hartshorn KL (2009) Interactions of alpha-, beta-, and theta-defensins with influenza A virus and surfactant protein D. J Immunol 182(12):7878–7887. https://doi.org/10.4049/jimmunol.0804049

Epand RM, Walker C, Epand RF, Magarvey NA (2016) Molecular mechanisms of membrane targeting antibiotics. Biochim Biophys Acta 1858(5):980–987. https://doi.org/10.1016/j.bbamem.2015.10.018

Gabay JE, Scott RW, Campanelli D, Griffith J, Wilde C, Marra MN, Seeger M, Nathan CF (1989) Antibiotic proteins of human polymorphonuclear leukocytes. Proc Natl Acad Sci U S A 86(14):5610–5614. https://doi.org/10.1073/pnas.86.14.5610

Ganz T, Selsted ME, Szklarek D, Harwig SS, Daher K, Bainton DF, Lehrer RI (1985) Defensins. Natural peptide antibiotics of human neutrophils. J Clin Invest 76(4):1427–1435. https://doi.org/10.1172/JCI112120

Ghosh SK, Gerken TA, Schneider KM, Feng Z, McCormick TS, Weinberg A (2007) Quantification of human beta-defensin-2 and -3 in body fluids: application for studies of innate immunity. Clin Chem 53(4):757–765. https://doi.org/10.1373/clinchem.2006.081430

Guilhelmelli F, Vilela N, Albuquerque P, Derengowski L, Silva-Pereira I, Kyaw CM (2013) Antibiotic development challenges: the various mechanisms of action of antimicrobial peptides and of bacterial resistance. Front Microbiol 4:353. https://doi.org/10.3389/fmicb.2013.00353

Gulyaev AE, Gelperina SE, Skidan IN, Antropov AS, Kivman GY, Kreuter J (1999) Significant transport of doxorubicin into the brain with polysorbate 80-coated nanoparticles. Pharm Res 16(10):1564–1569. https://doi.org/10.1023/A:1018983904537

Jin G, Weinberg A (2019) Human antimicrobial peptides and cancer. Semin Cell Dev Biol 88:156–162. https://doi.org/10.1016/j.semcdb.2018.04.006

Krivorotova T, Cirkovas A, Maciulyte S, Staneviciene R, Budriene S, Serviene E, Sereikaite J, Nisin-loaded pectin nanoparticles for food preservation, Food Hydrocoll. 54 (2016) 49–56. https://doi.org/10.1016/J.FOODHYD.2015.09.015

Klotman ME, Rapista A, Teleshova N, Micsenyi A, Jarvis GA, Lu W, Porter E, Chang TL (2008) Neisseria gonorrhoeae-induced human defensins 5 and 6 increase HIV infectivity: role in enhanced transmission. J Immunol 180(9):6176–6185. https://doi.org/10.4049/jimmunol.180.9.6176

Lehrer RI, Lu W (2012) α-Defensins in human innate immunity. Immunol Rev 245(1):84–112. https://doi.org/10.1111/j.1600-065X.2011.01082.x

Lehrer RI, Cole AM, Selsted ME (2012) θ-Defensins: cyclic peptides with endless potential. J Biol Chem 287(32):27014–27019. https://doi.org/10.1074/jbc.R112.346098

Lei R, Hou J, Chen Q, Yuan W, Cheng B, Sun Y, Jin Y, Ge L, Ben-Sasson SA, Chen J, Wang H, Lu W, Fang X (2018) Self-assembling myristoylated human α-defensin 5 as a next-generation nanobiotics potentiates therapeutic efficacy in bacterial infection. ACS Nano 12(6):5284–5296. https://doi.org/10.1021/acsnano.7b09109

Lei J, Sun L, Huang S, Zhu C, Li P, He J, Mackey V, Coy DH, He Q (2019) The antimicrobial peptides and their potential clinical applications. Am J Transl Res 11(7):3919–3931

Li Y, Shi J (2014) Hollow-structured mesoporous materials: chemical synthesis, functionalization and applications. Adv Mater 26(20):3176–3205. https://doi.org/10.1002/adma.201305319

Li X, Robinson SM, Gupta A, Saha K, Jiang Z, Moyano DF, Sahar A, Riley MA, Rotello VM (2014) Functional gold nanoparticles as potent antimicrobial agents against multi-drug-resistant bacteria. ACS Nano 8(10):10682–10686. https://doi.org/10.1021/nn5042625

Li W, Li Y, Sun P, Zhang N, Zhao Y, Qin S, Zhao Y (2020) Antimicrobial peptide-modified silver nanoparticles for enhancing the antibacterial efficacy. RSC Adv 10(64):38746–38754. https://doi.org/10.1039/D0RA05640E

Liang Z, Liu Y, Sun X, Lin J, Yao J, Song Y, Li M, Liu T, Zhou X (2020) Immunoregulatory and antimicrobial activity of bovine neutrophil β-defensin-5-loaded PLGA nanoparticles against Mycobacterium bovis. Pharmaceutics 12(12):1172. https://doi.org/10.3390/pharmaceutics12121172

Lima R, Del Fiol FS, Balcão VM (2019) Prospects for the use of new technologies to combat multidrug-resistant bacteria. Front Pharmacol 10. https://doi.org/10.3389/fphar.2019.00692

Magiorakos A-P, Srinivasan A, Carey RB, Carmeli Y, Falagas ME, Giske CG, Harbarth S, Hindler JF, Kahlmeter G, Olsson-Liljequist B, Paterson DL, Rice LB, Stelling J, Struelens

MJ, Vatopoulos A, Weber JT, Monnet DL (2012) Multidrug-resistant, extensively drug-resistant and pandrug-resistant bacteria: an international expert proposal for interim standard definitions for acquired resistance. Clin Microbiol Infect 18(3):268–281. https://doi.org/10.1111/j.1469-0691.2011.03570.x

Mirza AZ, Siddiqui FA (2014) Nanomedicine and drug delivery: a mini review. Int Nano Lett 4(1):94. https://doi.org/10.1007/s40089-014-0094-7

Nguyen LT, Haney EF, Vogel HJ (2011) The expanding scope of antimicrobial peptide structures and their modes of action. Trends Biotechnol 29(9):464–472. https://doi.org/10.1016/j.tibtech.2011.05.001

Nikitina IG, Bukurova YA, Krasnov GS, Grineva EN, Karpov VL, Lisitsyn NA, Beresten SF (2012) Structure and function of enteric α-defensins in norm and pathology. Mol Biol 46(1):27–33. https://doi.org/10.1134/S0026893311060094

Pace BT, Lackner AA, Porter E, Pahar B (2017) The role of defensins in HIV pathogenesis. Mediat Inflamm 2017:5186904. https://doi.org/10.1155/2017/5186904

Pachón-Ibáñez ME, Smani Y, Pachón J, Sánchez-Céspedes J (2017) Perspectives for clinical use of engineered human host defense antimicrobial peptides. FEMS Microbiol Rev 41(3):323–342. https://doi.org/10.1093/femsre/fux012

Piers KL, Brown MH, Hancock RE (1993) Recombinant DNA procedures for producing small antimicrobial cationic peptides in bacteria. Gene 134(1):7–13. https://doi.org/10.1016/0378-1119(93)90168-3

Pitkethly MJ (2004) Nanomaterials—the driving force. Mater Today 7(12, Suppl):20–29. https://doi.org/10.1016/S1369-7021(04)00627-3

Pizzolato-Cezar LR, Okuda-Shinagawa NM, Machini MT (2019) Combinatory therapy antimicrobial peptide-antibiotic to minimize the ongoing rise of resistance. Front Microbiol 10. https://doi.org/10.3389/fmicb.2019.01703

Porter E, Yang H, Yavagal S, Preza GC, Murillo O, Lima H, Greene S, Mahoozi L, Klein-Patel M, Diamond G, Gulati S, Ganz T, Rice PA, Quayle AJ (2005) Distinct defensin profiles in Neisseria gonorrhoeae and Chlamydia trachomatis urethritis reveal novel epithelial cell-neutrophil interactions. Infect Immun 73(8):4823–4833. https://doi.org/10.1128/IAI.73.8.4823-4833.2005

Rai A, Pinto S, Velho TR, Ferreira AF, Moita C, Trivedi U, Evangelista M, Comune M, Rumbaugh KP, Simões PN, Moita L, Ferreira L (2016) One-step synthesis of high-density peptide-conjugated gold nanoparticles with antimicrobial efficacy in a systemic infection model. Biomaterials 85:99–110. https://doi.org/10.1016/j.biomaterials.2016.01.051

Rozek A, Powers J-PS, Friedrich CL, Hancock REW (2003) Structure-based design of an indolicidin peptide analogue with increased protease stability. Biochemistry 42(48):14130–14138. https://doi.org/10.1021/bi035643g

Salouti M, Mirzaei F, Shapouri R, Ahangari A, Synergistic antibacterial activity of plant peptide MBP-1 and silver nanoparticles combination on healing of infected wound due to staphylococcus aureus, Jundishapur J. Microbiol. 9 (2016). https://doi.org/10.5812/JJM.27997.

Sánchez-López E, Gomes D, Esteruelas G, Bonilla L, Lopez-Machado AL, Galindo R, Cano A, Espina M, Ettcheto M, Camins A, Silva AM, Durazzo A, Santini A, Garcia ML, Souto EB (2020) Metal-based nanoparticles as antimicrobial agents: an overview. Nanomaterials 10(2). https://doi.org/10.3390/nano10020292

Sandreschi S, Piras AM, Batoni G, Chiellini F (2016) Perspectives on polymeric nanostructures for the therapeutic application of antimicrobial peptides. Nanomedicine 11(13):1729–1744. https://doi.org/10.2217/nnm-2016-0057

Selsted ME, Harwig SS, Ganz T, Schilling JW, Lehrer RI (1985) Primary structures of three human neutrophil defensins. J Clin Invest 76(4):1436–1439. https://doi.org/10.1172/JCI112121

Tai KP, Kamdar K, Yamaki J, Le VV, Tran D, Tran P, Selsted ME, Ouellette AJ, Wong-Beringer A (2015) Microbicidal effects of α- and θ-defensins against antibiotic-resistant Staphylococcus aureus and Pseudomonas aeruginosa. Innate Immun 21(1):17–29. https://doi.org/10.1177/1753425913514784

Tang YQ, Yuan J, Osapay G, Osapay K, Tran D, Miller CJ, Ouellette AJ, Selsted ME (1999) A cyclic antimicrobial peptide produced in primate leukocytes by the ligation of two truncated alpha-defensins. Science 286(5439):498–502. https://doi.org/10.1126/science.286.5439.498

Tecle T, Tripathi S, Hartshorn KL (2010) Review: defensins and cathelicidins in lung immunity. Innate Immun 16(3):151–159. https://doi.org/10.1177/1753425910365734

Teixeira MC, Carbone C, Sousa MC, Espina M, Garcia ML, Sanchez-Lopez E, Souto EB (2020) Nanomedicines for the delivery of antimicrobial peptides (AMPs). Nano 10(3). https://doi.org/10.3390/nano10030560

Teles RM, Graeber TG, Krutzik SR, Montoya D, Schenk M, Lee DJ, Komisopoulou E, Kelly-Scumpia K, Chun R, Iyer SS, Sarno EN, Rea TH, Hewison M, Adams JS, Popper SJ, Relman DA, Stenger S, Bloom BR, Cheng G, Modlin RL (2013) Type I interferon suppresses type II interferon-triggered human anti-mycobacterial responses. Science 339(6126):1448–1453. https://doi.org/10.1126/science.1233665

Unger T, Oren Z, Shai Y (2001) The effect of cyclization of magainin 2 and melittin analogues on structure, function, and model membrane interactions: implication to their mode of action. Biochemistry 40:6388–6397. https://doi.org/10.1021/bi0026066

Valore EV, Ganz T (1997) Laboratory production of antimicrobial peptides in native conformation. Methods Mol Biol 78:115–131. https://doi.org/10.1385/0-89603-408-9:115

Wang G (2014) Human antimicrobial peptides and proteins. Pharmaceuticals 7(5):545–594. https://doi.org/10.3390/ph7050545

Wang W, Mulakala C, Ward SC, Jung G, Luong H, Pham D, Waring AJ, Kaznessis Y, Lu W, Bradley KA, Lehrer RI (2006) Retrocyclins kill bacilli and germinating spores of Bacillus anthracis and inactivate anthrax lethal toxin. J Biol Chem 281(43):32755–32764. https://doi.org/10.1074/jbc.M603614200

Wang C, Zhao G, Wang S, Chen Y, Gong Y, Chen S, Xu Y, Hu M, Wang X, Zeng H, Wang A, Liu D, Su Y, Cheng T, Chen F, Wang J (2018) A simplified derivative of human defensin 5 with potent and efficient activity against multidrug-resistant Acinetobacter baumannii. Antimicrob Agents Chemother 62(2):e01504–e01517. https://doi.org/10.1128/AAC.01504-17

Water JJ, Smart S, Franzyk H, Foged C, Nielsen HM, Nanoparticle-mediated delivery of the antimicrobial peptide plectasin against Staphylococcus aureus in infected epithelial cells, Eur. J. Pharm. Biopharm. 92(2015) 65–73. https://doi.org/10.1016/J.EJPB.2015.02.009

Wright GD, Sutherland AD (2007) New strategies for combating multidrug-resistant bacteria. Trends Mol Med 13(6):260–267. https://doi.org/10.1016/j.molmed.2007.04.004

Wu Z, Ericksen B, Tucker K, Lubkowski J, Lu W (2004) Synthesis and characterization of human alpha-defensins 4-6. J Peptide Res 64(3):118–125. https://doi.org/10.1111/j.1399-3011.2004.00179.x

Wu C, Wu T, Fang Z, Zheng J, Xu S, Chen S, Hu Y, Ye X (2016) Formation, characterization and release kinetics of chitosan/γ-PGA encapsulated nisin nanoparticles. RSC Adv 6(52):46686–46695. https://doi.org/10.1039/C6RA06003J

Xu Z, Chen H, Yin X, Xu N, Cen P (2005a) High-level expression of soluble human beta-defensin-2 fused with green fluorescent protein in Escherichia coli cell-free system. Appl Biochem Biotechnol 127(1):53–62. https://doi.org/10.1385/abab:127:1:053

Xu Z, Wang F, Peng L, Fang X, Cen P (2005b) Expression of human beta-defensin-2 with multiple joined genes in Escherichia coli. Appl Biochem Biotechnol 120(1):1–13. https://doi.org/10.1385/abab:120:1:01

Yamaguchi Y, Ouchi Y (2012) Antimicrobial peptide defensin: identification of novel isoforms and the characterization of their physiological roles and their significance in the pathogenesis of diseases. Proc Jpn Acad B Phys Biol Sci 88(4):152–166. https://doi.org/10.2183/pjab.88.152

Yasin B, Wang W, Pang M, Cheshenko N, Hong T, Waring AJ, Herold BC, Wagar EA, Lehrer RI (2004) Theta defensins protect cells from infection by herpes simplex virus by inhibiting viral adhesion and entry. J Virol 78(10):5147–5156. https://doi.org/10.1128/jvi.78.10.5147-5156.2004

Yeaman MR, Yount NY (2003) Mechanisms of antimicrobial peptide action and resistance. Pharmacol Rev 55(1):27–55. https://doi.org/10.1124/pr.55.1.2

Zasloff M (2007) Antimicrobial peptides, innate immunity, and the normally sterile urinary tract. J Am Soc Nephrol 18(11):2810–2816. https://doi.org/10.1681/ASN.2007050611

Zhang L, Falla T, Wu M, Fidai S, Burian J, Kay W, Hancock RE (1998) Determinants of recombinant production of antimicrobial cationic peptides and creation of peptide variants in bacteria. Biochem Biophys Res Commun 247(3):674–680. https://doi.org/10.1006/bbrc.1998.8848

Zhang W, Li Y, Niu J, Chen Y (2013) Photogeneration of reactive oxygen species on uncoated silver, gold, nickel, and silicon nanoparticles and their antibacterial effects. Langmuir 29(15):4647–4651. https://doi.org/10.1021/la400500t

Zhao G, Chen Y, He Y, Chen F, Gong Y, Chen S, Xu Y, Su Y, Wang C, Wang J (2019) Succinylated casein-coated peptide-mesoporous silica nanoparticles as an antibiotic against intestinal bacterial infection. Biomater Sci 7(6):2440–2451. https://doi.org/10.1039/C9BM00003H

Zhong Z, Xu Z, Peng L, Huang L, Fang X, Cen P (2006) Tandem repeat mhBD2 gene enhance the soluble fusion expression of hBD2 in Escherichia coli. Appl Microbiol Biotechnol 71(5):661–667. https://doi.org/10.1007/s00253-005-0212-6

Chapter 20
Implementation of Diverse Nano-cargos to Disguise and Fight Multidrug Resistance

Tharun K. Bollu, Divya S. Parimi, Chandra S. Bhatt, Imran Uddin, and Anil K. Suresh

20.1 Introduction to Nano-vehicles

Antimicrobial resistance call for efficient ways to combat adamant pathogenic micro-organisms, which gradually develop resistance towards modern drugs that can no longer be effective with the same dose regimen and may even lead to fatal complications if untreated (World Health Organization (WHO) 2020). Nanotechnology is emerging as one of the promising tools to curb widespread infections caused by several multidrug-resistant microbes, especially metallic nanoparticles, as they possess intrinsic antibacterial properties (Suresh 2015). We shall discuss various nano-vehicles, which are being introduced as efficient antimicrobial adjuncts in combating multidrug resistance.

Nano-vehicles are supposedly pharmacologically inert and biocompatible materials chemically synthesized using inorganic and organic sources. Physically, these nanostructures are available in different dimensions of solid core and/or hollow structures ranging from sub-nanometre to few micrometres. Being devoid of any chemical reactivity and driven by their high surface-to-volume ratio, surface functionalization, charge-based drug and cellular binding, and high potential for targeted drug delivery, various nanostructures are employed as carriers of potent pharmaceutical drugs (Panyam and Labhasetwar 2003). Nanocarriers have been revolutionizing healthcare sectors, especially in clinical prognosis. These nanostructure paraphernalia allowed researchers and clinicians to defend prevailing pathogenic infections in the inertia of antimicrobial resistance (Baptista et al. 2018). In the interest of current chapter, the applicability of nano-vehicles is limited to antimicrobial resistance; however, several other applications of nano-vehicles in

T. K. Bollu · D. S. Parimi · C. S. Bhatt · I. Uddin · A. K. Suresh (✉)
Bio-Nanotechnology Laboratory, Department of Biotechnology, SRM University-AP, Amaravati, Andhra Pradesh, India
e-mail: anil.s@srmap.edu.in

N. Akhtar et al. (eds.), *Emerging Modalities in Mitigation of Antimicrobial Resistance*,
https://doi.org/10.1007/978-3-030-84126-3_20

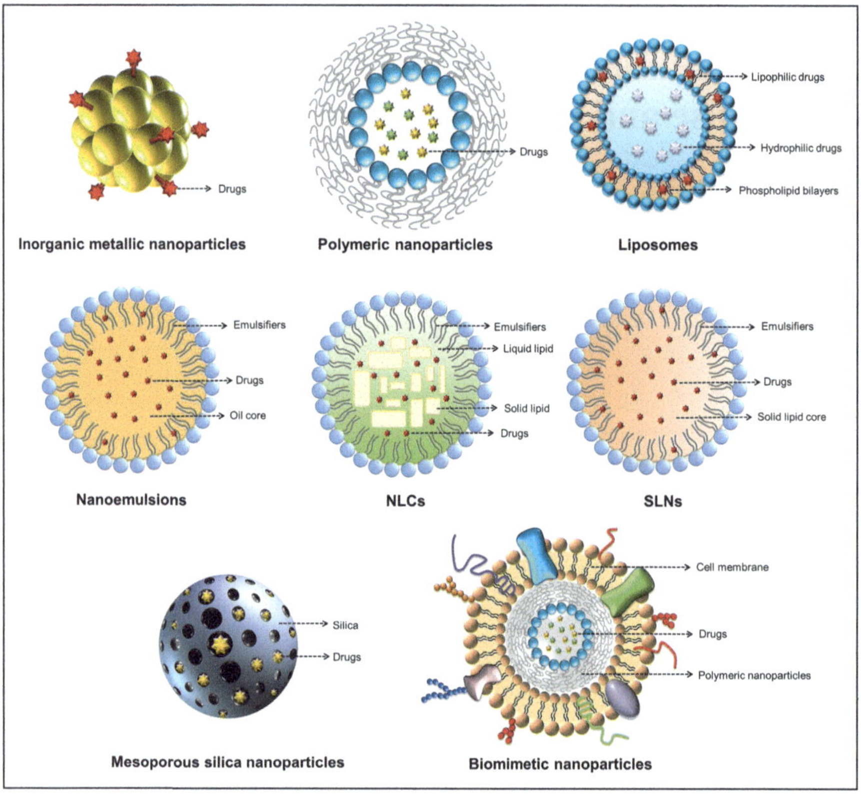

Fig. 20.1 Different types of nano-vehicles used in drug delivery. Reproduced from (Yeh et al. 2020) © 2020

biomedical fields are vastly reported. List of commonly used nanostructures to disguise and fight multidrug resistance is discussed in the below sections with relevant examples, and few selected nano-vectors are illustrated in Figs. 20.1 and 20.2.

20.1.1 Metallic Nanoparticle-Based Nano-cargos

Metallic nanoparticle is metal-based nanoparticles that are fabricated by reducing metal precursor salts that yield different core materials (gold, silver, platinum, palladium, zinc, iron), which fall in the size range from 1 to 100 nm. Such nano-scaled matter possesses exceptional intrinsic properties like high surface-to-volume ratio, chemical reactivity, catalytic activity, and physiochemical, optoelectronic and biological properties. These unique functionalities of nanoparticles driven by specific core, size and shape are leveraged in the fields of bio-medicine, food and agricultural, electronics, health care, industrial and textile (Anil Kumar and Khan 2010).

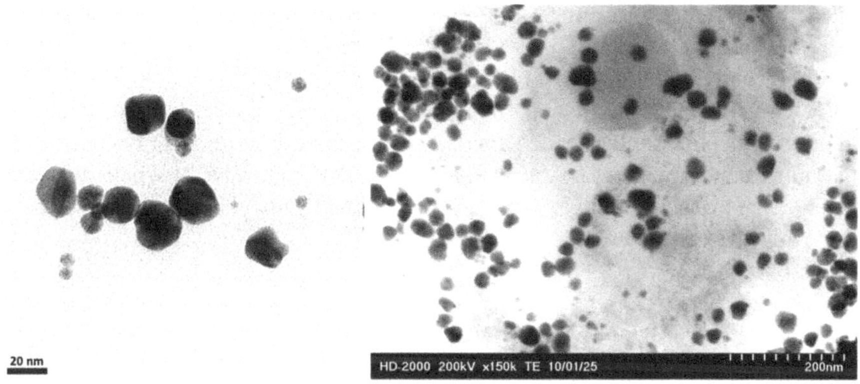

Fig. 20.2 Selected TEM images of metal nano-vehicles used in drug delivery. Transmission electron microscopy of metallic gold (**a**), silver (**b**) and silver sulphide (**c**) nanoparticles

Various metallic nanomaterials that are physiologically inert are employed as carrier platforms in accommodating antimicrobial peptides and antimicrobial entities onto their surface, which potentiates the bactericidal activity.

One among those is gold nanoparticles, which are well known for their biocompatible nature. Amenable surface functionalization property of gold nanoparticles explored handful of possibilities in tuning the nanoparticle surface by multiple functionalization intended to target desired application. For instance, regulating functional groups on gold nanoparticles surface with cationic monolayers could interfere with Gram-positive and Gram-negative bacteria. Xiaoning Li et.al synthesized gold nanoparticles by incorporating different chain lengths and aromatic groups to impart and understand the cationic properties and observed that synthesis using n-decane end linkage provided greater hydrophobicity and is effective against 11 clinical isolates of uropathogenic *E. coli* strains. Growth of all the selected *E. coli* strains was suppressed, which are otherwise resistant against seventeen different drugs depending on the strain used. These functionalized nanoparticles acted by disrupting the integrity of bacterial cell membrane to induce toxicity and were effective against superbugs like *Pseudomonas aeruginosa* and *Staphylococcus aureus* including the methicillin-resistant *S. aureus* (Li et al. 2014). In addition to surface functionalizations, gold nanoparticles were also used to conjugate antibiotic drugs like vancomycin and methicillin to potentiate the effect of antimicrobial efficacy. Mohammed Fayaz et al. compared the antibacterial activity of vancomycin alone and vancomycin-bound gold nanoparticles (VBGNPs) against *S. aureus* and *E. coli*. Compared to free vancomycin, VBGNPs non-specifically bound to transpeptidase of the glycopeptidyl precursors prominent on the vancomycin-resistant *S. aureus* to lyse the cell wall inducing toxicity (Fayaz et al. 2011). In another study, gold nanodots prepared by depositing antimicrobial peptide surfactin showed greater antimicrobial activity towards both non-multidrug and multidrug-resistant bacteria due to the synergistic effect of disrupting bacterial cell membrane. Using these antimicrobial peptide-loaded gold nanodots as dressing material for methicillin-resistant *S. aureus* wound healing showed rapid healing in rats (Chen et al. 2015).

Significantly, several other bimetallic nanoparticles such as Au/Ag, Au/Pt and Au/Fe_3O_4 and metal oxide nanoparticles such as graphene oxide, TiO_2 and CuO are reported to deploy antibacterial activities against multidrug-resistant bacteria with their intrinsic mechanisms (Baptista et al. 2018). Different ways by which different metallic nanoparticles interact with bacterial cell and pathogenic biofilms are illustrated in Fig. 20.3.

20.1.2 Carbon-Based Nano-vehicles

Nanostructures are made of pure carbon atoms arranged in different allotropic fashions, characterized by sp^2 bonding in the core (Bianco et al. 2005). Carbon dots, fullerenes, carbon nanotubes (single-walled/multiwalled), graphite, graphene and graphene oxide are different classifications of carbon nanoparticles based upon their dimension. Desired carbon nanostructures can be easily synthesized by microwave

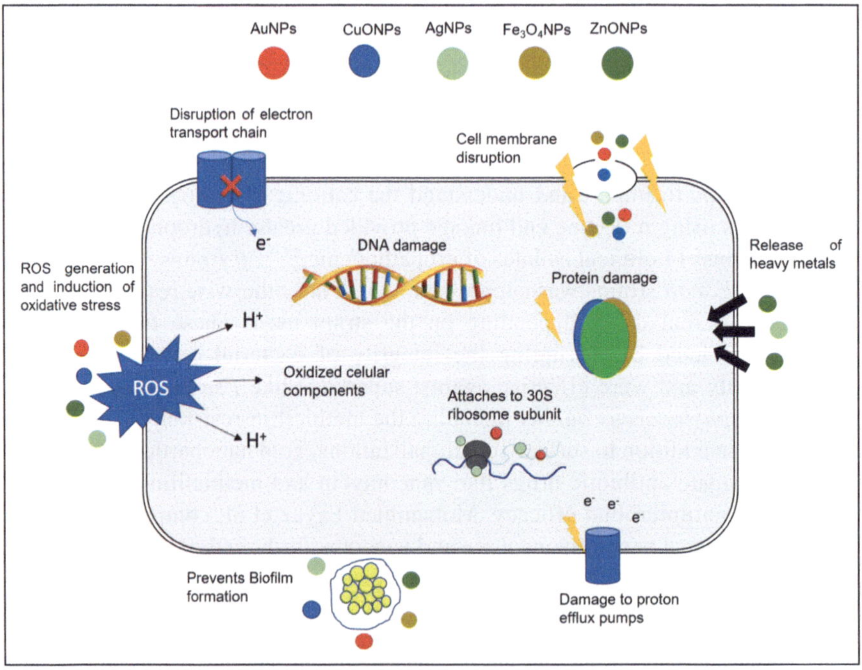

Fig. 20.3 Illustration of different mechanisms by which nanoparticles can elucidate toxic responses to bacterial cells. DNA—deoxyribonucleic acid; ROS—reactive oxygen species; AuNPs—gold NPs; CuONPs—copper oxide NPs; AgNPs—silver NPs; Fe_3O_4NPs—iron oxide NPs; ZnONPs—zinc oxide NPs. Reproduced from (Baptista et al. 2018) Copyright © 2018

irradiation, pyrolysis, carbonization and chemical vapour deposition. Due to the arrangement of carbon atoms in different dimensions, each structure exhibits unique electrical, thermal conductivities, high stability, and mechanical and optical properties (Baddour and Briens 2005). Therefore, carbon-based nanoparticles find applications in varied fields like biomedical engineering, drug delivery and cancer therapies.

Carbon being biocompatible and greater surface areas provided by multiple nanostructures opened up scope to utilize them as nano-vehicles in drug delivery. Very recently, Tak H. Kim reported that carbon nanoparticles conjugated with antibiotic tetracycline improved its efficacy on antimicrobial-resistant *Klebsiella pneumoniae* by tenfold. In contrary to the earlier methods, tetracycline-CNP conjugates are functional in inhibiting the process of drug extrusion, an inherent efflux mechanism of drug-resistant bacterial strains to protect itself and further generations from the antibiotic threat. Tetracycline when used alone showed less inhibitory action due to active drug efflux mechanism by bacteria, thereby escaping from the clutches of free antibiotic moieties. Whereas covalent bonding of tetracycline with CNPs disguises bacterial cell to misinterpret the tetracycline-CNP complex to tetracycline drug moieties and instantly initiate drug efflux mechanism, the tetracycline-CNP complex thus appeared in the vicinity of cytoplasmic membrane, thereby impairing the efflux system subsequently resulting in cell toxicity mediated by tetracycline-bound complex. These disguised mechanisms decipher the novel possibilities of implementing nanostructure for efficient drug deliveries (Kim et al. 2020).

Carbon nanotubes are made from either single-walled or multi-walled carbon atoms forming long hollow tubes by rolling of single or multiple graphene sheets. The aspect ratios of carbon nanotubes are as high as 132,000,000:1; i.e., length of carbon tube is usually greater than its diameter (Wang et al. 2009). These single lattice carbon nanotube structures have great scope to accommodate drug moieties because of their large surface areas and hollow channels. Afeefah Khazi-syed et al. demonstrated that suspending single-walled carbon nanotubes (SWCNTs) in doxycycline and methicillin water suspensions led to the non-covalent binding of drug onto the nanotube surface, which promoted the antibacterial effect in methicillin-resistant *Staphylococcus epidermidis*. SWCNT-methicillin complex was reported to exhibit 40-fold enhancement in inhibition of bacterial colony development. SWCNT internalization was likely due to their cell membrane permeability, thereby releasing methicillin to induce toxicity. On the other hand, SWCNTs also protect drug moieties from degradation (Khazi-Syed et al. 2019). Similarly, Mohyeddin Assali et al. reported the functionalization of carbon nanotubes with high payloads of ciprofloxacin, the most effective antibiotic of fluoroquinolones. Antibacterial activity of these functionalized SWCNTs was intensified by 16-fold against *Staphylococcus aureus* and *Pseudomonas aeruginosa* and by eightfold against *E. coli* when compared to free ciprofloxacin activity. High surface areas of SWCNTs favour abundant ciprofloxacin charging, and multiple nanotubes are exposed with bacteria thereupon augmenting the amount of ciprofloxacin interaction with the bacteria. On the other hand, these functionalized SWCNTs cripple the efflux pump defensive mechanism of bacteria used to gain resistance to antibiotics (Assali et al. 2017).

20.1.3 Liposomes-Based Nano-assemblies

Liposomes are small vesicular structures composed of lipid bilayer assemblages, enclosing aqueous cavities (Cevc 2012). The amphiphilic property of phospholipids gives an arrangement of hydrophilic heads facing the surface and hydrophobic tails towards the core-forming unilayer oligo and/or multi-lamellar bilayers. Liposomes can be prepared using natural or synthetic lipid molecules in situ, by techniques like rehydration of dried lipid films, sonication, micro-emulsification, membrane extrusion and solvent dispersion. Liposomes are widely used for incorporating nutrients and pharmaceutical drugs either in the aqueous core or in the phospholipid bilayers (Karmali and Chaudhuri 2007). Nanotechnology-driven liposomes are fabricated within 100–200 diameters that can efficiently penetrate infectious biofilms and reach the bacteria. The fusogenic property of liposomes, i.e., the ability of liposomal bilayer to directly fuse with bacterial cell membrane and expel highly potent antimicrobial drugs, provides great advantage in liposomal antimicrobial carriers over free antimicrobials (Scriboni et al. 2019). Either strains of bacteria that are resistant to selected antibiotics showed susceptibility to such drug encapsulation strategies. Liposomal encapsulation of drug also merits for conserving antibiotics from bacterial enzyme degradation. In the interest of infection control, and liposomal properties, these can be classified into natural, anionic, cationic, zwitterionic and fusogenic liposomes, which are capable of delivering selective agonistic interactions on pathogenic bacteria. Anne-Sophie et.al demonstrated the encapsulation of antibiotic tobramycin in anionic liposomes (negatively charged), which affirmed selective delivery of antibiotic to *Burkholderia cepacia* complex, a causative organism of lung infections that has intrinsic resistance to several antibiotics. Moreover, electrostatic interactions arising between negatively charged lipids and tobramycin anionic liposomes depicted increased loading of tobramycin by eight times, when compared to neutral liposomes, implying the abundance of antibiotic concentrations available at the proximity of pathogenic biofilms (Messiaen et al. 2013). Similarly, combinations of antibiotics such as incorporating both hydrophobic and hydrophilic drugs in the lipid bilayers were also employed for efficient antibacterial activity and drug delivery. Readers may further refer Wang et al. (2009) who compiled different antibiotics encapsulate using liposomal carriers.

20.1.4 Liquid Crystalline Particle-Based Nano-carriers

These are also made up of lipid bilayers but are curled to develop two- and three-dimensional structures bridged by water molecules; therefore, they are intermediate forms between solid and liquid phases. Thermotropics and lyotropics are the two types of liquid crystals classified based on the crystallinity exhibited as a function of temperature and concentration, respectively (Mo et al. 2017). Thermotropics are employed in monitor display screens, whereas lyotropics find applications in drug delivery owing to their good solubility, withstanding enzyme degradation and good absorption by human tissues. Liquid crystals are formed by suspending any amphiphilic groups in suitable solvents, where the amphiphilic molecules aggregate to form a monolayer exposing hydrophilic heads outward by protecting hydrophobic tails in the core rounded up to form micelles. Upon availability of more such amphiphilic solute groups, the liquid crystal pattern extends to form phases of hexagonal, cubic and lamellar arrangements (Mo et al. 2017). Desired liquid crystal forms are used to dissolve the drugs in liquid matrix so that they are steered towards the selective pathogenic sites for targeted drug delivery. Chelsea R. Thorn et al. recently performed an interesting research availing drug delivery from monoolein liquid crystal nanoparticle triggered by bacterial lipase. Addressing the limitation of using ciprofloxacin, these nanoparticles showed 82-fold and sevenfold higher release of rifampicin and alginate lyase, respectively, by diffusing monoolein liquid crystal nanoparticle through bacterial lipase. Rifampicin is a drug of choice to treat predominant MRSA infections, while alginate lyase actively degrades bacterial secretions that form biofilm. The non-regiopecific lipase activity actively digested the cubic structure to lamellar, thereby provoking the antibiotic release. Thus, such MO-LCNPs find efficient applications in infections caused by *Pseudomonas aeruginosa* and *Staphylococcus aureus* that secrete lipase as a part of natural living process (Thorn et al. 2020). Another recent study performed by Xiangfeng Lai et al. illustrates the use of phytantriol-based cubosomes against polymyxin-resistant Gram-negative bacteria (*Acinetobacter baumannii*, *Klebsiella pneumoniae* and *Pseudomonas aeruginosa*) for improved efficacy. Two model systems of Gram-negative bacteria, one with lipopolysaccharide membrane and the other lacking lipopolysaccharide membrane, were interacted with cubosomes. Good number of cubosomes are adsorbed on the surface of lipopolysaccharide-deficient model, which leads to rearrangement of bacterial cell membrane and penetration of cubosomes into bilayer systems, thereby exchanging lipids and subsequently disrupting the membrane (Lai et al. 2020).

20.1.5 Dendritic System Nano-assemblages

Dendrimers are highly branched polymeric molecules, which are nanometre in size obtained by controlled synthesis pathways. Dendrimers are usually monodispersed, symmetric and spherical in nature having multiple end groups. Polyamidoamine (PAMAM), polyglycerol and carbosilane are some of the common dendrimers (Martin-Serrano et al. 2019; Abedi-Gaballu et al. 2018; Yang and Zimmerman 2015; Jiménez et al. 2012). The dendritic end groups can be functionalized with selective functional groups to impart desired physiological/pharmacological activity that can have clinical significance. Reinaldo G. Bellini et al. studied the association of anti-tuberculosis drug rifampicin with a fourth-generation poly(amidoamine) (G4-PAMAM) dendrimer and evaluated the activity in neutral and acidic pH conditions. The complex was stable in neutral pH but instantly expelled the contents in acidic pH. This controlled behaviour of the complex demonstrated the selective release of the drug in acidic conditions, which is efficient against mycobacterium that usually resides in the acidic region of macrophage. The stability of complex in physiological pH and drug triggering behaviour by the change in pH credits the successful fabrications of dendrimers to deliver antibiotics without suffering the bacterial resistance (Bellini et al. 2015).

20.1.6 Polymer-Based Nano-matrices

Polymers are formed by long repetitive chains of smaller molecules (monomer). DNA, cellulose, chitosan and proteins are the examples of natural polymers, whereas polystyrene and PEG are synthetic in origin formed by the process of polymerization. Length of polymers ranges from sub-nanometre to few micrometres and can be microstructured into linear, branched, unbranched and cross-linked structures by the physical arrangement of monomer molecules (Martin-Serrano et al. 2019). Depending upon the structure, polymers possess unique physiochemical properties like tensile strength, diffusability, crystallization and melting transitions, ionic bond formations and electrical insulations. Therefore, polymers are tremendously used in packaging, manufacturing industries, synthetic fibres and medical applications. Polymer nanoparticles are being used extensively in biomedicine due to their biocompatibility/biodegradable nature. Antibacterial polymeric nanosystems have become phenomenal in tackling multidrug-resistant bacteria and their infectious biofilms through effective detection and binding abilities (Faya et al. 2018). Antibiotic encapsulations in these polymeric systems such as nanogels, vesicles and polymeric micelles serve as effective drug vehicles (Ding et al. 2019). Such encapsulations when selectively designed are capable in delivering multimodal therapies by responding to bacterial-infected environments such as alteration in pH and overexpression of enzymes. Abilities of certain MDR bacteria to produce hydrolytic enzymes such as penicillin G, amidase and B-lactamase, which console antibiotic activity of potent antibiotics such as penicillin, cephalosporins, cephamycins and carbapenams, are rate limiting the efficiencies of antibiotics. Yamin et al. synthesized bacterial enzyme-responsive polymer vesicles with amphiphilic diblock copolymer containing enzyme cleavable linkages (hydrophobic) and PEG blocks (hydrophilic), which can allow encapsulation of desired hydrophobic antibiotics, viz. deprotonated quinupristin/dalfopristin in hydrophobic bilayers and hydrophilic antibiotics, like gentamicin, vancomycin and parasin I in aqueous layers. Such antibiotic-loaded polymer vesicles experience structural rearrangement to expel the antibiotics to curb the growth of MRSA (Li et al. 2016). Similarly, antibody-loaded polymeric nanogels and antibody-conjugated polymeric nanosystems with series of hyperbranched antibiotics like gentamicin, tobramycin and neomycin exhibited potent antibacterial efficiency. Amjed et al. successfully sorbed antibiotic penicillin G onto fluorescent polystyrene nanoparticles. Even at lowest concentrations used the antibacterial activity was enhanced against Gram-positive and Gram-negative strains of bacteria tested which also includes MDR strains that are specifically resistant to penicillin G (Alabresm et al. 2020).

20.1.7 Hydrogel-Based Nano-networks

Hydrogels are basically cross-linked polymer chain networks and are hydrophilic in nature. They are highly absorbent containing about 90% of water (Narayanaswamy and Torchilin 2019; Hu et al. 2019). Hydrogels are chemically composed of polyvinyl alcohols, polyethylene glycol, sodium polyacrylate and acrylate polymers. Modifying the polymer concentration and cross-linking concentration provides great mechanical properties to hydrogels, thus making that available for various applications. Hydrogels containing antibiotics are promising technology in dealing with drug resistance by designing combination of synthetic antimicrobial polymers with specific antibiotics. Hydrogels containing poly(N-hydroxyethyl acrylamide)/salicylate were developed by Zhao et al., which possess antimicrobial activities against *E. coli* and *S. epidermidis*. Salicylate is a natural antibacterial compound extracted from plants, and abundant loading of salicylate (up to 80%) hindered the growth of bacteria by 98%. The release of salicylate from hydrogel complex was attributed through diffusion process as the function of concentration gradient of salicylate from hydrogels to biological fluids (Zhao et al. 2013). Similarly, Tan et al. also introduced seaweed-containing PVA/PVP hydrogels, which are highly active against clinically relevant pathogenic bacteria like *S. aureus*, *E. cloacae* and *C. perfringens*. Hydrogel containing antibiotics, antimicrobial metal nanoparticles, antimicrobial polymers and peptides are few other strategies being developed in this area (Ng et al. 2014).

20.1.8 Cyclodextrin-Based Nano-vesicles

Dextrins are cyclic oligosaccharides composed of macrocyclic rings of glucose subunits linked by alpha-1,4 glycosidic bonds giving a hollow structure. Cyclodextrins are usually produced from starch through enzyme conversion. Based on the number of glucose subunits involved in the ring creating a cone shape, cyclodextrins are classified as α-alpha (6), β-beta (7) and γ-gamma (8) cyclodextrins and easily form complexes with selective compounds owing to their hydrophilic interior and hydrophobic exterior surfaces (Crini 2014). Therefore, cyclodextrins are widely chosen as vehicles in delivering variety of drugs due to the high solubility and stabilities conferred by cyclodextrins to the selected drugs. Incorporating antibiotics in the cavities of cyclodextrins enhances their solubility, efficiency and reduced dosage regimen, thus counteracting antibiotic resistance. Yucai Wei et al. recently synthesized enrofloxacin/florfenicol-loaded cyclodextrin metal-organic framework, which showed increased bactericidal activity on

S. aureus and *E. coli* compared to free antibiotics. Antibiotics were encapsulated in the porous structures of the framework, thus increasing the solubility and bioavailability of antibiotics. Moreover, the antibiotic release was sustained for longer times, continually up to 4 hours demonstrating higher and longer inhibition capabilities of bacteria (Wei et al. 2021). Overcoming the limitation of reversible/virustatic activity of sulfonated cyclodextrins, an antiviral for HIV, Samuel T Jones et al. further modified the cyclodextrins with mercaptoundecane sulfonic acids that conferred efficient virucidal molecules effective against different viruses like herpes simplex virus (HSV), respiratory syncytial virus (RSV), dengue virus and Zika virus (Jones et al. 2020).

20.1.9 Aptamer-Based Nano-templates

Aptamers are monomers of nucleic acids such as RNA, single strands of DNA or peptides forming short chains, which are capable of binding to target moieties. Aptamers are formed in vitro by the process called systemic evolution of ligands by exponential enrichment (SELEX) (Mallikaratchy 2017). Through this process, a number of aptamers can be developed that selectively bind to the target, therefore supporting both the clinical and research needs. Their smaller size, ease of modification and in vitro synthesis afford great merits to encourage their use in biomedicine, biosensing, imaging and target drug delivery. Nanoparticle-conjugated aptamers are majorly used in drug delivery because their high selectivity and nanoparticles protect nucleic acid from enzymatic digestion (Ding et al. 2017; Poolsup and Kim 2017). The stability of antimicrobial peptides can be increased in association with nanoparticle conjugates. Boeun Lee et al. demonstrated such study by loading an antimicrobial peptide $HPA3P^{His}$ onto gold nanoparticle-DNA aptamer selective against *Vibrio vulnificus* that is resistant to antibiotic regimens. The developed conjugate was functional in intracellular delivery of $HPA3P^{His}$, thereby interfering with the pathogenic bacteria besides protecting the antimicrobial peptide from proteolytic degradation. Intravenous injection of the conjugate in *V. vulnificus*-infected mice showed 100% antibacterial activity with selective targeting and 0% toxicity to the mammalian cells. Thus, such efficient intracellular uptake mechanism also minimizes the antibiotic dose regimen besides being highly effective (Lee et al. 2017). Fan Chen et al. identified a single aptamer (NK2) from the SELEX process, which has high affinity towards highly virulent strain Mycobacterium tuberculosis (H37Rv). The aptamer NK2 upon selectively binding to H37Rv provokes the CD4+T cells to produce IFN-γ, thereby significantly lowering the bacterial abundance (Chen et al. 2007). Similarly, aptamers can thus be used further upon drug conjugations and chemical modifications to treat several bacterial and viral infections.

The list of such nano-vehicles may become exhaustive in the near future due to extended research progress simultaneous evident by efficient treatment modalities in combating the pathogenic infections caused by antimicrobial-resistant bacteria.

As the nano-vehicle-based drug delivery is still in infancy, efforts are needed to fasten various nano-vehicles for their optimal therapeutic use.

20.2 Limitations on the Use of Nano-carriers

From the pool of metallic nanoparticles available for varied clinical applications, selection of highly biocompatible core metals is imperative to ensure treatment compliance. Majority of metal nanoparticles (except gold) used as either carriers alone or incorporated in suitable encapsulations tend to dissolute into respective ions in aqueous atmospheres, which interferes with the healthy host cells and may hamper the active cell metabolisms. Physiochemical properties of nanoparticles like size, shape and surface charge influence the stability of nanoparticles when exposed to different physiological conditions like pH, enzymes and secretions, which may prone to aggregate nanoparticles if they are instable. Peptide molecules such as antimicrobial peptides are also limited to short life span and poor permeability (Lee et al. 2017). Therefore, peptides may get denature before they could reach the target. Hydrogel-based nano-carriers through bio-inert suffer the limitations of oxidative degradation such as in PEG-based hydrogels and poly-HEMA-based hydrogels known to cause local tissue dehydration as these hydrogels have low hydration levels, thus driving the water diffusion through concentration gradient. Nonetheless, bacterial-specific and bacterial-targeted deliverable capability of several nano-carries addressing subtle limitations as evidenced from careful monitoring is highly needed to ensure host cell compliance and optimal nano-carrier utilization.

20.3 Conclusions

In summary, while introducing the readers on the emerging of nano-based carriers 'nano-vectors' for targeting microbial killing. The emergence of various forms of nano-vectors that have evolved to address the needs of current therapeutics to decrease the after-effects of drugs by attaining high payloads is discussed one by one. Choice of nano-cargos to dictate a selective biomedical implication is discussed as a crucial factor, as not all nanoparticles can be used for all applications. Desired nanoparticle has to be designed depending on the objective and nanoparticle properties to adopt the same. Interaction of these nano-cargos with various bacterial systems to combat multidrug resistance and the various mechanisms by which the bactericidal action is induced is discussed. Several mean by which nanoparticles can pose threat to a microorganism by inducing cell membrane, cell organelle and DNA damage are discussed. Along with cellular response to nanoparticle, stress in terms of ROD-induced oxidative stress is highlighted. Finally, limitations with regard to the exiting nano-carriers are discussed with a scope to design novel nano-carriers with unique properties.

References

Abedi-Gaballu F, Dehghan G, Ghaffari M, Yekta R, Abbaspour-Ravasjani S, Baradaran B, Hamblin MR (2018) PAMAM dendrimers as efficient drug and gene delivery nanosystems for cancer therapy. Appl Mater Today 12:177–190

Alabresm A, Chen YP, Wichter-Chandler S, Lead J, Benicewicz BC, Decho AW (2020) Nanoparticles as antibiotic-delivery vehicles (ADVs) overcome resistance by MRSA and other MDR bacterial pathogens: the grenade hypothesis. J Glob Antimicrob Resist 22:811–817

Anil Kumar S, Khan MI (2010) Heterofunctional nanomaterials: fabrication, properties and applications in nanobiotechnology. J Nanosci Nanotechnol 10(7):4124–4134

Assali M, Zaid AN, Abdallah F, Almasri M, Khayyat R (2017) Single-walled carbon nanotubes-ciprofloxacin nanoantibiotics-strategy to improve ciprofloxacin antibacterial activity. Int J Nanomedicine 12:6647–6659

Baddour CE, Briens C (2005) Carbon nanotube synthesis: a review. Int J Chem React Eng 3(1)

Baptista PV, McCusker MP, Carvalho A, Ferreira DA, Mohan NM, Martins M, Fernandes AR (2018) Nano-strategies to fight multidrug resistant bacteria—"A Battle of the Titans". Front Microbiol 9:1441

Bellini RG, Guimarães AP, Pacheco MAC, Dias DM, Furtado VR, de Alencastro RB, Horta BAC (2015) Association of the anti-tuberculosis drug rifampicin with a PAMAM dendrmer. J Mol Graph Model 60:34–42

Bianco A, Hoebeke J, Kostarelos K, Prato M, Partidos CD (2005) Carbon nanotubes: on the road to deliver. Curr Drug Deliv 2(3):253–259

Cevc G (2012) Rational design of new product candidates: the next generation of highly deformable bilayer vesicles for noninvasive, targeted therapy. J Control Release 160(2):135–146

Chen F, Zhou J, Luo F, Mohammed AB, Zhang XL (2007) Aptamer from whole-bacterium SELEX as new therapeutic reagent against virulent Mycobacterium tuberculosis. Biochem Biophys Res Commun 357(3):743–748

Chen WY, Chang HY, Lu JK, Huang YC, Harroun SG, Tseng YT et al (2015) Self-assembly of antimicrobial peptides on gold nanodots: against multidrug-resistant bacteria and wound-healing application. Adv Funct Mater 25(46):7189–7199

Crini G (2014) A history of cyclodextrins. Chem Rev 114(21):10940–10975

Ding F, Gao Y, He X (2017) Recent progresses in biomedical applications of aptamer-functionalized systems. Bioorg Med Chem Lett 27(18):4256–4269

Ding X, Wang A, Tong W, Xu FJ (2019) Biodegradable antibacterial polymeric nanosystems: a new hope to cope with multidrug-resistant bacteria. Small 15(20):1900999

Faya M, Kalhapure RS, Kumalo HM, Waddad AY, Omolo C, Govender T (2018) Conjugates and nano-delivery of antimicrobial peptides for enhancing therapeutic activity. J Drug Deliv Sci Technol 44:153–171

Fayaz AM, Girilal M, Mahdy SA, Somsundar SS, Venkatesan R, Kalaichelvan PT (2011) Vancomycin bound biogenic gold nanoparticles: a different perspective for development of anti VRSA agents. Process Biochem 46(3):636–641

Hu W, Wang Z, Xiao Y, Zhang S, Wang J (2019) Advances in crosslinking strategies of biomedical hydrogels. Biomater Sci 7(3):843–855

Jiménez JL, Gómez R, Briz V, Madrid R, Bryszewsk M, De La Mata FJ, Munoz-Fernandez MA (2012) Carbosilane dendrimers as carriers of siRNA. J Drug Deliv Sci Technol 22(1):75–82

Jones ST, Cagno V, Janeček M, Ortiz D, Gasilova N, Piret J, Tapparel C (2020) Modified cyclodextrins as broad-spectrum antivirals. Sci Adv 6(5):eaax9318

Karmali PP, Chaudhuri A (2007) Cationic liposomes as non-viral carriers of gene medicines: resolved issues, open questions, and future promises. Med Res Rev 27(5):696–722

Khazi-Syed A, Hasan MT, Campbell E, Gonzalez-Rodriguez R, Naumov AV (2019) Single-walled carbon nanotube-assisted antibiotic delivery and imaging in S. epidermidis strains addressing antibiotic resistance. Nano 9(12):1685

Kim TH, Raiz A, Unni AD, Murhekar S, Donose BC, Floetenmeyer M et al (2020) Combating antibiotic-resistant gram-negative bacteria strains with tetracycline-conjugated carbon nanoparticles. Adv Biosyst 4(9):2000074

Lai X, Ding Y, Wu CM, Chen X, Jiang JH, Hsu HY et al (2020) Phytantriol-based cubosome formulation as an antimicrobial against lipopolysaccharide-deficient gram-negative bacteria. ACS Appl Mater Interfaces 12(40):44485–44498

Lee B, Park J, Ryu M, Kim S, Joo M, Yeom, J. H.,& Bae, J. (2017) Antimicrobial peptide-loaded gold nanoparticle-DNA aptamer conjugates as highly effective antibacterial therapeutics against Vibrio vulnificus. Sci Rep 7(1):1–10

Li X, Robinson SM, Gupta A, Saha K, Jiang Z, Moyano DF et al (2014) Functional gold nanoparticles as potent antimicrobial agents against multi-drug-resistant bacteria. ACS Nano 8(10):10682–10686

Li Y, Liu G, Wang X, Hu J, Liu S (2016) Enzyme-responsive polymeric vesicles for bacterial-strain-selective delivery of antimicrobial agents. Angew Chem Int Ed Engl 55(5):1760–1764

Mallikaratchy P (2017) Evolution of complex target SELEX to identify aptamers against mammalian cell-surface antigens. Molecules 22(2):215

Martin-Serrano Á, Gómez R, Ortega P, de la Mata FJ (2019) Nanosystems as vehicles for the delivery of antimicrobial peptides (AMPs). Pharmaceutics 11(9):448

Messiaen AS, Forier K, Nelis H, Braeckmans K, Coenye T (2013) Transport of nanoparticles and tobramycin-loaded liposomes in Burkholderia cepacia complex biofilms. PLoS One 8(11):e79220

Mo J, Milleret G, Nagaraj M (2017) Liquid crystal nanoparticles for commercial drug delivery. Liq Cryst Rev 5(2):69–85

Narayanaswamy R, Torchilin VP (2019) Hydrogels and their applications in targeted drug delivery. Molecules 24(3):603

Ng VW, Chan JM, Sardon H, Ono RJ, García JM, Yang YY, Hedrick JL (2014) Antimicrobial hydrogels: a new weapon in the arsenal against multidrug-resistant infections. Adv Drug Deliv Rev 78:46–62

Panyam J, Labhasetwar V (2003) Biodegradable nanoparticles for drug and gene delivery to cells and tissue. Adv Drug Deliv Rev 55(3):329–347

Poolsup S, Kim CY (2017) Therapeutic applications of synthetic nucleic acid aptamers. Curr Opin Biotechnol 48:180–186

Scriboni AB, Couto VM, Ribeiro LNDM, Freires IA, Groppo FC, De Paula E, Cogo-Müller K (2019) Fusogenic liposomes increase the antimicrobial activity of vancomycin against staphylococcus aureus biofilm. Front Pharmacol 10:1401

Suresh AK (2015) Co-relating metallic nanoparticle characteristics and bacterial toxicity. Springer, Cham

Thorn CR, Clulow AJ, Boyd BJ, Prestidge CA, Thomas N (2020) Bacterial lipase triggers the release of antibiotics from digestible liquid crystal nanoparticles. J Control Release 319:168–182

Wang X, Li Q, Xie J, Jin Z, Wang J, Li Y et al (2009) Fabrication of ultralong and electrically uniform single-walled carbon nanotubes on clean substrates. Nano Lett 9(9):3137–3141

Wei Y, Chen C, Zhai S, Tan M, Zhao J, Zhu X, Dai T (2021) Enrofloxacin/florfenicol loaded cyclodextrin metal-organic-framework for drug delivery and controlled release. Drug Deliv 28(1):372–379

World Health Organization (WHO) (2020) Antibiotic resistance. Available online at: https://www.who.int/news-room/fact-sheets/detail/antibiotic-resistance

Yang SK, Zimmerman SC (2015) Water-soluble polyglycerol dendrimers with two orthogonally reactive core functional groups for one-pot functionalization. Macromolecules 48(8):2504–2508

Yeh YC, Huang TH, Yang SC, Chen CC, Fang JY (2020) Nano-based drug delivery or targeting to eradicate bacteria for infection mitigation: a review of recent advances. Front Chem 8:286

Zhao C, Li X, Li L, Cheng G, Gong X, Zheng J (2013) Dual functionality of antimicrobial and antifouling of poly (N-hydroxyethyl acrylamide)/salicylate hydrogels. Langmuir 29(5):1517–1524

Chapter 21
Silver Nanoparticles as Potent Multidrug-Resistant Incorporants in Biomedicine

Imran Uddin, Divya S. Parimi, Tarun K. Bollu, Chandra S. Bhatt, and Anil K. Suresh

21.1 Introduction

Currently, there is inadequacy in the development of new antibiotics to tackle the emergent antimicrobial resistance. Continuous rise in the emergence of resistant strains has not outweighed by the readiness of novel therapeutic agents for several reasons (Rawson et al. 2020). Primarily, policymakers should practice to elude the recommendation of new antibiotics until they are obligatory for resistance development. Conversely, society wishes that pharmaceutical companies should design and develop innovative drugs with no negative impacts on health. Furthermore, the durability of antibiotics being short-term; companies cannot make an unrelenting revenue, and high screening and development cost of the design, monitoring onus for novel drugs makes it difficult to meet the demand of affordable antibiotics (Wilson et al. 2020).

Since earlier decades, noble metal nanoparticles exhibiting unique physical, chemical, and biological properties due to their size in the nano regime have elicited tremendous interest. Metal nanoparticles are attracting a great deal of attention because of their potential for achieving specific processes and selectivity, especially in biomedical and pharmaceutical applications. Among the various inorganic nanoparticles, silver nanoparticles are employed extensively since ancient times, to fight bacterial infections and control spoilage of food.

I. Uddin · D. S. Parimi · T. K. Bollu · C. S. Bhatt · A. K. Suresh (✉)
Bio-Nanotechnology Laboratory, Department of Biological Sciences, SRM University-AP, Amaravati, Andhra Pradesh, India
e-mail: anil.s@srmap.edu.in

N. Akhtar et al. (eds.), *Emerging Modalities in Mitigation of Antimicrobial Resistance*,
https://doi.org/10.1007/978-3-030-84126-3_21

21.2 Silver Nanoparticles as an Alternative Therapeutic to Disguise Multidrug-Resistant Microbes

Resistance of a microorganism to an antimicrobial drug that was originally effective for treatment of infections when becomes nonfunctional due to continuous exposure is termed as antimicrobial resistance. Resistant microorganisms (including bacteria, fungi, viruses, and parasites) can eventually withstand attack by antimicrobial drugs over time, and so the standard treatment becomes inefficient and the infections persist increasing the risk of patient threat and further spread (Kingwell 2016). The evolution of resistant strains is a natural phenomenon that occurs when microorganisms replicate themselves erroneously or when resistant traits are exchanged. The use and misuse of antimicrobial drugs accelerate the emergence of drug-resistant strains. With continuous use, antibiotics became usual which on onehand act as a curable agent against disease but also have a profound impact on the life of bacteria, which results in the development of resistive nature against antibiotics and chemotherapeutic agents. Poor infection control practices, inadequate sanitary conditions, and inappropriate food handling encourage the further spread of antimicrobial resistance (Laws et al. 2019).

Multidrug resistance is a natural phenomenon of insensitivity of microbes like bacteria, fungi, viruses, and parasites to the administered antimicrobial drug which are structurally different and have different molecular targets, which leads to ineffective and spreading of infection. As per the survey performed across various regions of the globe, it has been found that a number of antibiotic-resistant species have been evolved showing inhibitory effects against antibiotics with increased mortality and hence referred as "superbugs". These superbugs obstruct the drug action by intensifying the spreading of resistant microbes. Also, the difference in resistance profiles of bacterial and fungal pathogens is responsible for the impact on the efficacy of antimicrobial agents. Therefore, multidrug resistance is considered a threat to human health in the modern era (Gray and Wenzel 2020).

The action of antimicrobial agents on the microbe could hinder the metabolic pathway of nucleotide amalgamation, where a number of alleles (allele count) are inflated, relative to the expected mutation which inhibits the DNA/RNA synthesis and hence disrupts the cell membrane. With the development of multidrug resistance nature in microorganisms, the effectiveness of drug decreases as the organisms gain the tendency of chromosomal mutations or exchange of extrachromosomal DNA elements through conjugation which causes alteration in the composition of cell membrane resulting in decreased permeability and uptake of drugs by the microbial cell (Klemm et al. 2018). Another mode of multidrug resistance is the artificial expression of drug-targeted enzymes which generate alternate target molecules and cause interference in protein synthesis which affects the drug approach to the targeted site. Additional causatives of multidrug resistance are enzymatic degradation, conformational changes, certain point mutations, and chemical transformation of antimicrobials. Chemo resistance during the course of cancer treatment is also a major problem, which has emerged as multidrug resistance. When antibiotics were misused for the cure of bacterial infection, it caused a genetic modification in the bacteria, which made bacteria antibiotic-resistant (Cole 2014).

As far as the clinical resistance is concerned, it arises when the concentration of antimicrobial drugs such that it poses higher probability of therapeutic failure. Some diseases which cause human death because of multidrug resistance are tuberculosis (Srinivasan et al. 2020), pneumonia (Cillóniz et al. 2019), HIV (Günthard et al. 2019), influenza (Guk et al. 2020), malaria (Ross and Fidock 2019), and several yeast infections. Immunocompromised conditions like HIV, diabetes, severe burns are key factors responsible for the spread of multidrug resistance. This is because the old conventional drugs have been replaced by novel drugs with more treatment time resulting in the increment of treatment cost and economic burden for the patient. With the introduction of nanotechnology, it becomes possible to utilize unique physicochemical properties of metallic nanoparticles which arise due to higher surface to volume ratio, in various fields of medical science (Zhao et al. 2020). Considering the antimicrobial potential of silver nanoparticles (AgNPs) against broad spectrum of microorganisms, they have been employed as curable agents against multidrug resistance and nonmultidrug resistance strains, because of their microbicidal activity which leads to dysfunctions and structural disruptions of cells. Also, a number of functionalization/conjugation of AgNPs with commercial antibiotic drugs is performed that is greatly going on to improve the bactericidal activity of AgNPs leading to the development of a new class of nano-based nanobiotics against multidrug resistance.

21.3 Impact of AgNPs in Biomedicine

Although modern medicine has advanced at an astonishing pace and helped increase life expectancy, a remedy for incurable/untreatable diseases, epidemics and pandemics suffer from disadvantages, such as side effects, drug administered deaths, deal with health side effects and added sickness than promoting health and can be overcome by nanomedicines. A wide range of NPs, including metal NPs, silica NPs, lipid-based NMs, dendrimers, polymers, nanogels, carbon-based NMs, graphene oxides are developed to enhance the pharmacological and therapeutic properties of conventional drugs (Giner-Casares et al. 2016). Metal NPs are considered as promising candidates for biomedical research due to their exceptional physiochemical properties, large surface area, photothermal potential, and electrostatic charge. In future, medicines based on nanotechnology have significant potential to increase the pharmaceutical market, targeted drug delivery systems, and bioimaging. AgNPs are used in different biomedical applications, viz. drug delivery systems, catheters, dental applications, bone healing, wound healing, and other medical applications (Li et al. 2016).

AgNPs used in biomaterials like endodontics, periodontics, and implant dentistry prevent biofilm formation, micro leakage, and secondary caries. Further, they enhance the mechanical properties of restorative material and improve the overall bond strength between dentin and biomaterial. Also during the root canal therapy, complete elimination of microorganisms is not possible; hence AgNPs are used as irritants in place of sodium hypochlorite, ethylene diamine tetraacetic acid (EDTA) for having a complete clean sterile canal. Lotfi et al. showed that at lower concentrations, AgNPs show effective irrigation due to its deeper penetrating antimicrobial effect. Although the application of AgNPs in orthodontics is still at its primitive states, they have been incorporated in cement for orthodontic cases to prevent white spot lesions (Corrêa et al. 2015). AgNPs find application in different fields of dentistry, like implantology dental prostheses, restorative and endodontic dentistry. The profound characteristics of AgNPs hold a conspicuous place in nanomaterial-related restorative, regenerative, and multifunctional biomedicine (Nguyen and Hiorth 2015). Silver has shown promising results in caries prophylaxis in the form of nano-silver diamine fluoride (SDF), the disadvantage associated is tooth staining. As the particle size reduces, antibacterial effect enhanced due to increased reacting surface area. Similarly, in nano regime AgNPs have large contact surface area which further enhances the antimicrobial property (Santos et al. 2013).

AgNP-based nanosystems in particular have been assessed as appropriate transporters of several therapeutic molecules comprising antioxidant (Jiang et al. 2018), anti-inflammatory (Muhammad et al. 2016), anticancer (Al-Obaidi et al. 2018), and antimicrobial (Venugopal et al. 2017) substances. Due to their inherent bactericidal potential, AgNPs have grabbed distinctive consideration, and are well appraise as efficacious antibacterial drug-delivery systems, functioning both passive as well as active carriers for nano drugs in antimicrobial drugs delivery.

For the procurement of innovative and improved efficiency of drug delivery systems in response to different factors (optical, thermal, and pH) intonations to spot infectious, inflammatory, and malignant disorders, hybrid molecular units comprising of AgNPs were efficaciously chosen, owing to their incomparable biocompatibility, sturdy absorption characteristics, and its low toxicity. In biomedical devices like catheters, AgNPs play pivotal role to prevent device like infection (Thomas et al. 2015). Further, AgNPs-modified CVCs have shown very well inhibition against gram-positive as well as gram-negative bacterial biofilm. The attachment of AgNPs with bacterial cells is dependent on surface area to volume ratio hence bactericidal effects are size-dependent. Silver-treated catheters epitomize a possible strategy for decreasing dialysis-associated infections in patients enduring peritoneal catheters; though antimicrobial efficacy and method of procuring Ag^{+} may vary (Aflori 2014). It has been observed that hydrophilic surface properties of polymer matrix occur due to infused AgNPs which lead to retardation of biofilm formation and protein and electrolyte deposition accountable for coating and attachment of microorganisms onto the surface (Samuel and Guggenbichler 2004).

Some biocomposites comprising of silver-like AquacelTM (ConvaTec), PolyMemSilverTM (Aspen), ActicoatTM and BactigrasTM (Smith & Nephew), and TegadermTM (3 M) have been approved by Food and Drug Administration (FDA). Along with these commercial products, encouraging results have been witnessed in regard to the amalgamation of AgNPs with natural biomaterials for improved wound-healing treatment, like modified cotton fabrics, chitosan, bacterial cellulose, and sodium alginate (Emam et al. 2015). The AgNPs and Ag^+ carriers also find applications in treating the delinquent wound healing process of diabetic patients as wounds in these patients are escorted by various secondary infections thereby helping diabetic patients in wound healing at early stages. The enhanced antibacterial effects shown by AgNPs have gained special interest as their application in treatment of wounds, coating of medical devices, though their safety and biocompatibility need to be thoroughly studied (Rigo et al. 2013).

On comparing the effects of different nanoparticles, it has been observed that AgNPs increase the process of differentiation in preosteoblast cells (MC3T3-1) by following tissue mineralization like bone (Qing et al. 2018). Today, silver-coated prostheses present an alternative in the prophylaxis of tumor-related infections along with general trauma-related infections. However, no clinical studies associating with the lasting clinical effect of nano silver-coated implants used for revision arthroplasty are not reported (Brennan et al. 2015). The self-repairing ability of bones becomes limited during bacterial infection when occurs in bone defects. When compared to typical antibiotics, silver in nano dimension exhibits greater antibacterial effects of broad spectrum. Moreover, the resistance in bacteria against AgNPs is an infrequent phenomenon, therefore highlighting antibacterial mechanisms of nanosilver is essential. In antibiotic-resistant bacteria (methicillin-resistant *Staphylococcus aureus*), AgNPs have shown inhibiting potential in development of biofilm (Lu et al. 2016). In techniques like bone-replacement, doping materials are used as AgNPs in artificial as well as bioinspired bone scaffolds. The inimitable physiochemical and biofunctional characteristics like antiplatelet, antiangiogenesis, anti-inflammatory, antiviral, antifungal, and antibacterial activities, have made AgNPs a promising tool for implementing in biomedical applications. In fact, AgNPs were explicitly studied for their encouraging anticancer property. It was observed in different human cancerous cell lines (U251 glioblastoma cells, MDA-MB-231 breast cells, and IMR-90 lung fibroblasts) (Thapa et al. 2017). AgNPs are readily taken up by the mammalian cells, penetrate and cellularly localize *via* energy-dependent internalization pathways. Another enticing characteristic of AgNPs is precise fluorescence behavior which helps in detection and diagnosis applications (Mattea et al. 2017; Burdus et al. 2018).

Cardiovascular diseases (CVDs) are one of the reasons for global human death, with an history of greater than 17.7 million deaths in the year 2015. Prosthetic silicone heart valve having elemental silver was first silver-grafted cardiovascular medical device to evade bacterial infection and to reduce inflammation (Jiang et al. 2017).

The human eye is a multifarious organ, with extensive vascularization and innervation which are generally subjected to microbial infections. Compounds which have nanosilver have shown advancement of therapies even for the sensitive portions of our system the "eye infection". Coating of AgNPs with calcium indicators has reduced the damage in cells of the retina and may perhaps be experimentally suitable for retinal imaging in animal mouse models (Weng et al. 2017). The antibacterial effects of AgNPs are important facets for ocular applications. The broad-spectrum properties of AgNPs make them promising candidates not merely for antimicrobial effects, but also in multidrug resistance confronting tactics (Pind'akova et al. 2017).

21.4 Silver NPs as a Potent Antimicrobial Therapeutic Against Multidrug Resistance

Metal-based nanoparticles are effective against pathogens because of their nonspecific bacterial poisonousness nature along with their small size and hence exhibit a range of antibacterial action toward gram-positive and gram-negative bacteria. It has been used since ancient time not only for jewellery but also because of its antiseptic property specifically for open wounds and burns. AgNPs have an effective bactericidal nature against a broad spectrum of bacteria and can inhibit biofilm during wound healing. It is used as a substitute for conventional antibiotics like antifungal, antiviral, and anti-inflammatory. Considering the antibacterial effect of AgNPs that it exhibits by various modes of action in which silver ions interrupt DNA replication in bacterial cells, thus causing denaturation and rupture of cell organelles *via* oxidation (Marambio-Jones and Hoek 2010). In this type of cell lysis, AgNPs having a size ranging from 1 to 10 nm with minimum inhibitory concentration ranging from 0.003 to 0.5 mg/mL cause cell inhibition for microbes *Fusobacterium nucleatum, Streptococcus mutans,* and *Actinomycesoris* (Lara et al. 2010). Sondi et al., in 2004, showed that AgNPs cause 100% inhibition of *Escherichia coli* when MIC is 50–60 μg cm^{-3} (Sondi and Salopek-Sondi 2004). A mechanistic approach of interaction of AgNPs with antibiotics where amikacin-functionalized AgNPs revealed enhanced antibacterial activity against *E. coli* than normal antibiotics. Interaction of AgNPs possibly induces conformational changes and folding characteristics of amikacin molecules (Uddin 2018).

The size, shape, and surface properties of AgNPs and its interaction with cells can be either a physical or chemical transformation (Fig. 21.1). The physical interactions include disruption of membranes, membrane activity, transport processes, protein conformation/folding, and protein aggregation/fibrillation. Simultaneously, chemical interactions include the production of ROS, oxidative damage through catalysis, lipid peroxidation, and disturbance of cell membrane transport activity. Both physical and chemical processes cause cell death. The production of ROS is considered a critical chemical factor of causing cell cytotoxicity. However, the physicochemical properties of AgNPs are strongly responsible for the change in membrane morphology and stability. AgNPs are also potent enough to affect mitochondrial function, leading to apoptosis. Generally, the movement of NPs ends up in lysosomes where the cell could either digest or excrete out, based on its physicochemical properties. NPs can interact with protein corona because both are of the same size magnitude and compatible electron transfer. It is believed that ROS is responsible for DNA damage.

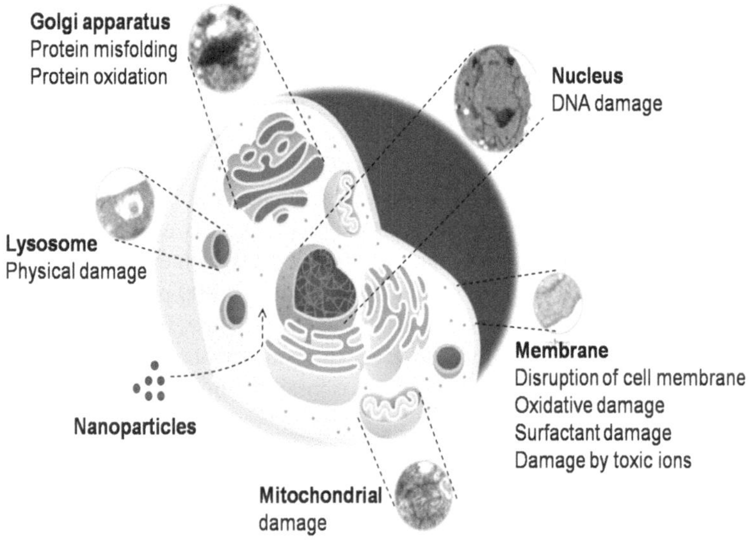

Fig. 21.1 Various modes of interaction of nanoparticles with cells

21.5 Mechanistics of AgNPs on Multidrug-Resistant Bacteria in Biomedicine

The mode of action of AgNPs in case of multidrug-resistant bacteria (MDRB) differs greatly by which outmoded antibiotics act, and therefore resistance with regards to antibiotics can be overcome by replacing conventional antibiotics with nanoparticles. Though the antimicrobial effects of AgNPs have been comprehensively investigated, their bactericidal mode is still ambiguous. Some studies indicated that AgNPs attack to the cell wall; penetrate through it, causing structural damage in cell membrane, increasing permeability, uncontrolled transport *via* cytoplasmic membrane leading to cell death.

Kim and coworkers reported that generation of free radicals might be attributed to antibacterial mechanisms leading to membrane damage (Kim et al. 2007). Morones and coworkers anticipated the idea that silver ions have affinity for thiol groups in most of the phosphorus-containing bases and enzymes, asserting the impairment caused by interactions of AgNPs with DNA (Morones et al. 2005). This contact can inhibit DNA replication and cell division, eventually causing cell death. Shrivastava and coworkers suggested that AgNPs inhibit bacterial growth by modifying the phosphotyrosine residue of putative bacterial peptides which further can affect cellular signaling (Shrivastava et al. 2007). Hwang et al. studied mechanistic aspect of toxicity of AgNPs and the silver ions on stress-specific bioluminescent bacteria. It was observed that ROS produce due to movement of silver ions in the cells and lead to membrane damage by AgNPs (Hwang et al. 2008). Klueh and coworkers proposed inhibition of bacterial growth occurs through silver atoms which restrict enzyme activity by binding to thiol groups of the plasma membrane components and alter the function of components of cell membrane, which is essential energy generation and ion transport. Silver induces the compilation of disulfide bonds in between oxygen and thiol groups (R-S-S-R) within the cell. The disulphide alters the shape and structure of enzymes thus affecting the functioning of enzymes. He also put forward that bactericidal effects of the AgNPs could be attributed to the Ag^+ ions, which make way into the cell and intercalate with pyrimidine and purine bases of DNA thereby disturbing the hydrogen bonding between the two antiparallel strands ultimately leading to denaturation of DNA (Klueh et al. 2000).

It has been observed that Ag^+ at the concentration of 900 ppb, reacts with cells alters the functioning of enzymes (Succinyl Co-A synthetase, fructose bisphosphate aldolase, maltose transporter (MalK)). The translation of protein was restricted through binding of Ag^+ ions with the 30S ribosomal subunit which disables the ribosome complex formation.

The TCA cycle enzyme, succinyl coenzyme A synthetase was down-regulated on exposure of Ag^+ ions. Therefore, the proteins, key players of cells, are altered by the application of AgNPs, resulting in bacterial cell death (Yamanaka et al. 2005).

21.6 Factors Influencing the Bactericidal Effect of AgNPs

Microbes are unlikely to develop resistance against silver, as they do against conventional and narrow-target antibiotics, because the metal attacks a broad range of targets in the organisms, which means that they would have to develop a host of mutations simultaneously to protect themselves. There are following factors that can significantly influence the bactericidal effect.

21.6.1 Nanoparticle Species

While the properties of core particle species can be a great role in dictating the potential causative in microbial toxicity, relative toxicity of one species to another is not simple as factors such as particle size, shape, stabilizers, and syntheses can influence toxicity. Suresh et al. suggested that, as nanoparticles size decreases or change in its curvature occurs which further leads to the reactivity and directly influences the enhancement of toxicity (Suresh et al. 2013). Surface coating of the nanomaterials can also have enhanced the microbial toxicity. Rigid controlling of physical characteristics is highly challenging, and various syntheses may pertain to differences in surface coating and contaminant toxic after reaction leftovers. Moreover, diverse encapping processes may involve additives, surfactants, and chemicals that are not fully eliminated from the reaction. For example, C60 that was initially toxic was later demonstrated to be safe, and it is the remnant of tetrahydrofuran that is used in the synthesis of C60 was the toxicity culprit. Likewise, despite of their use of using Ag-resistant *E. coli* in the studies, presence of formaldehyde was demonstrated for the killing of Ag-resistant bacterium (Marambio-Jones and Hoek 2010). Therefore, the properties of nanoparticles may also rely in par on other leftovers present in the reaction formulation and the coatings associated with the nanoparticles.

Evidence supporting the role of silver ions upon its dissolution from parent AgNPs as the reason toxicity is unfolding. For example, the antibacterial potential of 9 nm spheres of AgNPs against silver-resistant bacterium has been attributed to chemisorbed Ag^+ ions that are formed under extreme oxygen-sensitive conditions (Levard et al. 2012).

21.6.2 Particle Appearance and Morphology

Differences in the iterated antimicrobial potential of AgNPs also correlate to particle's size and/or shape distribution. Decrease in the size of the particle leads to increase in the ratio of surface area to mass thereby altering to the physicochemical properties such as surface atom reactivity, electronic and optical properties. Such a phenomena result in clumping of particles and can significantly influence the particle reactivity and cell-binding characteristics (Rai et al. 2012). The change in properties of nanoparticles is largely governed by their small size rather than their bulk counterparts. The properties of materials change as their size approaches the nanoscale and as the percentage of atoms at the surface of a material becomes significant. The smaller the size, greater the surface area to volume ratio, hence greater is the bactericidal effect of AgNPs. Since the particles in nano dimensions have greater surface reactivity, thus the antibacterial effects of silver are size-dependent. AgNPs undergo a size-dependent interaction with human immunodeficiency virus type 1, preferably *via* binding to gp120 glycoprotein knobs (Raimondi et al. 2005). The morphology of nanoparticles plays a key role on antibacterial effect, Pal and coworkers investigated the effect of different morphology (spherical, rod, and triangular) nanoparticles synthesized by chemical route against *E. coli* at various concentrations. It has been observed that triangular shape nanoparticles were shown more inhibition than spherical nanoparticles followed by rod shaped against *E. coli* (Pal et al. 2007). In Fig. 21.2, UV-Visible absorption spectra of silver nitrate ($AgNO_3$) and after synthesizing AgNPs are shown. Further, crystalline nature AgNPs explored through XRD pattern and surface morphology are also shown in TEM micrograph.

21.6.3 Concentration

Morones and coworkers studied interactive effects of AgNPs of various concentrations on bacterial cells (*E. coli*) at mid-log period of bacterial growth. Further dose-dependent activity of AgNPs is also measured on both the *Gram* + as well as

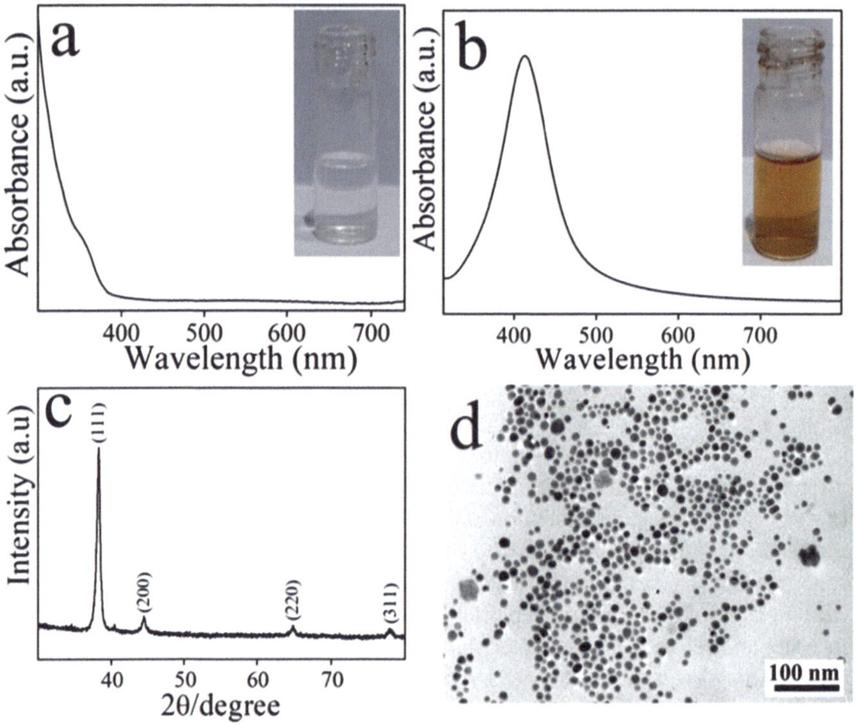

Fig. 21.2 UV-Visible spectra of (**a**) precursor ($AgNO_3$) as-synthesized AgNPs, (**b**) AgNPs showing their morphological distribution, (**c**) powder XRD-pattern of AgNPs demonstrating their crystallinity, (**d**) TEM image

Gram-bacteria. It has been observed that the action of AgNPs is dose-dependent and AgNPs depict conspicuous antibacterial effects against gram-negative bacteria in comparison to the gram-positive ones.

21.6.4 Stabilizing Agents

Silver nanoparticles are encamped through stabilizing agent and its determinant in dictating mechanistic approach and environmental as well as biological destiny of nanoparticles, as the surface is often primary approach of interaction. Charge on nanoparticles influenced by surface capping, which in turn can affect the binding of the nanoparticle to the cellular surface. Figure 21.3 shows the transmission electron microscopy image analysis of mode of interaction of AgNPs with bacteria. The binding of the nanoparticles to the surface can further induce stress,

and localize potential toxic impacts. Additionally, the surface coating can influence the release of metallic ions from the nanoparticles in reaction medium

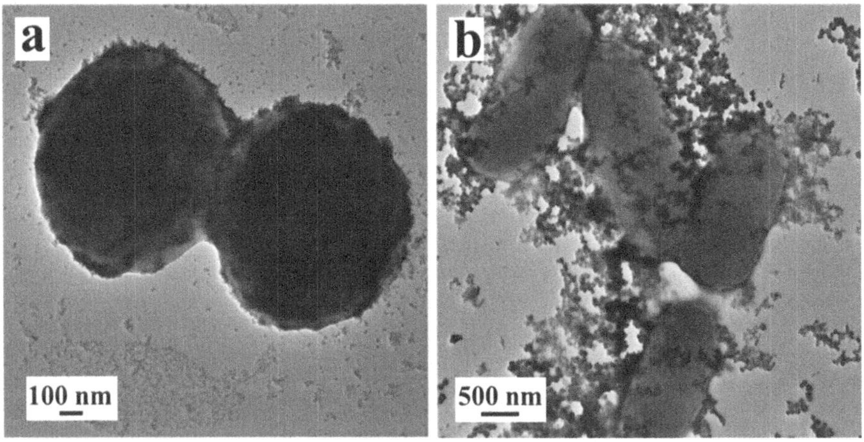

Fig. 21.3 Transmission electron microscopy images of interaction of AgNPs with bacteria

or aqueous environments. Suresh et al. while demonstrating the role of various surface coatings on AgNP use three types of encapping agents; peptide-coated biogenic, colloidal, and oleate (Suresh et al. 2010). The surface features are also strongly influenced by environmental parameters such as pH, ionic strength, presence of organic molecules, and temperature. Consequently, the surface coatings can be difficult to prescribe and can strongly influence the apparent toxicity of the nanoparticle. For example, the sensitivity of *N. europaea* to Ag^{+} ions in comparison with 20 and 80 nm diameter AgNPs under varied physiological parameters showed that seemingly minor factors like the sequence of addition of particles and the presence or absence of medium constituents can have huge influence on the stability, reactivity, and toxicity of the nanoparticles including the subject microorganism.

21.7 Conclusions

Silver nanoparticles have been proven as potent therapeutic alternative to disguise multidrug- resistant microorganisms, which effectively inhibit various pathogenic bacteria, fungi, and viruses either by membrane destruction, ROS generation, DNA damage, enzyme inactivation or protein denaturation. AgNPs had shown unique properties and therapeutic action, exhibit different antibacterial effects against gram-positive and gram-negative bacteria which depend upon particle appearance and morphology, dose and nanoparticle species. This chapter summarizes AgNPs and cell interaction in bacteria through which they inhibit multidrug resistance in bacteria.

References

Aflori M (2014) Surface characterization of peritoneal dialysis catheter containing AgNPs. Rev Roum Chim 59(6–7):523–526

Al-Obaidi H, Kalgudi R, Zariwala MG (2018) Fabrication of inhaled hybrid silver/ciprofloxacin nanoparticles with synergetic effect against Pseudomonas aeruginosa. Eur J Pharm Biopharm 128:27–35

Brennan SA, NíFhoghlú C, Devitt BM, O'mahony FJ, Brabazon D, Walsh A (2015) AgNPs and their orthopaedic applications. Bone Joint J 97(5):582–589

Burdus AC, Gherasim O, Grumezescu AM, Mogoanta L, Ficai A, Ecaterina Andronescu E (2018) Biomedical applications of silver nanoparticles: an up-to-date overview. Nano 8(9):681

Cillóniz C, Dominedò C, Torres A (2019) Multidrug resistant gram-negative bacteria in community-acquired pneumonia. Crit Care 23(1):1–9

Cole SP (2014) Targeting multidrug resistance protein 1 (MRP1, ABCC1): past, present, and future. Annu Rev Pharmacol Toxicol 54:95–117

Corrêa JM, Mori M, Sanches HL, Cruz ADD, Poiate E, Poiate IAVP (2015) AgNPs in dental biomaterials. Int J Biomater 2015:485275

Emam HE, Saleh NH, Nagy KS, Zahran MK (2015) Functionalization of medical cotton by direct incorporation of AgNPs. Int J Biol Macromol 78:249–256

Giner-Casares JJ, Henriksen-Lacey M, Coronado-Puchau M, Liz-Marzán LM (2016) Inorganic nanoparticles for biomedicine: where materials scientists meet medical research. Mater Today 19(1):19–28

Gray DA, Wenzel M (2020) Multitarget approaches against multiresistant superbugs. ACS Infect Dis 6(6):1346–1365

Guk K, Kim H, Lee M, Choi YA, Hwang SG, Han G et al (2020) Development of A4 antibody for detection of neuraminidase I223R/H275Y-associated antiviral multidrug-resistant influenza virus. Nat Commun 11(1):1–11

Günthard HF, Calvez V, Paredes R, Pillay D, Shafer RW, Wensing AM et al (2019) Human immunodeficiency virus drug resistance: 2018 recommendations of the International Antiviral Society–USA panel. Clin Infect Dis 68(2):177–187

Hwang ET, Lee JH, Chae YJ, Kim YS, Kim BC, Sang BI, Gu MB (2008) Analysis of the toxic mode of action of AgNPs using stress-specific bioluminescent bacteria. Small 4(6):746–750

Jiang W, Rutherford D, Vuong T, Liu H (2017) Nanomaterials for treating cardiovascular diseases: a review. Bioact Mater 2:185–198

Jiang Q, Yu S, Li X, Ma C, Li A (2018) Evaluation of local anesthetic effects of Lidocaine-Ibuprofen ionic liquid stabilized AgNPs in Male Swiss mice. J Photochem Photobiol B Biol 178:367–370

Kim JS, Kuk E, Yu KN, Kim JH, Park SJ, Lee HJ, Cho MH (2007) Antimicrobial effects of AgNPs. Nanomedicine 3(1):95–101

Kingwell K (2016) Antimicrobial drugs: multifaceted approach to multidrug resistance. Nat Rev Drug Discov 15(11):750

Klemm EJ, Wong VK, Dougan G (2018) Emergence of dominant multidrug-resistant bacterial clades: lessons from history and whole-genome sequencing. Proc Natl Acad Sci 115(51):12872–12877

Klueh U, Wagner V, Kelly S, Johnson A, Bryers JD (2000) Efficacy of silver-coated fabric to prevent bacterial colonization and subsequent device-based biofilm formation. J Biomed Mater Res 53(6):621–631

Lara HH, Ixtepan-Turrent L, Garza-Treviño EN, Rodriguez-Padilla C (2010) PVP-coated AgNPs block the transmission of cell-free and cell-associated HIV-1 in human cervical culture. J Nanobiotechnol 8(1):15

Laws M, Shaaban A, Rahman KM (2019) Antibiotic resistance breakers: current approaches and future directions. FEMS Microbiol Rev 43(5):490–516

Levard C, Hotze EM, Lowry GV, Brown GE Jr (2012) Environmental transformations of silver nanoparticles: impact on stability and toxicity. Environ Sci Technol 46(13):6900–6914
Li R, Chen J, Cesario TC, Wang X, Yuan JS, Rentzepis PM (2016) Synergistic reaction of silver nitrate, AgNPs, and methylene blue against bacteria. Proc Natl Acad Sci 113(48):13612–13617
Lu H, Liu Y, Guo J, Wu H, Wang J, Wu G (2016) Biomaterials with antibacterial and osteoinductive properties to repair infected bone defects. Int J Mol Sci 17(3):334
Marambio-Jones C, Hoek EM (2010) A review of the antibacterial effects of silver nanomaterials and potential implications for human health and the environment. J Nanopart Res 12(5):1531–1551
Mattea F, Vedelago J, Malano F, Gomez C, Strumia MC, Valente M (2017) AgNPs in X-ray biomedical applications. Radiat Phys Chem 130:442–450
Morones JR, Elechiguerra JL, Camacho A, Holt K, Kouri JB, Ramírez JT, Yacaman MJ (2005) The bactericidal effect of AgNPs. Nanotechnology 16(10):2346
Muhammad Z, Raza A, Ghafoor S, Naeem A, Naz SS, Riaz S et al (2016) PEG capped methotrexate AgNPs for efficient anticancer activity and biocompatibility. Eur J Pharm Sci 91:251–255
Nguyen S, Hiorth M (2015) Advanced drug delivery systems for local treatment of the oral cavity. Ther Deliv 6(5):595–608
Pal S, Tak YK, Song JM (2007) Does the antibacterial activity of AgNPs depend on the shape of the nanoparticle? A study of the gram-negative bacterium Escherichia coli. Appl Environ Microbiol 73(6):1712–1720
Pinďakova L, Kasparkova V, Kejlova K, Dvorakova M, Krsek D, Jirova D, Kašparová L (2017) Behaviour of AgNPs in simulated saliva and gastrointestinal fluids. Int J Pharm 527(1–2):12–20
Qing T, Mahmood M, Zheng Y, Biris AS, Shi L, Casciano DA (2018) A genomic characterization of the influence of AgNPs on bone differentiation in MC3T3-E1 cells. J Appl Toxicol 38(2):172–179
Rai MK, Deshmukh SD, Ingle AP, Gade AK (2012) Silver nanoparticles: the powerful nanoweapon against multidrug-resistant bacteria. J Appl Microbiol 112:841–852
Raimondi F, Scherer GG, Kötz R, Wokaun A (2005) Nanoparticles in energy technology: examples from electrochemistry and catalysis. Angew Chem Int Ed 44(15):2190–2209
Rawson TM, Ming D, Ahmad R, Moore LS, Holmes AH (2020) Antimicrobial use, drug-resistant infections and COVID-19. Nat Rev Microbiol 18(8):409–410
Rigo C, Ferroni L, Tocco I, Roman M, Munivrana I, Gardin C, Zavan B (2013) Active AgNPs for wound healing. Int J Mol Sci 14(3):4817–4840
Ross LS, Fidock DA (2019) Elucidating mechanisms of drug-resistant Plasmodium falciparum. Cell Host Microbe 26(1):35–47
Samuel U, Guggenbichler JP (2004) Prevention of catheter-related infections: the potential of a new nano-silver impregnated catheter. Int J Antimicrob Agents 23:75–78
Santos VE, Targino AGR, Flores MAP, Pessoa HDLF, Galembeck A, Rosenblatt A (2013) Antimicrobial activity of AgNPs in treating dental caries. Revista da Faculdade de Odontologia-UPF 18(3)
Shrivastava S, Bera T, Roy A, Singh G, Ramachandrarao P, Dash D (2007) Characterization of enhanced antibacterial effects of novel AgNPs. Nanotechnology 18(22):225103
Sondi I, Salopek-Sondi B (2004) AgNPs as antimicrobial agent: a case study on E. coli as a model for Gram-negative bacteria. J Colloid Interface Sci 275(1):177–182
Srinivasan V, Ha VT, Vinh DN, Thai PV, Ha D, Lan NH et al (2020) Sources of multi-drug resistance in patients with previous isoniazid-resistant tuberculosis identified using whole genome sequencing: a longitudinal cohort study. Clin Infect Dis 71(10):e532–e539
Suresh AK, Pelletier DA, Wang W, Moon JW, Gu B, Mortensen NP, Doktycz MJ (2010) Silver nanocrystallites: biofabrication using Shewanella oneidensis, and an evaluation of their comparative toxicity on gram-negative and gram-positive bacteria. Environ Sci Technol 44(13):5210–5215
Suresh AK, Pelletier DA, Doktycz MJ (2013) Relating nanomaterial properties and microbial toxicity. Nanoscale 5(2):463–474

Thapa RK, Kim JH, Jeong JH, Shin BS, Choi HG, Yong CS, Kim JO (2017) Silver nanoparticle-embedded graphene oxide-methotrexate for targeted cancer treatment. Colloids Surf B: Biointerfaces 153:95–103

Thomas R, Soumya KR, Mathew J, Radhakrishnan EK (2015) Inhibitory effect of silver nanoparticle fabricated urinary catheter on colonization efficiency of Coagulase Negative Staphylococci. J Photochem Photobiol B Biol 149:68–77

Uddin I (2018) Mechanistic approach to study conjugation of nanoparticles for biomedical applications. Spectrochim Acta A Mol Biomol Spectros 202:238–243

Venugopal K, Rather HA, Rajagopal K, Shanthi MP, Sheriff K, Illiyas M et al (2017) Synthesis of AgNPs (Ag NPs) for anticancer activities (MCF 7 breast and A549 lung cell lines) of the crude extract of Syzygium aromaticum. J Photochem Photobiol B Biol 167:282–289

Weng Y, Liu J, Jin S, Guo W, Liang X, Hu Z (2017) Nanotechnology-based strategies for treatment of ocular disease. Acta Pharm Sin B 7(3):281–291

Wilson DN, Hauryliuk V, Atkinson GC, O'Neill AJ (2020) Target protection as a key antibiotic resistance mechanism. Nat Rev Microbiol 18(11):637–648

Yamanaka M, Hara K, Kudo J (2005) Bactericidal actions of a silver ion solution on Escherichia coli, studied by energy-filtering transmission electron microscopy and proteomic analysis. Appl Environ Microbiol 71(11):7589–7593

Zhao X, Jia Y, Dong R, Deng J, Tang H, Hu F, Jiang X (2020) Bimetallic nanoparticles against multi-drug-resistant bacteria. Chem Commun 56(74):10918–10921

Chapter 22
Role of Gold Nanoparticles Against Multidrug Resistance (MDR) Bacteria: An Emerging Therapeutic Revolution

Kaushik Kumar Bharadwaj, Bijuli Rabha, Bhabesh Kumar Choudhury, Aditi Das, Lydia Islary, Dorothy Bhattacharjya, Monoswi Chakraborty, Debabrat Baishya, and Arabinda Ghosh

22.1 Introduction

Infectious diseases are increasing continuously in our day to day life and this reason has now become a serious threat globally. This has led to concern of increase in the development of multidrug resistance bacteria which has now become a serious nationwide burning topic in the public health sector. Large numbers of antibiotics are used against enormous number of both gram-positive and gram-negative bacteria in humans, animals, and fish resulting in the selection of pathogenic bacteria resistant to antibiotics and antimicrobials. Multidrug resistance bacteria may arise due to, first, accumulation of multiple genes each coding for resistance to a single drug inside a cell/R plasmids. Second, due to increased gene expression that codes for multidrug efflux pumps. Third, due to altered physiological state, which is the occurrence of "persister" cells, i.e., it was observed that high concentration of antibiotics does not kill a bacterial population, leaving behind a persister population which is genetically identical to the susceptible cells (Nikaido 2009). Hence, the use of nanostructure material or nanoparticle has been developed as an emerging technique to overcome the threat of multidrug resistance bacteria. Nanoparticles have a size varying within the range of 1-100 nm. Nanomaterials are used as potential candidates in disease diagnostics, drug delivery system, and medical imaging due to their improvised

K. K. Bharadwaj · B. Rabha · A. Das · L. Islary · D. Bhattacharjya · D. Baishya
Department of Bioengineering and Technology, Gauhati University Institute of Science and Technology, Guwahati, Assam, India

B. K. Choudhury
Department of Chemistry, Gauhati University, Guwahati, Assam, India

M. Chakraborty · A. Ghosh (✉)
Microbiology Division, Department of Botany, Gauhati University, Guwahati, Assam, India
e-mail: dra.ghosh@gauhati.ac.in

N. Akhtar et al. (eds.), *Emerging Modalities in Mitigation of Antimicrobial Resistance*,
https://doi.org/10.1007/978-3-030-84126-3_22

characteristic such as high surface area to volume ratio and small size effect (Okkeh et al. 2021). Different types of nanoparticles can be used which may be metallic, organic or carbon nanotubes. They can be used to assist the delivery of drugs, inhibit the formation of biofilm and activity of bio efflux pumps, and thus circumvent antimicrobial activity by themselves. Thus, nanoparticles furnish us a promising solution by acting as a carrier for antibiotic and antimicrobial compound (Baptista et al. 2018). In context to this, gold nanoparticles can be used as tool against multidrug resistance bacteria. However, gold nanoparticles offer excellent bactericidal activity against MDR gram-positive and gram-negative bacteria unlike other free drugs. One of the most extraordinary properties of AuNPs is their capacity to transform light into heat under laser irradiation. Thus, this property can be exploited to develop photothermal nanovectors to destroy MDR bacteria at a molecular level. Gold nanoparticles are considered as inert and nontoxic and can be synthesized by chemical synthesis based on the reduction of chloroauric acid by citrate (Baptista et al. 2018). These AuNPs enhance the inhibitory effects of antibiotics when conjugated with ampicillin, kanamycin, and streptomycin. They lower the minimum inhibitory concentrations of the antibiotic counterparts against the bacteria. Since they act at nanoscale range; hence, they can interfere with the molecular pathway, interact with bacterial cells, and regulate cell membrane penetration. Fluoroquinolone antibiotics and AuNPs can be used for the treatment of multidrug-resistant *E. coli* bacterial strains (Lee et al. 2019). Gold nanoparticles exhibit optical properties which are determined through Plasmon resonance, associated with the excitation of electrons and localized from the visible to the infrared (IR) region, depending on the particle size, shape, and structure (Dykman and Khlebtsov 2011). They act as a metal precursor, reducing agent, and a stabilizer. Certain in vitro researches have condemned that AuNPs once incorporated within a cell may give rise to endogenous reactive oxygen species (ROS) which may later result to DNA damage, cell death, and eventually cell cycle arrest. Furthermore, it decreases the viability of human cells about 100% at a concentration of 0.1 ppm to less than 40% at a concentration of 10 ppm resulting in dose-related toxicity. But due to the natural characteristic of gold nanoparticles such as biocompatible nature, ease of surface functionalization, optical properties, and antibacterial efficacies, they are widely used to fight against both multidrug resistance gram-positive and gram-negative bacteria (Okkeh et al. 2021). Antimicrobial actions of Cephalosporin type antibiotic (cefaclor) conjugated with AuNPs interact with the outer peptidoglycan layer of the bacteria where it has been observed that cefaclor enhance the membrane porosity and AuNPs create holes in cell walls. This interaction resulted into hindered DNA unwinding and transcription and resulted into high cefaclor concentration in the bacteria. Antibiotic named Cefotaxime when conjugated with AuNPs generates extended spectrum beta-lactamase which resulted into the deposition of high cefotaxime concentration in the bacterial cell. Furthermore, AuNPs when conjugated with methylene blue reduce the bacterial survival capacity enhancing the generations of ROS other than singlet oxygen (Shaikh et al. 2019).

22.2 Antimicrobial Mechanism of Gold Nanoparticles

Of late, the bacterial drug resistance has been greatly curtailing the effective antibiotic options and is consequently posing perils globally. Therefore, novel methodologies are highly sought-after and are straightaway required to significantly enhance the antimicrobial activity. However, associating antibiotics to nanoparticles is bringing about a promising scope by increasing overall drug avidity, bioavailability, and internalization by mammalian cells (Singh et al. 2017). Ag, Cu, and Zn ions along with zinc and copper oxides are some of the nanoscale compounds that have proven exalted antibacterial efficacy. The common consensus with regard to antibacterial activity is that the gold nanoparticle, AuNP-drug conjugates demonstrate stronger antibacterial activity than specific nanoparticles and drugs when administered separately.

Gold particles at the nanoscale (AuNPs) have demonstrated uniquely useful antimicrobial capabilities. Being the metal with bare minimum activity, gold exhibits steady chemical properties, nontoxicity, and excellent biological compatibility (Hammer and Norskov 1995). AuNPs are multivalent, and their substantial surface area provides AuNPs with a plethora of specific binding sites for the targeted bacterial cell (Dreaden et al. 2012). AuNPs have been shown to have a wide range of antibacterial action against gram-negative and gram-positive bacteria (Slavin et al. 2017). In contrast to traditional antibiotics, AuNPs do not easily acquire drug resistance (Yang et al. 2018) as they are shown to target a variety of components (DNA and protein) in bacteria, thereby posing perils for the pathogen to evolve and evade all the damage. The bactericidal characteristics of AuNPs are directly proportional to their size, dispersibility, and surface modification. By the means of electrostatic adsorption on the bacterial surface, reactive oxygen species (ROS) damage, and membrane damage to proteins and DNA, the gold nanoparticles induce death and bring about cell lysis. AuNPs are cytotoxicant-free and have a high degree of biocompatibility (Hammer and Norskov 1995). Currently, the majority of antibacterial investigations using gold nanoparticles have been conducted on traditional bacteria, with just a fraction on multidrug-resistant bacteria. This is therefore a critical step in paving the path for future study. Antibacterial activity is quantified in terms of the minimum inhibitory concentration (MIC), the number of bacteria reduced, the zone of inhibition, the rate of bacteria inactivation, and the proportion of bacterial survival (Amini et al. 2019). Due to the disparate units, it is tedious to conduct impartial comparisons across research. As a result, this becomes one of the many elements of nanoscience that needs standardization. Additionally, coalescing theoretical research and practical application is a tremendous task. Numerous problems remain unresolved. The selectivity and location of action of nanoparticles on the cytomembrane, as well as their transmembrane activity, are of primary importance (Gu et al. 2021). In general, when nanoparticles come into contact with bacteria, they induce oxidative stress mechanisms, enzymatic inhibition, protein deactivation, and gene expression alterations. The antibacterial effect of natural products (NPs) is mediated by oxidative stress caused by reactive oxygen species (ROS) (Rudramurthy et al. 2016;

Dwivedi et al. 2014). ROS are naturally occurring byproducts of cellular oxidative metabolism and play a significant role in the regulation of cell survival and death, as well as cell signaling and differentiation. ROS are produced during aerobic respiration in bacteria, and their formation is precisely controlled by the cell's antioxidant machinery (Li et al. 2012). Excessive generation of ROS disrupts redox equilibrium, resulting in oxidative stress that damages membrane lipids and substantially changes DNA and protein structure (Dwivedi et al. 2014). While endogenous antioxidants may neutralize O_2- and H_2O_2, singlet oxygen (1[O2]) and •OH have been found to cause acute microbial mortality (Zaidi et al. 2017). Gold NPs may generate distinctive ROS, such as hydroxyl radical (·OH) and hydrogen peroxide (H_2O_2) (Wang et al. 2017). In this way, the amount of ROS produced by NPs is determined by the chemical composition of the nanoparticles. Due to their high surface-to-volume ratio, metallic NPs have demonstrated significant antibacterial effectiveness and are presently being explored for the treatment of bacterial infections. An increased generation of ROS, including free radicals, is generally associated with an elevated ratio. It was shown that under UV irradiation, ROS production and metal ion release greatly increased the antibacterial activity of uncoated AuNPs in aqueous solution (365 nm). Antibacterial activity of AuNP against *E. coli, Salmonella typhi, Pseudomonas aeruginosa,* and *Klebsiella pneumoniae* was ascribed to oxidative stress caused by increased intracellular ROS generation (Umamaheswari et al. 2014). Studies evaluating the gold nanoparticles and laser-AuNP combination therapy against *C. pseudotuberculosis* suggested that the mode of action is directly linked to ROS production, which explains why an increase in oxidative stress in microbial cells in the form of vacuole formation is a sign of potent activity. This effect was enhanced by AuNPs-laser, resulting in a dramatic degradation of the integrity of bacterial cell membrane owing to laser light. As a corollary, AuNPs' antibacterial activity is increased by at least onefold. Metal ions produced by metal oxides are absorbed by cells and reach the intracellular compartment, where they interact with functional groups on proteins and nucleic acids, such as carboxyl (–COOH) groups, amino (–NH), and mercapto (–SH) groups (Wang et al. 2017). This kind of interaction changes the cell structure, inhibits enzyme activity, and disrupts bacteria's regular physiological functions. Copper oxide (CuO) nanoparticles have been demonstrated to substantially affect protein expression and to prevent denitrification. Additionally, it has been shown that CuONPs alter the proteins that are involved in nitrogen metabolism, and electron transport. The interaction of gold–superparamagnetic iron oxide nanoparticles with bacterial proteins influences the metabolism and redox system of bacteria (Niemirowicz et al. 2014). Nanoparticles may also enter the bacterial cell through absorption, and thereby releasing metal ions into the extracellular environment. It may also bind to the bacterial cell membrane's negatively charged functional groups. Silver ions (produced from silver NPs) deposited on the cell membrane, for example, have been shown to cause protein coagulation (Jung et al. 2008).

The bactericidal action of graphene oxide/Cu/Ag NPs against *E. coli, P. aeruginosa, K. pneumoniae, S. aureus*, and Methicillin-resistant *S. aureus* (MRSA) was recognized by Jankauskaite and colleagues as a potential synergy between various damaging routes (Jankauskaitė et al. 2018). The NPs' interaction with the cell wall is nonoxidative.

Bacterial cell walls and membranes serve as protective barriers against deterioration in the environment. Different adsorption pathways for NPs are provided by cell membrane components (Lesniak et al. 2013). The cell wall of the gram-negative bacteria is made up of lipoproteins, phospholipids, and lipid polysaccharides (LPS) which form a barrier that enables only specific macromolecules to get through (Zaidi et al. 2017). On the surface of their cell walls, gram-positive bacteria have a strong negative charge. Because teichoic acid is only produced in gram-positive bacteria, LPS binds NPs by providing the negatively charged sections of gram-negative bacteria's cell walls; NPs are distributed throughout the phosphate chain. Gram-positive bacteria have a stronger antibacterial effect as a result (Wang et al. 2017). Yu and colleagues produced hydroxyapatite whisker (HAPw)/zinc oxide (ZnO) NPs and tested them for antibacterial activity against *S. aureus, E. coli,* and *Streptococcus mutans*. According to the researchers, the components and structure of the bacterial cell wall affect antibiotic effectiveness.

The antibiotic action of these NPs may be increased in gram-positive bacteria, and certain components may inhibit NPs from sticking to the bacterial cell barrier. A plethora of excellent research on the antimicrobial activity and toxicity of gold nanoparticles have been reported in recent decades. AuNPs are thus reckoned as composite innovations with attributes that are distinctive from those of conventional materials, and they have enormous potential for using antimicrobial agent against MDR bacteria.

22.3 Combination/Conjugation Strategies of Gold Nanoparticles

Vancomycin when conjugated with AuNPs acts as a reducing and capping agent and thus reduced the antibacterial capacity. Similarly, when conjugated with colistin reduces the minimum inhibitory concentration providing improved efficacy. Conjugation in this case can be achieved by both covalent and noncovalent interaction. It have synergistic effect against *E.coli, S. aureus*, *E. faecium*, *A. baumannii*, and *P. aeruginosa* when conjugated with fluoroquinolone, polymyxin B, ciprofloxacin, ceftazidime, ampicillin, clindamycin, or erythromycin. They also induce reversal of antimicrobial resistance and boost antimicrobial effects when conjugated with antibiotics against vancomycin-resistant *Enterococcus* and methicillin-resistant *S. aureus* (Okkeh et al. 2021; Baptista et al. 2018). AuNPs when conjugated with antimicrobial peptides increase the stability of the peptide and are protected from protease degradation and have the property of water solubility. They can functionally fold itself into amphiphilic α-helical structure and thus reduce its antimicrobial activity. One such example of conjugation is antimicrobial peptide (AMP), $HPA3P^{His}$, when loaded onto a gold nanoparticle-DNA aptamer

(AuNP-Apt) conjugate (AuNP-Apt-HPA3P^{His}) is an effective therapeutic tool against *V. vulnificus* infection. Photoluminescent gold nanodots (AuNDs) when conjugated with surfactin, AuNP when conjugated with Cecropin melittin Magainin-1 Tet-20, and AuNPs when conjugated with antiapoptotic peptide show several healing features in the human body. Antibiotics such as ciprofloxacin when conjugated with NPs result into antimicrobial effect against MDR microorganisms. Peptide named 1018 K6 when conjugated to AuNPs can maintain its antimicrobial activity by folding into a functionally relevant α-helix structure when in the presence of a membranous environment. Cecropin-melittin (CM-SH) has antibacterial properties and thus when conjugated in the surface of AuNPs through Au-S bond, resulted in higher antimicrobial activity and higher stability (Lee et al. 2017; Teixeira et al. 2020). Drugs such as kanamycin are known to be broad spectrum antibiotic. This drug when conjugated with AuNP develops antibacterial activity against multidrug bacterial strains. It provides improved efficacy against both resistant and susceptible grám-positive –negative bacterial strains. kan-AuNP reduces minimum inhibitory concentration in comparison to free kanamycin when tested. They result into bacterial cell lysis. Prior to this AuNPs when conjugated with ampicillin, streptomycin also reduces the minimum inhibitory concentration compared to the free drug counterparts. In addition to this, we also have vancomycin-resistant based on mutation or cell membrane when conjugated with gold NPs. Another example, AuNPs and AuNPs-laser combined therapy against *C. pseudotuberculosis* and the mechanism of action are related with ROS production that results in vacuole formation as a potent activity. Conjugation AuNPs with Carbapenems (Imipenem (Ipm) and Meropenem (Mem)) as a delivering strategy against carbapenem-resistant gram-negative bacteria resulted in significant increase in antibacterial activity against all the tested isolates which were previously isolated from an infected individual. Cefotaxime-conjugated AuNPs are used to target drug-resistant CTX-M-producing bacteria. Concanavalin-A (ConA) directed dextran capped AuNPs (GNPDEX-ConA) when conjugated with methylene blue (MB) (MB@GNPDEX-ConA) and photodynamic therapy (PDT) enhanced the efficacy and selectivity of MB induced killing of MDR microorganisms. Glucosamine-gold nanoparticle-graphene oxide and UV-irradiated GlcN-AuNP-GO were evaluated against *E. coli* and *E. faecalis* (Payne et al. 2016). AuNPs when cross-linked with collagen, gelatin or chitosan provide great efficacy and easy binding with macromolecules. AuNPs inhibit the ATP synthase activity and the multidrug-resistant bacteria are killed by non-ROS-dependent pathway. Flavonoid when coated with gold nanoparticles on colonization of *Enterococcus faecalis* resulted in higher reduction of bacterial counts both in in vitro and in vivo. Trisodium citrate (TSC) and chlorogenic acid (CGA) can synthesize gold nanoparticles (CGA-AuNPs) with different particle sizes indicating antibacterial activity. Green synthesis of dispersed cuboid gold nanoparticles using algae indicated that the synthesized gold nanoparticles had certain potential for certain pathogens like *E.coli* and *Staphylococcus aureus*. Stems of *Tinospora cordifolia* can synthesize gold nanoparticles (AuNPs) and showed greater effect on the biofilm of *Pseudomonas aeruginosa* PAO1. The aqueous extract of C. sativa stem can be used to synthesize gold nanoparticles without adding other reducing, stabilizing or capping agents and upon conjugation it showed bactericidal effects against *Pseudomonas aeruginosa* and *Escherichia coli*. Flavonoid glycerin can be used as a reducing capping agent and upon conjugation with synthetic gold nanoparticles showed antibacterial efficacy against opportunistic MDR-resistant bacterial pathogens. Green synthesis of gold-chitosan hybrid nanoparticles (Au-CS hNPs) showed antibacterial activity (Singh et al. 2020; Riaz et al. 2020; Su et al. 2020). This is an advancing technology to combat against multidrug resistance bacteria where RNA is used to target DNA recognition and the cas enzymes thus destroy the nucleic acids. They enable specific removal of microorganisms and enhance the antimicrobial property by targeting specific resistance gen, undesirable polymorphism, biofilm formation, and virulence. Thus, they can enable the selective and specific removal of microorganisms. Carbapenem-resistant *K. pneumonia* resulted in efficiency of genome editing which facilitates resistance to carbapenems (Lima et al. 2019). These conjugates have shown increased efficacy and enhanced synergistic action by increasing the concentration of drugs at the target site. They reduce the toxicity at cellular level by reducing reactive oxygen species generation. Gold nanoparticles conjugated with aminoglycosides showed efficient antibacterial agents against *Staphylococcus aureus*, *Micrococcus luteus*, *E. coli,* and *Pseudomonas aeruginosa*. A hydrophobic cationic-conjugated gold nanoparticle reduces the minimum inhibitory concentration (MIC) of fluoroquinolone against multidrug-resistant bacteria by 8–16 times. Carbapenem when conjugated with gold nanoparticles resulted in antibacterial activity against MDR bacteria such as Klebsiella *pneumonia*, *Proteus mirabilis,* and *Acinetobacter baumannii.* Imipenem-conjugated gold nanoparticles indicated fourfold decrease in the MIC. Meropenem upon conjugation decreased the MIC by threefold. Mannose-substituted gold metal nanoparticles resulted in high efficiency of antibiotics capped AuNPs compared to free antibiotics against MDR bacteria. Gold NPs conjugated with cephalosporin showed high efficacy of AuNPs-cefaclor against *E. coli* and *S. aureus* (Masri et al. 2019). These conjugates inhibit the biofilm formation, trigger both innate and adaptive host immune response, generate reactive oxygen species, and induce intracellular effects, thus can be greatly used against multidrug-resistant bacteria. They are often used as a vector to carry antimicrobial moiety, hence improving biocidal property. Green synthetic AuNPs can be used as effective nanoantibiotics to combat biofilm-related infections caused by *Pseudomonas aeruginosa*. Combination of synthetic thiol ligands with p-mercaptobenzoic acid adapted Au- nanoparticles, and this conjugate showed antibacterial effects against MDR bacteria. They delayed the development of resistance compared to the antibiotic alone by affecting the transcription process for a wide number of genes. An amino-substituted pyrimidines-conjugated Au nanoparticle had antibacterial activities against MDR bacteria through chelating agents such as Mg^{2+} or Ca^{2+} and thus suppresses the ATPase activities. Combination of gold nanoparticles with 4-Dimethyl aminopyridinium propylthioacetate showed antibacterial effects against *E. coli* (Baptista et al. 2018).

22.4 Different Variants of Gold Nanostructures in Combating MDR

Gold Nanoparticles (AuNPs) Gold nanoparticles are small structures having size of about 0.1–100 nm. Among nanomaterials of other metals, gold nanoparticles have garnered much attention as they are more stable in comparison to the others (Hu et al. 2020). The methods for their synthesis include chemical, physical, and photochemical reduction of gold salts, microwave irradiation, etc. (Chen et al. 2008). Their application ranges from electronics to therapeutic drug deliveries and biosensors. The growing threat of microbial antibiotic resistance calls for novel therapeutic techniques and the use of gold nanoparticles in this scenario is a promising field (Tao 2018). Li et al. (2014) showed that the surface modification of gold nanoparticles displayed high antimicrobial effect against multidrug-resistant gram-positive and gram-negative uropathogens, with very low toxicity to mammalian cells. Khan et al. (2017) developed a complex of Methylene blue and Concanavalin-A (ConA) directed dextran-capped gold nanoparticles (MB@GNPDEX-ConA) that showed efficient light-activated antibacterial potential against a wide range of MDR bacteria like *Escherichia coli* and *Enterobacter cloacae,* with low toxicity to mammalian cells, thus giving a prospective for photodynamic therapy against such MDR bacteria.

Gold Nanorods (AuNR) Gold nanorods are gold nanoparticles that are given an elongated rod-like shape rather than the conventional spherical one. They have been preferred more than the other gold nanoparticles owing to their exceptional optical properties (Pérez-Juste et al. 2005). Due to their elongated shape, the plasmon frequency position increases longitudinally which gives them the advantage of having absorptions along wavelengths ranging from visible to infrared and also gives high quantum yield of fluorescent molecules (Stone et al. 2011). They have been synthesized by methods like seed-mediated chemical growth, electrochemical method, photochemical reduction, lithographic methods, catalytic methods, template methods, etc. (Huang et al. 2009). Because of their optical properties, they have been used in areas like biological sensing and imaging, tracking, drug delivery, photothermal therapies of cancer cells, etc. (Stone et al. 2011). Studies have also shown the potential use of these gold nanorods as antimicrobial agents. Prepared AuNRs with surface charge transformable properties showed compatibility with mammalian cells while displaying net positive charge in blood stream, while they changed to net positive charge in acidic fluids, corresponding to microbial infections. They were able to show that these nanorods had a potent photothermal antibacterial efficiency (Qiao et al. 2020). Shokri et al. (2015) showed in a similar way that when AuNRs were conjugated with antiprotein A antibody and used it as a targeting agent against methicillin-resistant *Staphylococcus aureus* with the application of photothermal therapy. The results showed a significant reduction in the bacterial cell viability both in vitro (82%) and in vivo (73%). Chen et al. (2020) developed a drug delivery system against *S. aureus* by integrating AuNRs, having photothermal potential and BF2b, an antimicrobial peptide. The results confirmed that AuNRs and BF2b conjugate can be used as a potential tool against drug-resistant bacteria.

Gold Nanoclusters (AuNC) AuNCs contain several 100 gold atoms, having size not more than 2 nm. They are a new form of gold nanoparticles that have attracted the attention of many because of their exceptional properties like high biocompatibility, high catalytic activity, long fluorescence lifetime, etc. (Cui et al. 2020). They differ from other larger AuNPs in the fact that they possess fluorescence emission in the near-infrared to visible region, but no surface plasmon resonance absorption in the visible region (Zheng et al. 2017) and in the catalytic activity. These characteristics make them a potent candidate for various applications in the field of fluorescence sensing, bioimaging, drug delivery system, etc. As such, their potential against multidrug-resistant microbes has also been explored. Li et al. (2020) developed highly water-soluble and positively charged AuNCs by using (11-mercaptoundecyl)-NNN-trimethylammonium bromide (MUTAB) and found that they possess antimicrobial activity against a wide range of microbes including the multidrug-resistant ones, causing bacterial cell membrane and DNA damage, with very low toxicity to mammalian cells and no apparent drug resistance. In another study by Zheng et al. (2019), AuNCs were conjugated with daptomycin, an antimicrobial peptide to obtain an antibacterial hybrid with high potential against *Staphylococcus aureus*. The resulting conjugate showed efficiency in conferring bacterial DNA damage and also reducing the drug resistance ability of the bacteria. Also, the fluorescence of the resulting hybrid structure was found to be enhanced.

Gold Nanostar (AuNS) Gold nanostars are a type of polyhedral gold nanostructures having a central core and several uneven arms of various lengths (Jiang et al. 2013). They have gained popularity due to their unique optical properties, like their localized surface plasmon resonance can be manipulated to the near infrared region. This makes them an exceptional candidate for application in various fields, like bioimaging, sensing, nanomedicine, etc. (Niu et al. 2015). The antimicrobial properties of these gold nanostars have also been observed. Rovati et al. (2019) prepared antimicrobial surfaces using gold nanostars by functionalizing them with a monolayer of thiols, which did not change their antibacterial property, but instead improved their stability. An antimicrobial film was prepared by Rossi et al. (2020) by conjugating gold nanorods and gold nanostars with crystal violet, a photosensitizer dye. The one involving gold nanostars showed efficient antimicrobial activity against gram-negative bacteria within 4 h of exposure of light due to the reactive oxygen species generated by the particles. This study thus showed the possibility of developing antimicrobial surfaces of objects that come in contact frequently.

Gold Nanocages (AuNCG) Gold nanocages are a class of gold nanoparticles with hollow interior and porous walls size less than 100 nm. As with the other nanostructures, the localized surface plasmon resonance of these nanocages can also be altered throughout the visible and IR region. These properties make them suitable for use in various biomedical applications including cancer therapy (Skrabalak 2008). Wu et al. (2019) developed a conjugate of gold and silver silica-coated nanocages to study their antibacterial effects. Consequently, this conjugate showed high efficiency against a broad range of gram-positive (*Staphylococcus aureus*) and gram-negative bacteria (*Escherechia coli*) by the synergistic action of the released silver ions and the near IR-induced hyperthermia. The study was demonstrated in vivo as well, on a rat model with wound infection.

22.5 Antimicrobial Drug Resistance Mechanisms in Bacteria

Resistance to antimicrobial agents has been turned into a foremost trace of morbidity and mortality worldwide. With the discovery of antibiotics in the 1900s, the healthcare community thought that the conflict with contagious diseases was won. However, it was soon discovered that a lot of bacteria hold befall resilient to many microbial agents. The comprehensive effect of antibiotic resistance is potentially devastating, threatening to regular earlier improvements against certain infectious diseases to the preantibiotic epoch. Even though nearly all antibiotic-resistant bacteria firstly emerged in hospitals, drug-resistant strains are seemly further ordinary in the community. Studies have disclosed that bacteria acquire resistance mechanisms by two routes, either by mutation or by means of directives provided by their genetic material (Munita and Arias 2016). A proper summary of antimicrobial resistance must perceive between the intrinsic antimicrobial resistance that characteristic bacteria may possess during planktonic growth and the recalcitrance to antimicrobial penetration accessible to bacteria in their protective biofilm mode of growth, familiar to the majority of bacterial infections. Over 300 years back, Antoni van Leeuwenhoek described that vinegar, as an obsolete antimicrobial, killed merely bacteria that inhabited at the outside of a bacterial layer affixed to a surface, what known as "biofilm" today (Donlan and Costerton 2002). Thereupon, biofilm formation can be categorized as a focal implement of antibiotic resistance along with other intrinsic mechanisms such as enzymatic inactivation of antimicrobials, target modification, immunity and bypass, and efflux (Yu et al. 2020).

Biofilm Formation Biofilms are complex architectural colonies of microorganisms within a self-composed extracellular polymeric matrix containing polysaccharides, extracellular DNA, structural proteins cell debris, and nucleic acids (Kassinger and van Hoek 2020). The occurrence of a biofilm involves attachment of cells to a surface, multiplication, maturation, and assembly of polymeric matrix, and colonization of new surfaces. To produce a protective extracellular matrix or "slime", the bacteria will afterward start to guise a monolayer. Initially, matrix formation is dominated by eDNA, and assumed over shortly by polysaccharides and structural proteins. During these stages, microcolonies are formed with significant growth and quorum sensing. The biofilm enlarges in a three-dimensional approach and the attachment is now irrevocable. In the last stage, some cells get detached from the mature biofilm and dispersed into the environment as planktonic cells potentially repeat the cycle of biofilm formation (Kostakioti et al. 2013) (Fig. 22.1).

The biofilm mode of growth confers on the coupled organisms a perceptible decline in antimicrobial vulnerability. The consequence on responsiveness may be intrinsic or acquired (Donlan 2001). Biofilms of selected strains, adhering keenly to hydrophobic silicon rubber, bent an extensive EPS matrix and therewith became not as much of susceptible to gentamycin (Muszanska et al. 2012). However, these strains merely feebly adhered to hydrophilic, polymer brush-coated silicon rubber, parting the bacteria in the notion of being in a planktonic state and suitably had alike responsiveness to gentamycin as planktonic bacteria. In a biofilm, depth-dependent gradients exist, toward the foundation of a biofilm that grounds a stumpy metabolic activity, coupled with an advanced tolerance to antibiotics. These at a low level metabolically dynamic or metabolically sluggish bacterium in a biofilm mode of augmentation are in a so-called "dormant" state. They remain genetically unchanged, but are phenotypic variants. Persisters having 1% of a biofilm population, with their antibiotic tolerance frequently related to chronic infections as they be capable of recoil regrowing when antibiotic pressure is relieved (Cohen et al. 2013). Antibiotic-resistant bacteria have genetically altered to deter antibiotic activity and grow in the poise of antibiotics. Generally, a biofilm has heteroresitance phenomenon, housing subpopulations of bacteria which are liable and tenacious to an antibiotic (El-Halfawy and Valvano 2015). Biofilm research is being paid new awareness due to pervasive allocation of biofilms in diseases and their resistance to plentiful antimicrobial

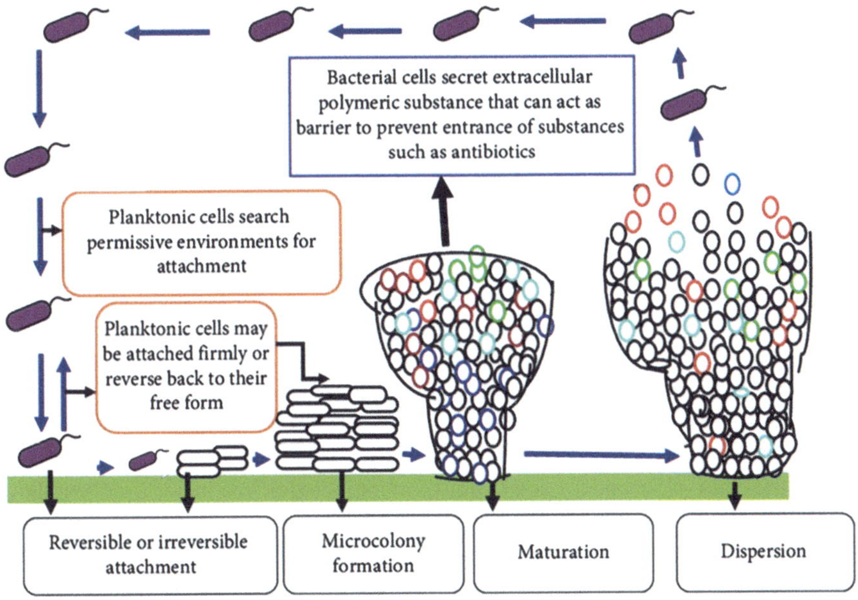

Fig. 22.1 Biofilm formation and structure (Adapted with own modification from Abebe (2020))

treatments. Even after understanding the basic structure and development of this protective shield of bacteria, underlying processes responsible for the transition from planktonic to sessile cells are up in the air.

Enzymatic Inactivation of Antimicrobials Countless bacterial strains own the aptitude to ooze enzymes to protect themselves against antibiotics or other antimicrobials (Egorov et al. 2018). After secretion, these defensive enzymes will add an acetyl or phosphate group maintaining the specificity with the antibiotic which will inactivate it by blocking a specific molecular site as well as induce structural changes to the molecule. These alterations can prevent the antibiotic molecules from binding to specific proteins by steric hindrance, forthwith generating antibiotic resistance. Thousands of enzymes have been identified till date inactivating different classes of antibiotics such as β-lactams, rifampicins, etc. For example, β-lactamase enzyme destroyed the core of β-lactam antibiotic to inactivate serine residues or Zn^{2+}, i.e., the active sites (Tooke et al. 2019).

Target Modification Alteration in the target sites of antibiotics is a frequent means of resistance (Lambert 2005). Different antibiotics have different cellular components including genetic material, cytoplasmic membrane, etc. to kill a bacterium. Antibiotics bind to their targets with lofty resemblance to disapprovingly shape the typical functioning of the target. However, after achieving antibiotic resistance, target sites are capable of being modified to lower the affinity of the antibiotic which makes it unable to interact with its target. The main reason behind changes in target site is spontaneous mutation of a bacterial gene on the chromosome and selection in the manifestation of the antibiotic. For example, mutations in DNA gyrase and RNA polymerase, generating resistance to the quinolones and rifamycins, respectively. In some different cases, attainment of resistance may interest transmission of resistance genes from other organisms through genetic exchange mechanisms including conjugation, transduction, or transformation, e.g., modification of peptidoglycan precursors, etc.

Efflux Pump These are transport proteins that help in the extrusion of substrates which are potentially toxic, such as drugs or compounds or chemicals from the cellular interior to the external environment (Sharma et al. 2019). These proteins are present in both gram-positive and -negative bacteria, and also in the major part of eukaryotic organisms. Effective efflux from prokaryotic as well as eukaryotic cells fanatically modulates the endeavor of a large quantity of antibiotics. In the group of primary active transporters, pumps may be specialized for one substrate or antibiotics of multiple classes, which can be coupled with MDR (multiple drug resistance). Again, the class of secondary active transporters includes five major families; they are- MF (major facilitator), RND (resistance-nodulation-division), SMR (small multidrug resistance), MATE (multidrug and toxic efflux), and ABC (ATP binding cassette). Each of these families consumes the proton motive force as an energy source, except ABC family which hustles the export of substrates through ATP hydrolysis. There are several different efflux pumps which indicate their ancestral origins have been studied in entire bacterial genome. These systems help to pump out the solids out of the cell. Efflux pumps set aside the bacteria to police their internal environment by removing deadly substances together with metabolites, antimicrobial agents, and quorum sensing signal molecules. Efflux pumps may be produced by a single component or by manifold components, with the later life form brings into being exclusively in gram-negative bacteria. Clinically, germane efflux systems in gram-negative bacteria mainly fit into the RND super family and are typically tranquil of a cytoplasmic membrane pump, an outer membrane protein channel, and a periplasmic protein (Routh et al. 2011). The efflux pump systems found most frequently in gram-positive bacteria such as Bmr and Blt in *Bacillus subtitis* and NorA in Staphylococcus aureus; involve MFS and ABC transporters (Kumar et al. 2013). Efflux pumps are prominent in provisions of high efficiency of drug extrusion as well as broad substrate specificities. It is well known that in quorum sensing regulation which regulates the expression of several virulence factors, and also the biofilm differentiation, efflux pumps have been involved. For example, in *Pseudomonas aeruginosa*, QS-dependent lecA::lux expression is found to be down-regulated by a mutation in a probable RND-like efflux pump transporter. Posterior studies have also demonstrated that QS is partly dependent upon efflux. Later studies comprise in addition demonstrated that QS is fairly dependent upon efflux. Hence, an active transport through efflux pumps is necessary for the signal molecule 3OC12-HSL in case of *P. aeruginosa* to diffuse across the cell membrane. Consequently, an increase in efflux pump activity may increase extrusion or intrusion of QS molecule which might have numerous impacts on biofilm formation.

22.6 Intracellular Uptake of Gold Nanostructures

The entry of the NPs into the cells represents the initial step of drug delivery process. Endocytosis is one of the key pathways for cellular uptake of nanoparticles (Behzadi et al. 2017). Recent studies have revealed that cellular uptake in case of GNPs depends upon the size and surface charge of the NPs. In case of AuNPs uptake, shape also plays a vital role (Manzanares and Ceña 2020). Between the spherical AuNPs and rod- or bar-shaped AuNPs, spherical ones are easily absorbed by the cells and rod- or bar-shaped are further easily extruded from the cells (Chithrani and Chan 2007). Additionally, star- and rod-shaped AuNPs taken up by the cells through CME with a significant involvement of CVME for the later in manifestation of FBS, period in its absence, star- and rod-shaped AuNPs are absorbed by macropinocytosis and following an independent pathway respectively (Ding et al. 2018). Again, 15-45 nm spherical AuNPs go through CME process in presence of FBS, whereas follow macropinocytosis in lack of FBS. According to several other studies, 15-30 nm AuNPs, form a complex with DNA, taken up by CME (Wang et al. 2019a, b). However, due to its larger size, 80 nm AuNPs are endocytosed by macropinocytosis. Again, in agreement with the other studies which indicate mainly for nearly all gold nanostructures, the major uptake pathways considered to be receptor-mediated endocytosis

(RME). It is also called as clathrin-mediated endocytosis which forms vesicles, allowing the adsorption of receptor-specific substances only. Once surface proteins (ligands) adsorb on AuNPs, network with receptors on the cell membrane, RME occurs. The ligand-receptor correlation has been exploited in applications to convalesce targeting of NPs. The plasma membrane possesses several specific receptors which play the major role in helping GNPs to get through the outer surface casing of the cell. Inside the cell plasma, a protein is present known as clathrin protein which guides and plays the leading role, the whole process of endocytosis and vesicle formation, respectively. This cytoplasmic protein has a trimeric structure with three clathrin heavy chains and three light chains (Kirchhausen et al. 2014). After adsorbing by the surface proteins, the gold nanostructures first bind to their specific receptors present on the plasma membrane. Some adaptor proteins are also present in the cytoplasm which helps the clathrin protein to get attached with the receptors. Adaptor proteins are made up of four components- AP-I, AP-II, AP-III, AP-IV. The attachment of the clathrin protein with the receptors via adaptor proteins results in the formation of an invagination. As the process is mediated by clathrin protein, it is known as "clathrin-mediated pit." This pit takes the form of a vesicle with the help of dynamin, a GTPase enzyme which converts GTP into GDP, inorganic phosphate, and energy. The vesicle formed is called as "clathrin-coated vesicle" (Fig. 22.2).

$$GTP \rightarrow GDP + iP + \text{energy}$$

This is now followed by the uncoating process. The bond between the adaptor protein and clathrin protein breaks and both go to their former places to repeat the

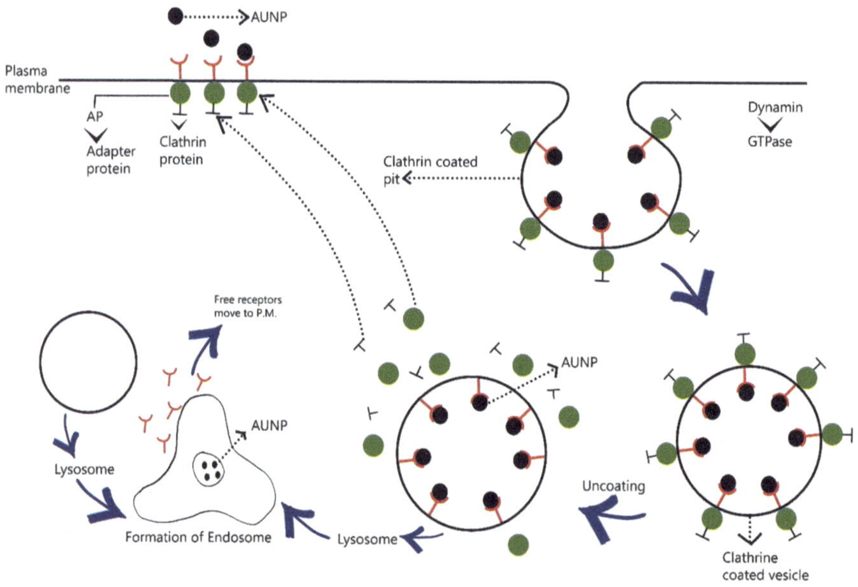

Fig. 22.2 Intracellular uptake of gold nanostructures

cycle. After this, fusion of the vesicle with the lysosome occurs, leading to the formation of secondary lysosome or endosome. This results in the digestion of the gold nanostructures. The receptors get free and take up the initial specific positions (Popova et al. 2013).

22.6.1 *Pharmacokinetic and Pharmacodynamic Characteristics of AuNPs*

Nanoparticles are found as the most effective tool in diagnosis as well in therapeutics because of their adaptable character (Kang et al. 2015). As a drug delivery system, nanoparticles provide a hopeful deal to achieve a drug formulation with modified pharmacokinetic and pharmacodynamic properties (Marcato 2014).

Half-life: The half-life of gold nanoparticles (AuNPs) is inversely proportional to the particle size, which means half-life increases when size of the particle decreases.

Absorption: It is very much necessary to classify the precise physicochemical characteristics which consent to most uptake and growth in secondary target organs of nanoparticlulate drug delivery system. A very less percentage of AuNPs, i.e., about 0.37–0.01% for large size, are absorbed orally. After the oral administration, absorption of AuNPs across intestinal barriers and accumulation in secondary target organs are regulated by its size and the surface charge. In rats, negatively charged AuNPs have a higher absorption than positive particles (0.37% vs. 0.14% respectively; 2.8 nm). Again, inhalation absorption of gold nanoparticles is 0.06–5.5%.

Tissue distribution: After venous injection, 51.3–96.9% and 2–11.4% of AuNPs are distributed to the liver and spleen, respectively. It crosses the blood-brain barrier (BBB) to a lower degree. Gold nanoparticles travel across the human placental barrier in restricted amounts. The placental conveying of AuNPs depends equally on embryonic/placental maturation and size and/or surface composition and modification. However, distribution of these particles to tissues remains for a long time, i.e., more than 6 months.

Metabolism: AuNPs get degraded within the endosomal compartments in case of mammalian cells.

Elimination: The elimination of metallic NPs through urinary and biliary pathways is normally low leading to their enduring accrual in the liver and spleen. Also, NPs exhibit prolonged tissue retention as they do not go through biodegradation into biologically benign components which lead to amplified toxicity. In case of AuNPs-

1. Renal elimination is low: After parenteral injection in rats, 9% of 1-4 nm AuNPs are eliminated within 24 hours.
2. If the size of the particle is less than the threshold value of 5.5 nm, then renal elimination is more efficient than biliary excretion.
3. Again, if the particle size is 13 nm, then biliary excretion is higher than urinary excretion.

22.7 Novel Therapeutic Approach of Gold Nanostructures for Combating MDR: Phage Therapy, Combined Ablation Photothermal Therapy, Photodynamic Therapy

22.7.1 Phage Therapy

Phage therapy is a controversial against bacterial infections till a decade ago when penicillin was discovered in 1928 by Alexander Flaming (Lin et al. 2017; Gaynes 2017). The term phage indicates the bacteriophage or bacteria-specific viruses in short which were found in use to against pathogens to treat them. The phages have the ability to disrupt bacterial metabolism pushing bacterium to lyse. D'Herelle used phages to treat dysentery in the year 1919 under the supervision of Professor Victor-Henri Hutinel, which is supposed to be first attempt to use phages in therapy (Sulakvelidze et al. 2001). The drug resistance bacteria are currently a global threat. It is seen that number of infections increase which become impossible to treat having high morbidity and mortality. In such situation, the use of bacteriophages which can kill bacteria is studied as strategy to combat antibiotic resistance (Altamirano and Barr 2019). The cause of antibiotic resistance can be considered both as natural as well as man-made. It is observed that in microbial population, genes conferring resistant to modern antibiotics are exist. The clinical and industrial overuse of antibiotic is regarded as one of the major cause that accelerates the increase in number of drug-resistant bacteria. The current use of phage therapy has been improved using the knowledge of phage biology, genetics, immunology, and pharmacology (Altamirano and Barr 2019). The utility of phage therapy is analyzed using in vivo animal models. The *P. aeruginosa* phage therapy is very popular now. The mouse model for administration of *P. aeruginosa* is found successful between 80% and 100% when administered intraperitoneally or intranasally. Phage therapy is also found very successful to treat ulcerative lesions and chronic otitis in catfish and dogs (Pires et al. 2015).

Phage therapy is applied for treatment of infected postoperative wounds in cancer patients too which helped in faster cleaning of wounds and healing without deforming scars in comparison to patients treated with antibiotics. It is also mentioned that highly specific bacteriophages are more effective than poly-specific phage cocktails. On the other hand, bacteria create colony very fast on burned skin. They produce a biofilm which is resistant to antibiotics. Again such patients often suffer from sepsis, intoxication, changes in microbiota, and lymphopenia. Here also phage therapy is potential to prevent sepsis (Morozova et al. 2018). The phage therapy is successful to eliminate multiple drug-resistant *P. aeruginosa* or successful skin graft. The report also says that phage therapy might be alternative to human ulcers where antibiotics are common practice.

22.7.2 Combined Ablation Photothermal Therapy

Photothermal ablation therapy employs agents that can absorb near infrared light (NIR) then convert it to heat to ablate cancer cells which are considered to be a highly efficient as anticancer strategy (Wang et al. 2019a, b). The ideal photothermal ablation is able to eradicate primary tumor as well as induce antitumor immunity by activating the immune system that prevents tumor recurrence. The thermal ablation has advantage over conventional methods like flexibility, minimal invasiveness, and cost-effective (Ashikbayeva et al. 2019). The search of therapeutic agents for photothermal therapy is a challenge. However, the gold nanoparticles (GNPs) high biocompatibility and its ability to be injected intravenously accumulate in cancer cells efficiently because of its high permeability and retention effect. The GNPs that are available in various forms including nanospheres, nanoshells, nanorods, nanocages, nanostars, etc. are found to be useful in photothermal treatment (Li et al. 2014). The functionalized GNPs are also effective against both gram-negative and gram-positive uropathogens as well as multidrug-resistant bacteria (Kang et al. 2020). GNPs have unique ability to convert NIR into heat energy which can induce hyperthermia due to surface plasmon resonance (SPR). Hence it is useful for photothermal ablation (Kang et al. 2020). Hybrid magnetic gold nanoparticles (HNPs) are found effective as catalyst to treat colorectal cancer liver metastases (CRLM) using photothermal sensitization (Yang 2017). The photothermal ablation (PTA) is able to perform in three basic modes such as (a) using only light, (b) light along with molecule for photothermal conversion, and (c) light along with metal nanostructures for photothermal conversion. Due to high efficiency of metal nanostructures in photothermal conversion, it is a good choice as photothermal agents. However, they are not easy to deliver in the cancer sites as well as possess potential toxicity. Noble metals like silver, gold nanoparticles are biocompatible with low toxicity. So they are most discussed as PTA agents. To be suitable PTA agent, metal nanostructure must possess (a) small size, (b) spherical or nearly spherical shape, and (c) strong, narrow NIR absorption. For intracellular uptake, the nanostructure shall be in optimal size range that minimizes friction too. Smaller sizes also help in tunable SPR effects which instead help in strong NIR absorption (Nadejda and Zhong 2009).

22.7.3 Photodynamic Therapy

The photodynamic therapy (PDT) is a technique that requires light, photosensitizer, and tissue oxygen (Jelveh and Chithrani 2011). On irradiation of photosensitizer, it reacts with ground state molecular oxygen ($_3O_2$) to form excited singlet oxygen ($^1O_2$) (Stacey and Pope 2013) and other reactive oxygen species (ROS) that can destroy target tissues. In the process, the photosensitizers are excited with light of suitable wavelength to generate ROS which help in inducing apoptosis or necrosis. To get the effective results, the photosensitizer is administered in the target sites followed by activation by light. The PDT got its regulatory approval using hematoporphyrin derivative to treat cancer in advanced stage of lung, digestive tract, genitourinary tract in Netherland, Japan, Germany, USA, and Canada in 1993 (Dougherty et al. 1998). Recently gold nanoparticles (GNPs) are found to be preferable delivery agents for PDT drugs. GNPs-based delivery offers enhanced permeability and retention (EPR) effect providing accumulation of drugs in tumor sites. Russell and coworkers first reported the self-assembly of GNPs with photosensitizers (Hone et al. 2002). Various studies have revealed that GNPs possess low toxicity and biocompatibility that make it applicable for biomedical applications. More often GNPs are combined with photosensitizer as has advantages like enhanced production of singlet oxygen as well as other ROS. Unlike PTA agents, the use of GNPs is not limited to spherical-shaped nanoparticles but other nanomaterials like nanorods, nanostars, nanoclusters are also reported (Calavia et al. 2018). It is emphasized that we need to monitor the levels of tissue oxygenation during PDT for clear understanding of therapeutic application. As photochemical consumption and vascular inhibition deplete the oxygen levels, it is believed that PDT efficacy also reduced accordingly.

22.8 Future Perspectives and Conclusions

The nanomedicine era has hardly begun. Nanotechnology has a ton of potential in terms of health and biological science. Due to their physicochemical properties, gold nanoparticles are developed as therapeutic and diagnostic agents both in vitro and in vivo among the various inorganic nanoparticles available today. Because of their possible applications in a variety of fields, including chemical, biomedical, and optical fields, AuNPs have taken a big hit. They are widely acknowledged as biosafety and have been actively used as drug carriers, cosmetics, surgical fillings, and antimicrobial materials (Selvaraj et al. 2010). Although there are many AuNPs antibacterial-based methods, and the majority of them appear to be successful, a plethora of drug researches have investigated AuNPs as drug carriers, while the side effects are under consideration. Until AuNPs may be used in clinical environments, their toxicity to species must be thoroughly understood. AuNPs have been shown to have cytotoxic effects both in vitro and in vivo, with various toxic effects identified and associated with AuNPs size, shape, doses, sampling points, surface coating and functionalization, and cell lines or animal models. AuNPs have demonstrated an array of effectiveness in a number of fields, but they are still in the initial phages of development and are not yet suitable for use in theranostics. The constraint to translational research on these systems in the future will be a dearth of deeper understanding of the interactions between AuNPs, biomacromolecules and the immune system, homogeneity of material preparations, and a lack of information about pertinent biomarkers, along with the biocompatibility concerns. Green AuNPs synthesis, multimodality bioimaging analyses, and theranostics are all pertinent areas of research that are of interest.

Gold nanoparticles interact with bacteria in a variety of ways, allowing them to bypass their resistance mechanisms. Researchers have discovered that these resistant bacteria are often vulnerable against these new therapeutic techniques, ranging from mechanical stress caused by gold nanoparticles acting on bacterial cell walls to hyperthermia and ROS produced by photoactivation. The mechanisms of action of the various antibacterial gold nanoparticles are still being investigated and understood, but they are providing new hope in an era when bacterial resistance continues to rise despite almost any other available treatment choice. Long-term biosafety and experimental observations in vivo and in vitro are critical for transferring this novel therapeutics into clinical practice and guiding their development. Another field, where we believe there is room for progress, is the standardization of experiments in order to obtain comparable studies. We expect to see secure and successful nano-gold-based therapeutics in the foreseeable future, where gold's properties are thoroughly used to eliminate the world's most deadly superbugs forthwith.

References

Abebe, GM (2020) The role of bacterial biofilm in antibiotic resistance and food contamination. Int J Microbiol, Volume 2020, Article ID 1705814

Altamirano FLG, Barr J (2019) Phage therapy in the postbiotic era. Clin Microbiol Rev 32:e00066-18

Amini A, Kamali M, Amini B, Najafi A, Narmani A, Hasani L et al (2019) Bio-barcode technology for detection of Staphylococcus aureus protein a based on gold and iron nanoparticles. Int J Biol Macromol 124:1256–1263

Ashikbayeva Z, Tosi D, Balmassov D, Schena E, Saccomandi P, Inglexakis V (2019) Application of nanoparticles and nanomaterials in thermal ablation therapy of cancer. Nano 9:1195

Baptista PV, Mc Cusker MP, Carvalho A, Ferreira DA, Mohan NM, Martins M, Fernandes AR (2018) Nano-strategies to fight multidrug-resistant bacteria—"A Battle of the Titans". Front Microbiol 9:1441

Behzadi S et al (2017) Cellular uptake of nanoparticles: journey inside the cell. Chem Soc Rev 46(14):4218–4244

Calavia PG, Bruce G, Gracia L, Russell DA (2018) Russel, Photosensitiser-gold nanoparticle conjugates for photodynamic therapy of cancer. Photochem Photobiol Sci 17(11):1534–1552

Chen PC, Mwakwari SC, Oyelere AK (2008) Gold nanoparticles: from nanomedicine to nanosensing. Nanotechnol Sci Appl 1:45–65

Chen J et al (2020) Integration of antimicrobial peptides and gold nanorods for bimodal antibacterial applications. Biomater Sci 8:4447–4457

Chithrani DB, Chan WCW (2007) Elucidating the mechanism of cellular uptake and removal of protein-coated gold nanoparticles of different sizes and shapes. Nano Lett 7(6):1542–1550

Cohen NR, Lobritz MA, Collins JJ (2013) Microbial persistence and the road to drug resistance. Cell Host Microbe 13(6):632–642

Cui H et al (2020) Development of gold nanoclusters: from preparation to applications in the field of biomedicine. J Mater Chem C 8:14312–14333

Ding L, Yao C, Yin X, Li C, Huang Y, Wu M, Wang B, Guo X, Wang Y, Wu M (2018) Size, shape, and protein corona determine cellular uptake and removal mechanisms of gold nanoparticles. Small 14(42)

Donlan RM (2001) Biofilm formation: a clinically relevant microbiological process. Clin Infect Dis 33(8):1387–1392

Donlan RM, Costerton JW (2002) Biofilms: survival mechanisms of clinically relevant microorganisms. Clin Microbiol Rev 15(2):167–193

Dougherty TJ, Gomer CJ, Henderson BW, Jori G, Kessel D, Korbelik M, Moan J, Peng Q (1998) Photodynamic therapy. J Natl Cancer Inst 90:889–905

Dreaden EC, Alkilany AM, Huang X, Murphy CJ, El-Sayed MA (2012) The golden age: gold nanoparticles for biomedicine. Chem Soc Rev 41(7):2740–2779

Dwivedi S, Wahab R, Khan F, Mishra YK, Musarrat J, Al-Khedhairy AA (2014) Reactive oxygen species mediated bacterial biofilm inhibition via zinc oxide nanoparticles and their statistical determination. PLoS One 9(11):e111289

Dykman LA, Khlebtsov NG (2011) Gold nanoparticles in biology and medicine: recent advances and prospects. Acta Nat 3(2):34–55

Egorov AM, Ulyashova MM, Rubtsova MY (2018) Bacterial enzymes and antibiotic resistance. Acta-Naturae 10(4):33–48

El-Halfawy OM, Valvano MA (2015) Antimicrobial heteroresistance: an emerging field in need of clarity. Clin Microbiol Rev 28(1):191–207

Gaynes R (2017) The discovery of penicillin- New insights after more than 75 years of clinical use. Emerg Infect Dis 23:849–853

Gu Y, Chen C, Mao Z, Bachman H, Becker R, Rufo J, Huang TJ (2021) Acoustofluidic centrifuge for nanoparticle enrichment and separation. Sci Adv 7(1):eabc0467

Hammer B, Norskov JK (1995) Why gold is the noblest of all the metals. Nature 376(6537):238–240

Hone DC, Walker PI, Evans-Gowing R, FitzGerald S, Beeby A, Chambrier I, Cook M, Russell DA (2002) Generation of cytotoxic singlet oxygen via phthalocyanine stabilized gold nanoparticles: a potential delivery vehicle for photodynamic therapy. Langmuir 18(8):2985–2987

Hu X, Zhang Y, Ding T, Liu J, Zhao H (2020) Multifunctional gold nanoparticles: a novel nanomaterial for various medical applications and biological activities. Front Bioeng Biotechnol 8:990

Huang X, Neretina S, El-Sayed MA (2009) Gold nanorods: from synthesis and properties to biological and biomedical applications. Adv Mater 21(48):4880–4910

Jankauskaitė V, Lozovskis P, Valeika V, Vitkauskienė A (2018) Graphene oxide and metal particles nanocomposites for inhibition of pathogenic bacteria strains. In: Proceedings of the 7th international conference on advanced material and systems, pp 117–122

Jelveh S, Chithrani DB (2011) Gold nanostructures as a platform for combinational therapy in future cancer therapeutics. Cancers 3:1082–1110

Jiang L, Tang Y, Liow C, Wu J, Sun Y, Jiang Y, Dong Z, Li S, Dravid VP, Chen X (2013) Synthesis of fivefold stellate polyhedral gold nanoparticles with {110}-facets via a seed-mediated growth method. Small 9(5):705–710

Jung HS, Lee JK, Lee S, Hong KS, Shin H (2008) Acid adsorption on TiO2 nanoparticles-an electrochemical properties study. J Phys Chem C 112(22):8476–8480

Kang H, Mintri S, Menon AV, Lee HY, Choi HS, Kim J (2015) Pharmacokinetics, pharmacodynamics and toxicology of theranostic nanoparticles. Nanoscale 7(45):18848–18862

Kang MS, Lee SY, Kim KS, Han D (2020) State of the art biocompatible gold nanoparticles for cancer theragnosis. Pharmaceutics 12:701

Kassinger SJ, van Hoek ML (2020) Biofilm architecture: an emerging synthetic biology target. Synthetic and systems. Biotechnology 13:5(1)–5(10

Khan S, Khan SN, Meena R, Dar AM, Pal R, Khan AU (2017) Photoinactivation of multidrug-resistant bacteria by monomeric methylene blue-conjugated gold nanoparticles. J Photochem Photobiol B Biol 174:150–161

Kirchhausen T, Owen D, Harrison SC (2014) Molecular structure, function, and dynamics of Clathrin-mediated membrane traffic. Cold Spring Harb Perspect Biol 6(5):a016725

Kostakioti M, Hadjifrangiskou M, Hultgren SJ (2013) Bacterial biofilms: development, dispersal, and therapeutic strategies in the dawn of the postantibiotic era. Cold Spring Harb Persepect Med 3(4):a010306

Kumar S et al (2013) Modulation of bacterial multidrug resistance efflux pumps of the major facilitator superfamily. Int J Bacteriol 2013:204141

Lambert PA (2005) Bacterial resistance to antibiotics: modified target sites. Adv Drug Deliv Rev 57(10):1471–1485

Lee B, Park J, Ryu M, Kim S, Joo M, Yeom JH, Kim S, Park Y, Lee K, Bae J (2017) Antimicrobial peptide-loaded gold nanoparticle-DNA aptamer conjugates as highly effective antibacterial therapeutics against *Vibrio vulnificus*. Sci Rep 7(1):1–10

Lee NY, Ko WC, Hsueh PR (2019) Nanoparticles in the treatment of infections caused by multidrug-resistant organisms. Front Pharmacol 10:1153

Lesniak A, Salvati A, Santos-Martinez MJ, Radomski MW, Dawson KA, Åberg C (2013) Nanoparticle adhesion to the cell membrane and its effect on nanoparticle uptake efficiency. J Am Chem Soc 135(4):1438–1444

Li Z, Barnes JC, Bosoy A, Stoddart JF, Zink JI (2012) Mesoporous silica nanoparticles in biomedical applications. Chem Soc Rev 41(7):2590–2605. 2985–2987

Li X, Robinson SM, Gupta A, Saha K, Jiang Z, Moyano DF, Sahar A, Riley MA, Rotello M (2014) Functional gold nanoparticles as potent antimicrobial agents against multi-drug-resistant bacteria. ACS Nano 8:10682–10686

Li Y, Zhen J, Tian Q, Shen C, Zhang L, Yang K, Shang L (2020) One step synthesis of positively charged gold nanoclusters as effective antimicrobial nanoagents against multidrug-resistant bacteria and biofilms. J Colloid Interface Sci 569:235–243. https://doi.org/10.1016/j.jcis.2020.02.084 PMID: 32113021

Lima R, Del Fiol FS, Balcão VM (2019) Prospects for the use of new technologies to combat multidrug-resistant bacteria. Front Pharmacol 10:692

Lin DM, Koskella B, Lin HC (2017) Phage therapy: an alternative to antibiotics in the age of multidrug resistance. World J Gastrointest Pharmacol Ther 8(3):162–173

Manzanares D, Ceña V (2020) Endocytosis: the nanoparticle and submicron. Pharmaceutics 12(4):371

Marcato PD (2014) Pharmacokinetics and pharmacodynamics of nanomaterials. In: Duran N, Guterres S, Alves O (eds) Nanotoxicology, nanomedicine and nanotoxicology. Springer, New York, NY

Masri A, Anwar A, Khan NA, Siddiqui R (2019) The use of nanomedicine for targeted therapy against bacterial infections. Antibiotics 8(4):260

Morozova VV, Vlassov VV, Tikunova NV (2018) Applications of bacteriophages in the treatment of localized infections in humans. Front Microbiol 9:1696

Munita JM, Arias CA (2016) Mechanisms of antibiotic resistance. ASM Sci

Muszanska AK, Nejadnik MR, Chen Y, van den Heuvel ER, Busscher HJ, van der Mei HC, Norde W (2012) Bacterial adhesion forces with substratum surfaces and the susceptibility of biofilms to antibiotics. Antimicrob Agents Chemother 56(9):4961–4964

Nadejda R, Zhong Z (2009) Photothermal ablation therapy for cancer based on metal nanostructures. Sci China Ser B-Chem 52:1559–1575

Niemirowicz K, Swiecicka I, Wilczewska AZ, Misztalewska I, Kalska-Szostko B, Bienias K et al (2014) Gold-functionalized magnetic nanoparticles restrict growth of Pseudomonas aeruginosa. Int J Nanomedicine 9:2217

Nikaido H (2009) Multidrug resistance in bacteria. Annu Rev Biochem 78:119–146

Niu W, Chua YAA, Zhang W, Huang H, Lu X (2015) Highly symmetric gold nanostars: crystallographic control and surface-enhanced raman scattering property. J Am Chem Soc 137(33):10460–10463

Okkeh M, Bloise N, Restivo E, De Vita L, Pallavicini P, Visai L (2021) Gold nanoparticles: can they be the next magic bullet for multidrug-resistant bacteria? Nano 11(2):312

Payne JN, Waghwani HK, Connor MG, Hamilton W, Tockstein S, Moolani H, Chavda F, Badwaik V, Lawrenz MB, Dakshinamurthy R (2016) Novel synthesis of kanamycin-conjugated gold nanoparticles with potent antibacterial activity. Front Microbiol 7:607

Pérez-Juste J, Pastoriza-Santosa I, Liz-Marzán LM, Mulvaney P (2005) Gold nanorods: synthesis, characterization and applications. Coord Chem Rev 249:1870–1901

Pires DP, Boas DV, Silankovra S, Azeredo J (2015) Phage therapy: a step forward in the treatment of pseudomonas aeruginosa infections. J Virol 89:7449–7456

Popova NV, Deyev IE, Petrenko AG (2013) Clathrin-mediated endocytosis and adaptor proteins. Acta Nat 5(3):62–73

Qiao Z, Yao Y, Song S, Yin M, Yang M, Yan D, Yanga L, Luo J (2020) Gold nanorods with surface charge switchable activities for enhanced photothermal killing of bacteria and eradication of biofilm. J Mater Chem B 8:3138–3149

Riaz S, Fatima Rana N, Hussain I, Tanweer T, Nawaz A, Menaa F, Janjua HA, Alam T, Batool A, Naeem A, Hameed M (2020) Effect of flavonoid-coated gold nanoparticles on bacterial colonization in mice organs. Nano 10(9):1769

Rossi F, Khoo EH, Su X, Thanh NTK (2020) Study of the effect of anisotropic gold nanoparticles on plasmonic coupling with a photosensitizer for antimicrobial film. ACS Appl Bio Mater 3(1):315–326

Routh MD, Zalucki Y, Long F, Zhang Q, Shafer WM, Yu EW (2011) Efflux pumps of the resistance-nodulation-division family: a perspective of their structure, function and regulation in gram-negative bacteria. Adv Enzymol Relat Areas Mol Biol 77:109–146

Rovati D, Albini B, Galinetto P, Grisoli P, Bassi B, Pallavicini P, Dacarro G, Taglietti A (2019) High stability thiol-coated gold nanostars monolayers with photo-thermal antibacterial activity and wettability control. Nano 9(9):1288

Rudramurthy GR, Swamy MK, Sinniah UR, Ghasemzadeh A (2016) Nanoparticles: alternatives against drug-resistant pathogenic microbes. Molecules 21(7):836

Selvaraj V, Grace AN, Alagar M, Hamerton I (2010) Antimicrobial and anticancer efficacy of antineoplastic agent capped gold nanoparticles. J Biomed Nanotechnol 6(2):129–137

Shaikh S, Nazam N, Rizvi SMD, Ahmad K, Baig MH, Lee EJ, Choi I (2019) Mechanistic insights into the antimicrobial actions of metallic nanoparticles and their implications for multidrug resistance. Int J Mol Sci 20(10):2468

Sharma A, Gupta VK, Pathania R (2019) Efflux pump inhibitors for bacterial pathogens: from bench to bedside. Indian J Med Res 149(2):129–145

Shokri R, Salouti M, Zanjani RS (2015) Anti protein A antibody-gold nanorods conjugate: a targeting agent for selective killing of methicillin-resistant *Staphylococcus aureus* using photothermal therapy method. J Microbiol 53(2):116–121

Singh R, Patil S, Singh N, Gupta S (2017) Dual functionality nano bioconjugates targeting intracellular bacteria in cancer cells with enhanced antimicrobial activity. Sci Rep 7(1):1–10

Singh A, Gautam PK, Verma A, Singh V, Shivapriya PM, Shivalkar S, Sahoo AK, Samanta SK (2020) Green synthesis of metallic nanoparticles as effective alternatives to treat antibiotics-resistant bacterial infections: a review. Biotechnol Rep 25:e00427

Skrabalak SE (2008) Gold nanocages: synthesis, properties, and applications. Acc Chem Res 41(12):1587–1595

Slavin YN, Asnis J, Häfeli UO, Bach H (2017) Metal nanoparticles: understanding the mechanisms behind antibacterial activity. J Nanobiotechnol 15(1):1–20

Stacey OJ, Pope SJA (2013) New avenues in the design and potential application of metal complexes for photodynamic therapy. RSC Adv 3:25550

Stone J, Jackson S, Wright D (2011) Biological applications of gold. WIREs Nanomed Nanobiotechnol 3:100–109

Su C, Huang K, Li HH, Lu YG, Zheng DL (2020) Antibacterial properties of functionalized gold nanoparticles and their application in oral biology. J Nanomater 2020:5616379

Sulakvelidze A, Alavidze Z, Glenn Morris J Jr (2001) Bacteriophage therapy. Antimicrob Agents Chemother 45:649–659

Tao C (2018) Antimicrobial activity and toxicity of gold nanoparticles: research progress, challenges and prospects. Lett Appl Microbiol 67(6):537–543

Teixeira MC, Carbone C, Sousa MC, Espina M, Garcia ML, Sanchez-Lopez E, Souto EB (2020) Nanomedicines for the delivery of antimicrobial peptides (amps). Nano 10(3):560

Tooke CL, Hinchliffe P, Bragginton EC, Colenso CK, Hirvonen VH, Takebayashi Y, Spencer J (2019) β-Lactamases and β-lactamase inhibitors in the 21st century. J Mol Biol 431(18):3472–3500

Umamaheswari K, Baskar R, Chandru K, Rajendiran N, Chandirasekar S (2014) Antibacterial activity of gold nanoparticles and their toxicity assessment. BMC Infect Dis 14(3):1–1

Wang L, Hu C, Shao L (2017) The antimicrobial activity of nanoparticles: present situation and prospects for the future. Int J Nanomedicine 12:1227

Wang H, Chen B, He M, Li X, Chen P, Hu B (2019a) Study on uptake of gold nanoparticles by single cells using droplet microfluidic chip-inductively coupled plasma mass spectrometry. Talanta 200:398–407

Wang X, Li J, Kawazoe N, Chen G (2019b) Photothermal ablation of cancer cells by albumin-modified gold nanorods and activation of dendritic cells. Materials 12:31

Wu S, Li A, Zhao X, Zhang C, Yu B, Zhao N, Xu F-J (2019) Silica-coated golsilver nanocages as photothermal antibacterial agents for combined anti-infective therapy. ACS Appl Mater Interfaces 11:19

Yang X (2017) Enhancing photothermal ablation of colorectal liver metastases with targeted hybrid nanoparticles. Radiology 285:691–701

Yang T, Zhao YL, Tong Y, Jiao ZB, Wei J, Cai JX et al (2018) Multicomponent intermetallic nanoparticles and superb mechanical behaviors of complex alloys. Science 362(6417):933–937

Yu T, Jiang G, Gao R, Chen G, Ren Y, Liu J (2020) Circumventing antimicrobial-resistance and preventing its development in novel, bacterial infection-control strategies Informa. Expert Opin Drug Deliv 17(8):1151–1164

Zaidi S, Misba L, Khan AU (2017) Nano-therapeutics: a revolution in infection control in post antibiotic era. Nanomedicine 13(7):2281–2301

Zheng Y, Lai L, Liu W, Jiang H, Wang X (2017) Recent advances in biomedical applications of fluorescent gold nanoclusters. Adv Colloid Interf Sci 242:1–16

Zheng Y, Liu W, Chen Y, Li C, Jiang H, Wang X (2019) Conjugating gold nanoclusters and antimicrobial peptides: from aggregation-induced emission to antibacterial synergy. J Colloid Interface Sci 546:1–10

Chapter 23
Carbon Nanoparticles: A Potential Cost-Effective Approach to Counter Antimicrobial Resistance

Afroza Khanam and Firdous Ahmad Khan

23.1 Introduction

Antimicrobial resistance (AMR) is one of the most important problems the world is facing at the moment. The World Health Organization (WHO) has declared AMR as "one of the top 10 global public health threats facing humanity" (WHO n.d.). Infectious diseases are a major cause of deaths worldwide and the emergence of multidrug resistant pathogens results in increased death rates and significantly higher therapeutic costs (Dadgostar 2019). The list of these pathogens is ever-growing and no longer limited to specific geographical locations or healthcare settings (Prestinaci et al. 2015). It has been predicted that by 2050, 300 million additional deaths would occur and an extra US$100 trillion would be incurred as a result of this issue (Dong et al. 2020). To address this problem, there is a global push to search for new antimicrobial agents that are resistance-free, and ideally cost-effective and of natural origin.

Nanoparticles have emerged as one of the effective alternatives to traditional antimicrobials. Apart from being effective, nanoparticles offer several advantages over other antimicrobials. Nanoparticles are relatively safer and less vulnerable to development of resistance due to their multifaceted actions and the inability of microbes to build up multiple gene mutations simultaneously (Plank 2009; Ruddaraju et al. 2020). Some examples of nanoparticles that have been tested for their antimicrobial actions include silver, silver oxide, titanium dioxide, zinc oxide, gold, calcium oxide, silica, copper oxide, magnesium oxide, and carbon-based nanomaterials (Azam et al. 2012; Besinis et al. 2014; Emami-Karvani and Chehrazi 2011; Usman et al. 2013; Chen et al. 2013; Pal et al. 2007; Zarei et al. 2014; Cataldo and Da Ros 2008). Besides being cost-effective and relatively easy to synthesize,

A. Khanam (✉) · F. A. Khan
Department of Large Animal Medicine and Surgery, School of Veterinary Medicine, St. George's University, True Blue, Grenada, West Indies

N. Akhtar et al. (eds.), *Emerging Modalities in Mitigation of Antimicrobial Resistance*,
https://doi.org/10.1007/978-3-030-84126-3_23

carbon nanoparticles have a high surface-to-volume ratio and unique chemical and physical properties that make them attractive potential alternatives to traditional antimicrobials (Maleki Dizaj et al. 2015).

23.2 Antibacterial Activity of Carbon Nanoparticles

Carbon nanoparticles have been shown to have broad-spectrum antimicrobial activities (Table 23.1). In contrast to metal nanoparticles that can potentially accumulate in animal tissues and cause cytotoxicity, carbon nanoparticles, especially those derived from natural sources, are safer alternatives to conventional antibiotics. Carbon nanoparticles derived from date pits have been tested for antibacterial activity against *Staphylococcus epidermidis*, *Serratia marcescens*, *Mycobacterium smegmatis*, and *Escherichia coli* (*E. coli*). The nanoparticles inhibited the growth of all four bacteria at a concentration of 0.5 mg/mL (Altaikyzy et al. 2018). Similarly, carbon nanoparticles isolated from kitchen soot (derived from wood and coconut husk) were found to be effective against *Proteus refrigere*, *Pseudomonas aeruginosa*, *Staphylococcus aureus* (*S. aureus*), and *Streptococcus haemolyticus* (*S. haemolyticus*) (Varghese et al. 2013). Another study by the same group demonstrated that CNPs and CNPs encapsulated in functionally modified β-cyclodextrin effectively inhibited the growth of *S. aureus*, *S. haemolyticus*, *Klebsiella pneumoniae* (*K. pneumoniae*), and *E. coli* (Varghese and Kuriakose 2016). In yet another study by a different group, carbon nanoparticles derived from butter and mustard oil were shown to be effective against *K. pneumoniae* (Mohanty et al. 2007).

Another group of carbon nanoparticles, known as carbon dots (CDots; dimensions less than 10 nm) have been shown to have antimicrobial activities (Al Awak et al. 2017; Meziani et al. 2016). It is believed that as the size of carbon particles reduces, their antimicrobial activities increase on account of their larger surface area per unit volume and high extent of atoms in the surface and close to surface layers (Dong et al. 2020). Since the effects of CDots are mediated primarily through photodynamic inactivation (PDI), the use of CDots is a minimally invasive and safe option suited for multiple applications (Maisch 2009; Yin et al. 2015). Carbon dots functionalized with 2,2-(ethylenedioxy)bis(ethylamine) effectively inhibited growth of *E. coli* both in suspensions and on culture plates (Meziani et al. 2016). Other microbes that were shown to be susceptible to the antimicrobial actions of CDots include *S. aureus*, *Bacillus subtilis*, and *methicillin-resistant S. aureus* (MRSA) (Li et al. 2018; Hou et al. 2017). In another study, CDots embedded in polydimethylsiloxane and polyurethane inhibited the growth of *S. aureus*, *E. coli*, and *K. pneumoniae* (Kovacova et al. 2018). Efforts to enhance the antimicrobial activity of CDots by surface modification to increase their affinity to bind with bacteria have met with success. Examples include spermidine-bound CDots (Li et al. 2016) and hybrid CDots with metal and metal oxides (Yan et al. 2019; Liu et al. 2017; Zhang et al. 2018; De et al. 2015). Another approach that has been tested included a combination of CDots and

Table 23.1 A summary of published studies on the antibacterial activity of carbon-based nanoparticles

Type of nanoparticles	Carbon source	Effective against microorganisms	References
Carbon nanoparticles	Date pits	*Staphylococcus epidermidis*, *Serratia marcescens*, *Mycobacterium smegmatis*, and *E. coli*	Altaikyzy et al. (2018)
Carbon nanoparticles	Kitchen soot (wood and coconut husk)	*Proteus refrigere*, *Pseudomonas aeruginosa (P. aeruginosa)*, *S. aureus*, and *S. haemolyticus*	Varghese et al. (2013)
Carbon nanoparticles and carbon nanoparticles encapsulated in β-cyclodextrin with {5-[4-(dimethylamino) benzylidene]-4-oxo-2-thioxo-1, 3-thiazolidin-3-yl} acetic acid	Kitchen soot	*S. aureus*, *S. haemolyticus*, *K. pneumoniae*, and *E. coli*	Varghese and Kuriakose (2016)
Carbon nanoparticles	Butter and mustard oil	*K. pneumoniae*	Mohanty et al. (2007)
Carbon dots functionalized with 2,2-(ethylenedioxy) bis(ethylamine)	Commercially supplied carbon nanopowder	*E. coli*	Meziani et al. (2016)
Carbon dots	Vitamin C	*S. aureus*, *Bacillus subtilis*, *Bacillus sp. WL-6*, *E. coli*, and ampicillin-resistant *E. coli*	Li et al. (2018)
Graphitic carbon dots	Ciprofloxacin hydrochloride	*S. aureus* and *E. coli*	Hou et al. (2017)
Carbon dots capped with polyurethane	Polyoxyethylene-polyoxypropylene-polyoxyethylene, Pluronic 68	*S. aureus* and *E. coli*	Kovacova et al. (2018)
Spermidine-capped carbon quantum dots	Ammonium citrate	*E. coli*, *S. aureus*, *B. subtilis*, *P. aeruginosa*, and methicillin-resistant *S. aureus* (MRSA)	Li et al. (2016)
Carbon dot decorated titanium dioxide	Graphite rods	*S. aureus* and *E. coli*	Yan et al. (2019)

(continued)

Table 23.1 (continued)

Type of nanoparticles	Carbon source	Effective against microorganisms	References
ZnO/graphene quantum dot nanocomposites	Citric acid	*S. aureus* and *E. coli*	Liu et al. (2017)
Carbon dots decorated $Na_2W_4O_{13}$ composite with WO_3	Ethanol	*E. coli*	Zhang et al. (2018)
OMMT-carbon dot reduced Cu_2O nanohybrid	Banana	*S. aureus*, *Bacillus subtilis*, *K. pneumoniae*, and *P. aeruginosa*	De et al. (2015)
2, 2′-(ethylenedioxy) bis(ethylamine) (EDA) capped carbon dots	Carbon nanopowders	*E.coli* and *Bacillus subtilis*	Dong et al. (2017)
Carbon dots functionalized with ampicillin	Citric acid	*E. coli*	Jijie et al. (2018)
Carbon dots as carrier of ciprofloxacin hydrochloride	Gum arabic	*Bacillus subtilis*, *S. aureus*, *E coli*, and *P. aeruginosa*	Thakur et al. (2014)
Carbon dots	Tamarind	*K. pneumoniae*, *E. coli*, *S. aureus*, and *P. aeruginosa*	Jhonsi et al. (2018)
Carbon dots	Vitamin C	*S. aureus* and *E. coli*	Li et al. (2018)

other antiseptic agents. A synergistic effect was observed between CDots and hydrogen peroxide but not between CDots and sodium carbonate or acetic acid (Dong et al. 2017). Carbon dots have also been used as carriers of antibiotics to improve their stability and facilitate sustained release to minimize the development of antibiotic resistance (Jijie et al. 2018; Thakur et al. 2014).

23.3 Antifungal Activity of Carbon Nanoparticles

Fungal infections are, in general, more time-consuming and challenging to treat due to reasons including the difficulties with fungal culture and sensitivity and a limited availability of antifungal agents. This issue is further compounded by the problem of antifungal resistance. Carbon nanoparticles have been demonstrated to have antifungal activity (Table 23.2). Studies have shown CDots to be effective against *Candida albicans* (*C. albicans*) (Jhonsi et al. 2018), and *Pyricularia grisea* and *Rhizoctonia solani* (Li et al. 2018). Apart from their antibacterial activity, some of the hybrid CDots have also been shown to have antifungal activity against *C. albicans* (Priyadarshini et al. 2018; Gao et al. 2019) and yeasts (Thakur et al. 2014). Further studies are required to investigate the activity of various types of carbon nanoparticles against different fungal species.

Table 23.2 A summary of published studies on the antifungal and antiviral activities of carbon-based nanoparticles

Type of nanoparticles	Carbon source	Effective against microorganisms	References
OMMT-carbon dot reduced Cu_2O nanohybrid	Banana	*Candida albicans*	De et al. (2015)
Carbon dots	Tamarind	*Candida albicans*	Jhonsi et al. (2018)
Carbon dots	Vitamin C	*Rhizoctonia solani* and *Pyricularia grisea*	Li et al. (2018)
Carbon dot and gold encapsulated carbon dot	Citric acid and PEG	*Candida albicans*	Priyadarshini et al. (2018)
Nitrogen and chloride co-doped carbon dots		*Candida albicans*	Gao et al. (2019)
Benzoxazine monomer-derived carbon dots	Benzoxazine monomer	Flaviviruses (Japanese encephalitis, Zika, and dengue viruses) and nonenveloped viruses (porcine parvovirus and adenovirus-associated virus)	Huang et al. (2019)
Nitrogen and chloride co-doped carbon dots	Citric acid, Choline chloride, and urea	*Candida albicans*	Khanam et al. (2013)

23.4 Antiviral Activity of Carbon Nanoparticles

Although there are relatively fewer studies on the antiviral effects of carbon nanoparticles, the available evidence suggests that carbon nanoparticles also have some antiviral activity (Table 23.2). It has been reported that CDots can block infection of cells with certain viruses, including porcine parvovirus, Japanese encephalitis, Zika, and dengue viruses (Huang et al. 2019). More research is warranted to evaluate if carbon nanoparticles can block infection of cells with different viruses or inhibit viral replication inside the body.

23.5 Synthesis of Carbon Nanoparticles and their Mechanism of Action

Compared to other nanoparticles, carbon nanoparticles are relatively easy and inexpensive to synthesize (Fig. 23.1). Carbon nanoparticles can be collected from sources such as candle soot in a simple, facile, and low-cost manner (Khanam et al. 2013). Other researchers have used different sources such as date pits (Altaikyzy et al. 2018), wood and coconut husks (Varghese et al. 2013), butter and mustard oil (Mohanty et al.

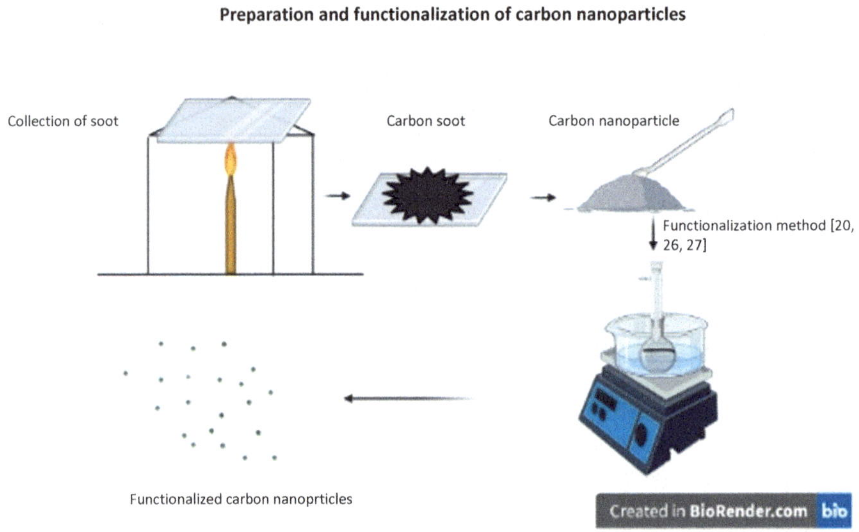

Fig. 23.1 A schematic diagram of the synthesis and functionalization of carbon nanoparticles

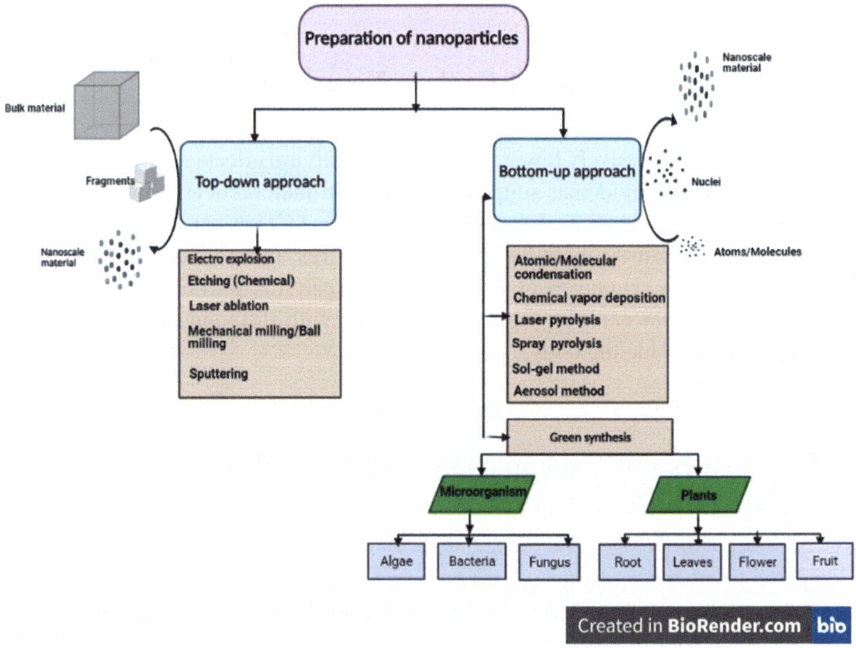

Fig. 23.2 Top-down and bottom-up approaches used for synthesis of carbon dots

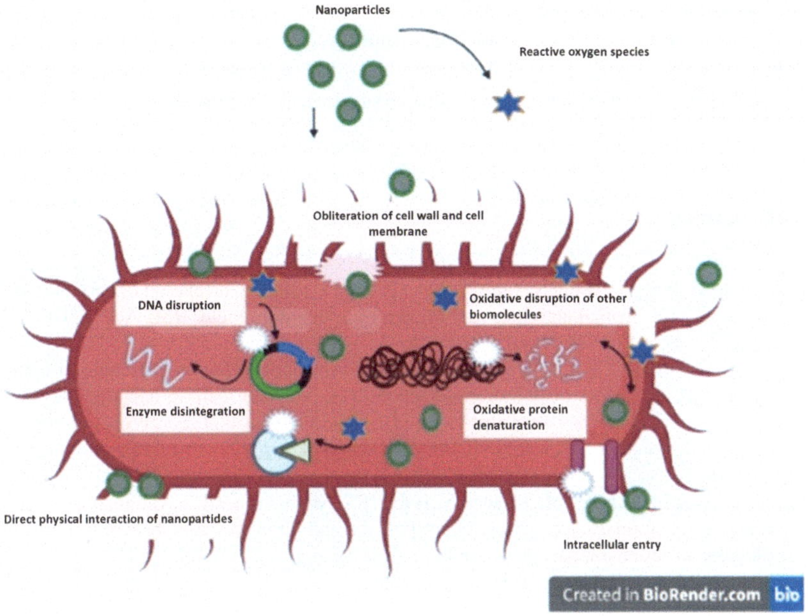

Fig. 23.3 Mechanisms of antimicrobial action of carbon-based nanoparticles

2007). Carbon dots can be prepared using either a top-down approach or a bottom-up approach (Sharma and Das 2019). The top-down approach follows fragmenting the bigger carbon structures into nanoscale material through electrochemical oxidation, discharge, chemical oxidation, and ultrasonic processes (Lim et al. 2015). This approach has some limitations including difficult reaction conditions, prolonged reaction time, and costly reagents (Wang and Hu 2014). The bottom-up approach involves transforming smaller carbon structure into carbon dots using thermal decomposition, ultrasonic treatment, carbonization, microwave synthesis, pyrolysis, solvothermal, and hydrothermal treatment methods (Fig. 23.2). Different mechanisms of action have been proposed for the antimicrobial activity of carbon-based nanoparticles (Fig. 23.3) including oxidative stress and direct interaction with the cell membrane and intracellular structures (Maleki Dizaj et al. 2015; Varghese et al. 2013).

23.6 Conclusions

Owing to the broad-spectrum antimicrobial activity of carbon nanoparticles and the ease with which they can be synthesized and functionalized, this class of nanoparticles offers a potentially viable alternative to traditional antimicrobials against which resitance is observed. Carbon nanoparticles are also a potentially good option for treatment of multidrug resistant infections in developing countries where the high cost of some traditional antimicrobial agents makes them an unaffordable or impractical option. Clinical studies are warranted in future to confirm the antimicrobial activity and safety of carbon nanoparticles in live animals.

References

Al Awak MM, Wang P, Wang S, Tang Y, Sun YP, Yang L (2017) Correlation of carbon dots' light-activated antimicrobial activities and fluorescence quantum yield. RSC Adv 7:30177–30184. https://doi.org/10.1039/C7RA05397E

Altaikyzy A, Fan H, Xie Y (2018) Effects of carbon nanoparticles on bacteria and application potential as biosensors of pollution detection of ocean water carbon nanoparticles as antibiotics and biosensors of ocean water. In: 2018 international conference on intelligent informatics and biomedical sciences (ICIIBMS), Bangkok, pp 177–180. https://doi.org/10.1109/ICIIBMS.2018.8550011

Azam A, Ahmed AS, Oves M, Khan MS, Habib SS, Memic A (2012) Antimicrobial activity of metal oxide nanoparticles against Gram-positive and Gram-negative bacteria: a comparative study. Int J Nanomedicine 7:6003–6009. https://doi.org/10.2147/IJN.S35347

Besinis A, De Peralta T, Handy RD (2014) The antibacterial effects of silver, titanium dioxide and silica dioxide nanoparticles compared to the dental disinfectant chlorhexidine on Streptococcus mutans using a suite of bioassays. Nanotoxicology 8:1–16. https://doi.org/10.3109/17435390.2012.742935

Cataldo F, Da Ros T (2008) Medicinal chemistry and pharmacological potential of fullerenes and carbon nanotubes. Springer, Trieste. https://doi.org/10.1007/978-1-4020-6845-4

Chen Q, Xue Y, Sun J (2013) Kupfer cell-mediated hepatic injury induced by silica nanoparticles in vitro and in vivo. Int J Nanomedicine 8:1129–1140. https://doi.org/10.2147/IJN.S42242

Dadgostar P (2019) Antimicrobial resistance: implications and costs. Infect Drug Resist 12:3903–3910. https://doi.org/10.2147/IDR.S234610

De B, Gupta K, Mandal M, Karak N (2015) Biocide immobilized OMMT-carbon dot reduced Cu_2O nanohybrid/hyperbranched epoxy nanocomposites: mechanical, thermal, antimicrobial and optical properties. Mater Sci Eng C Mater Biol Appl 56:74–83. https://doi.org/10.1016/j.msec.2015.06.023

Dong X, Awak MA, Tomlinson N, Tang Y, Sun YP, Yang L (2017) Antibacterial effects of carbon dots in combination with other antimicrobial reagents. PLoS One 12:e0185324. https://doi.org/10.1371/journal.pone.0185324

Dong X, Liang W, Meziani MJ, Sun YP, Yang L (2020) Carbon dots as potent antimicrobial agents. Theranostics 10:671–686. https://doi.org/10.7150/thno.39863

Emami-Karvani Z, Chehrazi P (2011) Antibacterial activity of ZnO nanoparticle on gram-positive and gram-negative bacteria. Afr J Microbiol Res 5:1368–1373. https://doi.org/10.5897/AJMR10.159

Gao Z, Li X, Shi L, Yang Y (2019) Deep eutectic solvents-derived carbon dots for detection of mercury (II), photocatalytic antifungal activity and fluorescent labeling for C. albicans. Spectrochim Acta A Mol Biomol Spectrosc 220:117080. https://doi.org/10.1016/j.saa.2019.04.072

Hou P, Yang T, Liu H, Li YF, Huang CZ (2017) An active structure preservation method for developing functional graphitic carbon dots as an effective antibacterial agent and a sensitive pH and Al(iii) nanosensor. Nanoscale 9:17334–17341. https://doi.org/10.1039/c7nr05539k

Huang S, Gu J, Ye J, Fang B, Wan S, Wang C, Ashraf U, Li Q, Wang X, Shao L, Song Y, Zheng X, Cao F, Cao S (2019) Benzoxazine monomer-derived carbon dots as a broad-spectrum agent to block viral infectivity. J Colloid Interface Sci 542:198–206. https://doi.org/10.1016/j.jcis.2019.02.010

Jhonsi MA, Ananth DA, Nambirajan G, Sivasudha T, Yamini R, Bera S, Kathiravan A (2018) Antimicrobial activity, cytotoxicity and DNA binding studies of carbon dots. Spectrochim Acta A Mol Biomol Spectrosc 196:295–302. https://doi.org/10.1016/j.saa.2018.02.030

Jijie R, Barras A, Bouckaert J, Dumitrascu N, Szunerits S, Boukherroub R (2018) Enhanced antibacterial activity of carbon dots functionalized with ampicillin combined with visible light triggered photodynamic effects. Colloids Surf B Biointerfaces 170:347–354. https://doi.org/10.1016/j.colsurfb.2018.06.040

Khanam A, Tripathi SK, Roy D, Nasim M (2013) A facile and novel synthetic method for the preparation of hydroxyl capped fluorescent carbon nanoparticles. Colloids Surf B Biointerfaces 102:63–69. https://doi.org/10.1016/j.colsurfb.2012.08.016

Kovacova M, Markovic ZM, Humpolicek P, Micusik M, Svajdlenkova H, Kleinova A et al (2018) Carbon quantum dots modified polyurethane nanocomposite as effective photocatalytic and antibacterial agents. ACS Biomater Sci Eng 4:3983–3993. https://doi.org/10.1021/acsbiomaterials.8b00582

Li YJ, Harroun SG, Su YC, Huang CF, Unnikrishnan B, Lin HJ, Lin CH, Huang CC (2016) Synthesis of self-assembled spermidine-carbon quantum dots effective against multidrug-resistant bacteria. Adv Healthc Mater 5:2545–2554. https://doi.org/10.1002/adhm.201600297

Li H, Huang J, Song Y, Zhang M, Wang H, Lu F, Huang H, Liu Y, Dai X, Gu Z, Yang Z, Zhou R, Kang Z (2018) Degradable carbon dots with broad-spectrum antibacterial activity. ACS Appl Mater Interfaces 10:26936–26946. https://doi.org/10.1021/acsami.8b08832

Lim SY, Shen W, Gao Z (2015) Carbon quantum dots and their applications. Chem Soc Rev 44:362–381. https://doi.org/10.1039/C4CS00269E

Liu J, Rojas-Andrade MD, Chata G, Peng Y, Roseman G, Lu JE, Millhauser GL, Saltikov C, Chen S (2017) Photo-enhanced antibacterial activity of ZnO/graphene quantum dot nanocomposites. Nanoscale 10:158–166. https://doi.org/10.1039/c7nr07367d

Maisch T (2009) A new strategy to destroy antibiotic-resistant microorganisms: antimicrobial photodynamic treatment. Mini Rev Med Chem 9:974–983. https://doi.org/10.2174/138955709788681582

Maleki Dizaj S, Mennati A, Jafari S, Khezri K, Adibkia K (2015) Antimicrobial activity of carbon-based nanoparticles. Adv Pharm Bull 5:19–23. https://doi.org/10.5681/apb.2015.003

Meziani MJ, Dong X, Zhu L, Jones LP, LeCroy GE, Yang F, Wang S, Wang P, Zhao Y, Yang L, Tripp RA, Sun YP (2016) Visible-light-activated bactericidal functions of carbon "quantum" dots. ACS Appl Mater Interfaces 8:10761–10766. https://doi.org/10.1021/acsami.6b01765

Mohanty B, Verma AK, Claesson P, Bohidar HB (2007) Physical and anti-microbial characteristics of carbon nanoparticles prepared from lamp soot. Nanotechnology 18:445102. https://doi.org/10.1088/0957-4484/18/44/445102

Pal S, Tak YK, Song JM (2007) Does the antibacterial activity of silver nanoparticles depend on the shape of the nanoparticle? A study of the gram-negative bacterium Escherichia coli. Appl Environ Microb 73:1712–1720. https://doi.org/10.1128/AEM.02218-06

Plank C (2009) Nanomedicine: silence the target. Nat Nanotechnol 4:544–545. https://doi.org/10.1038/nnano.2009.251

Prestinaci F, Pezzotti P, Pantosti A (2015) Antimicrobial resistance: a global multifaceted phenomenon. Pathog Glob Health 109:309–318. https://doi.org/10.1179/2047773215Y.0000000030

Priyadarshini E, Rawat K, Prasad T, Bohidar HB (2018) Antifungal efficacy of Au@ carbon dots nanoconjugates against opportunistic fungal pathogen, Candida albicans. Colloids Surf B Biointerf 163:355–361. https://doi.org/10.1016/j.colsurfb.2018.01.006

Ruddaraju LK, Pammi SVN, Guntuku GS, Padavala VS, Kolapalli VRM (2020) A review on antibacterials to combat resistance: from ancient era of plants and metals to present and future perspectives of green nano technological combinations. Asian J Pharm Sci 15:42–59. https://doi.org/10.1016/j.ajps.2019.03.002

Sharma A, Das J (2019) Small molecules derived carbon dots: synthesis and applications in sensing, catalysis, imaging, and biomedicine. J Nanobiotechnol 17:92. https://doi.org/10.1186/s12951-019-0525-8

Thakur M, Pandey S, Mewada A, Patil V, Khade M, Goshi E, Sharon M (2014) Antibiotic conjugated fluorescent carbon dots as a theranostic agent for controlled drug release, bioimaging, and enhanced antimicrobial activity. J Drug Deliv 2014:282193. https://doi.org/10.1155/2014/282193

Usman MS, El Zowalaty ME, Shameli K, Zainuddin N, Salama M, Ibrahim NA (2013) Synthesis, characterization, and antimicrobial properties of copper nanoparticles. Int J Nanomedicine 8:4467–4479. https://doi.org/10.2147/IJN.S50837

Varghese S, Kuriakose S (2016) Synthesis, characterisation and antibacterial studies of carbon nanoparticle and carbon nanoparticle encapsulated in functionally modified β-cyclodextrin with {5-[4-(dimethylamino) benzylidene]-4-oxo-2-thioxo-1, 3-thiazolidin-3-yl} acetic acid. Supramol Chem 28:293–304. https://doi.org/10.1080/10610278.2015.1107064

Varghese S, Kuriakose S, Jose S (2013) Antimicrobial activity of carbon nanoparticles isolated from natural sources against pathogenic Gram-negative and Gram-positive bacteria. J Nanosci 2013:457865. https://doi.org/10.1155/2013/457865

Wang Y, Hu A (2014) Carbon quantum dots: synthesis, properties and applications. J Mater Chem C 2:6921–6939. https://doi.org/10.1039/C4TC00988F

WHO (n.d.) Antimicrobial resistance. In: The World Health Organization Newsroom Factsheets. https://www.who.int/publications-detail-redirect/global-action-plan-on-antimicrobial-resistance. Accessed 25 Dec 2020

Yan YY, Kuang WC, Shi LJ, Ye XL, Yang YH, Xie XB et al (2019) Carbon quantum dot-decorated TiO2 for fast and sustainable antibacterial properties under visible-light. J Alloy Compd 777:234–243. https://doi.org/10.1016/j.jallcom.2018.10.191

Yin R, Agrawal T, Khan U, Gupta GK, Rai V, Huang YY, Hamblin MR (2015) Antimicrobial photodynamic inactivation in nanomedicine: small light strides against bad bugs. Nanomedicine (Lond) 10:2379–2404. https://doi.org/10.2217/nnm.15.67

Zarei M, Jamnejad A, Khajehali E (2014) Antibacterial effect of silver nanoparticles against four foodborne pathogens. Jundishapur J Microb 7:E8720. https://doi.org/10.5812/jjm.8720

Zhang J, Liu X, Wang X, Mu L, Yuan M, Liu B, Shi H (2018) Carbon dots-decorated $Na_2W_4O_{13}$ composite with WO_3 for highly efficient photocatalytic antibacterial activity. J Hazard Mater 359:1–8. https://doi.org/10.1016/j.jhazmat.2018.06.072

Chapter 24
Antimicrobial Interfaces as Augmentative Strategy Against Antimicrobial Resistance

Sunny Dholpuria, Rohit Kumar, Niraj Kumar Jha, Shipra Jha, and Sharad Agrawal

24.1 Introduction

Advancement of lifesaving drugs has effectively decreased the mortality rate in human population, but overuse of such drugs (mainly antibiotics) has increased the threat of antimicrobial resistance globally (McEwen and Collignon 2018). In recent years, we have seen antimicrobial resistance enormously emerges at global scale and is spreading from country to country at faster rate than we thought. Resistant microbes which are also known as superbugs are endemic in many parts of the world (Christaki et al. 2020) and uncontrolled use of antimicrobial drugs is majorly responsible for the consequences of antimicrobial threat (Zaman et al. 2017). The presence of microorganisms like bacteria, fungi, moulds, and viruses in our immediate surrounding causes undesirable and serious health problems. In order to combat such health issues, the patient relies on antibiotics, and excessive consumptions of antibiotics sometimes result in secondary infections and antimicrobial resistance is one of them. Antimicrobial resistance (AMR) is not a recent phenomenon, but it has become a critical health issue in recent times due to the excessive use of antibiotics in lot many sectors like animal farming (Xu et al. 2020a), agriculture (Suriyasathaporn et al. 2012), sewage treatment plants (Tambosi et al. 2010), and

S. Dholpuria (✉)
Department of Life Sciences, J.C. Bose University of Science and Technology, YMCA, Faridabad, Haryana, India
e-mail: sunnydholpuria@jcboseust.ac.in

R. Kumar · S. Jha · S. Agrawal
Department of Life Sciences, School of Basic Science and Research (SBSR), Sharda University, Greater Noida, India

N. K. Jha
Department of Biotechnology, School of Engineering and Technology (SET), Sharda University, Greater Noida, India

N. Akhtar et al. (eds.), *Emerging Modalities in Mitigation of Antimicrobial Resistance*,
https://doi.org/10.1007/978-3-030-84126-3_24

many more. In year 2017, World Health Organization (WHO) published a list of 12 bacteria stating them as threat of AMR to human health (Duval et al. 2019). The causes and consequences of antimicrobial resistance in different countries are very complex and may develop through the practices of healthcare toward the use of various types of antimicrobials (Ayukekbong et al. 2017). Although the use of drugs and antibiotics in healthcare and clinics is necessary but the research community is falling behind in the process of making new drugs to overcome the threat of newly discovered antimicrobial-resistant species (Moo et al. 2020). Poultry, which is one of the most widespread food industries across the world, has raised a major concern due to the direct/indirect involvement in creating AMR among human population (Xu et al. 2020a; Alam et al. 2019) and in order to minimize these adverse effects, certain rules need to be followed on the usage of antibiotics and antimicrobials in animal livestock (Lhermie et al. 2017). Contrary to this, effluents from hospitals containing high level of antibiotics and various other toxic chemical residues challenge the microbial population and confer the resistance property against them (Rowe et al. 2017). Likewise, faulty agriculture practices especially in irrigation further enhance the problems of AMR for the human population (Hong et al. 2020). Various action plans toward antimicrobial stewardship, improvisation in antibiotics, and techniques for the identification of drug-resistant microbes are promising toward combating the threat of AMR (Jansen et al. 2018). The aim of antimicrobial stewardship programs is to implement antimicrobial-targeted interventions to reduce inappropriate antimicrobial prescribing in order to prevent the development of antimicrobial resistance (Cole et al. 2019). We can use vaccines to control infections over a long time without becoming antiquated because vaccines work prophylactically (Tagliabue and Rappuoli 2018). All these steps are being advised to minimize the use of antibiotics in order to overcome the adverse effects from them.

But still there are significant areas such as animal and food industries where the use of antibiotics could not be reduced to great extent in order to fulfill the needs of overgrowing human population. This is a major concern especially in developing and/or overpopulated countries like India, China, and Bangladesh where agriculture and animal food sector use the antibiotics in large amount thereby causing serious health issues to the people (Van Boeckel et al. 2015). For the prevention and cure of such human diseases, there is unimaginable use of antibiotics and chemicals in daily life. Majorly these kinds of infections can be avoided by taking proper precautions but still the problems exist due to population, pollution, poor facilities of wastewater treatment, and many more. These problems are commonly found associated with the occupation of the individual especially health workers in hospitals and people associated with cleanliness, making them more prone to infections. The presence and growth of microorganisms on their clothes cause serious health problems and also deteriorate the quality of textiles fabrics (Ravindra and Murugesh Babu 2016). A hospital in China has observed methicillin-resistant *Staphylococcus epidermis* in the hand samples of nurses and doctors (Xu et al. 2020b). Likewise, in Saudi Arabia similar thing has been observed among hospital as well as nonhospital personnel for the presence of antibiotic resistance strains of *Staphylococcus aureus*.

Along with this the equipment used in such professions and some other work place is also the source of causing serious health issues to the individual coming in direct or indirect contact with them. As these infections are mostly bacterial, this consequently leads to the consumption of antibiotics and thereby causing concern for AMR. By keeping these things in account, the production of antimicrobial textiles and antimicrobial coating materials seems to be quite promising in reducing the problem of AMR. The use of antimicrobials in textiles such as silver- and zinc-based nanoparticles, silane quaternary ammonium compounds, and chitosan imparts the antibacterial properties to the textiles products for combating the threat of AMR. Antimicrobial's coating of healthcare devices and surfaces of surgical equipment with electrochemical deposition polymer nanoparticles are also found very effective for fighting against bacterial infections. Antimicrobial textiles and coated textiles possess effective potential to reduce the microbial growth and it also enhances the quality and durability of fabrics. This chapter talks about advancements in antimicrobial textiles and coating materials along with their applications and importance in present world.

24.2 Antimicrobial Fabrics in Textiles

Working in a microbial environment such as hospitals, laboratories, and wastewater treatment plants confers higher chances of bacterial infections than any other place. This risk of infection can increase several folds if the textiles used as garments and dressing materials do not offer hindrance to microbial attachment and colonization as they are in direct contact with the body of individuals. Also, microbial growth on fabrics can lead to deterioration of its quality. This necessitates the development of textile fibers which not only offer greater degree of antimicrobial activity but also have more durability. Various approaches to develop such fabric have been considered so far which include treatment of fabrics with reagents as well as naturally derived fibers.

Having a healthy lifestyle is one of the primary requirements for well-being of our body, but high prevalence of pathogenic microbes in our surroundings has created many problems (Emam et al. 2019). The use of antimicrobials in textiles intends to provide protection against microbial invasions (Kramer et al. 2006) as well as such antimicrobial impregnations in textiles have also been found to be quite beneficial for improving the quality of fabrics and reducing potential health risks, e.g., cytotoxicity, bad odor, sensitization, and skin allergies (Kramer et al. 2006). The implementation of antimicrobials in textiles has attracted the world with a conflict of interest aiming to reduce the infective microbial load, unpleasant odor, and enhancing the life span of textile fabrics. Recent advancements include the use of less soluble or nonleaching antimicrobials which have been found to be advantageous in overcoming the toxic effect on aquatic life due to their leaching. Another such advancement is the development of antimicrobial textiles which are effective against a broad-spectrum of pathogenic microorganisms, but do not harm the nonpathogenic microbes present on the skin of wearer. There is a huge market of antimicrobial textiles in healthcare, food packaging industries, and home furnishing owing to their wide applications. In comparison of traditional textiles, the global markets of present textile approaches are becoming highly competitive. Many technologies have been developed to prevent microbial growth but implementation of these techniques in textiles is not as per current requirements (Hiremath et al. 2018).

Various types of antimicrobials agents are used in textiles depending upon their different biocidal effects and target mechanisms against microorganisms. These different mechanisms can be (Morais et al. 2016):

- Inhibiting cell wall synthesis and targeting monomer polymerization
- Blocking of cross polymer linking beta-lactam by some antimicrobial agents
- Inhibition of protein synthesis and nucleic acid synthesis (DNA/RNA)

The eco-friendly attribute of antimicrobial chitosan has facilitated its applications in textiles. The use of nanomaterials such as silver and zinc nanoparticles as antimicrobials in textiles is advantageous due to their physiological as well as biological properties. Green synthesis and environment friendly approaches in nanomaterials are very reliable due to their antibacterial properties. Recent advances in field of microbicidal polymeric ammonium compounds and their activeness against broad-spectrum range of microorganisms have attracted the concern of textile industries for various applications.

24.3 Imparting Antimicrobial Property by Chemical Treatments

Treating fabrics such as cotton and polyester with certain specific chemicals and use of different scientific techniques for the same have been the prime focus of interest in research involving the development of antimicrobial textiles for varied applications. Such innovations not only increase the microbe fighting ability of textiles, but also contribute to the enhancement of their quality and durability. The finished antimicrobial fabrics can be applied to multiple industries to prevent infections caused by various microbes and improve public health (Chen et al. 2019). In one such research, covalent bonding of novel tert-butoxylcarbonyl (BOC)-protected guanidine which contains an isocyanate group was performed on the surface of cotton fabric. The BOC group was later on deprotected through the process of acidification. The fabric thus developed, showed 88.5% and 99.9% antibacterial activity for *E. coli* and *S. aureus,* respectively (Cao et al. 2020) which is much greater in comparison to the normal untreated cotton fabric. Another finding on cotton fabric was carried out using the process of heterogenous transesterification and the acetoacetyl group was attached directly on the surface of cotton fabric for imparting dual function of hydrophobicity and antibacterial properties. The cotton-acac modified with gentamicin and octadecyl amine (Rong et al. 2019) and the results were very promising which led to improvement in overall bacterial growth inhibition.

Polyester fabrics are also one of the popular choices for textiles. However, polyester does not offer adequate protection against microbial activity by themselves. Therefore, in order to be used as textiles in medical sector, it becomes important that polyesters undergo certain kind of modifications. One such innovation involves the modification of micro denier polyesters with sericin which is used to be extracted from yellow silk cocoons. The novel fabric thus developed shows outstanding effects against both bacteria as well as fungi (Rajalakshmi et al. 2018). Apart from this many a time blending of two fabrics in varying proportions can also increase the antimicrobial activity of the synthesized textile by several folds. Such blended mixtures can even give rise to a large number of knitted fabrics. Polyester silver nanocomposite fibers when blended with normal polyester fibers in varying proportions in terms of yarn count and knitting machine gauge can produce up to 27 types of knitted fabrics which provide good antibacterial activity (about 65–99%) against both *S. aureus* and *E. coli* bacteria (Khude et al. 2020).

Such innovations involving chemical treatment of fabrics and fibers seem to be quite promising in the future for the development of alternative approaches for textiles offering antimicrobial protection to be used in healthcare centers and laboratories and in places where the risk of microbial infection is quite high.

24.4 Developing Antimicrobial Textiles Using Eco-Friendly Approach

Chemical treatment of textiles has been proven beneficial in enhancing their antimicrobial property largely. However, if such enhancements can be made using naturally derived products and using eco-friendly procedures, then such approaches are always the center of attention. The idea to produce antimicrobial fabrics by modification with reagents from natural sources is the most preferred of all and has shifted the focus of new research, as the developed fibers are also biodegradable.

In a study on bamboo-derived fibers, it was found that natural bamboo fibers showed about 8–95% reduction in the growth of *K. pneumoniae* and about 3–50% in the growth of *S. aureus* (Prang Rocky and Thompson 2020). However, when the bamboo fibers are regenerated, it reduces its antibacterial properties significantly (Ali et al. 2021). Many a time substances derived from natural sources can also be utilized due to their natural antagonistic effects toward microbial growth. There are many such substances around us, which can show promising results. Casein, a naturally occurring product can be harnessed for the development of antimicrobial textiles. Casein derived from bovine milk, after being subjected to acid hydrolysis in aqueous media, is blended into PP matrix in the melt phase by extrusion. After processing this blend by the melt spinning process, production of multifilament takes place that can be further woven into textiles. Such textiles show excellent antibacterial activity against both gram-positive as well as gram-negative bacteria (Belkhir et al. 2021). These results are also promising in the perspective for preventing the wastage of dairy products and even their recycling. Another such product is natural pine/spruce rosin that has also been proven beneficial in improving the antibacterial properties of textiles. Using procedures like intrusion compounding and melt spinning, a mixture of rosin with polyethylene, polypropylene, polylactic acid, polyamide, and corn starch-based polymer can produce fibers for wide applications like filters and medical textiles. Polyethylene with about 10% rosin content had significant antibacterial effects against *E. coli* (Kanerva et al. 2019).

The antimicrobial activity of neem (*Azadirachta indica*) leaves and bark extracts implemented with cotton fabrics was found to be effective against gram-negative bacteria. Neem extract was applied on medical textiles by direct application and microencapsulation. It was found that durability of textiles given by direct method to washing was poor but application by encapsulation method showed good durability after washing (Radhika 2019). These modifications of cotton fabrics with neem extracts enhance the properties of fabric such as elongation, tensile strength, and antimicrobial performance. In recent studies, it was found that eucalyptus oil has different medical application and also property of cleansing agent. It shows antimicrobial and antifungal properties. Eucalyptus shows the antimicrobial properties against *E. coli* by applying on wool and cotton fabrics. But modified fabrics lose its antimicrobial property after every wash cycle. In woolen clothes its durability is high due to low washing period of woolen clothes is limited (Radhika 2019). Polyethylene terephthalate (PET) has been widely used for treatment of fabrics due to properties like easy processability and quick drying. Finishing of fabrics with PET exhibits durable broad-spectrum antimicrobial activities against gram-negative, gram-positive bacteria and fungi (Chen et al. 2019).

Betulin is a natural compound that can be obtained from the outer bark of birch. In an experiment, carboxyl-functionalized cellulose was prepared after oxidation and later on by esterification was attached covalently to betulin which led to the improvement in the antibacterial property by 99% removal and growth inhibition of bacteria (Huang et al. 2019). Similarly, lysozyme is a natural enzyme, which is known to show efficient antimicrobial preventive properties. Harnessing lysozyme from chicken egg white and its immobilization onto wool fibers by cross-linking has produced fibers having better antibacterial properties and even its durability (Yang et al. 2020).

Use of biopolymers and eco-friendly natural antimicrobials for the development of textile materials for medical sectors and food industries maintains health and hygiene of products. Various methods researched and reported the fabrics such as cotton, wool, and linen finish with antimicrobials to be effective against various types of bacteria, viruses, and fungi (Kost et al. 2020).

When a person is injured, the growth of microbes on the dressing materials can slow down the process of healing of wound, which necessitates the development of a biodegradable alternative having antimicrobial properties. In a recent approach, quercetin-loaded polylactide-based fibers have been developed through electrospinning and the obtained woven fabric showed good effects against *S. aureus*, *K. pneumonia,* and *E. coli* (Rigo et al. 2018).

24.5 Surface Modification of Textiles and Appliances with Antimicrobial Coating

The process of formation of biofilms and bacterial colonies on the surface of different kinds of medical devices, food processing equipment, and appliances related to the daily use can cause serious infections as well as unwanted side effects. So, in order to improve the sterility of such surfaces or medical devices like catheters, scissors, and other equipment used in surgery which comes in direct contact with the patient body, there is a huge need of antimicrobial coatings on them. Nanoscience-based solutions used as antimicrobial coating on surfaces are very reliable due to their surface modification properties that provide microbial resistance. Antimicrobial surfaces developed using nanotechnology include hydrogel films with temporal release of active agents (Langowska et al. 2014) and nanoreactors which again are catalytically active compartments to produce reactive agents (Li et al. 2018). The surface coating of materials with QAC is also an effective and promising method for the formation of microbe-free appliances used in food industries, wastewater treatment plants, and health care (Jiao et al. 2017).

In addition to these, contact killing and antimicrobial properties of copper nanoparticles (Maghimaa and Alharbi 2020) are also effective tools which can be used for coating of surfaces to make effective against microbes. Various properties like antifungal and antibacterial properties make copper nanoparticle suitable for use in various healthcare devices as coating agent. All these properties of different elements and compounds have made promising role toward fighting against microbial infections.

24.6 Silver and Zinc Oxide Nanomaterials as Antimicrobial Agent for Natural and Synthetic Textiles

The research in the field of nanomaterials has significantly increased in past few years due to their extensive biological, magnetic, optic, and catalytic properties (Ozin and Arsenault 2015). Developments in nanoscience research have flourished much more in recent times and it is now considered in every technical research for developing new products and similar to various applications, it can also help to tackle problems like AMR and related speculations (Aarthi et al. 2018). Green synthesis of nanoparticles has emerged as an alternative approach, as it is cost-effective and even minimizes the consumption of hazardous chemicals (Ozin and Arsenault 2015). The ongoing researches about the use of antimicrobials in textiles highlight their effectiveness in resisting the growth of microbes. Owing to their antimicrobial applications, the use of silver nanoparticles (AgNPs) has significantly increased in textiles and numerous studies focus on exploring the greenway synthesis of AgNPs. It has been found that after treating with AgNPs, the fabrics like cotton, wool, and linen show effective activity against bacteria like *E. coli* and *S. aureus*. In addition to this, the use of AgNPs instead of toxic chemicals is also much more cost-efficient and eco-friendlier (Nateghi and Shateri-Khalilabad 2015). Due to these advantages, the demand of antimicrobial fabrics such as cotton, wool, and polyester has significantly increased. Also, by enhancing the durability, nanomaterials used as antimicrobials in textiles seem to be quite promising. Impregnation of textiles with silver nanoparticles shows reduction in growth of microbes like *E. coli* to an extent of about 99% and it is not found to have any harmful and toxic effects on wearer and environment (Fiedot-Toboła et al. 2018). Such modified textiles products are effective against both gram-positive and gram-negative bacteria present on surfaces (Sirelkhatim et al. 2015). Despite all these advantages, incorporation of silver nanoparticles in textiles requires post surface treatments that affect the durability of fabric after several washes. In addition to AgNPs, antimicrobial activity of zinc oxide (ZnO) nanoparticles has also gained significant attention worldwide in developing antimicrobial textiles. Zinc oxide is a biosafe material that possesses photo-oxidizing and photocatalysis activity on biological species (Sharma et al. 2019). Also, ZnO is known to reduce the occurrence of foodborne diseases which has led to an increase use of zinc oxide as antimicrobial in the food packaging industries. However, in some cases, the excessive use of zinc oxide nanoparticles also found to increase the defects on modified surfaces. Incorporation of zinc oxide nanoparticles in textiles and on surface coatings is effective against the broad-spectrum of pathogenic microorganisms. Although, the use of silver nanoparticles is preferred over zinc oxides as in a study on linen fabrics the incorporation of silver nanoparticle resulted 100% reduction in the growth of microorganisms (Hasan et al. 2020) and therefore considered as better approach. Deposition of antimicrobials on nylon fabric surface in terms of excellent coloration and antibacterial properties has been developed by chitosan-mediated silver nanoparticles synthesis (Kloos and Musselwhite 1975). Use of silver nanoparticles as antimicrobials shows promising effect against wide range of microbes with very low toxicity and used in textiles as an antimicrobial agent to prevent infections which in long run helps to reduce the instances of AMR arising from irresponsible and unnecessary consumption of antibiotics (Fig. 24.1).

24.7 Quaternary Ammonium Compounds (QACs) as Antimicrobials

The quaternary ammonium compounds consist of 12–18 carbon atoms, a long chain of alkyl group, and a compatible anion. The mechanism of action of QACs against microbes includes process like inhibition of nucleic acid (DNA and RNA) and damages to the cell membrane of bacteria; along with this it also inhibits the multiplication of bacteria. These compounds possess greater degree of antimicrobial and anticarrier activities, and are very effective against bacteria like *Staphylococcus*

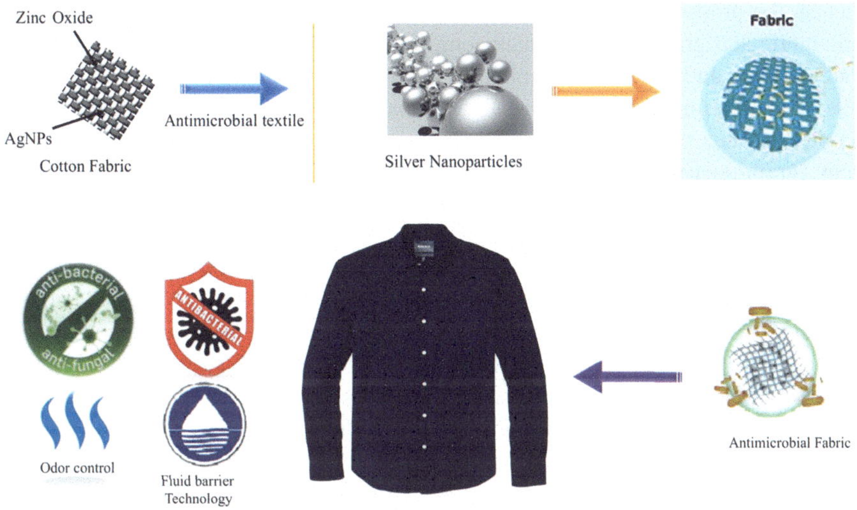

Fig. 24.1 Zinc oxide and silver nanoparticle-based antimicrobial fabrics

aureus, Streptococcus and *Bacillus* species (Kennedy et al. 1995). Newly developed QACs are found to possess virucidal activity and may be used as antimicrobials (Żywicka et al. 2018). Various bacterial membranes have been modified with these QAC compounds and these have been used in wastewater treatment plants because of their effective antimicrobial properties (Żywicka et al. 2018; Elena and Miri 2018). The covalent surface grafting of QACs seems to be quite promising in the formation of effective antimicrobial materials that could be used in medical sector, food and textile industries (Caschera et al. 2019). In textile industries, QACs containing long alkyl chains (12–18 C atoms) are most widely used for the treatment of fabrics to make them resistant to microbes. The antimicrobial action of these compounds depends on the length of alkyl chain, presence of perfluorinated groups, and the cationic ammonium groups. These QACs used as antimicrobial agents are very effective against gram-positive bacteria, gram-negative bacteria, viruses, and also some fungi. The use of QAC in textiles showed 99.99% effectiveness against *E. coli* and *S. aureus* (Khan et al. 2015).

24.8 Natural Polymeric and Graphene Oxide Nanocomposites with Antimicrobial Activities

Graphene, an atomically thin carbonaceous material, has been a popular research topic in the recent past, due to its mechanical, electrical, and antimicrobial properties. Reduction of graphite oxides offers a wide range of possibilities for synthesizing graphene nanocomposites using very easy methodologies (Konwar et al. 2016).

Hence, the production of antimicrobial compounds with impregnations of natural polymeric and graphene oxide nanocomposites exhibits a considerable antimicrobial activities and key tool in combating AMR problems instead of toxic chemicals. It is noted that impregnation of iron oxide-coated graphene nanocomposites on different appliances shows significant antimicrobial activity against bacterial pathogens (gram-positive and gram-negative) such as *E. coli, S. aureus,* and *Candida albicans* (Zhang et al. 2016a). The natural polymeric nanocomposites have received a lot of attention from the past decades due to their various antimicrobial activities and that is why natural polymers such as chitosan, dextran, rubber, and fibrin are enormously used in textiles instead of synthetic ones due to their elastic and antimicrobial properties. Antibacterial activities of natural green polymeric nanocomposites make them useful for having their applications in packaging materials, textiles, and surface coating (Kowalczyk et al. 2017). These polymers maintain the stability of fabrics such as cotton and also impart the treated surface with new functional properties such as UV resistance. But the main problem with respect to implementation of these natural polymers to textile fabrics is the reduction in durability of fabrics. To overcome this issue, use of inorganic nanoparticles with improved finishing is being applied on fabrics. Graphene and its derivatives such as graphene oxides exhibit excellent antimicrobial and mechanical properties. The best antistatic properties can be obtained by coating the material with 1.5% of graphene provided the surface resistance and volume resistance should be of order $10^{\wedge 5}$ and $10^{\wedge 3}$ Ω, respectively and these kinds of properties make the fabric suitable in acquiring antimicrobial properties (Knetsch and Koole 2011).

24.9 Antimicrobial Coatings on Medical Devices

Antimicrobials are commonly added to paint formulation, food processing equipment, medical devices to maintain the coatings against microbial attacks and to provide protection against fungal growth. Applications of nanomaterials implemented as coating agents have been used to prevent adhesion and subsequent biofilm formation (Wei et al. 2019). Recent advances in context to threat of AMR have led to developing new antimicrobial coating materials. Bacterial colonization and biofilm formation on medical devices potentially cause life threatening infections and hazardous impacts on patient's health. The microbes present in biofilm that develops on medical devices are very difficult to control using conventional antimicrobial agents as consequences of intrinsic resistance mechanisms. Several approaches such as hydrophilic coatings, bacteria-repellent coatings, and hydrogels-loaded antibiotics seem to be promising in their ability to prevent microbial infections associated with medical devices (Zhang and Wagner 2017). The considerable way to prevent life threatening infections caused by microorganisms is to modify the surfaces of medical devices in such a way that no microbial attachment on these devices could occur. These approaches of complete surface treatment require modifications with mostly hydrophilic polymeric surface coatings (Wei et al. 2019).

Hydrophilic coatings, when applied to medical equipment confer self-cleaning, antimicrobial, and anticorrosive properties (Mehta et al. 2010). Hydrophilic polymer coating evaluates that this method has significant capabilities linked with devices used in clinical settings (Zakrzewski et al. 2021).

The surface modifications of orthodontic appliances materials have provided an effective antimicrobial and mechanical characteristic during orthodontic treatment. Nanotechnology and its application in the field of orthodontics including antimicrobial properties enhancement is leading to overcome the problems like enamel demineralization and friction during orthodontic movements (Ryu et al. 2012). The use of orthodontic braces without any antimicrobial coatings significantly increases the risk of dental plaque and microbial infections. Nanoparticles such as silver platinum have a larger surface area to volume ratio that helps in providing more surface area for antimicrobial activities (Ryu et al. 2012). Silver-platinum (Ag-Pt) alloys, which have high degree of biocompatibility, excellent resistance to sterilization conditions, and antibacterial properties to different bacteria, are associated with long-term antibacterial efficiency (Yemmireddy and Hung 2017). Due to the biofilm formation and colonization of bacteria, the enamel decalcification occurs and this ends up to orthodontic treatments. Organic acids produced by the adherent bacteria can cause enamel demineralization around a bracket, which can deteriorate the outcomes of treatment (Yemmireddy and Hung 2017). The release of silver ions from the Ag-Pt coating of brackets provides a resistance against *Streptococcus mutans* and reduces the problems of tooth decay. Various bacterial infections can be reduced by use of fluoride and antibiotics but in addition to these, methods such as coating of Ag-Pt alloys has proved to be quite effective against bacterial infections in post dental treatments.

24.10 Antimicrobial Photocatalytic Coating

Improper cleaning and microbial association with surfaces cause life threatening infections and hazardous impacts on society. Several approaches such as cleaning, sanitation, and disinfection of surfaces of more equipment have been used to address current and emerging microbial safety challenges. Advanced oxidation processes involving photocatalytic nanomaterials coatings have shown great significant effects on a wide range of microorganisms and chemical contaminants (Wei et al. 2014). Copper-containing coatings can provide significant protection against wide range of microbial strain but due to the poor corrosion efficacy and high cost, the use of titanium dioxide (TiO_2) is preferably more active against microbes. These photocatalytic coatings show excellent properties against microbes in outdoor environment (Jaksik et al. 2018). Normally, the reduction in number of species of bacteria takes approximately 5 h and same numbers of bacteria get eliminated in 2 h with implementation of TiO_2 coating (Jaksik et al. 2018). In this coating, photocatalyst is used as ingredients to remove the microbes from the surfaces. The active ingredient is titanium dioxide which does not dissolve in water and used for coating of various surfaces. The surfaces when implemented by this coating have shown antibacterial activity against *E. coli* and *S. aureus* with 99.5% reduction in bacterial population. Photocatalytic coating surfaces happen to be free from biofilm formation and colonization of bacteria for a long time and this decreases the burden of healthcare (Zhang et al. 2016b).

In textiles, TiO_2 is often used in development of antimicrobial-efficient cotton fibers. Spray of TiO_2 emulsion with polymer generative additives, on the cotton fibers, along with heating up to 100 °C generates the TiO_2 -coated fabric surfaces. Coating of fibers with TiO_2 uniformly across the entire surface of the fiber without any cracking enhances the antimicrobial and nondegrading properties of fabrics. The observation and testing of TiO_2-imparting cotton fabrics confirmed with antimicrobial activity, which further invent the application of photocatalyst coating in the field of textiles and antimicrobial coating materials.

24.11 Antimicrobial Durability and Cost Efficiency

For an ideal and effective antimicrobial, the durability and its cost efficiency are the most significant factors among other factors. Durability of antimicrobials used in textiles mainly depends on implementation method and subjected to various treatments such as drying and washing. This challenge is further more complicated by the evaluation of durability that is more complex and time-consuming than that on antimicrobial activity (Xu et al. 2018). A large number of nanoparticles have been used for their antimicrobial properties but implementation of nanoparticles with textiles creates an aspect of durability too. The covalently attachment of L-cysteine to the fibers such as cotton and adhesion of silver nanoparticles to the surface of fibers significantly increase durability of fabrics with great antibacterial activity (Ramya and Subapriya 2012). Cost efficiency is also an influencing factor for the use of antimicrobials agents. Synthesis of nanoparticles such as silver, copper, and their implementation with textiles is a costly process (Aly and Albutti 2014). Due to this problem, antimicrobial textiles and antimicrobial coating materials are not for every consumer of the society. In context of textiles, use of herbal-synthesized nanoparticles approaches the long-term investment toward the public health and maintains the cost efficiency of antimicrobials. New innovative techniques to combating AMR with the use of significant antimicrobials imparting with textiles overcome the problems of durability and cost effectiveness in recent time.

24.12 Environmental Impacts of Antimicrobials

The demand for antimicrobial fabrics, antimicrobial coating materials significantly increases in recent time due to its combating efficiency for AMR and health-related potency. But high use of antimicrobials induces a negative impact on human health and environment. Wastes discharged from textile industries, comprising of silver nanoparticle residues (used as antimicrobials) as well as wastewater, which drains after washing of antimicrobial fabrics, are acting as a potential threat to the aquatic ecosystem. Therefore, specific intervention measures for the use of antimicrobials in textiles and coating materials should be established and implemented to mitigate negative impact on the environment (Singer et al. 2016). Environmental regulatory laws should be implemented significantly for the global AMR action plans (Morais et al. 2016). In May 2015, the 68th world health assembly adopted the global action plan on AMR. The goal of the global action plan is to ensure, for as long as possible, continuity of successful treatment and prevention of infectious diseases with effective and safe medicines that are quality-assured. Any antimicrobial used in textiles and coating of surfaces must be nontoxic to human health and environment. Due to the emergence of AMR, antimicrobials, which are synthesized from plants, have been extensively used in field of textiles to reduce the negative impact on environmental problems. The main significance of herbal antimicrobials is that they do not exhibit toxic effects on healthy microbes present on the wearer's skin. The ideal antimicrobial is which produces efficient effect at low dose and negative impact on environment. Besides the "green technologies", the usages of green modified antimicrobial techniques such as chitosan, sericin, and alginate are regenerable and have very low environmental impacts (Uddin 2014). An increasing consumption of antimicrobial finishing fibrous products and regulation for the control of such products provided reasons to appreciate the environment concern associated with finished textiles (Sayed and Korgaonkar 2020). The long-term benefits and potential use of antimicrobials significantly balance the both implementation in textiles and impact on environment. In future, more efforts require increasing the use of green techniques in implementation of antimicrobials to textiles and coating materials.

24.13 Conclusions

Antimicrobial fabrics and coating materials like silver oxide, zinc oxide, quaternary ammonium compound, and graphene look quite promising in providing protection against microbial growth on clothes and thereby directly minimize the risks of infections from them. Along with this, materials like silver-platinum alloy and titanium dioxide confer a good antimicrobial property when used as a coating material especially for medical equipment. All these things seem to be the need of present and future world in combating the problems associated with microbes as well as with the use of antimicrobial medicines. However, leaching of antimicrobial coating materials in the environment sometimes confers AMR to the microbes as well as the cost and durability of antimicrobial textiles are also major concerns. But if we look at the future commercial value and requirements of antimicrobial textiles and coating materials, then it seems like an extensive research needs to be done in overcoming the problems associated with them, and more prominent and eco-friendly compounds should be explored or made along with cost effectiveness of the end products.

References

Aarthi C, Govindarajan M, Rajaraman P, Alharbi NS, Kadaikunnan S, Khaled JM, Mothana RA, Siddiqui NA, Benelli G (2018) Eco-friendly and cost-effective Ag nanocrystals fabricated using the leaf extract of Habenaria plantaginea: toxicity on six mosquito vectors and four non-target species. Environ Sci Pollut Res 25:10317–10327

Alam MU, Rahman M, Islam MA, Asaduzzaman M, Sarker S, Rousham E, Unicomb L (2019) Human exposure to antimicrobial resistance from poultry production: assessing hygiene and waste-disposal practices in Bangladesh. Int J Hyg Environ Health 222:1068–1076

Ali M, Mahmood AH, Hussain S, Ahmed F (2021) An investigation into the antibacterial properties of bamboo/cotton blended fabric and potential limitations of the test method AATCC 147. J Nat Fibers 18:51–58

Aly SM, Albutti A (2014) Antimicrobials use in aquaculture and their public health impact. J Aquac Res Dev 5:1

Ayukekbong JA, Ntemgwa M, Atabe AN (2017) The threat of antimicrobial resistance in developing countries: causes and control strategies. Antimicrob Resist Infect Control 6:47

Belkhir K, Pillon C, Cayla A, Campagne C (2021) Antibacterial textile based on hydrolyzed milk casein. Materials 14:251

Cao Y, Gu J, Wang S, Zhang Z, Yu H, Li J, Chen S (2020) Guanidine-functionalized cotton fabrics for achieving permanent antibacterial activity without compromising their physicochemical properties and cytocompatibility. Cellulose 27:6027–6036

Caschera A, Mistry KB, Bedard J, Ronan E, Syed MA, Khan AU, Lough AJ, Wolfaardt G, Foucher DA (2019) Surface-attached sulfonamide containing quaternary ammonium antimicrobials for textiles and plastics. RSC Adv 9:3140–3150

Chen S, Zhang S, Galluzzi M, Li F, Zhang X, Yang X, Liu X, Cai X, Zhu X, Du B, Li J (2019) Insight into multifunctional polyester fabrics finished by one-step eco-friendly strategy. Chem Eng J 358:634–642

Christaki E, Marcou M, Tofarides A (2020) Antimicrobial resistance in bacteria: mechanisms, evolution, and persistence. J Mol Evol 88:26–40

Cole KA, Rivard KR, Dumkow LE (2019) Antimicrobial stewardship interventions to combat antibiotic resistance: an update on targeted strategies. Curr Infect Dis Rep 21:33

Duval RE, Grare M, Demoré B (2019) Fight against antimicrobial resistance: we always need new antibacterials but for right bacteria. Molecules 24:3152

Elena P, Miri K (2018) Formation of contact active antimicrobial surfaces by covalent grafting of quaternary ammonium compounds. Colloids Surf B: Biointerfaces 169:195–205

Emam HE, Ahmed HB, Gomaa E, Helal MH, Abdelhameed RM (2019) Doping of silver vanadate and silver tungstate nanoparticles for enhancement the photocatalytic activity of MIL-125-NH2 in dye degradation. J Photochem Photobiol A Chem 383:111986

Fiedot-Toboła M, Ciesielska M, Maliszewska I, Rac-Rumijowska O, Suchorska-Woźniak P, Teterycz H, Bryjak M (2018) Deposition of zinc oxide on different polymer textiles and their antibacterial properties. Materials 11:707

Hasan KF, Wang H, Mahmud S, Jahid MA, Islam M, Jin W, Genyang C (2020) Colorful and antibacterial nylon fabric via in-situ biosynthesis of chitosan-mediated nanosilver. J Mater Res Technol 9:16135–16145

Hiremath L, Kumar SN, Sukanya P (2018) Development of antimicrobial smart textiles fabricated with magnetite nano particles obtained through green synthesis. Mater Today Proc 5:21030–21039

Hong PY, Mantilla-Calderon D, Wang C (2020) Metagenomics as a tool to monitor reclaimed-water quality. Appl Environ Microbiol 86

Huang T, Chen C, Li D, Ek M (2019) Hydrophobic and antibacterial textile fibers prepared by covalently attaching betulin to cellulose. Cellulose 26(1):665–677

Jaksik J, Tran P, Galvez V, Martinez I, Ortiz D, Ly A, McEntee M, Durke EM, Aishee ST, Cua M, Touhami A (2018) Advanced cotton fibers exhibit efficient photocatalytic self-cleaning and antimicrobial activity. J Photochem Photobiol A Chem 365:77–85

Jansen KU, Knirsch C, Anderson AS (2018) The role of vaccines in preventing bacterial antimicrobial resistance. Nat Med 24:10–19

Jiao Y, Niu LN, Ma S, Li J, Tay FR, Chen JH (2017) Quaternary ammonium-based biomedical materials: state-of-the-art, toxicological aspects and antimicrobial resistance. Prog Polym Sci 71:53–90

Kanerva M, Puolakka A, Takala TM, Elert AM, Mylläri V, Jönkkäri I, Sarlin E, Seitsonen J, Ruokolainen J, Saris P, Vuorinen J (2019) Antibacterial polymer fibers by rosin compounding and melt-spinning. Mater Today Commun 20:100527

Kennedy MA, Mellon VS, Caldwell G, Potgieter LN (1995) Virucidal efficacy of the newer quaternary ammonium compounds. J Am Anim Hosp Assoc 31:254–258

Khan M, Tahir MN, Adil SF, Khan HU, Siddiqui MRH, Al-warthan AA, Tremel W (2015) Graphene-based metal and metal oxide nanocomposites: synthesis, properties and their applications. J Mater Chem A 3:18753–18808

Khude P, Majumdar A, Butola BS (2020) Development of antibacterial knitted fabrics from polyester-silver nanocomposite fibers. Indian J Fibre Text Res 45:419–425

Kloos WE, Musselwhite MS (1975) Distribution and persistence of Staphylococcus and Micrococcus species and other aerobic bacteria on human skin. Appl Microbiol 30:381–395

Knetsch ML, Koole LH (2011) New strategies in the development of antimicrobial coatings: the example of increasing usage of silver and silver nanoparticles. Polymers 3:340–366

Konwar A, Kalita S, Kotoky J, Chowdhury D (2016) Chitosan–iron oxide-coated graphene oxide nanocomposite hydrogel: a robust and soft antimicrobial biofilm. ACS Appl Mater Interfaces 8:20625–20634

Kost B, Svyntkivska M, Brzeziński M, Makowski T, Piorkowska E, Rajkowska K, Kunicka-Styczyńska A, Biela T (2020) PLA/β-CD-based fibers loaded with quercetin as potential antibacterial dressing materials. Colloids Surf B: Biointerfaces 190:110949

Kowalczyk D, Fortuniak W, Mizerska U, Kaminska I, Makowski T, Brzezinski S, Piorkowska E (2017) Modification of cotton fabric with graphene and reduced graphene oxide using sol–gel method. Cellulose 24:4057–4068

Kramer A, Guggenbichler P, Heldt P, Jünger M, Ladwig A, Thierbach H, Weber U, Daeschlein G (2006) Hygienic relevance and risk assessment of antimicrobial-impregnated textiles. Biofunct Text Skin 33:78–109

Langowska K, Kowal J, Palivan CG, Meier W (2014) A general strategy for creating self-defending surfaces for controlled drug production for long periods of time. J Mater Chem B 2:4684–4693

Lhermie G, Gröhn YT, Raboisson D (2017) Addressing antimicrobial resistance: an overview of priority actions to prevent suboptimal antimicrobial use in food-animal production. Front Microbiol 7:2114

Li Y, Pi QM, You HH, Li JQ, Wang PC, Yang X, Wu Y (2018) A smart multi-functional coating based on anti-pathogen micelles tethered with copper nanoparticles via a biosynthesis method using l-vitamin C. RSC Adv 8:18272–18283

Maghimaa M, Alharbi SA (2020) Green synthesis of silver nanoparticles from Curcuma longa L. and coating on the cotton fabrics for antimicrobial applications and wound healing activity. J Photochem Photobiol B Biol 204:111806

McEwen SA, Collignon PJ (2018) Antimicrobial resistance: a one health perspective. In: Antimicrobial resistance in bacteria from livestock and companion animals, pp 521–547

Mehta RI, Mehta RI, Solis OE, Jahan R, Salamon N, Tobis JM, Yong WH, Vinters HV, Fishbein MC (2010) Hydrophilic polymer emboli: an under-recognized iatrogenic cause of ischemia and infarct. Mod Pathol 23:921–930

Moo CL, Yang SK, Yusoff K, Ajat M, Thomas W, Abushelaibi A, Lim SHE, Lai KS (2020) Mechanisms of antimicrobial resistance (AMR) and alternative approaches to overcome AMR. Curr Drug Discov Technol 17:430–447

Morais DS, Guedes RM, Lopes MA (2016) Antimicrobial approaches for textiles: from research to market. Materials 9:498

Nateghi MR, Shateri-Khalilabad M (2015) Silver nanowire-functionalized cotton fabric. Carbohydr Polym 117:160–168

Ozin GA, Arsenault A (2015) Nanochemistry: a chemical approach to nanomaterials. Royal Society of Chemistry

Prang Rocky B, Thompson AJ (2020) Investigation and comparison of antibacterial property of bamboo plants, natural bamboo fibers and commercial bamboo viscose textiles. J Text Inst:1–12

Radhika D (2019) Review study on antimicrobial finishes on textiles–plant extracts and their application. Int Res J Eng Technol 6:3581–3588

Rajalakshmi M, Uddandrao VS, Saravanan G, Vadivukkarasi S, Koushik CV (2018) Biomodification of cotton and micro-denier polyester with sericin to develop potent antibacterial and antifungal textile products. J Inst Eng Ser E 99:119–127

Ramya M, Subapriya MS (2012) Green synthesis of silver nanoparticles. Int J Pharm Med Biol Sci 1:54–61

Ravindra KB, Murugesh Babu K (2016) Study of antimicrobial properties of fabrics treated with Ocimum sanctum L (tulsi) extract as a natural active agent. J Nat Fibers 13:619–627

Rigo S, Cai C, Gunkel-Grabole G, Maurizi L, Zhang X, Xu J, Palivan CG (2018) Nanoscience-based strategies to engineer antimicrobial surfaces. Adv Sci 5:1700892

Rong L, Liu H, Wang B, Mao Z, Xu H, Zhang L, Zhong Y, Feng X, Sui X (2019) Durable antibacterial and hydrophobic cotton fabrics utilizing enamine bonds. Carbohydr Polym 211:173–180

Rowe WP, Baker-Austin C, Verner-Jeffreys DW, Ryan JJ, Micallef C, Maskell DJ, Pearce GP (2017) Overexpression of antibiotic resistance genes in hospital effluents over time. J Antimicrob Chemother 72:1617–1623

Ryu HS, Bae IH, Lee KG, Hwang HS, Lee KH, Koh JT, Cho JH (2012) Antibacterial effect of silver-platinum coating for orthodontic appliances. Angle Orthod 82:151–157

Sayed U, Korgaonkar S (2020) Sustainable finishing of cotton fabric using eco-friendly natural products. Int J Adv Sci Eng 7:1589–1593

Sharma A, Kadam S, Mathur P, Sheikh J (2019) Re-using henna natural dyeing wastewater for coloration and multifunctional finishing of linen fabric. Sustain Chem Pharm 11:17–22

Singer AC, Shaw H, Rhodes V, Hart A (2016) Review of antimicrobial resistance in the environment and its relevance to environmental regulators. Front Microbiol 7:1728

Sirelkhatim A, Mahmud S, Seeni A, Kaus NHM, Ann LC, Bakhori SKM, Hasan H, Mohamad D (2015) Review on zinc oxide nanoparticles: antibacterial activity and toxicity mechanism. Nano-micro Lett 7:219–242

Suriyasathaporn W, Chupia V, Sing-Lah T, Wongsawan K, Mektrirat R, Chaisri W (2012) Increases of antibiotic resistance in excessive use of antibiotics in smallholder dairy farms in northern Thailand. Asian Australas J Anim Sci 25:1322

Tagliabue A, Rappuoli R (2018) Changing priorities in vaccinology: antibiotic resistance moving to the top. Front Immunol 9:1068

Tambosi JL, Yamanaka LY, José HJ, Moreira RDFPM, Schröder HF (2010) Recent research data on the removal of pharmaceuticals from sewage treatment plants (STP). Quím Nova 33:411–420

Uddin F (2014) Environmental concerns in antimicrobial finishing of textiles. Int J Text Sci 3:15–20

Van Boeckel TP, Brower C, Gilbert M, Grenfell BT, Levin SA, Robinson TP, Teillant A, Laxminarayan R (2015) Global trends in antimicrobial use in food animals. Proc Natl Acad Sci U S A 112:5649–5654

Wei X, Yang Z, Tay SL, Gao W (2014) Photocatalytic TiO_2 nanoparticles enhanced polymer antimicrobial coating. Appl Surf Sci 290:274–279

Wei T, Yu Q, Chen H (2019) Responsive and synergistic antibacterial coatings: fighting against bacteria in a smart and effective way. Adv Healthc Mater 8:1801381

Xu Q, Ke X, Cai D, Zhang Y, Fu F, Endo T, Liu X (2018) Silver-based, single-sided antibacterial cotton fabrics with improved durability via an l-cysteine binding effect. Cellulose 25:2129–2141

Xu J, Sangthong R, McNeil E, Tang R, Chongsuvivatwong V (2020a) Antibiotic use in chicken farms in northwestern China. Antimicrob Resist Infect Control 9:1–9

Xu Z, Cave R, Chen L, Yangkyi T, Liu Y, Li K, Meng G, Niu K, Zhang W, Tang N, Shen J (2020b) Antibiotic resistance and molecular characteristics of methicillin-resistant Staphylococcus epidermidis recovered from hospital personnel in China. J Glob Antimicrob Resist 22:195–201

Yang W, Zhang N, Wang Q, Wang P, Yu Y (2020) Development of an eco-friendly antibacterial textile: lysozyme immobilization on wool fabric. Bioprocess Biosyst Eng 43:1639–1648

Yemmireddy VK, Hung YC (2017) Using photocatalyst metal oxides as antimicrobial surface coatings to ensure food safety—opportunities and challenges. Compr Rev Food Sci Food Saf 16:617–631

Zakrzewski W, Dobrzynski M, Dobrzynski W, Zawadzka-Knefel A, Janecki M, Kurek K, Lubojanski A, Szymonowicz M, Rybak Z, Wiglusz RJ (2021) Nanomaterials application in orthodontics. Nanomaterials 11:337

Zaman SB, Hussain MA, Nye R, Mehta V, Mamun KT, Hossain N (2017) A review on antibiotic resistance: alarm bells are ringing. Cureus 9

Zhang Z, Wagner VE (eds) (2017) Antimicrobial coatings and modifications on medical devices. Springer, Berlin

Zhang J, Kong Q, Yang L, Wang DY (2016a) Few layered $Co(OH)_2$ ultrathin nanosheet-based polyurethane nanocomposites with reduced fire hazard: from eco-friendly flame retardance to sustainable recycling. Green Chem 18:3066–3074

Zhang Y, Xu Q, Fu F, Liu X (2016b) Durable antimicrobial cotton textiles modified with inorganic nanoparticles. Cellulose 23:2791–2808

Żywicka A, Fijałkowski K, Junka AF, Grzesiak J, El Fray M (2018) Modification of bacterial cellulose with quaternary ammonium compounds based on fatty acids and amino acids and the effect on antimicrobial activity. Biomacromolecules 19:1528–1538

Chapter 25
Nanostrategies Against Rising Antimicrobial Resistance (AMR)-Metallic Nanoparticles as Nanoweapon

Purnima Sharma, Dinesh Goyal, and Bhupendra Chudasama

25.1 Introduction

The term infectivity attributes to the ability of microorganisms (bacteria, parasites, virus, and fungi) to enter and multiply in the body of host causing infections. Infectious diseases have caused millions of deaths and strained the global healthcare system with an unprecedented load. Microbial infections cause nearly nine million deaths annually, which account for 41% of the global disease burden, measured in terms of Disability-Adjusted Life Years (DALYS) (GDB 2015; Hessling et al. 2017). Microbial diseases with their estimated mortality from year 1918 to 2020 are presented in Table 25.1. Infections with most pathogens trigger the immune response inside the host, which inactivates or kills the disease-causing pathogen as first line of defense. However, if the host is unable to do so then external help in the form of antibiotics is required to treat the infectious disease (Pankey and Sabath 2004).

P. Sharma
Department of Biotechnology, Thapar Institute of Engineering and Technology (Deemed University), Patiala, Punjab, India

School of Physics and Materials Science, Thapar Institute of Engineering and Technology (Deemed University), Patiala, Punjab, India

D. Goyal
Department of Biotechnology, Thapar Institute of Engineering and Technology (Deemed University), Patiala, Punjab, India

B. Chudasama (✉)
School of Physics and Materials Science, Thapar Institute of Engineering and Technology (Deemed University), Patiala, Punjab, India

Thapar-VT Center of Excellence in Emerging Materials (CEEMS), Thapar Institute of Engineering and Technology (Deemed University), Patiala, Punjab, India
e-mail: bnchudasama@thapar.edu

N. Akhtar et al. (eds.), *Emerging Modalities in Mitigation of Antimicrobial Resistance*,
https://doi.org/10.1007/978-3-030-84126-3_25

Table 25.1 Global mortality due to infectious diseases (1918–2020) (Whitworth 2018; Hackett 2018; Bloom and Cadarette 2019; WHO 2019; International Association for Medical Assistance 2019; Coronavirus Outbreak 2021)

Year	Pathogen	Geographic region	Cases/Mortality
1918–1920	Influenza (Spanish flu)	Worldwide	500 million cases and 30 to 100 million deaths
1957–1958	Influenza (Asian flu)	Worldwide	One to two million deaths
1968–1969	Influenza (Hong Kong flu)	Worldwide	500,000 to two million deaths
1960-present	HIV/AIDS	Worldwide, primarily Africa	70 million cases and 35 million deaths
1961-present	Cholera	Worldwide	1.4 to four million annual cases and 21,000 to 143,000 annual deaths
1974	Smallpox	India	130,000 cases and 26,000 deaths
1994	Plague	India	693 suspected cases and 56 deaths
2002–2003	SARS	Originated in China, spread to 37 countries	8098 cases and 774 deaths
2009	Influenza (swine flu)	Worldwide	284,000 deaths
2014–2016	Ebola	West Africa, primarily Guinea, Liberia, and Sierra Leone	28,600 cases and 11,325 deaths
2015-present	Zika	The Americas, Primarily Brazil	Unknown number of cases
2016	Dengue	Worldwide	100 million cases and 38,000 deaths
2017	Plague	Madagascar	2417 cases and 209 deaths
2017–2018	Cholera	Worldwide	1.2 million cases and 2627 deaths
2018	Listeria	South Africa	1060 cases and 216 deaths
2017–2018	Hepatitis A	USA	10,582 cases
2019–2020	Corona virus	Worldwide	155, 192, 083 cases and 3, 244, 581 deaths till 05/05/2021

The word "antibiotic" hails from the word "antibiosis" (meaning against life). Antibiotics are chemical compounds that kill or inhibit the growth of microorganisms. Antibiotics are classified as antibacterial, antifungal, and antiviral agents depending on their target group. However, all antibacterial compounds are commonly referred as antibiotics (Etebu and Arikekpar 2016). For decades, antibiotics have been used to treat various diseases. They are also used in various medical procedures like organ transplant and chemotherapy (Etebu and Arikekpar 2016). Broad-spectrum antibiotics target both the gram-negative and gram-positive bacteria while narrow-spectrum antibiotics only target one of them (Etebu and Arikekpar 2016). Antibiotics are either bactericidal (lethal to bacteria) or bacteriostatic (causing growth inhibition of bacteria) (Kohanski et al. 2010). They cause cell deaths or the cessation of growth by targeting vital physiology and biochemistry of bacterial cells. Major targets of antibiotics are: bacterial cell walls, cell membranes, DNA and RNA synthesis, protein synthesis, and folic acid (vitamin B9) metabolism (Kohanski et al. 2010). For example, the β-lactam class of antibiotics such as penicillin, cephalosporins, and carbapenem blocks the synthesis of the bacterial cell wall, while antibiotics like tetracycline, aminoglycoside, and macrolide target bacterial ribosomes. Antibacterial action of antibiotics may result in irreversible disruption of the cell integrity by interacting with target molecules through irreversible binding via covalent interaction with enzyme or cellular structure of cells (Kapoor et al. 2017).

Antibiotics have been used extensively for the controlled treatment of infectious diseases and in various surgical procedures like organ transplant and chemotherapy (Wright 2010). Excessive and uncontrolled use of antibiotics has resulted into development of Antimicrobial Resistance (AMR) in both intrahospital and nonhospital strains (Allahverdiyev et al. 2011). Development of antibiotic-resistant strains possesses a serious global threat to the disease management.

Center for Disease Control and Prevention (CDC), World Economic Forum, Infectious Diseases Society of America, and World Health Organization (WHO) have declared antimicrobial resistance (AMR) as one of the "biggest threats to public health globally" (Allahverdiyev et al. 2011; Spellberg et al. 2016). In European Union alone, an estimated 25,000 deaths per annum was caused by AMR (Padiyara et al. 2018). Antibiotic-resistant pathogen-associated hospital-acquired infections (HAIs) cause 99,000 deaths per annum in US (Aslam et al. 2018).

Many intrahospital pathogens have developed resistance against multiple antibiotics resulting in decrease in the efficiency of treatment. Decline in the effectiveness of existing antibiotics has caused higher morbidity and mortality in patients who are infected with multidrug-resistant (MDR) strains (Davies and Davies 2010; Spellberg and Gilbert 2014; González-Candelas et al. 2017). Deaths due to MDR alone are nearly 7,00,000 per annum (Betts et al. 2018). In past two decades, hospitals have witnessed increase in MDR-acquired infections due to the production of β-lactamase which is responsible for development of drug resistance against third generation antibiotics (Blair et al. 2014). Following mechanisms are responsible for antibiotic resistance in microorganisms:

1. Increased efflux of drugs and their reduced uptake.
2. Development of modifying enzymes/drug degrading inside microorganisms.
3. Enzymatic breakdown of antibiotics.
4. Alterations in antibiotic targets wherein new biosynthetic machinery is engaged to alter cell-wall structure.
5. Formation of protective biofilm around the bacteria that prevent its exposure to antibiotics.

Injection of antibiotics as a growth promoter in livestock has increased in recent years resulting in greater MDR strains in animals. Antibiotic consumption in livestock is expected to rise up to 67% by 2030 (Van Boeckel et al. 2015). This could also affect human health, because mobile genetic elements (MGEs) from resistant bacteria make their ways from animals to humans by various means (Molbak 2004; Humphrey et al. 2005).

Development of new and unconventional antibiotics is the need of the hour to overcome challenges put forward by multiple drug resistance (MDR). Metal nanoparticles owing to their high surface-to-volume ratio exhibit enhanced chemical and biological activities (Khurana et al. 2014). Nanoparticles impart toxicity on microorganisms (Table 25.2) by disturbing permeability and respiration of cell walls, inhibit enzymes, deactivate proteins, electrolyte imbalance, interfering with metabolic pathways of cells, damage to the cell membrane, reacting with sulfur-containing proteins and phosphorus-containing compounds such as DNA, induce oxidative stress by generating reactive oxygen species (ROS), generation of free radicals, modify gene expression levels, damage biomolecules like proteins, lipids, and DNA, and prevent biofilm formation (Hemeg 2017; Shaikha et al. 2019; Martinez et al. 2019). Due to the different target sites of metal nanoparticles, the probability of a microorganism developing resistance against metal nanoparticles is very low (Chopra 2007; Harikumar and

Table 25.2 Antibacterial effects of metal nanoparticles on different bacteria (Hemeg 2017)

Nanoparticles	Target bacteria	Antibacterial mechanism
Silver (Ag)	*Acinetobacter baumannii*, *S. typhi*, *Vibrio cholerae*, *B. subtilis*, *S. aureus* (multiple drug resistance), *Escherichia coli*, *Streptococcus pyogenes*, *Pseudomonas aeruginosa*, coagulase-negative *Staphylococcus epidermis*, *Enterococcus faecalis*, *Klebsiella pneumoniae*, *Listeria monocytogenes*, *Proteus mirabilis*, *Micrococcus luteus*	ROS generation, lipid peroxidation, inhibition of cytochromes of ETC, bacterial membrane disintegration, inhibition of cell wall synthesis, increase in membrane permeability, dissipation of proton gradient resulting in lysis, adhesion to cell surface causing lipid and protein damage, ribosome destabilization, intercalation between DNA bases, disruption of biofilms
Copper (Cu)	*Escherichia coli*, *Bacillus subtilis*, *Proteus vulgaris*	ROS generation, disorganization of membrane, inhibition of DNA replication, dissipation of cell membrane potential, lipid peroxidation, protein oxidation, DNA degradation
Zinc oxide (ZnO)	*S. aureus*, *E. coli*, *P. aeruginosa*, *B. subtilis*, *Stenotrophomonas acidaminiphila*, methicillin-resistant *Streptococcus agalactiae*, *E. coli*, *Klebsiella oxytoca*, *S. aureus*, *S. pyogenes*, methicillin-resistant *Staphylococcus aureus*	ROS production, disruption of membrane, adsorption to cell surface, lipids and protein damage, inhibition of microbial biofilm formation

Aravind 2016). Among metals, gold, silver, and copper are known for their antiseptic properties since ancient times (Vimbela et al. 2017). The ligand-based metal (copper complexes) and metal-based formulations of inorganic nanoparticles (barium sulfate) have also showed higher antibacterial activities (Lobana et al. 2020; Sooch et al. 2020).

It has been found that metal nanoparticles exhibit significant toxicity even at low dose levels, where their bulk counter parts were otherwise biocompatible (Morones et al. 2005; Giannousi et al. 2014; Azam and El-Said 2014; Suker and Albadran 2015; Chakraborty and Basu 2017). High active surface area and smaller size of nanoparticles provide them greater biochemical reactivity and membrane permeability (Wang et al. 2017).

Nanoparticle–antibiotic conjugates are promising nanoplatforms to combat bacterial resistance. They help in lowering the dosage by synergistic effect and reduce noxiousness (Sekhon 2010; Pissuwana et al. 2011). Further, these conjugates can increase the concentration of antibiotics at the antibiotic–microorganism contact site (Li et al. 2005). Decrease in the overall drug concentration in the conjugated system makes them less toxic and improves their dosage limits (Chen et al. 2009). Possible mechanisms leading to the synergistic effects in antibacterial activities of metal nanoparticles–antibiotic conjugates are summarized in Table 25.3.

Table 25.3 Mechanisms of antibacterial activities of nanoparticles–antibiotic conjugates on different bacteria (Li et al. 2005; Birla et al. 2009; Banoee et al. 2010; Selvarani 2010; Mandava et al. 2017)

Nanoparticle	Antibiotics	Bacteria	Mechanism
Silver	Kanamycin, tetracycline, streptomycin, ampicillin, amoxicillin, gentamicin, vancomycin, erythromycin	*S. aureus, E. coli, Micrococcus luteus, Salmonella typhi, P. aeruginosa*	Increased antibacterial action of nanoparticles and antibiotics. Ampicillin destroys cell wall and increases penetration of silver nanoparticles Silver nanoparticles act as antibiotic carrier and facilitate approach of hydrophobic antibiotic drugs into the bacterial cell membrane
Zinc oxide	15 antibiotics from different groups	*S. aureus, E. coli*	Increased activity of ciprofloxacin in the presence of ZnO nanoparticles by influencing activity of membrane
Copper	Tetracycline, streptomycin, gentamycin, amoxicillin, ampicillin, ciprofloxacin	*E. coli, M. luteus, Salmonella typhi, Streptococcus pyrogenes, S. mutans*	Increased activity of nanoparticles in the presence of antibiotics either by penetrating membrane or by ROS formation

25.2 Silver Nanoparticles

Silver has historically been widely known due to its antibacterial properties (Chang et al. 2006). Moreover, different salts of silver (Ag) and their by-products are also used as bactericidal agents (Melaiye and Youngs 2005). However, nanosilver greatly enhances the functionality of the material which further increased its applications in various healthcare products including wound dressings, bandages, lotions, ointments, water purifications, and medical devices (Thomas et al. 2007), and also used as protective agents for HIV patients (Shameli et al. 2012). Due to smaller size and greater surface to volume ratio, the silver nanoparticles (AgNPs) hold an extensive contact area with the microbes which makes them a robust bactericidal material. AgNPs showed a broad spectrum of antibacterial properties against a wide range of gram-positive and gram-negative bacteria. The AgNPs have capability to interrupt with cellular functions, including the cell permeation and potentially disrupts the cellular components by reacting with the sulfur-mediated proteins and phosphorus-mediated complexes like deoxyribonucleic acid (DNA) (Pal et al. 2007; Raffi et al. 2008; Ahamed et al. 2010). AgNPs can also interact with the respiratory enzyme system, thereby generating reactive oxygen species (ROS) (hydrogen peroxide (H_2O_2), hydroxyl (OH^-), and superoxide (O_2^-) radicals) that induce oxidative stress which further damages the proteins and nucleic acids (Gurunathan et al. 2018). Therefore, AgNPs have the potential to be widely used as antibacterial agent against drug resistance bacteria such as MRSA. It is estimated that, AgNPs as an antimicrobial agent have an annual demand of 3125 tons/year for medicine and 2800 tons/year in the field of food, hygiene, and water purification (Swathy et al. 2014).

AgNPs have shown enhanced antimicrobial potential in number of studies. Work done by Sondi and Sondi (2004) and Morones et al. (2005) showed that the AgNPs accumulate on the cell wall of *E. coli* that leads to formation of pits. These pits caused a loss of outer membrane integrity, resulting in the release of LPS molecules and membrane proteins and eventually caused cell death. AgNPs can also release silver ions which lead to cell damage. Pal et al. (2007) demonstrated that the surface area to volume ratio of AgNPs and the crystallographic surface structures are important factors that determine the antibacterial activity of AgNPs against gram-negative bacterium *E. coli*. Energy-filtering transmission electron microscopy images revealed that AgNPs caused changes in cell membrane of *E.coli*, resulting in cell death. Studies on AgNPs antibacterial activity against gram-positive *S. aureus* and gram-negative *E. coli* have shown that the growth in case of *E. coli* was inhibited at low AgNPs concentrations than in case of *S. aureus*, and indicated the higher antibacterial activity of AgNPs against gram-negative than gram-positive (Kim et al. 2007). Transmission electron microscopy has revealed that after a few minutes of contact with AgNPs, the cell membrane of *E. coli* cells gets completely disrupted at concentrations as low as 60 μg/mL and resulted into inhibition of bacterial cell growth and multiplication (Raffi et al. 2008). Ayala-Nunez et al. (2009) reported a dose-dependent antimicrobial activity of AgNPs against methicillin-resistant *Staphylococcus aureus* MRSA and non-MRSA by minimum inhibitory concentration (MIC), minimum bactericidal concentration (MBC). They found that AgNPs are effective bactericidal agents against both MRSA and non-MRSA regardless of the resistance mechanisms that confer importance to these bacteria as an emergent pathogen.

The antibacterial activity and antibacterial mechanism of AgNPs that were studied by Li et al. (2010) on E. coli (ATCC 8739) were investigated by evaluating growth pattern, cell membrane permeability, and morphology following treatment with nanoparticles. The results indicated that 10 μg/mL of AgNPs could completely inhibit the growth of *E. coli* (10^7 cfu/mL cells in liquid medium). AgNPs resulted in the leakage of reducing sugars and proteins by destroying the permeability of the bacterial membranes and induced the respiratory chain dehydrogenases into inactive state resulted in cell decomposition and death eventually. Whereas AgNPs at 50 μg/mL lead to pits and gaps formation as observed by transmission electron microscopy and scanning electron microscopy. They concluded that AgNPs caused structural damage of bacterial cell membrane and suppress the activity of some membranous enzymes which caused bacterial death (Li et al. 2010).

Uniform linoleic acid-capped AgNPs of 12 nm showed antibacterial activities against *S. basillus, S. aureus, and P. aureginosa*. They demonstrated that linoleic acid-capped AgNPs can be used as effective antibacterial agent (Das et al. 2011). Colloidal AgNPs synthesized by green route with controlled size and high stability showed high antimicrobial and bactericidal activity against *E. coli* and *S. aureus* (Dehnavi et al. 2012). In other study, the biologically synthesized AgNPs were shown higher antibacterial potential against gram-negative bacterium *S. typhi* in combination with two standard antibiotics (ampicillin and gentamycin). They showed that higher enhancing effect was observed for ampicillin in comparison to gentamicin due to increased potency of ampicillin-mediated cell wall lysis with combination of AgNPs (Rajawat and Qureshi 2012). This suggested that AgNPs must be increasing the local concentration of antibiotics at the site of action and thus helped in improving the potency of antibiotics. AgNPs showed higher antibacterial potential against gram-negative and gram-positive with decreasing particle size. Lu et al. (2013) prepared AgNPs of different sizes (~5, 15, and 55 nm) by reduction method and found that AgNPs (5 nm) exhibited the highest antibacterial activity against five anaerobic oral pathogenic bacteria *S. mutans, S. sanguis, S. mitis, A. actinomycetemcomitans, F. nuceatum,* and one aerobic bacteria *E. coli*, indicated potential used of AgNPs in oral antibacterial materials in the form of dental restorative material, dental implants, caries inhibitory solution and mouthwash. Similarly AgNPs synthesized with various sizes, i.e., 5, 7, 10, 15, 20, 30, 50, 63, 85, and 100 nm showed bacteriostatic/bactericidal effect in size and dose-dependent manner as determined by the minimum inhibitory concentration (MIC) and minimum bactericidal concentration (MBC) against *E. coli and S. aureus*. AgNPs with less than 10 nm size showed higher antibacterial efficacy whereas AgNPs of 5 nm mediated the fastest bactericidal activity against all the tested strains as revealed through delayed bacterial growth kinetics, corresponding MIC/MBC values, and disk diffusion tests.

The surface charge of AgNPs showed different effects on its antibacterial efficacy. Positively charged AgNPs exhibit the highest bactericidal activity against gram-positive (*S. aureus, Streptococcus mutants,* and *Streptococcus pyogenes*) and gram-negative bacteria (*E. coli* and *P. vulgaris*). The negatively charged AgNPs showed lesser activity while neutral AgNPs show intermediate antibacterial activity against tested strains (Abbaszadegan et al. 2015). The interaction of AgNPs with DNA may cause shearing or denaturation of the DNA and interruption in cell division also (Hsueh et al. 2015; Kumar et al. 2016). AgNPs (10 nm) are attached to the cell wall of cholera and interrupting the cell permeability that leads to bacterial cell death (Gahlawat et al. 2016). AgNPs of different shapes and sizes prepared by solution-based chemical reduction routes showed in vitro antibacterial properties against two types of gram-negative bacteria *P. aeruginosa* and *E. coli* as indicated by Kirby–Bauer disk diffusion susceptibility method. It was noticed that the smallest-sized spherical AgNPs showed better antibacterial activity against both bacterial strains as compared to the triangular and larger spherical-shaped AgNPs (Raza et al. 2016). Similar study was done by Acharya et al. (2018) in which they found bactericidal effects of both spherical and rod-shaped AgNPs against gram-positive (*S. aureus, B. subtilis*) and gram-negative (*E. coli, K. pneumoniae* AWD5, P. *aeruginosa*) bacterial strains.

Liao et al. (2019) reported the antimicrobial effect of AgNPs on clinical isolates of resistant *P. aeruginosa* through determining its minimal inhibitory concentration (MIC) and minimal bactericidal concentration (MBC) values, alterations of morphology, and structure by transmission electron microscopy (TEM) and also the analyzed differentially expressed proteins (superoxide dismutase (SOD), catalase (CAT), and peroxidase (POD)). Results revealed that AgNPs had highly bactericidal effect on the drug-resistant *P. aeruginosa* with the MIC range of 1.406–5.625 μg/mL and the MBC range of 2.813–5.625 μg/mL. TEM images showed that AgNPs could enter in bacterial cell and impaired its morphology and structure. The proteomics analyses revealed AgNPs-treated bacteria have high levels of SOD, CAT, and POD. The apoptosis rate of AgNP-treated bacteria was remarkably higher than that of the untreated bacteria. Shu et al. (2020) reported the antibacterial activity of shape-controlled and well-dispersed AgNPs (13.8 nm) which was synthesized by using green method in which biomolecules of reductive amino acids, alpha-linolenic acid, and carbohydrates in yeast extract played a significant role. The AgNPs in combination with antibiotic (ampicillin) reversed the resistance in ampicillin-resistant *E. coli* cells. They found that surface coatings of AgNPs enhanced their affinity toward the bacterial membrane and increased the permeability of the cell wall which further changed the peptidoglycan configuration and finally led to the apoptosis of bacteria. Mahmod et al. (2021) showed antibacterial activity of green synthesized AgNPs against gram-positive (*S. aureus, S. mutans,* and *S. epidermidis*) and gram-negative (*K. pneumoniae* and *P. aeruginosa*) bacteria. They found that gram-negative bacteria were more sensitive than gram-positive bacteria. Results indicated that synthesized AgNPs might be used to develop new antibacterial drug for combating against various diseases.

25.3 Copper Nanoparticles

Copper nanoparticles (CuNPs) have gained increasing attention due to their unique physical and chemical antimicrobial properties (Cioffi et al. 2005). CuNPs have many industrial applications: in gas sensors, high temperature superconductors, solar cells, and wood preservatives (Salvadori et al. 2014). Copper metal is used as an antimicrobial agent since ancient times. Compounds such as $CuSO_4$ and $Cu(OH)_2$ containing copper are used as the inorganic antibacterial agents (Raffi et al. 2010). Copper is a trace element as well as essential micronutrient, which is necessary for growth and maintenance of bone, connective tissue, brain, and heart (Azizi et al. 2017). Deficiency of copper can lead to anemia and improper fetal development during pregnancy. It plays a vital role in oxygen transport and iron homeostasis (Habibovic and Barralet 2011; Vimbela et al. 2017). In 2008, Environmental Protection Agency of US approved copper in medicinal products for human usage (Azizi et al. 2017). CuNPs also help in collagen cross-linking and in bone matrix formation (Klaine et al. 2008). CuNPs and their complexes are now used as antibacterial, antiviral, antifungal, and antifouling agents (Ventola 2015). The antibacterial effect of CuNPs on the bacterial cell involves multiple pathways, including adhesion to gram-negative bacterial cell wall due to electrostatic interaction, effecting protein structure of cell membrane, release of cytoplasmic contents, denaturation of the intracellular proteins, and interaction with phosphorus- and sulfur-containing compounds like DNA, loss of cellular functions, and reactive oxygen species (Raffi et al. 2010).

Antibacterial activities of copper nanoparticles (CuNPs)-based polymer metal nanocomposites were reported by Cioffi et al. (2005). Synthesized nanocomposite was capable of releasing metal species in a controlled manner that inhibits the growth of the organism. The biostatic activities of composites were correlated to the nanoparticle loading. Yoon et al. (2007) studied the antimicrobial activity of AgNPs and CuNPs on *E. coli* and *B. subtilis*. They observed that CuNPs exhibited higher antibacterial activities against gram-positive, *B. subtilis* while AgNPs were more active against gram-negative, *E. coli*. Ruparelia et al. (2008) studied the antibacterial activity of AgNPs and CuNPs on *E . coli, S. aureus and B. subtilis. They reported that AgNPs exhibited higher antibacterial activity against E. coli and S. aureus, while CuNPs were more bactericidal to B. subtilis. CuNPs-loaded* fibers by borohydride induced were tested their antibacterial activities against *E. coli*. These copper nanoparticles loaded fibers can be used in burn/wound dressing (Mary et al. 2009).

A study conducted by Raffi et al. (2010) evaluated the antibacterial activity of CuNPs in liquid and solid medium against gram-negative *E. coli*. They observed formation of cavities/pits in the bacterial cell wall after their interaction with CuNPs. Rispoli et al. (2010) studied the antimicrobial activity of CuNPs against *E. coli* and evaluated effects of pH, temperature, aeration rate, concentration of nanoparticles, and concentration of bacteria. It was concluded that the antibacterial activities of CuNPs not only depend on primary effect of test parameters, but also on the interactive effect of these parameters.

Mullite-based CuNPs were synthesized and their antibacterial activities against pathogenic strains: *E. coli, S. aureus*, Methicillin-resistant *Staphylococcus aureus* (MRSA), and *Shigella flexneri* were reported. They reported that 70 μg/mL of CuNPs were able to kill all tested microorganisms (Bagchi et al. 2012). Chatterjee et al. (2012) evaluated the antibacterial activity of CuNPs against *E. coli*. They observed increase in the number of filamentous bacteria with increase in the concentration of CuNPs indicating that the antibacterial activity of CuNPs was because of the growth of filamentous bacteria. The synergistic effects of chitosan-CuNPs on gram-positive bacteria (methicillin-resistant *S. aureus* (MRSA) and *Bacillus subtilis*), gram-negative bacteria (*Salmonella choleraesuis* and *Pseudomonas aeruginosa*), and yeast (*Candida albicans*) with nystatin (for yeast), ampicillin (for gram-negative bacteria), and streptomycin (for gram-positive bacteria) were evaluated. They have concluded that CuNPs in combination with antibiotics exhibit higher antibacterial activities against tested bacteria (Usman et al. 2012).

The antibacterial activities of CuNPs and copper oxide nanoparticles on *E. coli, Streptococcus pyogenes, Pseudomonas aeruginosa,* and *S. aureus* showed that CuNPs have higher antibacterial activities as compared to their oxides, indicated CuNPs as potential antibacterial agents than oxide (Karthik and Geetha 2013). Usman et al. (2013) observed that chitosan-coated copper nanoparticles have higher antibacterial activities against gram-negative bacteria than the gram-positive bacteria. Similarly, Shiv and Jong-Whan (2014) evaluated antibacterial activity of CuNPs against gram-positive (*Listeria monocytogenes*) and gram-negative bacteria (*E. coli*). They observed that gram-positive bacteria exhibit higher susceptibility toward CuNPs than the gram-negative bacteria. Figueroa et al. (2004) investigated antibacterial effects of copper, nickel, and bimetallic Cu-Ni nanoparticles on dental pathogens (*S. aureus, E. coli,* and *Streptococcus mutans*). They observed that CuNPs exhibited bactericidal effects while Ni and bimetallic Cu-Ni nanoparticles only show bacteriostatic effects on tested microorganisms.

Kruk et al. (2015) synthesized CuNPs by the reduction of copper ions with hydrazine in SDS aqueous solution and evaluated their antimicrobial activities against standard and clinical strains of methicillin-resistant *S. aureus (MRSA)* and *Candida* species (*C. albicans* and *C. parapsilosis*). Synthesized CuNPs exhibited strong antibacterial and antifungal activities. The antibacterial activities of CuNPs both against standard and clinical strains were stronger than their antifungal activities. CuNPs loaded with regenerated bacterial cellulose (RC) showed antibacterial activity against *S. aureus* (ATCC 6538), *B. subtilis* (ATCC 9372), *C. albicans* (CMCC(F)98,001), *E. coli* (ATCC 25922), and *P. aeruginosa* (ATCC 27853) (Shao et al. 2016). The synergistic effects of metal nanoparticles (silver and copper) with antibiotics (tetracycline and kanamycin) were studied on environmental friendly bacteria (*B. subtilis* and *P. fluorescens*). The antimicrobial activity of tetracycline was improved by 286–346% and 0–28% with silver and CuNPs, respectively whereas the improvement was 154–289% for silver and 3–20% for CuNPs with kanamycin (Khurana et al. 2016).

Ebrahimi et al. (2017) reported the antibacterial activities of CuNPs synthesized by green route against gram-positive (*S. aureus* and *B. cereus*) and gram-negative

(*E. coli* and *K. pneumonia*) bacteria. They observed that gram-positive strains were more susceptible to copper nanoparticles as compared to the gram-negative bacteria. In other study, synergistic effects of CuNPs with antibiotics ampicillin, amoxicillin, gentamicin, and ciprofloxacin were studied against bacterial strains *E. coli, S. typii, M. luteus,* and *S. mutans* (Mandava et al. 2017). Amongst the tested antibiotics, CuNPs-Ampicillin conjugates have highest synergistic antibacterial activities. Zia et al. (2018) synthesized AgNPs and CuNPs by chemical reduction method and studied their antibacterial activity against common human pathogenic bacteria (*S. aureus* and *E. coli*). They observed dose-dependent antibacterial activity of nanoparticles. Both silver and copper nanoparticles exhibited similar antibacterial activities against *S. aureus*, while *E. coli* was more susceptible to AgNPs. Selvarani (2010) investigated the synergistic effects of CuNPs with commercial antibiotics (tetracycline, rifampicin, chloramphenicol, vancomycin, gentamycin, streptomycin, kanamycin, tobramycin, penicillin, and ampicillin) against *B. cereus, E. coli, P. aeruginosa,* and *S. aureus*. Strong synergy of CuNPs was observed with all tested antibiotics.

Mesoporous silica (MSN)-maleamic and MSN-maleamic-Cu nanoparticles were synthesized and their antimicrobial activities on gram-positive and gram-negative bacteria were tested. MICs of *E. coli* and *S. aureus* were 125 μg/mL and 250 μg/mL, respectively. ROS levels in MSN-maleamic-Cu-treated *E. coli* were 40% higher, while in *S. aureus*, it was 30% higher as compared to controls, indicating that ROS plays a vital role in their antibacterial activities (Diana et al. 2019). Jayarambabu et al. (2019) synthesized CuNPs by using green synthesis route and evaluated their antibacterial activity on gram-negative (*E. coli*) and gram-positive (*B. subtilis*) bacteria. They have also observed that gram-positive strains were more susceptible to CuNPs as compared to the gram-negative strains. Kaur et al. (2019) reported synergistic activity of CuNPs with erythromycin, azithromycin, and norfloxacin against gram-positive and gram-negative bacterial strains. In case of antibiotics, the test bacteria (*Klebsiella* sp. and *Pseudomonas* sp.) showed resistance at all tested concentrations. In conjugation with CuNPs, the antibacterial activity of all antibiotics increased suggesting that combination of antibiotics with nanoparticles has significant synergistic antibacterial effects on *Klebsiella* sp. and *Pseudomonas* sp.

Green synthesized CuNPs showed antibacterial activity against *E. coli, Klebsiella pneumoniae, S. aureus*, and *E. faecalis*. They observed antibacterial activities of CuNPs against these pathogens. They observed higher antibacterial activities mainly due to CuNPs than CPE (Das et al. 2020). Yaqub et al. (2020) prepared CuNPs by chemical and biological synthesis and evaluated their synergistic effects with doxycycline against *E.coli* and *P. aeruginosa.* They reported that chemically synthesized CuNPs showed higher antimicrobial activity with doxycycline as compared to green synthesized copper nanoparticles. Jessop et al. (2021) studied antibacterial activities of hybrid composite (metallic copper nanoparticles (CuNPs) and novel cationic π-conjugated polyelectrolyte (CPE)) against gram-negative (*E. coli* and *Salmonella enteritidis*) and gram-positive bacteria (*S. aureus* and *Enterococcus faecalis*).

25.4 Gold Nanoparticles

For many decades, gold has been used in many medicines. Robert Koch was first to explore the biocidal potential of gold (Glišić and Djuran 2014). However, nanoparticles generally show good biocompatibility, strong adsorption ability, and applications in medical imaging, drug delivery owing to their surface, quantum size, and small-size effects (Mukherjee et al. 2016). Gold nanoparticles (AuNPs) have many applications starting from engineering to medicine field in past years (Patra et al. 2015). AuNPs have gained increasing attention due to their optical and electrical properties which further depend on their shape and size of nanoparticles (Verissimo et al. 2016). The biocompatibility of AuNPs made them a potential candidate to be used in the treatment of arthritis and cancer (Jain et al. 2006). Functionalized AuNPs with different biomolecules such as drugs, genes, peptides, and other targeting ligands have been exploited for a wide range of biomedical applications, not only in research and development sector but also in various commercially viable point of care systems viz. cytosensors, immunosensors, drug delivery, cancer imaging, aptasensing, and most advanced theranostics devices (Burygin et al. 2009; Mahato et al. 2019). AuNPs can also sense endocytosis, tumor metabolites, and receptors in cells, under dark-field light-scattering microscopy (Dykman and Khlebtsov 2011). Some AuNPs-based diagnostic kits are under clinical trials. FDA also approvals for AuNP-based in vitro diagnostic systems and clinical trials of AuNPs as cancer and cardiovascular treatments (Mieszawska et al. 2013). AuNPs have been used in the development of quantification of blood glucose, biosensors, toxic metals, disease markers, and insecticides (Liu and Lu 2003) and they also have the potential to degrade and detoxify some toxic pollutants (Lopez et al. 2004; Hernández et al. 2006).

AuNPs with a strongly bound capping (poly-allylamine hydrochloride) and a weakly bound capping agent (citrate) could directly contact with the bacterial (*E. coli*) cell membrane due to its positively charged nature mediated by strong electrostatic attraction to the negatively charged bilayer. The cationic particles are moderately toxic, whereas anionic particles are quite nontoxic in nature (Goodman et al. 2004). Polysiloxane polymers containing embedded methylene blue and AuNPs were shown significant antimicrobial activity against methicillin-resistant *S. aureus* and *E. coli*. The bacterial cell death was due to the light-induced production of singlet oxygen and reactive oxygen species induced by AuNPs (Perni et al. 2009). Antibiotic (gentamicin)–AuNP (15-nm) conjugates showed high antibacterial activity against tested *E. coli* K12 by the agar-well-diffusion method. AuNPs themselves do not have any antimicrobial activity, they may act as drug carriers by increasing surface area and carrying a lot of drug on its surface which further increased antibiotic concentration at the site of bacterium–particle contact and enhanced antibacterial property of antibiotics (Burygin et al. 2009). Amino-substituted pyrimidines-capped AuNPs showed antibacterial activities against multidrug-resistant clinical isolates of *E.coli* and *P. aeruginosa*. The antibacterial mechanism involved targeting the bacteria cell via multiple targets, destabilizing cell membranes, binding to nucleic acids, inhibiting protein synthesis, and inducing the leakage of cytoplasmic contents, suggested these NPs could probably bypass the resistance mechanisms employed by bacteria to combat conventional antibiotics (Zhao et al. 2010).

Ampicillin-functionalized AuNP (Brown et al. 2012) was functioned as effective broad-spectrum antibacterial agent against gram-negative and gram-positive bacteria. Functionalized AuNP mainly utilizes ampicillin's ability to permeate the outer membrane of bacteria and then nanoparticles enter the bacteria to achieve their antibacterial effects. This combined antibacterial effect can inhibit the growth of multiple-antibiotic-resistant isolates of *P. aeruginosa, E. aerogenes,* and Methicillin-Resistant *S. aureus*. Similarly, dextrose-coated gold nanoparticles (dGNPs) showed bacteriostatic and bactericidal effects against gram-negative (*E. coli*) and gram-positive (*S. epidermidis*) (Badwaik et al. 2012). The antibacterial action of dGNPs caused via disruption of the cell membrane leading to possible leakage of the cytoplasmic contents including nucleic acids. Tiwari et al. (2011) investigated the antibacterial and antifungal activities of 5-fluorouracil-functionalized AuNPs against *Micrococcus luteus, S. aureus, P. aeruginosa, E. coli, Aspergillus fumigates* (*A. fumigates*), and *Aspergillus niger* (*A. niger*). Results showed that AuNPs showed more activity on gram-negative bacteria than gram-positive due to membrane structure difference. AuNPs also showed antifungal activity against tested fungus also. Moreover, according to Wani et al. (2013), AuNPs may act both at the plasma membrane and in the cytoplasm. AuNPs inhibited H + -ATPase-mediated proton pumping and that might alter the normal cell conformation that leads to loss of activity in *Candida* sp.

Lima et al. (2013) reported the antimicrobial effect of AuNPs (5 nm) against *E. coli* and *Salmonella typhi*. They found that AuNPs reduced 90–95% growth of *E. coli* and *S. typhi*. The main factors that influenced the biocidal properties were the roughness and the dispersion of the AuNPs in the medium. The antibacterial efficacy of drugs (acridine derivatives: 9-aminoacridine hydrochloride hydrate (9AA-HCl), acridine yellow (AY), acridine orange (AO), and proflavine (Pro))-coated citrate-stabilized AuNPs studied against gram-positive (*B. subtilis*) and gram-negative bacteria (*E.coli*). AuNPs conjugated 9AA-HCl and AO resulted severe alterations in the cell wall structure that leads to leakage of cellular contents and finally cell death (Mitraa et al. 2014).

Patra and Baek (2015) reported that AuNPs with the standard antibiotics (kanamycin and rifampicin) showed higher antibacterial activity against five food-borne pathogens and demonstrated high synergistic activity. Results showed that AuNPs have antioxidant and antiprotease inhibitory activity also. The antibacterial activities of AuNP–kanamycin conjugates and kanamycin against gram-positive (*S. epidermidis*) and the gram-negative (*Enterobacter aerogenes*) concluded that the MIC values of the conjugate were significantly lower than that of free kanamycin. Similar result was concluded by Rattanata et al. (2016), against the food-borne pathogenic bacterial species *Plesiomonas shigelloides* and *Shigella flexneri* by gallic acid and AuNP–gallic acid. AuNP–gallic acid altered lipids, proteins, and nucleic acids of the bacterial cell membrane of treated bacteria. AuNP–levofloxacin conjugates (27.2 nm) were more efficient than levofloxacin alone and improved the antibacterial efficacy against *S. aureus*, *E. coli,* and *P. aeruginosa* by 1.94, 2.89, and 1.46 times, respectively (Bagga et al. 2017).

The antibacterial activity of biosynthesized *Nigella arvensis* (NA)-conjugated gold nanoparticles (NA-AuNPs) was determined by well diffusion method against *S. epidermidis*, *B. subtilis*, *S. aureus*, *E. coli*, *Serratia marcescens,* and *P. aeruginosa.* They concluded that NA-AuNPs can be considered as promising antibacterial agents in biomedical applications (Chahardoli et al. 2018). The peptide/triclosan-comodified gold nanoparticles showed better antibacterial effect than pure gold nanoparticles (Wang et al. 2019). Sun et al. (2019) observed that AuNPs cofunctionalized with both bovine serum albumin (BSA) and 4,6-diamino-2-pyrimidinethiol (DAPT) can generate conjugates (Au-DAPT-BSA) with progressive antimicrobial activities against gram-negative bacteria and gram-positive bacteria. Au-DAPT-BSA induced no drug resistance and strengthening the nanomaterial-based bactericides like AuNPs against multiple drug-resistant bacteria (MDR).

The green synthesized AuNPs with average size of 19.45 nm showed antimicrobial potential against a number of pathogenic microorganisms. The green synthesized AuNPs exhibited higher antimicrobial activity toward gram-positive bacteria and gram-negative bacteria (Dongaa et al. 2020). Biogenic AuNPs were synthesized from Brazilian Red *Propolis* hydroethanolic extract and its fractions which showed great potential to produce AuNPs in size range of 8–15 nm. These biogenic AuNPs showed antimicrobial, antifungal activity, and high cytotoxicity at low concentrations suggested that these biogenic AuNPs act as promising candidate in nanomedicine (Botteon et al. 2021).

25.5 Conclusions

Due to the development of multidrug-resistance (MDR) in microorganisms, there is an urgent need for development of new antimicrobial agents that will overcome the threats put up by MDR. Metal nanoparticles owing to their high surface-to-volume ratio exhibit enhanced chemical and biological activities that make them a potential candidate for development of antimicrobial drugs. There are various metal nanoparticles like gold, silver, and copper which possess enhanced antibacterial and cytotoxic activities. Because of the unique capability to target multiple structural sites of organisms randomly, organisms are unlikely to develop resistance against them unlike conventional antibiotics.

References

Abbaszadegan A, Ghahramani Y, Gholami A, Hemmateenejad B, Dorostkar S, Nabavizadeh M, Shargh H (2015) The effect of charge at the surface of silver nanoparticles on antimicrobial activity against gram-positive and gram-negative bacteria: a preliminary study. J Nanomater 16:53

Acharya D, Singha KM, Pandey P, Mohanta B, Rajkumari J, Singha LP (2018) Shape dependent physical mutilation and lethal effects of silver nanoparticles on bacteria. Sci Rep 8:201

Ahamed M, Alsalhi MS, Siddiqui M (2010) Silver nanoparticle applications and human health. Clin Chim Acta 41:1841

Allahverdiyev AM, Volodymyrivna KK, Abamor ES, Bagirova M, Rafailovich M (2011) Coping with antibiotic resistance: combining nanoparticles with antibiotics and other antimicrobial agents. Expert Rev Anti Infect Ther 9:1035

Aslam B, Wang W, Arshad MI, Khurshid M, Muzammil S, Rasool MH, Nisar MA, Alvi RF, Aslam MA, Qamar MV, Salamat MKF, Baloch Z (2018) Antibiotic resistance: a rundown of a global crisis. Infect Drug Resist 11:1645

Ayala-Nunez NV, Villegas HHL, Turrent LCI, Padilla CR (2009) Silver Nanoparticles Toxicity and Bactericidal Effect Against Methicillin-Resistant *Staphylococcus aureus*: Nanoscale Does Matter. J Nanobiotechnol 5:2

Azam ES, El-Said WA (2014) Synthesis of copper/nickel nanoparticles using newly synthesized Schiff-base metals complexes and their cytotoxicity/catalytic activities. Bioorg Chem 57:5

Azizi M, Hedayatollah G, Fatemeh Y, Fariba D, Hojjat A (2017) Cytotoxic effect of albumin coated copper nanoparticle on human breast cancer cells of MDA-MB 231. PLoS ONE 12: e0188639

Badwaik VD, Vangala LM, Pender DS, Willis CB, Aguilar ZP, Gonzalez MS, Paripelly R, Dakshinamurthy R (2012) Size-dependent antimicrobial properties of sugar-encapsulated gold nanoparticles synthesized by a green method. Nanoscale Res Lett 7:623

Bagchi B, Dey S, Bhandary S, Das S, Bhattacharya A, Basu R, Nandy P (2012) Antimicrobial efficacy and biocompatibility study of copper nanoparticle adsorbed mullite aggregates. Mater Sci Eng C 32:1897

Bagga P, Hussain Siddiqui H, Akhtar J, Mahmood T, Zahera M, Sajid Khan M (2017) Gold nanoparticles conjugated levofloxacin: for improved antibacterial activity over levofloxacin alone. Curr Drug Deliv 14:1114

Banoee M, Seif S, Nazari ZE, Fesharaki PJ, Shahverdi HR, Moballegh A, Moghaddam KM, Shahverdi AR (2010) Zno nanoparticles enhanced antibacterial activity of ciprofloxacin against *Staphylococcus aureus* and *Escherichia coli*. J Biomed Mater Res Pt B Appl Biomater 93:557

Betts JW, Hornsey M, La Ragione RM (2018) Novel Antibacterials: Alternatives to Traditional Antibiotics. Adv Microb Physiol 73:123

Birla SS, Tiwari VV, Gade AK, Ingle AP, Yadav AP, Rai MK (2009) Fabrication of silver nanoparticles by Phoma glomerata and its combined effect against Escherichia coli, *Pseudomonas aeruginosa* and *Staphylococcus aureus*. Lett Appl Microbiol 48:173

Blair JM, Webber MA, Baylay AJ, Ogbolu DO, Piddock LJ (2014) Molecular mechanisms of antibiotic resistance. Nat Rev Microbiol 13:42

Bloom DE, Cadarette D (2019) Infectious Disease Threats in the Twenty-First Century: Strengthening the Global Response. Front Immunol 10:549

Botteon CEA, Silva LB, Ccana-Ccapatinta GV, Silva TS, Ambrosio SR, Veneziani RCS, Bastos JK, Marcato PD (2021) Biosynthesis and characterization of gold nanoparticles using Brazilian red propolis and evaluation of its antimicrobial and anticancer activities. Sci Rep 11:1974

Brown AN, Smith K, Samuels TA, Lu J, Obare SO, Scott ME (2012) Nanoparticles functionalized with ampicillin destroy multiple-antibiotic-resistant isolates of *Pseudomonas aeruginosa* and *Enterobacter aerogenes* and methicillin-resistant *Staphylococcus aureus*. Appl Environ Microbiol 78:2768

Burygin GL, Khlebtsov BN, Shantrokha AN, Dykman LA, Bogatyrev VA, Khlebtsov NG (2009) On the enhanced antibacterial activity of antibiotics mixed with gold nanoparticles. Nanoscale Res Lett 4:794

Chahardoli A, Karimi N, Sadeghi F, Fattahi A (2018) Green approach for synthesis of gold nanoparticles from Nigella arvensis leaf extract and evaluation of their antibacterial, antioxidant, cytotoxicity and catalytic activities. Artificial cells. Nanomed Biotechnol 46:579

Chakraborty R, Basu T (2017) Metallic copper nanoparticlesinduce apoptosis in a human skin melanoma A-375 cell line. Nanotechnology 28:105101

Chang TT, Gish RG, de Man R, Gadano A, Sollano J, Chao YC, Lok AS, Han KH, Goodman Z, Zhu J, Cross A, DeHertogh D, Wilber R, Colonno E, Apelian D (2006) A comparison of entecavir and lamivudine for HBeAg-positive chronic hepatitis B. N Engl J Med 354:1001

Chatterjee AK, Sarkar RK, Chattopadhyay AP, Aich P, Chakraborty R, Basu T (2012) A simple robust method for synthesis of metallic copper nanoparticles of high antibacterial potency against *E. coli*. J Nanotechnol 23:085103

Chen F, Shi Z, Neoh KG, Kang ET (2009) Antioxidant and antibacterial activities of eugenol and carvacrol-grafted chitosan nanoparticles. Biotechnol Bioeng 104:30

Chopra I (2007) The increasing use of silver-based products as antimicrobial agents: a useful development or a cause for concern? J Antimicrob Chemother 59:578.

Cioffi N, Torsi L, Ditaranto N, Giuseppina T, Lina G, Luigia S, Teresa BZ, Maria DA, Giorgio PZ, Enrico T (2005) Copper nanoparticle/polymer composites with antifungal and bacteriostatic properties. Chem Mater 17:5255

Coronavirus Outbreak. https://www.worldometers.info/coronavirus/. Accessed 5 May 2021

Das R, Gang S, Siddhartha S, Nath S (2011) Preparation and antibacterial activity of silver nanoparticles. J Biomater Nanobiotechnol 2:472

Das PE, Abu Yousef IA, Majalawieh AF, Narasimhan A, Poltronieri P (2020) Green synthesis of encapsulated copper nanoparticles using a hydroalcoholic extract of *Moringa oleifera* leaves and assessment of their antioxidant and antimicrobial activities. Molecules 25:555

Davies J, Davies D (2010) Origins and evolution of antibiotic resistance. Microbiol Mol Biol Rev 74:417

Dehnavi AS, Raisi A, Aroujalian A (2012) Control size and stability of colloidal silver nanoparticles with antibacterial activity prepared by a green synthesis method. Synth React Inorg Met Org Chem 43:543

Diana DG, Perla AK, Sanjiv P, Antonio RD, Paulina PL, Santiaqo GR (2019) Preparation and study of the antibacterial applications and oxidative stress induction of copper maleamate-functionalized mesoporous silica nanoparticles. Pharmaceutics 11:30

Donga S, Bhadu GR, Chandaa S (2020) Antimicrobial, antioxidant and anticancer activities of gold nanoparticles green synthesized using *Mangifera indica* seed aqueous extract. Artifical cells. Nanomed Biotechnol 48:1315

Dykman L, Khlebtsov N (2011) Gold nanoparticles in biology and medicine: recent advances and prospects. Acta Natura 3:34

Ebrahimi K, Shiravand S, Mahmoudvand H (2017) Biosynthesis of copper nanoparticles using aqueous extract of Capparis spinosa fruit and investigation of its antibacterial activity. Marmara Pharmaceut J 21:866

Etebu E, Arikekpar I (2016) Antibiotics: Classification and mechanisms of action with emphasis on molecular perspectives. Int J Appl Microbiol Biotechnol Res 4:90

Figueroa LA, Morales-Luckie RA, Scougall-Vilchis RJ, Olea-Mejia OF (2004) Synthesis, characterization and antibacterial activity of copper, nickel and bimetallic Cu–Ni nanoparticles for potential use in dental materials. Prog Nat Sci Mater Int 24:321

Gahlawat G, Shikha S, Chaddha BS, Chaudhuri SR, Mayilraj S, Choudhury AR (2016) Microbial glycolipoprotein-capped silver nanoparticles as emerging antibacterial agents against cholera. Microbial Cell Fact 15:25

GDB (2015) Diseases and injury incidence and prevalence collaborates Global regional, and national incidence, prevalence, and years lived with disability for 310 diseases and injuries, 1990-2015: a systematic analysis for the Global Burden of Disease Study. Lancet 388:1459

Giannousi K, Lafazanis K, Arvanitidis J, Pantazaki A, Samara CD (2014) Hydrothermal synthesis of copper based nanoparticles: antimicrobial screening and interaction with DNA. J Inorg Chem 133:24

Glišić BĐ, Djuran MI (2014) Gold complexes as antimicrobial agents: an overview of different biological activities in relation to the oxidation state of the gold ion and the ligand structure. Dalton Trans 43:5950

González-Candelas F, Comas I, Martínez JL, Galán JC, Baquero F (2017) The evolution of antibiotic resistance. Genetics and evolution of infectious diseases. Elsevier, London, 257

Goodman C, McCusker C, Yilmaz T, Rotello V (2004) Toxicity of gold nanoparticles functionalized with cationic and anionic side chains. Bioconjug Chem 15:897

Gurunathan S, Choi YZ, Kim JH (2018) Antibacterial Efficacy of Silver Nanoparticles on Endometritis Caused by *Prevotella melaninogenica* and *Arcanobacterum pyogenes* in Dairy Cattle. Int J Mol Sci 19:1210

Habibovic P and JE Barralet (2011) Bioinorganics and Biomaterials: Bone repair. Acta Biomater 7:3013

Hackett DW. Hepatitis A outbreak reaches 10,582 cases during 2018. Precis vaccines. Accessed 31 Dec 2018

Harikumar PS, Aravind A (2016) Antibacterial Activity of Copper Nanoparticles and Copper Nanocomposites against *Escherichia Coli* Bacteria. Int J Sci 5:83

Hemeg HA (2017) Nanomaterials for alternative antibacterial therapy. Int J Nanomed 12:8211

Hernández J, Solla-Gullón J, Herrero E, Aldaz A, Feliu JM (2006) Methanol oxidation on gold nanoparticles in alkaline media: Unusual electrocatalytic activity. Electrochem Acta 52:1662

Hessling M, Feiertag J, Hoenes K (2017) Pathogens Provoking Most Deaths Worldwide: A Review. Biosci Biotechnol Res Commun 10:3

Hsueh YH, Lin KS, Ke WJ, Hsieh CT, Chiang CL, Tzou DY, Liu ST (2015) The antimicrobial properties of silver nanoparticles in *Bacillus subtilis* are mediated by released Ag+ ions. PLoS One 10:e0144306

Humphrey TJ, Jorgensen F, Frost JA, Wadda H, Domingue G, Elviss NC, Griggs DJ, Piddock LJ (2005) Prevalence and subtypes of ciprofloxacin-resistant Campylobacter spp. in commercial poultry flocks before, during, and after treatment with fluoroquinolones. Antimicrob Agents Chemother 49:690

International Association for Medical Assistance for Travelers. United States of America recommended vaccinations: Hepatitis A http://www.iamat.org/country/united-states-of-america/risk/hepatitis-a. Accessed 3 Jan 2019

Jain PK, Lee KS, Sayed IH, Sayed MA (2006) Calculated absorption and scattering properties of gold nanoparticles of different size, shape, and composition: applications in biological imaging and biomedicine. J Phys Chem B 110:7238

Jayarambabu N, Allam A, Rao TV, Kumar RR (2019) Green synthesis of Cu nanoparticles using Curcuma longa extract and their application in antimicrobial activity. Mater Lett 259:126813

Jessop IA, Perez YP, Jachura A, Nunez H, Saldies C, Isaacs M, Tundidor-Camba A, Terraza CA, Duran IA, Camarada M, Carcamo-Vega JJ (2021) New hybrid copper nanoparticles/conjugated polyelectrolyte composite with antibacterial activity. Polymers 13:401

Kapoor G, Saigal S, Elongavan A (2017) Action and resistance mechanisms of antibiotics: A guide for clinicians. J Anesthesiol Clin Pharmacol 33:300

Karthik AD, Geetha K (2013) Synthesis of Copper Precursor, Copper and its oxide Nanoparticles by Green Chemical Reduction Method and its Antimicrobial Activity. J Appl Pharmaceut Sci 3:16

Kaur P, Nene AG, Sharma D, Somani PR, Tuli HS (2019) Synergistic effect of copper nanoparticles and antibiotics to enhance antibacterial potential. BioMater Technol 1:33

Khurana C, Vala AK, Andhariya N, Pandey OP, Chudasama B (2014) Antibacterial activity of silver: the role of hydrodynamic particle size at nanoscale. J Biomed Mater Res A 102A:3361

Khurana C, Sharma P, Pandey OP, Chudasama B (2016) Synergistic Effect of Metal Nanoparticles on the Antimicrobial Activities of Antibiotics against Biorecycling Microbes. J Mater Sci Technol 32:524

Kim JS, Kuk E, Yu KN, Kim JH, Park SJ, Lee HJ, Kim SH, Park YK, Park YH, Hwang CY, Kim YK, Lee YS, Jeong DH, Cho MH (2007) Antimicrobial effects of silver nanoparticles. Nanomedicine 3:95

Klaine SJ, Alvarez PJJ, Batley GE, Fernandes TF, Habdy RD, Lyon DY, Mahendra S, McLaughlin MJ, Lead JR (2008) Nanomaterials in the environment: Behaviour, fate, bioavailability and effects. Environ Toxicol Chem 27:1825

Kohanski MA, Dwyer DJ, Collins JJ (2010) How antibiotics kill bacteria: from targets to networks. Nat Rev 8:423

Kruk T, Szczepanowicz K, Stefanska J, Socha RP, Warszynski P (2015) Synthesis and antimicrobial activity of monodisperse copper nanoparticles. Colloids Surf B Biointerfaces 128:17

Kumar N, Das S, Jyoti A, Kaushik S (2016) Synergistic effect of silver nanoparticles with doxycycline against *Klebsiella Pneumonia*. Int J Pharm Pharm Sci 8:183–186

Li P, Li J, Wu C, Wu Q, Li J (2005) Synergistic antibacterial effects of β-lactam antibiotic combined with silver nanoparticles. Nanotechnology 16:1912

Li WR, Xie XB, Shi QS, Zeng HY, Ou-Yang YS, Chen YB (2010) Antibacterial activity and mechanism of silver nanoparticles on *Escherichia coli*. Appl Microbiol Biotechnol 85:1115

Liao S, Zhang Y, Pan X, Zhu F, Jiang C, Liu Q, Cheng Z, Dai G, Wu G, Wang L, Chen L (2019) Antibacterial activity and mechanism of silver nanoparticles against multidrug-resistant *Pseudomonas aeruginos*a. Int J Nanomed 14:1469

Lima E, Guerra R, Lara V, Guzmán A (2013) Gold nanoparticles as efficient antimicrobial agents for Escherichia coli and *Salmonella typhi*. Chem Cent J 7:1

Liu J, Lu YA (2003) A colorimetric lead biosensor using DNAzyme-directed assembly of gold nanoparticles. J Am Chem Soc 125:6642

Lobana TS, Kaushal M, Bala R, Nim L, Paul K, Arora DS, Bhatia A, Arora S, Jasinski JP (2020) Di-2-pyridylketone-N 1-substituted thiosemicarbazone derivatives of copper(II): Biosafe antimicrobial potential and high anticancer activity against immortalized L6 rat skeletal muscle cells. J Inorg Biochem 212:111205

Lopez N, Janssens T, Clausen B, Xu Y, Mavrikakis M, Bligaard T, Nørskov JK (2004) On the origin of the catalytic activity of gold nanoparticles for low-temperature CO oxidation. J Catal 223:232

Lu Z, Rong K, Li J, Yang H, Chen R (2013) Size-dependent antibacterial activities of silver nanoparticles against oral anaerobic pathogenic bacteria. J Mater Sci Mater Med 24:1465

Mahato K, Nagpal S, Shah MA, Srivastava A, Maurya PK, Roy S, Jaiswal A, Singh R, Chandra P (2019) Gold nanoparticle surface engineering strategies and their applications in biomedicine and diagnostics. 3 Biotech 9:57

Mahmod R, Junayed A, Bhowmick C, Sompa SA, Sultana T, Akter T, Abedin MZ, Zubair MA, Islam MA, Moga MR, Khatun M, Sikder MA (2021) Antibacterial activity of silver nanoparticles synthesized from leaf and flower extracts of *Galinsoga formosa*. J Adv Biotechnol Exp Therapeut 4:178

Mandava K, Kadimcharla K, Kessara NR, Sumayya NF, Parthyusha B, Batchu UR (2017) Green synthesis of stable copper nanoparticles and synergistic activity with antibiotics. Indian J Pharm Sci 24:136

Martinez NN, Orozco MFS, Martinez-Castanon GA, Mendez FT, Ruiz F (2019) Molecular mechanisms of bacterial resistance to metal and metal oxide nanoparticles. Int J Mol Sci 20:2808

Mary G, Baypai SK, Chand N (2009) Copper (II) ions and copper nanoparticles-loaded chemically modified cotton cellulose fibers with fair antibacterial properties. J Appl Polym Sci 113:757

Melaiye A, Youngs WJ (2005) Silver and its application as an antimicrobial agent. Expert Opin Ther Pat 15:125

Mieszawska AJ, Mulder WJ, Fayad ZA, Cormode DP (2013) Multifunctional gold nanoparticles for diagnosis and therapy of disease. Mol Pharm 10:831

Mitraa P, Chakraborty PK, Sahab P, Rayc P, Basu S (2014) Antibacterial efficacy of acridine derivatives conjugated with gold nanoparticles. Int J Pharmaceut 473:636

Molbak K (2004) Spread of resistant bacteria and resistance genes from animals to humans--the public health consequences. J Veter Med B Infect Dis Public Health 51:364

Morones JR, Elechiguerra JL, Camacho A, Holit K, Kouri JB, Ramirez JT, Yacaman MJ (2005) The bactericidal effect of silver nanoparticles. J Nanotechnol 16:2346

Mukherjee S, Sau S, Madhuri D, Bollu VS, Madhusudana K, Sreedhar B, Banerjee R, Patra CR (2016) Green synthesis and characterization of monodispersed gold nanoparticles: toxicity study, delivery of doxorubicin and its bio-distribution in mouse model. J Biomed Nanotechnol 12:165

Pal S, Tak YK, Song JM (2007) Does the antibacterial activity of silver nanoparticles depend on the shape of the nanoparticle? A study of the Gram-negative bacterium *Escherichia coli*. Appl Environ Microbiol 73:1712
Pankey GA, Sabath LD (2004) Clinical relevance of bacteriostatic versus bactericidal mechanisms of action in the treatment of Gram-positive bacterial infections. Clin Infect Dis 38:864
Padiyara P, Inoue H, Sprenger M (2018) Global Governance Mechanisms to Address Antimicrobial Resistance. Infect Dis Res Treat 11:1
Patra JK, Baek KH (2015) Novel green synthesis of gold nanoparticles using Citrullus lanatus rind and investigation of proteasome inhibitory activity, antibacterial, and antioxidant potential. Int J Nanomed 10:7253
Patra JM, Panda SS, Dhal NK (2015) A review on green synthesis of gold nanoparticles. Int J Pharma Biosci 6:251
Perni S, Piccirillo C, Pratten J, Prokopovich P, Chrzanowski W, Parkin IP, Wilson M (2009) The antimicrobial properties of light-activated polymers containing methylene blue and gold nanoparticles. Biomaterials 30:89
Pissuwana D, Niddomea T, Cortiec MB (2011) The forthcoming applications of gold nanoparticles in drug and gene delivery systems. J Control Release 149:65
Raffi M, Hussain F, Bhatti T, Akhter J, Hameed A, Hasan M (2008) Antibacterial Characterization of Silver Nanoparticles against *E. Coli* ATCC-15224. J Mater Sci Technol 24:192
Raffi M, Mehrwan S, Bhatti TM, Akhtar JI, Hameed A, Yanwar W, Ul-Hasan MM (2010) Investigations into the antibacterial behavior of copper nanoparticles against *Escherichia coli*. Ann Microbiol 60:75
Rajawat S, Qureshi MS (2012) Comparative Study on Bactericidal Effect of Silver Nanoparticles, Synthesized Using Green Technology, in Combination with Antibiotics on *Salmonella Typhi*. J Biomed Nanotechnol 3:480
Rattanata N, Klaynongsruang S, Leelayuwat C, Limpaiboon T, Lulitanond A, Boonsiri P, Chio-Srichan S, Soontaranon S, Rugmai S, Daduang J (2016) Gallic acid conjugated with gold nanoparticles: antibacterial activity and mechanism of action on foodborne pathogens. Int J Nanomedicine 11:3347
Raza MA, Kanwal Z, Rauf A, Sabri AN, Riaz S, Naseem S (2016) Size- and Shape-Dependent Antibacterial Studies of Silver Nanoparticles Synthesized by Wet Chemical Routes. Nanomaterials 6:74
Rispoli F, Angelov A, Badia D, Kumar A, Seal S, Shah V (2010) Understanding the toxicity of aggregated zero valent copper nanoparticles against *Escherichia coli*. J Hazard Mater 180:212
Ruparelia JP, Chatterjee AK, Duttagupta SP, Mukherji S (2008) Strain specificity in antimicrobial activity of silver and copper nanoparticles. Acta Biomater 4:707
Salvadori MR, Ando RA, do Nascimento CA, Corrêa B (2014) Intracellular Biosynthesis and Removal of Copper Nanoparticles by Dead Biomass of Yeast Isolated from the Wastewater of a Mine in the Brazilian Amazonia. PLoS One 9:e87968
Sekhon BP (2010) Metalloantibiotics and antibiotic mimics - an overview. J Pharmaceut Educ Res 1:1
Selvarani M (2010) Investigation of the synergistic antibacterial action of copper nanoparticles on certain antibiotics against human pathogens. Int J Pharm Pharmaceut Sci 10:83
Shaikha S, Nazam N, Rizvi SMD, Ahmad K, Baig MH, Lee EJ, Choi I (2019) Mechanistic insights into the antimicrobial actions of metallic nanoparticles and their implications for multidrug resistance. Int J Mol Sci 20:2468
Shameli K, Ahmad MB, Jazayeri SD, Shabanzadeh P, Sangpour P, Jahangrian H, Gharayetri Y (2012) Investigation of antibacterial properties silver nanoparticles prepared via green method. Chem Cent J 6:73.
Shao W, Wang S, Wu J, Huang M, Liu H, Min H (2016) Synthesis and antimicrobial activity of copper nanoparticle loaded regenerated bacterial cellulose membranes. RSC Adv 6:65879
Shiv S, Jong-Whan R (2014) Effect of copper salts and reducing agents on characteristics and antimicrobial activity of copper nanoparticles. Mater Lett 132:307

Shu M, He F, Li Z, Lu X, Ma Y, Zhou Z, Yang Z, Gao F, Zeng M (2020) Biosynthesis and antibacterial activity of silver nanoparticles using yeast extract as reducing and capping agents. Nanoscale Res Lett 15:14

Sondi I, Sondi BS (2004) Silver nanoparticles as antimicrobial agent: a case study on *E. coli* as a model for Gram-negative bacteria. J Colloid Interface Sci 275:177

Sooch BS, Mann MK, Sharma M (2020) Metal-Doped Barium Sulphate Nanoparticles Decorated with Gelatin as Antibacterial Agents. J Cluster Sci 32:1141

Spellberg B, Gilbert DN (2014) The Future of antibiotics and resistance: A tribute to a career of leadership by John Bartlett. Clin Infect Dis 59:71

Spellberg B, Srinivasan A, Chambers HF (2016) New Societal Approaches to Empowering Antibiotic Stewardship. J Am Med Assoc 315:1229

Suker DS, Albadran RM (2015) Cytotoxic Effects of Titanium Dioxide Nanoparticles on Rat Embryo Fibroblast REF-3 Cell Line *in vitro*. Eur J Exp Biol 3:354

Sun Z, Zheng W, Zhu G, Lian J, Wang J, Hui P, He S, Chen W, Jiang X (2019) Albumin broadens the antibacterial capabilities of nonantibiotic small molecule-capped gold nanoparticles. ACS Appl Mater Interfaces 11:45381

Swathy J, Sankar MU, Chaudhary A, Aigal S, Pradeep T (2014) Antimicrobial silver: An unprecedented anion effect. Sci Rep 4:7161

Thomas V, Yallapu MM, Sreedhar B, Bajpai S (2007) A versatile strategy to fabricate hydrogel-silver nanocomposites and investigation of their antimicrobial activity. J Colloid Interface Sci 315:389

Tiwari PM, Vig K, Dennis VA, Singh SR (2011) Functionalized gold nanoparticles and their biomedical applications. Nanomaterials 1:31

Usman MS, Irabim NA, Shameli K, Zainuddin N, Yunus WMW (2012) Copper Nanoparticles Mediated by Chitosan: Synthesis and Characterization via Chemical Methods. Molecules 17:14928

Usman SM, Modamed ZEE, Kamyer S, Norhazlin Z, Mahamed S, Nor Azoma I (2013) Synthesis, characterization, and antimicrobial properties of copper nanoparticles. Int J Nanomedicine 8:4467

Van Boeckel TP, Brower C, Gilbert M, Grenfell BT, Levin SA, Robinson TP, Teillent A, Laxminnarayan R (2015) Global trends in antimicrobial use in food animals. Proc Natl Acad Sci 112:5649

Ventola CL (2015) The antibiotic resistance crisis:part1: causes and threats. P T 40:277

Verissimo TV, Santos NT, Silva JR, Azevedo RB, Gomes AJ, Lunardi CN (2016) *In vitro* cytotoxicity and phototoxicity of surface-modified gold nanoparticles associated with neutral red as a potential drug delivery system in phototherapy. Mater Sci Eng C 65:199

Vimbela GV, Ngo SM, Fraze C, Yang L, Stout D (2017) Antibacterial properties and toxicity from metallic nanomaterials. Int J Nanomed 12:3941

Wang L, Chen H, Lonqquan S (2017) The antimicrobial activity of nanoparticles: present situation and prospects for the future. Int J Nanomed 12:1227

Wang J, Cai X, Yang L, Huang F, Liu J (2019) Antibacterial peptide/triclosan dual functionalized gold nanoparticles for enhanced antibacterial effect. Biomed Eng Clin Med 3:252

Wani IA, Ahmad T, Manzoor N (2013) Size and shape dependant antifungal activity of gold nanoparticles: a case study of *Candida*. Colloids Surf B Biointerfaces 101:162

Whitworth J (2018) Food safety news. Accessed 4 September

World Health Organization (2019) Outbreak update—Cholera in Yemen, 20 December 2018. http://www.emro.who.int/pandemic-epidemic-diseases/cholera/outbreak-update-cholera-in-yemen-20-december-2018.html. Accessed 3 Jan 2019

Wright GD (2010) Q&A: Antibiotic resistance: where does it come from and what can we do about it? BMC Biol 8:123

Yaqub A, Malkani N, Shabbir A, Ditta AS, Tanvir F, Ali S, Naz M, Kazmi SAR, Ullah R (2020) Novel biosynthesis of copper nanoparticles using *Zingiber* and *Allium* sp. with synergic effect of doxycycline for anticancer and bactericidal activity. Curr Microbiol 77:2287

Yoon KY, Hoon Byeon J, Park JH, Hwang J (2007) Susceptibility constants of *Escherichia coli* and *Bacillus subtilis* to silver and copper nanoparticles. Sci Total Environ 373:572

Zhao Y, Tian Y, Cui Y, Liu W, Ma W, Jiang X (2010) Small molecule-capped gold nanoparticles as potent antibacterial agents that target gram-negative bacteria. J Am Chem Soc 132:12349

Zia R, Raiz M, Farooq N, Qamar A, Aujum S (2018) Antibacterial activity of Ag and Cu nanoparticles synthesized by chemical reduction method: a comparative analysis. Mater Res Expr 5:075012

Index

N. Akhtar et al. (eds.), *Emerging Modalities in Mitigation of Antimicrobial Resistance*,
https://doi.org/10.1007/978-3-030-84126-3

The manufacturer's authorised representative in the EU is Springer Nature Customer Service Centre GmbH, Europaplatz 3, 69115 Heidelberg, Germany. If you have any concerns regarding our products, please contact ProductSafety@springernature.com

Printed and bound by CPI Group (UK) Ltd, Croydon, CR0 4YY
30/09/2026
02225699-0016